Blood Cell Biochemistry

Volume 1
Erythroid Cells

Blood Cell Biochemistry

Series Editor
J. R. Harris, *North East Thames Regional Transfusion Centre, Brentwood, Essex, England*

Volume 1 **Erythroid Cells**
Edited by J. R. Harris

Volume 2 **Megakaryocytes, Platelets, Macrophages, and Eosinophils**
Edited by J. R. Harris

Volume 3 **Lymphocytes and Granulocytes**
Edited by J. R. Harris

Blood Cell Biochemistry

Volume 1
Erythroid Cells

Edited by
J. R. Harris
North East Thames Regional Transfusion Centre
Brentwood, Essex, England

Plenum Press • New York and London

Library of Congress Cataloging-in-Publication Data

Erythroid cells / edited by J.R. Harris.
p. cm. -- (Blood cell biochemistry ; v. 1)
Includes bibliographical references.
Includes index.
ISBN 0-306-43462-8
1. Erythrocytes. 2. Erythrocyte membrances. 3. Erythropoiesis--Molecular aspects. 4. Membrane proteins. I. Harris, James R. II. Series.
[DNLM: 1. Biochemistry. 2. Erythrocytes--physiology. WH 150 E734]
QP96.E783 1990
612.1'11--dc20
DNLM/DLC
for Library of Congress 90-7254

A Division of Plenum Publishing Corporation
233 Spring Street, New York, N.Y. 10013

Printed in the United States of America

Contributors

Henry R. V. Arnstein Division of Biomolecular Sciences, Department of Biochemistry, King's College London, London WC2R 2LS, England

Grzegorz Bartosz Laboratory of Biophysics of Development and Aging, Department of Biophysics, University of Lodz, 90-237 Lodz, Poland

Sophia A. Bonanou-Tzedaki Division of Biomolecular Sciences, Department of Biochemistry, King's College London, London WC2R 2LS, England

Augusta Brovelli Department of Biochemistry, Faculty of Science, University of Pavia, 27100 Pavia, Italy

Z. Ioav Cabantchik Department of Biological Chemistry, Institute of Life Sciences, Hebrew University, Jerusalem, Israel 91904; and Laboratory of Cell Biology and Genetics, National Institute of Diabetes and Kidney Diseases, National Institutes of Health, Betheseda, Maryland 20892, USA

Jean-Pierre Cartron Unité INSERM U76, Institut National de Transfusion Sanguine, 75015 Paris, France

Yves Colin Unité INSERM U76, Institut National de Transfusion Sanguine, 75015 Paris, France

Bernhard Deuticke Department of Physiology, Medical Faculty, Technical University, D-5100 Aachen, Federal Republic of Germany

Ofer Eidelman Department of Biological Chemistry, Institute of Life Sciences, Hebrew University, Jerusalem, Israel 91904

Klas Forsbeck Research and Development, Kabi, S-112 87 Stockholm, Sweden

Minoru Fukuda La Jolla Cancer Research Foundation, La Jolla, California 92037, USA

Reinhard Grebe Department of Physiology, Medical Faculty, Technical University, D-5100 Aachen, Federal Republic of Germany

Cees W. M. Haest Department of Physiology, Medical Faculty, Technical University, D-5100 Aachen, Federal Republic of Germany

J. Robin Harris North East Thames Regional Transfusion Centre, Brentwood, Essex CM15 8DP, England

Marie-Jeanne Heynen Department of Hematology, University Hospital, B-3000 Leuven, Belgium

Shinichi Kudo La Jolla Cancer Research Foundation, La Jolla, California 92037, USA

Hans U. Lutz Laboratory for Biochemistry, Swiss Federal Institute of Technology, ETH-Zentrum, CH-8092 Zurich, Switzerland

Makoto Nakao Department of Biochemistry, Tokyo Medical and Dental University School of Medicine, Tokyo, Japan

Greg B. Ralston Department of Biochemistry, University of Sydney, Sydney, New South Wales 2006, Australia

Samuel M. Rapoport Institute of Biochemistry, Humboldt University of Berlin, School of Medicine (Charité), DDR-1040 Berlin, German Democratic Republic

Catherine Rice-Evans Department of Biochemistry and Chemistry, Royal Free Hospital School of Medicine, London NW3 2PF, England

Stephen T. Sawyer Division of Hematology, Department of Medicine, Vanderbilt University School of Medicine, Nashville, Tennessee 37232, USA

Tankred Schewe Institute of Biochemistry, Humboldt University of Berlin, School of Medicine (Charité), DDR-1040 Berlin, German Democratic Republic

Marilyn J. Telen Division of Hematology/Oncology, Department of Medicine, and Immunohematology Laboratory, Transfusion Service, Duke University Medical Center, Durham, North Carolina 27710, USA

Bernd-Joachim Thiele Institute of Biochemistry, Humboldt University of Berlin, School of Medicine (Charité), DDR-1040 Berlin, German Democratic Republic

Preface

This series of books, devoted to aspects of blood cell biochemistry, development, immunology, and ultrastructure, has evolved and separated from the long-established Plenum series *Subcellular Biochemistry*. It is the intention of these volumes to draw together related areas of investigation and to provide, in the fullness of time, complete coverage of this rapidly advancing important biomedical discipline. Both fundamental and medically applied topics, dealing with normal and pathological cells, will be included.

This, the first volume of the series, contains a diverse collection of chapters, all of which relate to erythroid cells. The range of material included is extremely broad and the authors have used contrasting technical approaches, both within their personal experimental studies and within their manuscripts. This has led to the production of a very interesting compilation, which does, nevertheless, possess a strong overall thematic unity. As with all edited volumes, some topics of importance and interest are not included. This may be because of oversight on my part, as editor, or because the authors originally selected failed to submit their manuscript by the agreed-upon submission date. For these omissions I take full responsibility and trust that at least some of the topics omitted, for instance membrane cation transport systems, will be covered within a future volume of the series.

This book commences with two chapters of a developmental nature. The first, by Marie-Jeanne Heynen, deals with ultrastructural aspects of erythroid development, and the second, by Marilyn J. Telen, with the expression of erythroid membrane protein antigens. These are followed by two chapters that relate to the life-span and removal from the circulation of the erythrocyte. Grzegorz Bartosz has written on *in vivo* aging of the erythrocyte with emphasis upon membrane changes, and Hans U. Lutz has contributed a chapter on erythrocyte clearance that deals in depth with current ideas on the role of immunological aspects in erythrocyte removal. Then follows a group of seven chapters with a strong biochemical content. Sophia A. Bonanou-Tzedaki and Henry R. V. Arnstein have written on macromolecular synthesis and degradation during terminal erythroid cell development, and Samuel M. Rapoport and his colleagues discuss the maturational breakdown of mitochondria and other organelles in reticulocytes. Two chapters then relate to the red blood cell cytoskeleton. The first, by Makoto Nakao, discusses from a broad biochemical and ultrastructural stance the function and structure of the cytoskeleton, and

Greg B. Ralston then deals more specifically with spectrin–actin interactions. My own chapter on the biochemistry and transmission electron optical imaging of some oligomeric proteins and enzymes from erythroid cells emphasizes the strength of this technical combination for structural studies on high-molecular-weight proteins. The molecular genetics and biochemistry of the glycophorins are presented by Jean-Pierre Cartron and his colleagues; then follows a detailed discussion from Z. Ioav Cabantchik of the red blood cell anion transport system. The next two chapters, on erythroid receptors, strongly complement one another. Stephen T. Sawyer writes on the distribution and structure of the erythropoietin receptors and their role in receptor-mediated endocytosis, and Klas Forsbeck deals with the transferrin receptor and iron accumulation in erythroid cells. The final group of three chapters deals with pathological and drug-induced changes of the erythrocyte. Iron-mediated oxidative stress in erythrocytes is considered from a strongly biochemical stance by Catherine Rice-Evans, and her chapter is followed by the contribution of Augusta Brovelli on erythrocyte-mediated damage in hemolytic anemias. A detailed consideration of the action of drugs on the erythrocyte membrane from Bernhard Deuticke and his colleagues completes this exciting volume.

Having studied erythrocyte membrane proteins since the commencement of my research career in 1965, it has been a pleasant task to compile and edit this volume, which draws on so many interesting aspects of current studies on developing and mature erythroid cells. It is my hope that this book may be of widespread interest and use to those studying or wishing to know about erythroid cells.

J. R. Harris

Brentwood

Contents

Chapter 3
Erythrocyte Membrane Changes during Aging *in Vivo*
Grzegorz Bartosz

Chapter 9

Some High-Molecular-Weight Oligomeric Proteins and Enzymes of Reticulocytes and Erythrocytes

J. Robin Harris

Chapter 10
Molecular Genetics of Human Erythrocyte Sialoglycoproteins: Glycophorins A, B, C, and D
Jean-Pierre Cartron, Yves Colin, Shinichi Kudo, and Minoru Fukuda

Chapter 11
The Anion Transport System of Red Blood Cell Membranes
Z. Ioav Cabantchik

Chapter 12
Receptors for Erythropoietin: Distribution, Structure, and Role in Receptor-Mediated Endocytosis in Erythroid Cells
Stephen T. Sawyer

Chapter 1

Ultrastructural Changes during Erythroid Development

Marie-Jeanne Heynen

1. INTRODUCTION

The differentiation and maturation of erythroid cells are associated with biochemical (Koury *et al.*, 1987; Harrison, 1984), antigenic (Loken *et al.*, 1987), and morphological events (Castoldi and Beutler, 1988; Bessis *et al.*, 1986; Breton-Gorius and Reyes, 1976; Bessis, 1973; Tanaka and Goodman, 1972).

The proliferation of the erythrocyte precursors is realized predominantly in the morphologically unrecognizable erythroid compartment. Under the influence of erythropoietin, erythropoietin-sensitive progenitor cells differentiate into proerythroblasts (Quesenberry, 1986). The proerythroblast has a limited proliferation capability and undergoes marked morphological changes, which are associated with the synthesis of hemoglobin and other characteristic erythrocyte proteins (Harrison, 1984). In the human erythron model (Bessis *et al.*, 1986), one proerythroblast is assumed to yield 8 to 16 polychromatic erythroblasts by three to four succeeding cell divisions occurring within 70–72 hr. This has been confirmed recently by following the progenies of single proerythroblasts using cinematography (Furukawa *et al.*, 1987).

The ultrastructure of erythroid cells is quite similar in man (Castoldi and Beutler, 1988; Bessie *et al.*, 1986; Breton-Gorius and Reyes, 1976; Smetana *et al.*, 1975; Bessis, 1973; Tanaka and Goodman, 1972; Kent *et al.*, 1966) and in the mammalian experimental animals: dog (Simpson and Kling, 1967, 1968), mouse (Skutelsky and Danon, 1967; Orlic *et al.*, 1965), rat (Heynen *et al.*, 1985; Heynen and Verwilghen, 1982; Le Carpentier and Prenant, 1975; Ben-Ishay and Yoffey, 1972), guinea pig (Rosse and Trotter, 1974a,b), and rabbit (Gasko and Danon, 1972; Rifkind *et al.*, 1964; Grasso *et al.*, 1962).

Marie-Jeanne Heynen Department of Hematology, University Hospital, B-3000 Leuven, Belgium.

The ultrastructural changes during erythroid development in mammals will be described, particularly their quantitative aspects. In addition, the ultrastructural characteristics of defective maturation under some dyserythropoietic conditions (sideroblastic anemia, megaloblastic anemia, thalassemia) and of stress erythropoiesis will be outlined briefly.

2. STEM CELLS

The morphological and ultrastructural identification of the multipotent hemopoietic stem cell is complicated by their small number in the bone marrow. Several possible stem cell candidates have been proposed over the years by various authors (Wickramasinghe, 1975). The most convincing results came from the work of van Bekkum *et al.* (1971). These workers were able to prepare high concentrations of mouse and rat bone marrow pluripotent stem cells, as could be estimated from their ability to form spleen colonies (CFU-S). In these stem-cell-enriched preparations they found a morphological cell entity, which they considered to be the candidate stem cell. Although these candidate stem cells have roughly the same size as small lymphocytes, they could readily be distinguished from the latter by the presence of a nucleus with two large nucleoli and a finely dispersed chromatin pattern with only a few aggregates of condensed chromatin adjacent to the nuclear membrane. The cytoplasm was scant and devoid of most organelles. Only several small mitochondria, a few small vesicles, few or no polyribosomes, and numerous monoribosomes could be observed. A similar population of cells was identified in human and monkey bone marrow cultures by Dicke *et al.* (1973).

Attempts have been made to identify the erythroid progenitor cell (CFU-E). This was performed in experiments with rats (Ben-Ishay and Yoffey, 1972) and mice (Nijhof and Wierenga, 1983), whose erythropoiesis had been stimulated after prior inhibition. In the bone marrow of the rats, small lymphocyte-like cells and "transitional" cells are found in new erythroblastic islands before the appearance of proerythroblasts (Ben-Ishay and Yoffey, 1972). These "transitional cells" are thought to be the immediate precursors of proerythroblasts, since they were labelled with ^{55}Fe and synthesized hemoglobin in detectable amounts (Rosse and Trotter, 1974b). The immediate precursors of the proerythroblasts from mouse spleens (Nijhof and Wierenga, 1983) are cells that are easily distinguishable from the pluripotent stem cell described by van Bekkum *et al.* (1971) by the indented nucleus and the high density of ribosomes and mitochondria. They are indistinguishable from proerythroblasts, except for the high concentration of mitochondria in one area of the cytoplasm.

It is possible that the morphological heterogeneity within the stem cell compartments reflects differences in the cell cycle status rather than differences between cells of different potentials (van Bekkum *et al.*, 1979).

3. ERYTHROBLASTIC ISLANDS

The proliferation and maturation of erythroblasts occur in erythroblastic islands (Figure 1). They are an anatomical unit, consisting of one centrally located macrophage,

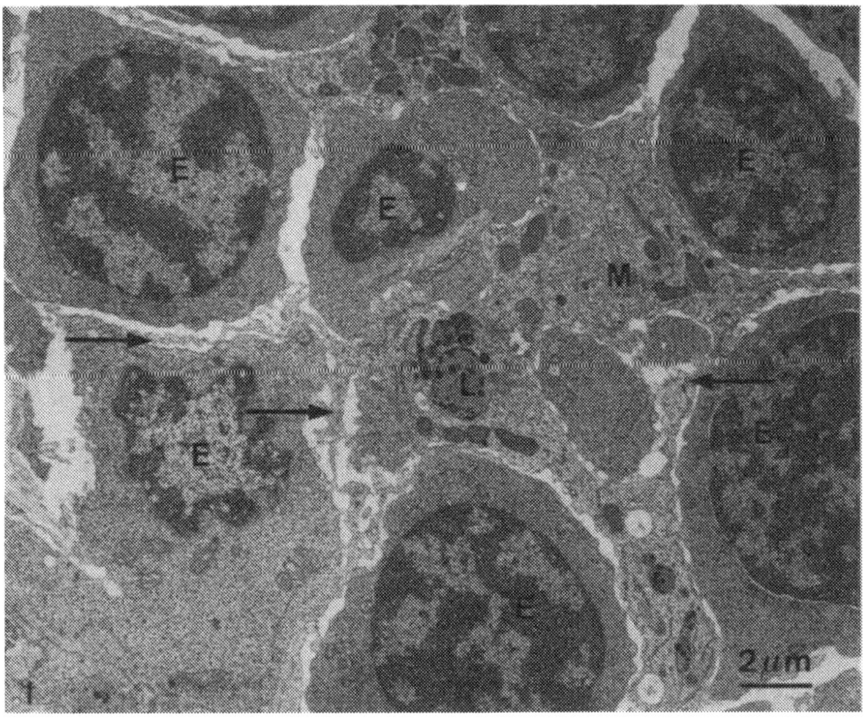

FIGURE 1. Erythroblastic island from the bone marrow of a normal Wistar rat. Several late erythroblasts (E) are localized in close contact with the cell body and cytoplasmic extensions (arrows) of the macrophage (M). Secondary lysosomes (L) can be observed in the macrophage cytoplasm.

surrounded by erythroblasts of all stages of maturation, which remain in contact with the cell body of the macrophage or its long cytoplasmic extensions throughout development (Bessis, 1973). In isolated erythroblastic islands it can be observed more readily that the erythroblasts are arranged in the form of concentric rings, with the outer cells being more mature than the inner (Le Carpentier and Prenant, 1973).

The functional significance of the relationship between macrophage and developing erythrocyte precursors has been the subject of many discussions (Bessis, 1973). The frequent finding of pyknotic nuclei and degraded nuclei inside the cytoplasm of the central macrophage indicates its role in the elimination of extruded erythroblast nuclei. Nevertheless, it has been shown that the presence of the central macrophage is not a prerequisite for the nuclear expulsion to occur (Simpson and Kling, 1967, Bessis, 1973). The central macrophage also has a role in the phagocytosis and degradation of defective erythroblasts and old erythrocytes (Ghadially, 1979). Since the description of rhopheocytosis, the role of the central macrophage in the transfer of Fe and ferritin to maturing red cell precursors has been considered (Bessis, 1973; Bessis and Breton-Gorius, 1962), but direct proof is still lacking.

From studies of return of erythropoiesis in erythropoietically depressed rebound marrow (Ben-Ishay and Yoffey, 1972), it appears that the central macrophage plays an essential role in the re-formation of the erythroblastic island. This raised the question whether the central macrophage may transmit growth factors to the developing erythroid cells. Recently it has been shown that macrophages derived from mouse bone marrow are functionally capable of producing and secreting hemopoietic growth factors (Rich, 1986) and expressing the erythropoietin gene (Rich *et al.*, 1988).

4. ERYTHROBLASTS AND RETICULOCYTES

In the red cell line, the proerythroblast is the earliest precursor that can be recognized on morphological grounds. With the electron microscope they are not easy to identify, for they are similar to other large undifferentiated cells (Figure 2a). Their association with the macrophage or with its projections and the rhopheocytosis of ferritin (Figure 2b) may be helpful for their identification. Early erythroid cells can also be identified by more elaborate methods, which are an indication of hemoglobin synthesis (Rosse and Trotter, 1974a). With the advent of monoclonal antibodies, antigens on the plasma membrane of erythroid cells (Loken *et al.*, 1987) may be used to identify early erythroid cells.

It is possible to recognize four classes of erythroblasts (Wickramasinghe, 1975; Bessis, 1973) in the light microscope. This morphological classification depends on subjective criteria, including nuclear and cellular size, chromatin texture and density, nucleolar morphology, number and size, and cytoplasmic staining.

By electron microscopy, thin sections of erythroid cells can be identified, using the criteria of light microscopy, as proerythroblasts (Figures 2a, 3a), basophilic erythroblasts (Figure 4a), polychromatophilic erythroblasts (Figure 5, 6), orthochromatic erythroblasts (Figure 7), and reticulocytes (Figure 8a). This procedure has been used in most investigations on normal and pathological erythropoiesis (Castoldi and Beutler, 1988; Bessis *et al.*, 1986; Fresco, 1981; Breton-Gorius and Reyes, 1976; Bessis, 1973).

When erythrocyte precursors have to be classified in subsequent maturation stages for a quantitative analysis of their cell organelles, a division, according to their chromatin pattern (Rosse and Trotter, 1974a), into early (Figures 2a, 3b), intermediate (Figures 4a, 5), and late erythroblasts (Figures 6, 7) is more suitable (Tablin and Weiss, 1985; Heynen and Verwilghen, 1982).

Mammalian reticulocytes (Figure 8a) do not contain a nucleus, but they still have mitochondria, endoplasmic reticulum, centrioles, remnants of the Golgi apparatus, and polyribosomes. These polyribosomes are actively synthesizing hemoglobin, since one-third of the erythrocyte hemoglobin is formed in the reticulocyte.

4.1. Maturation of the Nucleus

During erythroid maturation, there is a marked condensation of the chromatin of the interphase nucleus, which involves a packing of chromatin fibrils into heterochromatin. Parallel to these morphological changes, the nucleus undergoes a decrease of RNA and DNA synthesis (Yataganas *et al.*, 1970). In erythrocyte precursors, which can be identified only by the presence of ferritin (Figure 2) or by the autoradiographic detection of ^{55}Fe incorporation and the pseudoperoxidase reaction of hemoglobin (Rosse and Trotter,

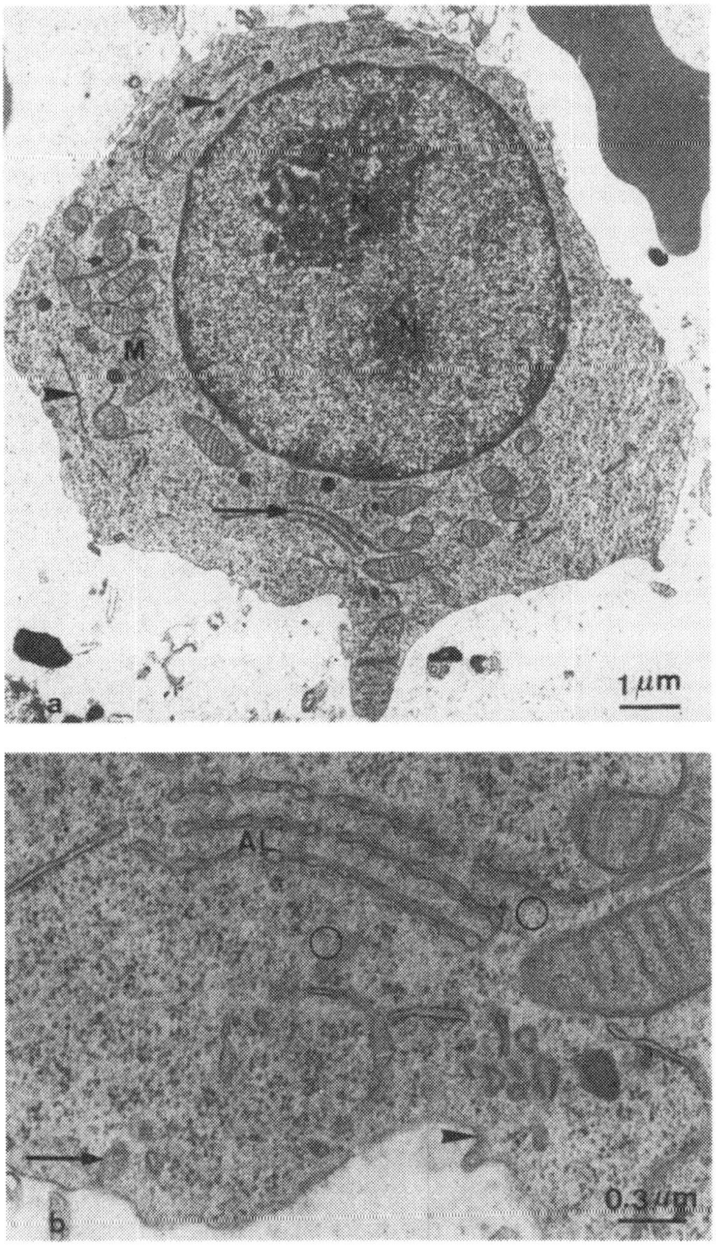

FIGURE 2. A proerythroblast from the bone marrow of a normal human control. (a) The nuclear chromatin is finely dispersed, and small aggregates of heterochromatin are localized along the nuclear membrane and the nucleoli (N). The cytoplasm contains numerous polyribosomes, large mitochondrial profiles (M), rigid strands of RER (arrowheads), and annulate lamellae (arrow). (b) Part of (a) showing the annulate lamellae (AL) and ferritin attached to the plasma membrane (arrowhead), inside a rhopheocytotic vesicle (arrow), and free in the cytoplasm (circle).

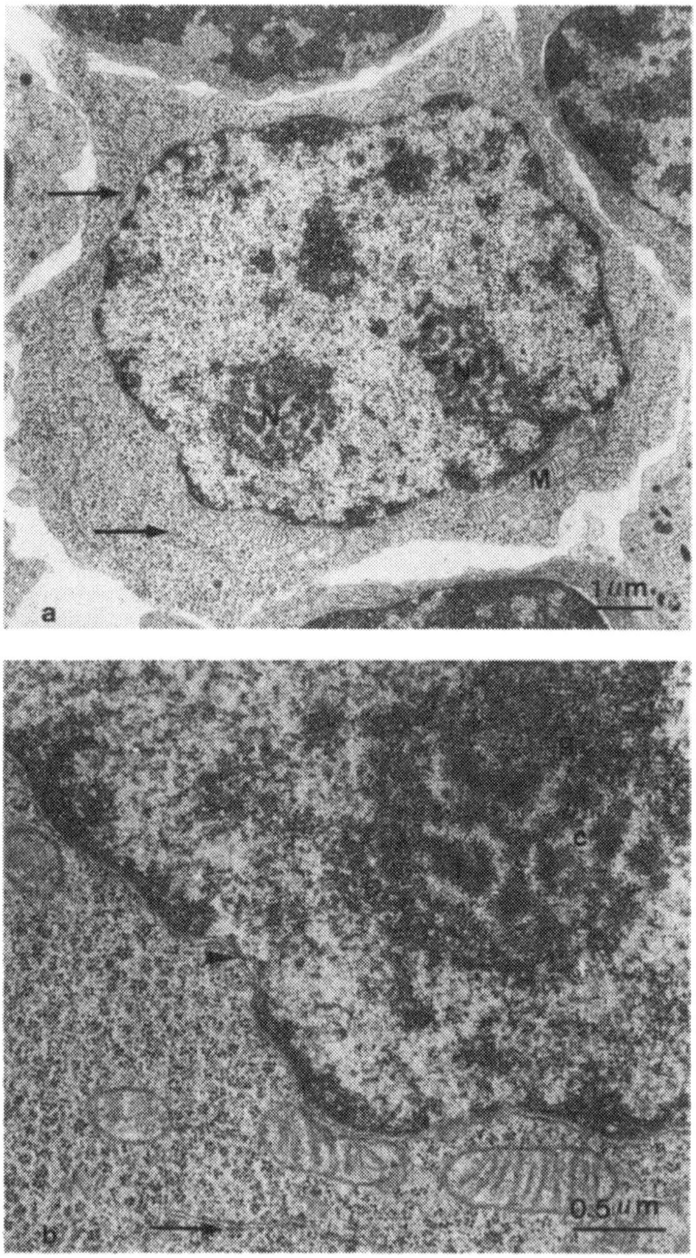

FIGURE 3. A proerythroblast from the bone marrow of a normal Wistar rat. (a) The nucleus is predominantly euchromatic. Aggregates of heterochromatin are associated with the two nucleoli (N) and the nuclear membrane. The cytoplasm contains numerous polyribosomes, large mitochrondrial profiles (M), and long strands of RER (arrows). (b) Part of (a) showing a nucleolus, composed of granular (g) and fibrillar (f) components and fibrillar centers (c). Heterochromatin is attached to the nucleolus and to the nuclear membrane with interruptions at the nuclear pore complexes (arrowhead). The cytoplasm contains a high density of polyribosomes. Ribosomes attached to the RER membranes occur in irregularly distributed patches (arrow).

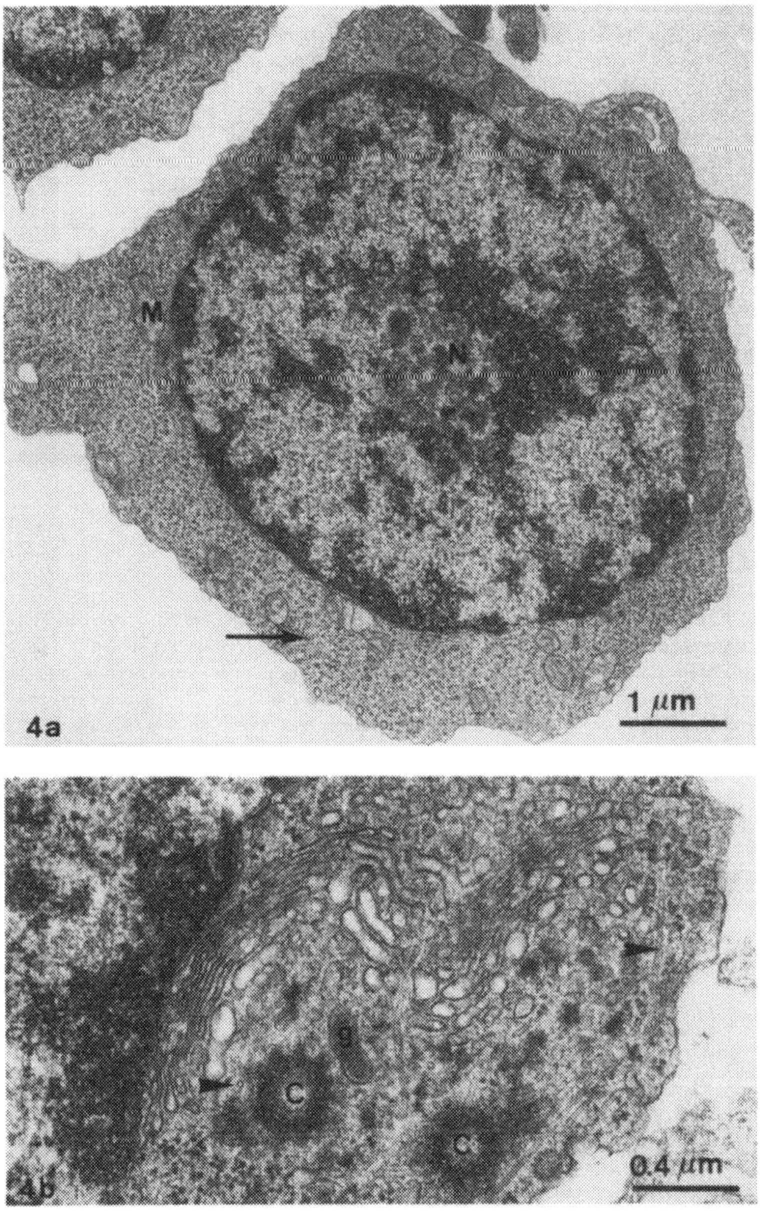

FIGURE 4. Basophilic erythroblasts from the bone marrow of a normal Wistar rat. (a) The clumping of the nuclear chromatin is increased relative to that of the proerythroblast shown in Figure 3. An important fraction of the heterochromatin occurs free from the nuclear membrane and the large, complex nucleolus (N). The cytoplasm contains a high density of polyribosomes, several mitochrondrial profiles (M), and a few strands of RER (arrow). (b) Part of a basophilic erythroblast showing a Golgi area with two Golgi stacks, each composed of several cisternae. The *trans* face of the Golgi stack is oriented toward the centrioles (C). Numerous smooth and fuzzy-coated vesicles, microtubules (arrowheads), and a granule (g) can be seen.

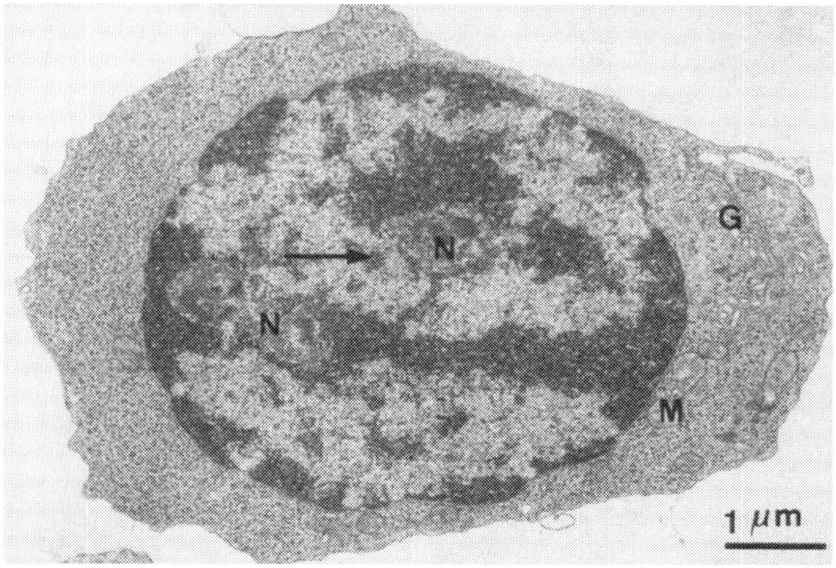

FIGURE 5. A polychromatophilic erythroblast from the bone marrow of a normal Wistar rat. The clumping of the nuclear chromatin is increased relative to that of the basophilic erythroblast shown in Figure 4a. Small aggregates of chromatin (arrow) still remain distributed in the euchromatin area. According to the chromatin pattern, this erythroblast will be classified as an "intermediate erythroblast." Two nucleoli (N) are present. The cytoplasm contains a high density of polyribosomes, several mitochondrial profiles (M), and a Golgi zone (G).

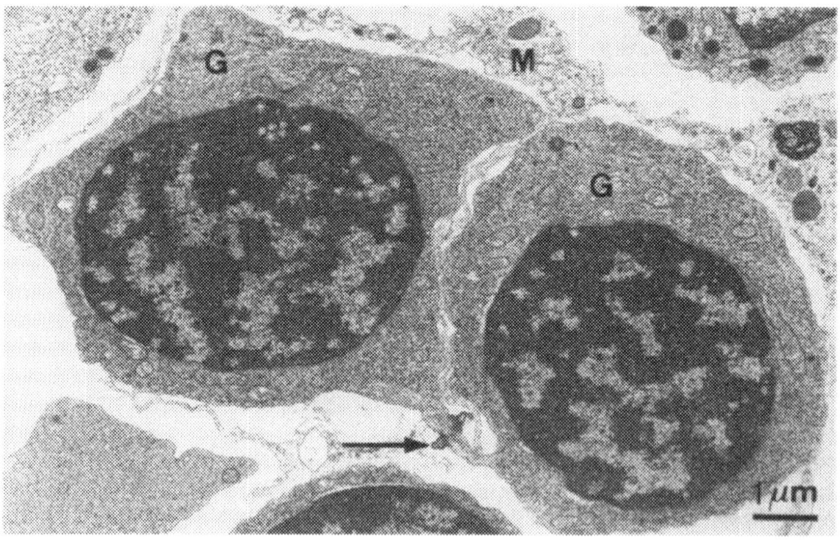

FIGURE 6. Two late polychromatophilic erythroblasts from the bone marrow of a normal Wistar rat. The two cells are connected by a cell bridge, composed of microtubules and a midbody (arrow). The nuclear chromatin is condensed into large heterochromatin clumps. According to the chromatin pattern, these cells are classified as "late erythroblasts" and it is assumed that they terminated their last cell division. The cytoplasm contains a high density of polyribosomes, several mitochondria, and a Golgi apparatus (G). Due to the accumulation of hemoglobin, the electron density of the cytoplasm is higher in the erythroblasts than in the macrophage (M).

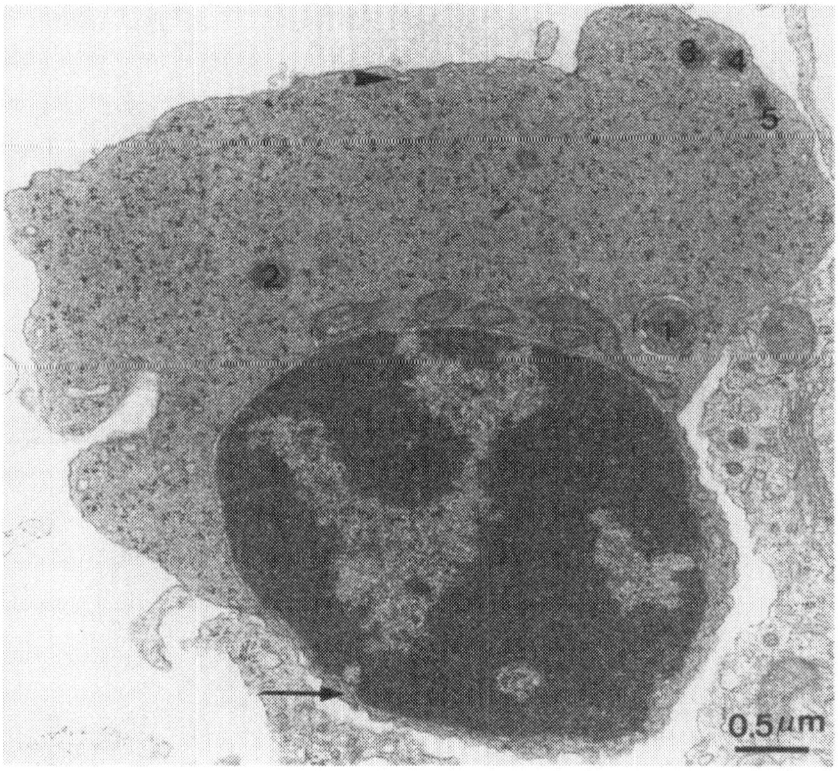

FIGURE 7. An orthochromatic erythroblast from the bone marrow of a normal Wistar rat. The nucleus is pyknotic, eccentric, and partly surrounded by a narrow ruffled rim of cytoplasm without organelles (arrow). Mitochondria are accumulated near the nucleus. Relative to previous maturation stages, the density and clustering of the ribosomes have decreased. The electron density of the cytoplasm is increased due to the accumulation of hemoglobin. Five autophagosomes are visible: (1) containing cytoplasm and a mitochondrial profile, (2–5) containing degraded material. Short strands of RER (arrowhead) are present.

1974a), the nuclear chromatin is finely dispersed with few peripheral aggregates of heterochromatin. In the first morphologically recognizable erythroblasts (Figure 3), the heterochromatin is increased relative to the earlier erythrocyte precursors, but it remains predominantly associated with the nucleolus and the nuclear envelope with interruptions at the nuclear pore complexes. During further maturation, the condensation of the nuclear chromatin increases and an increasing fraction of the heterochromatin occurs free from the nuclear envelope and the nucleolus (Figures 4a, 5). Finally, the chromatin is condensed in that it forms a coherent dense mass (Figure 7); these nuclei are described as pyknotic. The condensation of the nuclear chromatin, which is associated with a decrease of the nuclear protein (Spivak, 1975), results in a reduction of the nuclear size, although the nuclear DNA content remains diploid in the nondividing erythroblasts (Yataganas *et al.*, 1970).

Together with the chromatin condensation, nucleolar changes occur. The number and size of the nucleoli reach a maximum in the human basophilic erythroblasts and then gradually decrease (Smetana *et al.*, 1975). In the early erythroblast maturation stages with

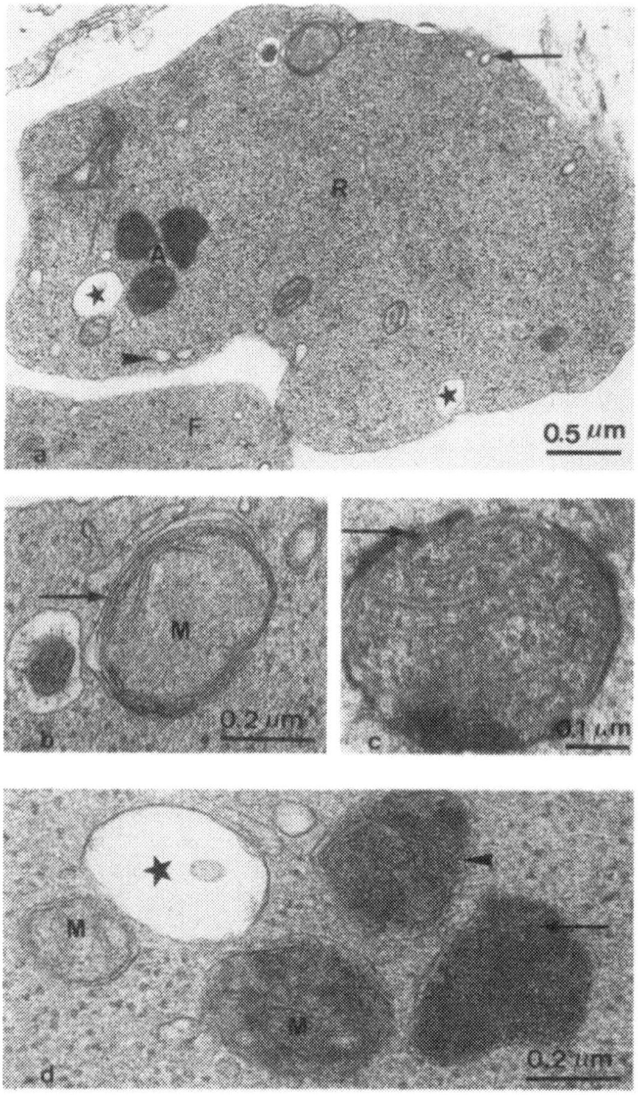

FIGURE 8. Reticulocytes from the bone marrow of a normal Wistar rat. (a) The density of ribosomes is higher in the reticulocyte (R) than in the reticulocyte fragment (F). Several mitochondria and three autophagosomes (A) are present. Numerous rhopheocytotic vesicles and fuzzy-coated vesicles (arrow) and large clear vacuoles (stars) can be seen as well as short strands of RER (arrowhead). (b) Part of reticulocyte R from (a) showing a mitochondrion (M) in close association with a strand of RER (arrow). (c) Part of a reticulocyte showing acid phosphatase reaction product (arrow) in the isolating cisterna around a mitochondrion. (d) Part of reticulocyte R from (a) showing an intact mitochondrion (M), a clear vacuole (star), and three autophagosomes containing a mitochondrion (M), membranous material, amorphous electron-dense material (arrow), and ferritinlike granules (arrowheads).

a large number of ribosomes in the cytoplasm, the fibrillar and granular nucleolar components form, together with the fibrillar centers, a complex network, designed as nucleolonema (Figures 2a, 3, 4a). These complex nucleoli are surrounded by a discontinuous layer of perinucleolar chromatin. In the advanced erythroblast maturation stages, which contain a reduced number of ribosomes in their cytoplasm, micronucleoli are present (Smetana *et al.*, 1975). In most micronucleoli the nucleolar components are segregated in clearly separated areas. These micronucleoli are thought to produce little if any preribosomal RNA. The decreasing number of ribosomes in erythroblasts containing micronucleoli supports such a conclusion.

After splenectomy, in hyposplenism, hemolytic and megaloblastic anemias, nuclear chromatin bound by a typical double membrane of the nuclear envelope can be found in reticulocytes and erythrocytes. These "Howell Jolly bodies" appear to arise from nuclear fragmentation (karyorrhexis) or incomplete expulsion of the nucleus (Bessis *et al.*, 1986; Fresco, 1981).

4.2. Maturation of the Cytoplasm

Although there is an increase of the cytoplasmic volume after each cell division, the daughter cells do not reach the dry mass (Yataganas *et al.*, 1970) nor the volume of the precursor cells (Koury *et al.*, 1988a). Hemoglobin accumulates in the cytoplasm of erythrocyte precursors in the course of maturation. This can be measured by means of scanning microspectrophotometry of individual cells (Yataganas *et al.*, 1970). A higher concentration of hemoglobin in the cytoplasm results in an increased electron density of the cytoplasm in conventional electron microscopic preparations (Figures 6, 7). The increased concentration of hemoglobin in the cytoplasm is even more apparent after peroxidase cytochemical staining (Figure 9a), which makes use of the pseudoperoxidase reaction of hemoglobin (Breton-Gorius, 1970). The electron density increases proportionally to the amount of hemoglobin and can be measured on electron microscopic negatives (Rosse and Trotter, 1974a).

While hemoglobin is accumulating in the cytoplasm, the density of the cell organelles (ribosomes, rough endoplasmic reticulum, Golgi apparatus, mitochondria) in the cytoplasm is decreasing. This is already visible on electron micrographs (Figures 7, 8a). Stereologic analysis of erythroblasts and reticulocytes of rat bone marrow (Figure 10) (Heynen and Verwilghen, 1982) gives a more accurate picture and shows that there is a significant reduction of the number of ribosomes and volume of mitochondria per unit volume cytoplasm, between the intermediate and late erythroblasts. This maturation step coincides largely with the last cell division, after which there is a shift from predominantly nonheme protein to mainly hemoglobin synthesis (Denton *et al.*, 1975). The decrease of the density of ribosomes and mitochondria per unit volume cytoplasm continues during the maturation of late erythroblasts, in which a renewal of ribosomes (Smetana *et al.*, 1975) and mitochondria (Nijhof and Wierenga, 1984) does not take place, while hemoglobin is still accumulating. After nuclear extrusion the decrease of the number of ribosomes and the volume of mitochondria per unit volume cytoplasm is accelerated by the maturational breakdown of these structures (Rapoport and Schewe, 1986; Heynen *et al.*, 1985) (see also Sections 4.3 and 4.4).

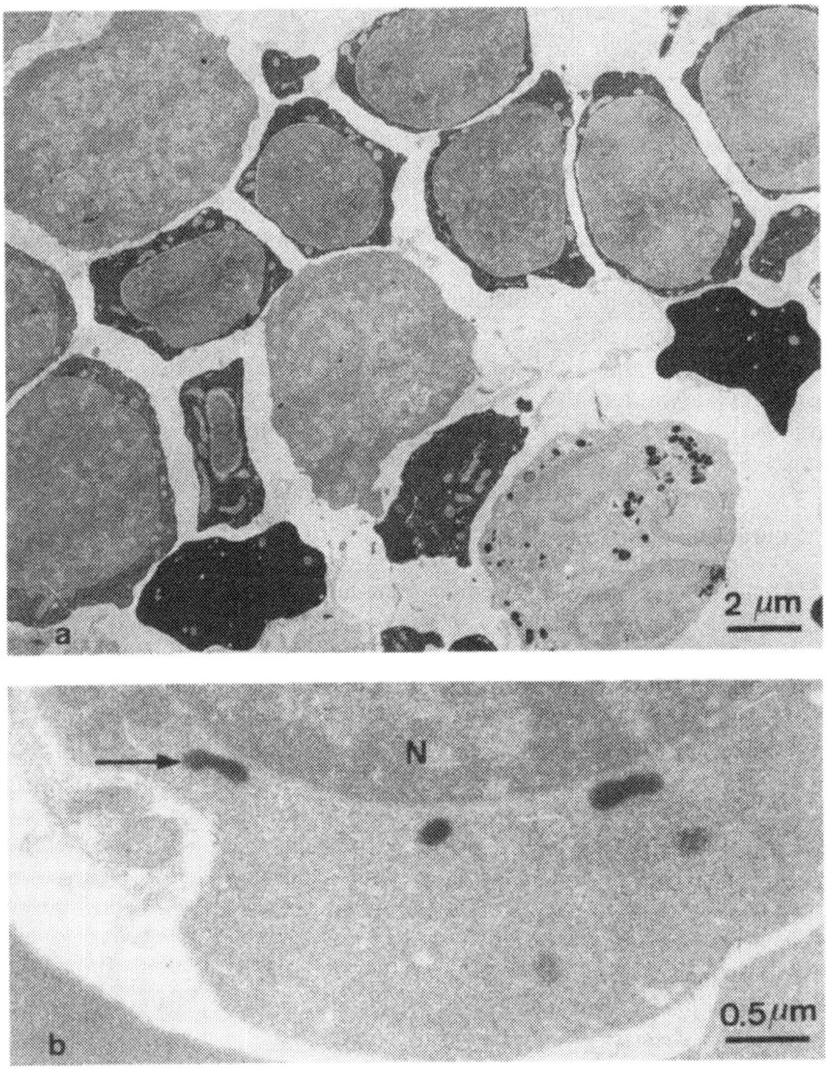

FIGURE 9. Pseudoperoxidase and catalase cytochemistry. (a) Part of the bone marrow of a normal Wistar rat, incubated in a medium for the demonstration of peroxidase activity. The intensity of the pseudoperoxidase reaction is related to the concentration of hemoglobin in the different maturation stages. (b) Part of a polychromatophilic erythroblast from a normal human control, incubated in a medium for the demonstration of catalase. Three catalase-positive microperoxisomes (arrow) can be seen in close proximity to the nucleus (N).

4.3. Mitochondria

Several mitochondrial profiles can be observed in the erythroblast and reticulocyte sections. In the early and intermediate erythroblasts, they do not have a preferential localization, but in the late erythroblasts with an eccentric nucleus, mitochondria are accumulated near the nucleus (Figure 7). In this area, vesicles and cisternae are concentrated as well.

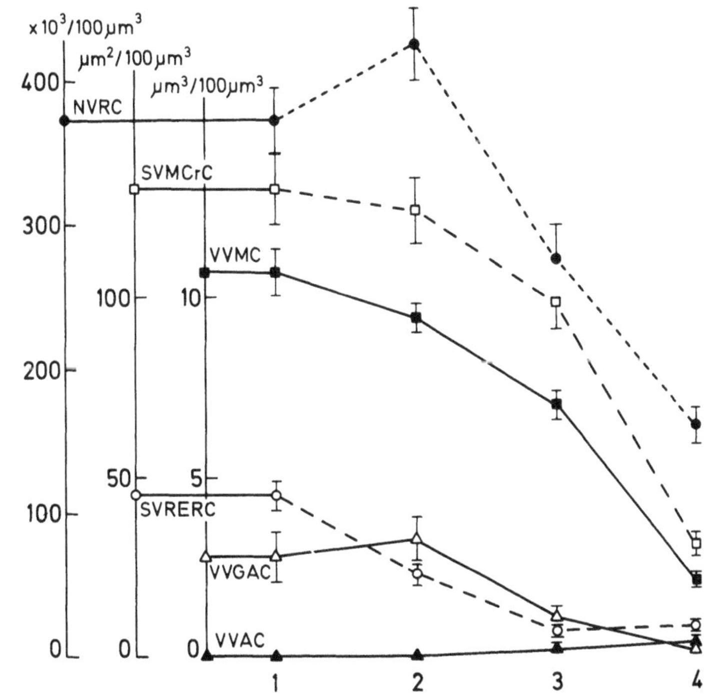

FIGURE 10. Stereologic parameters of early (1), intermediate (2), late (3) erythroblasts and reticulocytes (4) from the bone marrow of normal Wistar rats (data from Heynen and Verwilghen, 1982). Values are expressed as mean ± S.E.M. per 100 µm cytoplasm. NVRC, number of ribosomes; SVMCrC, surface area of cristae membranes; VVMC, volume of mitochondria; SVRERC, surface area of RER membranes; VVGAC, volume of Golgi area; VVAC, volume of autophagosomes.

The size of the mitochondrial profiles in sections of erythrocyte precursors decreases during the course of maturation.

The surface area of mitochondrial cristae per unit volume mitochondria does, however, not change significantly. The value of this mitochondrial parameter is slightly higher in late erythroblasts and reticulocytes than in proliferating erythroblasts (Heynen and Verwilghen, 1982). This may be interpreted as a sign of intactness of the mitochondria in the former, for degradation of mitochondria is associated with swelling and destruction of the cristae. Indeed, in erythroblasts and reticulocytes from well-prepared bone marrow samples of humans and rats, swollen mitochondria or mitochondria with other signs of degradation are not observed. In late erythroblasts with an eccentric nucleus (Figure 7) and in bone marrow reticulocytes (Figure 8) of normal rats, a selective autophagocytosis of mitochondria has been observed (Heynen *et al.*, 1985).

The sequestration of mitochondria inside a double membrane is brought about by the coalescence of small vesicles and cisternae of smooth and rough endoplasmic reticulum (Figure 8b). This has been observed in circulating reticulocytes of anemic dogs. Under these experimental conditions, fusion of the inclusions with the plasma membrane and subsequent extrusion of fairly intact mitochondria have been observed (Simpson and Kling, 1968). In normal rats, however, primary lysosomes, carrying acid hydrolases, fuse with the autophagic vacuoles and mitochondria are degraded inside the secondary lysosomes (Figure 8c,d).

It has been computed that autophagy can account for the elimination of mitochondria during maturation of rat reticulocytes in the bone marrow (Heynen *et al.*, 1985). Circulating reticulocytes do, however, still contain a few mitochondria. A fraction of the mitochondria in peripheral blood reticulocytes appears to be swollen, with destroyed cristae (Gasko and Danon, 1972; Schewe *et al.*, 1977). The fraction of free but degraded mitochondria increases with reticulocyte age and is correlated with decreasing aerobic metabolism (Gasko and Danon, 1972). A specific lipoxygenase has been described, which is responsible for the maturational breakdown of free reticulocyte mitochondria (Rapoport and Schewe, 1986; Schewe *et al.*, 1977).

It may be concluded that the maturational elimination of mitochondria in the rat is brought about by (1) autophagy in bone marrow erythroblasts and reticulocytes and (2) degradation in the cytosol in peripheral blood reticulocytes, but (3) expulsion of the mitochondria together with the nucleus has also been observed (Nijhoff and Wierenga (1984).

4.4. Ribosomes

The most abundant structures in erythroblasts and reticulocytes are ribosomes. They occur in clusters of polyribosomes. Electron microscopy of ribosomal fractions of lysed rabbit reticulocytes reveals mainly clusters of five ribosomes (Warner *et al.*, 1962). The polyribosomes in reticulocytes are thought to be predominantly involved in globin synthesis (Clissold *et al.*, 1977). It has been computed that the mRNA for a globin polypeptide chain is about the right length to hold together a five-unit polyribosome (Warner *et al.*, 1962). In ultrathin sections of erythrocyte precursors, however, ribosomes are observed singly or as clusters containing two to six ribosomes (Figure 3b) (Rifkind *et al.*, 1964). This is due to the fact that many of the polyribosomes are transected by the plane of the sectioning for the size of the ribosomal clusters is of the same order of magnitude as the section thickness. When one corrects the clustering of ribosomes for this sectioning effect (Perl, 1964), unclustered ribosomes account for less than 1% of the total ribosomal population in nucleated bone marrow erythroblasts of phenlhydrazine-treated rabbits and the peak increase of ribosomes is in clusters of four (Rifkind *et al.*, 1964). Uncorrected figures for the sectioning effect show that the clustering of ribosomes does not change before the nuclear extrusion in bone marrow erythroblasts of normal rats (Heynen and Verwilghen, 1982). During maturation of normal rat reticulocytes (Heynen and Verwilghen, 1982) and stress reticulocytes from rabbits *in vivo* and *in vitro* (Rifkind *et al.*, 1964), the number and size of polyribosomes decrease.

The reduction in number of ribosomes per unit volume cytoplasm has already started after the last cell division (Heynen and Verwilghen, 1982). This is thought to be due to the accumulation of hemoglobin in the cytoplasm without a concomitant production of new ribosomes (Smetana *et al.*, 1975). After the nuclear extrusion, the reduction of the number of ribosomes per unit volume cytoplasm continues (Heynen and Verwilghen, 1982) and the ribosome clustering decreases, as has already been discussed. It is thought that the degradation of ribosomal and soluble RNA is responsible for the reduction of ribosomes and ribosomes bound to mRNA after the nuclear extrusion.

"Basophilic stippling" seen in erythroblasts and reticulocytes of patients with thalassemia or lead intoxication is due to a spontaneous pathological aggregation of ribosomes. A similar aggregation of ribosomes, artifactually produced *in vitro* by vital stains, such as brilliant cresyl blue or new methylene blue, is responsible for the "reticulofilamentous substance" seen during the traditional method of counting reticulocytes (Jensen *et al.*, 1965).

4.5. Endoplasmic Reticulum

The endoplasmic reticulum is not prominent in erythrocyte precursors. It consists of cisternae and vesicles scattered as single elements throughout the cytoplasm of all maturation stages. It is more abundant in the proliferating erythroblasts (Figures 2, 3, 10), where it occurs mainly as long rapid strands of rough endoplasmic reticulum (RER). Patches of evenly spaced ribosomes are distributed in an asymmetrical and irregular way along the membranes (Figure 3b). In most erythropoietic cells the endoplasmic reticulum occurs in close proximity to mitochondria or to the plasma membrane (Grasso *et al.*, 1978).

After the last cell division, there is an important reduction of the amount of RER per unit volume cytoplasm (Heynen and Verwilghen, 1982). In the nonproliferating late erythroblasts and reticulocytes (Figures 7, 8a), the endoplasmic reticulum consists mainly of clear agranular cisternae and vesicles. Part of these elements originate from endocytosis of the plasma membrane, which brings about the remodeling of an irregularly shaped reticulocyte into a biconcave erythrocyte (Gasko and Danon, 1974). These vesicles may participate as well in the autophagocytosis of mitochondria, which starts together with the nuclear extrusion (Heynen *et al.*, 1985). It has been shown that the internalization of plasma membrane is followed by intracellular digestion or by exocytosis (Gasko and Danon, 1974). Extrusion of cytoplasmic vesicles in the form of "surface multivesicular structures" has been described in maturing rat reticulocytes (Seelig, 1972).

"Annulate lamellae," i.e., parallel, double membranes periodically joined by pores, with a structure similar to that of the nuclear pores, can be seen in proerythroblasts (Figure 2). These structures occur more frequently and in more advanced maturation stages in dyserythropoiesis (Verwilghen *et al.*, 1975).

4.6. Golgi Apparatus

The Golgi apparatus (Figure 4b) is well developed in erythroblasts. It consists of two stacks of five to six cisternae. A high concentration of microtubules, smooth, and "fuzzy-coated" vesicles is found in the Golgi zone.

The Golgi area, defined as the zone with the Golgi stacks, associated vesicles, and centrioles, occupies about 3% of the cytoplasmic volume in the proliferating erythroblasts of Wistar rats (Figure 10). After the last cell division, the Golgi area occupies only 1% of the late erythroblast cytoplasm and only 0.2% of the bone marrow reticulocyte cytoplasm (Heynen and Verwilghen, 1982).

Taking into account the role of the Golgi apparatus in cell biology (Rothman, 1985), one can assume that the Golgi apparatus of erythroid cells is involved in the modification and sorting, for specific destinations of membrane proteins and the lysosomal enzymes.

The acid phosphatase reaction product can be found in the Golgi cisternae, in small cytoplasmic vesicles (primary lysosomes), as well as in secondary lysosomes (Figure 8c).

4.7. Autophagosomes

Autophagosomes are membrane-enclosed bodies with a polymorphic content, mitochondria, cytoplasm, and degraded material (Figure 8d). Autophagosomes are found in circulating human reticulocytes and erythrocytes (Kent *et al.*, 1966). In normal rats, they are not found in proliferating erythroid cells, but after nuclear extrusion autophagosomes appear in the cytoplasm with a density of 0.4% (Heynen *et al.*, 1985). They are thought to be responsible for the elimination of mitochondria during reticulocyte maturation (see Section 4.3 and Chapter 6). Autophagosomes are more frequent under pathological conditions (Kent *et al.*, 1966).

When incubated for the cytochemical demonstration of acid phosphatase, the reaction product can be found in autophagic vacuoles of erythroid cells (Kent *et al.*, 1966; Heynen *et al.*, 1985). First, it appears in the cisternae of the isolating double membrane (Figure 8c). Subsequently, the inner membrane is degraded, leaving only the outer one. Thereafter, the acid phosphatase reaction product can be found over the content of the autophagic vacuole. The degradation proceeds until only membranous structures, amorphous electron-dense material, and ferritinlike particles (Figure 8d) are left. These particles, measuring approximately 5 to 6 nm, are considered to contain iron (Kent *et al.*, 1966). It is thought that the spleen plays a role in the elimination of autophagic vacuoles (Kent *et al.*, 1966).

4.8. Peroxisomes

When a cytochemical technique is used for the demonstration of catalase in erythroid cells (Breton-Gorius and Guichard, 1975), small catalase-positive granules are found in erythroblasts and reticulocytes. According to criteria defined by Novikoff and Novikoff (1973), they can be considered as microperoxisomes. Microperoxisomes are more frequent in the early erythroblasts and disappear in the reticulocytes (Breton-Gorius and Guichard, 1975). They are localized in close proximity to the nuclear membrane (Figure 9b).

5. EXTRUSION OF THE NUCLEUS

The extrusion of the nucleus is an exclusive characteristic of mammalian erythrocytes. The circulating red blood cells of the other vertebrates retain their pyknotic nuclei. The nuclear expulsion can be observed *in vitro* by means of phase-contrast cinematography, where it appears to be associated with violent movements and lasts from 10 to 30 min at 37°C (Bessis, 1973).

The onset of the enucleation is manifested morphologically by the nucleus leaving its central position in the cell to occupy an eccentric site close to a small area of the cell membrane. Progressively more of the nuclear membrane comes into close juxtaposition with the cell membrane (Figure 7). Skutelsky and Danon (1967) describe this process in mice as a penetration of the nucleus through a narrow passage into a pseudopodlike cytoplasmic protrusion, the nucleus being deformed in this process. The constriction passes over the entire nucleus, until the nucleus is completely surrounded by a narrow rim of cytoplasm, bordered by an undulating membrane. This constriction is also apparent in scanning electron micrographs of mouse erythroblasts (Repasky and Eckert, 1981). F-actin appears to be concentrated in this constriction (annular ring) between the extruding nucleus and the forming reticulocyte (Koury *et al.*, 1988b).

Koury *et al.* (1988b) argue that since in their experiments cytochalasin D blocked enucleation, the annular F-actin structure is essential for the extrusion of the nucleus. Inhibition of enucleation by microfilament-disrupting agents has also been observed under other experimental conditions (Repasky and Eckert, 1981). Chasis *et al.* (1988), however, report only a decreased motility of erythroblasts, arrested at the stage of nuclear extrusion, by cytochalasin B. The blockage of enucleation was realized in their experiment by the microtubule-disrupting agent, colchicine. Lemke and Linss (1984) and Koury *et al.* (1988b), however, did conclude from their experiments that microtubules are not essential for enucleation.

Although there are conflicting results concerning the role of cytoskeletal structures in enucleation, it appears that the annular ring finally separates the enveloped nucleus and the early reticulocyte. Cytoplasmic vesicles, concentrated near the nucleus, may also contribute to this separation. This was apparent in ultrastructural studies of circulating erythroblasts from anemic dogs (Simpson and Kling, 1967), where these cytoplasmic vesicles fuse with each other and with the cell membrane in such a way that the pyknotic enveloped nucleus and the future reticulocyte are separated.

Enucleation is connected with marked changes of the distribution of membrane proteins in the membrane surrounding the nucleus and in the plasma membrane of the reticulocyte domain (Patel and Lodish, 1987; Geiduschek and Singer, 1979; Skutelsky and Farquhar, 1976). Alteration of the membrane skeleton may explain the undulating appearance of the plasma membrane surrounding the nucleus (Figure 7).

6. ULTRASTRUCTURAL ASPECTS OF IRON METABOLISM

Iron is taken up into the erythrocyte precursors by means of receptor-mediated endocytosis of iron-loaded transferrin (Iacopetta and Morgan, 1983) (see Chapter 13). It was found during development of erythroid cells from rat and mouse fetal liver, the transferrin receptor level increases from about 300,000 per cell at the early normoblast stage to a maximum of about 800,000 per cell on intermediate normoblasts and decreases progressively to 105,000 in the circulating reticulocytes. The rate of iron uptake was found to correlate closely with the number of transferrin receptors (Iacopetta *et al.*, 1982).

The initial event of the receptor-mediated endocytosis of ferritransferrin is the clustering of receptor ligand complexes into clathrin-coated pits (Crichton and Charloteaux-Wauters, 1987). Coated pits can be observed in all morphologically recognizable erythrocyte precursors. This phenomenon has been described as rhopheocytosis (Bessis, 1973). The internalization of iron–transferrin complexes into erythroid cells has been visualized by means of electron microscopy using various labeling methods (Blight and Morgan,

1987; Parmley *et al.*, 1983; Takahashi and Tavassoli, 1983; Light and Morgan, 1982; Zaman *et al.*, 1980; Sullivan *et al.*, 1976). Egyed *et al.* (1986) found a good correlation between the rate of iron uptake and the coated pit count in rabbit reticulocytes.

In *in vitro* experiments, guinea pig reticulocytes take up exogenous ferritin via high-affinity receptors (Blight and Morgan, 1983) through the same pathway as transferrin, involving the same coated pits and vesicles (Blight and Morgan, 1987). In these *in vitro* experiments, it was found as well that iron from ferritin can be utilized for hemoglobin synthesis and competes directly with iron derived from plasma transferrin (Blight and Morgan, 1983).

Due to the large number of iron atoms in the core, single ferritin molecules are visible as black grains in the electron microscope (Figures 2b, 11a). *In vivo,* ferritin can be found attached to the outer cell membrane of the rhopheocytotic pits and inside fuzzy-coated cytoplasmic vesicles of erythroid cells from humans (Figures 2b, 11a) and guinea pigs (Bessis, 1973). In view of the intimate association of the central macrophage with the maturing erythroblasts, it has been proposed that iron used in hemoglobin synthesis can also be derived from the ferritin-iron pool of the central macrophage (Bessis, 1973). However, in contrast with the *in vitro* experiments (Blight and Morgan, 1987), the observation of ferritin in rhopheocytotic vesicles of erythrocyte precursors in erythroblastic islands is no absolute proof that the ferritin molecules pass from the central macrophage to the erythroblast (Bessis and Breton-Gorius, 1962).

In rat, mouse, and rabbit erythroid cells, ferritin molecules are not visible in rhopheocytotic vesicles. This correlates with the finding that the uptake of ferritin into rat reticulocytes is entirely nonspecific (Blight and Morgan, 1983).

It is thought that iron taken up into the proliferating erythroblasts and not used immediately for heme synthesis is stored in ferritin (Konijn *et al.*, 1979). In an individual section of normal human proerythroblasts, roughly 500 to 1000 ferritin molecules can be counted (Bessis, 1973). In proerythroblasts, small aggregates of ferritin can also be found. When maturation proceeds, more and larger ferritin aggregates, eventually sequestered into a membrane, are visible in the erythrocyte precursors (Figure 11a) (Bessis and Breton-Gorius, 1962).

Large ferritin aggregates, "siderosomes," are visible in the optical microscope as iron-positive granules after the Perl's reaction. They can be detected in 40 to 60% of normoblasts, which are designated as sideroblasts. Siderosomes can also be found in reticulocytes and erythrocytes, which are designated as siderocytes. Under normal conditions, erythrocytes do not contain siderosomes. It is thought that siderosomes are removed by the "pitting function" of the spleen (Kent *et al.*, 1966).

The existence of ferritin aggregates indicates that the entrance and formation of ferritin molecules exceeds the consumption at this particular stage of development (Bessis and Breton-Gorius, 1962) and in a particular organism. In hypochromic iron-deficiency anemia sideroblasts and siderocytes are not seen, but in conditions with iron overload their number increases.

Every electron-dense particle inside a siderosome is not ferritin (Figure 11a). Due to the degradation of the ferritin protein shell, iron-positive granules of several sorts are formed (Richter, 1984). In Romanowsky stained smears, siderosomes containing ferritin, degradation products of ferritin, and proteolytic enzymes are visible as Pappenheimer bodies.

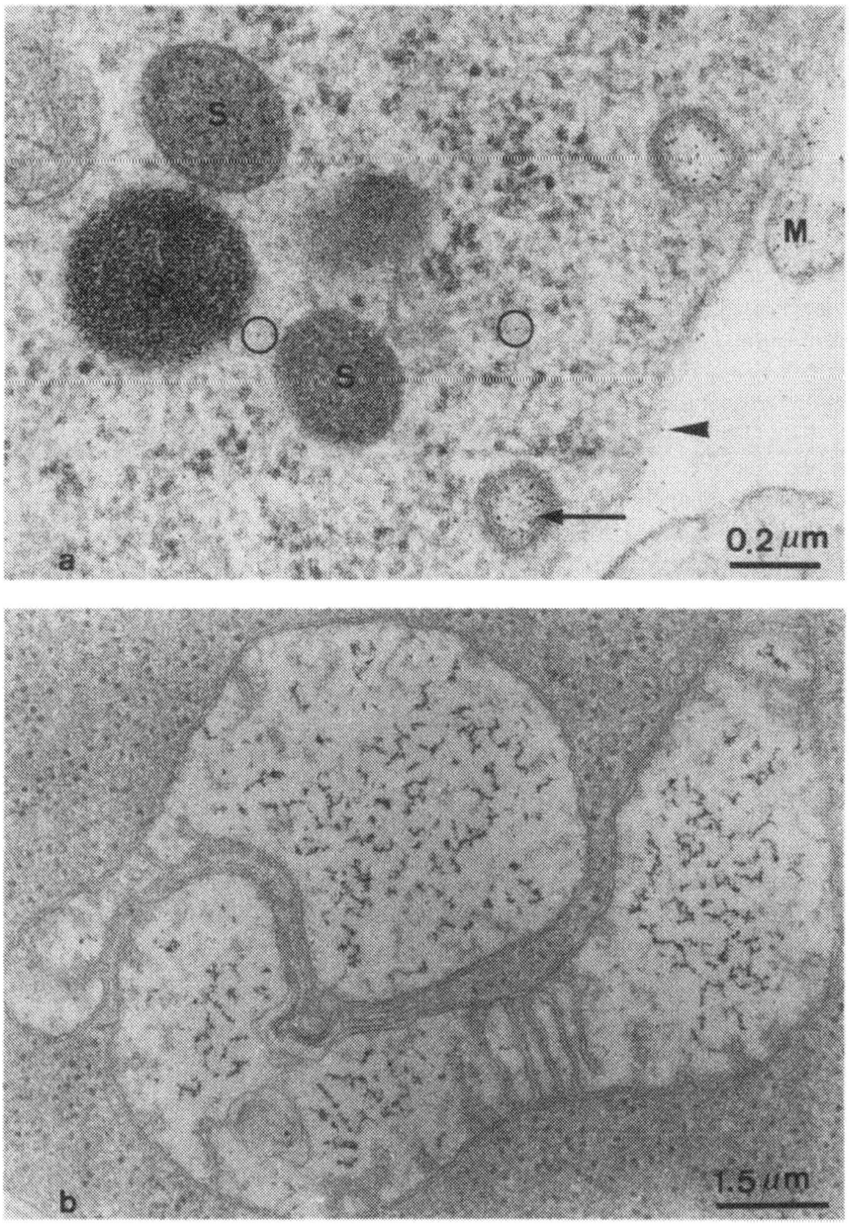

FIGURE 11. Iron in erythrocyte precursors. (a) Part of a basophilic erythroblast from a normal human control showing ferritin attached to the plasma membrane (arrowhead), inside a fuzzy-coated vesicle (arrow), and free in the cytoplasm of the erythroblast (circles), and of the associated macrophage (M). Ferritinlike granules are accumulated inside siderosomes (S). (b) A pathological siderocyte from the bone marrow of a Wistar rat with acute lead intoxication showing "ferruginous micelles," inside the swollen inner mitochondrial compartment.

7. DYSERYTHROPOIESIS

7.1. Ringed Sideroblasts

The final step of the heme synthetic pathway, i.e., the incorporation of iron into protoporphyrin, occurs in the mitochondria and iron has to be taken up into these organelles (Jones and Jones, 1968). Under normal conditions, iron cannot be detected in the mitochondria of mammals other than guinea pigs (Bessis and Breton-Gorius, 1962). However, in sideroblastic anemia (Jacobs, 1986; Bottomley, 1982), iron accumulates in the inner mitochondrial compartment in the form of "ferruginous micelles" (Bessis and Breton-Gorius, 1962). The intramitochondrial deposits of amorphous insoluble iron are morphologically different from ferritin (Figure 11b) (Bessis and Jensen, 1965). As seen by X-ray microanalysis, they contain iron and phosphorus (Heynen *et al.*, 1987; Grasso *et al.*, 1980; Ghadially *et al.*, 1979; Trump *et al.*, 1978) and are thought to be a mixture of iron phosphates (Grasso *et al.*, 1980).

Since the mitochondria are closely associated with the nucleus, the iron-positive granules, visible after Perl's staining, occur in a ring around the nucleus, which gives these cells the name "ringed sideroblasts" (Bowman, 1961).

In an experiment of acute lead intoxication of rats, iron-loaded mitochondria were only found in late erythroblasts and reticulocytes. It may be assumed that cells that are more active in heme synthesis, prior to the inhibition of heme synthesis by lead, accumulate more iron in their mitochondria than do proliferating erythroblasts, which are not so active in heme synthesis (Heynen *et al.*, 1987).

7.2. Megaloblasts

Under normal conditions, a synchronized pattern of nuclear and cytoplasmic maturation occurs. When there is defective DNA synthesis, megaloblasts are formed (Beck, 1986). In these cells an asynchronous maturation of nucleus and cytoplasm occurs.

Megaloblasts (Figure 12) are characterized ultrastructurally (Bessis *et al.*, 1986; Wickramasinghe, 1977) by: (1) an increase in cell and nuclear diameter at all stages of maturation, (2) a delay in chromatin condensation leading to a subnormal amount of heterochromatin, which is dispersed in fine clumps, in cells with advanced hemoglobinization, (3) one to four large nucleoli, and (4) numerous polyribosomes (which are responsible for the intense basophilia of megaloblasts on Romanowsky-stained smears). In addition, other nonspecific nuclear and cytoplasmic abnormalities may appear in the megaloblasts, which seem to be manifestations of the reaction of erythroblasts to cellular injury (Wickramasinghe, 1977; Fresco, 1981).

7.3. Beta-Thalassemia

The primary defect in homozygous beta-thalassemia is either a failure to synthesize beta chains or a severe depression in the rate of beta-chain synthesis (Weatherall, 1986). The unbalanced production of alpha chains results in the presence of multiple foci of precipitated alpha chains (Figure 13). Unlike Heinz bodies in phenylhydrazine-induced hemolytic anemia, alpha-chain precipitations are not attached to the plasma membrane

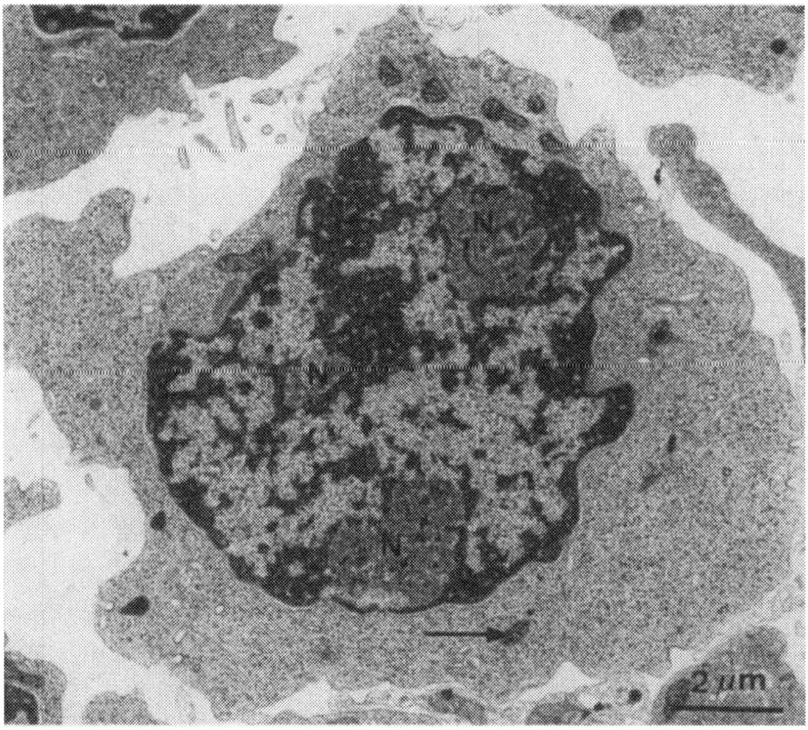

FIGURE 12. A megaloblast from the bone marrow of a patient with megaloblastic anemia. The nuclear chromatin is dispersed as fine clumps. Three large nucleoli (N) can be seen. The cytoplasm contains numerous polyribosomes and iron-loaded mitochondria (arrow).

(Polliack and Rachmilewitz, 1973). The proportion of cells containing precipitates and the extent of precipitation increase with erythroblast maturity (Wickramasinghe, 1975). In addition, other nonspecific abnormalities (Figure 13) may appear in the erythroblasts and reticulocytes (Fresco, 1981). The erythrocytes of beta-thalassemia are microcytic, hypochromic target cells.

8. STIMULATED ERYTHROPOIESIS

Experimental animals in which erythropoiesis has been stimulated by bleeding (Tablin and Weiss, 1985; Gasko and Danon, 1972, 1974) or phenylhydrazine-induced hemolysis (Blight and Morgan, 1983, 1987; Iacopetta and Morgan, 1983; Light and Morgan, 1982; Denton *et al.*, 1975; Le Carpentier and Prenant, 1975; Simpson and Kling, 1967, 1968; Rifkind *et al.*, 1964) have been used frequently to study multiple aspects of maturation of erythroid cells.

The intense erythropoietin response results in the production of "stress" reticulocytes, which may reach up to twice their normal size with a corresponding increase of

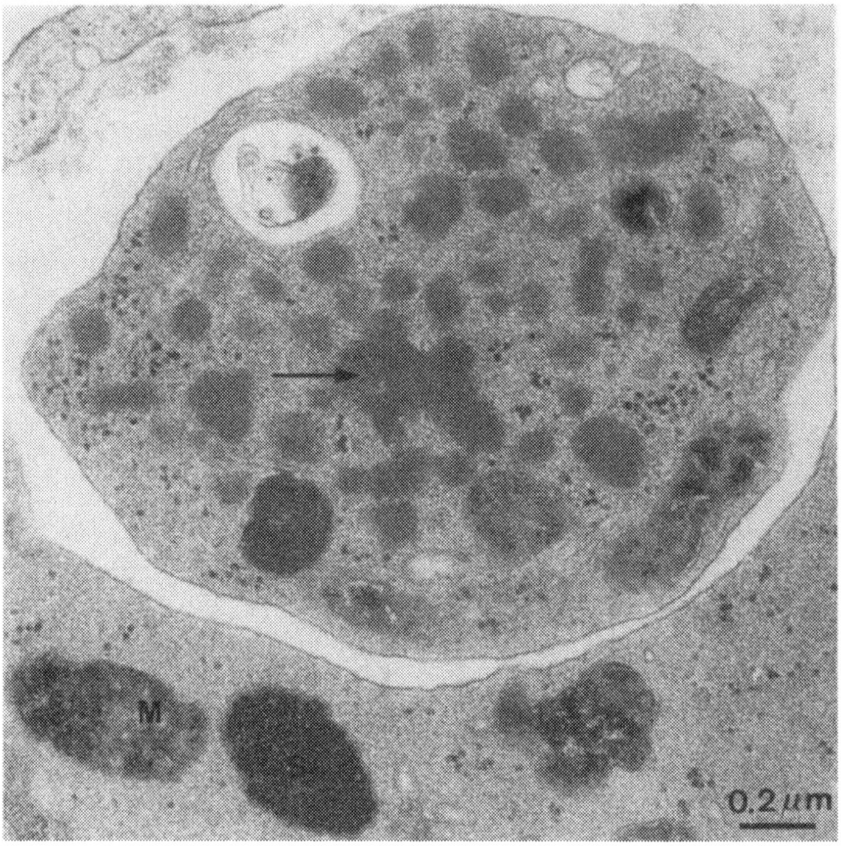

FIGURE 13. Part of two reticulocytes from the bone marrow of a patient with beta-thalassemia showing the alpha-chain precipitations (arrow) and the nonspecific accumulation of iron into the mitochondria (M) and a siderosome (S).

their hemoglobin content. These macroreticulocytes originate from macronormoblasts (Brecher *et al.*, 1975).

Morphometric analysis of erythrocyte precursors from equine bone marrow after bleeding (Tablin and Weiss, 1985) showed a significantly higher volume density of mitochondria and number of ribosomes in anemic than in normal animals at the reticulocyte stage. This has also been observed in erythrocyte precursors from phenlhydrazine-treated Wistar rats.

Whether the production of macroreticulocytes is due to a skipped terminal division or an early denucleation or some other phenomenon is not clear (Bessis *et al.*, 1986).

ACKNOWLEDGMENTS. I thank Professor Dr. R. L. Verwilghen and Dr. L. Van Hove for helpful criticism in reading the manuscript, and Catherine Mistiaen for typing the references.

9. REFERENCES

Beck, W. S., 1986, The megaloblastic anemias, in: *Hematology* (W. J. Williams, E. Beutler, A. J. Erslev, and M. A. Lichtman, eds.), pp. 434–465, McGraw-Hill, New York.

Ben-Ishay, Z., and Yoffey, J. M., 1972, Ultrastructural Studies of erythro-blastic islands of rat bone marrow. The resumption of erythropoiesis in erythropoietically depressed rebound marrow, *Lab. Invest.* **26**:637–647.

Bessis, M., 1973, *Living Blood Cells and Their Ultrastructure*, Springer, Berlin.

Bessis, M., and Breton-Gorius, J., 1962, Iron metabolism in the bone marrow as seen by electron microscopy: A critical review, *Blood* **19**:635–663.

Bessis, M., and Jensen, W. N., 1965, Sideroblastic anaemia, mitochondria and erythroblastic iron, *Br. J. Haematol.* **11**:49–51.

Bessis, M., Lessin, L. S., and Beutler, E., 1986, Morphology of the erythron, in: *Hematology* (W. J. Williams, E. Beutler, A. J. Erslev, and M. A. Lichtman, eds.), pp. 257–279, McGraw–Hill, New York.

Blight, G. D., and Morgan, E. H., 1983, Ferritin and iron uptakc by rcticulocytes, *Br. J. Haematol.* **55**:59–71.

Blight, G. D., and Morgan, E. H., 1987, Receptor-mediated endocytosis of transferring and ferritin by guinea-pig reticulocytes. Uptake by a common endocytic pathway, *Eur. J. Cell Biol.* **43**:260–265.

Bottomley, S. S., 1982, Sideroblastic anaemia, *Clin. Haematol.* **11**:389–409.

Bowman, W. D., 1961, Abnormal ("ringed") sideroblasts in various hematologic and non-hematologic disorders, *Blood* **18**:662–671.

Brecher, G., Prenant, M., Haley, J., and Bessis, M., 1975, Origin of stress macroreticulocytes from macronormoblasts, *Nouv. Rev. Fr. Hematol.* **15**:13–18.

Breton-Gorius, J., 1970, Utilisation de la diaminobenzidine pour la mise en évidence, au microscope électronique, de l'hémoglobine intracellulaire, *Nouv. Rev. Fr. Hematol.* **10**:243–256.

Breton-Gorius, J., and Guichard, J. 1975, Fine structural and cytochemical identification of microperoxisomes in developing human erythrocytic cells, *Am. J. Pathol.* **79**:523–536.

Breton-Gorius, J., and Reyes, F., 1976, Ultrastructure of human bone marrow cell maturation, *Int. Rev. Cytol.* **46**:251–321.

Castoldi, G. L., and Beutler, E., 1988, Erythrocytes, in: *Atlas of Blood Cells, Function and Pathology* (D. Zucker-Franklin, M. F. Greaves, C. E. Grossi, and A. M. Marmont, eds.), pp. 47–156, Edi-Ermes, Milan and Lea & Febiger, Philadelphia.

Chasis, J. A., Prenant, M., and Mohandas, N., 1988, Skeletal and cytoskeletal remodeling during reticulocyte maturation, *Blood* **72**:24a.

Clissold, P. M., Arnstein, H. R. V., and Chesterton, C. J., 1977, Quantitation of globin mRNA levels during erythroid development in the rabbit and discovery of a new β-related species in immature erythroblasts, *Cell* **11**:353–361.

Crichton, R. R., and Charloteaux-Wauters, M., 1987, Iron transport and storage, *Eur. J. Biochem.* **164**:485–506.

Denton, M. J., Spencer, N., and Arnstein, H. R. V., 1975, Biochemical and enzymic changes during erythrocyte differentiation. The significance of the final cell division, *Biochem. J.* **146**:205–211.

Dicke, K. A., van Noord, M. J., and van Bekkum, D. W., 1973, Attempts at morphological identification of the hemopoietic stem cell in rodents and primates, *Exp. Hematol.* **1**:36–45.

Egyed, A., Fodor, I., and Lelkes, G., 1986, Coated pit formation: A membrane function involved in the regulation of cellular iron uptake, *Br. J. Haematol.* **64**:263–269.

Fresco, R., 1981, Electron microscopy in the diagnosis of the bone marrow disorders of the erythroid series, *Semin. Hematol.* **18**:279–292.

Furukawa, T., Ikeda, H., Suda, J., Eguchi, M., Takaoka, T., and Suda, T., 1987, Cinemicrography of human erythroblasts—direct measurement of generation time and delineation of their pedigrees, *Blood Cells* **12**:531–539.

Gasko, O., and Danon, D., 1972, Deterioration and disappearance of mitochondria during reticulocyte maturation, *Exp. Cell Res.* **75**:159–169.

Gasko, O., and Danon, D., 1974, Endocytosis and exocytosis in membrane remodelling during reticulocyte maturation, *Br. J. Haematol.* **28**:463–470.

Geiduschek, J. B., and Singer, S. J., 1979, Molecular changes in the membranes of mouse erythroid cells accompanying differentiation, *Cell* **16**:149–163.

Ghadially, F. N., 1979, Haemorrhage and haemosiderin, *J. Submicrosc. Cytol.* **11**:271–291.

Ghadially, F. N., Lalonde, J.-M.A., and Mukherjee, T. M., 1979, Electron-probe X-ray analysis of intramitochondrial iron deposits in sideroblastic anaemia, *J. Submicrosc. Cytol.* **11**:503–510.

Grasso, J. A., Swift, H., and Ackerman, G. A., 1962, Observations on the development of erythrocytes in mammalian fetal liver, *J. Cell Biol.* **14**:235–254.

Grasso, J. A., Sullivan, A. L., and Chan, S.-C., 1978, Studies of the endoplasmic reticulum and plasma membrane-bound ribosomes in erythropoietic cells, *J. Cell Sci.* **31**:165–178.

Grasso, J. A., Myers, T. J., Hines, J. D., and Sullivan, A. L., 1980, Energy-dispersive X-ray analysis of the mitochondria of sideroblastic anaemia, *Br. J. Haematol.* **46**:57–72.

Harrison, P. R., 1984, Molecular analysis of erythropoiesis, *Exp. Cell Res.* **155**:321–344.

Heynen, M. J., and Verwilghen, R. L., 1982, A quantitative ultrastructural study of normal rat erythroblasts and reticulocytes, *Cell Tissue Res.* **224**:397–408.

Heynen, M. J., Tricot, G., and Verwilghen, R. L., 1985, Autophagy of mitochondria in rat bone marrow erythroid cells. Relation to nuclear extrusion, *Cell Tissue Res.* **239**:235–239.

Heynen, M. J., Zaman, Z., and Verwilghen, R. L., 1987, Effect of acute lead intoxication on the ultrastructure of rat erythroblasts and reticulocytes. Morphometric analysis and Röntgen micro-analysis, *Haematologia* **20**:131–153.

Iacopetta, B. J., and Morgan, E. H., 1983, The kinetics of transferrin endocytosis and iron uptake from transferrin in rabbit reticulocytes, *J. Biol. Chem.* **258**:9108–9115.

Iacopetta, B. J., Morgan, E. H., and Yeoh, G. C. T., 1982, Transferrin receptors and iron uptake during erythroid cell development, *Biochim. Biophys. Acta* **687**:204–210.

Jacobs, A., 1986, Primary acquired sideroblastic anaemia, *Br. J. Haematol.* **64**:415–418.

Jensen, W. N., Moreno, G. D., and Bessis, M., 1965, An electron microscopic description of basophilic stippling in red cells, *Blood* **25**:933–943.

Jones, M. S., and Jones, O. T. G., 1968, Evidence for the location of ferrochelatase on the inner membrane of rat liver mitochondria, *Biochem. Biophys. Res. Commun.* **31**:977–982.

Kent, G., Minick, O. T., Volini, F. I., and Orfei, E., 1966, Autophagic vacuoles in human red cells, *Am. J. Pathol.* **48**:831–858.

Konijn, A. M., Hershko, C., and Izak, G., 1979, Ferritin synthesis and iron uptake in developing erythroid cells, *Am. J. Hematol.* **6**:373–379.

Koury, M. J., Bondurant, M. C., and Atkinson, J. B., 1987, Erythropoietin control of terminal erythroid differentiation: Maintenance of cell viability, production of hemoglobin, and development of the erythrocyte membrane, *Blood Cells* **13**:217–226.

Koury, S. T., Koury, M. J., and Bondurant, M. C., 1988a, Morphological changes in erythroblasts during erythropoietin-induced terminal differentiation in vitro, *Exp. Hematol.* **16**:758–763.

Koury, S. T., Koury, M. J., and Bondurant, M. C., 1988b, Cytoskeletal function during the enucleation of mammalian erythroblasts, *Blood* **72**:29a.

Le Charpentier, Y., and Prenant, M., 1975, Isolement de l'ilot érythroblastique. Etude en microscopie optique et électronique é balayage, *Nouv. Rev. Fr. Hematol.* **15**:119–140.

Lemke, C., and Linss, W., 1984, Remarks on the role of microtubules in enucleating normoblasts, *Anat. Anz.* **156**:427–431.

Light, A., and Morgan, E. H., 1982, Transferrin endocytosis in reticulocytes: An electron microscope study using colloidal gold, *Scand. J. Haematol.* **28**:205–214.

Loken, M. R., Shah, V. O., Dattilio, K. L., and Civin, C. I., 1987, Flow cytometric analysis of human bone marrow: I. Normal erythroid development, *Blood* **69**:255–263.

Nijhof, W., and Wierenga, P. K., 1983, Isolation and characterization of the erythroid progenitor cell: CFU-E, *J. Cell Biol.* **96**:386–392.

Nijhof, W., and Wierenga, P. K., 1984, A new system for the study of erythroid cell differentiation, *Exp. Hematol.* **12**:115–120.

Novikoff, A. B., and Novikoff, P. M., 1973, Microperoxisomes, *J. Histochem. Cytochem,* **21**:963–966.

Orlic, D., Gordon, A. S., and Rhodin, J. A. G., 1965, An ultrastructural study of erythropoietin-induced red cell formation in mouse spleen, *J. Ultrastruct. Res.* **13**:516–542.

Parmley, R. T., Hajdu, I., Denys, F. R., 1983, Ultrastructural localization of the transferrin receptor and transferrin on marrow cell surfaces, *Br. J. Haematol.* **54**:633–641.

Patel, V. P., and Lodish, H. F., 1987, A fibronectin matrix is required for differentiation of murine erythroleukemia cells into reticulocytes, *J. Cell Biol.* **105:**3105–3118.

Perl, W., 1964, Correction of polyribosome distributions as observed in cell sections by electron microscopy, *J. Cell Biol.* **22:**613–621.

Polliack, A., and Rachmilewitz, F. A., 1973, Ultrastructural studies in β-thalassaemia major, *Br. J. Haematol.* **24:**319–326.

Quesenberry, P. J., 1986, Hemopoietic stem cells, in: *Hematology* (W. J. Williams, E. Beutler, A. J. Erslev, and M. A. Lichtman, eds.), pp. 129–143, McGraw–Hill, New York.

Rapoport, S. M., and Schewe, T., 1986, The maturational breakdown of mitochondria in reticulocytes, *Biochim. Biophys. Acta* **864:**471–495.

Repasky, E. A., and Eckert, B. S., 1981, The effect of cytochalasin B on the enucleation of erythroid cells in vitro, *Cell Tissue Res.* **221:**85–91.

Rich, I. N., 1986, A role for the macrophage in normal hemopoiesis. Functional capacity of bone-marrow-derived macrophages to release hemopoietic growth factors, *Exp. Hematol.* **14:**738–745.

Rich, I. N., Vogt, C., and Pentz, S., 1988, Erythropoietin gene expression in vitro and in vivo detected by in situ hybridization, *Blood Cells* **14:**505–520.

Richter, G. W., 1984, Studies of iron overload. Rat liver siderosome ferritin, *Lab. Invest.* **50:**26–35.

Rifkind, A. R., Danon, D., and Marks, P. A., 1964, Alterations in polyribosomes during erythroid cell maturation, *J. Cell Biol.* **22:**599–611.

Rosse, C., and Trotter, J. A., 1974a, A cytochemical and radioautographic analysis of erythropoiesis at the ultrastructural level, *Am. J. Anat.* **141:**41–72.

Rosse, C., and Trotter, J. A., 1974b, Cytochemical and radioautographic identification of cells induced to synthesize hemoglobin, *Blood* **43:**885–898.

Rothman, J. E., 1985, The compartmental organization of the Golgi apparatus, *Sci. Am.* **9:**84–95.

Schewe, T., Halangk, W., Hiebsch, C., and Rapoport, S., 1977, Degradation of mitochondria by cytosolic factors in reticulocytes, *Acta Biol. Med. Ger.* **36:**363–372.

Seelig, L. L., 1972, Surface multivesicular structures associated with maturing erythrocytes in rats, *Z. Zellforsch.* **133:**181–186.

Simpson, C. F., and Kling, J. M., 1967, The mechanism of denucleation in circulating erythroblasts, *J. Cell Biol.* **35:**237–245.

Simpson, C. F., and Kling, J. M., 1968, The mechanism of mitochondrial extrusion from phenylhydrazine-induced reticulocytes in the circulating blood, *J. Cell Biol.* **36:**103–109.

Skutelsky, E., and Danon, D., 1967, An electron microscopic study of nuclear elimination from late erythroblast, *J. Cell Biol.* **33:**625–635.

Skutelsky, E., and Farquhar, M. G., 1976, Variations in distribution of Con A receptor sites and anionic groups during red blood cell differentiation in the rat, *J. Cell Biol.* **71:**218–231.

Smetana, K., Gyorkey, F., Gyorkey, P., and Busch, H., 1975, Studies on nucleoli of maturing human erythroblasts, *Exp. Cell Res.* **91:**143–151.

Spivak, J. L., 1975, Chromosomal protein synthesis during erythropoiesis in the mouse spleen, *Exp. Cell Res.* **91:**253–262.

Sullivan, A. L., Grasso, J. A., and Weintraub, L. R., 1976, Micropinocytosis of transferrin by developing red cells: An electron microscopic study utilizing ferritin-conjugated transferrin and ferritin-conjugated antibodies to transferrin, *Blood* **47:**133–143.

Tablin, F., and Weiss, L., 1985, Equine bone marrow: A quantitative analysis of erythroid maturation, *Anat. Rec.* **213:**202–206.

Takahashi, K., and Tavassoli, M., 1983, Internalization of iron–transferrin complex by murine L1210 leukemia cells and rat reticulocytes demonstrated by a minibead probe, *J. Ultrastruct. Res.* **82:**314–321.

Tanaka, Y., and Goodman, J. R., 1972, *Electron Microscopy of Human Blood Cells,* Harper & Row, New York.

Trump, B. F., Berezesky, I. K., Rouben, B. A., Jiji, M., Mergner, W. J., and Bulger, R. E., 1978, Energy dispersive X-ray microanalysis of mitochondrial deposits in sideroblastic anemia, *Lab. Invest.* **39:**375–380.

van Bekkum, D. W., van Noord, M. J., Maat, B., and Dicke, K. A., 1971, Attempts at identification of hemopoietic stem cell in mouse, *Blood* **38:**547–558.

van Bekkum, D. W., van den Engh, G. J., Wagemaker, G., Bol, S. J. L., and Visser, J. W. M., 1979, Structural identity of the pluripotential hemopoietic stem cell, *Blood Cells* **5**:143–159.

Verwilghen, R. L., Broeckaert-Van Orshoven, A., and Heynen, M. J., 1975, Dyserythropoiesis and annulate lamellae, *Br. J. Haematol.* **30**:307–310.

Warner, J. R., Rich, A., and Hall, C. E., 1962, Electron microscope studies of ribosomal clusters synthesizing hemoglobin, *Science* **138**:1399–1403.

Weatherall, D. J., 1986, The thalassemias, in: *Hematology* (W. J. Williams, E. Beutler, A. J. Erslev, and M. A. Lichtman, eds.), pp. 493–521, McGraw-Hill, New York.

Wickramasinghe, S. N., 1975, *Human Bone Marrow,* Blackwell, Oxford.

Wickramasinghe, S. N., 1977, Pathophysiology of the proliferation and cytodifferentiation of megaloblasts, in: *Dyserythropoieses* (S. M. Lewis and R. L. Verwilghen, eds.), pp. 103–123, Academic Press, New York.

Yataganas, X., Gahrton, G., and Thorell, B., 1970, DNA, RNA and hemoglobin during erythroblast maturation, *Exp. Cell Res.* **62**:254–261.

Zaman, Z., Heynen, M. J., and Verwilghen, R. L., 1980, Studies on the mechanism of transferrin iron uptake by rat reticulocytes, *Biochim. Biophys. Acta* **632**:553–561.

Chapter 2

Expression of Human Erythrocyte Membrane Protein Antigens during Erythroid Differentiation

Marilyn J. Telen

Recent advances in hybridoma technology, fluorescence-activated cell sorting, and bone marrow culture techniques have led to investigations of the manner in which blood group antigens and other cell surface molecules are expressed during hematopoietic differentiation. These studies have demonstrated expression by erythroid progenitors and precursors of molecules or epitopes unique to the erythroid lineage as well as of other molecules whose expression is shared by precursor cells of other lineages. Some antigens are expressed by progenitor cells but lost during differentiation, while other antigens appear first during intermediate and late stages of erythropoiesis.

Understanding of the progression of cell surface antigen characteristics has also led to attempts to apply this knowledge to the characterization and identification of malignant hematopoietic cells that express antigenic markers of erythroid differentiation. Knowledge of the timing of expression of erythroid-specific molecules, such as the glycophorins, has thus helped identify erythroleukemias and study agents that affect differentiation of erythroleukemia cell lines.

1. OVERVIEW OF ERYTHROPOIESIS

Four cell stages have been described to precede the development of morphologically recognizable erythroid precursors. The earliest is the pluripotent stem cell, which can replicate itself as well as give rise to more differentiated cells (Nakahata and Ogawa,

Marilyn J. Telen Division of Hematology/Oncology, Department of Medicine, and Immunohematology Laboratory, Transfusion Service, Duke University Medical Center, Durham, North Carolina 27710, USA.

1982). Next is the CFU-GEMM, an early stem cell that has limited capacity for self-renewal; individual CFU-GEMM can give rise to a mixed population of granulocytic, erythroid, monocytic, and megakaryocytic colonies (Fauser and Messner, 1978). Further differentiation leads to the appearance of the committed stem cells: the BFU-E progenitor, followed by the CFU-E progenitor (Gordon *et al.*, 1985). These cells no longer have self-renewal capacity and give rise only to erythroid cells. The BFU-E in culture gives rise to large cellular aggregates (or bursts), each of which contains several colonies of erythroblasts; all of these cells arise from a single progenitor cell. Relatively primitive BFU-E can be distinguished from more differentiated BFU-E by the number of colonies they produce; less mature BFU-E have higher proliferative capacity. However, cell cycle activity of BFU-E is lower than in more differentiated progenitor cells, and the more primitive BFU-E take relatively longer to produce their maximum number of colonies than do more mature progenitor cells. CFU-E are more mature stem cells than BFU-E and give rise, *in vitro,* to single small colonies that appear during the first week in culture (as compared to 10–20 days for BFU-E of various degrees of maturity). There are more CFU-E than BFU-E in bone marrow, although BFU-E but not CFU-E may be found in peripheral blood. From these stem cells the morphologically recognizable erythroid precursor cells arise sequentially, namely the pronormoblast (erythroblast), basophilic normoblast, polychromatophilic normoblast, orthochromatic normoblast, nucleated red cell, and reticulocyte. The pronormoblast is a large cell with a large nucleus, often with prominent nucleoli; this cell is only slightly distinguishable from the myeloblast. Differentiation into a basophilic normoblast is accompanied by loss of nucleoli and decrease of the nucleus/cytoplasm ratio; the cytoplasm becomes more deeply basophilic due to a marked increase in ribosomes. The polychromatophilia of the next stage bears witness to the appearance of hemoglobin at this time; the nucleus has continued to contract. The more acidophilic orthochromatic normoblast is distinctly pink and contains a nucleus becoming increasingly pyknotic. The nucleated red cell's cytoplasm appears as acidophilic (or nearly so) as a mature erythrocyte, while its nucleus is a small, dense body ready for extrusion. The reticulocyte is still somewhat larger than a mature red cell and contains an abundance of ribosomes, which can be demonstrated by a variety of supravital stains. Thus, this differentiation process (summarized in Figure 1) entails as its three major characteristics gradual diminishment of cell size, involution and extrusion of the nucleus, and hemoglobinization of the cytoplasm.

2. TECHNIQUES USED TO STUDY EXPRESSION OF ANTIGENS DURING ERYTHROID DIFFERENTIATION

Investigation of the expression of various cell surface proteins and antigens by the normally small population of morphologically unrecognizable stem cells has been performed using both direct and indirect techniques. In general, the small number of stem cells in bone marrow has presented researchers with their most formidable obstacle. Another problem has been the development of antisera with appropriately restricted but also not overly fastidious specificities. Polyclonal heteroantisera have to their advantage the ability to recognize many epitopes on a single protein; however, these sera are often contaminated by antibody to extraneous proteins that may have been present in the

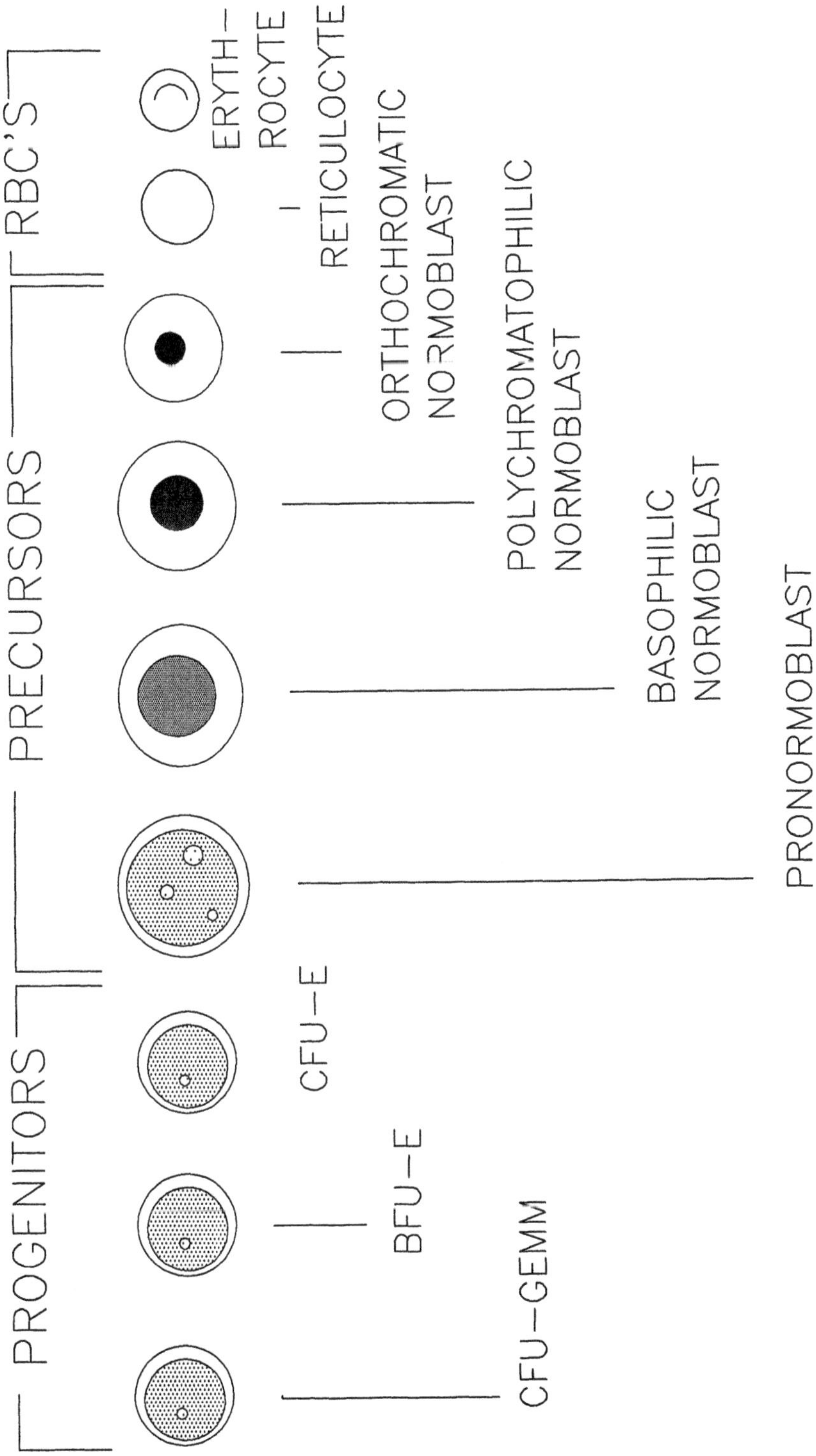

FIGURE 1. Schema of erythropoiesis: stem cells to precursors to circulating red blood cells.

immunogen. In addition, they may not be specific enough to recognize only the structures unique to a particular protein. Monoclonal antibodies, on the other hand, each recognize only one epitope. However, the particular epitope recognized may exist on a number of membrane structures. Furthermore, slight perturbation of the molecule of interest (e.g., loss of one sugar molecule) may render an antigen unreactive with a relevant monoclonal antibody.

Some investigators have used the ability of antibodies to induce complement-mediated cell lysis in order to selectively deplete mixed cell populations of progenitors that bear a particular surface antigen. In these systems, a mixed cell population comprising mononuclear bone marrow cells is incubated with specific antibody and complement; the surviving cells are then cultured to determine what types of progenitor cells remain. However, not all antibodies fix complement, and not all antigens are expressed in adequate density or appropriate topology for antibody-mediated complement fixation. A further unknown is the relative sensitivity or resistance to complement lysis of different cell types. And finally, an apparent effect on stem cells may reflect instead an effect on auxiliary cells that provide "help" to progenitor cell proliferation or differentiation.

Recent investigations have often relied largely on monoclonal antibodies and fluorescence-activated flow cytometry to sterilely separate antigen-positive and antigen-negative populations of cells. Advances in flow cytometry and computer imaging have also allowed the "sorting" of populations either with specific combinations of antigen expression using two-color fluorescence or with specific degrees of antigen expression (high versus low). However, these techniques detect low levels of expression poorly and may fail to detect up to 1000 molecules per cell (Civin and Loken, 1987). Furthermore, investigators often must determine "cutoff" points arbitrarily, depending on the characteristic "background" immunofluorescent staining of the cell population by a particular antibody. Another immunologic separation technique, called "panning," involves coating of plastic dishes with an antibody of interest and then manually separating adherent and nonadherent populations of cells. Manual separation can also be accomplished using antibody linked to magnetic beads. After either type of procedure, various populations of cells can then be further analyzed for their ability to bind selected antibodies as well as for their ability to give rise to different types of bone marrow cell colonies. These manual mechanical methods are also dependent on the presence of an adequate number of antigen sites and on the care with which the method is applied. Using analogous mechanical means to retard erythrocytes expressing less than 100 copies of acetylcholinesterase, investigators have shown that increased flow rate is sufficient to detach cells from the support matrix (Chow *et al.*, 1986). With all these methods, however, it is often problematic to determine whether absence of colonies typical of a particular progenitor is due to absence of the progenitor itself or to absence of accessory cells. However, this latter difficulty may frequently be overcome by the addition of soluble growth factors or "accessory cells" to the culture system.

Morphologically identifiable erythroid precursors have been somewhat easier to study, as cells can be sorted or panned according to their immunologic (antigenic) characteristics and then stained by routine (e.g., Wright) stains for direct identification and description. Early erythroid precursors grown *in vitro* have been studied directly by both immunologic and biochemical methods.

3. OVERVIEW OF ERYTHROCYTE MEMBRANE PROTEINS

Mature erythrocyte membranes bear on their surfaces a large number of proteins, among which some have been demonstrated to bear human blood group antigens. Characterized red cell membrane proteins include structural proteins, enzymes, complement regulatory proteins, and proteins whose functional roles are not yet known. The most abundant erythrocyte surface glycoprotein is band 3, the anion channel protein, of which approximately 1×10^6 copies exist on each cell (reviewed by Steck, 1978). This integral membrane protein by weight represents 25% of all erythrocyte membrane protein and allows the rapid transmembrane equilibration of Cl^- and HCO_3^-. Band 3 expresses the Ii blood group determinants and may also express ABH determinants. Absence of band 3 has not been described, although rare persons with abnormal band 3 protein have been identified (Kay *et al.*, 1988).

The next most abundant protein is glycophorin A, also called sialoglycoprotein α or MN glycophorin. Cells normally contain about 5×10^5 copies of this protein, which bears either the M or the N blood group antigen determinant at its amino terminus (reviewed by Anstee, 1986). Glycophorin A is a heavily glycosylated protein that bears approximately 60% of the sialic acid of the erythrocyte membrane. Antibodies may require the presence of most or all the normal complement of sialic acid (and other sugars) for recognition of M, N, and other glycophorin A antigens.

The other glycophorins occur in much lower copy numbers. Glycophorin B (sialoglycoprotein δ, SsU glycophorin) bears the Ss and U antigens, while glycophorin C (sialoglycoproteins β and γ) carries Gerbich determinants. Glycophorin B is partly homologous to glycophorin A and is encoded by genetic material close to and largely homologous with that encoding glycophorin A (Tate and Tanner, 1988). Glycophorin C, on the other hand, bears minimal homology to glycophorins A and B and is encoded by genetic material on a different chromosome (Colin *et al.*, 1986; Rahuel *et al.*, 1988; Mattei *et al.*, 1986).

Although band 3 and glycophorin A may carry oligosaccharide determinants expressed by other proteins or glycolipids, these proteins have generally been thought to be erythroid-specific. However, certain antibodies to glycophorins have also been shown to bind to some nonerythroid cells (Harvey *et al.*, 1988).

Another class of proteins either erythroid-specific or with erythroid-specific isoforms are the Rh proteins. Rh antigens (D, Cc, Ee) are currently thought to reside on two or three polymorphic proteins (reviewed by Issitt, 1988, 1989). These proteins are unusual in that they interact extensively with membrane lipid and lack glycosylation (Gahmberg, 1983). Again, although analogous proteins exist in other tissues, Rh-antigen-bearing proteins are thought to be specifically erythroid isoforms.

Numerous other classes of erythroid membrane proteins have been described and, to some extent, investigated concerning their expression during erythroid differentiation. Several other systems of blood group antigens are known to reside on proteins. These include the LW, Kell, Lutheran, Kidd, Duffy, Xg^1, and Ok^a antigens. In addition, numerous proteins belonging to the relatively newly described class of phosphatidylinositol-anchored membrane proteins exist on the erythrocyte surface. Several of these are complement regulatory proteins, such as decay-accelerating factor, homologous restriction factor

(C8 binding protein), and membrane inhibitor of reactive lysis (MIRL, p18). Decay-accelerating factor has been shown to bear the Cromer-related blood group determinants (Telen *et al.*, 1988). Preliminary evidence suggests that others of these proteins also express recognized blood group antigens (Telen and Moulds, unpublished data).

4. EXPRESSION OF MAJOR MEMBRANE GLYCOPROTEINS

Much of the original work examining markers of erythroid differentiation focused on glycophorin A and band 3, as these two proteins are expressed in high copy number on mature erythrocytes and were isolated earlier than many other erythroid membrane proteins. Polyclonal and monoclonal antibodies to these proteins have also been more readily available than antibodies to less abundant proteins.

4.1. Glycophorin A

Several investigatons have confirmed that glycophorin A is absent from committed erythroid stem cells (BFU-E and CFU-E) and appears along with the morphologically recognizable characteristics of erythroid differentiation (Robinson *et al.*, 1981; Gahmberg *et al.*, 1978). All bone marrow cells that express glycophorin A are morphologically recognizable erythroid precursors, although there has been some controversy as to whether glycophorin A is expressed by the earliest of such cells, the pronormoblast. Gahmberg *et al.* (1978) and Civin and Loken (1987), using polyclonal and monoclonal anti-glycophorin A antibodies, respectively, found that such antibodies did not react with pronormoblasts. However, Ekblom *et al.* (1985) and Robinson *et al.* (1981) were able to demonstrate that a polyclonal and monoclonal antiglycophorin, respectively, did react weakly with pronormoblasts. These conflicting results may have been due to the use of different techniques or to the use of different antibodies. Some researchers have also postulated that early erythroid cells express a glycophorin that is not fully glycosylated; thus, reactivity with monoclonal antibodies to the O-oligosaccharide-dependent M and N antigens occurred appreciably later than did expression of epitopes recognized by polyclonal antiglycophorin (Ekblom *et al.*, 1985). However, other investigators, using different techniques and different antibodies, have presented data suggesting that all glycophorin molecules expressed during normal erythroid differentiation are fully and normally glycosylated (Loken *et al.*, 1987). Direct biochemical characterization of the glycophorin molecules expressed by cells in various stages of maturation has not been performed. Thus, although it is certain that glycophorin A expression appears during the pronormoblast or basophilic normoblast stage, it remains controversial whether early expression of glycophorin occurs prior to the basophilic normoblast stage and whether the first glycophorin molecules expressed are fully glycosylated.

4.2. Glycophorin B

The expression of glycophorin B has not been investigated as fully as that of glycophorin A. Molecular genetic data obtained from the study of K562 erythroleukemia cells showed that the phorbol ester TPA coordinately down-regulates expression of both

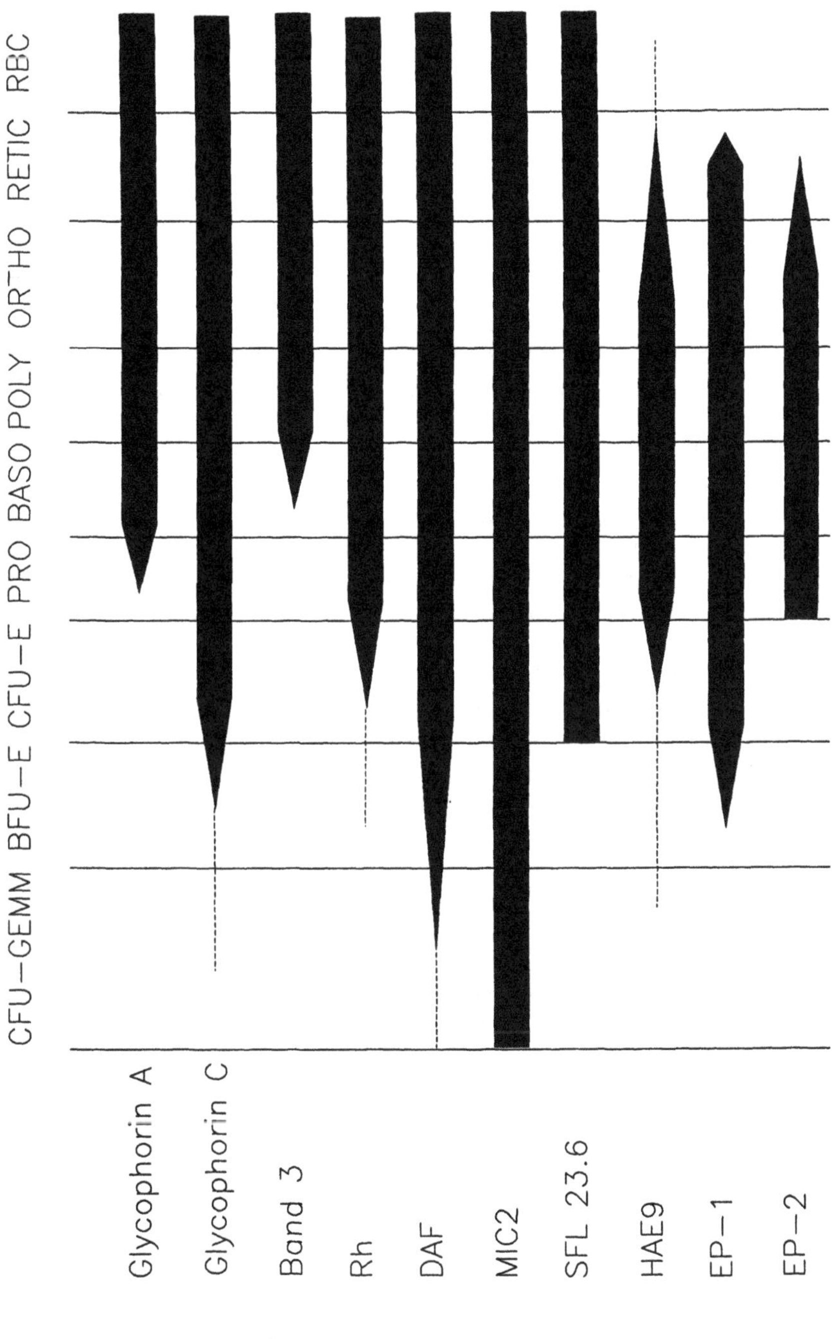

FIGURE 2. Expression of surface antigens during erythropoiesis.

glycophorin A and glycophorin B (Siebert and Fukuda, 1986). Also, it is known that the genes for these two proteins reside close to one another, making more feasible the coordinate regulation of their expression during normal erythroid differentiation. However, antibodies that recognize both glycophorins bind to more nucleated human bone marrow cells than do antibodies that recognize only glycophorin A (Telen *et al.*, 1987). These data could be explained by either the earlier appearance of glycophorin B or the detection of lower levels of expression of glycophorin A by simultaneous detection of expression of coordinately expressed glycophorin B.

4.3. Band 3

An early report of expression of band 3 by erythroid progenitors (Sieff *et al.*, 1982) based its conclusions on reactivity of a monoclonal antibody later shown to be related to the Rh blood group system and not reactive with band 3 (Ridgwell *et al.*, 1983). Like glycophorins A and B, band 3 is not expressed by committed erythroid stem cells but is instead expressed at about the time that hemoglobin synthesis occurs. Fukuda *et al.* (1980) demonstrated that erythroblasts grown *in vitro* expressed little band 3, as indicated by radiolabeling of cell membrane proteins and detection of protein with the molecular weight and labeling characteristics of band 3 in SDS–polyacrylamide gels. Thus, band 3 is most likely expressed just slightly later than glycophorin A. Work with monoclonal antibodies to the Wr^b blood group antigen, expression of which appears to depend on coordinate expression (and possible cell surface interaction) of band 3 and glycophorin A (Telen and Chasis, 1989), also supports this conclusion, as more bone marrow nucleated cells are reactive with antiglycophorin antibodies than with anti-Wr^b antibodies (Telen *et al.*, 1987). It has not been completely elucidated whether the band 3 expressed by early normoblasts differs substantially from that expressed by late normoblasts. Band 3 is known to bear Ii carbohydrate determinants (Fukuda and Fukuda, 1984), and bone marrow stress (e.g., marked reticulocytosis) is associated with increased expression of the i antigen, with concomitant decrease in the I antigen (O'Hara *et al.*, 1978). Expression of I versus i antigen appears to depend, at least in part, on the presence of an enzyme capable of converting a linear oligosaccharide structure into a branched one (Fukuda *et al.*, 1979). However, while it is clear that i expression decreases during erythropoiesis while expression of I increases (Vainchenker *et al.*, 1981; Sieff *et al.*, 1982), it is not clear how much of this change relates to changes in band 3 glycosylation and how much change may be due to loss of i from membrane structures other than band 3. Both I and i antigens reside on structures other than band 3, including membrane glycolipid (Fukuda and Fukuda, 1984). I expression increases during the transition from BFU-E to CFU-E and then to mature erythrocytes (Fukuda *et al.*, 1980; Sieff *et al.*, 1982), even though the appearance of band 3 is delayed until the normoblast stage. Thus, increase in I expression during the earlier stages of erythropoiesis must be due to changes involving non-band 3 structures.

5. EXPRESSION OF GLYCOPHORIN C

Glycophorin C is composed of two sialoglycoproteins of slightly different molecular weights (Anstee *et al.*, 1984). These proteins are most likely the product of a single gene, which is unrelated to the genes for glycophorins A and B (Colin *et al.*, 1989). Glycophorin C proteins bear the Gerbich blood group antigens (Reid *et al.*, 1987). Some investigators have found that monoclonal antibodies with Gerbich-related specificities bound to some CFU-GEMM, as well as to BFU-E and CFU-E (Greaves *et al.*, 1985). However, other workers have found later expression of glycophorin C, detectable only at the CFU-E stage (Bettaieb *et al.*, 1987). These latter investigators have pointed out, however, that the monoclonal antibody used in their study was carbohydrate-dependent and therefore might not recognize incompletely glycosylated protein present on early progenitors. Anti-glycophorin C antibodies bind to significantly more bone marrow cells than do antibodies to glycophorin A (Telen *et al.*, 1987). Also, panning nucleated bone marrow cells with anti-glycophorin C antibodies identifies small, undifferentiated cells as well as morphologically recognizable erythroid precursors. And finally, antibodies to glycophorin C bind to erythroleukemia cells far better than do antibodies to glycophorin A (Bettaieb *et al.*, 1987; Villeval *et al.*, 1986; Telen and Borowitz, 1990), suggesting that glycophorin C expression precedes expression of glycophorin A during normal erythroid differentiation.

A monoclonal antibody designated SFL 23.6 has been described as having a tissue distribution very similar to that of glycophorin C, although the target antigen of SFL 23.6 has not been identified (Gupta *et al.*, 1985). Antibody SFL 23.6 bound to most late erythroid progenitors (CFU-E), as well as to all erythroid cells of later maturational stages, including circulating erythrocytes. Reactivity with CFU-E was confirmed by these investigators using both cell sorting and complement-dependent cytotoxicity. Thus, it remains to be seen whether this antigen is expressed concurrently with but independently of glycophorin C.

6. EXPRESSION OF Rh ANTIGENS

The Rh antigens appear to reside on two or more nonglycosylated integral membrane proteins with molecular weights of 30–34 kDa. These proteins express frequently polymorphic epitopes as well as epitopes that are expressed by the vast majority of persons; more than 40 Rh antigens have been defined by human alloantisera. Studies with Rh antibodies have been hampered by the difficulty of producing murine hybridoma antibodies with specificities similar to human Rh antibodies and by the fact that antibodies to the Rh proteins are almost uniformly incapable of fixing complement, irrespective of antibody isotype. Nevertheless, some studies have been performed, although they have produced somewhat contradictory data. R6A monoclonal antibody recognizes a 34-kDa protein absent from Rh_{null} erythrocytes; this protein may bear the Ee or Cc antigens (Ridgwell *et al.*, 1983; Anstee, 1986). When bone marrow cells were separated by fluorescence-activated cell sorting into R6A-antigen-positive and -negative fractions, only 15% of BFU-E and 21% of CFU-E sorted into the positive fraction (Sieff *et al.*, 1982). Other investigators (Falkenburg *et al.*, 1985), using a monoclonal antibody with apparent Rh 29 specificity, found that 68% of CFU-E were recovered in the antigen-positive population obtained by immunofluorescence-activated cell sorting. However, only 2% of BFU-E were recovered in the Rh 29-positive cell population, suggesting that Rh 29 expression was acquired during the transition from early to late erythroid progenitor. Rearden and Masouredis (1977) also demonstrated that the ability to bind ^{125}I-labeled human anti-D increased with increasing maturation from pronormoblasts to later forms. However, one study (Rearden and Chiu, 1983) attempted to use cytotoxicity to determine whether the D antigen was on BFU-E and CFU-E and found that, although rabbit anti-human immunoglobulin was used to enhance cytotoxicity of human anti-D for erythroid progenitors, none was produced. Also, the presence of human anti-D did not inhibit *in vitro* growth of erythroid colonies.

Although this latter study is methodologically problematic, it does raise a question, again, as to the mechanism whereby one but not another Rh antigen might be expressed during erythroid differentiation. The various Rh proteins are likely to be encoded by either one gene or a large gene complex (Issitt, 1988), increasing the likelihood that expression of all Rh antigens are coordinately regulated. Second, Rh proteins are unusual in that they are not glycosylated (Gahmberg, 1983), making stage-specific glycosylation differences an unlikely cause of variable expression of Rh antigens, unlike glycophorin-related antigens. Thus, it would seem that the various Rh proteins and antigens are either expressed at different levels during different stages of differentiation or that additional elements influence the expression of these antigens. Again, although the absence of Rh antigens decreases expression of other antigens (including Fy5, LW, and glycophorin B) (reviewed in Issitt, 1989), no other proteins have been shown to be necessary for expression of Rh protein antigens. However, Rh proteins are known to have specific associations with both glycolipids and glycoproteins (Connor and Schroit, 1988; Moore and Green, 1987), the timing of whose expression is thus far unknown. And finally, some workers have suggested that certain membrane proteins undergo reorientation during erythrocyte maturation (Allen and Hoover, 1983); such a process could also explain differential expression of individual Rh antigens.

7. EXPRESSION OF OTHER BLOOD GROUP ANTIGENS RESIDENT ON PROTEINS

The proteins that bear Kell, Duffy, Lutheran, LW, and Kidd blood group antigens have been partially characterized, and monoclonal antibodies have been produced to some relevant specificities. However, expression of these antigens on morphologically identifiable erythroid precursors or on erythroid progenitors has for the most part not been studied. A monoclonal antibody to Lu^b has been used to identify the population of nucleated human bone marrow cells that express Lu^b. These studies, performed as part of the First International Workshop of Monoclonal Antibodies against Red Cells and Related Antigens, demonstrated that anti-Lu^b bound to relatively few cells, indicating acquisition of the antigen at about the time of acquisition of band 3 expression (i.e., during hemoglobinization).

The Ok^a antigen has recently been shown to reside on membrane proteins of 35–69 kDa expressed by erythrocytes and all peripheral blood mononuclear cells (Williams *et al.*, 1988). The epitope recognized by one Ok^a-related monoclonal antibody, Tral.85, appears to be expressed by both erythroid and myeloid progenitors; however, the antigen is lost during myeloid differentiation while it is expressed by erythroid cells at all stages of differentiation (including circulating erythrocytes) (Sieff and Greaves, 1983).

8. EXPRESSION OF COMPLEMENT REGULATORY PROTEINS

The erythrocyte bears on its surface numerous proteins that play a role in preventing lysis of the cell by activation of autologous complement. These proteins include decay-accelerating factor (DAF), complement receptor type I (CRI), C8 binding protein (homologous restriction factor), and membrane inhibitor of reactive lysis (MIRL). However, of these proteins, only one (DAF) has been studied with regard to expression during various stages of erythropoiesis. DAF is expressed by all circulating hematopoietic cells and has also recently been shown to be the protein that bears Cromer-related blood group antigens (Telen *et al.*, 1988). Studies of DAF expression have attempted to answer the question of whether expression of DAF is lost during differentiation of abnormal erythrocytes in the disease paroxysmal nocturnal hemoglobinuria (PNH), or whether clonally abnormal progenitors in this disease are incapable of expressing (or giving rise to cells capable of expressing) DAF. Moore *et al.* (1985) demonstrated that all BFU-E and CFU-E could be found in the DAF$^+$ fraction of normal bone marrow mononuclear cells sorted by fluorescence-activated flow cytometry. They also observed that, in bone marrow from patients with PNH, DAF$^+$ progenitors could give rise to DAF$^-$ progeny. Kanamaru *et al.* (1988) used similar methods, although employing a mixture of three monoclonal antibodies to DAF rather than polyclonal anti-DAF, and found that a majority of CFU-GEMM expressed some detectable DAF and that DAF expression increased with differentiation to BFU-E and CFU-E. These results were also supported by the authors' cytotoxicity studies, which demonstrated that the proportion of progenitors killed by anti-DAF antibodies and complement increased with increasing differentiation from CFU-GEMM to CFU-E. However, they also found that, in patients with PNH, significant numbers of erythroid progenitors were DAF$^-$, indicating loss of expression of this protein prior to differentiation. This concept is also supported by the work of Dessypris *et al.* (1983), who found that hematopoietic progenitors from patients with PNH were abnormally sensitive to complement-mediated lysis caused by the presence of either acidified serum or isotonic sucrose.

9. EXPRESSION OF THE MIC2 GENE PRODUCT

Many investigators have reported the production of monoclonal antibodies that recognize an erythrocyte and lymphocyte glycoprotein of approximately 32 kDa whose expression is linked to expression of the Xg^a blood group antigen (Goodfellow and Tippett, 1981). This protein is encoded by the MIC2 locus on the X chromosome and by the MIC2 Y locus on the Y chromosome (Goodfellow *et al.*, 1983, 1984; Darling *et al.*, 1986); indeed, this protein provided the first example of a functional gene mapping to the pseudoautosomal regions of the X and Y chromosomes (Goodfellow *et al.*, 1987). Many monoclonal antibodies that recognize this protein fail to react or react only weakly with Xg(a$^-$) red cells from female donors (Goodfellow and Tippett, 1981, Latron *et al.*, 1987). It is not known whether female Xg(a$^-$) donors' erythrocytes lack this protein or whether they express a polymorphic protein unreactive with these monoclonal antibodies. Also, there is conflicting evidence regarding whether the Xg^a antigen actually resides on this protein; one group of investigators (Latron *et al.*, 1987) found that purified MIC2 protein contained Xg^a activity, although other genetic evidence suggests that the XG locus is separate from MIC2 (Goodfellow *et al.*, 1987).

At least one antibody to this protein has been used to study expression of the MIC2 gene during erythropoiesis. Bodger *et al.* (1983) used both cell sorting methodology as well as arsanilated-antibody-directed cell cytotoxicity to determine expression of the MIC2 gene product. All CFU-GEMM, BFU-E, and CFU-E were separated by sorting into the antigen-positive population. Incubation of bone marrow cells with the arsanilated antibody and complement completely abrogated growth of mixed colonies (indicating effect on CFU-GEMM) and partially inhibited growth of BFU-E- and CFU-E-derived cells. These results are compatible with data that two antibodies to two different epitopes on this protein show similar patterns of reactivity with bone marrow cells as do antibodies to glycophorin C, except that a small number of lymphoid-appearing cells are also reactive (Telen, unpublished data).

10. ANTIGENS EXPRESSED BY PROGENITORS OR PRECURSORS BUT NOT BY MATURE ERYTHROCYTES

Erythroid progenitor cells express a number of antigens common to stem cells (reviewed by Sieff, 1986), and these are discussed elsewhere in this volume. However, several researchers have described antigens (or monoclonal antibodies to them) that appear to be unique to differentiating cells of erythroid lineage.

Monoclonal antibody HAE9, described by Mechetner *et al.* (1987), bound to all morphologically recognizable nucleated erythroid cells in bone marrow but not to mature circulating human erythrocytes. The antibody is also reactive with T-lymphocyte cell lines, but not with circulating leukocytes or malignant B-cell lines. The antigen was further characterized as resident on a 70-kDa protein. Cell sorting of bone marrow cells demonstrated that only a very small portion of CFU-GEMM and BFU-E expressed the HAE9 antigen, while 36% of CFU-E expressed it. Peak expression appeared during the pronormoblast stage (95% positive), and expression lessened slightly during differentiation to the orthochromatic normoblast (93% positive), while less than 2% of circulating erythrocytes were reactive. These investigators also presented evidence that the HAE9 antigen constitutes a uniquely human epitope of the previously described protein Ag-EB (Ievleva *et al.*, 1976), a protein present in the nucleated erythroid cells of many other mammals.

Yokochi *et al.* (1984) described two monoclonal antibodies that also appeared uniquely reactive with erythroid progenitors and/or precursors. EP-1 antibody reacted with 90% of BFU-E and CFU-E in cell sorting experiments and inhibited the growth of BFU-E and CFU-E in complement-dependent cytotoxicity assays. Morphologically identifiable nucleated erythroid precursors also reacted with EP-1. The EP-2 antigen, however, was present on a mean of only 2% of normal adult bone marrow mononuclear cells. Various studies of the timing of the appearance and disappearance of this antigen supported the conclusion that the EP-2 antigen was absent on erythroid progenitors and expressed only by early erythroblasts (pronormoblasts primarily). The investigators were unable to biochemically characterize the target antigens of the EP-1 and EP-2 antibodies.

11. APPLICATION OF ANTIGEN TYPING TO THE DIAGNOSIS OF ERYTHROLEUKEMIA

Overt erythroleukemia (M6-ANLL by the French–American–British classification) is a rare disorder diagnosed by morphological criteria, despite the fact that one-fourth to one-third of all nucleated bone marrow cells are erythroid. Typical presentation of the disease (Frydecka *et al.*, 1984) is a patient with anemia, circulating nucleated red blood cells, and bizarre megaloblastic erythrocytic precursors predominating in the marrow. As the disease progresses, cells with features more characteristic of myeloblasts become more prevalent in the bone marrow and peripheral blood, and the clinical picture becomes more typical of acute myeloblastic or myelomonocytic leukemia. While leukemias with the morphological features of erythroleukemia are found only rarely, there has been some evidence to suggest that there are an unknown number of leukemias of early erythroid derivation that cannot be identified by conventional morphological criteria. Some of these have been identified using antisera to various antigens unique to or characteristic of erythroid tissues.

The first efforts to use protein antigen markers to diagnose erythroleukemia employed various antisera to glycophorin A. However, while glycophorin A expression precedes appearance of hemoglobin slightly, it does not appear on committed stem cells. Therefore, it is not surprising that only a modest number of erythroid leukemias have been identified using anti-glycophorin A. Greaves *et al.* (1983), using a monoclonal antibody to glycophorin A, identified 12 cases of glycophorin A-positive leukemias of 724 leukemias screened. Approximately 3.5% of myeloid and 0.6% of lymphoid leukemic blast cell populations contained significant numbers of glycophorin A-positive cells. Also, as found by Liszka *et al.* (1983), antibodies that are able to detect only completely glycosylated glycophorin A may be useless in detecting protein expressed by relatively undifferentiated leukemic erythroblasts.

Studies using antibodies to glycophorin C to identify erythroleukemias have also been reported. Villeval *et al.* (1986) identified nine erythroleukemias among 120 cases of poorly differentiated acute leukemia using a variety of monoclonal and polyclonal antibodies. A monoclonal antibody to glycophorin A was reactive with two cases, while a monoclonal antibody to glycophorin C identified three cases, of which two were not positive with the monoclonal anti-glycophorin A. However, these investigators based their diagnoses of erythroleukemia on expression of two other proteins. Glycoprotein IV of platelets is absent on megakaryocytic stem cells but is expressed by CFU-E and erythroid precursors (Edelman *et al.*, 1986). All the leukemias determined to be erythroid by Villeval *et al.* were classified as erythroid on the basis of expression of this protein, along with—in most cases—cytoplasmic expression of carbonic anhydrase I (the erythroid-specific isoenzyme) (Villeval *et al.*, 1985; Walloch *et al.*, 1986). A second study (Telen and Borowitz, 1990) looked at expression of glycophorin C among 92 cases of leukemia, including 89 sequential cases studied prospectively. Among the cases studied prospectively, 5 had significant populations of cells expressing glycophorin C, while only 2 had glycophorin A-positive cell populations. Furthermore, in all cases, more cells expressed detectable glycophorin C than glycophorin A. Of the 5 cases with glycophorin C expression, 2 were morphologically erythroleukemias, while a third was erythroleukemic blast crisis of chronic myelogenous leukemia. The other 2 cases had been diagnosed as FAB-M2 leukemia and myelodysplasia with excess blasts. Thus, both studies support the concept that, since glycophorin C is expressed earlier during normal differentiation than is glycophorin A, the former might be a better marker for malignant cells arising from transformation of early erythroid progenitors.

Villeval *et al.* (1986) also looked at expression of other erythrocyte surface markers. A polyclonal antibody to band 3 bound to cells from only one erythroleukemia, which also strongly expressed glycophorins A and C. On the other hand, antibody R6A (related to the Rh complex) failed to demonstrate significant binding to any erythroleukemia cells.

Several other markers of early erythroid differentiation hold promise as diagnostic tools but have not yet been tested adequately. Andreasen and Olsson (1987) described a monoclonal antibody (K : 1-6F) that showed strong reactivity with erythroleukemia cells and with erythroid precursor cells in normal bone marrow. However, further clinical use of the antibody has not been reported. Another monoclonal antibody, RTF8X, was also reported to be a marker of the erythroid lineage (Sakurai *et al.*, 1988). This antibody does not bind to erythroid stem cells or circulating erythrocytes but does bind specifically to erythroblasts and to the erythroleukemia cell line K562. These investigators have suggested that this antibody recognizes a lineage-specific antigen whose expression is increased with malignant transformation. However, further clinical use of this antibody has not been reported. Likewise, clinical studies using antibodies SFL23.6 (Gupta *et al.*, 1985), EP-1, and EP-2 (Yokochi *et al.*, 1984) have not been reported.

12. SUMMARY

Although erythrocytes provided one of the earliest model systems with which to study membrane proteins, the number of well-characterized erythroid differentiation antigens is relatively small (summarized in Figure 2). Likewise, only a few of the more than 600 blood group antigens have been investigated as to the timing of their expression during erythropoiesis. Nevertheless, this field is being actively explored, and the next 5–10 years should see major advances in the biochemical characterization of protein-borne blood group antigens and in the understanding of the timing and regulation of their expression. Such progress should also contribute to our understanding of abnormal and malignant erythropoiesis as well.

ACKNOWLEDGMENTS. This work has been funded in part by grant CH-371 from the American Cancer Society and by grant R01 HL33572 from the National Institutes of Health. The author is the recipient of Research Career Development Award K04 HL02233 from the NIH. Dr. Wendell F. Rosse is acknowledged for insightful comments and for help in preparing the illustrations.

REFERENCES

Allen, R. W., and Hoover, B. A., 1983, Reorientation of membrane polypeptides during erythrocyte maturation, *Blood* **61:**803–806.

Andreasen, R. B., and Olsson, L., 1987, Identification and characterization of an antigen specific for normal erythroid precursor cells and its application in diagnosis of erythroleukemia, *Leuk. Res.* **11**:1045–1053.

Anstee, D. J., 1986, Red cell membrane antigens, in: *Red Cell Antigens and Antibodies* (G. Garratty, ed.), pp. 1–15, American Association of Blood Banks, Arlington, Va.

Anstee, D. J., Parsons, S. F., Ridgwell, K., Tanner, M. J. A., Merry, A. H., Thomson, E. E., Judson, P. A., Johnson, P., Bates, S., and Fraser, I. D., 1984, Two individuals with elliptocytic red cells apparently lack three minor erythrocyte membrane sialoglycoproteins, *Biochem. J.* **218**:615–619.

Bettaieb, A., Villeval, J. L., Kieffer, N., Farace, F., Edelman, L., Cartron, J. P., and Vainchenker, W., 1987, Early erythroid markers as probes for normal and leukaemic erythroid differentiation, *Ann. Inst. Pasteur/Immunol.* **138**:877–883.

Bodger, M. P., Izaguirre, C. A., Blacklock, H. A., and Hoffbrand, A. V., 1983, Surface antigenic determinants on human pluripotent and unipotent hematopoietic progenitor cells, *Blood* **61**:1006–1010.

Chow, F. L., Hall, S. E., Rosse, W. F., and Telen, M. J., 1986, Separation of the acetylcholinesterase-deficient red cells in paroxysmal nocturnal hemoglobinuria, *Blood* **67**:893–897.

Civin, C. I., and Loken, M. R., 1987, Cell surface antigens on human marrow cells. Dissection of hematopoietic development using monoclonal antibodies and multiparameter flow cytometry, *Int. J. Cell Cloning* **5**:267–288.

Colin, Y., Rahuel, C., London, J., Romeo, P. H., d'Auriol, L., Galibert, F., and Cartron, J. P., 1986, Isolation of cDNA clones and complete amino acid sequence of human erythrocyte glycophorin C, *J. Biol. Chem.* **261**:229–233.

Colin, Y., Kim, C. L. V., Tsapis, A., Clerge, M., d'Auriol, L., London, J., Galibert, F., and Cartron, J. P., 1989, Human erythrocyte glycophorin C, *J. Biol. Chem.* **264**:3773–3780.

Connor, J., and Schroit, A. J., 1988, Transbilayer movement of phosphatidylserine in erythrocytes: Inhibition of transport and preferential labeling of a 31,000-dalton protein by sulfhydryl reactive reagents, *Biochemistry* **27**:848–851.

Darling, S. M., Banting, G. S., Pym, B., Wolfe, J., and Goodfellow, P. N., 1986, Cloning an expressed gene shared by the human sex chromosomes, *Proc. Natl. Acad. Sci. USA* **83**:135–139.

Dessypris, E. N., Clark, D. A., McKee, L. C., Jr., and Krantz, S. B., 1983, Increased sensitivity to complement of erythroid and myeloid progenitors in paroxysmal nocturnal hemoglobinuria, *N. Engl. J. Med.* **309**:690–693.

Edelman, P., Vinci, G., Villeval, J. L., Vainchenker, W., Henri, A., Miglierina, R., Rouger, P., Reviron, J., Breton-Gorius, J., Sureau, C., and Edelman, L., 1986, A monoclonal antibody against an erythrocyte ontogenic antigen identifies fetal and adult erythroid progenitors, *Blood* **67**:56–63.

Ekblom, M., Gahmberg, C. G., and Andersson, L. C., 1985, Late expression of M and N antigens on glycophorin A during erythroid differentiation, *Blood* **66**:233–236.

Falkenburg, J. H., Fibbe, W. E., van der Vaart-Duinkerken, N., Nichols, M. E., Rubinstein, P., and Jansen, J., 1985, Human erythroid progenitor cells express rhesus antigens, *Blood* **66**:660–663.

Fauser, A. A., and Messner, H. A., 1978, Granuloerythropoietic colonies in human bone marrow, peripheral blood, and cord blood, *Blood* **52**:1243–1248.

Frydecka, I., Brodzka, W., Lawinska, B., and Sciborski, R., 1984, Clinical course and results of treatment of erythroleukemia, *Folia Haematol. (Leipzig)* **111**:282–289.

Fukuda, M., and Fukuda, M. N., 1984, Cell surface glycoproteins and carbohydrate antigens in development and differentiation of human erythroid cells in *The Biology of Glycoproteins* (R. J. Ivatt, ed.), pp. 183–234, Plenum Press, New York.

Fukuda, M., Fukuda, M. N., and Hakomori, S. I., 1979, Developmental change and genetic defect in the carbohydrate structure of band 3 glycoprotein of human erythrocyte membrane, *J. Biol. Chem.* **254**:3700–3703.

Fukuda, M., Fukuda, M. N., Papayannopoulou, T., and Hakomori, S., 1980, Membrane differentiation in human erythroid cells: Unique profiles of cell surface glycoproteins expressed in erythroblasts *in vitro* from three ontogenic stages of differentiation and maturation, *Proc. Natl. Acad. Sci. USA* **77**:3474–3478.

Gahmberg, C. G., 1983, Molecular characterization of the human red cell Rh_0(D) antigen, *Eur. Mol. Biol. Organ. J.* **2**:223–227.

Gahmberg, C. G., Jokinen, M., and Andersson, L. C., 1978, Expression of the major sialoglycoprotein (glycophorin) on erythroid cells in human bone marrow, *Blood* **52**:379–387.

Goodfellow, P. J., Darling, S., Banting, G., Pym, B., Mondello, C., and Goodfellow, P. N., 1987, Pseudoautosomal genes in man, *Development* **101**(Suppl.):119–125.

Goodfellow, P. N., and Tippett, P., 1981, A human quantitative polymorphism related to Xg blood groups, *Nature* **289:**404–405.

Goodfellow, P. N., Banting, G., Sheer, D., Ropers, H. H., Caine, A., Ferguson-Smith, M. A., Povey, S., and Voss, R., 1983, Genetic evidence that a Y-linked gene in man is homologous to a gene on the X chromosome, *Nature* **302:**346–349.

Goodfellow, P. N., Pym, B., Mohandas, T., and Shapiro, L. J., 1984, The cell surface antigen locus, MIC2X, escapes X-inactivation, *Am. J. Hum. Genet.* **36:**777–782.

Gordon, M. Y., Barrett, A. J., and Gordon-Smith, E. C., 1985, *Bone Marrow Disorders: The Biological Basis of Clinical Problems,* pp. 20–59, Blackwell, Oxford.

Greaves, M. F., Sieff, C., and Edwards, P. A. W., 1983, Monoclonal antiglycophorin as a probe for erythroleukemias, *Blood* **61:**645–651.

Greaves, M. F., Katz, F. E., Myers, C. D., Davies, L., and Sieff, C., 1985, Selective expression of cell surface antigens on human haematopoietic progenitor cells, in: *Hematopoietic Stem Cell Physiology* (E. P. Cronkite, N. Dainiak, R. P. McCaffrey, J. Palek, and P. J. Quesenberry, eds.), pp. 301–315, Liss, New York.

Gupta, A. D., Samoszuk, M. K., Papayannopoulou, T., and Stamatoyannopoulos, G., 1985, SFL 23.6: A monoclonal antibody reactive with CFU-E, erythroblasts, and erythrocytes, *Blood* **66:**522–526.

Harvey, J., Parsons, S. F., Anstee, D. J., and Bradley, B. A., 1988, Evidence for the occurrence of human erythrocyte membrane sialoglycoproteins in human kidney endothelial cells, *Vox Sang.* **55:**104–108.

Ievleva, E. S., Endelhardt, N. V., and Abelev, G. I., 1976, Specific antigen of murine erythroblasts, *Int. J. Cancer* **17:**798–805.

Issitt, P. D., 1988, Genetics of the Rh blood group system: Some current concepts, *Med. Lab. Sci.* **395:**404.

Issitt, P. D., 1989, The Rh blood group system, 1988: Eight new antigens in nine years and some observations on the biochemistry and genetics of the system, *Transf. Med. Rev.* **3:**1–12.

Kanamaru, A., Okuda, K., Ueda, E., Kitani, T., Kinoshita, T., and Nagai, T., 1988, Different distribution of decay-accelerating factor on hematopoietic progenitors from normal individuals and patients with paroxysmal nocturnal hemoglobinuria, *Blood* **72:**507–511.

Kay, M. M., Bosman, G. J., and Lawrence, C., 1988, Functional topography of band 3: Specific structural alteration linked to functional aberrations in human erythrocytes, *Proc. Natl. Acad. Sci. USA* **85:**492–496.

Latron, F., Blanchard, D., and Cartron, J. P., 1987, Immunochemical characterization of the human blood cell membrane glycoprotein recognized by monoclonal antibody 12E7, *Biochem. J.* **247:**757–764.

Liszka, K., Majdic, O., Bettelheim, P., and Knapp, W., 1983, Glycophorin A expression in malignant hematopoiesis, *Am. J. Hematol.* **15:**219–226.

Loken, M. R., Civin, C. I., Bigbee, W. L., Langlois, R. G., and Jensen, R. H., 1987, Coordinate glycosylation and cell surface expression of glycophorin during normal human erythropoiesis, *Blood* **70:**1959–1961.

Mattei, M. G., Colin, Y., Le Van Kim, C., Mattei, J. F., and Cartron, J. P., 1986, Localization of the gene for human erythrocyte glycophorin C to chromosome 2, q14–q21, *Hum. Genet.* **74:**420–422.

Mechetner, E. B., Tonevitsky, A. G., Ievleva, E. S., and Rozinova, E. N., 1987, Identification of a human erythroid cell surface antigen by monoclonal antibody HAE9, *Exp. Hematol.* **15:**355–359.

Moore, J. G., Frank, M. M., Muller-Eberhard, H. J., and Young, N. S., 1985, Decay-accelerating factor is present on paroxysmal nocturnal hemoglobinuria erythroid progenitors and lost during erythropoiesis in vitro, *J. Exp. Med.* **162:**1182–1192.

Moore, S., and Green, C., 1987, The identification of specific rhesus polypeptide/blood-group-ABH active glycoprotein complexes in the human red cell membrane, *Biochem. J.* **244:**735–741.

Nakahata, T., and Ogawa, M., 1982, Identification in culture of a class of hemopoietic colony-forming units with extensive capability to self-renew and generate multipotential hemopoietic colonies, *Proc. Natl. Acad. Sci. USA* **79:**3843–3847.

O'Hara, C. J., Shumkak, K. H., and Price, G. B., 1978, The i antigen on hyman myeloid progenitors, *Clin. Immunol. Immunopathol.* **10:**420–426.

Rahuel, C., London, J., d'Auriol, L., Mattei, M. G., Tournamille, C., Skrzynia, C., Levouc, Y., Galibert, F., and Cartron, J. P., 1988, Characterization of cDNA clones for human glycophorin A, *Eur. J. Biochem.* **172:**147–153.

Rearden, A., and Chiu, P., 1983, Lack of rhesus antigen expression by human committed erythroid progenitors, *Blood* **61:**525–529.

Rearden, A., and Masouredis, S. P., 1977, Blood group D antigen content of nucleated red cell precursors, *Blood* **50:**981–986.

Reid, M. E., Anstee, D. J., Tanner, M. J. A., Ridgwell, K., and Nurse, G. T., 1987, Structural relationships between human erythrocyte sialoglycoproteins β and γ and abnormal sialoglycoproteins found in certain rare human erythrocyte variants lacking Gerbich blood-group antigen(s), *Biochem. J.* **244:**123–128.

Ridgwell, K., Roberts, S. J., Tanner, M. J. A., and Anstee, D. J., 1983, Absence of two membrane proteins containing extracellular thiol groups in Rh_{null} human erythrocytes, *Biochem. J.* **213:**267–269.

Robinson, J., Sieff, C., Delia, D., Edwards, P. A. W., and Greaves, M., 1981, Expression of cell-surface HLA-DR, HLA-ABC and glycophorin during erythroid differentiation, *Nature* **289:**68–71.

Sakurai, T., Hara, H., and Nagai, K., 1988, Anti-K562 cell monoclonal antibody RTF8X recognizes tumor-associated antigen of erythroid lineage, *Blood* **72:**1487–1491.

Siebert, P. D., and Fukuda, M., 1986, Human glycophorin A and B are encoded by separate, single copy genes coordinately regulated by a tumor-promoting phorbol ester, *J. Biol. Chem.* **261:**12433–12436.

Sieff, C., 1986, Membrane antigen expression during hematopoietic differentiation, *CRC Crit. Rev. Oncol. Hematol.* **5:**1–36.

Sieff, C., and Greaves, M. F., 1983, Cell surface characteristics of hematopoietic cells, in: *Red Cell Membrane Glycoconjugates and Related Genetic Markers* (J. P. Cartron, P. Rouger, and C. Salmon, eds.), pp. 243ff, Librairie Arnette, Paris.

Sieff, C., Bicknell, D., Caine, G., Robinson, J., Lam, G., and Greaves, M. F., 1982, Changes in cell surface antigen expression during hematopoietic differentiation, *Blood* **60:**703–713.

Steck, T. L., 1978, The band 3 protein of the human red cell. A review, J. Supramol. 8:311–423.

Tate, C. G., and Tanner, M. J. A., 1988, Isolation of cDNA clones for human erythrocyte membrane sialoglycoproteins α and δ, *Biochem. J.* **254:**743–750.

Telen, M. J., and Borowitz, M. J., 1989, Glycophorin C as a marker of erythroid differentiation in non-lymphoid leukemias, manuscript in preparation.

Telen, M. J., and Chasis, J. A., 1989, Relationship of the human erythrocyte Wr^b antigen to an interaction between glycophorin A and band 3, submitted for publication.

Telen, M. J., Scearce, R. M., and Haynes, B. F., 1987, Human erythrocyte antigens. III. Characterization of a panel of murine monoclonal antibodies that react with human erythrocyte and erythroid precursor membranes, *Vox Sang.* **52:**236–243.

Telen, M. J., Hall, S. E., Green, A. M., Moulds, J. J., and Rosse, W. F., 1988, Identification of human erythrocyte blood group antigens on decay-accelerating factor (DAF) and an erythrocyte phenotype negative for DAF, *J. Exp. Med.* **167:**1993–1998.

Vainchenker, W., Testa, U., Rochant, H., Titeux, M., Henri, A., Bouguet, J., and Breton-Gorius, J., 1981, Cellular regulation of i and I antigen expression in human erythroblasts grown *in vitro, Stem Cells* **1:**97–110.

Villeval, J. L., Testa, U., Vinci, G., Tonthat, H., Bettaieb, A., Titeux, M., Cramer, P., Edelman, L., Rochant, H., Breton-Gorius, J., and Vainchenker, W., 1985, Carbonic anhydrase I is an early specific marker of normal human erythroid differentiation, *Blood* **66:**1162–1170.

Villeval, J. L., Cramer, P., Lemoine, F., Henri, A., Bettaieb, A., Bernaudin, F., Beuzard, Y., Berger, R., Flandrin, G., Breton-Gorius, J., and Vainchenker, W., 1986, Phenotype of early erythroblastic leukemias, *Blood* **68:**1167–1174.

Walloch, J., Frankel, S., Hrisinko, M. A., and Weil, S. C., 1986, Carbonic anhydrase: A marker for the erythroid phenotype in acute nonlymphocytic leukemia, *Blood* **68:**304–306.

Williams, B. P., Daniels, G. L., Pym, B., Sheer, D., Povey, S., Okubo, Y., Andrews, P. W., and Goodfellow, P. N., 1988, Biochemical and genetic analysis of the Ok^a blood group antigen, *Immunogenetics* **27:**322–329.

Yokochi, T., Brice, M., Rabinovitch, P. S., Papayannopoulou, T., and Stamatoyannopoulos, G., 1984, Monoclonal antibodies detecting antigenic determinants with restricted expression on erythroid cells: From the erythroid committed progenitor level to the mature erythroblast, *Blood* **63:**1376–1384.

Chapter 3

Erythrocyte Membrane Changes during Aging in Vivo

Grzegorz Bartosz

1. INTRODUCTION

The mammalian red blood cell is a specific but interesting object for research on aging. Devoid of a nucleus and other intracellular organelles, it lacks transcriptional and translational machinery. Although studies of this cell obviously cannot yield direct information on the DNA control of cellular aging, they do allow us to follow cell-age-related modifications of intracellular proteins and of the plasma membrane without the interference of the repair processes to the exchange of damaged macromolecules. Thus, the red blood cell is an ideal model for studies of membrane and protein aging in their natural cellular environment.

Another interesting problem in red blood cell aging is the mechanism of recognition and selective removal of senescent erythrocytes. Apart from its purely cognitive value, enabling understanding of cellular homeostasis in the body (perhaps tenable to other cells as well), this question is of practical importance for blood banking as knowledge of such a mechanism might lead to means of prolongation of the shelf life of fluid-supplemented erythrocyte concentrates and whole blood and open new therapeutic possibilities for the treatment of anemias.

Practical stimuli for studies of erythrocyte aging also include (1) the growing understanding that merely changes in the mean age of circulating erythrocyte populations may affect values of red blood cell parameters in routine hematological tests and (2) the recent trend to use only fractions of young erythrocytes (neocytes) instead of whole blood or unseparated erythrocytes for transfusion, aimed at a reduction of transfusion frequency

---

Grzegorz Bartosz Laboratory of Biophysics of Development and Aging, Department of Biophysics, University of Lodz, 90-237 Lodz, Poland.

and of the iron load of thalassemia patients (Bracey and McFarland, 1984; Picot *et al.*, 1987; Piomelli *et al.*, 1978).

Obviously, the combination of several factors such as the simplicity of mammalian red blood cells, the apparent ease of their age fractionation, and the practical implications of these investigations has stimulated numerous workers for several decades to study erythrocyte aging and to intensify these studies in recent years.

Among mammalian red blood cells, the life span of the human cell is most interesting, at least from a practical point of view because of the possible medical applications, so inevitably most studies have been performed on human erythrocytes. Therefore, if not otherwise stated, in what follows I will deal with results concerning human red cells.

2. HOW CAN THE AGING OF THE RED BLOOD CELL BE STUDIED?

A prerequisite for studies on alterations in erythrocyte properties during cellular aging is the desirability of comparing cell populations of various mean age. Several approaches have been employed for this purpose, none of them lacking drawbacks or disadvantages.

An ingenious early approach consisted of transfusion (in cattle *Bos taurus*) of red cells to a recipient whose own erythrocytes differed antigenically but who tolerated red cells of the donor. Then, by withdrawal of blood samples from the recipient at desired time intervals and lysing the recipient's erythrocytes with a specific antiserum, it was possible to study the donor's cells at various times after transfusion (Todd and White, 1911). An essentially similar method was used for human erythrocytes except that agglutination, rather than lysis, was employed for the removal of the recipient's cells (indirect differential agglutination; Ashby, 1919). A modification of this method enabled isolation of donor red cells by using an antiserum that agglutinated them but did not agglutinate recipient erythrocytes (direct differential agglutination). In this way, donor MM erythrocytes could be isolated from the blood of an NN recipient (Wiener, 1942) and donor Rh-positive cells could be recognized in an Rh-negative recipient using anti-Rh antibodies (Boorman *et al.*, 1942). However, technical difficulties limit the usefulness of this approach: the number of transfused cells is always small in relation to those of the recipient, and the separation of donor and recipient cells is not ideal, due, e.g., to cell coaggregation (Beutler, 1985b).

Another approach used for comparison of erythrocyte populations of various mean age is based on production of reticulocyte-rich blood, either by repeated bleeding or by oxidative (e.g., phenylhydrazine) stress. This is, however, related directly rather to the question of red cell maturation than senescence.

The vast majority of studies have been based on the separation of circulating erythrocytes into fractions of various mean age on the basis of differences in physical properties: osmotic fragility, surface charge density, and cell density.

Early workers shared the belief that older erythrocytes were more osmotically fragile than were younger ones (Chalfin, 1956; Levy *et al.*, 1959; Marks *et al.*, 1958a,b; Sass *et al.*, 1963; Simon and Topper, 1957), which was the premise for methods of age separation of red blood cells by serial osmotic hemolysis. Cells hemolyzing most easily, in solutions of highest tonicity, have been assumed to represent the oldest cells and those most resistant, the youngest (Bartosz and Desforges, 1967; Fornaini, 1967; Marks *et al.*, 1958a; Simon and Topper, 1957). However, more recent data have cast doubt on the relation between red cell aging and osmotic fragility (see Section 3.2), thus calling into question results of at least some studies employing those methods.

Separation of cells by preparative electrophoresis has been proposed on the basis of the decreasing surface charge density of aged erythrocytes (Gear, 1977), as indicated by earlier investigators (Danon and Marikovsky, 1961; Yaari, 1969). (Interestingly, these findings have not been confirmed by more recent studies; see Section 3.3.)

An interesting procedure for obtaining erythrocyte populations of increasing age, although painstaking and very demanding of experimental animals, is based on serial hypertransfusion. Half of a group of syngeneic animals are sacrificed and their blood transfused to the remaining animals to obtain a high hematocrit that suppresses erythropoiesis. This procedure is repeated every few days up to the time corresponding to nearly one red-cell life span, and produces after n hypertransfusions, $1/2^n$ of the original number of animals whose erythrocytes are appropriately aged. This procedure has been employed in rats (*Rattus rattus*) (Ganzoni *et al.*, 1971, 1976) and mice (*Mus musculus*) (Morrison *et al.*, 1983; Mueller *et al.*, 1987). In mice the survival half-time of erythrocytes obtained from a 56-day hypertransfused animal (estimated with ^{51}Cr) was less than 1 day (versus 15 days for the control) (Morrison *et al.*, 1983). Reservations concern the possibility that inhibition of hemopoiesis is less than complete and that such inhibition may alter red cell aging (Vaysse *et al.*, 1988b).

Most studies on the *in vivo* aging of red blood cells have consisted of comparison of the properties of erythrocytes separated by density based on the assumption that increasing red cell density is a function of their age. Erythrocytes have been separated into density (and presumably age) fractions by centrifugation of whole blood (O'Connell *et al.*, 1965) or packed erythrocytes (Gattegno *et al.*, 1975; Hall and Ellory, 1986; Prankerd, 1958) and separation of the top, lightest portion as being enriched in young cells and the bottom, densest portion as being enriched in old cells; by ultracentrifugation of a packed erythrocyte mass in an angle rotor (Murphy, 1973; by centrifugation through a series of phthalate ester mixtures of appropriate gravity (Brok *et al.*, 1966; Danon and Marikovsky, 1964, 1988; Joiner and Lauf, 1978; Takeshita *et al.*, 1983; Yaari, 1969); and by centrifugation through continuous or discontinuous density gradients of various substances like gum acacia (Kimura *et al.*, 1960), Stractan (arabinogalactan) (Corash *et al.*, 1974), bovine serum albumin (Baxter and Beeley, 1978; Bishop and Prentice, 1966; Leif and Vinograd, 1964; Piomelli *et al.*, 1967), dextran (Botscharova, 1973; Gross *et al.*, 1978), Ficoll–Triosil (Turner *et al.*, 1974), and, lately, Percoll (Branch *et al.*, 1983; Mackie *et al.*, 1987; Rennie *et al.*, 1979; Salvo *et al.*, 1982; Vettore *et al.*, 1980). Correlation between cell age and density has been demonstrated by monitoring the distribution of a radiolabel (such as ^{59}Fe or [^{14}C]glycine, which can be incorporated only into maturing cells) in the fractionated erythrocytes. ^{59}Fe radioactivity was found mostly (though not exclusively) in the lightest cell fractions soon after administration to a volunteer (or an animal) and the radioactivity peak moved down the cellular density distribution with time, as expected for a cell cohort whose density increases upon aging (Bennett and Kay, 1981; Bishop and Prentice, 1966; Piomelli *et al.*, 1967; ten Brinke and de Regt, 1970). In some cases, instead of a progressive shift of the radioactivity peak in the direction of increasing cell density, the distribution of label became gradually more homogeneous, radioactivity being increasingly equally found in all density fractions (Bartosz *et al.*, 1982b; Luthra *et al.*, 1979), supposedly due to the reutilization of hemoglobin iron.

The validity of age fractionation of erythrocytes by density has recently been questioned on the basis of observations made during hypertransfusion experiments. In mice, even after a 56-day hypertransfusion the red blood cells exhibit a broad range of densities, which suggests that the densest erythrocytes do not represent a homogeneous population of the oldest cells (Mueller *et al.*, 1987).

Moreover, it cannot be excluded that the increase in density (or any other change in the physical properties of the erythrocyte) is not a continuous function of age and that the cell is instead subject to rapid alterations just before its elimination (Clark, 1986, 1988).

In spite of these reservations, it can apparently be assumed that comparison of erythrocyte fractions of various densities allows one to interpret the progressive change of erythrocyte properties. However, one must be aware that this approach may not permit detection of phenomena occurring during the very last (and perhaps most interesting) period of erythrocyte survival, and that comparison of middle-density fractions with the lightest, reticulocyte-rich fractions may reflect changes occurring during red cell maturation rather than senescence (Beutler, 1985a,b).

In any case, the above-mentioned reservations seem to preclude quantitative conclusions on the relationships between red blood cell age and cellular properties on the basis of the lightest and densest cells, even assuming that an ideal cell separation according to density has been achieved.

A more recent technique, counterflow centrifugation (elutriation), is based on erythrocyte separation by cell size instead of density (Sanderson and Bird, 1977; Sanderson *et al.*, 1976; van der Vegt *et al.*, 1985b). It has been claimed to yield a superior age separation of erythrocytes compared to density fractionation (van der Vegt *et al.*, 1985a). This claim has been questioned (Vaysse *et al.*, 1988a,b), especially in mice where red blood cell fractions obtained by counterflow centrifugation did not differ in their survival times (Vaysse *et al.*, 1988a). Comparison of density-based and size-based fractionation of erythrocytes suggests that the pattern of age modification of these two parameters is not the same. Perhaps a combination of these two techniques may be superior to either used alone (van der Vegt *et al.*, 1985a).

Recently, some other procedures have been proposed for age separation of erythrocytes and isolation of senescent cells. Red blood cells of different ages show differences in their affinity for alkyl-Sepharose 6 MB, so that cell chromatography on *n*-propylamine-, *n*-hexylamine, *n*-octylamine-, or *n*-decylamine-Sepharose 6 MB columns might be used for age separation of erythrocytes (Shinozuka *et al.*, 1986), but evidence suggests that this technique is not superior to those used previously. A promising idea employed recently for the study of red cell aging *in vivo* is biotinylation of erythrocytes with *N*-hydroxysuccinimidbiotin. This procedure does not influence red blood cell survival; the covalently bound biotin remains on the cell surface and the biotinylated cells can be recovered from the circulation at any time by affinity chromatography on avidin-coated beads (Suzuki and Dale, 1987, 1988). Some results using this technique—e.g., an apparent increase in cellular ATP concentration by about 75% and a lack of activity changes of several glycolytic enzymes in rabbit (*Oryctolagus cuniculus*) erythrocytes 50–60 days after reinfusion into the circulation (Suzuki and Dale, 1988)—differ considerably from those obtained with density-fractionated cells.

Undoubtedly, new methods for the isolation of senescent cells are still needed. There is a consensus that the content of IgG on the surface of red blood cells increases with age (see Section 4.10). A method of cell separation based on the amount of bound IgG (e.g., in a cell sorter) might be of value, although technically difficult due to the low amounts of IgG present.

3. AGE-RELATED CHANGES IN MEMBRANE-DEPENDENT PHYSICAL PROPERTIES OF RED BLOOD CELLS

3.1. Cell Volume, Surface, and Shape

With one exception (Wilton, 1966), all studies demonstrate a volume reduction during red blood cell aging (see Table I). Hypertransfusion experiments on rat erythrocytes showed that the mean cellular volume decreased in a linear fashion with time, a 13–14% reduction being reached by day 40–45 of aging (Ganzoni *et al.*, 1971). However, analogous studies in mice did not show a progressive decrease in cell density with time (Morrison *et al.*, 1983).

Two factors appear to contribute to the decrease in cell volume with aging. One is a net electrolyte loss. Older cells contain progressively less potassium, the reported content of sodium being constant or increasing too slightly to compensate for the substantial fall in potassium content (Bartosz *et al.*, 1981a; Bernstein, 1959; Borun, 1964, Clark, 1985; Cohen *et al.*, 1976). The decrease in electrolyte content brings about osmotic water loss, leading to progressive cellular dehydration and increase in density.

The other factor, suggested by many authors, may be microvesiculation, i.e., release of small membrane fragments encapsulating a portion of the cytoplasm. Release of

Table I
Volume, Area, Area/Volume Ratio, and Diameter of Density-Separated Human Erythrocytes

	Cell fractions		Reference
	Top 10%	*Bottom 10%*	
Volume (fl)	91 ± 4	84 ± 4[a]	Tillman *et al.* (1980)
	Top[b]	*Bottom*[b]	
	111 ± 13%	99 ± 14%[c]	Nash and Wyard (1980)
Area (μm²)	141 ± 7%	130 ± 8%	
Area/volume	1.27 ± 5%	1.32 ± 7%	
Diameter (μm)	8.3 ± 5%	7.9 ± 5%	
	Top 5%	*Bottom 5%*	
Volume (fl)	95 ± 4	79 ± 3[a]	Linderkamp and Meiselman (1982)
	98 ± 13	78 ± 12[c]	

[a]Measured in a Coulter-type counter.
[b]Fraction size not specified.
[c]Derived from micropipette measurements.

microvesicles has been demonstrated during blood storage, metabolic depletion of erythrocytes and their damage by various agents, e.g., heat (Greenwalt and Dumaswala, 1988; Lutz *et al.*, 1977; Snyder *et al.*, 1985; Weed and Reed, 1966). Young (light) erythrocytes were found to shed microvesicles in significantly higher amounts under blood bank conditions (Greenwalt and Dumaswala, 1988) and after metabolic depletion (Snyder *et al.*, 1985) than old (dense) cells. Blood does not contain appreciable amounts of microvesicles although small amounts of erythrocyte-derived microvesicles from fresh normal blood have been reported (Greenwalt *et al.*, 1980; Dumaswala and Greenwalt, 1984). In vivo microvesiculation may be connected with red cell remodeling in the spleen (Halbhuber *et al.*, 1986).

Apparently, both suggested means of volume loss should be invoked to account for the observed pattern of volume and surface loss. Microvesiculation alone would not lead to cellular dehydration and would result only in a substantial loss of the surface to volume ratio of the cells. It would also bring about a considerable loss of cellular hemoglobin content, which is consistently reported by various authors to remain fairly constant during the life span of the red cell. On the other hand, electrolyte loss alone would lead to an increase in the ratio of cell surface to cell volume.

Although most authors agree that erythrocyte size decreases during aging (except for the unknown behavior of the oldest cells, just prior to their removal from the circulation), the changes in red cell shape are a matter of controversy. Some authors report that erythrocytes become more spherical as they age, based on observations by light microscopy (Ogiso *et al.*, 1985; Shiga *et al.*, 1979), while others have found no shape alterations (Bartosz *et al.*, 1981b; Tillman *et al.*, 1980).

Results of micropipette measurements indicate that sphericity, defined as 4.84(volume)$^{2/3}$/area, increases by less than 2% for the densest erythrocytes compared to the lightest, in both humans and rabbits; this would mean that red cells do not become spherical as they approach senescence (Nash and Wyard, 1982). However, it has been argued that since the cells become smaller, the lack of change in the surface-to-volume ratio is indicative of an increasingly spherical shape (van Oss, 1982).

3.2. Hemolytic Fragility

One would expect intuitively that red cell aging is accompanied by a decrease in its stability expressed by an increased sensitivity of its membrane to stressing factors. It is indeed generally true, although as perhaps for any general statement in biology, it cannot be taken for granted in all cases.

Early reports indicated an increased osmotic fragility of the denser, and presumably older, erythrocytes in hypotonic NaCl solutions (Chalfin, 1956; Hoffman, 1958; Levy *et al.*, 1959; Marks and Johnson, 1958; Marks *et al.*, 1958a,b; Rahman *et al.*, 1973; Sass *et al.*, 1963; Simon and Topper, 1957). On this basis, serial osmotic hemolysis has been proposed as a method for red cell separation according to age (see Section 2). However, even the earliest papers devoted to this question are not concordant. It has been noted (Cruz *et al.*, 1941) that new erythrocytes, formed after repeated bleeding, are less resistant to hypotonic hemolysis, and some authors (e.g., Prankerd, 1958) were unable to confirm the difference in osmotic fragility between the densest and lightest erythrocytes. Even now, a definite answer to the question of whether older red cells are more prone to hypotonic hemolysis is still equivocal. Part of the problem may lie in methodological and interspecies differences. For example, older (denser) bovine erythrocytes do not show an increased osmotic fragility (Bartosz, 1981d); indeed, they may even show a decreased osmotic fragility (Mosior *et al.*, 1984). Human erythrocytes separated by counterflow centrifugation were reported to show a progressive increase in osmotic fragility with decreasing cell size (van der Vegt *et al.*, 1985a), but several authors found no such differences between various density fractions of these cells (Cohen *et al.*, 1976; Luthra *et al.*, 1979), yet others found a higher osmotic fragility for the denser cells (Malachi *et al.*, 1986). A thorough study (Rifkind *et al.*, 1983) demonstrated that the major effect of erythrocyte aging is to produce a more heterogeneous distribution of osmotic fragility with no significant alterations of the midpoint in the osmotic fragility curve. Interestingly, the authors demonstrated that graded osmotic hemolysis is not an effective means for age fractionation of erythrocytes since the highest aspartate aminotransferase activity and the lowest hemoglobin A_{Ic} content were found in the cell fraction showing intermediate osmotic fragility.

No differences were reported in the lysis time in isosmotic glycerol solution for human erythrocyte fractions of various densities (Cohen *et al.*, 1976). I have observed a decrease in the glycerol lysis time and in the time for lysis in isoosmotic NH_4Cl solution for bovine red cells with increasing density (Bartosz, 1982b). (A reason for the species differences may be the membrane transport of glycerol by simple diffusion in bovine red cells and by facilitated diffusion in human erythrocytes.)

Comparison of the temperature dependence of the heat-induced lysis of sheep (*Ovis aries*) erythrocytes separated by density (top 10% versus bottom 10%) showed that the hemolysis can be described by a power function of time:

$$- (1/N)\ (dN/dt) = A\ t \exp(n - 1) \tag{1}$$

where N is the cell number, dN/dt the rate of hemolysis, and t the time. The Arrhenius-type plot of A was different for young (lightest) and old (densest) erythrocytes. For young erythrocytes the plot was linear in the temperature range of 42–56°C, indicating an activation enthalpy of hemolysis of 132 kcal/mol. Old cells were more fragile and on the Arrhenius plot showed a transition centered at about 50°C, with an activation enthalpy of 62 kcal/mol below and 83 kcal/mol above the transition point (Juckett and Rosenberg, 1982).

Older (denser) human erythrocytes were reported to be more prone to hemolysis in 0.172 M Tris-HCl buffer (37°C, pH 7.6) (Cohen *et al.*, 1976), to mechanical hemolysis (Danon, 1968), and to hemolysis induced by chlorpromazine and clemastine. Also, their osmotic fragility was affected more by 50 μM chlorpromazine compared to the young (light) cells (Ogiso *et al.*, 1985). On the other hand, young (light) red cells showed higher hemolysis in a histidine buffer containing a high (40 mM) concentration of Ca^{2+} (Luthra *et al.*, 1979) and their osmotic fragility was increased to a greater extent by parathyroid hormone (Malachi *et al.*, 1986).

Old (dense) human red cells were more prone to hemolysis induced by 10 mM H_2O_2 (Powers and Thurnham, 1980). Incubation in a system of xanthine oxidase and hypoxanthine generating reactive oxygen species ($O_2^{-}\cdot$, H_2O_2, $\cdot OH$) brought about a much greater lysis of old (dense) than of young (light) human (Walls *et al.*, 1976) and rat

(Gershon *et al.*, 1988) red cells. I have found that bovine red cells showed an age (density)-dependent increase in the hemolytic sensitivity to the "severe" oxidative stress induced by higher concentrations of H_2O_2, ascorbate (10 mM), and higher doses of gamma radiation (e.g., 500 Gy), while there was no significant augmentation of sensitivity to "mild" oxidative stress provoked by lower concentrations of ascorbate (3 mM) or lower doses of ionizing radiation (e.g., 100 Gy; Bartosz, 1981c).

3.3. Surface Charge

Early studies showed that the densest (oldest) human erythrocytes have an up to 30% lower electrophoretic mobility than the lightest (youngest) cells (Danon and Marikovsky, 1961; Yaari, 1969), indicating a loss of surface charge density. However, a more recent reinvestigation of this question did not confirm these findings: it was found using conventional electrophoresis, streak width electrophoresis, and electrophoretic light scattering that electrophoretic mobility is the same for the 5–7% lightest and the 2–10% densest human erythrocytes, separated by density according to two different methods. It implies that the effective net number of negative charges per unit area of membrane surface remains constant, even though the total number of charged groups may be decreased, e.g., by membrane loss (Luner *et al.*, 1977; Seaman *et al.*, 1977).

The main determinants of the negative surface charge of the mammalian erythrocyte are the sialic acid residues. The content of sialic acid per cell has been reported to decrease by 10–15% in the densest red cell fractions obtainable in several mammalian species (Bartosz *et al.*, 1984; Baxter and Beeley, 1978; Gattegno *et al.*, 1975; Greenwalt *et al.*, 1971), but the question as to whether the sialic acid content per membrane protein also decreases or remains constant is still debatable (see also Section 4.9).

However, age-related differences in surface charge properties of erythrocytes were also detected by cell partitioning in the two-phase system of dextran and poly(ethylene glycol) under conditions where a potential difference between the phases makes surface charge the main determinant of the partition. Initially, such differences were reported for rat but not human erythrocytes (Walter *et al.*, 1981) and ascribed to the ganglioside-bound sialic acid residues, which are fairly abundant in rat but not in human erythrocytes (Walter *et al.*, 1983). Subsequent precise studies of the cell distribution pattern during countercurrent separation in a two-phase system revealed a small but detectable change in the partition coefficient between the phases for aging human erythrocytes (Walter and Krob, 1983).

It is possible that although the net surface charge density determining the electrophoretic mobility of the cells does not change, there is an alteration in the spatial distribution of surface charges in the course of erythrocyte aging. Young erythrocytes show an increased label density upon incubation with positively charged colloidal particles such as iron hydroxide or cationized ferritin (Danon *et al.*, 1972; Marikovsky and Danon, 1969). Ferritin particles are about 12 nm in diameter and could potentially cover an area of 24 nm^2 on the cell surface. The glycophorin molecule, the main erythrocyte membrane receptor for ferritin, is much smaller, so the decreased ferritin binding by older erythrocytes is most likely due to steric hindrance brought about by aggregation of glycophorin (Seaman, 1983).

3.4. Cell Deformability

The red cell during its continuous travel through the circulation is forced countless times to pass capillaries that are smaller than its own diameter. The human erythrocyte has a diameter of about 8 μm and traverses microcapillaries that are 3 to 4 μm in diameter (Braasch, 1971). Therefore, the ability to reversibly change shape is of primary physiological importance for the erythrocyte and many studies have been devoted to the effect of red cell age on its deformability.

Decreased deformability of aged (densest) erythrocytes has been found using the following techniques: measurement of filtration time, initially through paper filters (Teitel, 1977) and, more recently, through polycarbonate filters (Goebel and Lanser, 1983; Tillman *et al.*, 1980); the ease of cell aspiration into micropipettes (LaCelle *et al.*, 1973a; Linderkamp and Meiselman, 1982; Nash and Meiselman, 1983; Nash and Wyard, 1980; Sutera *et al.*, 1985); the extent of deformation under the influence of centrifugal force (Corry and Meiselman, 1978); viscometry (Ogiso *et al.*, 1985; Tillman *et al.*, 1980; Usami *et al.*, 1971); microscopic measurement of cell deformation in a cone-plane viscometer (Rheoscope) (Sutera *et al.*, 1985); and ektacytometry (Clark *et al.*, 1983).

Both extrinsic and intrinsic factors determine red cell deformability. Extrinsic factors include cell size, morphology, and area-to-volume ratio; intrinsic ones consist of membrane mechanical properties and cytoplasmic viscosity (Nash and Meiselman, 1981).

Of the extrinsic factors, the most important seem to be the excess surface area fraction and minimum cylindrical diameter of a cell. Excess surface area is usually expressed as the excess surface area index (ESI), defined as the ratio of the actual cell surface area to the area of a sphere of the cell's volume:

$$\mathrm{ESI} = S/(36\pi)^{1/3}\, V^{2/3} \tag{2}$$

where S and V denote cell surface and volume, respectively.

Available data on the cell-age-related changes of ESI are contradictory: the excess surface area fraction was reported both to decrease substantially in old erythrocytes, by 25% (5% densest versus 5% lightest cells) (Linderkamp and Meiselman, 1982) and even by 48–65% (10% densest versus 10% lightest cells) (Sutera *et al.*, 1985), and even to show no significant changes (Nash and Wyard, 1980).

The minimum cylindrical diameter (MCD) is a measure of the smallest cylindrical opening through which a cell can pass without altering its surface area or volume. MCD was reported not to increase in old (dense) erythrocytes (Nash and Wyard, 1980).

While the role of extrinsic determinants in the diminution of deformability of old erythrocytes is still disputable, that of intrinsic determinants seems at present less controversial. Both alterations in membrane properties and increase in the hemoglobin concentration may contribute to decreased deformability of aged red cells. Viscosity of hemoglobin solutions is not a linear function of the protein concentration and increases very steeply above physiological concentrations. Sutera *et al.* (1985) calculated from their data indicating an 11% decrease in cell volume between the lightest 10% and densest 10% red cells that the intracellular hemoglobin viscosity should increase by 61% (from about 8 mPa sec to about 12.9 mPa sec). Williams and Morris (1980) predict a much higher increase during red cell aging, from about 9 mPa sec for young cells with an intracellular hemoglobin concentration of 31.7 g/dl to some 54 mPa sec in old cells with a hemoglobin concentration of 37.5 g/dl.

Kinetic studies of erythrocyte shape recovery after an abrupt cessation of shearing stress by turning off the viscometer or cutting off the pressure applied to a micropipette allowed for a further insight into the nature of age-dependent alterations in mechanical properties of the red cell membrane. For example, the data from rheoscope measurements could be fitted by a three-parametric equation:

$$(L - L_f)/(L_i - L_f) = \exp(-t/t_c) \tag{3}$$

where L is the instantaneous length corresponding to time t, L_i is the initial and L_f the final length (or diameter) of a discoidal cell, and t_c is the characteristic recovery time, a function of the viscoelastic properties of the erythrocyte membrane. Assuming that the red cell membrane can be described by the Kelvin–Voigt model of a viscoelastic solid, charcterized by the presence of elastic and viscous elements acting in parallel, t_c can be expressed as

$$t_c = n/u \tag{4}$$

with n being the membrane surface viscosity and u the two-dimensional (surface) shear elastic modulus.

Both micropipette and Rheoscope measurements have demonstrated that during aging of human erythrocytes the main change concerns membrane surface viscosity, the membrane shear elastic modulus showing only slight, if any, alteration (Table II). Thus, the red blood cell membrane does not stiffen but becomes more viscous during *in vivo* aging (Nash *et al.*, 1988).

3.5. Adhesiveness

Of the other surface properties that might be relevant for cell physiology and pathology, erythrocyte adhesiveness (especially to endothelial cells) should be considered.

Table II
Characteristic Recovery Time, Membrane Surface Viscosity, and Membrane Shear Elastic Modulus of Density-Separated Human Erythrocytes

t_c (msec)	$10^4\ n$ (dyn sec/cm)	$10^3\ u$ (dyn/cm)	Reference
127/152			Sutera *et al.* (1985)[a]
99/148	5.4/9.5	5.5/6.4	Linderkamp and Meiselman (1982)[b]
130/190	7.2/12.2		Nash and Meiselman (1983)[c]

[a]Top 10%/bottom 10% cells, rheoscope.
[b]Top 5%/bottom 5% cells, micropipette.
[c]Top 5%/bottom 5% cells, micropipette.

When this property was addressed experimentally, it was found that the oldest (densest) red blood cells had a higher adhesiveness to endothelial cells than the youngest (lightest). Although the effect of cell aging could be mimicked to some extent by neuraminidase treatment, removal of sialic acid from the cell surface would have to exceed 80% in order to obtain the adhesiveness of naturally aged cells, indicating that sialic acid loss alone is not the determining factor in this aging effect (Dhermy *et al.*, 1987).

3.6. Membrane Transport

In an attempt to explain the differences in cation content, several investigators have studied monovalent cation transport in density-separated erythrocytes. The results have been conflicting to some extent. Various authors have reported a decreased uptake of $^{42}K^{+}$ and $^{24}Na^{+}$ in old (densest) cells (Bernstein, 1959), no difference in $^{42}K^{+}$ and $^{24}Na^{+}$ uptake in density-separated cells (Borun, 1964), and a 20% higher K^{+} uptake (per liter cells) in old cells (Tosteson *et al.*, 1955). A more recent study has indicated that the active K^{+} transport is higher by about 37% in the youngest (lightest) 20% of erythrocytes (Joiner and Lauf, 1978). Measurements of [^{3}H]ouabain binding to density-separated cells suggest that this decrease in the transport rate is mainly, if not entirely, due to the ~33% loss in the number of ouabain-binding sites, i.e., the Na^{+}/K^{+} pump (Hentschel *et al.*, 1986; Joiner and Lauf, 1978) without any significant change in the turnover number of the pump. Such a loss might occur by microvesiculation. The lack of significant cell-age-related changes in the activity of Na^{+}/K^{+}-ATPase as expressed per milligram membrane protein (Cohen *et al.*, 1976) seems to confirm that scenario. Old red cells were found to have a 1.8-fold increased rate of passive (ouabain-insensitive) K^{+} influx (Joiner and Lauf, 1978) and an increased rate of Li^{+}/K^{+} cotransport (Hentschel *et al.*, 1986).

The Cl^{-}-dependent, furosemide-sensitive K^{+} efflux induced by *N*-ethylmalemide was decreased in old (dense) compared to young (light) erythrocytes (Berkowitz *et al.*, 1987).

Ca^{2+}-ATPase activity, reflecting the Ca^{+} pump activity, has been found to decrease with red cell aging (Hanahan and Ekholm, 1978; Luthra and Kim, 1980). The diminution of the erythrocyte's content of calmodulin may significantly contribute to this effect (Monzon *et al.*, 1982; Warth *et al.*, 1983). However, a recent study did not find any difference in the rate of Ca^{2+} efflux between the 1–2% densest cells and lighter, mature cells, after cell loading with calcium using the ionophore A23187 (Clark, 1985).

Studies of anion transport have also led to discrepant results although all anions studied are transported by the same protein channel (the band 3 protein of the erythrocyte membrane or capnophorin). No change in the rate of Cl^{-} transport was found (Białas, 1984). The transport rate of sulfate was reported to increase in the oldest (densest) cells (Zanner and Galey, 1985) while another group found a substantial decrease in the maximal sulfate flux and an increase in the Michaelis constant for this anion (Bosman and Kay, 1988; Kay *et al.*, 1986). Our studies pointed to a decrease in the transport rate of phosphate in human and bovine red cells, and a decreased transport rate, with no significant alterations in the activation energy for a spin-labeled analogue of phosphate (TEMPO-phosphate) in human cells (Bartosz *et al.*, 1987b).

The hydrophobic nonelectrolyte spin label 2,2,6,6-tetramethylpiperidine-1-oxide (TEMPO), which apparently penetrates the membrane by simple diffusion through its lipid phase, was found to permeate to denser red cells more slowly, probably due to the decrease in their lipid fluidity (Bartosz, 1981b).

Oxidized glutathione (GSSG) has been shown to be actively transported out of erythrocytes by two systems (GSSG-ATPases), one of high affinity and low capacity, apparently more relevant physiologically, and another of low affinity and high capacity. Comparison of the lightest (youngest) 20% and the densest (oldest) 20% of human erythrocytes showed a small decrease in the activity of the high-affinity transport system (measured with 20 μM GSSG) and lack of change in the activity of the low-affinity system (measured with 5 mM GSSG; Kondo *et al.*, 1981).

3.7. Transmembrane Potential

In bovine erythrocytes, we have found a decrease in the magnitude of the negative (with respect to the outside) transmembrane electric potential, by about 1.7 mV (from the value of −8.5 mV) between the 20% lightest and the 20% densest cells (Grzelinska and Bartosz, 1988). This membrane depolarization may affect electrolyte transport across the cell membrane.

4. CHANGES IN MEMBRANE COMPOSITION AND ORGANIZATION

4.1. Lipid Composition

Older erythrocytes contain less lipid (expressed either per cell or per milligram membrane protein) but their lipid composition is not subject to dramatic changes. Comparisons of the lipid content of the youngest and oldest red cell fractions obtained by various authors are given in Table III. The most marked differences take place between the fraction of lowest density and the next fraction suggesting that the most marked lipid loss

Table III
Lipid Content of the Youngest and Oldest Erythrocyte Fractions Obtained by Various Authors

Cell fraction	Total lipid	Phospholipids	Cholesterol	Reference
Lightest 10%[a]	4.79 ± 0.85	3.00 ± 0.36	1.12 ± 0.09	Westerman *et*
Densest 10%	4.09 ± 0.54	2.56 ± 0.32	0.94 ± 0.09	*al.* (1963)
Lightest 14%[a]	5.33	2.84	1.19	van Gastel *et al.*
Densest 14%	4.00	2.35	0.99	(1965)
Lightest 10%	4.15 ± 0.32[a]	3.75 ± 0.24[b]	1.05 ± 0.03[a]	Winterbourn and
Densest 20%	3.40 ± 0.05	3.30 ± 0.05	0.90 ± 0.03	Batt (1970)
100–105 fl[c]		4.61 ± 0.01[b]	3.33 ± 0.10[b]	van der Vegt *et*
80–85 fl		3.89 ± 0.10	2.80 ± 0.16	*al.* (1985a)

[a]mg/10^{10} cells.
[b]μmol/10^{10} cells.
[c]Mean cell volume (cells separated by counterflow centrifugation).

occurs during reticulocyte maturation, although the loss continues progressively at a lower rate (Winterbourn and Batt, 1970).

These results, expressed per milligram erythrocyte membrane protein, indicate a significant decrease of the membrane lipid-to-protein ratio, most probably due to the release of protein-poor vesicles in the process of microvesiculation. The data shown, as well as other results (Bartosz, 1981b; Cohen *et al.*, 1976; Shiga *et al.*, 1979), point to a lack of significant changes in the cholesterol/phospholipid ratio although some authors have reported an increase of this ratio in the oldest mammalian red cells (Kamber *et al.*, 1984; Marin *et al.*, 1988). Marin *et al.* (1988) fractionated ovine erythrocytes by density and found the cholesterol/phospholipid molar ratio to be 1.01 ± 0.04 in cells of density $<$ 1.100 g/cm^3. This ratio did not change significantly up to a density of 1.115 g/cm^3, but increased up to 1.40 ± 0.49 in cells of density between 1.115 and 1.120 g/cm^3 and even up to 6.86 ± 2.08 in erythrocytes of density $>$ 1.130 g/cm^3.

No significant changes in phospholipid composition were reported for human red cells (van Gastel *et al.*, 1965; Winterbourn and Batt, 1970), although a recent report indicates an increase in the phosphatidylcholine content and a decrease in the phosphatidylserine + phosphatidylinositol content (Jain, 1988) and the occurrence of changes

Table IV
Phospholipid Composition of Red Cells Separated by Density

Phospholipid (% total)	Human cells (Winterbourn and Batt, 1970)	
	Lightest 20%	Densest 20%
PE[a]	29	30.5
PC + PS	46	43.3
SM	24	25
Others	1	1
	Human cells (Jain, 1988)	
	1.043 g/cm^3 $< d <$ 1.053 g/cm^3	$d >$ 1.111 g/cm^3
PE	31.8 ± 0.6	32.2 ± 0.2
PC	27.8 ± 0.8	31.2 ± 0.3
PS + PI	16.1 ± 0.1	12.9 ± 0.5
SM	24.1 ± 0.4	23.0 ± 0.3
	Ovine cells (Marin *et al.*, 1988)[b]	
	$d <$ 1.100 g/cm^3	1.115 g/cm^3 $< d <$ 1.120 g/cm^3
PE	33.1 ± 0.1	38.1 ± 7.4
PS	16.0 ± 0.9	19.7 ± 0.5
SM	45.7 ± 4.5	38.1 ± 3.8
PC	1.0	0.6
PI	2.5	0.5
PA	1.6	2.6
DAG	ND	0.9

[a]PE, phosphatidylethanolamine; PS, phosphatidylserine; PC, phosphatidylcholine; PI, phosphatidylinositol; PA, phosphatidic acid; SM, sphingomyelin; DAG, diacylglycerol.
[b]Shown on a poster.

(especially of an increase in the phosphatidylserine/sphingomyelin ratio) was recently reported for ovine cells (Marin *et al.*, 1988; Table IV).

No gross change in the composition of the fatty acid residues was found to accompany cell aging, except for a small increase in the content of linoleic acid residues and a slight decrease in the content of arachidonic acid residues (Ogiso *et al.*, 1985; Phillips *et al.*, 1969; Shiga *et al.*, 1979; van Gastel *et al.*, 1965). According to van Gastel *et al.* (1965), the content of linoleic acid ($C_{18:2}$) increases from 15.2% for the 14% lightest cells to 17.9% for the 14% densest cells, which is accompanied by a decrease in the content of arachidonic acid ($C_{20:4}$) from 19.2% to 16.3%. The data of Ogiso *et al.* also indicate a decrease in the content of arachidonic acid, from 13.9 ± 1.1% in the 10% lightest cells to 11.4 ± 1.6% in the 10% densest cells. This effect contributes to the decrease in the ratio of unsaturated to saturated fatty acid residues from 1.55 ± 0.20 to 1.39 ± 0.20.

4.2. Accumulation of Lipid Peroxidation Products

There has been a long-standing controversy as to whether accumulation of lipid peroxidation products accompanies erythrocyte aging. The method most commonly used for the estimation of lipid peroxidation products, based on a color reaction between (mainly) malondialdehyde and thiobarbituric acid, was reported by most (though not all; Gershon *et al.*, 1988) authors to show no increased concentration in aged erythrocytes (Jain, 1988; Phillips *et al.*, 1969; Takeuchi *et al.*, 1985). However, this cannot be taken as evidence against accumulation of lipid peroxidation products during the red cell life span, since malondialdehyde is a reactive compound and reacts with amino groups of membrane components (leading to lipid and protein cross-linking) or diffuses out of the membrane, and thus this fraction is not available to react with thiobarbituric acid.

Nevertheless, other approaches have allowed the demonstration that lipid peroxidation products accumulate progressively in the red cell membrane: the lipid-soluble fluorescence of emission wavelength of about 455 nm—characteristic for Schiff-type adducts, presumably between the aldehyde groups of the final products of lipid peroxidation and the amino groups of phosphatidylserine and phosphatidylethanolamine—is increased in the densest (oldest) human (Jain, 1988), rat (Gershon *et al.*, 1988; Jain and Hochstein, 1980), and bovine (Bartosz *et al.*, 1981b) red cell fractions. With increasing cell density, a new ninhydrin-negative fraction appeared and increased with increasing cell age on thin-layer chromatograms of membrane phospholipids, of motility intermediate to that of phosphatidylethanolamine and phosphatidylserine. A similar lipid spot was formed when erythrocytes were treated with malondialdehyde. In the lightest human red cells ($d < 1.053$ g/cm^3), this fraction constituted 0.2 ± 0.1% of the total phospholipid, while in the densest fraction ($d > 1.111$ g/cm^3), it increased up to 1.5 ± 0.1% of the total phospholipid (Jain, 1988).

4.3. Lipid Organization and Fluidity

Studies of phospholipase A_2 accessibility of membrane phospholipids has shown an increased susceptibility in aged erythrocytes, indicative of a decreased phospholipid asymmetry (Shukla and Hanahan, 1982). The susceptibility of membrane cholesterol to cholesterol oxidase was reported to increase significantly in older red cells (Rifkind *et al.*,

1983). Measurements of fluorescence anisotropy revealed that the plasma membrane fluidity of reticulocytes is higher than that of mature erythrocytes (Katsumoto *et al.*, 1977). Further studies of human (Shiga *et al.*, 1979) and bovine erythrocytes (Bartosz, 1981b) demonstrated that this process of membrane rigidification proceeds progressively during the red cell life span. Electron spin resonance measurements using spin-labeled analogues of fatty acids with the paramagnetic nitroxide group attached at different positions of the polymethylene chain revealed that in human and bovine red cells, both the order parameter (an inverse measure of the amplitude of movement) and the rotational correlation time (an inverse measure of the rate movement) of the probes increased with increasing cell density (Bartosz, 1981b; Shiga *et al.*, 1979). Measurements of the fluorescence anisotropy of the hydrophobic probe 1,6-diphenylhexatriene (DPH), located in the midzone of the membrane lipid bilayer, and of the more hydrophilic probe 1-aminonaphthalene-8-sulfonate, located close to the membrane polar surface, showed that fluorescence anisotropy (a measure of membrane microviscosity) increases with density (Bartosz *et al.*, 1981b). The increase in membrane microviscosity, albeit small (about 18% between the lightest 17% and the densest 17% of bovine red cells as measured with DPH), may affect the enzymatic or transporting functions of membrane proteins. According to the hypothesis of membrane-fluidity-dependent vertical translocation of membrane proteins, an increase in membrane rigidity may enhance the exposure of membrane proteins to the extramembranous environment (Borochov and Shinitzky, 1976). In studies in which a decrease in membrane fluidity was reported, no significant increase in the membrane cholesterol/phospholipid ratio was found (Bartosz *et al.*, 1981b; Shiga *et al.*, 1979), so that membrane rigidification had to be accounted for by other mechanisms. The small differences in the phospholipid and fatty acid composition of membrane lipids might contribute to this effect, but the main factor seems to be the decrease in the lipid/protein ratio of the membranes since much smaller fluidity changes were found in liposomes prepared from lipids extracted from cells of various ages than in whole cell membranes (Bartosz *et al.*, 1981b).

4.4. Protein Composition

The protein content of the erythrocyte membrane was reported to remain constant (Shiga *et al.*, 1979) and to decrease (Cohen *et al.*, 1976; Wilson and Peterson, 1986) during erythrocyte aging. A small decrease in the membrane protein content (per cell) would be compatible with the postulated occurrence of a vesiculation process, resulting in the release of protein-poor vesicles from the membrane.

Apart from the increase in high-molecular-weight protein aggregates (see Section 4.5), no gross changes can be seen concerning membrane protein composition of erythrocyte fractions of various ages. An increased binding of cytoplasmic proteins to membranes of older erythrocytes was reported by Kadlubowski (1979), and the amount of immunorective myosin was found to decrease (Matovcik *et al.*, 1986).

Studies of hypertransfused mouse erythrocytes showed a progressive increase in the content of the membrane protein band 4.1a at the expense of band 4.1b. While in erythrocytes of control mice band 4.1b predominated, comprising about 70% of the total band 4.1a + band 4.b, after 55 days of hypertransfusion band 4.1a constituted 54% of the total (Morrison *et al.*, 1983; Mueller *et al.*, 1987). This 4.1b/4.1a conversion did not appear to have any significant effect on the extractability of spectrin. An increase in the Triton-insoluble membrane-associated globin was also found in aged erythrocytes of hypertransfused mice (Mueller *et al.*, 1987).

A similar change in the 4.1a/4.1b ratio seemingly takes place during aging of the human erythrocyte, for band 4.1a is deficient in red cells of patients with a variety of hemolytic anemias (Sauberman *et al.*, 1979) and in transient erythroblastopenia of childhood, a relatively common childhood disorder characterized by temporary cessation of erythroid cell production. As the reticulocyte level rises during recovery, band 4.1b becomes the predominant form of band 4.1 in these patients (Ravindranath *et al.*, 1987). Comparison of the 4.1a/4.1b ratio in various animal species has shown a correlation with the mean erythrocyte life span (i.e., the mean red cell age in the circulation; Inaba and Maede, 1988). The relationship between bands 4.1a and 4.1b during erythrocyte aging appears to be a product–substrate relationship but the nature of modification increasing the apparent molecular weight of the protein by about 2000 is unknown.

4.5. Covalent Protein Modifications

The red cell, in which the proteins do not turn over, is a good model for the study of progressive age-related modifications of membrane proteins. Four such modifications studied more extensively are: covalent aggregation, proteolytic degradation, glycation, and carboxyl methylation. All have been proposed to constitute a "biological clock" terminating red cell life by enabling selective recognition of senescent erythrocytes.

An increase in the content of high-molecular-weight covalently linked protein aggregates is observed in human, bovine, and rat older (denser) red cell fractions (Bartosz *et al.*, 1982a; Jain, 1988; Kadlubowski, 1979; Pfeffer and Swislocki, 1982). At least part of these aggregates are not disulfide cross-links, since they are not reducible by dithiothreitol or β-mercaptoethanol. The appearance of a high-molecular-weight (~ 255,000) band, representing a hemoglobin monomer (mainly α chain) adduct to spectrin band 1 in old (dense) erythrocytes, was also reported (Snyder *et al.*, 1983).

It has been suggested that transglutaminase (EC2.3.2.13) found in erythrocyte membranes may contribute to membrane protein cross-linking during red cell aging (Lorand *et al.*, 1976) but the latent activity of this enzyme cannot be unmasked unless the Ca^{2+} concentration inside the cell is raised to levels now believed to be completely unphysiological (Greenwalt *et al.*, 1980). Another source of these cross-links may be the reactions of malondialdehyde and other products of lipid peroxidation with protein amine groups. *In vitro* incubation of erythrocytes with malondialdehyde unequivocally resulted in formation of similar aggregates (Jain and Hochstein, 1980).

The lack of gross changes in the electrophoretic pattern of membrane proteins of various erythrocyte age fractions may suggest that proteolysis of membrane components does not take place in mature red cells, despite the presence of an array of proteolytic enzymes. However, it seems to be rather a question of subtlety of methodological approach. Kay was the first to demonstrate accumulation of the band 3 protein degradation products in denser (older) erythrocytes, localizing degradation products by immunoblotting and suggesting one of these products to be a "senescent cell antigen" (Bosman and Kay, 1988; Kay, 1984b). A recent study confirming the red cell-age-related degradation of the band 3 protein using a monoclonal antibody (Czerwiński *et al.*, 1988) pointed out difficulties of artifact production when studying fine alterations of membrane proteins, since only by working with whole cells lysed by heating in a detergent (SDS) solution were they able to avoid proteolytic artifacts, observed even in membranes prepared in the presence of protease inhibitors. All of these studies showed a progressive accumulation during red cell aging of the band 3 protein degradation products in the regions of 60, 40, and 20 kDa, not detectable in conventional Coomassie brilliant blue-stained electrophoretograms of membrane proteins.

Accumulation of oxidative damage of membrane proteins during erythrocyte aging has been inferred from a slight decrease in the content of membrane protein -SH groups and an increase in the content of methionine sulfoxide residues (0.45 versus 0.27 nmol/mg hemoglobin when the 13% densest and the 13% lightest human erythrocytes were compared; Brovelli *et al.*, 1983).

Like all proteins in contact with reducing sugars, particularly glucose, red cell membrane proteins undergo progressive nonenzymatic glycosylation (glycation) (Bartosz, 1982a; Schleicher *et al.*, 1981; Sutera *et al.*, 1985; Williamson *et al.*, 1985). When comparing the difference in the level of glycation between the lightest 10% and the densest 10% of human erythrocytes, Sutera *et al.* (1985) found a higher increase for membrane protein (66%) than for hemoglobin (35%), possibly due to the contact of higher glucose concentrations with membrane proteins than with hemoglobin. The physiological significance of erythrocyte membrane protein glycation is rather unclear. It has been suggested to result in a decreased red cell deformability, but the unchanged shape recovery time of diabetic relative to normal erythrocytes does not support this claim (Williamson *et al.*, 1985). The presence of glycated proteins in membranes of mouse and human erythrocytes promotes antibody-independent phagocytosis by homologous monocytes and macrophages (Vlassara *et al.*, 1987), but it is doubtful that glycation of membrane proteins may be a senescence signal for physiologic removal of erythrocytes. It may be noted, hoever, that mouse erythrocytes glycated *in vitro* have a shortened survival *in vivo* (Vlassara *et al.*, 1987) and that the red cell life span is somewhat decreased in diabetes (Peterson *et al.*, 1977).

Another covalent modification of erythrocyte membrane proteins is the transfer of the methyl group of *S*-adenosylmethionine to carboxylic acid residues to form methyl esters. The reaction is catalyzed by the enzyme protein carboxyl methyltransferase (*S*-adenosyl-L-methionine:protein *O*-methyltransferase, EC 2.1.1.24). In the erythrocyte membrane the main targets for the carboxyl methylation are D-aspartate residues formed by racemization of L-aspartate. The racemization rate of amino acid residues of erythrocyte membrane proteins is unexpectedly high, the times required to racemize 1% of the residues ($T_{1\%}$) being about 39 days for aspartate, asparagine, and isoaspartate residues in human red cell membrane proteins. This racemization rate is higher for intrinsic than for extrinsic membrane proteins ($T_{1\%}$ of 34 and 140 days, respectively), and is much higher than that for cytosolic erythrocyte proteins ($T_{1\%}$ of about 620 days) and one or two orders of magnitude higher than for lens and tooth enamel or dentin proteins (Barber and Clarke, 1983; Brunauer and Clarke, 1986; Lowenson and Clarke, 1988). A protein such as band 3 would contain about two racemized Asx residues in a 120-day human red cell. Comparison of the 5% lightest and the 10% densest human erythrocytes showed that the rate of carboxyl methylation of membrane proteins increases about 3-fold in older erythrocytes, the stoichiometry (residues bound per protein molecule) increasing 4.6-fold for band 2.1 protein, 4.1-fold for band 3 protein, and 3.1-fold for band 4.1 protein (Barber and Clarke, 1983). The physiological function of carboxyl methylation of erythrocyte membrane proteins remains a mystery. It might be a first step in the repair of racemized amino acids, enabling recovery of the L configuration, or a signal for proteolysis. The suggestion that protein carboxyl methylation may be a "biological clock" for erythrocytes came, among others, from the observation that in the rat a mutation resulting in a low content of erythrocyte 2,3-diphosphoglycerate and decreased red cell life span leads also to an increased rate of red cell protein carboxyl methylation (Brewer *et al.*, 1985).

Phosphorylation, a reversible modification of membrane proteins, whose physiological significance is also largely obscure, was reported to be altered with erythrocyte age. The rate of protein phosphorylation in isolated membranes by both cAMP-dependent and -independent protein kinases decreased with increasing density (age) of human erythrocytes (Fairbanks *et al.*, 1983). The main reason may be an altered protein structure. Isolated spectrin from dense (aged) rat erythrocytes was up to 10-fold less able than spectrin from light (young) cells to accept phosphate groups in the presence of erythrocyte protein kinase, while spectrin from young cells was phosphorylated to the same extent by the enzyme preparations isolated from young and old cells (O'Connell and Swislocki, 1983).

4.6. Protein Organization

Data on alterations in protein organization in the membrane accompanying red cell aging are scarce. In bovine erythrocytes, altered electron spin resonance spectra of membrane-bound maleimide spin label demonstrate changes in the conformation of some proteins (Bartosz, 1981b). Increased anisotropy of membrane tryptophan fluorescence indicates an enhanced protein immobilization (Bartosz *et al.*, 1981b). Increased quenching constants of membrane protein tyrosine and tryptophan fluorescence by polar quenchers (acrylamide, iodide) suggest an increased accessibility of these amino acids to the extramembranous environment, typical of protein denaturation and perhaps contributing to enhanced interactions of cytoplasmic proteins with the membrane in older erythrocytes (Grzelinska and Bartosz, 1989). Proteins in membranes of denser (older) erythrocytes are also more susceptible to degradation by exogenous proteases (Gaczynska *et al.*, 1986) and more prone to thermal denaturation (Bartosz *et al.*, 1987a). Enhanced cross-linking of membrane proteins (mainly spectrin and the band 3 protein) with Cu^{2+}-*o*-phenanthroline is evidence for increased noncovalent membrane protein aggregation (Gaczynska and Bartosz, 1986). However, in contrast to the above biochemical data for bovine erythrocytes suggesting a considerable increase in protein aggregation, a freeze-fracture electron microscopic study did not reveal any difference between human red cells of different densities (ages) in the size or distribution of intramembranous particles, thought to represent clusters of integral membrane proteins (Fischbeck *et al.*, 1982).

The number and affinity of ankyrin-binding sites on the inner face of the red cell membrane (residing on the N-terminal segments of band 3 molecules) decrease in dense (old) human erythrocytes (Kay *et al.*, 1988).

Table V
Membrane Enzymatic Activities of Erythrocytes of Various Age Fractions

Enzyme/species[a]	Activity change	Reference
Acetylcholinesterase (EC 3.1.1.7)		
H $L_{1/8}$ vs. $D_{1/8}$	Decrease by 21%	Kadlubowski and Agutter (1977)
H	Decrease, then increase	Galbraith and Watts (1980)
Rb $L_{1/10}$ vs. $D_{1/10}$	Decrease by 57%	Au and Chan (1983)
B $L_{1/5}$ vs. $D_{1/5}$	Decrease by 9% (V_{max})	Grzelinska *et al.* (1983)
Adenylate cyclase (both basal and NaF-stimulated; EC 4.6.1.1)		
H, R	Decrease	Pfeffer and Swislocki (1976)
Protein kinase (both basal and cAMP-stimulated; EC 2.7.1.37)		
R	Decrease	Pfeffer and Swislocki (1976)
Na^+ K^+-ATPase (EC 3.6.1.3)		
H	No change	Kadlubowski and Agutter (1977)
H	No change	Cohen *et al.* (1976)
H	Decrease	Kamber *et al.* (1984)
Ca^{2+}/Mg^{2+}-ATPase (EC 3.6.1.3)		
Rb	Decrease	Au and Chan (1983)
H	No consistent change	Cohen *et al.* (1976)
H	No change but altered calmodulin effect	Ekholm *et al.* (1981)
NADH : ferricyanide reductase		
H	No change	Kadlubowski and Agutter (1977)
Alkaline phosphate (EC 3.1.1.1)		
H $L_{1/8}$ vs. $D_{1/8}$	Decrease by 90%	Kadlubowski and Agutter (1977)
(Ca^{2+})-polyphosphoinositide phosphodiesterase (EC 3.1.4.11)		
H $L_{1/10}$ vs. $D_{1/10}$	Decrease by 57%	Palmer (1985)
Phosphatidylinositol-4-phosphate phosphatase		
H $L_{1/10}$ vs. $D_{1/10}$	Decrease by 12%	Palmer (1985)

[a]L, lightest cells; D, densest cells; B, bovine; H, human; R, rat; Rb, rabbit.

4.7. Membrane Enzymes

The majority of red cell enzymes show a progressive decrease during red cell aging and membranous enzymes generally follow that pattern, although detailed data from various studies are often divergent, as exemplified in Table V.

Apart from the specific activity, other enzyme properties may be subject to age-related alterations. The apparent Michaelis constant of acetylcholinesterase was found to be unchanged (Galbraith and Watts, 1980; Kadlubowski and Agutter, 1977) or increased (Grzelinska *et al.*, 1983) and the activation of the enzyme by Lubrol WX to be decreased

(Galbraith and Watts, 1980) with increasing cell density (age). Essentially similarly, the degree of crypticity (percent activity in the absence of a nonionic surfactant) of NADH:ferricyanide reductase was decreased. Lower activities of pyruvate kinase (2.7.1.40) and purine nucleoside phosphorylase (2.4.2.1) were associated with membranes of older erythrocytes while for adenylate kinase (2.7.4.3) this activity did not change substantially (Kadlubowski and Agutter, 1977). Glyceraldehyde 3-phosphate dehydrogenase (1.2.1.12) was detected only in the cytoplasm of light (young) human erythrocytes, while it was mostly membrane-bound in dense (old) cells (Bosman and Kay, 1988; Kay *et al.*, 1988).

4.8. Receptors

The insulin-binding capacity of erythrocytes was found to decrease with increasing cell density (age), the decrease being apparently exponential (Baumann and MacCart, 1984; Dons *et al.*, 1981; Kosmakos *et al.*, 1980), due to the loss in the amount of both high-affinity and low-affinity insulin receptors. Vesiculation does not seem to be the main reason for the receptor loss since the relative decline in the amount of insulin receptors is about fivefold higher than the relative total protein loss from the membrane (Wilson and Peterson, 1986). Also, the activity of complement receptor type 1 was reported to decrease with increasing red cell density (Ripoche and Sim, 1986).

4.9. Carbohydrates

Most of the studies in this field have concerned sialic acid, due to the hypothesis that desialylation is responsible for the recognition of senescent erythrocytes (see Section 6). The majority of workers have found a slight but significant (about 10–15%) age-related decrease in the amount of sialic acid residues per cell (Bartosz *et al.*, 1984; Baxter and Beeley, 1978; Greenwalt *et al.*, 1980).

It is debatable, however, whether the sialic acid content of membrane glycoproteins (the number of residues per molecule) decreases since both a 12–18% decrease of the sialic acid content per glycophorin molecule between light (young) and middle-density (middle-age) cells (Platt and Norwig, 1980) and a lack of difference between young and old human erythrocytes in this respect were reported (Lutz and Fehr, 1979).

Apart from sialic acid, the content of other carbohydrates also decreases with increasing red cell density (age). Decreased content of neutral hexoses, fucose, mannose, galactose, galactosamine, glucose, glucosamine, and 2-acetamido-2-deoxyglucose residues (per cell) was reported (Baxter and Beeley, 1978; Bladier *et al.*, 1980; Gattegno *et al.*, 1976, 1981a,b). No significant differences were noted when the results were expressed as the number of residues per 100 carbohydrate residues, suggesting a homogeneous decrease of various carbohydrate residues during erythrocyte aging (Bladier *et al.*, 1980). These results could be accounted for by microvesiculation, although a recent lectin binding study showed a decrease in the number of receptor sites for a variety of lectins (per unit surface area of red cells) indicative of a decrease in the content of α-mannosyl, β(α)-galactosyl, α(β)-*N*-acetylgalactosyl, galactosyl-β-1-3(4)-*N*-acetylglucosyl, galactosyl-β-1-4-*N*-acetylglucosyl-1-2-mannosyl, and/or galactosyl-β-1-3-*N*-acetylgalactosyl residues, without any significant change in the density of receptors for *Triticum vulgaris* lectin (Shinozuka *et al.*, 1988a). Similarly, immunological studies point to small decreases in the amounts of B and H but not A, M, and N antigens during *in vivo* aging of human erythrocytes (Shinozuka *et al.*, 1988b). The authors explain it as due to the loss of glycoconjugates containing most but not all carbohydrate residues from the erythrocyte membrane. The number of galactose oxidase (1.1.3.9) and sialyl transferase (2.4.99.1) receptor sites was reported to be higher in old (dense) than in young (light) erythrocytes, apparently for the same reason (Aminoff, 1988; Aminoff *et al.*, 1981).

4.10. Surface-Bound Immunoglobulins and Complement Components

An increase in the amount of cell-bound immunoglobulin was found with increasing red cell age. Most workers have reported a significant increase in bound IgG with increasing density of human (Bosman and Kay, 1988; Green *et al.*, 1985; Halbhuber *et al.*, 1986; Kay, 1978; Kay *et al.*, 1988; Khansari, 1988; Magnani *et al.*, 1988; Smalley and Tucker, 1983), rat (Gershon *et al.*, 1988; Glass *et al.*, 1983; Kay *et al.*, 1986), and bovine erythrocytes (Bartosz *et al.*, 1982b). One group reported that in the densest 0.4% human red cells, apart from IgG-positive cells (98%), also IgM-positive (30%) and IgA-positive cells (18%), while Kay (1978) found IgG (IgG1 and IgG3) but not IgM or IgA on dense human erythrocytes.

Data on the number of IgG molecules present on erythrocytes of the highest density fractions differ between various studies, from 50–100 (Kay, 1978) to about 600 (Halbhuber *et al.*, 1986; Smalley and Tucker, 1983). The latter authors found a low amount of IgG bound to erythrocytes of low- and intermediate-density fractions and over 600 IgG molecules in the 15% densest cells, with some (presumably the oldest) cells of this fraction having much higher amounts than others.

The increase in cell-bound IgG during erythrocyte aging was confirmed in mice using the hypertransfusion protocol (Singer *et al.*, 1986).

The specificity of IgG bound to senescent erythrocytes has been a subject of considerable controversy (for a detailed discussion see Chapter 4), various authors finding them directed against α-galactosyl residues (Alderman *et al.*, 1980; Galili *et al.*, 1983, 1984; Jancik and Schauer, 1974), β-galactosyl residues (Bladier *et al.*, 1984; Vaysse *et al.*, 1986), the band 3 protein (reacting more avidly with its aggregates; Drenckhahn, 1988; Low *et al.*, 1985; Lutz and Stringaro-Wipf, 1983; Schlüter and Drenckhahn, 1986) and Band 3 degradation products (Kay, 1984).

In addition to immunoglobulins, more C3b complement fragments (Freedman, 1984) and more C3b–IgG complexes (Lutz *et al.*, 1987) were found on older than on younger human red cells.

5. HYPOTHESES ON THE MECHANISM OF RED CELL MEMBRANE AGING

The mammalian red cell is a unique model for studies of the aging of the plasma membrane and its constituents. Several possible basic processes responsible for the aging of the erythrocyte membrane have been hypothesized. Although this question is very far from consensus, recalling these hypotheses seems worthwhile.

5.1. Mechanical Fatigue

It had been noticed some time ago that red cells in mammalian species as different as man, dog, and cat undergo approximately the same number of circulations during their lifetime. This observation gave basis to the hypothesis that *in vivo* longevity of erythrocytes may be determined by their mechanical fatigue during circulation (Allison, 1960). Such a statement brings a further question concerning the molecular mechanisms that might underlie their "mechanical fatigue." In each circulation passage, the erythrocyte undergoes a swelling and shrinking cycle (Li and Li, 1983), so that the denaturation of some membrane protein(s) during each functional cycle is a possibility. However, membrane protein damage independent of the swelling and shrinking but due to the release of superoxide radical anion in each cycle should also be considered (see Section 5.5).

5.2. ATP Depletion and Calcium Accumulation

It has been suggested that a progressive decline in the capacity for ATP generation, due to inactivation of ATP-producing enzymes such as pyruvate kinase, hexokinase (2.7.1.1), and glucose 6-phosphate dehydrogenase (1.1.1.49), determines the functional deterioration of the red cell (Chapman and Schaumburg, 1967; Lichtman, 1975; Marks *et al.*, 1958b; Piomelli, and Wyss, 1971).

The progressive ATP deficiency might bring about, among other things, a failure of ATP-driven pumps, especially the Ca^{2+} pump. This would lead to an elevation of the intracellular Ca^{2+} concentration and its consequences, including K^+ loss due to the Gardos effect (Ca^{2+}-induced increase in the red cell membrane K^+ permeability; Gardos, 1959), which might account for the decrease in cellular water, increase in cell density, and (also via a direct effect of calcium on the membrane) the loss of deformability (LaCelle *et al.*, 1973a,b; Weed, 1970), as well as activation of proteolysis and transglutaminase (Lorand *et al.*, 1976).

This hypothesis could account for several phenomena of erythrocyte aging. However, recent data do not follow this prediction (Clark, 1988). In particular, no significant diminution of the capacity to extrude Ca^{2+} was found in aged (dense) human erythrocytes (Clark, 1985), and rabbit red cells in which the Ca^{2+} level was considerably increased with the A23187 ionophore were found to have near-normal survival in the circulation (Bookchin *et al.*, 1985).

5.3. Decrease of Redox Potential

According to this hypothesis, the red cell deteriorates because its "reductive potential" decreases and the cell is no longer able to protect itself against oxidative reactions (Fornaini *et al.*, 1970; Prankerd, 1958). This concept is convergent with a more recent view presented in Section 5.5.

5.4. Nonspecific Reactions of Membrane Proteins

Protein components of the erythrocyte membrane, like other proteins, are subjected to various nonspecific mono- and bimolecular reactions (Bartosz, 1981a) including racemization, glycation, deamidation, proteolysis, and "spontaneous denaturation," i.e., structural and functional alterations without detectable chemical alterations (Rothstein, 1979). This class of reactions encompasses those with reactive oxygen species discussed below (Section 5.5). All these reactions could lead to a progressive functional deterioration of the erythrocyte membrane.

5.5. Reactive Oxygen Species

The recent interest in the biological and pathogenic roles of oxygen free radicals and other reactive oxygen species (ROS) has stimulated the idea that erythrocyte aging is due to reactions of ROS with red cell (membrane) constituents. Four groups of arguments may speak in favor of this hypothesis.

First, ROS are indeed produced inside erythrocytes and can act on them from the outside (e.g., released by stimulated phagocytes). A fraction of hemoglobin-bound oxygen is released as the superoxide radical anion $O_2^{-}\cdot$ with accompanying hemoglobin oxidation. Every day about 3% of the hemoglobin is oxidized in this way to methemoglobin and reduced back by the methemoglobin reductase system (Carrell *et al.*, 1975). But the actual $O_2^{-}\cdot$ production in erythrocytes seems to be higher than corresponding to that estimate; measurements with various techniques indicate that $O_2^{-}\cdot$ production ranges from 10^2–10^3 (Hebbel *et al.*, 1982; Scarpa *et al.*, 1985; Schacter, 1986) up to about 10^6 superoxide radicals per erythrocyte per second (Müller *et al.*, 1983). Erythrocytes of patients with Fanconi anemia (Scarpa *et al.*, 1985) or with sickle-cell anemia or trait (Hebbel *et al.*, 1982; Schacter, 1986), of shorter life span, generate more $O_2^{-}\cdot$. The superoxide radical anion is apparently not a very dangerous species by itself, but upon decomposition it forms H_2O_2 and, in cooperation with the latter, the extremely reactive hydroxyl radical $\cdot OH$ is able to damage proteins (inactivate enzymes) and initiate lipid peroxidation (Halliwell and Gutteridge, 1986; Hebbel, 1986).

Second, the red cell is endowed with protective enzymes against ROS: superoxide dismutase (1.15.1.1), catalase (1.11.1.6), glutathione peroxidase (1.11.1.9), as well as glutathione reductase (1.6.4.2) and pentose phosphate cycle enzymes, auxiliary in this respect, to regenerate reduced glutathione (Figure 1), but the activities of these enzymes decrease during erythrocyte aging (Abraham *et al.*, 1978; Bartosz and Bartkowiak, 1981; Bartosz *et al.*, 1978; Fornaini *et al.*, 1970; Galbraith and Watts, 1980; Gershon *et al.*, 1988; Glass and Gershon, 1981; Ogiso *et al.*, 1985; Spooner *et al.*, 1979; Takeuchi *et al.*, 1985), which, at least partly, may be accounted for by their reactions with ROS, especially other than their own substrates. Some investigators also found decreases in the concentrations of low-molecular-weight ROS scavengers, such as reduced glutathione and tocopherol, in aged erythrocytes (Abraham *et al.*, 1978; Ogiso *et al.*, 1985). Therefore, not only is the erythrocyte permanently exposed to a ROS flux, but also its defensive capacities against them are impaired with increasing time.

Third, the effects of ROS on erythrocytes are in many respects reminiscent of red cell (especially membrane) changes observed during cell aging: lipid peroxidation, decreased deformability and survival in the circulation (Jain *et al.*, 1983), membrane protein aggregation (Pfeffer and Swislocki, 1982; Snyder *et al.*, 1981; Yamaguchi *et al.*, 1983), inactivation of membranous enzymes (O'Malley *et al.*, 1966; Pfeffer and Swislocki, 1982; Shalev *et al.*, 1981), decreased membrane lipid fluidity (Rice-Evans and Hochstein,

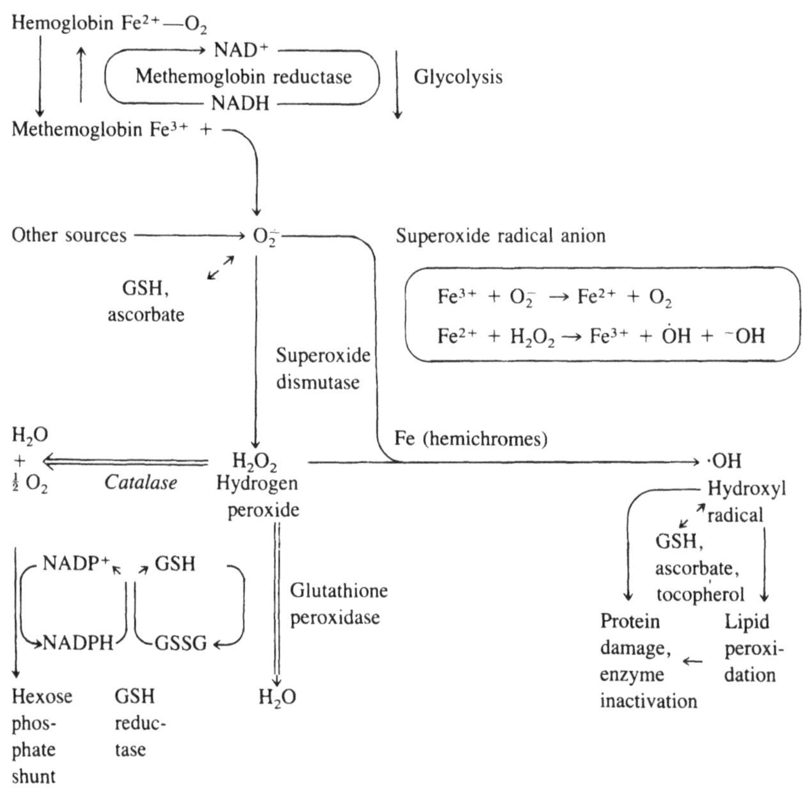

FIGURE 1. Reactive oxygen species and main antioxidants and antioxidative enzymes in the erythrocyte.

1981; Yamaguchi *et al.*, 1983), increased membrane permeability for K^+ (Maridonneau *et al.*, 1983; Orringer, 1984), and increased IgG binding (Bartosz *et al.*, 1982b).

Fourth, erythrocytes of vitamin E-deficient animals (Marvin *et al.*, 1963) and humans (Farrell *et al.*, 1977) have decreased life span (Kay *et al.*, 1986), increased K^+ permeability (Harm and Deamer, 1977), and their biochemical characteristics are similar to those of senescent red cells (Kay *et al.*, 1986).

The question of the molecular mechanisms of red cell aging certainly requires further studies, but it seems at present that the ROS hypothesis is worthy of serious consideration in this respect.

6. POSSIBLE MECHANISMS OF RECOGNITION OF SENESCENT ERYTHROCYTES

Several main hypotheses have been put forward in order to account for the selective recognition and removal of senescent erythrocytes. They imply that old erythrocytes are eliminated from the circulation due to:

1. Decreased surface charge density (Danon *et al.*, 1971; Marikovsky *et al.*, 1966)
2. Decreased deformability (Weed, 1970; Weed *et al.*, 1969; Williams and Morris, 1980)
3. Decreased lipid asymmetry of the erythrocyte membrane, especially exposure of phosphatidylserine in the outer leaflet (McEvoy *et al.*, 1986; Schroit *et al.*, 1985; Tanaka and Schroit, 1983)
4. Exposure of (a) residue(s) (e.g., β-D-galactosyl residues upon desialylation of membrane proteins) that is recognized by a lectin-like receptor on macrophages (Aminoff, 1988; Aminoff *et al.*, 1981; Küster and Schauer, 1981; Schlepper-Schäfer *et al.*, 1983)
5. Exposure of a "senescent cell antigen" and binding of an IgG autoantibody (Alderman *et al.*, 1980; Kay, 1978; see Chapter 4).

Although the "autoantibody hypothesis" seems to be currently most popular, in the opinion of many investigators, no hypothesis can be assumed as unequivocally verified or disproven. For example, though no electrophoretically detectable differences in the surface charge density of human erythrocytes were confirmed (see Section 3.3), it can be argued that perhaps not the average but rather the local surface charge density is important in erythrocyte interaction with phagocytic cells (Danon and Marikovsky, 1988).

The problem is far from clear, as indicated by the variety of hypotheses proposed. It is not certain whether the recognition of senescent erythrocytes results (as most researchers tacitly assume) from a progressive accumulation of changes, or is made possible by a sudden preterminal event. Similarly, it is still an open question whether erythrocyte recognition occurs by the same mechanism in various species (Clark, 1986, 1988). Finally, it may also be true that the elimination of functionally impaired senescent erythrocytes is too important to be left to a single mechanism (Beutler, in the discussion of Clark, 1986). Although one mechanism may be of predominant physiological importance, others may become more important or even dominant in pathology or when cells are subjected to various experimental treatments. So, although studied for many years, the question of red cell aging still poses basic questions. Answers to these questions may help in understanding tissue homeostasis, aging of other cells, and of the organism.

REFERENCES

Abraham, E. C., Taylor, J. F., and Lang, C. A., 1978, Influence of mouse age and erythrocyte age on glutathione metabolism, *Biochem. J.* **174:**819–825.

Alderman, E. M., Fudenberg, H. H., and Lovins, R. E., 1980, Binding of immunoglobulin classes to subpopulations of human red blood cells separated by density-gradient centrifugation, *Blood* **55:**817–822.

Allison, A. C., 1960, Turnover of erythrocytes and plasma proteins in mammals, *Nature* **180:**37–40.

Aminoff, D., 1988, The role of sialoglycoconjugates in the aging and sequestration of red cells from circulation, *Blood Cells* **14:**229–247.

Aminoff, D., Ghalambor, M. A., and Heinrich, C. J., 1981, GOST, galactose oxidase and sialyl transferase, substrate and receptor sites in erythrocyte senescence, in: *Erythrocyte Membranes. 2. Recent Clinical and Experimental Advances* (W. C. Kruckenberg, J. W. Eaton, and G. J. Brewers, eds.), pp. 267–278, Liss, New York.

Ashby, W., 1919, The determination of the length of life of transfused blood corpuscles in man, *J. Exp. Med.* **29:**267–281.

Au, K. S., and Chan, K. C., 1983, Variations in ($Ca^{2+}Mg^{2+}$)-ATPase, its inhibitor protein and calmodulin of density (age) separated rabbit erythrocytes, *Biochim. Biophys. Acta* **761:**291–295.

Barber, J. R., and Clarke, S., 1983, Membrane protein carboxyl methylation increases with human erythrocyte age, *J. Biol. Chem.* **258:**1189–1196.

Bartos, H. R., and Desforges, J. F., 1967, Enzymes as erythrocyte age reference standards, *Am. J. Med. Sci.* **254:**862–865.

Bartosz, G., 1981a, Non-specific reactions: Molecular basis for ageing, *J. Theor. Biol.* **91:**233–235.

Bartosz, G., 1981b, Aging of the erythrocyte. IV. Spin-label studies of membrane lipid, proteins and permeability, *Biochim. Biophys. Acta* **644:**69–73.

Bartosz, G., 1981c, Aging of the erythrocyte. VIII. Sensitivity to oxidant factors, *Acta Biol. Med. Ger.* **40:**985–989.

Bartosz, G., 1981d, Bovine erythrocyte membrane: Does not act as a molecular sieve or allow for hemolytic fractionation of red cells according to age, *Comp. Biochem. Physiol.* **68A:**273–275.

Bartosz, G., 1982a, Aging of the erythrocyte. XI. Membrane glycosylation, *Biochem. Med.* **27:**398–400.

Bartosz, G., 1982b, Aging of the erythrocyte. XV. Isoosmotic lysis times, *Experientia* **38:**1484–1485.

Bartosz, G., and Bartkowiak, A., 1981, Aging of the erythrocyte. II. Activities of peroxide-detoxifying enzymes, *Experientia* **37:**722.

Bartosz, G., Tannert, C., Fried, R., and Leyko, W., 1978, Superoxide dismutase activity decreases during erythrocyte aging, *Experientia* **34:**1464.

Bartosz, G., Swierczynski, B., and Gondko, R., 1981a, Aging of the erythrocyte. III. Cation content, *Experientia* **37:**723.

Bartosz, G., Szabo, G., Szöllösi, J., Szöllösi, J., and Damjanovich, S., 1981b, Aging of the erythrocyte. IX. Fluorescence studies on changes in membrane properties, *Mech. Ageing Dev.* **16:**265–274.

Bartosz, G., Soszynski, M., and Wasilewski, A., 1982a, Aging of the erythrocyte, XIX. Composition of membrane proteins, *Mech. Ageing Dev.* **19:** 45–52.

Bartosz, G., Soszynski, M., and Wasilewski, A., 1982b, Aging of the erythrocyte. XVII. Binding of autologous immunoglobulin G, *Mech. Ageing Dev.* **20:**223–232.

Bartosz, G., Grzelinska, E., and Bartkowiak, A., 1984, Aging of the erythrocyte. XIX. Decrease in surface charge density of bovine erythrocytes, *Mech. Ageing Dev.* **24:**1–7.

Bartosz, G., Christ, G., Bosse, H., Stephan, R., and Gärtner, H., 1987a, Thermal lability of membrane proteins of age separated erythrocytes as studied by electron spin resonance spin label technique, *Z. Naturforsch.* **42C:**1343–1344.

Bartosz, G., Gaczynka, M., Grzelinska, E., Soszynski, M., Michalak, W., and Gondko, R., 1987b, Aged erythrocytes exhibit decreased anion exchange, *Mech. Ageing. Dev.* **39:**245–250.

Baumann, G., and MacCart, J. G., 1984, Kinetics of cell age-dependent decline of insulin receptors in human red cells, *Am. J. Physiol.* **247:**E667–E674.

Baustad, B., and Nafstad, I., 1972, Haematological response to vitamin E in piglets, *Br. J. Nutr.* **28:**183–191.

Baxter, A., and Beeley, J. G., 1978, Surface carbohydrates of aged erythrocytes, *Biochem. Biophys. Res. Commun.* **83:**466–471.

Bennett, G. D., and Kay, M. M. B., 1981. Homeostatic removal of senescent murine erythrocytes by splenic macrophages, *Exp. Hematol.* **9:**297–307.

Berkowitz, L. R., Walstad, D., and Orringer, E. P., 1987, Effect of N-ethylmaleimide on K transport in density-separated human red blood cells, *Am. J. Physiol.* **253:**C7–C12.

Bernstein, R. E., 1959, Alterations in metabolic energetics and cation transport during aging of red cells, *J. Clin. Invest.* **38:**1572–1586.

Beutler, E., 1985a, How do red cell enzymes age? A new perspective, *Br. J. Haematol.* **61:**377–384.

Beutler, E., 1985b, Biphasic loss of red cell enzyme activity during in vivo ageing, in: *Cellular and Molecular Aspects of Ageing: The Red Cell as a Model* (J. W. Eaton, D. K. Konzen, and J. G. White, eds.), pp. 317–329, New York.

Białas, W. A., 1984, Alteration of Cl-transport in erythrocytes from patients with Huntington's disease, *Gen. Physiol. Biophys.* **3:**403–411.

Bishop, C., and Prentice, T. C., 1966, Separation of rabbit red cells by density in a bovine serum albumin gradient and correlation of red cell density with cell age after in vivo labeling with ^{59}Fe, *J. Cell. Physiol.* **67:**197–207.

Bladier, D., Gattegno, L., Fabia, F., Perret, G., and Cornillot, P., 1980, Individual variations of the seven carbohydrate components of human erythrocyte membrane during aging in vivo, *Carbohydr. Res.* **83:**371–376.

Bladier, D., Vassy, R., Perret, G., Gattegno, L., and Cornillot, P., 1984, Evidence for the participation of

glycoconjugates in the recognition of human old red blood cells by autologous macrophages, *IRCS Med. Sci.* **12**:889–890.
Bookchin, R. M., Roth, E. F., Jr., and Lew, V. L., 1985, Near-normal circulatory survival of rabbit red cells exposed to high levels of Ca and iconophore in vitro, *Blood* **66**:220–223.
Boorman, K. E., Dodd, B. E., and Mollison, P. L., 1942, The clinical significance of the Rh factor, *Br. Med. J.* **2**:535–538, 569–572.
Borochov, H., and Shinitzky, M., 1976, Vertical displacement of membrane proteins mediated by changes in microviscosity, *Proc. Natl. Acad. Sci. USA* **73**:4526–4530.
Borun, R. E., 1964, Some differences in erythrocyte composition and uptake of radioactive potassium, sodium, chromate and triiodothyronine associated with in vivo aging, *J. Lab. Clin. Med.* **62**:263–278.
Bosman, G.J.C.G.M., and Kay, M.M.B., 1988, Erythrocyte aging: A comparison of model systems for simulating cellular aging in vitro, *Blood Cells* **14**:19–35.
Botscharova, L., 1973, Eine Methode zur Berechnung des Sedimentationsverhaltens von Partiken in linearen Dextran-Dichtegradienten und ihre Anwendung auf die Trennung roter Blutzellen nach der Sedimentationsgeschwindigkeit, *Acta Biol. Med. Ger.* **30**:1–12.
Braasch, D., 1971, Red cell deformability and capillary blood flow, *Physiol. Rev.* **51**:679–701.
Bracey, A. M., and McFarland, F., 1984, Harvest of young red cells on an automated cell separator, *Transfusion* **24**:330–332.
Branch, D. R., Sy Siok Hian, A. L., Carlson, F., Maslow, W. C., and Petz, L. D., 1983, Erythrocyte age-fractionation using a Percoll™–Renographin^{R} density gradient: Application to autologous red cell antigen determinations in recently transfused patients, *Am. J. Clin. Pathol.* **80**:453–458.
Brewer, G. J., Mueller, G. A., Brewer, L. F., and Dick, R. D., 1985, A search for the primary biochemical effect of the *Dpg* gene; does this gene influence cellular aging? in: *Cellular and Molecular Aspects of Ageing: The Red Cell as a Model* (J. W. Eaton, D. K. Konzen, and J. G. White, eds.), pp. 403–426. Liss, New York.
Brok, F., Ramot, B., Zwang, E., and Danon, D., 1966, Enzyme activities in human blood cells of different age groups, *Isr. J. Med. Sci.* **2**:291–296.
Brovelli, A., Seppi, C., Pallavicini, G., and Balduini, C., 1983, Membrane processes during "in vivo" aging of human erythrocytes, *Biomed. Biochim. Acta* **42**:S122–S126.
Brunauer, L. S., and Clarke, S., 1986, Age-dependent accumulation of protein residues which can be hydrolyzed to D-aspartic acid in human erythrocytes, *J. Biol. Chem.* **261**:12538–12543.
Carrell, R. W., Winterbourn, C. C., and Rachmilewitz, E. A., 1975, Activated oxygen and haemolysis, *Br. J. Haematol.* **30**:259–264.
Chalfin, D., 1956, Differences between young and mature rabbit erythrocytes, *J. Cell. Comp. Physiol.* **47**:215–239.
Chapman, R. G., and Schaumburg, L., 1967, Glycolysis and glycolytic enzyme activity of aging red cells in man, *Br. J. Haematol.* **13**:665–678.
Clark, M. R., 1985, Selected ionic and metabolic characteristics of human red cell populations separated on Stractan density gradients, in: *Cellular and Molecular Aspects of Ageing: The Red Cell as a Model* (J. W. Eaton, D. K. Konzen, and J. G. White, eds.), pp. 381–386, Liss, New York.
Clark, M. R., 1986, Why does the normal red cell die? *Blood Cells* **12**:99–102.
Clark, M. R., 1988, Senescence of red blood cells: Progress and problems, *Physiol. Rev.* **68**:503–554.
Clark, M. R., Mohandas, N., and Shohet, S. B., 1983, Osmotic gradient ektacytometry: Comprehensive characterization of red cell volume and surface maintenance, *Blood* **61**:899–910.
Cohen, N. S., Eckholm, J. E., Luthra, M. G., and Hanahan, D. J., 1976, Biochemical characterization of density-separated human erythrocytes, *Biochim. Biophys. Acta* **419**:229–242.
Corash, L. M., Piomelli, S., Chen, H. C., Seaman, G. V. F., and Gross, E., 1974, Separation of erythrocytes according to age on a simplified density gradient, *J. Lab. Clin. Med.* **84**:147–151.
Corry, W. D., and Meiselman, H. J., 1978, Centrifugal method of determining red cell deformability, *Blood* **51**:693–701.
Cruz, W. O., Hahn, P. F., Bale, W. F., and Balfour, W. M., 1941, The effect of age on the susceptibility of the erythrocyte to hypotonic salt solutions, *Am. J. Med. Sci.* **202**:157–163.
Czerwinski, M., Wasniowska, K., Steuden, I., Duk, M., Wiedłocha, A., and Lisowska, E., 1988, Degradation of the human erythrocyte membrane band 3 studied with the monoclonal antibody directed against an epitope on the cytoplasmic fragment of band 3, *Eur. J. Biochem.* **174**:647–654.

Danon, D., 1968, Reversible deformability and mechanical fragility as a function of red cell age, in: *Hemorheology* (A. L. Copley, ed.), pp. 497–504, Pergamon Press, Elmsford, N.Y.

Danon, D., and Marikovsky, Y., 1961, Difference de charge electrique de surface entre erythrocytes jeunes et ages, *C. R. Acad. Sci.* **253:**1271–1272.

Danon, D., and Marikovsky, Y., 1964, Determination of density distribution of red cell population, *J. Lab. Clin. Med.* **64:**668–674.

Danon, D., and Marikovsky, Y., 1988, The aging of the red blood cell. A multifactor process, *Blood Cells* **14:**7–15.

Danon, D., Marikovsky, Y., and Skutelsky, E., 1971, The sequestration of old red cells and expulsed nuclei, in: *Red Cell Structure and Metabolism* (B. Ramot, ed.), pp. 23–38, Academic Press, New York.

Danon, D., Goldstein, L., Marikovsky, Y., and Skutelsky, E., 1972, Use of cationized ferritin as label of negative charges on cell surfaces, *J. Ultrastruct. Res.* **38:**500–510.

Dhermy, D., Simeon, J., Wautier, M.-P., Boivin, P., and Wautier, J.-L., 1987, Role of membrane sialic acid content in the adhesiveness of aged erythrocytes to human cultured endothelial cells, *Biochim. Biophys. Acta* **904:**201–206.

Dons, R. F., Corash, L. M., and Gorden, P., 1981, The insulin receptor is an age-dependent integral component of the human erythrocyte membrane, *J. Biol. Chem.* **256:**2982–2987.

Drenckhahn, D., 1988, Removal of old and abnormal red blood cells from circulation: Mechanical and immunologic mechanism, in: *Blood Cells, Rheology, and Aging* (D. Platt, ed.), pp. 62–76, Springer, Berlin.

Dumaswala, U. J., and Greenwalt, T. J., 1984, Human erythrocytes shed exocytic vesicles in vivo, *Transfusion* **24:**490–492.

Ekholm, J. E., Shukla, S. D., and Hanahan, D. J., 1981, Change in cytosolic calmodulin activity of density (age) separated human erythrocytes towards membrane $Ca^{2+}Mg^{2+}$ ATPase, *Biochem. Biophys. Res. Commun.* **103:**407–413.

Fairbanks, G., Palek, J., Dino, J. E., and Liu, P. A., 1983, Protein kinase and membrane protein phosphorylation in normal and abnormal human erythrocytes: Variation related to mean cell age, *Blood* **61:**850–857.

Farrell, P. M., Bieri, J. G., Fratantoni, J. F., Wood, R. E., and di Sant'Agnese, P. A., 1977, The occurrence and effects of human vitamin E deficiency. A study in patients with cystic fibrosis, *J. Clin. Invest.* **60:**233–241.

Fischbeck, K. H., Bonilla, E., and Schotland, D. L., 1982, Freeze-fracture characterization of "young" and "old" human erythrocytes, *Biochim. Biophys. Acta* **685:**207–210.

Fornaini, G., 1967, Biochemical modifications during the life span of the erythrocyte, *Ital. J. Biochem.* **16:**258–301.

Fornaini, G., Dacha, M., Fazi, A., Gargano, M., and Schiavo, E., 1970, Relationship between age and properties of human erythrocyte glutathione reductase, *Ital. J. Biochem.* **19:**345–360.

Freedman, J., 1984, Membrane-bound immunoglobulins and complement components on young and old red cells, *Transfusion* **24:**477–481.

Gaczynska, M., and Bartosz, G., 1986, Crosslinking of membrane proteins during erythrocyte ageing, *Int. J. Biochem.* **18:**377–382.

Gaczynska, M., Rosin, J., Soszynski, M., and Bartosz, G., 1986, Proteolytic susceptibility of membrane proteins during erythrocyte aging, *Mech. Ageing Dev.* **35:**109–121.

Galbraith, D. A., and Watts, D. C., 1980, Changes in some cytoplasmic enzymes from red cells fractionated into age groups by centrifugation in Ficoll–Triosil gradients. Comparison of normal humans with Duchenne muscular dystrophy, *Biochem. J.* **191:**63–70.

Galbraith, D. A., and Watts, D. C., 1981, Human erythrocyte acetyl cholinesterase in relation to cell age, *Biochem. J.* **195:**221–228.

Galili, U., Korkesh, A., Kahane, I., and Rachmilewitz, E. A., 1983, Demonstration of a natural antigalactosyl IgG antibody on thalassemic red blood cells. *Blood* **61:**1258–1264.

Galili, U., Rachmilewitz, E. A., Peleg, A., and Flechner, I., 1984, A unique natural human IgG antibody with anti-alpha-galactosyl specificity, *J. Exp. Med.* **160:**1519–1531.

Ganzoni, A. M., Oakes, R., and Hillman, R. S., 1971, Red cell aging in vivo, *J. Clin. Inv. Med.* **50:**1373–1378.

Ganzoni, A. M., Barras, P., and Mart, H. R., 1976, Red cell ageing and death, *Vox Sang,* **30:**161–174.

Gardos, G., 1959, The role of calcium in the potassium permeability of human erythrocytes, *Acta Physiol. Acad. Sci. Hung.* **15:**121–125.

Gattegno, L., Bladier, D., and Cornillot, P., 1975, Ageing in vivo and neuraminidase treatment of rabbit erythrocytes: influence on half-life as assessed by 51Cr labelling, *Hoppe-Seyler's Z. Physiol. Chem.* **356**:391–397.

Gattegno, L., Bladier, D., Garnier, M., and Cornillot, P., 1976, Changes in carbohydrate content of surface membranes of human erythrocytes during ageing, *Carbohydr. Res.* **52**:197–208.

Gattegno, L., Perret, G., Fabia, F., Bladier, D., and Cornillot, P., 1981a, In vivo ageing of human erythrocytes and cell-surface labeling by D-galactose oxidase and sodium borotritide, *Carbohydr. Res.* **95**:283–290.

Gattegno, L., Perret, G., Fabia, F., and Cornillot, P., 1981b, Decrease of carbohydrate in membrane glycoproteins during human erythrocyte ageing in vivo, *Mech. Ageing Dev.* **16**:205–219.

Gear, A. R. L., 1977, Age-dependent separation of erythrocyte by preparative electrophoresis, *J. Lab. Clin. Med.* **90**:744–753.

Gershon, D., Glass, G. A., and Gershon, H., 1988, The effect of host and cell age on the rat erythrocyte: Biochemical aspects, in: *Blood Cells, Rheology, and Aging* (D. Platt, ed.), pp. 42–50, Springer, Berlin.

Glass, G. A., and Gershon, D., 1981, Enzymatic changes in rat erythrocytes with increasing cell and donor age: Loss of superoxide dismutase activity associated with increases in catalytically defective forms, *Biochem. Biophys. Res. Commun.* **103**:1245–1253.

Glass, G. A., Gershon, H., and Gershon, D., 1983, The effect of donor and cell age on several characteristics of rat erythrocytes, *Exp. Hematol.* **11**:987–995.

Goebel, K. M., and Lanser, K. G., 1983, Biorheological and metabolic dysfunctions of density-fractionated erythrocytes in diabetics with peripheral vascular disease, *Biomed. Biochim. Acta* **42**:102–106.

Green, G. A., Rehn, M. M., and Kalra, W. K., 1985, Cell-bound autologous immunoglobulin in erythrocyte subpopulations from patients with sickle cell disease, *Blood* **65**:1127–1133.

Greenwalt, T. J., and Dumaswala, U. J., 1988, Effect of red cell age on vesiculation in vitro, *Br. J. Haematol.* **68**:465–467.

Greenwalt, T. J., Steane, E. A., and Pine, N. E., 1971, Changes in erythrocyte surface antigens with aging in vivo, in: *Glycoproteins of Blood Cells and Plasma* (G. A. Jamieson and T. J. Greenwalt, eds.), pp. 235–244, Lippincott, Philadelphia.

Greenwalt, T. J., Steane, E. A., Lau, F. O., and Sweeney-Hammond, K., 1980, Aging of the human erythrocyte, in: *Immunobiology of the Erythrocyte* (S. G. Sandler, J. Nusbacher, and M. S. Schanfield, eds.), pp. 195–212, Liss, New York.

Gross, J., Staak, R., and Syllm-Rapoport, I., 1978, Veränderungen der anorganischen Pyrophosphatase während Reifung und Altern von roten Blutzellen des Neugeborenen, *Acta Biol. Med. Ger.* **37**:403–408.

Grzelinska, E., and Bartosz, G., 1988, Membrane potential decreases during erythrocyte aging, *Cell Biol. Int. Rep.* **12**:497.

Grzelinska, E., and Bartosz, G., 1989, Effect of cell age on the quenching of erythrocyte membrane protein fluorescence, *Cytobios* **57**:149–154.

Grzelinska, E., Bartosz, G., and Bartkowiak, A., 1983, Aging of the erythrocyte. XVIII. Changes in kinetic properties of acetylcholinesterase, *Enzyme* **30**:95–98.

Halbhuber, K.-J., Linss, W., Zimmermann, N., Oehring, H., and Pätzold, L., 1986, Cytochemical and cell-biological investigations of the signal function of the erythrocyte plasmalemma–The membrane structure as code for cell life span, *Acta Histochem.* **33**:S23–S44.

Hall, A. C., and Ellory, J. C., 1986, Evidence for the presence of volume-sensitive KCl transport in "young" human red cells, *Biochim. Biophys. Acta* **858**:317–320.

Halliwell, B., and Gutteridge, J. M. C., 1986, Oxygen free radicals and iron in relation to biology and medicine: Some problems and concepts, *Arch. Biochem. Biophys.* **246**:501–514.

Hanahan, D. J., and Ekholm, J. E., 1978, The expression of optimum ATPase activities in human erythrocytes. A comparison of different lytic procedures, *Arch. Biochem. Biophys.* **187**:170–179.

Harm, W., and Deamer, D. W., 1977, Altered potassium permeability in vitamin E-deficient rat erythrocytes, *Physiol. Chem. Phys.* **9**:501–512.

Hebbel, R. P., 1986, Autoxidation and the sickle erythrocyte membrane: A possible model of iron decompartmentalization, in: *Free Radicals, Aging, and Degenerative Diseases*, pp. 395–424, Liss, New York.

Hebbel, R. P., Eaton, J. W., Balasingam, M., and Steinberg, M. H., 1982, Spontaneous oxygen radical generation by sickle erythrocytes, *J. Clin. Invest.* **70**:1253–1259.

Hentschel, W. M., Wu, L. L., Tobin, G. O., Anstall, H. B., Smith, J. B., Williams, R. R., and Ash, K. O., 1986, Erythrocyte cation transport activities as a function of cell age, *Clin. Chim. Acta* **157**:33–44.

Hoffman, J. F., 1958, On the relationship of certain erythrocyte characteristics to their physiological age, *J. Cell. Comp. Physiol.* **51**:415–423.

Inaba, M., and Maede, Y., 1988, Correlation between protein 4.1a/4.1b ratio and erythrocyte life span, *Biochim. Biophys. Acta* **944**:256–264.

Jain, S. K., 1988, Evidence for membrane lipid peroxidation during the in vivo aging of human erythrocytes, *Biochim. Biophys. Acta* **937**:205–210.

Jain, S. K., and Hochstein, P., 1980, Polymerization of membrane components in aging red blood cells, *Biochem. Biophys. Res. Commun.* **92**:247–254.

Jain, S. K., Mohandas, N., Clark, M. R., and Shohet, S. B., 1983, The effect of malonyldialdehyde, a product of lipid peroxidation on the deformability, dehydration and ^{51}Cr-survival of erythrocytes, *Br. J. Haematol.* **53**:247–255.

Jancik, J., and Schauer, R., 1974, Sialic acid—a determinant of the life-time of erythrocytes, *Hoppe-Seyler's Z. Physiol. Chem.* **355**:395–400.

Joiner, C. H., and Lauf, P. K., 1978, Ouabain binding and potassium transport in young and old populations of human red cells, *Membr. Biochem.* **1**:187–202.

Juckett, D. A., and Rosenberg, B., 1982, The kinetics and thermodynamics of lysis of young and old sheep red blood cells, *Mech. Ageing Dev.* **18**:33–45.

Kadlubowski, M., 1979, The effect of in vivo aging of the human erythrocyte on the protein of the plasma membrane: A characterization, *Int. J. Biochem.* **9**:67–88.

Kadlubowski, M., and Agutter, P. S., 1977, Changes in the activities of some membrane-associated enzymes during in vivo ageing of the normal human erythrocyte, *Br. J. Haematol.* **37**:111–125.

Kamber, E., Poyiagi, A., and Deliconstantinos, G., 1984, Modifications in the activity of membrane-bound enzymes during in vivo ageing of human and rabbit erythrocytes, *Comp. Biochem. Physiol.* **77B**:95–99.

Katsumoto, Y., Tanaka, F., Hagihara, M., and Yagi, K., 1977, Changes in membrane fluidity of erythrocytes during cell maturation, *Biochem. Biophys. Res. Commun.* **78**:609–614.

Kay, M.M.B., 1978, Role of physiologic autoantibody in the removal of senescent human red cells, *J. Supramol. Struct.* **9**:555–567.

Kay, M.M.B., 1984a, Localization of senescent cell antigen on band 3, *Proc. Natl. Acad. Sci. USA* **81**:5753–5757.

Kay, M.M.B., 1984b, Band 3, the predominant transmembrane polypeptide, undergoes proteolytic degradation as cells age, *Monogr. Dev. Biol.* **17**:245–253.

Kay, M.M.B., Goodman, S., Whitfield, C., Wong, P., Zaki, L., and Rudloff, V., 1984, The senescent cell antigen is immunologically related to band 3, *Proc. Natl. Acad. Sci. USA* **80**:1631–1635.

Kay, M.M.B., Bosman, G.J.C.G.M., Shapiro, S. S., Bendich, A., and Bassel, P., 1986, Oxidation as a possible mechanism of cellular aging: Vitamin E deficiency causes premature aging and IgG binding to erythrocytes, *Proc. Natl. Acad. Sci. USA* **83**:2463–2467.

Kay, M.M.B., Bosman, G.J.C.G.M., Johnson, G. J., and Beth, A. H., 1988, Band-3 polymers and aggregates, and hemoglobin precipitates in red cell aging, *Blood Cells* **14**:275–289.

Khansari, N., 1988, Mechanism for elimination of senescent red blood cells from circulation, in: *Blood Cells, Rheology, and Aging* (D. Platt, ed.), pp. 77–89, Springer, Berlin.

Kimura, R. E., Suzuki, T., and Kinoshita, Y., 1960, Separation of reticulocytes by means of multi-layer centrifugation, *Nature* **188**:1201–1202.

Kondo, T., Dale, G. L., and Beutler, E., 1981, Studies on glutathione transport utilizing inside-out vesicles prepared from human erythrocytes, *Biochim. Biophys. Acta* **645**:132–136.

Kosmakos, F. C., Nagulesparan, M., and Bennett, P. H., 1980, Insulin binding to erythrocytes: A negative correlation with red cell age, *J. Clin. Endocrinol. Metab.* **51**:46–50.

Küster, J. M., and Schauer, R., 1981, Phagocytosis of sialidase-treated rat erythrocytes: Evidence for a two-step mechanism, *Hoppe-Seyler's Z. Physiol. Chem.* **362**:1507–1514.

LaCelle, P. L., Kirkpatrick, F. H., and Udkow, M., 1973a, Relation of altered deformability, ATP, DPG and Ca^{++} concentration in senescent erythrocytes, in: *Erythrocytes, Thrombocytes, Leukocytes. Recent Advances in Membrane and Metabolic Research* (E. Gerlach, K. Moser, E. Deutsch, and W. Wilmanns, eds.), pp. 49–52, Thieme, Stuttgart.

LaCelle, P. L., Kirkpatrick, F. H., Udkow, M. D., and Arkin, B., 1973b. Membrane fragmentation and Ca-membrane interaction: Potential mechanisms of shape change in the senescent red cell, in: *Red Cell Shape* (M. Bessis, R. I. Weed, and P. F. Leblond, eds.), pp. 69–78, Springer, Berlin.

Leif, R. C., and Vinograd, J., 1964, The distribution of buoyant density of human erythrocytes in bovine serum albumin solutions, *Proc. Natl. Acad. Sci. USA* **51**:520–528.

Levy, L. M., Walter, H., and Sass, M. D., 1959, Enzymes and radioactivity in erythrocytes of different age, *Nature* **184**:643–644.

Li, C. K. N., and Li, E. K. H., 1983, Mechanical fatigue as a possible determinant of in vivo longevity of red blood cells, *IEEE Trans. Biomed. Eng.* **30**:226–227.

Lichtman, M. A., 1975, Does ATP decrease exponentially during red cell aging? *Nouv. Rev. Fr. Hematol.* **15**:625–632.

Linderkamp, O., and Meiselman, H. J., 1982, Geometric, osmotic, and membrane mechanical properties of density-separated human red cells, *Blood* **59**:1121–1127.

Lorand, L., Weissmann, L. B., Epel, D. L., and Bruner-Lorand, J., 1976, Role of the intrinsic transglutaminase in the Ca^{2+}-mediated crosslinking of erythrocyte proteins, *Proc. Natl. Acad. Sci. USA* **73**:4479–4481.

Low, P. S., Waugh, S. M., Zinke, K., and Drenckhahn, D., 1985, The role of hemoglobin denaturation and band 3 clustering in red blood cell aging, *Science* **227**:531–533.

Lowenson, J., and Clarke, S., 1988, Does the chemical instability of aspartyl and asparaginyl residues in proteins contribute to erythrocyte aging? The role of protein carboxyl methylation reactions, *Blood Cells* **14**:103–117.

Luner, S. J., Szklarek, D., Knox, R. J., Seaman, G. V. F., Josefovicz, J. Y., and Ware, B. R., 1977, Red cell charge is not a function of cell age, *Nature* **269**:719–721.

Luthra, M. G., and Kim, H. D., 1980, (Ca^{2+}-Mg^{2+})-ATPase of density separated human red cells: Effect of calcium and a soluble cytoplasmic activator (calmodulin), *Biochim. Biophys. Acta* **600**:480–488.

Luthra, M. G., Friedman, J. M., and Sears, D. A., 1979, Studies of density fractions of normal human erythrocytes labeled with iron-59 in vivo, *J. Lab. Clin. Med.* **94**:879–896.

Lutz, H. U., and Fehr, J., 1979, Total sialic acid content of glycophorins during senescence of human red blood cells, *J. Biol. Chem.* **254**:11177–11180.

Lutz, H. U., and Stringaro-Wipf, G., 1983, Senescent red cell-bound IgG is attached to band 3 protein, *Biomed. Biochim. Acta* **42**:S117–S121.

Lutz, H. U., Liu, S. C., and Palek, J., 1977, Release of spectrin-free vesicles from human erythrocytes during ATP depletion. I. Characterization of spectrin-free vesicles, *J. Cell Biol.* **73**:548–560.

Lutz, H. U., Stammler, P., Furter, C., and Fasler, S., 1987, "Anti-Bande-3"-Antikörper aktivieren Komplement über den alternativen Weg, *Schweiz. Med. Wochenschr.* **117**:1821–1824.

McEvoy, L., Williamson, P., and Schlegel, R. A., 1986, Membrane phospholipid asymmetry as a determinant of erythrocyte recognition by macrophages, *Proc. Natl. Acad. Sci. USA* **83**:3311–3315.

Mackie, L. H., Frank, R. S., and Hochmuth, R. M., 1987, Erythrocyte density separation on discontinuous "Percoll" gradient, *Biorheology* **24**:227–230.

Magnani, M., Papa, S., Rossi, L., Vitale, M., Fornaini, G., and Manzoni, F. A., 1988, Membrane-bound immunoglobulins increase during red blood cell aging, *Acta Haematol.* **79**:127–132.

Malachi, T., Bogin, E., Gafter, U., and Levi, J., 1986, Parathyroid hormone effect on the fragility of human young and old red blood cells in uremia, *Nephron* **42**:52–57.

Maridonneau, J., Braquet, P., and Garay, R. P., 1983, Na^+ and K^+ transport damage induced by oxygen free radicals in human red cell membranes, *J. Biol. Chem.* **258**:3107–3113.

Marikovsky, Y., and Danon, D., 1969, Electron microscope analysis of young and old red blood cells stained with colloidal iron for surface charge evaluation, *J. Cell Biol.* **43**:1–7.

Marikovsky, Y., Danon, D., and Katchalsky, A., 1966, Agglutination by polylysine of young and old red blood cells, *Biochim. Biophys. Acta* **124**:154–159.

Marin, M. S., Sanchez-Yagüe, J., Caberaz, J. A., and Llanillo, M., 1988, Phospholipid composition and aminophospholipid topology in erythrocyte plasma membranes of different ages, 14th International Congress of Biochemistry, Prague, Abstracts, Thursday: 674.

Marks, P. A., and Johnson, A. B., 1958, Relationship between the age of human erythrocytes and their osmotic resistance: A basis for separating young and old erythrocytes, *J. Clin. Invest.* **37**:1542–1548.

Marks, P. A., Johnson, A. B., and Hirschberg, E., 1958a, Effect of age on the enzyme activity in erythrocytes, *Proc. Natl. Acad. Sci. USA* **44**:529–536.

Marks, P. A., Johnson, A. B., Hirschberg, E., and Banks, J., 1958b, Studies on the mechanism of aging of human red blood cells, *Ann. N.Y. Acad. Sci.* **75**:95–105.

Matovcik, L. M., Gröschel-Stewart, U., and Schrier, S. L., 1986, Myosin in adult and neonatal human erythrocyte membranes, *Blood* **67**:1668–1674.

Monzon, C. M., Penniston, J. T., Fairbanks, V. F., and Burgert, E. O., Jr., 1982, Erythrocyte calmodulin correlates with red cell age, *Br. J. Haematol.* **51**:261–264.

Morrison, M., Jackson, C. W., Mueller, T. J., Huang, T., Dockter, M. E., Walker, W. S., Singer, J. A., and Edwards, H. H., 1983, Does cell density correlate with red cell age? *Biomed. Biochim. Acta* **42:**S107–S111.

Mosior, M., Gomułkiewicz, J., Bobrowska, M., Komorowska, M., and Koter, M., 1984, Effect of phosphate ions on osmotic fragility and membrane fluidity of bovine erythrocytes, *Stud. Biophys.* **99:**117–126.

Mueller, T. J., Jackson, C. W., Dockter, M. E., and Morrison, M., 1987, Membrane skeletal alterations during in vivo mouse red cell aging. Increase in the band 4.1a : 4.1b ratio, *J. Clin. Invest.* **79:**492–499.

Müller, M., Dumdey, R., and Rapoport, S., 1983, Superoxide radicals in the metabolism of the red cell, *Biomed. Biochim. Acta* **42:**S297–S301.

Murphy, J. R., 1973, Influence of temperature and method of centrifugation on the separation of erythrocytes, *J. Lab. Clin. Med.* **82:**334–341.

Nash, G. B., and Meiselman, H. J., 1981, Red cell ageing: Changes in deformability and other possible determinants of in vivo survival, *Microcirculation* **1:**255–284.

Nash, G. B., and Meiselman, H. J., 1983, Red cells and ghost viscoelasticity. Effects of hemoglobin concentration and in vivo aging, *Biophys. J.* **43:**63–73.

Nash, G. B., and Wyard, S. J., 1980, Changes in surface area and volume measured by micropipette aspiration for erythrocytes ageing in vivo, *Biorheology* **17:**479–484.

Nash, G. B., and Wyard, S. J., 1982, Shape of ageing erythrocytes, *Biorheology* **19:**727.

Nash, G. B., Linderkamp, O., Pfafferoth, C., and Meiselman, H. J., 1988, Changes in red cell mechanisms during in vivo aging: Possible influence on removal of senescent cells, in: *Blood Cells, Rheology, and Aging* (D. Platt, ed.), pp. 99–112, Springer, Berlin.

O'Connell, D. J., Caruso, C. J., and Sass, M. D., 1965, Separation of erythrocytes of different ages, *Clin. Chem.* (Winston-Salem, N.C.) **11:**771–781.

O'Connell, M. A., and Swislocki, N. I., 1983, Spectrin phosphorylation in senescent rat erythrocytes, *Mech. Ageing Dev.* **22:**51–70.

Ogiso, T., Iwaki, M., Takagi, T., Hirai, I., and Kashiyama, T., 1985, Increased sensitivity of aged erythrocytes to drugs and age-related loss of cell components, *Chem. Pharm. Bull.* **33:**5404–5412.

O'Malley, B. W., Engel, C. E., Meriwether, W. D., and Zirkle, L. G., Jr., 1966, Inhibition of erythrocyte acetylcholinesterase by peroxides, *Biochemistry* **5:**40–45.

Orringer, E. P., 1984, A further characterization of the selective K movements observed in human red blood cells following acetylphenylhydrazine exposure, *Am. J. Hematol.* **16:**355–366.

Palmer, F.B.St.C., 1985, Polyphosphoinositide metabolism in aging human erythrocytes, *Can. J. Biochem. Cell Biol.* **63:**927–931.

Peterson, C. M., Jones, R. L., Koenig, R. J., Melvin, E. T., and Lehrman, M. L., 1977, Reversible hematologic sequelae of diabetes mellitus, *Ann. Intern. Med.* **86:**425–429.

Pfeffer, S. R., and Swislocki, N. I., 1976, Age-related decline in the activities of erythrocyte membrane adenylate cyclase and protein kinase, *Arch. Biochem. Biophys.* **177:**117–122.

Pfeffer, S. R., and Swislocki, N. I., 1982, Role of peroxidation in erythrocyte aging, *Mech. Ageing Dev.* **18:**355–367.

Phillips, G. B., Dodge, J. T., and Howe, C., 1969, The effect of aging of human red cells in vivo on their fatty acid composition, *Lipids* **4:**544–549.

Picot, C., Girot, R., Loutounda, J., Mattlinger, B., Maier-Redelsperger, M., Feo, C., Chevalier, A., and Barritault, L., 1987, Enrichment of blood units with young red cells (neocytes) with the IBM 2991 cell washer, *Eur. J. Haematol.* **39:**214–220.

Piomelli, S., and Wyss, S. R., 1971, Metabolic death of the red blood cell, *Blood* **38:**832.

Piomelli, S., Lurinsky, G., and Wasserman, L. R., 1967, The mechanism of red cell aging. I. Relationship between cell age and specific gravity evaluated by ultracentrifugation in a discontinuous density gradient, *J. Lab. Clin. Med.* **69:**659–674.

Piomelli, S., Seaman, C., Reibman, J., Tytun, A., Graziano, J., Tabachnik, N., and Corash, L., 1978, Separation of younger red cells with improved survival in vivo: An approach to chronic transfusion therapy, *Proc. Natl. Acad. Sci. USA* **75:**3474–3478.

Platt, D., and Norwig, P., 1980, Biochemical studies of membrane glycoproteins during red cell aging, *Mech. Ageing Dev.* **14:**119–126.

Powers, H. J., and Thurnham, D. I., 1980, Effect of cell age on the malondialdehyde formation in erythrocytes in vitro, *Biochem. Soc. Trans.* **8:**195–196.

Prankerd, T. A. J., 1958, Ageing of red cells, *J. Physiol.* (London) **143:**325–331.

Rahman, Y. E., Elson, D. L., and Cerny, E. A., 1973, Studies on the mechanism of erythrocyte aging and destruction. I. Separation of rat erythrocytes according to age by Ficoll gradient centrifugation, *Mech. Ageing Dev.* **2:**141–150.

Ravindranath, Y., Brohn, F., and Johnson, R. M., 1987, Erythrocyte age-dependent changes of membrane protein 4.1: Studies in transient erythroblastopenia, *Pediatr. Res.* **21:**275–278.

Rennie, C. M., Thompson, S., Parker, A. C., and Maddy, A., 1979, Human erythrocyte fractionation in "Percoll" density gradients, *Clin. Chim. Acta* **98:**119–125.

Rice-Evans, C., and Hochstein, P., 1981, Alterations in erythrocyte membrane fluidity by phenylhydrazine-induced peroxidation of lipids, *Biochem. Biophys. Res. Commun.* **100:**1537–1542.

Rifkind, M., Araki, K., and Hadley, E. C., 1983, The relationship between the osmotic fragility of human erythrocytes and cell age, *Arch. Biochem. Biophys.* **222:**582–589.

Ripoche, J., and Sim, R. B., 1986, Loss of complement receptor type 1 (CR1) on ageing of erythrocytes. Studies of proteolytic release of the receptor, *Biochem. J.* **235:**815–821.

Rothstein, M., 1979, The formation of altered enzymes in aging animals, *Mech. Ageing Dev.* **9:**197–202.

Salvo, G., Caprari, P., Samoggia, P., Mariani, G., and Salvati, A. M., 1982, Human erythrocyte separation according to age on a discontinuous Percoll density gradient, *Clin. Chim. Acta* **122:**293–300.

Sanderson, R. J., and Bird, K. W., 1977, Separation by counterflow centrifugation, *Methods Cell Biol.* **15:**1–14.

Sanderson, R. J., Bird, K. E., Palmer, N. F., and Brenman, J., 1976, Design principles for a counterflow centrifugation cell separation chamber, *Anal. Biochem.* **71:**615–622.

Sass, M. D., Levy, L. M., and Walter, H., 1963, Characteristics of erythrocytes of different ages. II. Enzyme activity and osmotic fragility, *Can. J. Biochem. Physiol.* **41:**2287–2296.

Sauberman, N., Fortier, N. L., Fairbanks, G. F., O'Connor, R. J., and Snyder, L. M., 1979, Red cell membrane in hemolytic disease. Studies on variables affecting electrophoretic analysis, *Biochim. Biophys. Acta* **556:**292–313.

Scarpa, M., Rigo, A., Mono, F., Isacchi, G., Novelli, G., and Dallapiccola, B., 1985, Increased rate of superoxide ion generation in Fanconi anemia erythrocytes, *Biochem. Biophys. Res. Commun.* **130:**127–132.

Schacter, L. P., 1986, Generation of superoxide anion and hydrogen peroxide by erythrocytes from individuals with sickle cell trait or normal hemoglobin, *Eur. J. Clin. Invest.* **16:**204–210.

Schleicher, E., Scheller, L., and Wieland, D. H., 1981, Quantitation of lysine-bound glucose of normal and diabetic erythrocyte membranes by HPLC analysis of furosine [E-N(L-furoylmethyl)-L-lysine], *Biochem. Biophys. Res. Commun.* **99:**1011–1019.

Schlepper-Schäfer, J., Kolb-Bachofen, V., and Kolb, H., 1983, Identification of a receptor for senescent erythrocytes on liver macrophages, *Biochem. Biophys. Res. Commun.* **115:**551–559.

Schlüter, K., and Drenckhahn, D., 1986, Co-clustering of denatured hemoglobin with band 3: Its role in binding of autoantibodies against band 3 to abnormal and aged erythrocytes, *Proc. Natl. Acad. Sci. USA* **83:**6131–6141.

Schroit, A. J., Madsen, J. W., and Tanaka, Y., 1985, In vivo recognition and clearance of red blood cells containing phosphatidylserine in their plasma membranes, *J. Biol. Chem.* **260:**5131–5138.

Seaman, G.V.F., 1983, Electrochemical properties of the peripheral zone of erythrocytes, *Ann. N.Y. Acad. Sci.* **416:**176–189.

Seaman, G.V.F., Knox, R. J., Nordt, F. J., and Regan, D. H., 1977, Red cell aging. I. Surface charge density and sialic acid content of density-fractionated human erythrocytes, *Blood* **50:**1001–1011.

Shalev, O., Leida, M. N., Hebbel, R. P., Jacob, H. S., and Eaton, J. W., 1981, Abnormal erythrocyte calcium homeostasis in oxidant-induced hemolytic disease, *Blood* **58:**1232–1235.

Shiga, T., Maeda, N., Suda, T., Kon, K., and Sekiya, M., 1979, The decreased membrane fluidity of in vivo aged, human erythrocytes. A spin label study, *Biochim. Biophys. Acta* **553:**84–95.

Shinozuka, T., Takei, S., and Watanabe H., 1986, Affinity of young and old human erythrocytes for alkyl-Sepharose 6 MB gels, *J. Chromatogr.* **375:**380–385.

Shinozuka, T., Takei, S., Yanagida, J. I., Watanabe, H., and Ohkuma, S., 1988a, Comparative study on the main membrane-surface sialoglycopeptides released from young and old human erythrocytes with trypsin, *Comp. Biochem. Physiol.* **89B:**309–315.

Shinozuka, T., Takei, S., Yanagida, J. I., Watanabe, H., and Ohkuma, S., 1988b, Number and distribution density of ABH and MN antigen sites on young and old human erythrocyte surfaces, *Life Sci.* **43:**683–689.

Shukla, S. D., and Hanahan, D. J., 1981, Identification of domains of phosphatidylcholine in human erythrocyte plasma membranes, *J. Biol. Chem.* **257:**2908–2911.

Simon, E. R., and Topper, Y. L., 1957, Fractionation of human erythrocytes on the basis of their age, *Nature* **459:**1211–1212.

Singer, J. A., Jennings, L. K., Jackson, C. W., Dockter, M. E., Morrison, M., and Walker, W. S., 1986, Erythrocyte homeostasis: Antibody-mediated recognition of the senescent state by macrophages, *Proc. Natl. Acad. Sci. USA* **83:**5498–5501.

Smalley, C. E., and Tucker, E. M., 1983, Blood group A antigen site distribution and immunoglobulin binding in relation to red cell age, *Br. J. Haematol.* **54:**209–219.

Snyder, L. M., Sauberman, N., Condara, H., Dolan, J., Jacobs, J., Szymanski, I., and Fortier, N. L., 1981, Red cell membrane response to hydrogen-peroxide sensitivity in hereditary xerocytosis and in other abnormal red cells, *Br. J. Haematol.* **48:**435–444.

Snyder, L. M., Leb, L., Piotrowski, J., Sauberman, N., Liu, S. C., and Fortier, N. L., 1983, Irreversible spectrin–hemoglobin crosslinking in vivo: A marker for red cell senescence, *Br. J. Haematol.* **53:**379–384.

Snyder, L. M., Fairbanks, G., (Piotrowski) Trainor, J., Fortier, N. L., Jacobs, J. B., and Leb, L., 1985, Properties and characterization of vesicles released by young and old human red cells, *Br. J. Haematol.* **59:**513–522.

Spooner, R. J., Percy, R. A., and Rumley, A. G., 1979, The effect of erythrocyte ageing on some vitamin and mineral dependent enzymes, *Clin. Biochem.* **12:**289–290.

Sutera, S. P., Gardner, R. A., Boylan, C. W., Carroll, G. L., Chang, K. C., Marvel, J. S., Kilo, C., Gonen, B., and Williamson, J. R., 1985, Age-related changes in deformability of human erythrocytes, *Blood* **65:**275–282.

Suzuki, T., and Dale, G. L., 1987, Biotinylated erythrocytes: In vivo survival and in vitro recovery, *Blood* **70:**791–795.

Suzuki, T., and Dale, G. L., 1988, Senescent erythrocytes: Isolation of in vivo aged cells and their biochemical characteristics, *Proc. Natl. Acad. Sci. USA* **85:**1647–1651.

Takeshita, M., Tamura, M., Yubisui, T., and Yoneyama, Y., 1983, Exponential decay of cytochrome b_5 and cytochrome b_5 reductase during senescence of erythrocytes: Relation to the increased methemoglobin content, *J. Biochem.* **93:**931–934.

Takeuchi, N., Shishino, K., Bando, S., Murase, M., Go, S., and Uchida, K., 1985, Aging change of riboflavin concentration and glutathione reductase activity in erythrocytes, *Arch. Gerontol. Geriatr.* **4:**205–210.

Tanaka, Y., and Schroit, A. J., 1983, Insertion of a fluorescent phosphatidylserine into the plasma membrane of red blood cells. Recognition by autologous macrophages, *J. Biol. Chem.* **258:**11335–113343.

Teitel, P., 1977, Basic principles of the filterability test and analysis of erythrocyte flow behaviour, *Blood Cells* **3:**55–70.

ten Brinke, M., and de Regt, J., 1970, ^{51}Cr-half life time of heavy and light human erythrocytes, *Scand. J. Hematol.* **7:**36–41.

Tillman, W., Levin, C., Prindull, G., and Schröter, W., 1980, Rheological properties of young and aged human erythrocytes, *Klin. Wochenschr.* **58:**569–574.

Todd, C., and White, R. G., 1911, On the fate of red blood corpuscles when injected into the circulation of an animal of the same species: With a new method for the determination of the total volume of blood, *Proc. R. Soc. London Ser. B* **84:**255–259.

Tosteson, D. C., Carlsen, E., and Dunham, E. T., 1955, The effect of sickling on ion transport. I. Effect of sickling on potassium transport, *J. Gen. Physiol.* **39:**31–53.

Turner, B. M., Fisher, R. A., and Harris, H., 1974, The stage related loss of activity of four enzymes in the human erythrocyte, *Clin. Chim. Acta* **50:**85–95.

Usami, S., Chein, S., and Gregersen, M. I., 1971, Viscometric behaviour of young and aged erythrocytes, in: *Theoretical and Clinical Hemorheology* (H. H. Hartert and A. L. Copley, eds.), pp. 266–270, Springer, Berlin.

van der Vegt, S. G. L., Ruben, A. M. T., Werre, J. M., de Gier, J., and Staal, G. E. J., 1985a, Membrane characteristics and osmotic fragility of red cells, fractionated with anglehead centrifugation and counterflow centrifugation, *Br. J. Haematol.* **61:**405–413.

van der Vegt, S. G. L., Ruben, A. M. T., Werre, J. M., Palsma, D. M. H., Verhofen, C. W., de Gier, J., and Staal, G. E. J., 1985b, Counterflow centrifugation of red cell populations: A cell age related separation technique, *Br. J. Haematol.* **61:**393–403.

van Gastel, C., van den Berg, D., de Gier, J., and van Deenen, L. L. M., 1965, Some lipid characteristics of normal red blood cells of different age, *Br. J. Haematol.* **11:**193–199.

van Oss, C. J., 1982, Shape of aging erythrocytes, *Biorheology* **19:**725.

Vaysse, J., Gattegno, L., Bladier, D., and Aminoff, D., 1986, Adhesion and erythrophagocytosis of human senescent erythrocytes by autologous monocytes and their inhibition by beta-galactosyl derivatives, *Proc. Natl. Acad. Sci. USA* **83:**1339–1343.

Vaysse, J., Vassy, R., Eclache, V., Bladier, D., Gattegno, L., and Pilardeau, P., 1988a, Does red blood cell size correlate with red blood cell age in mouse? *Mech. Ageing Dev.* **44:**265–276.

Vaysse, J., Vassy, R., Eclache, V., Gattegno, L., Bladier, D., and Pilardeau, P., 1988b, Some characteristics of human red blood cells separated according to their size: A comparison with density-fractionated red blood cells, *Am. J. Hematol.* **28:**232–238.

Vettore, L., de Matteis, M. C., and Zampini, P., 1980, A new density gradient system for the separation of human red blood cells, *Am. J. Hematol.* **8:**291–297.

Vlassara, H., Valinsky, J., Brownlee, M., Cerami, C., Nishimoto, S., and Cerami, A., 1987, Advanced glycosylation endproducts on erythrocyte cell surface induce receptor-mediated phagocytosis by macrophages. A model for turnover of aging cells, *J. Exp. Med.* **166:**539–549.

Walls, R., Kumar, K. S., and Hochstein, P., 1976, Aging of human erythrocytes. Differential sensitivity of young and old erythrocytes to hemolysis induced by peroxide in the presence of thyroxine, *Arch. Biochem. Biophys.* **174:**463–468.

Walter, H., and Krob, E. J., 1983, Detection of surface differences between two closely related cell populations by partitioning isotopically labeled mixed cell populations in two-polymer aqueous phases. I. Human red blood cell subpopulations, *Cell Biophys.* **5:**301–306.

Walter, H., Krob, E. J., and Ascher, G. S., 1981, Aging of erythrocytes results in altered red cell surface properties in the rat, but not in the human. Studies by partitioning in two-polymer aqueous phase systems, *Biochim. Biophys. Acta* **641:**202–215.

Walter, H., Tamblyn, C. H., Krob, E. J., and Seaman, G. V. F., 1983, The effect of neuraminidase on the relative surface charge-associated properties of rat red blood cells of different ages, *Biochim. Biophys. Acta* **734:**368–372.

Warth, J. A., Brewer, G. J., Gnegy, M. E., Treisman, G., and Near, K., 1983, Calmodulin level in whole blood correlates with the percentage of reticulocytes, *Am. J. Hematol.* **15:**153–157.

Weed, R. I., 1970, The importance of erythrocyte deformability, *Am. J. Med.* **49:**147–150.

Weed, R. I., and Reed, C. F., 1966, Membrane alterations leading to red cell destruction, *Am. J. Med.* **41:**681–698.

Weed, R. I., LaCelle, P. L., and Merrill, E. W., 1969, Metabolic dependence of red cell deformability, *J. Clin. Invest.* **48:**795–809.

Westerman, M. P., Pierce, L. E., and Jensen, W. N., 1963, Erythrocyte lipids: A comparison of normal young and normal old populations, *J. Lab. Clin. Med.* **62:**394–400.

Wiener, A. S., 1942, Hemolytic transfusion reactions. I. Diagnosis, with special reference to the method of differential agglutination, *Am. J. Clin. Pathol.* **12:**189–199.

Williams, A. R., and Morris, D. R., 1980, The internal viscosity of the human erythrocyte may determine its lifespan in vivo, *Scand. J. Haematol.* **24:**57–62.

Williamson, J. R., Gardner, R. A., Boylan, C. W., Carroll, G. L., Chang, K. I., Marvel, J. S., Gonen, B., Kilo, C., Tran-Son-Tay, R., and Sutera, S. P., 1985, Microrheologic investigation of erythrocyte deformability in diabetes mellitus, *Blood* **65:**283–288.

Wilson, C., and Peterson, S. W., 1986, Insulin receptor processing as a function of erythrocyte age. A kinetic model for down-regulation, *J. Biol. Chem.* **261:**2123–2128.

Wilton, A., 1966, An attempt to separate erythrocytes according to age by a new type of centrifuge, *Acta Haematol.* **35:**163–175.

Winterbourn, C. C., and Batt, R. D., 1970, Lipid composition of human red cells of different ages, *Biochim. Biophys. Acta* **202:**1–8.

Yaari, A., 1969, Mobility of human red blood cells of different age groups in an electric field, *Blood* **33:**159–163.

Yamaguchi, T., Fujita, Y., Kuroki, S., Ohtsuka, K., and Kimoto, E., 1983, A study on the reaction of human erythrocytes with hydrogen peroxide, *J. Biochem.* **94:**379–386.

Zanner, M. A., and Galey, W. R., 1985, Aged human erythrocytes exhibit increased anion exchange, *Biochim. Biophys. Acta* **818:**310–315.

Chapter 4

Erythrocyte Clearance

Hans U. Lutz

1. INTRODUCTION

1.1. Tissue Homeostasis

Maintenance of tissue function includes disposal of nonfunctional, senescent, and aberrant cells. Removal of these cells is a tissue homeostatic process that occurs in one way or another in any multicellular organism with somatic and germline tissues. Vertebrates with a constant body weight and tissues of mesenchymal and mesodermal origin had to acquire tools to eliminate cells by means other than mummification or release to the surroundings. Our knowledge of specific clearance mechanisms is very limited. Our ignorance is in part due to the reliability with which the clearance systems work. There is to my knowledge no known disease or metabolic error that can be traced to an improper clearance system. In contrast to this, a large number of diseases are initiated by an overreaction of a particular clearance mechanism. Consider, for example, the various autoimmune diseases where excessive clearance is mediated either by cellular autoimmunity or by autoantibodies, phagocytes, and in part by complement. Thus, the humoral and cellular defense systems are the primary and definitely the most versatile tools for a selective clearance also of senescent and altered self. The recent insights into the generation of diversity among antibodies (Tonegawa, 1985; Alt *et al.*, 1987), the immunological network (Jerne, 1984; Bottomly, 1984; Hooper, 1987), the persistence of autoreactive B cells in ontogeny (Bona, 1988; Benner *et al.*, 1982), the existence of naturally occurring autoantibodies in healthy subjects (Lutz and Wipf, 1982; Guilbert *et al.*, 1982; Lutz *et al.*, 1984), and the understanding of complement regulation (Müller-Eberhard and Schreiber, 1980; Fearon and Wong, 1983; Schifferli *et al.*, 1986) have paved a road into the complexity of tissue homeostasis and clearance.

Hans U. Lutz Laboratory for Biochemistry, Swiss Federal Institute of Technology, ETH-Zentrum, CH-8092 Zurich, Switzerland.

In reviewing this field I will concentrate on clearance of red blood cells (red cells) since their senescence is not perturbed by ongoing protein synthesis. My review is not comprehensive as I will emphasize immunological aspects for which there exist well-established parallels in clearance of invading organisms.

1.2. Previously Reviewed Aspects and an Outline

In preparing this review I have tried not to duplicate previous ones. For example, characterization of senescent and young red cells has been discussed in two recent reviews (Clark and Shohet, 1985; Clark, 1988). Clark and Shohet (1985) describe the proposed mechanisms of red cell aging in great detail and concentrate their critique on density separation techniques as a way to isolate young and old red cell subpopulations. A critical evaluation of desialylation and vesicle release to account for a lowered surface-to-volume ratio to old red cells has been published by Bocci (1981). Determination of red cell-associated IgG (Schreiber, 1985), shortened red cell survival in anemia (Frank, 1985), and drug-induced immune hemolytic anemia (Petz and Branch, 1985) have been reviewed recently. Clearance mechanisms for red cells are the most controversial of the subjects listed and information regarding this topic is scattered over many years, journals, and groups. This situation renders reviewing difficult, in particular evaluation of the data that were obtained by constantly changing methods. Hence, my approach was to introduce a section on "methodological problems." This serves two purposes: it refers to experimental standard techniques, and it lists a number of reasons for the restriction to immunological aspects of clearance. It is followed by a chronology of the efforts aimed at elucidating the mechanism(s) of red cell removal. For each hypothesis a few key papers are cited. Emphasis is put on the role of naturally occurring antibodies that stimulate alternative complement pathway C3b deposition. This subject is embedded in a section in which I review aspects of complement that have become relevant to understanding red cell clearance. It also focuses on clinical aspects related to red cell clearance. I continue by reviewing data on the regulation of naturally occurring antibodies and discuss the relationship between cellular age and body age.

1.3. General Aspects in Understanding Red Cell Clearance in Humans

Unequivocal evidence against a random removal of human red cells from the circulation has been available since Ashby's differential agglutination studies (for review see Frank, 1985): he followed the survival of transfused red cells carrying antigens not normally present in the recipient. The data were compatible with a selective clearance of red cells aged roughly 120 days. Thus, almost all red cells withstand a period of 120 days in circulation and are then suddenly cleared by phagocytosis and to a small extent by intravascular lysis.

Their uniform life span suggested the existence of an endogenous "clock" that is turned on by the loss of the nucleus and the endoplasmic reticulum. As red cells can no longer replace damaged or modified proteins, they lose several enzymatic activities (Beutler, 1985a,b; Cohen *et al.*, 1976; Pfeffer and Swislocki, 1976; Kadlubowski and Agutter, 1977), as they age. Another hypothesis has put emphasis on a "wearing out" process that might be mediated by constant shearing of red cells, oxidative stress, or hydrolytic enzymes within plasma. The process as a whole offers a number of scientifically exciting questions: What mechanism allows a selective recognition of old cells? Which of the many changes in an aging red cell generates a signal of age on the surface of the cell? How is this signal recognized and transformed into a reaction against self? What makes the red cell so uniquely stable to survive for 120 days as a hemoglobin-filled bag? None of these questions has been answered unequivocally, despite the fact that they have been tackled many times from a variety of viewpoints. It is the aim of this review to evaluate the partial answers given so far.

2. METHODOLOGICAL PROBLEMS

2.1. Separation of Young from Old Red Cells

The major problem in studying properties of young and old red cells is a reliable method to isolate red cell subpopulations. Bleeding and hypertransfusion (Singer *et al.*, 1986; Walker *et al.*, 1984) as well as in vivo aging of tagged cells (Suzuki and Dale, 1987) have successfully been applied in animals to generate red cell populations enriched in young or old cells, respectively. As these techniques are not applicable to humans, other means had to be introduced. Graded osmotic hemolysis was one of them, but it failed to yield membranes from distinct subpopulations, since differences in osmotic fragility are small among mature red cells (for details see Clark and Shohet, 1985). In contrast, density fractionation was more promising. The red cell buoyant density increases with cell age due to a loss of potassium and accompanying water (Astrup, 1974; Hoffman, 1958; Lee *et al.*, 1984). Studies on red cell subpopulations became quite popular after Murphy (1973) reported on self-orientation of red cells according to density during high-speed centrifugation of packed red cells in angle rotors. The simplicity of the technique was tempting. It has been widely used with variable success. Far more powerful than centrifugation of whole blood is separation of red cells on preformed density gradients (Piomelli, 1988) containing dextran (Schmidt *et al.*, 1975), stractan (Corash *et al.*, 1974; Piomelli *et al.*, 1978), serum albumin (Piomelli *et al.*, 1967; Schmidt *et al.*, 1975), or on self-forming Percoll gradients (Lutz and Fehr, 1979; Vettore *et al.*, 1980). Red cells separated on density gradients showed the least contamination by cells of other density (Lutz, 1988). Their functional properties differed, with the light fraction representing a mixture of reticulocytes and young mature cells and the dense fraction representing predominantly old red cells. This was best illustrated by labeling a cohort of red cells ex vivo with [^{14}C]glycine (Piomelli *et al.*, 1967): label was incorporated only in the youngest cells that yet contained the endoplasmic reticulum. Fractionation of red cells on density gradients after reinfusion showed the labeled red cells to represent the lightest cells 1 day after reinfusion and the heaviest third of cells 60 days later. There is little doubt about old cells being denser than young mature ones. It is still controversial, however, whether the two phenomena do in fact correlate throughout the life span of red cells. Doubts have been raised against a correlation of red cell density with cell age (Beutler, 1985a; Clark and Shohet, 1985). These doubts were mainly based on radioactive iron distribution among red cells that were fractionated by whole blood centrifugation according to Murphy (Luthra *et al.*, 1979). Since released iron can be reused and since the applied separation technique did not yield good separations of subpopulations when compared to any type of density gradient centrifugation, the doubts are not well grounded. The Murphy method suffers from a number of drawbacks: cells are virtually packed in little plasma or serum. This generates an enormous viscosity, which hinders red cells from reaching their isopycnic point within the tube (Lutz, 1988). High temperature and long centrifugation times favor hemolysis and damage by hydrolytic enzymes (Brovelli *et al.*, 1987; Lutz, 1987).

2.2. Removal of White from Red Cells

Many studies have been carried out with red cells obtained from anticoagulated blood by removal of the buffy coat and by washing the pelleted cells. It was, however, noticed by several investigators (Nakao, Engelfriet, Beutler) that this procedure did not completely eliminate white cell contamination. In particular, the remaining granulocytes release most of their proteases within 24 hr of cold storage (Jochum *et al.*, 1984; Duswald *et al.*, 1985) and are the primary cause of artifacts. Beutler *et al.* (1976) optimized filtration of heparinized blood through mixtures of α- and microcystalline cellulose, as a procedure to eliminate white cells. This approach has since been used widely and works best with heparinized blood. Omission of this step is the most likely source of errors and of inconsistent results (for discussion see Bocci, 1981; Lutz, 1987).

2.3. Opsonization of Red Cells

Washed red cells were added as such or following modification without serum contact to phagocytes in media. This type of assay led, for example, to the suggestion that senescent human red cells contain newly exposed β-galactose residues that are recognized by lectinlike receptors on phagocytes (Vaysse *et al.*, 1986). Similar types of experiments suggested that nonenzymatic glycosylation modified red cell surface components to the extent that cells became recognized by phagocytes (Vlassara *et al.*, 1987). Another type of approach was the addition of small amounts of eluted or purified antibodies to washed red cells and analysis of their subsequent phagocytosis (Kay, 1978; Galili *et al.*, 1986). While all approaches revealed interesting binding properties of phagocytes, they cannot mimic in vivo conditions.

A red cell in blood is surrounded by a serum solution containing 65 mg/ml protein of which 15 mg/ml are immunoglobulins. Hence, a phagocyte in vivo never encounters a "naked" red cell. Whatever modification a red cell goes through, it is in contact with whole serum and may thus be opsonized prior to reaching the phagocyte. Likewise, addition of a few micrograms of an antibody to washed red cells creates a completely unphysiologic situation. This was already noticed 20 years ago by studying binding of whole human IgG to red blood cells: $2–4 \times 10^5$ IgG bound per red cell (Grob *et al.*, 1967; Frommel *et al.*, 1967). Binding was via both Fab and Fc portions, was partially inhibited by albumin, and increased in low-ionic-strength media (Frommel *et al.*, 1967). The fraction of IgG that bound to red cells at low ionic strength and was eluted by isotonic salt solutions even received a name (erythrophilic IgG) (Fidalgo *et al.*, 1967).

Considering these findings, experiments aimed at demonstrating the role of certain modifications or of a given antibody ought to show their effectiveness in the presence of normal serum or diluted serum. While this approach will reduce effective binding of a potential antibody to red cells (Kurlander and Rosse, 1979; Jungi and Barandun, 1985), it reveals potentially relevant aspects of the target/effector cell interactions. At the same time, opsonization in whole or diluted serum allows one to study the involvement of the major humoral effector system, complement. This was largely neglected in the field of red cell aging.

2.4. Preparation of Membranes and Electrophoresis of Membrane Proteins

Comparison of membrane properties often requires lysis of red cells and membrane preparations. Any such preparation should be free of proteases. This is of primary importance since proteases from residual granulocytes as well as red cell-inherent, acid (Tökés and Chambers, 1975; Pontremoli *et al.*, 1979; Murakami *et al.*, 1979), neutral, SH-dependent (Pontremoli *et al.*, 1985; Hatanaka *et al.*, 1984; Rieder *et al.*, 1986), and ATP-dependent (Rieder *et al.*, 1986; Au, 1987) proteases are primarily found among the light fractions on density gradients (Rieder *et al.*, 1986; Brovelli *et al.*, 1987). Thus, at some stage during or after lysis (in the cold), proteases should be inhibited. Various protocols have been suggested, but to my knowledge no study has rigorously compared the effect of the various protocols. Our experience is that hemolysis of red cells at a volume ratio of 1/30 in 5 mM phosphate and 1 mM EDTA (pH 7.4) and a subsequent wash in hemolysis buffer containing diisopropylfluorophosphate (DFP) (0.4 mM) (modified from Hargreaves *et al.*, 1980) yielded reproducible results and at the same time allowed safe disposal of the supernate that yet contained DFP after the second centrifugation.

Analysis of membrane proteins by SDS PAGE requires a number of precautions, when not carried out immediately after membrane preparation. Samples that should maintain their content of SH groups and their actual extent of disulfide bonding should be alkylated in the presence of 1% SDS and 5 mM *N*-ethylmaleimide (NEM) (Lutz, 1984). The detergent renders sites accessible and NEM blocks free SH groups and thus prevents subsequent disulfide bond formation during freezing and thawing. Subsequent electrophoresis of these samples on SDS PAGE may be carried out with or without reduction, provided the concentration of added reducing agent exceeds that of NEM. Incubation with reducing agent for 30 min at 37°C is generally superior to boiling, which tends to generate protein aggregates. More important, however, is subsequent alkylation of all samples (1) to prevent reoxidation of reduced material during electrophoresis and (2) to run reduced and unreduced samples side by side without interference (Lutz *et al.*, 1987a).

Based on these arguments I will put emphasis on studies that in one way or another considered the above-mentioned methodological problems.

3. A HISTORICAL VIEW ON STUDIES AIMED AT ELUCIDATING THE MECHANISM(S) OF RED CELL CLEARANCE

3.1. Fragmentation and Desialylation

Fragmentation of red cells by shearing of vesicles was long thought to eventually render cells spherical and thus less deformable (Weed and Reed, 1966). Reduced deformability was considered to prolong the time a red cell remained exposed at high hematocrit to the hypoglycemic conditions within the spleen's orifices. The sum of these adverse effects should have allowed adherent phagocytes to clear the less deformable cells preferentially. Splenectomy, however, eliminated the adverse effects accompanying spherocytosis—which can be considered a model for this type of mechanism—but had little effect on the life span of red cells. This implied that clearance occurred within the entire reticuloendothelial system and was not confined to the spleen and its stress-imposing capillaries. In addition, the loss of deformability (Nash and Meiselman, 1981) was mainly due to the loss of potassium and cell water (Lee *et al.*, 1984; Astrup, 1974) rather than a loss of membrane, in the form of vesicles (Dumaswala and Greenwalt, 1984), which accounts for less than 5–10% of the surface as judged from lipid loss (Winterbourn and Batt, 1970). Another type of wearing out process was suggested by Danon and Marikovsky (1962), who reported that senescent red cells have lost 10–15% of their surface charge. The negative charges on red cells are due to sialic acid residues, which are located almost exclusively on sialoglycoproteins, since red cells contain very few gangliosides (Schultz *et al.*, 1984; Watanabe *et al.*, 1979). A loss of charge can occur either by release of sialic acid via glycosidases or by proteolysis of sialoglycoproteins. Either process was compatible with the idea that it is not the red cell that determines the time of its removal, but the length of time it circulates and is exposed to exogenous modifiers.

Artificial removal of sialic acid by neuraminidase (Durocher *et al.*, 1975) and modification of sialic acid by periodate (Durocher *et al.*, 1976) indeed resulted in increased in vivo clearance in rats. Reduction of oxidized sialic acid with borohydride reversed the effect of periodate (Durocher *et al.*, 1976). While desialylation suggested recognition via galactose-specific lectinlike components on macrophages, the modifications of sialic acid called for another mechanism. The latter phenomenon did not receive enough credit, although it appears interpretable in view of more recent findings concerning the effect of sialic acid on complement (Michalek *et al.*, 1988) (see also Section 4.3). A selective removal of sialic acid and exposure of penultimate galactose residues was a more attractive hypothesis with many facets in common with recognition of desialylated serum glycoproteins by the asialoglycoprotein receptor on hepatocytes (Ashwell and Harford, 1982). Although the hypothesis appeared reasonable, it had one major disadvantage: it assumed that red cell aging correlated with desialylation to an extent as suggested by Danon and comparable to that achievable by neuraminidase treatment. The use of more sophisticated separation techniques later allowed the demonstration that young and old red cells had the same surface charge (Luner *et al.*, 1977). Direct quantitation of sialic acid content and reference to the number of sialoglycoproteins further demonstrated identical numbers of sialic acid residues (± 1.5%) per sialoglycoprotein (Lutz and Fehr, 1979). These findings left room for the loss of some sialic acid molecules along with membrane in the form of vesicles (Lutz and Fehr, 1979; Lutz, 1978; Dumaswala and Greenwalt, 1984), but rendered desialylation and exposure of penultimate galactose residues an unlikely process. The findings did not, however, exclude a selective loss of less than 1.5% of sialic acid molecules. Others confirmed that young and old red cells have the same number of sialic acid residues per sialoglycoprotein (Miyahara and Spiro, 1984), unaltered surface charge (Walter *et al.*, 1981), and increased rather than reduced reactivity with wheat germ agglutinin (Shinozuka *et al.*, 1988). Nevertheless, some proponents continue to claim that sialoglycoproteins are desialylated as human red cells age (Vaysse *et al.*, 1986).

3.2. Naturally Occurring Antibodies on Senescent Red Cells

New input came in the early 1970s when DeHeer and Edington (1974) and Cunningham (1974) described spontaneous generation of antiself reacting antibodies from spleen cells of mice. They verified the presence of these antibodies by a guinea pig complement-dependent lysis in a plaque-forming assay. Kay (1975, 1978) presented evidence for the selective recognition of senescent and in vitro-stored but not of young human red cells by autologous or homologous IgG. This suggested alterations within red cells that expose a signal on the surface of senescent red cells, a signal that is recognized by certain IgGs that mediate phagocytosis. The inherent principle called for a red-cell-associated endogenous clock to determine the time of clearance. Kay's findings were poorly accepted initially, since the existence of antibodies reacting with self was not established and contradicted the dogma (horror autotoxicus) that antiself reactive lymphocytes were eliminated during ontogeny. Rapidly growing knowledge in understanding the immune network (for reviews see Bretscher and Cohn, 1970; Jerne, 1984; Bottomly, 1984) put her findings into a new perspective. Likewise, direct demonstration of a number of naturally occurring antibodies to autologous and homologous red cell proteins and purification of antispectrin (Lutz and Wipf, 1982) and later of anti-band 3 antibodies (Lutz *et al.*, 1984) from IgG of individuals and from pooled IgG established their existence.

In the meantime, several laboratories attempted to identify the specificity of red-cell-associated IgG. Kay (1981) concluded from a protein that coeluted with red-cell-associated IgG that it had an apparent molecular mass of 62 kDa and served as senescent antigen. Fudenberg's group (Alderman *et al.*, 1981) claimed IgG to bind to asialoglycophorin. Their results were questioned, since further washing removed the antibodies and because the immunosorbent bound glycophorin (Kay and Bennett, 1982). We suggested that red-cell-associated IgG was bound to oligomerized band 3 protein, since protein A added to detergent extracts of surface-^{125}I-iodinated young and old red cells precipitated not only IgG but also label associated with band 3 protein and band 3 oligomers primarily from extracts of old cells (Lutz, 1981; Lutz and Stringaro-Wipf, 1983). This finding was unexpected, since common knowledge suggested exposure of a minor protein rather than modification of the most abundant protein. Later, the discrepancy between Kay's data and ours was eliminated, since Kay (1984) found that IgG eluted from red cells bound primarily to band 3 protein rather than the 62-kDa protein on immunoblots of membrane proteins. The difference that remained concerned the mechanism of exposure of antigenic sites. Kay suggested that a cytoplasmic cleavage of band 3 protein exposed the 62-kDa protein on senescent red cells, which thereby were recognized by anti-band 3 antibodies. We had good reasons to assume that the 62-kDa protein originated from preparations not well protected from proteolysis (Lutz, 1981). Our results favored another mechanism of exposure, namely increased binding of anti-band 3 antibodies to band 3 oligomers on senescent red cells. This interpretation was based on detection of cross-linked band 3 protein on old red cells (Lutz and Stringaro-Wipf, 1983), on a correlation between exoplasmic cross-linkability of band 3 protein (Schweizer *et al.*, 1982) and binding of autologous IgG (Mueller and Lutz, 1983), as well as on enhanced binding of anti-band 3 to band 3 oligomers on immunoblots (Lutz *et al.*, 1984). This suggested that band 3 protein oligomerization favored binding of anti-band 3 presumably by lowering dissociation of bivalently bound antibody.

The controversy as to whether antigenic band 3 protein is exposed by a topologic rearrangement or whether a neoantigen is "produced" by proteolytic cleavage of band 3 protein will remain unsettled, since it is extremely difficult to pinpoint the few hundred molecules that carry antibodies among the 10^6 molecules of band 3 protein. A slight increase in the number of band 3-derived 60-kDa protein in membranes of old cells is not sufficient evidence to call this the modification that renders red cells old (Czerwinski *et al.*, 1988; Morrison *et al.*, 1985), in particular, because the 60-kDa fragment was not formed by a cytoplasmic, but rather an exoplasmic cleavage (Morrison *et al.*, 1985) similar to that obtained by chymotrypsin. If generation of this fragment were sufficient to expose an antigenic site, chymotrypsin treatment should have increased binding of anti-band 3 antibodies. It did not (Kay and Goodman, 1984).

Anti-band 3 antibodies were not, however, the only antibodies that were claimed to be primarily associated with senescent red cells. Galili and co-workers reported on antibodies from normal human IgG that specifically bind to α-galactosyl residues (αGal) (Galili *et al.*, 1983, 1984), a type of carbohydrate that is very scarce on human red cells (Galili *et al.*, 1987b). Anti-αGal antibodies were described as naturally occurring antibodies, although their concentration in serum was about 500 to 1000 times that of naturally occurring antispectrin or anti-band 3 antibodies (Galili *et al.*, 1984). Studies on anti-αGal antibodies in several species later revealed that these antibodies are acquired antibodies not present in animals containing large amounts of αGal on their red cell surface (Galili *et al.*, 1987a). Anti-αGal antibodies could be eluted by galactose or melibiose from thalassemic (Galili *et al.*, 1983), from normal young, and in larger quantities from senescent red cells (Galili *et al.*, 1986). As a consequence of this, phagocytosis of eluted red cells was less efficient for both young and old red cells and supplementation of eluted antibodies increased phagocytosis not only of old but also of young red cells by a factor of 2 to 3 (Figure 3 in Galili *et al.*, 1986). Hence, these antibodies stimulated the extent of phagocytosis, but were not responsible for a selective recognition of senescent red cells. In agreement with this, extensive washing of young and old red cells depleted red cells of anti-αGal, but did not eliminate selective phagocytosis of senescent red cells (Kay and Bosman, 1985). Thus, anti-band 3 antibodies appeared more likely to selectively recognize senescent red cells.

Studies of Low and Drenckhahn (Low *et al.*, 1985) appeared to support a functional role of anti-band 3 antibodies: treatment of red cells with 10 mM phenylhydrazine not only produced Heinz bodies but also aggregated band 3 protein and at the same time resulted in clustered binding of IgG to the red cell surface at sites superimposable on band 3 oligomers. Nevertheless, their data were not compatible with this explanation, since most of the antibodies eluted from phenylhydrazine-treated red cells had in fact anti-spectrin antibody specificity (see p. 61397 in Schlüter and Drenckhahn, 1986). This is not surprising, since the high concentration of phenylhydrazine caused partial lysis and liberated spectrin, which formed immune complexes with antispectrin antibodies in serum. Thus, Low and Drenckhahn probably demonstrated binding of C3b-solubilized immune complexes to complement receptor type 1 (CR1) of red cells. This explanation is highly probable, since the 1000 CR1 molecules per red cell are clustered in a few patches (Paccaud *et al.*, 1988). The chances of one of these clusters being superimposed on a band 3 protein cluster is likely, since band 3 is 1000 times more prevalent. On the other hand, eluates from Heinz body-containing red cells of patients with unstable hemoglobin variants indeed revealed anti-band 3 antibodies (Schlüter and Drenckhahn, 1986), but less anti-IgG reactivity on intact cells than after phenylhydrazine treatment (Low *et al.*, 1985). Antispectrin antibodies were also found associated with in vivo aged red cells from thalassemic patients (Wiener *et al.*, 1986). Whether these antispectrin antibodies had an opsonizing effect and were involved in accelerated clearance of thalassemic red cells is questionable since antispectrin antibodies do not cross-react with surface proteins (Lutz *et al.*, 1987b). They could have induced an innocent bystander opsonization (see Section 4.7), but except for this their direct role in opsonizing oxidatively stressed and thalassemic red cells is unlikely, since spectrin is not accessible, neither on thalassemic nor on old red cells.

3.3. Cell-Bound IgG and Complement Components

The functional role of red-cell-associated, naturally occurring antibodies remains unclear. The reason for this uncertainty is mainly due to the low number of red-cell-associated IgGs. This value had earlier been determined on normal red cells by several groups, and was at the most a few hundred IgG molecules on senescent and less than one third of this on young red cells (Szymanski *et al.*, 1980; Yam *et al.*, 1982; Mueller and Lutz, 1983; Kay, 1978; Green *et al.*, 1985; Freedman, 1984; Wegner *et al.*, 1984). Cell-bound IgG was higher on sickle red cells (Hebbel and Miller, 1984) and sickling under nitrogen enhanced the amount of cell-associated IgG two-fold (Green and Kalra, 1988). None of the investigations on the mechanism of clearance addressed the quantitative aspect, though it is relevant with respect to opsonization. Several studies on induced, red-cell-specific antibodies have clearly shown that a few hundred IgG antibodies per cell is far from sufficient to induce phagocytosis by Fc receptors alone (Ehlenberger and Nussenzweig, 1977; van Loghem, 1978; Kurlander and Rosse, 1979). On the other hand, a few hundred IgG molecules strongly stimulated phagocytosis, if the target cells also carried C3b (Ehlenberger and Nussenzweig, 1977; Kurlander and Rosse, 1979). In view of these well-established phenomena, it was unlikely that the few hundred IgGs on senescent red cells were able to induce efficient phagocytosis via Fc receptors alone. They could, however, be efficient mediators of phagocytosis, if accompanied by C3b deposition.

Complement deposition on old red cells was, however, considered unlikely, since normal direct Coombs tests with anti-C3 reagents remained negative with whole normal red cell populations and even with dense red cells (for review see Frank, 1985). On the other hand, specific Coombs reagents revealed 50–150 C3d fragments per human red cell (Freedman and Massey, 1979; Chaplin *et al.*, 1981). More sensitive assays allowed Freedman (1984) to detect complement and in particular C3 fragments primarily on senescent red cells. In fact, the ratio of its content (old/young) was even higher than that for cell-bound IgG.

Another line of research, aimed at improving red cell storage for transfusion, also put emphasis on complement: storage of red cells in media depleted of plasma prolonged the effective use of conserved blood by almost a factor of 2 (Högman *et al.*, 1983). Thus, plasma factors have a deleterious effect on cold-stored red cells. The most likely factors that lower posttransfusion survival of whole blood stored in CPD adenine are components of the complement system which are diluted out in plasma-depleted red cells by the optimal additive solution. Szymanski *et al.* (1985) showed a negative correlation between red cell survival and red-cell-associated C3b or iC3b by carefully determining residual anti-C3c antibody concentrations after equilibrium binding of anti-C3c to stored red cells. Detection of C3b or iC3b on stored red cells is of particular relevance. It implies that C3b deposition was not due to binding of immune complexes carrying C3b to CR1 of red cells. Such a binding is transient and peaks at less than 3 min (Ross *et al.*, 1985) to 5 min (Horgan *et al.*, 1984), because CR1-bound C3b is inactivated to C3d and dissociates from the receptor (Taylor *et al.*, 1985; Horgan *et al.*, 1984), which has hardly any affinity for C3d,g (Fearon, 1984). Instead, C3b deposited on stored cells remained bound for days as C3b or iC3b. Hence, it was covalently bound either to a red cell membrane acceptor or to cell-associated, naturally occurring antibodies (for more information see Chapter 4).

3.4. Oxidative Damage in Cellular Aging and as a Tool to Accelerate Aging

One approach to study red cell aging is to enrich those cells from the circulation that are about to be cleared. This will never allow one to pick the cells that are just in the process of being phagocytized and will yield cells that are not only old but also opsonized by the surrounding plasma. Thus, various attempts have been made to mimic the aging process by *in vitro* modifications, to study dose response and opsonization.

Oxidative damage by a variety of oxidants results in peroxidation of lipids (Jain, 1988) and "nonenzymatic mixed function oxidation" of proteins (Oliver *et al.*, 1987). Peroxidation of lipids generates aldehydes among which malondialdehyde is the most common. The latter reacts with nucleophiles, primarily aminolipids (Jain, 1988) and proteins. Malondialdehyde is also capable of cross-linking proteins, although direct proof for this has not yet been given. Oxidative damage also yields protein oligomers by other routes: by disulfide bond formation, when the damage depleted cellular glutathione (Jain and Hochstein, 1980; Coetzer and Zail, 1980); and by transglutaminase, when permeability barriers become impaired and cellular Ca^{2+} is increased, as was shown for unstable hemoglobins (Lorand *et al.*, 1987).

Senescent red cells in fact reveal some of the signs of oxidative damage. They contain increased amounts (+50%) of high-molecular-weight oligomers consisting mainly of spectrin (Jain and Hochstein, 1980). Their content of aminophospholipid adducts with malondialdehyde is seven-fold higher (Jain, 1988) than in young red cells. The number of borohydride-accessible groups on proteins is increased by 16% on old cells (Oliver *et al.*, 1987). Most of these changes are small and their immediate impact on red cell life span and clearance is not obvious. More stringent correlations come from studying model systems for aging, e.g., vitamin E-deficient rats (Kay *et al.*, 1986), and from oxidation-sensitive red cells due to a glutathione deficiency (Arese, 1982): high-molecular-weight protein oligomers, for example, are more prominent in glucose 6-phosphate dehydrogenase-deficient red cells (Allen *et al.*, 1978) and further increase in numbers during a challenge with oxidant as in favic crisis (Turrini *et al.*, 1985). While protein oligomers are normally disulfide-bonded, in favic crisis some are resistant to reduction and may be cross-linked by transglutaminase (Turrini *et al.*, 1985). This is likely, since intracellular calcium is tenfold higher than in control cells (DeFlora *et al.*, 1985).

In vitro modification of red cells with various oxidants allowed study of the dose response of these agents and the extent of opsonization, phagocytosis, and clearance. Phenylhydrazine, H_2O_2, and several peroxides generated the expected oxidative damage

(Snyder *et al.*, 1988; Waugh and Low, 1985) and resulted in enhanced phagocytosis of red cells (Hebbel and Miller, 1984; Magnani *et al.*, 1986). Treatment of red cells with increasing concentrations of malondialdehyde stimulated, and subsequent opsonization with serum further enhanced, their phagocytosis (Hebbel and Miller, 1984).Oxidative damage also caused unexpected changes: it perturbed phospholipid asymmetry and exposed phosphatidylserine (PS) on the outer monolayer (Jain, 1985). PS is almost exclusively located on the cytoplasmic side in normal cells, since an ATP-dependent translocator maintains asymmetry for PS (Devaux, 1988). Artificially incorporated, chemically modified PS remained in the outer monolayer and mediated recognition of these cells by phagocytes *in vitro* and induced *in vivo* clearance of these cells (McEvoy *et al.*, 1986; Schroit *et al.*, 1985). At present it is not known whether clearance of PS-dotted red cells occurs by direct binding to phagocytes or whether it involves naturally occurring antibodies to phospholipids, or even other factors. Though elegant, the model may not bear on red cell senescence, since the chemically modified PS was not translocated to the inner monolayer.

Diamide is yet another agent that may perturb phospholipid asymmetry. It selectively oxidizes SH groups (Kosower *et al.* 1969) and penetrates red cell membranes where it causes protein oligomerization by disulfide bond formation (Haest *et al.*, 1977). At the high concentration of 5 mM it increased the accessibility of aminophospholipids to phospholipase A_2, but had no effect on aminogroup labeling (Franck *et al.*, 1986). This suggested that diamide destabilized the membrane, but did not cause aminophospholipids to flip spontaneously to the outer monolayer.

Diamide has pronounced effects on red cell survival at considerably lower concentrations than those required to produce marked depletion of glutathione and destabilization of the membrane. It accelerated *in vivo* clearance of dog red cells (Johnson *et al.*, 1980) at concentrations below 400 μM. It also cross-linked band 3 protein on intact human red cells in a dose-dependent way from 20 to 1000 μM (Lutz *et al.*, 1987a). The amount of oligomerized protein was, however, 1% or less and was verified by enrichment of oligomerized proteins. Diamide-treated human red cells (20–200 μM) were indeed phagocytized by adherent monocytic phagocytes (Bussolino *et al.*, 1987), and required opsonization with whole serum (Lutz *et al.*, 1987a). Phagocytosis was primarily dependent on active complement. In addition, the extent of phagocytosis was stimulated by supplementing whole serum with 20–100 times the physiologic concentration of naturally occurring anti-band 3 antibodies (Lutz *et al.*, 1987a). The effect of anti-band 3 antibodies was specific, since it could not be mimicked by increasing the concentration of another naurally occurring antibody with specificity for the cytoskeletal protein spectrin. Furthermore, the enhancing effect of anti-band 3 antibodies on phagocytosis was mediated by an anti-band-3-dependent stimulation of C3b deposition.

The most puzzling finding was that anti-band 3 antibodies exerted their stimulatory effect on C3b deposition despite a very small increase in the amount of red-cell-associated anti-band 3 (increment: <20 IgG per cell, when measured in complement-inactivated serum) (Lutz *et al.*, 1987a). The low extent of opsonization by IgG and the significant deposition of C3b, however, were also observed for various types of anemias (Petz and Branch, 1985; Schreiber, 1985). It became obvious that anti-band 3 antibodies have some unusual properties. To unravel their unique properties, we characterized cell-associated IgG, complement components, and complexes of the two on in vivo aged human red cells.

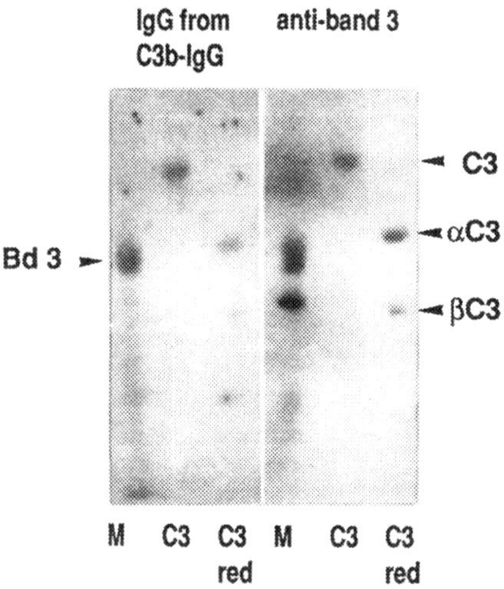

FIGURE 1. Binding of ^{125}I-iodinated anti-band 3 antibodies and IgG that was cleaved off C3b–IgG complexes from senescent red cells, to blotted membrane proteins and to C3. Anti-band 3 antibodies were purified from Sandoglobulin and labeled with ^{125}I (Lutz *et al.*, 1984). C3b–IgG complexes from senescent red cells (Lutz and Fehr, 1979; Lutz *et al.*, 1984) were enriched by gel filtration in buffer containing SDS (Lutz *et al.*, 1987c). Complexes were treated with 1 M hydroxylamine at pH 10 and the liberated IgG was then purified on immobilized protein A and ^{125}I-iodinated (Fasler, 1989). Red cell membrane proteins and purified C3 were electrophoresed on SDS PAGE and blotted (Lutz *et al.*, 1988). Blots were incubated with labeled IgG from C3b–IgG complexes or with anti-band 3 antibodies. Shown are autoradiographs of the immunoblots. M, red cell membranes; C3, unreduced C3; C3 red, reduced C3; Bd3, band 3 protein.

Senescent red cells did not only contain anti-band 3 IgG, but also covalently linked complexes containing C3b or iC3b bound to IgG (Lutz *et al.*, 1987c, 1988). Moreover, IgG liberated from such complexes by hydroxylamine treatment and purification on protein A Sepharose revealed exclusive binding to band 3 protein on immunoblots (Fasler, 1989; Fasler and Lutz, unpublished) (Figure 1). Thus, anti-band 3 antibodies mediated C3b deposition to senescent red cells, as was expected from their stimulation of C3b deposition to oxidatively stressed red cells. Since a detailed understanding of their role requires knowledge of recent breakthroughs in the complement field, I will discuss it while reviewing aspects of complement.

4. THE ROLE OF NATURALLY OCCURRING ANTIBODIES AND OF COMPLEMENT IN TISSUE HOMEOSTASIS

4.1. General Principles in Antibody- and Complement-Mediated Opsonization

Binding of antibodies to red cells does not itself damage the cell. Bound antibodies can, however, mediate adherence of the target red cell to phagocytes, which upon this contact engulf the target cell. The effectiveness of immune adherence is greatly increased when target red cells are opsonized with antibodies and complement. The concerted opsonization with antibodies and complement simultaneously activates receptors for IgG (FcR) and complement C3b (CR1) on phagocytes. Thus, deposition of C3 on target cells in the form of C3b represents the crucial event in opsonization by complement. C3 is not only the most abundant of the complement proteins, it is also the central component (Figure 2). It can be activated by both the classical and the alternative complement pathway (for reviews on complement see Ross, 1986; Fearon and Wong, 1983; Müler-Eberhard and Schreiber, 1980).

Activation means proteolytic cleavage of a 9-kDa fragment (C3a) from the α polypeptide of C3 and generation of the activated molecule, C3b (Figure 2). Cleavage intro-

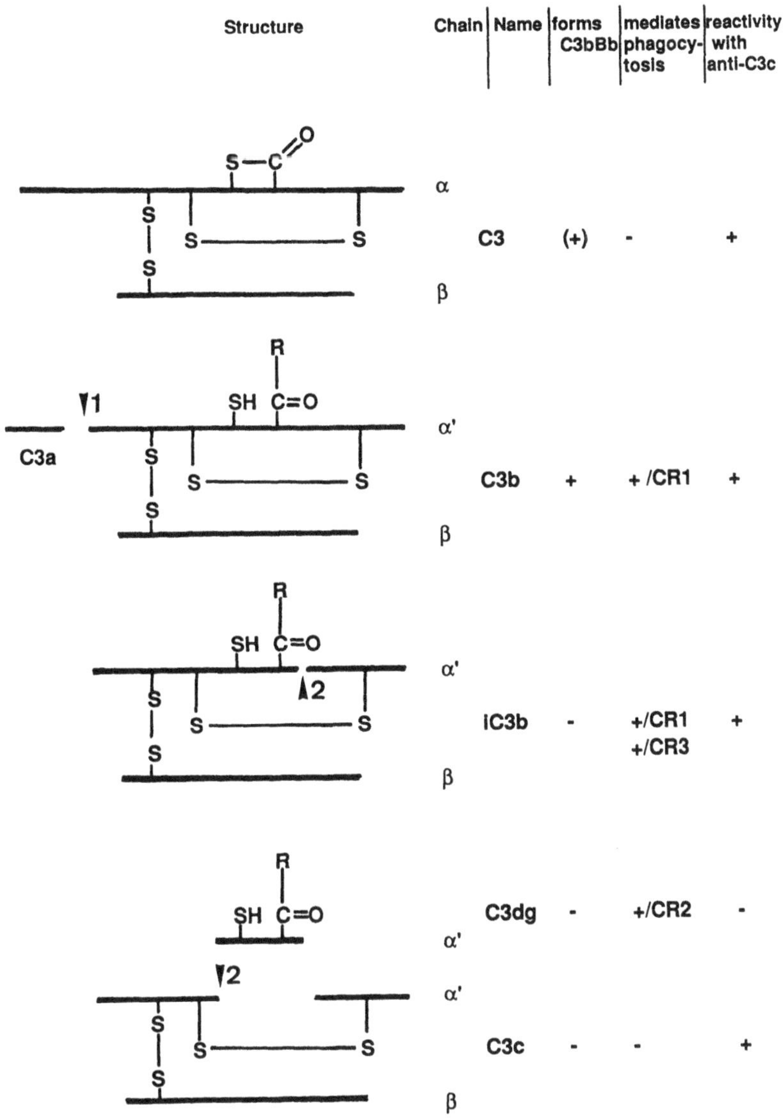

FIGURE 2. The structures of C3 and its fragments, and their major properties. The simplified structures of C3 and its fragments are shown together with some of their properties: their ability to form a convertase ("forms C3bBb"); whether they mediate phagocytosis and by which receptor ("mediates phagocytosis") and the reactivity of monoclonal anti-C3c antibodies with the individual fragment ("reactivity with anti-C3c"). 1, cleavage site of C3 convertases; 2, cleavage sites for factor I; R, protein–OH.

duces mechanical (conformational) stress in the α' polypeptide of C3b. This opens the thioester bond within α' subunit and creates a highly reactive species [nascent C3b (nC3b), containing a carbonium ion with a half-life of 60 μsec] that forms mainly fluid-phase C3b–OH (C3b) with water (for review see Müller-Eberhard and Schreiber, 1980; Capel *et al.*, 1978; Law *et al.*, 1979) (Figure 2). The hypothetical carbonium ion of nC3b

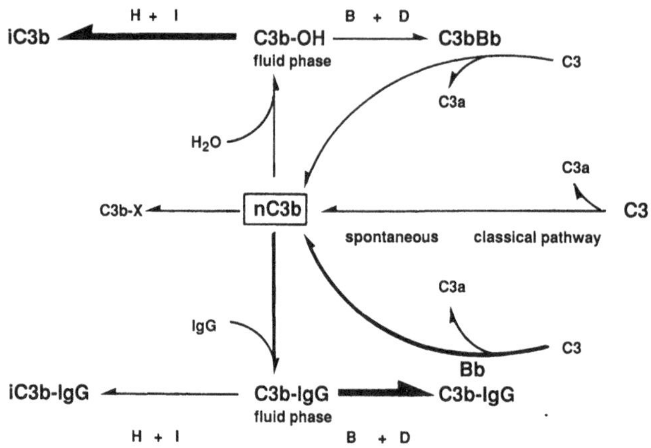

FIGURE 3. The alternative complement pathway. The scheme in the upper half illustrates the well-established amplification loop of the alternative pathway without showing the convertase-stabilizing role of properdin (for the meaning of the various abbreviations, see the text). The lower half is a mirrorlike duplication of the upper half with the following exceptions: C3b–OH represents the species that mediates the loop in the upper half, while C3b–IgG substitutes for C3b–OH in the lower half. In this way the preferred reaction that an intermediate will undergo can be indicated by the thickness of the arrows. Experimental evidence for this preference is discussed in the text. Thus, it is the lower part of the scheme that is more efficient in activation of the alternative pathway.

reacts primarily with water, since its concentration is highest. Theoretically, it should react far better with NH_2 groups. Inhibition studies with fluid-phase nucleophiles suggested the carbonium ion to react preferentially with tyrosine–OH, presumably because it is located in a hydrophobic pocket (Takata *et al.*, 1984). In the absence of serum proteins, ester bonds are also possible with OH groups of carbohydrates (Law *et al.*, 1979). Amide bonds are formed rarely (Takata *et al.*, 1984) with amino groups of proteins (Gadd and Reid, 1981). The ability of C3b to form covalent bonds with other proteins and cell surfaces (C3b–X) is a key process in many respects.

C3b–OH can associate with factor B from serum. The heterodimer (C3bB) is converted into an enzyme by an active but specific serum protease, factor D. Factor D cleaves B to Bb and Ba, when B is bound to C3b. Thereby, the heterodimer, C3bBb, has gained proteolytic specificity for C3 with the catalytic site being located in Bb. C3bBb represents the alternative C3 convertase, also called the amplification convertase, since it is responsible for the positive feedback loop (Figure 3). The positive feedback has much in common with a chain reaction that can eventually result in depletion of serum C3. Thus, the system needs to be regulated (for review see Mollnes and Lachmann, 1988). Regulation occurs primarily by modulation of the C3b concentration.

Alternative Pathway C3b Generation

1. By inactivation of C3b: C3b is inactivated by proteolytic cleavage (Davis *et al.*, 1984). A serum protease, factor I, acts as a specific protease and factor H binds to

C3b and enhances I-dependent proteolysis of C3b to several C3 fragments that no longer participate in complement activation (Figure 3).

Alternative pathway C3b deposition is further reduced by

2. Binding of C3b and of C3b–X to C3b receptors
3. Dissociation of active C3 convertases (C3bBb)

Much is known about the cellular receptors for C3b that prevent C3b from forming an alternative convertase and accelerate dissociation of a preformed C3bBb complex. These cellular receptors for C3b and for later forms of activated complement components are primarily responsible for the remarkable resistance of red cells to complement lysis. Hence, on normal red cells the complement cascade is interrupted at the C3b stage. The mechanisms that prevent complement lysis have been elucidated (see Section 4.8). Far less clearly understood are the steps that initiate and activate complement deposition.

Initiation

The classical pathway of complement deposition is initiated by antibodies that are bound to antigens. Initiation depends on the presence of IgG antibody clusters. This provides sufficient binding sites for the first component of complement, to assemble the classical C3 convertase (reviewed in a comprehensive manner by Ross, 1986). Activation of complement by the classical pathway results in low numbers of deposited C3b per IgG antibody bound to target.

Alternative pathway activation of complement is primarily independent of antibodies, but is dependent on the availability of not well defined surface properties (for review see Müller-Eberhard and Schreiber, 1980). The alternative pathway does not need initiation, since the few molecules of C3b that are generated spontaneously (tick over of the alternative pathway) keep it initiated at all times. It requires activation and this is achieved by raising the number of active C3 convertases.

Activation of the Alternative Pathway

Stabilization of C3 convertase:

- Serum contains a stabilizer of the alternative convertase (properdin) (Smith *et al.*, 1984; Farries *et al.*, 1987). It binds to C3b and stabilizes a C3bBb complex.
- Sera from some patients with glomerulonephritis contain IgG autoantibodies, called nephritic factors, that stabilize C3 convertases by binding to Bb (Daha and VanEs, 1981). This extends the half-life of the enzyme.

C3b inactivation is slowed down:

- Alternative pathway C3b deposition is increased when C3b inactivation is impaired, as is the case when C3b is covalently bound to IgG, forming C3b–IgG (see Section 4.2).
- Finally, some IgG antibodies are known to stimulate alternative pathway apparently without nephritic factor activity via a so far unknown mechanism (see Section 4.2).

4.2. The Role of IgG Antibodies in Stimulating Alternative Pathway C3b Deposition

The effect of complement on a cell type that rarely allows assembly of the entire cascade remains inefficient without concomitant amplification of C3b deposition by the alternative pathway. This puts emphasis on a pathway that is primarily independent of antibodies, except that some IgG antibodies—so far exclusively naturally occurring antibodies—stimulate alternative pathway C3b deposition. Schenkein and Ruddy (1981a,b) found antibodies to zymosan in human serum that stimulated alternative pathway C3b deposition to zymosan. Polhill *et al.* (1978) reported a human IgG fraction that was purified by adsorption to isolated rabbit red cell membranes. It stimulated alternative pathway lysis of rabbit red cells in diluted heterologous serum. A similar type of antibody was found for mouse red cells (Moore *et al.*, 1981). Likewise, sera from hypogammaglobulinemic patients were less efficient than normal sera in stimulating alternative pathway lysis of rabbit red cells (Polhill *et al.*, 1978). Hence, these antibodies stimulated alternative pathway lysis when the species restriction in assembly of the membrane attack complex was overcome.

In an autologous or allogeneic system, such antibodies can at best stimulate alternative pathway C3b deposition. Naturally occurring anti-band 3 antibodies from human serum IgG indeed stimulated C3b deposition on oxidatively stressed red cells in the presence of a 500- to 1000-fold excess of autologous or allogeneic IgG molecules (Lutz *et al.*, 1987a). The same antibodies are primarily bound to senescent red cells (Lutz and Stringaro-Wipf, 1983; Kay, 1984) as free antibodies and as C3b–IgG complexes (Lutz *et al.*, 1987c, 1988). The mechanism of complement stimulation remained unclear. Ruddy and collaborators suggested that IgG or its $(F(ab')_2$ fragment might provide its own C3 acceptor surface for assembly of the alternative pathway C3 convertase. Their view implied a unique site on these naturally occurring antibodies that is exposed upon binding to antigen. We suggested a mechanism (Lutz *et al.*, 1989) that extends this view by demonstrating an affinity between C3 and those IgG molecules that stimulate the alternative pathway. The hypothesis is heavily dependent on recent breakthroughs in understanding regulation of the alternative pathway and differences in IgG binding to native C3 for naturally occurring antibodies.

Fluid-phase C3b (C3b–OH) that is generated by either pathway or spontaneously can nucleate a new amplification convertase (C3bBb) (Figure 3). C3b-OH is, however, a poor stimulator as it is rapidly inactivated by the concerted action of serum factors H and I (for review see Müler-Eberhard and Schreiber, 1980). Thus, the preferred reaction is toward iC3b (Figure 3, indicated by the thickness of the arrows). C3b–OH is, however, not the only product of nC3b, it can also covalently bind to proteins (C3b–X) that differ from C3b–OH in several aspects. Among these complexes, C3b–IgG has unique properties as has been elucidated by Fries and colaborators:

1. C3b–IgG complexes are 30 times less sensitive to factor H- and I-induced inactivation than C3b (Fries *et al.*, 1984).
2. C3b–IgG and C3b can form equally well a C3 convertase (C3bBb or Bb–C3b–IgG) (Fries *et al.*, 1987).

Increased stability of C3b–IgG toward inactivation and unaltered ability to form a C3 convertase make formation of a C3 convertase the preferred reaction of C3b–IgG, in comparison to C3b–OH (Figure 3 lower half, indicated by the thickness of the arrows). It is obvious that formation of C3b–IgG complexes will inevitably stimulate alternative pathway C3b deposition.

The question we tried to tackle was why certain IgGs form C3b–IgG complexes more easily than others. IgG is the second most abundant component in serum and thus is a likely substrate for nascent C3b. In addition, Kulics *et al.* (1983) showed a weak binding of whole IgG to C3/C3b that was mediated by the Fab and the Fc portions of the IgG molecule. IgG binding was higher to C3 than to C3b and therefore the binding was not due to an immunoconglutinin. Immunoconglutinins have been described as reacting primarily with C3b and stabilizing a preformed C3 convertase (Durand and Burge, 1984). The experiments of Kulics *et al.* (1983) did not address the question of whether the observed binding represented a weak interaction of all IgG molecules with C3 or whether this affinity was unequally spread among various IgG molecules (subclasses, specific antibodies). In studying naturally occurring antispectrin and anti-band 3 antibodies we found significant differences in their binding to C3 on immunoblots (Lutz *et al.*, 1988), RIA (Lutz *et al.*, 1989), and ELISA (unpublished). Binding to native C3 (RIA and ELISA) was

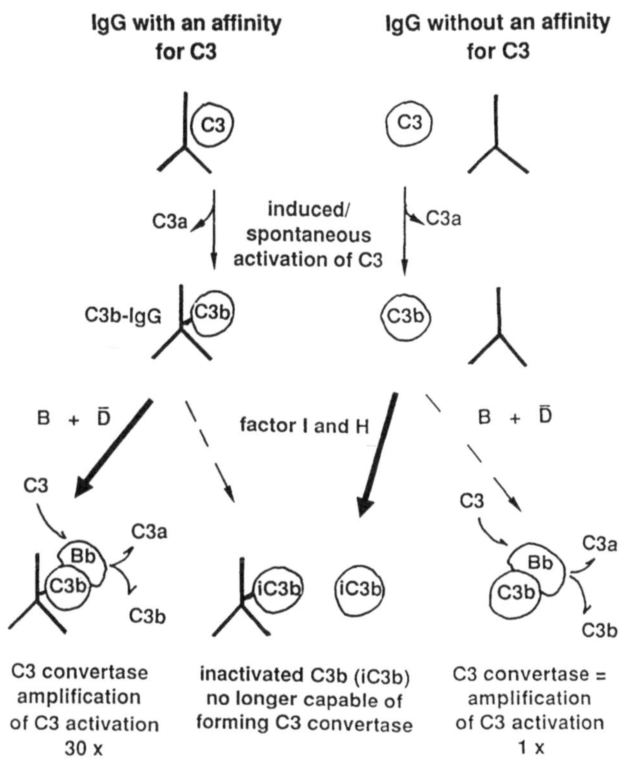

FIGURE 4. Schematic depiction of the hypothesis that IgG with an affinity for C3 can stimulate alternative complement pathway C3b deposition. The figure illustrates that those IgG molecules with an affinity for C3 are on average more often in the vicinity of C3 than those IgG molecules without this affinity. Thus, the chance that an IgG molecule with an affinity for C3 will become covalently bound to C3b is higher, simply because the short half-life of nascent C3b prevents it from choosing a partner. Once a C3b–IgG complex is formed, its preferred reaction is formation of a convertase as illustrated in Figure 3.

at least 10-fold higher for anti-band 3 than for antispectrin antibodies. Preliminary experiments further showed that trypsin-induced formation of covalently linked C3b–IgG complexes from antibodies with an affinity for C3 was possible at IgG-to-C3 ratios 10- to 100-fold lower than required for whole IgG. These findings suggest the following mechanism (Figure 4): IgG with an affinity for C3 is more often in the vicinity of C3 in serum than is IgG without this property. Thus, its chances of forming a C3b–IgG complex with nascent C3b are higher than for other IgGs, since nascent C3b with its short half-life has little chance to diffuse and find partners other than those in the vicinity (H_2O or a protein with affinity to intact C3). It remains to be elucidated whether this is a general principle and whether the IgG/C3 binding is mediated by the Fd region of IgG or involves other regions as well.

4.3. Erythrophilic IgG and Complement Deposition

Erythrophilic immunoglobulin represents the fraction of serum immunoglobulin that binds at low ionic strength to human red cells and can be eluted by isotonic NaCl (Fidalgo *et al.*, 1967). This immunoglobulin fraction is of immediate interest to the discussion of anti-band 3 and its ability to stimulate alternative pathway C3b deposition (Lutz *et al.*, 1987a, 1988), since anti-band 3 antibodies are among erythrophilic IgG as shown by Victoria *et al.* (1982). In addition, erythrophilic IgG contains other IgGs, some of which bind to polypeptides smaller than band 3 protein (Victoria *et al.*, 1982), as we could confirm with ^{125}I-iodinated erythrophilic IgG on immunoblots from red cell membranes (Lutz *et al.*, unpublished). Binding of erythrophilic IgG to red cells appears to depend on (1) charge effects and (2) reversible aggregation of IgG. Binding decreases steeply above pH 7.0 and below pH 5.0. At pH 7.0 binding reached up to 1.5 mg IgG/10^{10} red cells. It is much reduced on cells depleted of negative charges (Hsu *et al.*, 1979). Thus, low-ionic-strength conditions increases binding for IgG nonspecifically by charge–charge interactions, and apparently enhance specific binding of antibodies with a weak affinity (Szymanski *et al.*, 1981), presumably by forming IgG aggregates.

Moreover, binding of erythrophilic IgG from whole serum to red cells is accompanied by activation of complement and deposition of C4b/C3b (Hsu *et al.*, 1979; Szymanski *et al.*, 1981; Chaplin *et al.*, 1980). In low-ionic-strength serum containing Mg^{2+}, the alternative pathway is activated with almost exclusive C3b deposition (Chaplin *et al.*, 1980). Thus, binding of erythrophilic IgG at low ionic strength stimulated alternative pathway C3b deposition on red cells. Note that C3b deposited by the alternative pathway at low ionic strength showed a marked resistance to inactivation by factors I and H (Chaplin *et al.*, 1980), which was very similar indeed to that reported for C3b–IgG complexes (Fries *et al.*, 1984). Elution of bound IgG was incomplete when deposited at low ionic strength with active complement (Chaplin *et al.*, 1980; Szymanski *et al.*, 1981). This suggested that complement deposition generated complexes containing IgG that remain cell-associated and are probably C3b–IgG complexes containing IgG with an affinity for red cells.

4.4. Red-Cell-Associated IgG is the Major C3b Acceptor

Human red cells have receptors for activated C3b and its fragments (for review see Fearon, 1984; Schifferli *et al.*, 1986). Their role is to bind C3b-containing immune complexes (IC) and to aid in their clearance. On the other hand, when C3 is activated on the red cell surface or on serum IC, but close to the surface of red cells, C3b can form a covalent complex with surface components of the red cell. These components are called C3b acceptors (Ross, 1986). Cell surface carbohydrates had been suggested as acceptors, and Parker *et al.* (1984) reported that glycophorin A was the acceptor for nascent C3b when generated by alternative pathway convertase.

We came to a different conclusion by studying C3b–X complexes from SDS-solubilized membranes of young and old normal red cells. C3b–IgG or C3b–glycophorin complexes are larger than the biggest red cell membrane protein, spectrin band 1. Thus, gel filtration of SDS-solubilized membranes allowed us to enrich such complexes. Immunoblots of the collected material revealed complexes that reacted with labeled anti-IgG and anti-C3c antibodies (Lutz *et al.*, 1987c, 1988). Immunoblots from reduced samples and comparison with artificially generated C3b–IgG complexes (Fries *et al.*, 1984) showed the complexes to contain C3b–IgG and to be more abundant on senescent than young cells (Fasler, 1989; Fasler and Lutz, in preparation). The high-molecular-weight proteins revealed little anti-C3-reactive material that was not also reactive with anti-IgG. This suggested that the majority of covalently bound C3b on red cells was not bound to a red cell membrane protein, like glycophorin A, but rather to red-cell-associated IgG. These experiments left little doubt about C3b–IgG being involved in opsonization of in vivo-aged red cells. On the other hand, our findings are in contradiction to those of Parker *et al.* (1984) who reported that α′ C3b was bound to glycophorin A. Their conclusion was based on finding off-diagonal spots of glycophorin A on two-dimensional SDS PAGE from red cells incubated with alternative convertase and run with reduction in the first and hydroxylamine treatment in the second dimension. Since glycophorins are well known to aggregate and disaggregate in SDS (Futhmayr, 1978; Tuech and Morrison, 1974), Parker and colleagues' results are not conclusive without appropriate controls. We had earlier studied labeling of glycophorins on intact red cells and always found high-molecular-weight aggregates without having to activate the alternative pathway (Figure 2 in Schweizer *et al.*, 1982). Furthermore, in studying exoplasmic cross-linkability of glycoproteins (Schweizer *et al.*, 1982) we ran two-dimensional control gels with reduction in both dimensions. These gels always revealed off-diagonal spots from glycophorin A similar to those published by Parker *et al.* (Figure 5) and thus confirmed its reversible aggregation and disaggregation, also in the absence of complement activation. This shed some doubts on the identification of glycophorin A as the major acceptor for C3b on red cells.

Several other phenomena are difficult to conceive with glycophorin A as the C3b acceptor:

1. Sialic acid molecules that are primarily located on glycophorin A (Marchesi and Furthmayr, 1976) are known to inhibit activation of the alternative pathway. The mode of inhibition has recently been clarified for sialic acid on gangliosides. It increases binding of factor H to surface-associated C3b and thereby promotes inactivation (Michalek *et al.*, 1988). If sialic acid bound to protein were similarly active in mediating factor H binding, C3b bound covalently to glycophorin would have been degraded. Hence, it is unlikely that C3b specifically associated with the protein that carries the majority of sialic acids.

2. Artificial coating of red cells with C3d and C4d was obtained by incubating red cells in low-ionic-strength serum. Bound C3b was cleaved with small amounts of trypsin to release C3c/C4c. Removal of C3c/C4c generated red cells that survived equally well as control cells upon autotransfusion (Chaplin *et al.*, 1983). The immediate conclusion was

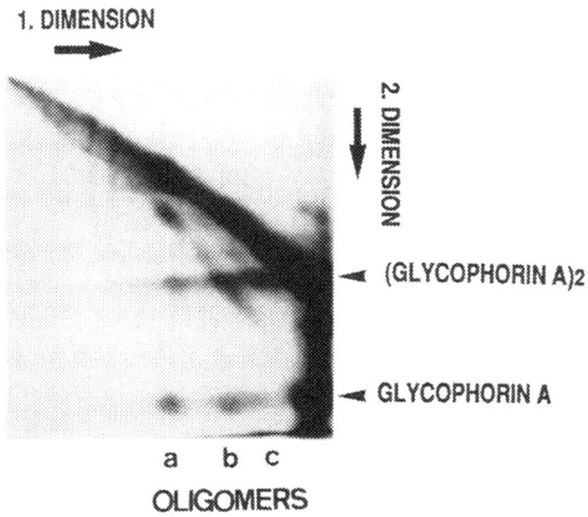

FIGURE 5. Reversible aggregation of glycophorin A generates a pattern on two-dimensional SDS PAGE reminiscent of a two-dimensional SDS PAGE obtained from surface ^{125}I-iodinated red cells on which alternative pathway C3b deposition was activated and membranes therefrom run with reduction in the first and with hydroxylamine in the second dimension (see Figure 4 in Parker *et al.*, 1984). Sialic acids on intact red cells were mildly oxidized with periodate and labeled (Schweizer *et al.*, 1982). Electrophoresis was with reduction in both dimensions. If a mixture of proteins were forming none or stable oligomers, their two-dimensional electropherogram should yield a straight line, normally a diagonal. Here we expected a diagonal and a vertical spread of polypeptides at the right side, since the acrylamide concentration was 4.5% in the first, and 8% in the second dimension. Glycophorin that formed oligomers (a–c) in the first dimension dissociated during electrophoresis in the second dimension and formed off-diagonal spots. Migration of oligomer *a* may correspond to that of bc2, bc3 and that of *b* to bc4 in the nomenclature used by Parker *et al.* (1984).

that red cells carrying C3d are not phagocytized. This is in contrast to the current knowledge on the distribution of C3d-specific receptors (CR2). Lymphocytes and monocytes contain CR2 (Rask *et al.*, 1988) and the latter mediate phagocytosis of red cells coated with C3d and IgG (Ehlengerger and Nussenzweig, 1977). On monocytes the number of CR2 increases, however, from 7000 to over 100,000 upon activation (Inada *et al.*, 1983). The authors' interpretation was that red-cell-associated C3d is unstable (Chaplin *et al.*, 1983), since 90% of it was lost within 6 days after autotransfusion. If C3d were bound to glycophorin, the ester bond would have had an approximate half-life of 22 days at 4°C in buffer (Fries *et al.*, 1984) and thus roughly 3–4 days at 37°C. This could not account for the rapid release observed. A 90% loss is, however, conceivable, if C3d were covalently bound to cell-associated components (like erythrophilic IgG), since erythrophilic IgG is involved in C3b deposition at low ionic strength (for details see Section 4.3) and is displaced from the membrane during autotransfusion.

4.5 Two Types of C3b–IgG complexes and Two Modes of Association with Red Cells

Those C3b–IgG complexes that were found on aging red cells contained anti-band 3 antibodies (Figure 1). Additional data (Fasler, 1989; Fasler and Lutz, unpublished) also suggest formation of C3b–anti-band 3 complexes on oxidatively stressed red cells, which contain increased numbers of oligomerized band 3 proteins (Lutz *et al.*, 1987a). This implies that C3b–anti-band 3 complexes interact via Fab of IgG with band 3 protein on red cells. We lack direct evidence for this conclusion. It was based on a correlation between increased numbers of oligomerized band 3 protein on senescent red cells and increased C3b deposition as well as enhanced phagocytosis (Lutz *et al.*, 1987a). Another line of indirect evidence is also in favor of this explanation: pretreatment of red cells with monoclonal anti-CR1 antibodies did not inhibit phagocytosis of oxidatively stressed red cells that were opsonized by increasing concentrations of anti-band 3 antibodies in whole serum (Lutz *et al.*, 1988). This result suggests that those C3b–IgG complexes that mediated red cell phagocytosis were not bound to CR1 of red cells. Antigen-associated C3b–IgG complexes remained intact and kept on nucleating new C3 convertases, which resulted in anti-band-3-stimulated C3b deposition. This process was, however, limited, presumably by decay-accelerating factor (DAF).

The second type of binding, namely that of a C3b–IgG complex to CR1 involves the C3b portion of the complex. A C3b–IgG complex can no longer nucleate a C3 convertase, when bound to CR1 (for review see Fearon, 1984; Holers *et al.*, 1985). The difference between a C3b–IgG complex that binds via Fab of its IgG to clustered antigens on red cells and a complex that binds via C3b to CR1 must reside in the specificities of complexed IgG. It is obvious that any C3b–IgG complex can bind to CR1 while only those carrying IgG with a specificity for cell antigens can associate via Fab portions with the membrane. The two modes of association of C3b–IgG complexes with red cells differ in their affinities.

A methodological difficulty arises from the bifunctional binding properties of C3b–IgG: assume a C3b–IgG complex that is associated via Fab with antigens on one red cell. The C3b portion of the complex can yet be recognized by CR1 of an adjacent red cell (the extent to which this is possible remains to be determined). This bridging of cells may induce transient rouleaux formation. It can also severely lower the number of intact C3b–IgG complexes, since binding of a red cell with its C3b–IgG to CR1 on a second red cell is expected to stimulate factor I-dependent and CR1-mediated C3b cleavage (Malbran *et al.*, 1987).

4.6. C3b–IgG Complexes as Mediators of Efficient Phagocytosis

The synergistic effect of IgG and C3b in phagocytosis was thought to arise from a concomitant binding of FcR and CR1 on phagocytes (Ehlenberger and Nussenzweig, 1977; Kurlander and Rosse, 1979; Shaw and Griffin, 1981). Efficient deposition of both opsonins is dependent on activation of the alternative pathway as is known for antibody-dependent killing of gram-negative bacteria (Hill *et al.*, 1978). More recently, Fries *et al.* (1987) studied whether target-bound C3b and IgG were equally potent in mediating phagocytosis as covalently linked C3b–IgG complexes. At a low extent of opsonization by IgG antibodies, C3b–IgG complexes were at least twice as potent as C3b plus IgG. Furthermore, the association constant for C3b–IgG to phagocytes was 9×10^7 M^{-1} and thus fives times higher than for C3b alone (Fries *et al.*, 1987). Additional results were found for another bacterium opsonized with antibacterial IgG in serum: at least one fifth of bound C3b was covalently bound to IgG. Preformed C3b–IgG complexes were three- to fourfold more effective in inducing complement-mediated lysis than antibody and serum (Joiner *et al.*, 1985). Thus, C3b–IgG complexes play a central role in opsonization. The presence of antigen-associated C3b–IgG complexes on target cells greatly facilitated binding to FcR and CR1 on phagocytes, since binding was bivalent as evident from competition studies (Malbran *et al.*, 1987). Analogous binding studies have not been carried out with C3b–anti-band 3 antibodies and human red cells.

4.7. Fluid-Phase C3b–IgG Complexes, Innocent Bystander Opsonization, and a Positive Direct Coombs Test for Complement

Circulating immune complexes (IC) with foreign or self antigens are a potential hazard. IC activate alternative pathway C3b deposition (Fujita *et al.*, 1981). However, efficient binding of solubilized IC appears to also involve elements of the classical pathway (Sherwood and Virella, 1986; Jepsen *et al.*, 1986; Schifferli *et al.*, 1987). In the course of this activation, IC are solubilized by deposition of C3b to IgG of the IC (Fujita *et al.*, 1981; Taylor *et al.*, 1983; Edberg *et al.*, 1987) and to fluid-phase IgG (Jacobs and Reichlin, 1983). C3b–IgG complexes generated during IC solubilization can nucleate a new C3 convertase and can opsonize innocent bystanders.

Since IC solubilization goes along with substantial C3b–IgG complex formation with serum IgG (Jacobs and Reichlin, 1983), it is likely that also here those IgG with an affinity for C3 (like anti-band 3) are the predominant targets among fluid-phase IgG molecules. The consequence could be that any IC solubilization is followed by enhanced formation of, e.g., C3b–anti-band 3 complexes that may accelerate clearance of red cells. This implies that the presence of IC and their solubilization may generate red cells showing a positive direct Coombs test for complement and in certain cases accelerated removal of red cells or anemia. Direct evidence for such a causal relationship is not available, but several clinical findings reveal a parallelism between the presence of IC and anemia.

A positive direct Coombs test for complement is predominantly found in patients with drug-induced immune hemolytic anemia with little or no hemolysis. Those who also have a positive Coombs test for IgG frequently present with hemolysis (Worlledge, 1978; Petz and Branch, 1985).

Blood from patients with lymphomas or leukemias frequently contains IC (Serrou and Rosenfeld, 1981) and in 10% of the cases their red cells gave a positive direct Coombs tests for complement components (Worlledge, 1978; Solal-Celigny *et al.*, 1984; Rondeau *et al.*, 1983; Gibson and Wells, 1988). Red cells with a positive Coombs test for complement also contained antibodies in two thirds of the cases (Solal-Celigny *et al.*, 1984). Eluted antibodies had invariably anti-blood-group-I specificity. This seems of particular relevance, since blood group I/i is mainly located on the carbohydrate portion of band 3 protein (Childs *et al.*, 1979). Hence, antibodies associated with these cells had a kind of "anti-band 3 specificity." Note, however, that anti-band 3 antibodies from normal human IgG do not bind to the carbohydrate-rich, chymotryptic fragment of band 3 protein (Lutz *et al.*, 1984).

Positive Coombs test for complement have also been reported at a low frequency for normal blood donors (Worlledge, 1978; Schreiber, 1985; Masouredis *et al.*, 1987). Increased opsonization can arise from a hereditary inefficiency in clearance of red cells (Lawley *et al.*, 1981) or from innocent bystander opsonization induced by either infection or drug. Masouredis *et al.* (1987) recently presented evidence for yet another possibility: they eluted bound IgG and found some antibodies with anti-Rhesus specificity, and substances that apparently were able to agglutinate red cells that were sensitized with certain anti-D antibodies. Their conclusion was that normal donors with a positive Coombs test contain anti-Rhesus antibodies and corresponding antidiotypic antibodies. While this explanation is compatible with their findings, it is not the only possibility, since they also detected invariably C3 fragments on these red cells and found the amount of "antidiotypiclike" substances to increase on red cells upon storage for several weeks. Since storage of red cells correlates with increased C3b binding (Szymanski *et al.*, 1985) and since senescent red cells contain C3b–IgG complexes (Lutz *et al.*, 1988), eluates collected by Masouredis *et al.* (1987) must have also contained covalently linked C3b–IgG complexes, which were not separated from IgG. The presence of C3b–IgG or iC3b–IgG complexes—presumably also containing anti-band 3 antibodies—could have mediated agglutination by complement receptors of red cells that were coated with anti-D. The results suggest that it was not anti-D that was relevant, but some unknown factor, since agglutination only occurred with a minority of anti-D preparations. Hence, the presence of antiidiotypic antibodies on red cells with a positive direct Coombs test is not firmly established. This and the possible involvement of C3b–IgG complexes with antibodies that have an affinity for C3 needs further investigation. At present it seems quite probable that positive Coombs tests and eventual hemolysis can result from IC-induced C3b deposition on naturally occurring, fluid-phase IgGs that have an affinity for C3 on their Fd, and for a red cell component on their Fab portion.

4.8. Protection from Complement Attack

Activation of C3 by either a cell-bound or a fluid-phase convertase in the vicinity of a red cell surface can result in complement deposition and eventually in phagocytosis or lysis. A variety of mechanisms have evolved that protect blood cells quite efficiently from the adverse effects of complement. One of them, CR1 or the immune adherence receptor, was initially discovered by Fearon (1979). CR1 is an integral membrane protein with an apparent molecular mass from 195 to 250 kDa and exists in four allotypic variants (for review see Holers *et al.*, 1985) (Figure 6). Binding of C3b–IgG complexes to CR1 on blood cells prevents these complexes from nucleating a C3 convertase and initiates their inactivation (for review see Fearon, 1984; Fearon and Wong, 1983) (Figure 7). In contrast to the resistance of C3b–IgG toward fluid-phase inactivation, CR1-bound C3b–IgG and C3b are cleaved with similar efficiency by factor I (Fries *et al.*, 1985). Hence, binding to CR1 quickly stops amplification of C3b deposition. Red cells contain only 500–1000 CR1 per cell, but since they make up the majority of cells in blood they contain 80–90% of all CR1. This illustrates the relevance of red cell CR1. The efficiency of CR1 in protecting red cells from complement deposition is, however, limited by their low number, their clustered arrangement (Paccaud *et al.*, 1988; Arnaout *et al.*, 1983), and presumably their restricted lateral mobility within the plane of the membrane.

DAF (Nicholson-Weller *et al.*, 1982) represents another type of protecting device. It is a 70-kDa glycoprotein and is present in 2000–4000 copies per red cell (Kinoshita *et al.*, 1985) (Figure 6). It is an ectoprotein linked to the membrane via a carbohydrate chain that is bound to phosphatidylinositol (Davitz *et al.*, 1986). Its amino acid sequence is known (Medof *et al.*, 1987). DAF binds to red-cell-associated C3b, C3b–IgG and C4b and thereby prevents them from forming convertases (Kinoshita *et al.*, 1986a,b) (Figure 7). Likewise, it binds to assembled C3 convertases and accelerates dissociation of Bb (or 2a for classical convertase) from the convertases (Fujita *et al.*, 1987). Unlike CR1, it has no cofactor activity for factor I-mediated cleavage of C3b. Its potency to protect from C3b deposition may depend on its lateral mobility, which is expected to be similar to that of a lipid.

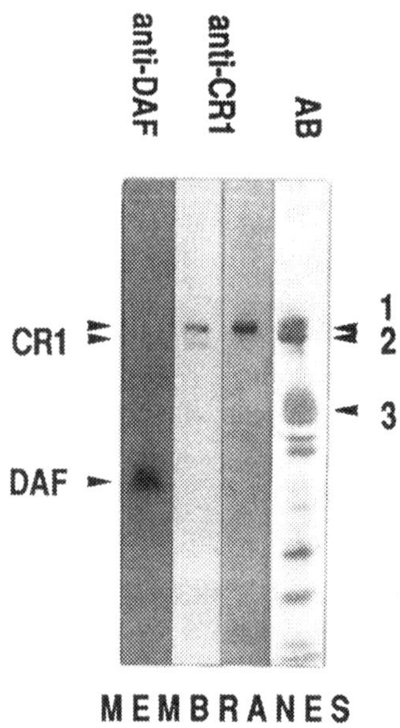

FIGURE 6. CR1 and DAF from human red cells. CR1 and DAF are shown by immunoblotting techniques using ^{125}I-iodinated, monoclonal anti-DAF (IA10, a gift from V. Nussenzweig, New York) and anti-CR1 (a gift from M. D. Kazatchkine, Paris). Red cell membrane proteins were electrophoresed without reduction and blotted onto nitrocellulose. A stained blot (AB) revealing the polypeptide pattern is shown together with autoradiographs from immunoblots: anti-CR1 binding is shown to red cell membrane proteins from two different donors, to illustrate the presence of either one or two anti-CR1-reactive bands.

Finally, there exists an integral membrane protein that is species specific and prevents formation of the membrane attack complex (Zalman *et al*, 1986; Schönemark *et al.*, 1986). It binds to C8/9 and thereby impairs formation of poly C9 and thus lysis.

4.9. Loss of CR1 and DAF from Red Cells in Disease and Cell Aging

The extent to which surrounding cells are opsonized by IC-induced activation of alternative pathway C3b deposition may depend on the degree to which the cells have lost protection against complement deposition.

Disease: The classical example with a deficiency in DAF (Nicholson-Weller *et al.*, 1983; Kinoshita *et al.*, 1985; Moore *et al.*, 1985) is paroxysmal nocturnal hemoglobinuria (PNH). The disease is heterogeneous and reveals three types of red cell subpopulations, which differ in their sensitivity to complement lysis. Type II and III cells are variably deficient in DAF (Moore *et al.*, 1985; Kinoshita *et al.*, 1985). The increased sensitivity of type III cells to binding of C8/9 and lysis is presumably due to an impairment of the C8/9 binding protein (Zalman *et al.*, 1986).

Deficiencies in protecting proteins are the result of defects and/or are acquired (for review see Kazatchkine *et al.*, 1987). Low numbers of CR1 on red cells in systemic lupus erythematosus (SLE) (Iida *et al.*, 1982; Walport *et al.*, 1985) correlated with disease activity and increased amounts of red-cell-bound C3d/C3dg, which served as a measure of previous IC-induced complement activation (Ross *et al.*, 1985). A similar correlation was found for PNH and for autoimmune hemolytic anemias with considerably higher numbers of red-cell-bound C3d (Ross *et al.*, 1985). Acquired deficiency of CR1 is either due to proteolytic stripping of CR1 or vesicle release. This is compatible with the occurrence of a serum form of CR1 ranging from 13 to 81 ng/ml (Yoon and Fearon, 1985). The latter is as active as red-cell-bound CR1 as a cofactor of C3b inactivation and has the same

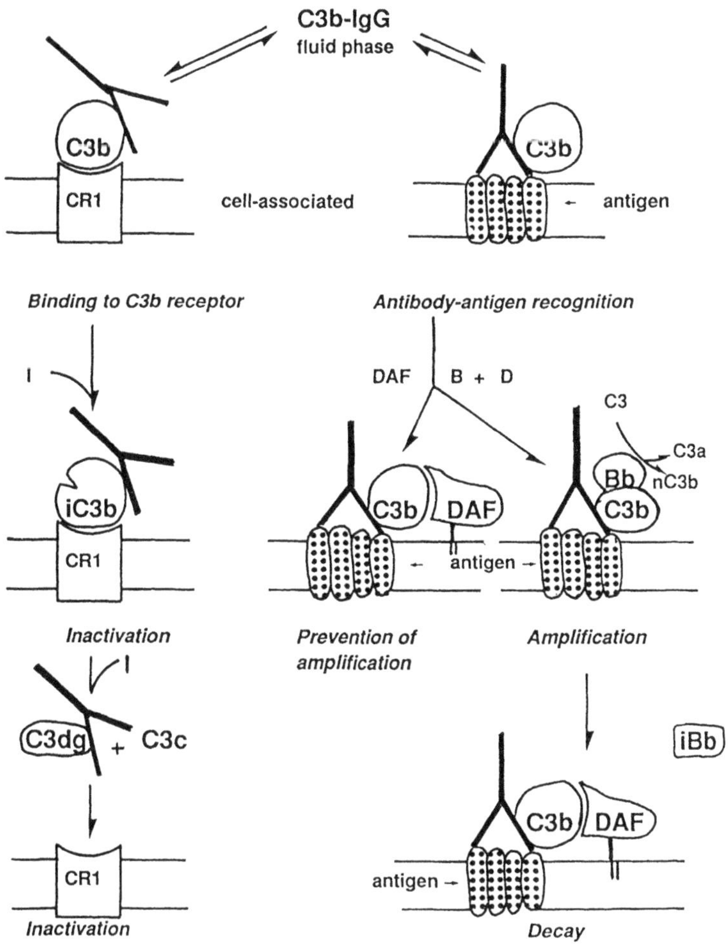

FIGURE 7. Protection from complement deposition by membrane-bound CR1 and DAF. Illustrated are two examples of how C3b–IgG complexes can be associated with red cells. The figure is self-explanatory, except for the abbreviations, which are explained in the text. A third type of C3b–IgG interaction with the red cell has been omitted, namely that of unspecifically bound IgG carrying C3b. This type of complex can also interact with either CR1 or DAF of the same red cell. Experimental evidence further suggests that membrane-associated C3b-IgG complexes can yet be recognized by CR1/DAF of the same cell (Kinoshita *et al.*, 1986a,b).

apparent molecular weight (Yoon and Fearon, 1985). Thus, a CR1 release by proteolysis is questionable.

Red cell aging: Ripoche and Sim (1986) studied the functional activity of CR1 on aging human red cells and found a decrease with cell age. This was later confirmed by others (Moldenhauer *et al.*, 1988). Hence, C3 convertases are more easily formed on senescent red cells than young red cells. This possibility is made even more probable, since we could also detect a 40–60% decrease in the number of DAF antigens with red cell age (Lutz *et al.*, unpublished). This increases the chances for opsonization either by an antibody that stimulates alternative pathway C3b deposition (anti-band 3) or by nascent

C3b originating during IC solubilization (innocent bystander mechanism). Thus, senescent red cells are less well protected against C3b deposition than young ones. Therefore, innocent bystander opsonization is not random, occurring preferentially on older cells. This may be the reason why drug-induced immune hemolytic anemias with an IC type of mechanism rarely cause severe anemia. Substantial hemolysis can occur in autoimmune hemolytic anemias, when the acquired antibody reacts with an antigen that is exposed on young and old red cells. Exceptions are known, where an autoantibody causing autoimmune hemolytic anemia binds preferentially to senescent red cells (Herron *et al.*, 1987).

Impaired protection from C3b deposition on senescent red cells evidently does not result in increased formation of membrane attack complexes, since these CR1- and DAF-deficient cells still circulate. This may be due to the presence of the species-specific, red-cell-associated C8/9 binding protein (Schönemark *et al.*, 1986; Zalman *et al.*, 1986), which selectively prevents formation of membrane attack complex with autologous complement.

The selective loss of 40–60% of CR1 and DAF from aging cells is not understood. While DAF may be lost by shearing, the integral membrane protein CR1 must disappear by other routes. It is conceivable that some CR1 are proteolytically degraded in the process of acting as a cofactor for factor-I-mediated C3b inactivation (Ross *et al.*, 1985). On the other hand, the clustered occurrence of CR1 (Paccaud *et al.*, 1988) would allow release of a substantial fraction of CR1 along with a very limited portion of the membrane by a type of exocytotic vesiculation. Senescent red cells have lost some membrane, but the potency to spontaneously release skeleton-free vesicles (Lutz *et al.*, 1977) decreases rapidly with cell age (Snyder *et al.*, 1985), while the loss of CR1 appears continuous throughout the life span of the cell (Ripoche and Sim, 1986; Moldenhauer *et al.*, 1988).

5. ON THE OCCURRENCE AND REGULATION OF NATURALLY OCCURRING ANTIBODIES IN ONTOGENY

The increasing relevance ascribed to naturally occurring antibodies as housekeeping antibodies in clearance of accidentally released, cellular constituents and of senescent red cells (Grabar, 1983; Lutz and Wipf, 1982; Guilbert *et al.*, 1982; Lutz, 1987) justifies a chapter on regulation of antiself reactivity. Autoreactivity is in fact a prerequisite in the selection of pre-B-cell clones. Selection of B cells occurs before antigen contact from a large variety of variable regions (V_H, V_L) that are encoded in the germline genes. One B cell with a certain idiotope will give rise to a counterpart with antiidiotypic properties (Jerne, 1984; Coutinho, 1980; Alt *et al.*, 1987). Since the antiidiotope represents the matching element to the idiotope of the first antibody or antibodylike structure on B cells, it is in Jerne's terms a partial mirror of an epitope of an antigen (Jerne, 1984). It is in fetal and early neonatal life when this network of pre-B cells is challenged with autoantigens and maternal IgGs. This challenge (Benner *et al.*, 1982) displaces the equilibrium within the network, requires regulation (Bottomly, 1984), and results in antiself much earlier than it results in anti-nonself.

5.1. Autoreactive B Cells Persist in Ontogeny

With regard to the B-cell compartment, a distinction between self and nonself does not exist, since at any time "nonself" must be present in a form of an internal image (Alt *et al.*, 1987). This explains why old individuals can yet develop autoimmune diseases and why infection with antiidiotypic antibodies can mimic an antigen and give rise to antibody production. Thus, the potential for autoreactivity persists through ontogeny.

Studies aimed at understanding generation of naturally occurring antibodies in germ-free animals have shown production of antibodies in the absence of exogenous stimuli (Benner *et al.*, 1982) and have revealed a size difference between those B cells that were naturally activated and those that were not (Portnoi *et al.*, 1988). Activation of B cells from the spleen was the least dependent on exogenous stimuli and was likewise independent of T cells (for review see Benner *et al.*, 1982). These B cells generated primarily, but not exclusively, antibodies of the IgM type. Forty to fifty percent of the active B cells secreted IgG of one of the four subclasses.

5.2. Regulation of Antiself Reactivity

The reason why certain B cells remain active and produce "background IgG," naturally occurring antibodies, or autoantibodies in disease remains open. Huge efforts in sequencing the genes of monoclonal antibodies from such B cells have not revealed a preferential use of certain gene families that encode for variable regions (Bona, 1988; Dighiero *et al.*, 1987). Thus, differentiation into antiself and antinonself reactivities has to occur by other means. Burnet (1959) suggested a clonal deletion. This theory was later considered to hold for most autoreactive, immunocompetent cells as a way to maintain self tolerance. This is not the case, at least not in general terms, since autoreactive T-helper cells specific for red cells persist in several species (Hooper, 1987) and may be involved in stimulating naturally occurring autoantibodies to red cell constituents and at the same time could mediate autoimmune reactivity to red cells in autoimmune hemolytic anemia. On the other hand, it has recently been shown that some autoreactive T-helper and T-suppressor cells are clonally deleted (MacDonald *et al.*, 1988). Hence, clonal deletion within the T-cell compartment may restrict autoreactivity to "background IgG" levels for some antigens. This negative selection can induce tolerance and thus allows self/nonself discrimination. A positive selection requires clonal selection and somatic mutations (Manser *et al.*, 1985) within the B-cell compartment and regulation by antigen-specific T cells and/or antiidiotypic antibodies (Solal-Celigny *et al.*, 1984; Valderrama *et al.*, 1988).

Autoreactive antibodies can exist at "background IgG" (Benner *et al.*, 1982) levels for most antigens and at elevated levels where T-suppressor cells are reduced (Moticka, 1983) or T-helper cells activated. In mice, overt production of autoreactive antibodies to red cells has been measured early in neonatal life (Miller and Calkins, 1988). Generation of T-suppressor cells followed with a delay by 2 weeks of age. It is possible that a similar process in humans leads to suppression of vividly autoreactive antibodies, but not of those reacting weakly, like anti-band 3 antibodies. This establishes persistence of self-reacting B cells, of red-cell-specific T-helper cells, and suggests control by T-suppressor cells. Indeed, production of naturally occurring antibodies was increased by eliminating and reduced by adding T-suppressor cells (Moticka, 1983). Experiments on mice also suggest naturally occurring antibodies to regulate immune response (Mahana *et al.*, 1988): injection of a naturally occurring antibody specific to trinitrophenylated BSA was inhibitory, while a similar injection with an antibody to a normally hidden antigen (myosin) stimulated immune response.

Nevertheless, it is not known whether the naturally occurring antibodies to human red cell constituents represent background IgG that are not under T-cell control or whether the actual titers are due to an active control by both sets of T cells or even antiidiotypic antibodies.

5.3. An Attempt to Group Naturally Occurring Antibodies

The scarce information on immunologic properties of naturally occurring antibodies favors their grouping on functional aspects. I have decided to group them into three types:

1. *Polyspecific antibodies*. These are mainly of the IgM type and show cross-reactivity with many unrelated antigens (Guilbert *et al.*, 1982; Seigneurin *et al.*, 1988) and appear to be exact transcripts of a particular germline gene for variable regions (Manser *et al.*, 1985; Baccala *et al.*, 1989).
2. *Specific antibodies that are mainly of the IgG class*. These antibodies show similar types of specificities as induced antibodies. Most of them are directed to normally hidden proteins (Lutz and Wipf, 1982; Lutz *et al.*, 1987b; Wiener *et al.*, 1986; Berti and Govoni, 1986). Since sequence conservation among hidden antigens is high, these antibodies show cross-reactivities with related antigens (Klinman *et al.*, 1988).
3. *Specific antibodies capable of stimulating alternative pathway C3b deposition*. This group of antibodies appears as specific as induced antibodies. They are directed against surface-exposed proteins/carbohydrates, but have a low affinity (Lutz *et al.* 1987a; Polhill *et al.*, 1978; Schenkein and Ruddy, 1981a,b; Okada *et al.*, 1983; Strejan *et al.*, 1981). In analogy to the properties of anti-band 3 antibodies, I suggest that these antibodies contain a binding site for C3 probably within the Fd region. It is the affinity for C3 that potentiates the effect of anti-band 3, since it increases the probability for C3b–anti-band 3 generation (see Section 4).

5.4 Cell Age and Body Age

Gershon and Gershon (1988) reported on accelerated aging of red cells in apparently healthy, elderly humans and in old rats, mice, and rabbits. These red cell populations contained more light and less dense cells than controls and showed a slight reticulocytosis. The authors explain this phenomenon by the cells' own impairment in full potential and by being exposed to an environment that is more hostile than the one of young subjects. All density-separated subpopulations had elevated numbers of red-cell-associated IgG, and young red cells, prominent among the light cells, had enzyme activities higher than old, but lower than young cells from controls. Accelerated cellular aging was also reported for mice by others (Magnani *et al.*, 1988). I think the host effect is detrimental, not via endothelial cells and oxidative damage, but via the immune system. Aged individuals have higher concentrations of IgG (predominantly IgG1 and IgG3; for review see Benner *et al.*, 1982). Increased IgG concentrations may originate from loss of suppression with body age and may comprise primarily autoreactive IgG (Meredith *et al.* 1979). This could result in increased opsonization as found for hypergammaglobulinemia (Heddle *et al.*, 1988). At the same time, the immune system of old individuals is impaired in reacting against a new challenge (Zharhary, 1988) and is more sluggish in reacting against newly introduced antigen as shown for the murine system (Albright *et al.*, 1988). In elderly humans, macrophage functions were less efficient than in controls when probed with anti-Rhesus antibody-coated autologous red cells (Melez *et al.*, 1988). This finding is not contradictory to accelerated clearance of aged red cells; instead, it suggests that accelerated aging of autologous red cells in elderly persons would be more obvious, if the reticuloendothelial system were to act as efficiently as in young persons. Thus, increased IgG on all red cell subpopulations from elderly humans may be indicative not only of elevated concentrations of naturally occurring anti-red-cell antibodies (presumably anti-band 3) but also of a higher threshold necessary for their clearance. Cross-transfusion experiments with density-separated red cells have not been carried out with human red cells, but could answer this point.

6. CONCLUSIONS AND OUTLOOK

In writing on "the role of naturally occurring antibodies and of complement in tissue homeostasis" I gathered information from various fields to evaluate a hypothesis. The hypothesis is as follows:

Naturally occurring antibodies with a low affinity to cell surface constituents have tissue homeostatic potential, if they possess an affinity for C3, presumably within the Fd region. Their binding to oligomerized antigen—a process that overcomes the low affinity by slowing down dissociation—activates alternative pathway C3b deposition in a similar fashion as an IC. The efficiency, however, by which bound antibody can promote C3b deposition is increased by its affinity for C3. This enhances the probability for the formation of C3b–IgG complexes, which are far better activators of the alternative pathway than C3b.

Experimental evidence supports some parts of the hypothesis, but as yet is not conclusive proof of it. It is best documented for anti-band 3 antibodies. Its verification or falsification may be more easily achieved with one of the other naturally occurring antibodies mentioned. The recent understanding of the species restriction of assembly of the membrane attack complex has given experimental support to the concept that the various types of naturally occurring IgG antibodies that stimulate the alternative pathway are similar. It will be of great interest to determine the specificities of the various naturally occurring antibodies having this potential. Of similar importance will be their affinity to C3, their cross-reactivities, and the precise location of the C3 binding site within the IgG.

The hypothesis has impacts that I could not foresee, but gradually became aware of. One of these impacts may be of clinical relevance:

Any generation of nascent C3b will also deposit C3b onto serum IgG. Thus, it is inevitable that C3b–IgG complex formation occurs among serum IgG, preferentially with IgG molecules possessing an affinity for C3. If the very same antibodies also have an affinity for cellular constituents, they can opsonize innocent bystanders, since binding of the complex is higher than for antibody alone.

I listed a number of clinical situations of transitory positive Coombs tests or even of transitory anemia that could be related to this phenomenon. There is just a hint that anti-band-3-like antibodies are involved in some cases. The coincidence calls for investigations of the clinical relevance of these naturally occurring antibodies. This, however, will not be trivial, since naturally occurring antibodies with an affinity for C3 have so far not been quantified in serum samples. The reason is simple: their binding to serum C3, which is in excess over presented antigen in ELISA, will inhibit binding to the antigen.

Another message from the compiled literature concerns the relevance of complement in clearance of red cells. This aspect has been neglected by many investigators in the field. Clinical data have always indicated the predominant role of complement in red cell clearance. Yet the connection between deposited C3 fragments and the few antibodies was not obvious. The evidence now available provides a preliminary explanation as to why there are only a few antibodies that can stimulate a significant deposition of C3b. The hypothesis is based on differences in affinities among IgG for C3. The crucial elements that support the concept have been worked out almost exclusively by Fries's group at the NIH in a careful study of the role of artificially generated C3b–IgG complexes.

Formulation of a hypothesis may be dangerous, it can be proved wrong, but it is a starting point for new approaches in the field, which are certainly needed.

ACKNOWLEDGMENTS. I thank Dr. S. Fasler, P. Stammler, and G. Mariani for reading the manuscript. This work was supported by the Swiss National Science Foundation, Bern, and the ETH.

7. REFERENCES

Albright, J. W., Matusewicz, N. M., and Albright, J. F., 1988, Aging of the murine immune system is reflected by declining ability to generate antibodies that promote elimination of *Trypanosoma musculi*, *J. Immunol.* **141:**1318–1325.

Alderman, E. M., Fudenberg, H. H., and Lovins, R. E., 1981, Isolation and characterization of an age-related antigen present on senescent human red blood cells, *Blood* **58:**341–349.

Allen, D. W., Johnson, G. J., Cadman, S., and Kaplan, M. E., 1978, Membrane polypeptide aggregates in glucose 6-phosphate dehydrogenase-deficient and in vitro aged red blood cells, *J. Lab. Clin. Med.* **91:**321–327.

Alt, F. W., Blackwell, T. K., and Yancopoulos, G. D., 1987, Development of the primary antibody repertoire, *Science* **238:**1079–1087.

Arese, P., 1982, Favism—A natural model for the study of hemolytic mechanisms, *Rev. Appl. Pure Pharm. Sci.* **3:**123–183.

Arnaout, M. A., Dana, N., Melamed, J., Medicus, R., and Colten, H. R., 1983, Low ionic strength or chemical cross-linking of monomeric C3b increases its binding affinity to the human complement C3b receptor, *Immunology* **48:**229–237.

Ashwell, G., and Harford, J., 1982, Carbohydrate-specific receptors of the liver, *Annu. Rev. Biochem.* **51:**531–554.

Astrup, J., 1974, Sodium and potassium in human red cell. Variations among centrifuged cells, *Scand. J. Clin. Lab. Invest.* **33:**231–237.

Au, K. S., 1987, Activation of erythrocyte membrane (Ca^{2+})-ATPase by calpain, *Biochim. Biophys. Acta* **905:**273–278.

Baccala, R., Quang, T. V., Gilbert, M., Ternynck, T., and Avrameas, S., 1989, Two murine natural polyreactive autoantibodies are encoded by nonmutated germ-line genes, *Proc. Natl. Acad. Sci. USA.* **86:**4624–4628.

Benner, R., van Oudenaren, A., Björklund, M., Ivars, F., and Holmberg, D., 1982, "Background" immunoglobulin production: Measurement, biological significance and regulation, *Immunol. today* **3:**243–249.

Berti, G., and Govoni, M., 1986, Rat antibodies to spectrin, the principal inner component of erythrocyte membrane. Immunochemical studies after SDS-gel electrophoresis, *Boll. Ist. Sieroter. Milan.* **65:**309–318.

Beutler, E., 1985a, How do red cell enzymes age? A new perspective, *Br. J. Haematol.* **61**:377–384.

Beutler, E., 1985b, Biphasic loss of red cell enzyme activity during in vivo aging, *Prog. Clin. Biol. Res.* **95**:317–329.

Beutler, E., West, C., and Blume, K-G., 1976, The removal of leukocytes and platelets from whole blood, *J. Lab Clin. Med.* **88**:328–333.

Bocci, V., 1981, Determinants of erythrocyte ageing: A reappraisal, *Br. J. Haematol.* **48**:515–522.

Bona, C. A., 1988, V Genes encoding autoantibodies: Molecular and phenotypic characteristics, *Annu. Rev. Immunol.* **6**:327–358.

Bottomly, K., 1984, 1984: All idiotypes are equal, but some are more equal than others, *Immunol. Rev.* **79**:45–61.

Bretscher, P., and Cohn, M., 1970, A theory of self-nonself discrimination, *Science* **168**:1042–1049.

Brovelli, A., Seppi, C., Bardoni, A., Balduini, C., and Lutz, H. U., 1987, Re-evaluation of the structural integrity of red cell glycoproteins during aging *in vivo* and nutrient deprivation, *Biochem. J.* **242**:115–121.

Burnet, F. M., 1959, *The Clonal Selection Theory of Acquired Immunity*, Cambridge University Press, London.

Bussolino, F., Turrini, F., and Arese, P., 1987, Measurement of phagocytosis utilizing ^{14}C-cyanate-labelled human red cells and monocytes, *Br. J. Haematol.* **66**:271–274.

Capel, P. J. A., Groeneboer, O., Grosveld, G., and Pondman, K. W., 1978, The binding of activated C3 to polysaccharides and immunoglobulins, *J. Immunol.* **121**:2566–2572.

Chaplin, H., Jr., Freedman, J., Massey, A., and Monroe, M. C., 1980, Characterization of red blood cells strongly coated *in vitro* by C3 via alternative pathway, *Transfusion* **20**:256–262.

Chaplin, H., Nasongkla, M., and Monroe, M. C., 1981, Quantitation of red blood cell-bound C3d in normal subjects and random hospitalized patients, *Br. J. Haematol* **48**:69–78.

Chaplin, H., Coleman, M. E., and Monroe, M. C., 1983, *In vivo* instability of red blood cell-bound C3d and C4d, *Blood* **62**:965–971.

Childs, R. A., Feizi, T., and Tonegawa, Y., 1979, Multiplicity of molecules carrying blood-group I antigen on erythrocyte membranes, *Biochem. J.* **181**:533–538.

Clark, M. R., 1988, Senescence of red blood cells: Progress and problems, *Physiol. Rev.* **68**:503–554.

Clark, M. R., and Shohet, S. B., 1985, Red cell senescence, *Clin. Haematol.* **14**:223–257.

Coetzer, T. L., and Zail, S., 1980, Membrane protein complexes in GSH-depleted red cells, *Blood* **56**:159–167.

Cohen, N. S., Ekholm, J. E., Luthra, M. G., and Hanahan, D. J., 1976, Biochemical characterization of density-separated human erythrocytes, *Biochim. Biophys. Acta* **419**:229–242.

Corash, L. M., Piomelli, S., Chen, H. C., Seaman, C., and Gross, E., 1974, Separation of erythrocytes according to age on a simplified density gradient, *J. Lab. Clin. Med.* **84**:147–151.

Coutinho, A., 1980,The self–nonself discrimination and the nature and acquisition of the antibody repertoire, *Ann. Immunol. Inst. Pasteur* **131D**:236–253.

Cunningham, A. J., 1974, Large numbers of cells in normal mice produce antibody components of isologous erythrocytes, *Nature* **252**:749–751.

Czerwinski, M., Wasniowska, K., Stueden, I., Duk, M., Wiedlocka, A., and Lisowska, E., 1988, Degradation of the human erythrocyte membrane band 3 studied with the monoclonal antibody directed against an epitope on the cytoplasmic fragment of band 3, *Eur. J. Biochem.* **174**:647–654.

Daha, M. R., and VanEs, L. A., 1981, Stabilization of homologous and heterologous cell-bound amplification convertases, C3bBb, by C3 nephritic factor, *Immunology* **43**:33–38.

Danon, D., and Marikovsky, Y., 1962, Différence de charge de surface entre erythrocytes jeunes et âgés, *C. R. Acad. Sci Ser. D* **253**:1271–1272.

Davis, A. E., III, Harrison, R. A., and Lachmann, P. J., 1984, Physiologic inactivation of fluid phase C3b: Isolation and structural analyses of C3c, C3d,g (α2D), and C3g, *J. Immunol.* **132**:1960–1966.

Davitz, M. A., Low, M. G., and Nussenzweig, V., 1986, Release of decay-accelerating factor (DAF) from the cell membrane by phosphatidylinositol-specific phospholipase C (PIPLC), *J. Exp. Med.* **163**:1150–1159.

DeFlora, A., Benatti, U., Forteleoni, G. G., and Meloni, T., 1985, Favism: Disordered erythrocyte calcium homeostasis, *Blood,* **66**:294–297.

DeHeer, D. H., and Edington, T. S., 1974, Identification of derepressed autoimmunocompetent B lymphoid cells in NZB mice, *Clin. Exp. Immunol.* **16**:431–443.

Devaux, P. F., 1988, Phospholipid flippases, *FEBS Lett.* **234**:8–12.

Dighiero, G., Poncet, P., Matthes, T., and Kaushik, A., 1987, Is autoantibody production related to particular B-cell subsets and variable region genes? *Pathol. Immunopathol. Res.* **6**:371–389.

Dumaswala, U. J., and Greenwalt, T. J., 1984, Human erythrocytes shed exocytic vesicles *in vivo, Transfusion* **24:**490–492.

Durand, C. G., and Burge, J. J., 1984, A new enzyme-linked immunosorbent assay ELISA for measuring immunoconglutinins directed against human complement C3: Findings in systemic lupus erythematosus, *J. Immunol. Methods* **73:**57–66.

Durocher, J. R., Payne, R. C., and Conrad, M. E., 1975, Role of sialic acid in erythrocyte survival, *Blood* **45:**11–19.

Durocher, J. R., Viner, C. L., Sims, G. A., and Werner, L. M., 1976, Effect of periodate on erythrocyte survival, *Blood* **48:**992–992.

Duswald, K.-H, Jochum, M., Schramm, W., and Fritz, H., 1985, Released granulocytic elastase: An indicator of pathobiochemical alterations in septicemia after abdominal surgery, *Surgery* **98:**892–899.

Edberg, J. C., Kujala, G. A., and Taylor, R. P., 1987, Rapid immune adherence reactivity of nascent, soluble antibody/DNA immune complexes in the circulation, *J. Immunol.* **139:**1240–1244.

Ehlenberger, A. G., and Nussenzweig, V., 1977, The role of membrane receptors for C3b and C3d in phagocytosis, *J. Exp. Med.* **145:**357–371.

Farries, T. C., Finch, J. T., Lachmann, P. J., and Harrison, R. A., 1987, Resolution and analysis of "native" and "activated" properdin, *Biochem. J.* **243:**507–517.

Fasler, S., 1989, Charakterisierung erythrozyten-assozierter IgG-und Komplementfaktoren, Dissertation ETH No. 8777, Zurich.

Fearon, D. T., 1979, Regulation of the amplification C3 convertase of human complement by an inhibitory protein isolated from human erythrocyte membrane, *Proc. Natl. Acad. Sci. USA* **76:**5867–5871.

Fearon, D. T., 1984, Cellular receptors for fragments of the third component of complement, *Immunol. Today* **5:**105–110.

Fearon, D. T., and Wong, W. W., 1983, Complement ligand–receptor interactions that mediate biological responses, *Annu. Rev. Immunol.* **1:**243–271.

Fidalgo, B. V., Katayama, Y., and Najjar, V. A., 1967, The physiological role of the lymphoid system. V. The binding of autologous (erythrophilic) globulin to human red blood cells, *Biochemistry* **6:**3378–3385.

Franck, P. F. H., Op den Kamp, J. A. F., Roelofsen, B., and van Deenen, L. L. M., 1986, Does diamide treatment of intact human erythrocytes cause a loss of phospholipid asymmetry? *Biochem. Biophys. Acta* **857:**127–130.

Frank, M. M., 1985, Assessment of shortened in vivo RBC survival and sites of RBC destruction: in: *Immune Hemolytic Anemias* (H. Chaplin, Jr., ed.), pp. 95–118, Churchill/Livingstone, Edinburgh.

Freedman, J., 1984, Membrane-bound immunoglobulins and complement components on young and old red cells, *Transfusion* **24:**477–481.

Freedman, J., and Massey, A., 1979, Complement components detected on normal red blood cells taken into EDTA and CPD, *Vox Sang.* **37:**1–8.

Fries, L. F., Gaither, T. A., Hammer, C. H., and Frank, M. M., 1984, C3b covalently bound to IgG demonstrates a reduced rate of inactivation by factors H and I, *J. Exp. Med.* **160:**1640–1655.

Fries, L. F., Prince, G. M., Gaither, T. A., and Frank, M. M., 1985, Factor I cofactor activity of CR1 overcomes the protective effect of IgG on covalently bound C3b residues, *J. Immunol.* **135:**2673–2679.

Fries, L. F., Siwik, S. A., Malbran, A., and Frank, M. M., 1987, Phagocytosis of target particles bearing C3b–IgG covalent complexes by human monocytes and polymorphonuclear leucocytes, *Immunology* **62:**45–51.

Frommel, D., Grob, P. J., Masouredis, S. P., and Isliker, H. C., 1967, Studies on the mechanism of immunoglobulin binding to red cells, *Immunology* **13:**501–508.

Fujita, T., Takata, Y., and Tamura, N., 1981, Solubilization of immune precipitates by six isolated alternative pathway proteins, *J. Exp. Med.* **154:**1743–1751.

Fujita, T., Inoue, T., Ogawa, K., Iida, K., and Tamura, N., 1987, The mechanism of action of decay-accelerating factor (DAF). DAF inhibits the assembly of C3-convertase by dissociating C2a and Bb, *J. Exp. Med.* **166:**1221–1228.

Furthmayr, H., 1978, Glycophorins A, B, and C: A family of sialoglycoproteins. Isolation and preliminary characterization of trypsin derived peptides, *J. Supramol. Struct.* **9:**79–95.

Gadd, K. J., and Reid, K. B. M., 1981, The binding of complement component C3 to antibody–antigen aggregates after activation of the alternative pathway in human serum, *Biochem. J.* **195:**471–480.

Galili, U., Korkesh, A., Kahane, I., and Rachmilewitz, E. A., 1983, Demonstration of a natural antigalactosyl IgG antibody on thalassemic red blood cells, *Blood* **61:**1258–1264.

Galili, U., Rachmilewitz, E. A., Peleg, A., and Flechner, I., 1984, A unique natural human IgG antibody with anti-αgalactosyl specificity, *J. Exp. Med.* **160:**1519–1531.

Galili, U., Clark, M. R., and Shohet, S. B., 1986, Excessive binding of natural anti-αgalactosyl immunoglobulin G to sickle erythrocytes may contribute to extravascular cell destruction, *J. Clin. Invest.* **77:**27–33.

Galili, U., Clark, M. R., Shohet, S. B., Buehler, J., and Macher, B. A., 1987a, Evolutionary relationship between the natural anti-Gal antibody and the Gal-α1-3Gal epitope in primates, *Proc. Natl. Acad. Sci. USA* **84:**1369–1373.

Galili, U., Basbaum, C. B., Shohet, S. B., Buehler, J., and Macher, B. A., 1987b, Identification of erythrocyte Gal α1-3Gal glycosphingolipids with a mouse monoclonal antibody, Gal-13, *J. Biol. Chem.* **262:**4683–4688.

Gershon, H., and Gershon, D., 1988, Altered enzyme function and premature sequestration of erythrocytes in aged individuals, *Blood Cells* **14:**93–98.

Gibson, J., and Wells, J., 1988, Positive direct antiglobulin test in acute leukemia at presentation, *Transfusion* **28:**398–398.

Grabar, P., 1983, Autoantibodies and the physiological role of immunoglobulins, *Immunol. Today* **4:**337–340.

Green, G. A., and Kalra, V. K., 1988, Sickling-induced binding of immunoglobulin to sickle erythrocytes, *Blood* **71:**636–639.

Green, G. A., Rehn, M. M., and Kalra, V. K., 1985, Cell-bound autologous immunoglobulin in erythrocyte subpopulations from patients with sickle cell disease, *Blood* **65:**1127–1133.

Grob, P. J., Frommel, D., Isliker, H. C., and Masouredis, S. P., 1967, Interaction of IgG and its fragments with red cells, *Immunology* **13:**489–499.

Guilbert, B., Digherio, G., and Avrameas, S., 1982, Naturally occurring antibodies against nine common antigens in human sera, *J. Immunol.* **128:**2779–2787.

Haest, C. W. M., Kamp, D., Plasa, G., and Deuticke, B., 1977, Intra- and intermolecular cross-linking of membrane proteins in intact erythrocytes and ghosts by SH-oxidizing agents, *Biochim. Biophys. Acta* **469:**226–230.

Hargreaves, W. R., Giedd, K. N., Verkleij, A., and Branton, D., 1980, Reassociation of ankyrin with band 3 in erythrocyte membranes and lipid vesicles, *J. Biol. Chem.* **255:**11965–11972.

Hatanaka, M., Yoshimura, N., Murakami, T., Kannagi, R., and Murachi, T., 1984, Evidence for membrane-associated calpain I in human erythrocytes. Detection by an immunoelectrophoretic blotting method using monospecific antibody, *Biochemistry* **23:**3272–3276.

Hebbel, R. P., and Miller, W. J., 1984, Phagocytosis of sickle erythrocytes: Immunologic and oxidative determinants of hemolytic anemia, *Blood* **64:**733–741.

Heddle, N. M., Kelton, J. G., Kurchyn, K. L., and Ali, M. A. M., 1988, Hypergammaglobulinemia can be associated with a positive direct antiglobulin test, a nonreactive eluate, and no evidence of hemolysis, *Transfusion* **28:**29–33.

Herron, R., Clark, M., and Smith, D. S., 1987, An autoantibody with activity dependent on red cell age in the serum of a patient with autoimmune hemolytic anemia and a negative direct antiglobulin test, *Vox Sang.* **52:**71–74.

Hill, W., Shears, A. L., and Hibbit, K. G., 1978, The requirements of specific antibody for the killing of *E. coli* by the alternative complement pathway in bovine serum, *Immunology* **34:**131–140.

Hoffman, J. F., 1958, On the relationship of certain erythrocyte characteristics to their physiological age, *J. Cell Comp. Physiol.* **51:**415–423.

Högman, C. F., Akerblom, O., Hedlund, K., Rosén, I., and Wiklund, L., 1983, Red cell suspensions in SAGM medium, *Vox Sang.* **45:**217–223.

Holers, V. M., Cole, J. L., Lublin, D. M., Seya, T., and Atkinson, J. P., 1985, Human C3b- and C4b-regulatory proteins: A new multi-gene family, *Immunol. Today* **6:**188–192.

Hooper, D. C., 1987, Self-tolerance for erythrocytes is not maintained by clonal deletion of T helper cells, *Immunol. Today* **8:**327–330.

Horgan, C., Burge, J., Crawford, L., and Taylor, R. P., 1984, The kinetics of ^{3}H-dsDNA/anti-DNA immune complex formation. Binding by red blood cells, and release into serum: Effect of DNA molecular weight and conditions of antibody excess, *J. Immunol.* **133:**2079–2084.

Hsu, T. C. S., Steinberg, J., LeDoux, R., and Sawitsky, A., 1979, The low ionic strength reaction of human blood: Relationship between the binding of serum immunoglobulin and complement to red blood cells and surface charge of the cells, *Br. J. Haematol.* **42:**403–415.

Iida, Mornaghi, R., and Nussenzweig, V., 1982, Complement receptor (CR1) deficiency in erythrocytes from patients with systemic lupus erythematosus, *J. Exp. Med.* **155:**1427–1438.

Inada, S., Brown, E. J., Gaither, T. A., Hammer, C. H., Takahashi, T., and Frank, M. M., 1983, C3d receptors are expressed on human monocytes after in vitro cultivation, *Proc. Natl. Acad. Sci. USA* **80:**2351–2355.

Jacobs, R. J., and Reichlin, M., 1983, Generation of low M. W., C3-bearing immunoglobulin in human serum, *J. Immunol.* **130:**2775–2781.

Jain, S. K., 1985, *In vivo* externalization of phosphatidylserine and phosphatidylethanolamine in the membrane bilayer and hypercoagulability by the lipid peroxidation of erythrocytes in rats, *J. Clin. Invest.* **76:**281–286.

Jain, S. K., 1988, Evidence for membrane lipid peroxidation during the *in vivo* aging of human erythrocytes, *Biochim. Biophys. Acta* **937:**205–210.

Jain, S. K., and Hochstein, P., 1980, Polymerization of membrane components in aging red blood cells, *Biochem. Biophys. Res. Commun.* **92:**247–254.

Jepsen, H. H., Svehag, S.-E., Jarlbaek, L., and Baatrup, G., 1986, Interaction of complement-solubilized immune complexes with CR1 receptors on human erythrocytes. The binding reaction, *Scand. J. Immunol.* **23:**65–73.

Jerne, N. K., 1984, Idiotypic networks and other preconceived ideas, *Immunol. Rev.* **79:**6–24.

Jochum, M., Mempel, W., Fritz, H., Dwenger, A., Lindena, J., Schweitzer, G., Barthels, M., and Trautschold, I., 1984, Release of granulocytic lysosomal enzymes and concentrations of plasma factors in stored blood, 18th Cong. Int. Soc. Blood Transf. pp. 169–169.

Johnson, G. J., Allen, D. W., Flynn, T. P., Finkel, B., and White, J. G., 1980, Decreased survival *in vivo* of diamide-incubated dog erythrocytes. A model of oxidant-induced hemolysis, *J. Clin. Invest.* **66:**955–961.

Joiner, K. A., Fries, L. F., Schmetz, M. A., and Frank, M. M., 1985, IgG bearing covalently bound C3b has enhanced bactericidal activity for Escherichia coli O111, *J. Exp. Med.* **162:**877–889.

Jungi, T. W., and Barandun, S., 1985, Estimation of the degree of opsonization of homologous erythrocytes by IgG for intravenous and intramuscular use, *Vox Sang.* **49:**9–19.

Kadlubowski, M., and Agutter, P. S., 1977, Changes in the activities of some membrane-associated enzymes during *in vivo* ageing of the normal human erythrocyte, *Br. J. Haematol.* **37:**111–125.

Kay, M. M. B., 1975, Mechanism of removal of senescent cells by human macrophages in situ, *Proc. Natl. Acad. Sci. USA* **72:**3521–3525.

Kay, M. M. B., 1978, Role of physiologic autoantibody in the removal of senescent human red cells, *J. Supramol. Struct.* **9:**555–567.

Kay, M. M. B, 1981, Isolation of the phagocytosis-inducing IgG-binding antigen on senescent somatic cells, *Nature* **289:**491–494.

Kay, M. M. B., 1984, Localization of senescent cell antigen on band 3, *Proc. Natl. Acad. Sci. USA* **81:**5753–5757.

Kay, M. M. B., and Bennett, G. D., 1982, To the editor, *Blood* **59:**1111–1112.

Kay, M. M. B., and Bosman, G. J. C. G. M., 1985, Naturally occurring human "antigalactosyl" IgG antibodies are heterophile antibodies recognizing blood-group-related substances, *Exp. Hematol.* **13:**1103–1112.

Kay, M. M. B., and Goodman, J. R., 1984, IgG antibodies do not bind to band 3 in intact erythrocytes: Enzymatic treatment of cells is required for IgG binding, *Biomed. Biochim. Acta* **43:**841–846.

Kay, M. M. B., Bosman, G. J. C. G. M., Shapiro, S. S., Bendich, A., and Bassel, P. S., 1986, Oxidation as a possible mechanism of cellular aging: Vitamin E deficiency causes premature aging and IgG binding to erythrocytes, *Proc. Natl. Acad. Sci. USA* **83:**2463–2467.

Kazatchkine, M. D., Jouvin, M. H., Wilson, J. G., Fischer, E., and Fisher, A., 1987, Human diseases associated with C3 receptor deficiencies, *Immunol. Lett.* **14:**191–195.

Kinoshita, T., Medof, M. E., Silber, R., and Nussenzweig, V., 1985, Distribution of decay-accelerating factor in the peripheral blood of normal individuals and patients with paroxysmal nocturnal hemoglobinuria, *J. Exp. Med.* **162:**75–92.

Kinoshita, T., Medof, M. E., and Nussenzweig, V., 1986a, Endogenous association of decay-accelerating factor (DAF) with C4b and C3b on cell membranes, *J. Biol. Chem.* **136:**3390–3395.

Kinoshita, T., Medof, M. E., Hong, K., and Nussenzweig, V., 1986b, Membrane-bound C4b interacts endogenously with complement receptor CR1 of human red cells, *J. Exp. Med.* **164:**1377–1388.

Klinman, D. M., Banks, S., Hartman, A., and Steinberg, A. D., 1988, Natural murine autoantibodies and conventional antibodies exhibit similar degrees of antigenic cross-reactivity, *J. Clin. Invest.* **82:**652–657.

Kosower, N. S., Kosower, E. M., Wertheim, B., and Correa, W. S., 1969, Diamide, a new reagent for the intracellular oxidation of glutathione to the disulfide, *Biochem. Biophys. Res. Commun.* **37:**593–596.

Kulics, J., Rajnavölgyi, E., Füst, G., and Gergely, J., 1983, Interaction of C3 and C3b with immunoglobulin G, *Mol. Immunol.* **20:**805–810.

Kurlander, R. J., and Rosse, W. F., 1979, Monocyte-mediated destruction in the presence of serum of red cells coated with antibody, *Blood* **54:**1131–1139.

Law, S. K., Lichtenberg, N. A., and Levine, R. P., 1979, Evidence for an ester linkage between the labile binding site of C3b and receptive surfaces, *J. Immunol.* **123:**1388–1394.

Lawley, T. J., Hall, R. P., Fauci, A. S., Katz, S. I., Hamburger, M. J., and Frank, M. M., 1981, Defective Fc receptor functions associated with the HLA B8/DRW3 haplotype, *N. Engl. J. Med.* **304:**185–192.

Lee, P., Kirk, R. G., and Hoffman, J. G., 1984, Interrelations among Na and K content, cell volume, and buoyant density in human red blood cell populations, *J. Membr. Biol.* **79:**119–126.

Lorand, L., Michalska, M., Murthy, S. N. P., Shohet, S. B., and Wilson, J., 1987, Cross-linked polymers in the red cell membranes of a patient with Hb-Köln disease, *Biochem. Biophys. Res. Commun.* **147:**602 697.

Low, P. S., Waugh, S. M., Zinke, K., and Drenckhahn, D., 1985, The role of hemoglobin denaturation and band 3 clustering, *Science* **227:**531–533.

Luner, S. J. Szklarek, D., Knox, R. J., Seaman, G. V. F., Josefowicz, J. Y., and Ware, B. R., 1977, Red cell charge is not a function of cell age, *Nature* **269:**719–721.

Luthra, M. G., Friedman, J. M., and Sears, D. A., 1979, Studies on density fractions of normal human erythrocytes labeled with iron-59 in vivo, *J. Lab. Clin. Med.* **94:**879–896.

Lutz, H. U., 1978, Vesicles isolated from ATP-depleted erythrocytes and out of thrombocyte-rich plasma, *J. Supramol. Struct.* **8:**375–389.

Lutz, H. U., 1981, Elimination alter Erythrozyten aus der Zirkulation: Freilegung eines zellalterspezifischen Antigens auf alternden Erythrozyten, *Schweiz. Med. Wochenschr.* **111:**1507–1517.

Lutz, H. U., 1984, A cyclic AMP-dependent phosphorylation of spectrin dimer, *FEBS Lett.* **169:**323–329.

Lutz, H. U., 1987, Red cell clearance (A review), *Biomed. Biochim. Acta* **46:**65–71.

Lutz, H. U., 1988, Commentary to "The relationship of red cell enzymes to cell life-span" by E. Beutler, *Blood Cells* **14:**76–80.

Lutz, H. U., and Fehr, J., 1979, Total sialic acid content of glycophorins during senescence of human red blood cells, *J. Biol. Chem.* **254:**11177–11180.

Lutz, H. U., and Stringaro-Wipf, G., 1983, Senescent red cell-bound IgG is attached to band 3 protein, *Biomed. Biochim. Acta* **42:**117–121.

Lutz, H. U., and Wipf, G., 1982, Naturally occurring autoantibodies to skeletal proteins from human red blood cells, *J. Immunol.* **128:**1695–1699.

Lutz, H. U., Liu, S.-C., and Palek, J., 1977, Release of spectrin-free vesicles from human erythrocytes during ATP depletion, *J. Cell Biol.* **73:**548–560.

Lutz, H. U., Flepp, R., and Stringaro-Wipf, G., 1984, Naturally occurring autoantibodies to exoplasmic and cryptic regions of band 3 protein of human red blood cells, *J. Immunol.* **133:**2610–2618.

Lutz, H. U., Bussolino, F., Flepp, R., Fasler, S., Stammler, P., Kazatchkine, M. D., and Arese, P., 1987a, Naturally occurring anti-band 3 antibodies and complement together mediate phagocytosis of oxidatively stressed human red blood cells, *Proc. Natl. Acad. Sci. USA* **84:**7368–7372.

Lutz, H. U., Flepp, R., Stammler, P., and Baccala, R., 1987b, Red cell associated, naturally occurring anti-spectrin *Clin. exp. Immunol.* **67:**674–676.

Lutz, H. U., Stammler, P., Furter, C., and Fasler, S., 1987c, "Anti-Bande 3"-Antikörper aktivieren Komplement über den alternativen Weg, *Schweiz. Med. Wochenschr.* **117:**1821–1824.

Lutz, H. U., Fasler, S., Stammler, P., Bussolino, F., and Arese, P., 1988, Naturally occurring anti-band 3 antibodies and complement in phagocytosis of oxidatively-stressed and in clearance of senescent red cells, *Blood Cells* **14:**175–195.

Lutz, H. U., Fasler, S., and Stammler, P., 1989, An affinity for complement C3 as a possible reason for the potency of naturally occurring antibodies in mediating tissue homeostasis, *Beitr. Infusionsther.* **24:**193–199.

MacDonald, H. R., Schneider, R., Lees, R. K., Howe, R. C., Acha-Orbea, H., Festenstein, H., Zinkernagel, R. M., and Hengartner, H., 1988, T-cell receptor Vβ use predicts reactively and tolerance to Mis-encoded antigens, *Nature* **332:**40–45.

McEvoy, L., Williamson, P., and Schlegel, R. A., 1986, Membrane phospholipid asymmetry as a determinant of erythrocyte recognition by macrophages, *Proc. Natl. Acad. Sci. USA* **83:**3311–3315.

Magnani, M., Stocchi, V., Cucchiarini, L., Chiarantini, L., and Fornaini, G., 1986, Red blood cell phagocytosis and lysis following oxidative damage by phenylhydrazine, *Cell Biochem. Funct.* **4:**263–269.

Magnani, M., Rossi, L., Stocchi, V., Cucchiarini, L., Piacentini, G., and Fornaini, G., 1988, Effect of age on some properties of mice erythrocytes, *Mech. Ageing Dev.* **42:**37–47.

Mahana, W., Guilbert, B., and Avrameas, S., 1988, Regulation of the humoral immune response by polyspecific natural autoantibodies, *Ann. Inst. Pasteur Paris* **139:**349–360.

Malbran, A., Frank, M. M., and Fries, L. F., 1987, Interactions of monomeric IgG bearing covalently bound C3b with polymorphonuclear leucocytes, *Immunology* **61:**15–20.

Manser, T., Wysocki, L. J., Gridley, T., Near, R. I., and Gefter, M. L., 1985, The molecular evolution of the immune response, *Immunol. Today* **6:**94–101.

Marchesi, V. T., and Furthmayr, H., 1976, The red cell membrane, *Annu. Rev. Biochem.* **45:**667–698.

Masouredis, S. P., Branks, M. J., and Victoria, E. J., 1987, Antiidiotypic IgG crossreactive with Rh alloantibodies in red cell autoimmunity, *Blood* **70:**710–715.

Medof, M. E., Lublin, D. M., Holers, V. M., Ayers, D. J., Getty, R. R., Leykam, J. F., Atkinson, J. P., and Tykocinski, M. L., 1987, Cloning and characterization of cDNAs encoding the complete sequence of decay-accelerating factor of human complement, *Proc. Natl. Acad. Sci. USA* **84:**2007–2011.

Melez, K. A., Fries, L. F., Bener, B. S., Quinn, T., and Frank, M. M., 1988, Decline in rates of clearance of IgG-sensitized erythrocytes with increasing age, *Blood* **71:**1726–1730.

Meredith, P. J., Kristie, J. A., and Walford, R. L., 1979, Aging increases expression of LPS-induced autoantibody-secreting B cells, *J. Immunol.* **123:**87–91.

Michalek, M. T., Bremer, E. G., and Mold, C., 1988, Effect of gangliosides on activation of the alternative pathway of human complement, *J. Immunol.* **140:**1581–1587.

Miller, R. D., and Calkins, C. E., 1988, Development of self-tolerance in normal mice. Appearance of suppressor cells that maintain adult self-tolerance follows the neonatal autoantibody response, *J. Immunol.* **141:**2206–2210.

Miyahara, K., and Spiro, M. J., 1984, Nonuniform loss of membrane glycoconjugates during in vivo aging of human erythrocytes: Studies of normal and diabetic red cell saccharides, *Arch. Biochem. Biophys.* **232:**310–322.

Moldenhauer, F., Botto, M., and Walport, M. J., 1988, The rate of loss of CR1 from ageing erythrocytes in vivo in normal subjects and SLE patients: No correlation with structural or numerical polymorphisms, *Clin. Exp. Immunol.* **72:**74–78.

Mollnes, T. E., and Lachmann, P. J., 1988, Regulation of complement, *Scand. J. Immunol.* **27:**127–142.

Moore, F. D., Fearon, D. T., and Austen, K. F., 1981, IgG on mouse erythrocytes augments activation of the human alternative complement pathway by enhancing deposition of C3b, *J. Immunol.* **126:**1805–1809.

Moore, J. G., Frank, M. M. Müller-Eberhard, H. J., and Young, N. S., 1985, Decay-accelerating factor is present on paroxysmal nocturnal hemoglobinuria erythroid progenitors and lost during erythropoiesis in vitro, *J. Exp. Med.* **162:**1182–1192.

Morrison, M., Grant, W., Smith, H. T., Mueller, T. J., and Hsu, L., 1985, Catabolism of the anion transport protein in human erythrocytes, *Biochemistry* **24:**6311–6315.

Moticka, E. J., 1983, Regulation of naturally occurring autoantibody secretion by a radiosensitive lymphocyte: Initial characterization and ontogeny, *Cell. Immunol.* **81:**36–44.

Mueller, H., and Lutz, H. U., 1983, Binding of autologous IgG to human red blood cells before and after ATP-depletion. Selective exposure of binding sites (autoantigens) on spectrin-free vesicles, *Biochim. Biophys. Acta* **729:**249–257.

Müler-Eberhard, H. J., and Schreiber, R. D., 1980, Molecular biology and chemistry of the alternative pathway of complement, *Adv. Immunol.* **29:**1–53.

Murakami, T., Suzuki, Y., and Murachi, T., 1979, An acid protease in human erythrocytes and its localization in the inner membrane, *Eur. J. Biochem.* **96:**221–227.

Murphy, J. R., 1973, Influence of temperature and method of centrifugation on the separation of erythrocytes, *J. Lab. Clin. Med.* **82:**334–341.

Nash, G. B., and Meiselman, H. J., 1981, Red cell ageing: Changes in deformability and other possible determinants of in vivo survival, *Microcirculation* **1:**255–284.

Nicholson-Weller, A., Burge, J., Fearon, D. T., Weller, P. F., and Austen, K. F., 1982, Isolation of a human erythrocyte membrane glycoprotein with decay-accelerating activity for C3 convertases of the complement system, *J. Immunol.* **129:**184–189.

Nicholson-Weller, A., March, J. P., Rosenfeld, S. I., and Austen, K. F., 1983, Affected erythrocytes of patients with paroxysmal nocturnal hemoglobinuria are deficient in the complement regulatory protein, decay accelerating factor, *Proc. Natl. Acad. Sci. USA* **80:**5066–5070.

Okada, N., Yasuda, T., Tsumita, T., and Okada, H., 1983, Activation of the alternative complement pathway by natural antibody to glycolipids in guinea-pig serum, *Immunology* **50:**75–84.

Oliver, C. N., Ahn, B-W., Moerman, E. J., Goldstein, S., and Stadtman, E. R., 1987, Age-related changes in oxidized proteins, *J. Biol. Chem.* **262:**5488–5491.

Paccaud, J.-P., Carpentier, J.-L., and Schifferli, J. A., 1988, Direct evidence for the clustered nature of complement receptors type 1 in the erythrocyte membrane, *J. Immunol.* **141:**3889–3894.

Parker, C. J., Soldato, C. M., and Rosse, W. F., 1984, Abnormality of glycophorin-α on paroxysmal nocturnal hemoglobinuria erythrocytes, *J. Clin. Invest.* **73:**1130–1143.

Petz, L. D., and Branch, D. R., 1985, Drug-induced immune hemolytic anemia, in: *Immune Hemolytic Anemias.* (H. Chaplin, Jr., ed.), pp. 47–94, Churchill/Livingstone, Edinburgh.

Pfeffer, S. R., and Swislocki, N. I., 1976, Age-related decline in the activities of erythrocyte membrane adenylate cyclase and protein kinase, *Arch. Biochem. Biophys.* **177:**117–122.

Piomelli, S., 1988, The relationship of red cell enzymes to red cell life-span by E. Beutler. Commentary, *Blood Cells* **14:**81–86.

Piomelli, S., Lurinsky, G., and Wasserman, L. R., 1967, The mechanism of red cell aging. I. Relationship between cell age and specific gravity evaluated by ultracentrifugation in a discontinuous density gradient, *J. Lab. Clin. Med.* **69:**659–674.

Piomelli, S., Seaman, C., Reigman, J., Tytun, A., Graziano, J., Tabachnik, N., and Corash, L., 1978, Separation of younger red cells with improved survival *in vivo*: An approach to chronic transfusion therapy, *Proc. Natl. Acad. Sci. USA* **75:**3474–3478.

Polhill, R. B., Jr., Newman, S. L., Pruitt, K. M., and Johnston, R. B., Jr., 1978, Kinetic assessment of alternative complement pathway activity in a hemolytic system. II. Influence of antibody on alternative pathway activation, *J. Immunol.* **121:**371–376.

Pontremoli, S., Salamino, F., Sparatore, B., Melloni, E., Morelli, A., Benatti, U., and DeFlora, A., 1979, Isolation and partial characterization of three acidic proteinases in erythrocyte membranes, *Biochem. J.* **181:**559–568.

Pontremoli, S., Melloni, E., Sparatore, B., Salamino, F., Michetti, M., Sacco, O., and Horecker, B. L., 1985, Role of phospholipids in the activation of the (Ca^{2+})-dependent neutral proteinase of human erythrocytes, *Biochem. Biophys. Res. Commun.* **129:**389–395.

Portnoi, D., Lundkvist, I., and Coutinho, A., 1988, Inverse correlation between the utilization of an idiotype in specific immune responses and its representation in pre-immune "natural" antibodies, *Eur. J. Immunol.* **18:**571–576.

Rask, R., Rasmussen, J. M., Hansen, H. V., Bysted, P., and Svehag, S.-E., 1988, Complement C3d,g/Epstein–Barr virus receptor density on human B-lymphocytes eliminated by immunoenzymatic assay and immunocytochemistry, *J. Clin. Lab. Immunol.* **25:**153–156.

Rieder, R. F., Ibrahim, A., and Etlinger, J. D., 1986, A soluble adenosine triphosphate-dependent proteolytic system in human peripheral red blood cells, *Blood* **67:**1293–1297.

Ripoche, J., and Sim, R. B., 1986, Loss of complement receptor type 1(CR1) on ageing of erythrocytes. Studies of proteolytic release of the receptor, *Biochem. J.* **235:**815–821.

Rondeau, E., Solal-Celigny, P., Dhermy, D., Vroclans, M., Brousse, N., Bernard, J. F., and Boivin, P., 1983, Immune disorders in agnogenic myeloid metaplasia: Relations to myelofibrosis, *Br. J. Haematol.* **53:**467–475.

Ross, G. D., 1986, *Immunobiology of the Complement System,* Academic Press, New York.

Ross, G. D., Yount, W. J., Walport, M. J., Winfield, J. B., Parker, C. J., Fuller, C. R., Taylor, R. P., Myones, B. L., and Lachmann, P. J., 1985, Disease-associated loss of erythrocyte complement receptors (CR1, C3b receptors) in patients with systemic lupus erythematosus and other disease involving autoantibodies and/or complement activation, *J. Immunol.* **135:**2005–2014.

Schenkein, H. A., and Ruddy, S., 1981a, The role of immunoglobulins in alternative complement pathway activation by zymosan. I. Human IgG with specificity for zymosan enhances alternative pathway activation by zymosan, *J. Immunol.* **1260:**7–10.

Schenkein, H. A., and Ruddy, S., 1981b, The role of immunoglobulins in alternative pathway activation by zymosan. II. The effect of IgG on the kinetics of the alternative pathway, *J. Immunol.* **126:**11–15.

Schifferli, J. A., Ng, Y. C., and Peters, D. K., 1986, The role of complement and its receptor in the elimination of immune complexes, *N. Engl. J. Med.* **315:**488–495.

Schifferli, J. A., Hauptmann, G., and Paccaud, J.-P., 1987, Complement-mediated adherence of immune complexes to human erythrocytes, *FEBS Lett.* **213:**415–418.

Schlüter, K., and Drenckhahn, D., 1986, Co-clustering of denatured hemoglobin with band 3: Its role in binding of autoantibodies against band 3 to abnormal and aged erythrocytes, *Proc. Natl. Acad. Sci. USA* **83:**6137–6141.

Schmidt, G., Gross, J., Moller, R., and Staak, R., 1975, Trennung von roten Blutzellen im isopyknischen Dextran und Albumindichtegradienten, *Acta Biol. Med. Ger.* **34:**1621–1630.

Schönemark, S., Rauterberg, E. W., Shin, M. L., Löke, S., Roelcke, D., and Hänsch, G. M., 1986, Homologous species restriction in lysis of human erythrocytes: A membrane-derived protein with C8-binding capacity functions as an inhibitor, *J. Immunol.* **136:**1772–1776.

Schreiber, A. D., 1985, Quantitation of RBC-bound immunoglobulin and complement components, in: *Immune Hemolytic Anemias* (H. Chaplin, Jr., ed.), pp. 155–175, Churchill/Livingstone, Edinburgh.

Schroit, A. J., Madsen, J. W., and Tanaka, Y., 1985, *In vivo* recognition and clearance of red blood cells containing phosphatidylserine in their plasma membranes, *J. Biol. Chem.* **260:**5131–5138.

Schultz, H., Mischer, O., and Schenk, H., 1984, Gehalt und Muster der Ganglioside von Membranen menschlicher Erythrozyten, *Biomed. Biochim. Acta* **43:**1127–1133.

Schweizer, E., Angst, W., and Lutz, H. U., 1982, Glycoprotein topology on intact human red blood cells reevaluated by cross-linking following amino-group supplementation, *Biochemistry* **21:**6807–6818.

Seigneurin, J. M., Guilbert, B., Bourgeat, M. J., and Avrameas, S., 1988, Polyspecific natural antibodies and autoantibodies secreted by human lymphocytes immortalized with Epstein–Barr virus, *Blood* **71:**581–585.

Serrou, B., and Rosenfeld, C., 1981, Immune complexes and plasma exchanges in cancer patients, in: *Human Cancer Immunology* (B. Serrou and C. Rosenfeld, eds.), pp. 1–342, Elsevier/North-Holland, Amsterdam.

Shaw, D. R., and Griffin, F. M., Jr., 1981, Phagocytosis requires repeated triggering of macrophage phagocytic receptors during particle ingestion, *Nature* **289:**409–411.

Sherwood, T. A., and Virella, G., 1986, The binding of immune complexes to human red cells: Complement requirements and fate of the RBC-bound IC after interaction with phagocytic cells, *Clin. Exp. Immunol.* **64:**195–204.

Shinozuka, T., Takei, S., Yanagida, J., Watanabe, H., and Ohkuma, S., 1988, Binding of lectins to "young" and "old" human erythrocytes, *Blut* **57:**117–123.

Singer, J. A., Jennings, L. K., Jackson, C. W., Dockter, M. E., Morrison, M., and Walker, W. S., 1986, Erythrocyte homeostasis: Antibody-mediated recognition of the senescent state by macrophages, *Proc. Natl. Acad. Sci. USA* **83:**5498–5501.

Smith, C. A., Pangburn, M. K., Vogel, C.-H., and Müller-Eberhard, H. J., 1984, Molecular architecture of human properdin, a positive regulator of the alternative pathway of complement, *J. Biol. Chem.* **259:**4582–4588.

Snyder, L. M., Fairbanks, G., Trainor, J., Fortier, N. L., Jacobs, J. B., and Leb, L., 1985, Properties and characterization of vesicles released by young and old human red cells, *Br. J. Haematol.* **54:**513–522.

Snyder, L. M., Fortier, N. L., Leb, L., McKenney, J., Trainor, J., Sheerin, H., and Mohandas, N., 1988, The role of membrane protein sulfhydryl groups in hydrogen peroxide-mediated membrane damage in human erythrocytes, *Biochim. Biophys. Acta* **937:**229–240.

Solal-Celigny, P., Vazeux, R., Vroclans, M., Amar, M., Herrera, A., Bernard, J. F., and Boivin, P., 1984, Positive Coombs test in acute leukaemia, *Br. J. Haematol.* **57:**563–569.

Strejan, G. H., Essani, K., and Surlan, D., 1981, Naturally occurring antibodies to liposomes. II. Specificity and electrophoretic pattern of rabbit antibodies reacting with sphingomyelin-containing liposomes, *J. Immunol.* **127:**160–165.

Suzuki, T., and Dale, G. L., 1987, Biotinylated erythrocytes: *In vivo* survival and *in vitro* recovery, *Blood* **70:**791–795.

Szymanski, I. O., Odgren, P. R., Fortier, N., and Snyder, L. M., 1980, Red blood cell associated IgG in normal and pathologic states, *Blood* **55:**48–54.

Szymanski, I. O., Keegan, M. G., and Odgren, P. R., 1981, Complement and Ig uptake by erythrocytes in low ionic strength solutions, *Vox Sang.* **41:**151–159.

Szymanski, I. O., Odgren, P. R., and Valeri, C. R., 1985, Relationship between the third component of human complement (C3) bound to stored preserved erythrocytes and their viability *in vivo*, *Vox Sang.* **49:**34–41.

Takata, Y., Tamura, N., and Fujita, T., 1984, Interaction of C3 with antigen–antibody complexes in the process of solubilization of immune precipitates, *J. Immunol.* **132:**2531–2537.

Taylor, R. P., Burge, J., Horgan, C., and Shasby, D. M., 1983, The complement-mediated binding of soluble antibody/dsDNA immune-complexes to human neutrophils, *J. Immunol.* **130:**2656–2662.

Taylor, R. P., Horgan, C., Hooper, M., and Burge, J., 1985, Dynamics of interaction between complement-fixing antibody/dsDNA immune complexes and erythrocytes. *In vitro* studies and potential general applications to clinical immune complex testing, *J. Clin. Invest.* **75:**102–111.

Tökés, Z. A., and Chambers, S. M., 1975, Proteolytic activity associated with human erythrocyte membranes. Self-digestion of isolated human erythrocyte membranes, *Biochim. Biophys. Acta* **389:**325–338.

Tonegawa, S., 1985, The molecules of the immune system, *Sci. Am.* **253:**104–113.

Tuech, J. K., and Morrison, M., 1974, Human erythrocyte membrane sialoglycoproteins: A study of interconversion, *Biochem. Biophys. Res. Commun.* **59:**352–360.

Turrini, F., Naitana, A., Mannuzzu, L., Pescarmona, G., and Arese, P., 1985, Increased red cell calcium adenosine triphosphate, and altered membrane proteins during Fava bean hemolysis in glucose 6-phosphate dehydrogenase-deficient (Mediterranean variant) individuals, *Blood* **66:**302–305.

Valderrama, R., Eggers, A. E., Revan, S., Moomjy, M., Frost, M., Pipia, P., and DiPaola, M., 1988, Idiotypic control of the immune response, *J. Neuroimmunol.* **20:**269–276.

van Loghem, E., 1978, Genetic studies on human immunoglobulins, in: *Handbook of Experimental Immunology,* Volume I (D. M. Weir, ed.), p. 16, Blackwell, Oxford.

Vaysse, J., Gattegno, L., Bladier, D., and Aminoff, D., 1986, Adhesion and erythrophagocytosis of human senescent erythrocytes by autologous monocytes and their inhibition by β-galactosyl derivatives, *Proc. Natl. Acad. Sci. USA* **83:**1339–1343.

Vettore, L., DeMatteis, M. C., and Zampini, P., 1980, A new density gradient system for the separation of human red blood cells, *Am. J. Hematol.* **8:**291–297.

Victoria, E. J., Mahan, L. C., and Masouredis, S. P., 1982, The IgG binding function of the normal red cell plasma membrane: Identification of integral polypeptides that bind IgG, *Br. J. Haematol.* **50:**101–110.

Vlassara, H., Valinsky, J., Brownlee, M., Cerami, C., Nishimoto, S., and Cerami, A., 1987, Advanced glycosylation endproducts on erythrocyte cell surface induce receptor-mediated phagocytosis by macrophages, *J. Exp. Med.* **166:**539–549.

Walker, W. S., Singer, J. A., Morrison, M., and Jackson, C. W., 1984, Preferential phagocytosis of *in vivo* aged murine red blood cells by a machrophage-like cell line, *Br. J. Haematol.* **58:**259–266.

Walport, M. J., Ross, G. D., Mackworth-Young, C., Watson, J. V., Hogg, N., and Lachmann, P. J., 1985, Family studies of erythrocyte complement receptor type 1 levels: Reduced in patients with SLE are acquired, not inherited, *Clin. Exp. Immunol.* **59:**547–554.

Walter, H., Krob, E. J., and Ascher, G. S., 1981, Aging of erythrocytes results in altered red cell surface properties in the rat, but not in the human. Studies by partitioning in two-polymer aqueous phase systems, *Biochim. Biophys. Acta* **641:**202–215.

Watanabe, K., Powell, M. E., and Hakomori, S.-I., 1979, Isolation and characterization of gangliosides with a new sialosyl linkage and core structure, *J. Biol. Chem.* **254:**8223–8229.

Waugh, S. M., and Low, P. S., 1985, Hemichrome binding to band 3: Nucleation of Heinz bodies on the erythrocyte membrane, *Biochemistry* **24:**34–39.

Weed, R. I., and Reed, C. F., 1966, Membrane alterations leading to red cell destruction, *Am. J. Med.* **41:**6881–698.

Wegner, G., Tanner, C., Maretzki, D., and Schössler, W., 1984, Immunoglobulin G binding to human erythrocytes, *Biomed. Biochim. Acta* **43:**179–186.

Wiener, E., Hughes-Jones, N. C., Irish, W. T., and Wickramasinghe, S. N., 1986, Elution of antispectrin antibodies from red cells in homozygous β-thalassemia, *Clin. Exp. Immunol.* **63:**680–686.

Winterbourn, C. C., and Batt, R. D., 1970, Lipid composition of human red cells of different ages, *Biochim. Biophys. Acta* **202:**1–8.

Worlledge, S. M., 1978, Annotation: The interpretation of a positive direct antiglobulin test, *Br. J. Haematol.* **39:**157–162.

Yam, P., Petz, L. D., and Spath, P., 1982, Detection of IgG sensitization of red cells with ^{125}I-staphylococcal protein A, *Am. J. Hematol.* **12:**337–346.

Yoon, S. H., and Fearon, D. T., 1985, Characterization of a soluble form of the C3b/C4b receptor (CR1) in human plasma, *J. Immunol.* **134:**3332–3338.
Zalman, L. S., Wood, L. M., and Müler-Eberhard, H. J., 1986, Isolation of a human erythrocyte membrane protein capable of inhibiting expression of homologous complement transmembrane channels, *Proc. Natl. Acad. Sci. USA* **83:**6975–6979.
Zharhary, D., 1988, Age-related changes in the capability of the bone marrow to generate B cells, *J. Immunol.* **141:**1863–1869.

Chapter 5

Macromolecular Synthesis and Degradation during Terminal Erythroid Cell Development

Sophia A. Bonanou-Tzedaki and Henry R. V. Arnstein

1. INTRODUCTION

In this chapter we have focused on the major changes in nucleic acids and proteins that occur during terminal erythroid cell development and have attempted to illustrate how these macromolecular changes account for the observed cellular changes. Before discussing these changes in any detail, we present a brief description of the main cellular and molecular events in erythropoiesis.

1.1 Outline of Erythroid Cell Development

In the adult mammal, erythropoiesis, the production of red blood cells from precursors in the bone marrow, is the end result of proliferation and differentiation of three distinct cell populations: pluripotent hematopoietic stem cells, committed erythroid precursor cells, and recognizable maturing erythroblasts (Paul, 1976). The production of committed erythroid precursors from the progeny of pluripotent stem cells is not well understood, but is believed to be influenced by various growth factors and cell–cell interactions that occur in the bone marrow microenvironment (Dexter, 1987). Committed erythroid precursors are a population of cells of defined age structure, burst-forming units (BFU-e) giving rise to colony-forming-unit erythroid cells (CFU-e). These erythroid cells are characterized by increasing responsiveness to erythropoietin (Gregory, 1976), a hormone secreted from the kidney in response to hypoxia. Erythropoietin increases the proportion of committed erythroid precursors entering the terminal maturation pathway and accelerates their rate of terminal differentiation (see review by Spivak, 1986). The

Sophia A. Bonanou-Tzedaki and Henry R. V. Arnstein Division of Biomolecular Sciences, Department of Biochemistry, King's College London, London WC2R 2LS, England.

Table I
Outline of Mammalian Erythroid Cell Development

Stage of erythroid cell development	Time (days)	Cellular changes	Molecular changes
1. Hemopoietic stem cells			
		"Commitment" to erythroid cell pathway	
2. Committed erythroid progenitors		Increasing responsiveness to erythropoietin ↓	Acquisition of erythropoietin receptors?
BFU-e, CFU-e	14		Globin genes acquire hypersensitive sites?
3. Morphologically identifiable cells			
Proerythroblast		Decrease in cell size ↓	Induction of globin gene transcription
		Increase in hemoglobin content ↓	Synthesis of spectrin, transferrin receptor, glycophorin
Basophilic erythroblast	2–3 (Proerythroblast to Orthochromatic erythroblast)		High DNA, RNA, and protein synthesis
Polychromatic erythroblast			Peak of globin gene transcription
			Peak synthesis of heme, transferrin receptor, spectrin, glycophorin
Orthochromatic erythroblast			Peak synthesis of globin, band 3
		Condensation and loss of nucleus	Cessation of DNA and RNA synthesis
			Loss of some surface receptors
Reticulocyte	2–3 (Orthochromatic erythroblast to Erythrocyte)		Peak activity of catalase, superoxide dismutase, carbonic anhydrase
			Synthesis of lipoxygenase
		Loss of cell organelles	Cessation of protein synthesis
Erythrocyte			Peak of hemoglobin content

first morphologically identifiable erythroid cell is the proerythroblast, a large cell more than 20 μm in diameter, with a large nucleus containing diffuse chromatin and basophilic cytoplasm (Bessis, 1973). Like the late CFU-e this cell is absolutely dependent on erythropoietin for survival and differentiation. In the presence of the hormone, the proerythroblast undergoes a limited number of cell divisions, occurring within 3–4 days, giving rise to basophilic, polychromatic, and orthochromatic erythroblasts (cells of decreasing size and increasing hemoglobin content) and finally, after the extrusion of the nucleus, to reticulocytes. These enucleated cells, approximately one-tenth the volume of the proerythroblasts, mature for 2–3 days mainly in the bone marrow before becoming erythrocytes.

An outline of the major cellular and biochemical changes that occur in mammalian erythroid cell development is shown in Table I. In avian or amphibian erythropoiesis the main difference is that the condensed nucleus is not extruded but is retained in the erythrocyte although it is no longer active in transcription.

1.2. Coordinate Changes in Macromolecular Synthesis and Degradation during Terminal Erythroid Cell Differentiation: An Overview

The production of mature red blood cells from actively proliferating erythroid cell precursors involves the acquisition of a series of specific proteins. These include the erythropoietin and transferrin receptors, glycophorin, spectrin, and hemoglobin, as well as the various enzymes required for the functioning of erythrocytes during their protracted life span. At the same time this process involves the elimination of a number of enzymes and other proteins involved in nucleic acid metabolism, oxidative phosphorylation, and protein synthesis and degradation, which in mammals culminates in the extrusion of the nucleus and the eventual loss of all subcellular organelles (mitochondria, lysosomes, ribosomes). The coordination of these molecular changes has been the subject of extensive research and will be discussed in detail later in this chapter. Here we only present a brief outline of the chronological sequence of these changes.

Expression of erythropoietin receptors and the acquisition of DNase I–nuclease hypersensitive sites next to erythroid-specific genes is already detectable at the CFU-e stage. Proerythroblasts have ongoing synthesis of transferrin receptors, spectrin, glycophorin, and carbonic anhydrase. The initial response of CFU-e/proerythroblast cells to erythropoietin is characterized by increased glucose uptake, DNA, RNA, and protein synthesis, and all these activities reach a peak at the basophilic erythroblast stage. DNA and histone synthesis ceases at the polychromatic/orthochromatic stage and RNA synthesis is no longer observed 12 hr later, at the late orthochromatic stage. Induction of globin gene transcription is first detectable in proerythroblasts and culminates in polychromatic erythroblasts. Globin synthesis is initiated without delay as soon as the mRNA has been transcribed, peaks at the orthochromatic erythroblast stage, and persists for a considerable time in reticulocytes. Maximal transferrin binding, iron uptake, and synthesis of heme biosynthetic enzymes and heme precede the peak of globin synthesis by 10–20 hr. Synthesis of spectrin, band 3, and other membrane proteins peak at different times during terminal erythroid cell maturation but their accumulation in the plasma membrane becomes prominent at the late orthochromatic/reticulocyte stage. Loss of transferrin receptors and other membrane proteins and a profound remodeling of the plasma membrane also occur in late orthochromatic erythroblasts, when the cell nucleus is extruded. During reticulocyte maturation, synthesis of lipoxygenase results in the loss of subcellular organelles and associated functions (Rapoport and Schewe, 1986) while the activity of certain erythrocyte-specific soluble enzymes (e.g., catalase, carbonic anhydrase, and superoxide dismutase) reaches a peak level. In going from reticulocytes to erythrocytes, there is an initial rapid decline in the activity of most soluble enzymes, which is followed by a prolonged slight decrease as the red cells age (Beutler, 1988). The difference in proteins synthesized by bone marrow erythroblasts and circulating reticulocytes is illustrated in Figure 1.

The preponderance of hemoglobin in the mature erythrocyte does not appear to

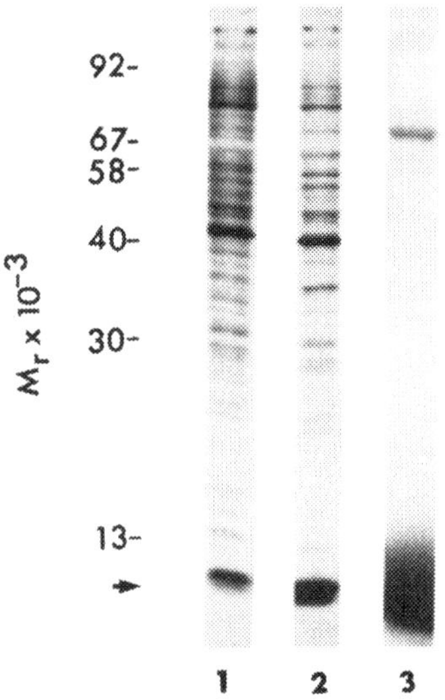

FIGURE 1. Changes in protein synthesis during rabbit erythroid cell development. Immature and mature bone marrow erythroid cells and circulating reticulocytes were labeled with [^{35}S]methionine and the radioactive cytoplasmic proteins were separated by SDS PAGE. Lane 1: proerythroblasts and basophilic erythroblasts; lane 2: polychromatic and orthochromatic cells; lane 3: reticulocytes. Globin is indicated by the arrow. (Results from authors' laboratory.)

involve enhanced rates of globin gene transcription (Clissold *et al.*, 1977) but can be accounted for by the very long half-life of globin mRNA (> 60 hr; Volloch and Housman, 1981), the stability of hemoglobin, the cessation of transcription at the time of nuclear condensation, and an approximately twofold reduction in cell volume at each of the preceding cell divisions. During active transcription, this reduction in cell volume greatly enhances the cellular concentration of mRNAs, whereas after transcription ceases in the terminal stages of erythroid cell development the degradation of the short-lived nonglobin mRNAs ensures that globin synthesis becomes predominant.

1.3. Control Mechanisms Operating during Erythroid Cell Differentiation

Eukaryotic gene expression is a complex process that may be regulated at numerous levels (Darnell, 1982), e.g., gene transcription, RNA processing, nucleocytoplasmic transport, mRNA degradation, translation, or protein turnover. The profound molecular changes that characterize the terminal erythroid cell differentiation program are brought about by the operation of all of these control mechanisms (Arnstein *et al.*, 1987). For example, analysis of erythropoietic diseases like β-thalassemia, a genetically heterogeneous anemia (Weatherall and Clegg, 1981), has demonstrated that the absence or reduced synthesis of adult β-globin can arise from defects at all levels of gene expression. In normal erythropoiesis the initial erythropoietin-dependent increase in proerythroblast cell proliferation is marked by intense transcription of previously expressed as well as new genes, with increased processing and transport to cytoplasm of certain mRNAs, leading to increased synthesis of transferrin receptors, heme biosynthetic enzymes, spectrin, and globin. After the final cell division at the orthochromatic erythroblast stage, RNA transcription ceases but changes in the synthesis and accumulation of globin, intracellular enzymes, and membrane proteins continue, due to the differential stability and selective degradation of various mRNAs and proteins. Finally, the synthesis of lipoxygenase from masked ribonucleoprotein complexes is under translational control in the reticulocyte.

Other control mechanisms may result from the action of various intermediates of erythroid cell differentiation as inhibitors or activators of cellular processes. For example, heme, besides its role as a structural component of hemoglobin, acts as an activator of globin gene transcription (Charnay and Maniatis, 1983) and of heme biosynthetic enzymes (Ibraham *et al.*, 1983), as a factor required for the prevention of the formation of an inhibitor of initiation of protein synthesis (Ochoa and De Haro, 1979), as a ribonuclease inhibitor (Burka, 1968), and as an inhibitor of proteolysis of ubiquinated proteins (Haas and Rose, 1981).

Control mechanisms that are not fully understood are also responsible for the differential expression of various globin genes within the same gene family (e.g., β_{major} and β_{minor} globin in the adult mouse and ζ or α and ϵ, γ, or β globin during human embryonic and fetal development). The precise macromolecular changes responsible for the dramatic switch in humans from fetal hemoglobin (HbF), containing $\alpha_2\gamma_2$, to adult hemoglobin (HbA), containing $\alpha_2\beta_2$, upon gestation remain to be elucidated. (For a recent review of the organization, structure, and function of the human globin genes, see Karlsson and Nienhuis, 1985.) In man the switch appears to be transcriptional in nature, in response to changes in the cell's environment (Anagnou *et al.*, 1985; Groudine *et al.*, 1986), but in the chick embryo the switch appears to be due to a change in the erythroid cell line (Fucci *et al.*, 1987).

1.4. Model Systems for the Study of Terminal Erythroid Cell Maturation

The paucity of committed erythroid cell progenitors and our inability to identify them morphologically, combined with the scarcity of erythropoietin-responsive cell lines has meant that, until recently, most studies of terminal erythroid cell differentiation have been performed either with relatively unhomogeneous cell populations, enriched in immature erythroblasts, or with virally transformed erythroleukemic cells. Depending on the model system used, minor variations have been encountered and the following brief description of the cell systems usually employed is therefore pertinent.

Cell populations enriched in proerythroblasts and erythroid progenitor cells have been obtained from chick embryos, from human, mouse, or rat fetal liver (an erythropoietic organ in the mammal), from mouse spleen, or from the bone marrow of animals in which the erythroid cell population was expanded by the experimental induction of hypoxia or anemia. These cells can be cultured in suspension or on semisolid media either immediately or after further enrichment of erythroid progenitor cells by velocity or isopycnic centrifugation in various media and removal of nonerythroid cells by immunological techniques. In our laboratory, anemic rabbit bone marrow erythroid cells are isolated by Percoll gradient density centrifugation (Harrison *et al.*, 1981) or are separated into fractions enriched in cells of a particular developmental stage by velocity sedimentation at unit gravity (Denton and Arnstein, 1973), or bovine serum albumin density centrifugation (Borsook *et al.*, 1969). Immature erythroblasts prepared in this way respond to erythropoietin by accelerated proliferation and differentiation (Bonanou-Tzedaki *et al.*, 1987). Hemin has also been reported to accelerate the proliferation (Porter *et al.*, 1979; Mayeux *et al.*, 1986b) and the differentiation of immature erythroblasts, and its mode of action may vary depending on the exact developmental stage of the cells (Bonanou-Tzedaki *et al.*, 1981). A variety of other substances, including β-adrenergic agonists (Brown and Adamson, 1977), prostaglandins (Mayeux *et al.*, 1986a), glucocorticoids (Gidari, 1981), insulin (Kurtz *et al.*, 1983), growth hormone (Golde *et al.*, 1977), thyroid hormones (Popovic *et al.*, 1977), sex hormones (Urabe *et al.*, 1979), IL3 (Goodman *et al.*, 1985), GM-CSF (Sieff *et al.*, 1985), and hematopoietic cell-derived activities (Linch and Nathan, 1984), can also affect the extent and rate of terminal erythroid cell development.

A system that has been extensively employed as a model for erythropoiesis is the Friend virus-transformed mouse erythroleukemia (MEL) cells. These cells (Friend *et al.*, 1971) appear to be arrested at the CFU-e or proerythroblast stage but can be maintained indefinitely in culture. Although they have lost the capacity to respond to erythropoietin, they can be induced to differentiate in a multistep process by the addition of various structurally diverse compounds, such as dimethylsulfoxide, butyric acid, aliphatic bisamides, hypoxanthine, actinomycin D, or hemin (see review by Marks and Rifkind, 1978). With all inducers, a lag period lasting from a few to many hours precedes the "commitment" to differentiation and the onset of terminal maturation. Although some erythroleukemic cell variants that respond to erythropoietin have been obtained (see references in Broudy *et al.*, 1988), the recent demonstration that proerythroblasts, isolated more than 95% pure from the spleens of mice infected with the anemia strain of the Friend virus (FVA), can terminally differentiate in response to erythropoietin (Koury *et al.*, 1984) has greatly facilitated studies of the hormonal regulation of terminal erythroid cell differentiation. Erythropoietin-responsive mutants of virally transformed avian erythroid cells have also been extensively employed (Graf and Beug, 1983). Other model systems often used include two human erythroleukemic cell lines that respond to some inducers, e.g., butyric acid or hemin, but not to erythropoietin: the K562 cells, which synthesize embryonic and fetal globins (Rutherford *et al.*, 1979a), and the HEL cells, which synthesize only fetal ($\alpha_2\gamma_2$) globin (Martin and Papayannopoulou, 1982).

2. CHANGES IN NUCLEIC ACID SYNTHESIS AND DEGRADATION

2.1. Overall Changes in DNA and RNA Synthesis

There are several differences in the pattern of nucleic acid synthesis during the terminal differentiation of immature erythroid cells derived from hematopoietic tissues and MEL cells treated with inducers.

In bone marrow erythroblasts, fractionated at different stages of development, synthesis of DNA continues at a constant rate per cell between the proerythroblasts and the late polychromatic erythroblasts and then declines precipitously in the orthochromatic cells (Denton and Arnstein, 1973) (Figure 2). The pattern of RNA synthesis is broadly similar, although there is a progressive decrease in the rate as the cells approach the final cell division. Addition of erythropoietin to hormone-responsive cells leads to a rapid increase in the synthesis of RNA (Gross and Goldwasser, 1969; Maniatis *et al.*, 1973; Bonanou-Tzedaki *et al.*, 1984; Bondurant *et al.*, 1985) and DNA (Gross and Goldwasser, 1970; Nijhof and Wierenga, 1983; Bonanou-Tzedaki *et al.*, 1987). This increase is due to the effects of the hormone on cell proliferation, peaks about 24 hr later (Bondurant *et al.*, 1985; Bonanou-Tzedaki *et al.*, 1987), and is followed by a decrease in nucleic acid synthesis as the cells mature. Most of the initial hormone-dependent increase in RNA

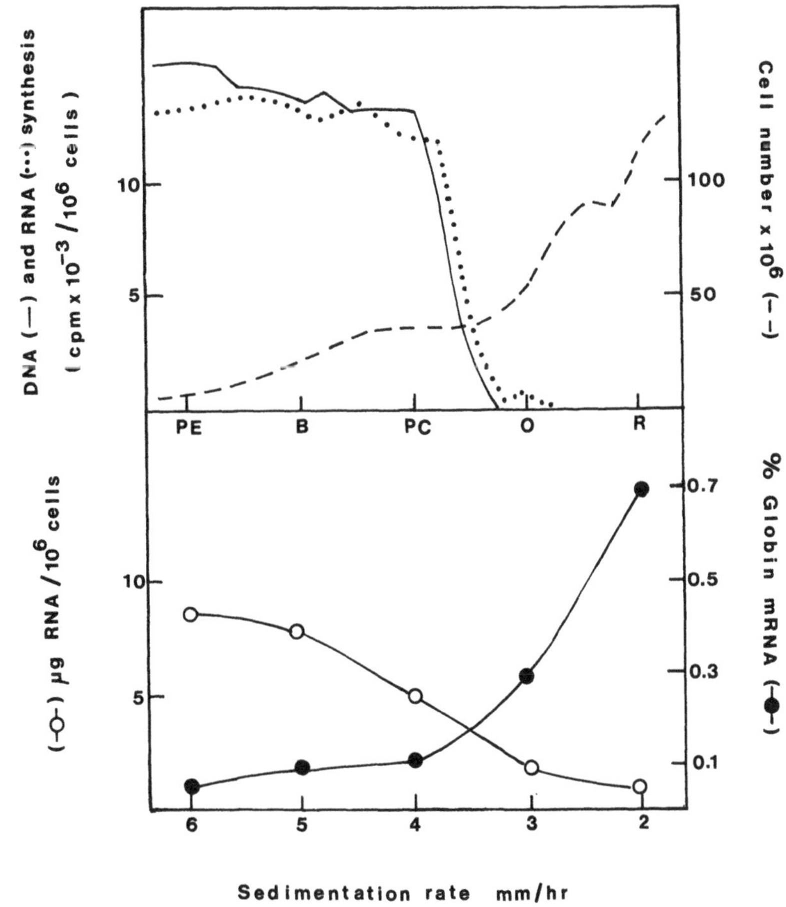

FIGURE 2. Changes in nucleic acid synthesis and content during rabbit erythroid cell development. Anemic rabbit bone marrow erythroid cells were separated by velocity sedimentation at unit gravity and 25 fractions were collected. DNA and RNA synthesis were determined in each fraction by the incorporation of [^{3}H]thymidine and [^{3}H]uridine into acid-insoluble radioactivity, the RNA content was measured by the orcinol reaction, and the percentage of total RNA present as globin mRNA was calculated by hybridization to globin cDNA. Different regions of the gradient contained predominantly the cell types indicated: PE, proerythroblasts; B, basophilic erythroblasts; PC, polychromatic erythroblasts; O, orthochromatic erythroblasts; R, reticulocytes. (Results adapted from Denton and Arnstein, 1973, and Clissold *et al.*, 1977.)

synthesis, which can be detected as early as 15 min (Gross and Goldwasser, 1969), is due to the increased synthesis of ribosomal RNA, since this accounts for more than 80% of all RNA synthesis, but transcription of mRNA sequences appears to be equally stimulated. Thus, a 5-hr incubation of immature rabbit erythroblasts with erythropoietin produced a 50% increase in both ribosomal, poly(A)$^+$, and globin mRNA synthesis whereas incubation with 50 μM hemin produced a preferential increase in the synthesis of poly(A)$^+$ and globin mRNA sequences (Bonanou-Tzedaki *et al.*, 1984).

When inducers are added to actively growing and dividing MEL cells, the initiation of their terminal differentiation program is characterized by an initial slight decrease in DNA synthesis, due to the transient delay of cells entering the S phase of the cell cycle (Marks and Rifkind, 1978; Reuben *et al.*, 1980), and DNA synthesis continues at about 75% of the rate of the control uninduced cultures until 50–60 hr, when it decreases as the cells undergo terminal cell division. There is evidence that structural changes in DNA accompany this process (Marks and Rifkind, 1978). During MEL cell differentiation, there is a progressive decrease in RNA content, and a decrease in the rate of total RNA synthesis has been detected as early as 1.5 hr. Production of ribosomal RNA is selectively reduced (Sherton and Kabat, 1976), and a reduction in the rates of accumulation and half lives of tRNA (Kleiman *et al.*, 1980) and of small nuclear RNAs (Kunzio and Goldstein, 1981) has been observed. It has been suggested that a reduction in the RNA content relative to the protein content of MEL cells is a prerequisite for the triggering of differentiation by inducers (Harel *et al.*, 1981) and it is interesting that inhibitors of RNA synthesis, e.g., actinomycin D, are potent inducers of differentiation (Terada *et al.*, 1978), whereas tumor promoters inhibit terminal cell maturation (Fibach *et al.*, 1979). However, it should be remembered that in MEL cells accumulation of hemoglobin and cessation of cell division can be independently regulated since induction with hemin increases globin gene expression without producing terminal cell division (Marks and Rifkind, 1978; Lowenhaupt and Lingrel, 1979).

2.2. Changes in Globin mRNA

All globin genes consist of three exons and two introns and the α-like and β-like genes are located in different chromosomes (Karlsson and Nienhuis, 1985). Transcriptional activation of the globin genes is associated with characteristic changes in chromatin in the vicinity of the α- and β-globin gene transcription units, involving altered nucleosome spacing, demethylation of certain DNA sequences, and the appearance of DNase I and S1-nuclease hypersensitive sites (Harrison, 1984; Yu and Smith, 1985; Benegra *et al.*, 1986). Such changes, although necessary, are not by themselves sufficient for initiation of transcription as they are already present in erythroid cell progenitors (CFU-e), in uninduced MEL cells, or in unexpressed adult globin genes during embryonic development. It has been demonstrated recently that erythroid cell-specific gene expression is regulated by a factor that binds to a core hexanucleotide DNA sequence in a stage-independent manner (Tsai *et al.*, 1989). There is increasing evidence that globin mRNA synthesis is developmentally controlled by erythroid cell-stage-specific *trans*-acting factors that interact with *cis*-controlling elements present in and around the α- and β-globin gene clusters and that the α- and β-globin genes are regulated differently (Wall *et al.*, 1988; Mantovani *et al.*, 1988; Kim *et al.*, 1988). Detailed description of the transcriptional regulation of globin genes or of the processing of the globin mRNA precursors (Bastos and Aviv, 1977) is beyond the scope of this chapter.

2.2.1. Globin mRNA Synthesis in Hematopoietic Cells

Induction of transcription of globin mRNA in response to erythropoietin occurs at the CFU-e/proerythroblast stage and globin mRNA becomes detectable within 2–4 hr of hormone addition (Nijhof *et al.*, 1987; Bondurant *et al.*, 1985). In erythroblasts isolated from FVA-infected mice, globin gene transcription reaches a maximum after 36 hr with erythropoietin and is at least 25 times the initial rate, which is virtually undetectable (Bondurant *et al.*, 1985). In bone marrow erythroblasts from anemic rabbits, hemin increases the relative rate of globin mRNA synthesis 5-fold, compared to a 2-fold increase produced by erythropoietin, after 4 6 hr incubation (Bonanou Tzedaki *et al.*, 1984), although transcription of other poly(A)$^+$ RNAs is also increased. In these isolated anemic rabbit bone marrow proerythroblasts, there is already some ongoing transcription of globin genes (Clissold *et al.*, 1977), presumably because of the previous exposure of the bone marrow cells to erythropoietin in the intact animals (Conkie *et al.*, 1975). The level of globin mRNA rises from approximately 7000 molecules/cell to approximately 10,000–15,000 molecules/cell in the basophilic/polychromatic erythroblast and reaches 20,000–25,000 molecules/cell after cell division stops. Since this increase in globin mRNA is accompanied by an approximately 2-fold decrease in cell volume at each division as proerythroblasts develop into orthochromatic cells, with a consequent 10-fold overall reduction in cell volume, the concentration of globin mRNA in the orthochromatic erythroblast and reticulocyte is at least 30-fold greater than that found in proerythroblasts, and in reticulocytes globin mRNA accounts for more than 90% of the poly(A)$^+$ RNA.

The amounts of α- and β-globin mRNA present in rabbit reticulocytes are not equal. There are two α- and one β-globin gene (Hunt, 1976) but unequal transcription results in only 1.3 (Mezl *et al.*, 1979) to 1.5 (Kazazian *et al.*, 1974) times more α than β mRNA. However, almost equimolar amounts of α- and β-globin are synthesized due to translational control (Lodish, 1976). (See Section 3.3.1.)

2.2.2. Globin mRNA Synthesis in MEL Cells

In MEL cells globin gene transcripts can be detected after a lag period of about 30 hr following addition of dimethylsulfoxide, but after only 4 hr when hemin is used as the inducer (Ross and Sautner, 1976). Hemin simultaneously initiates the rapid accumulation of both α- and β-globin mRNA whereas with dimethylsulfoxide, hexamethylene-bisacetamide, or butyric acid, β-globin mRNA sequences appear after 20–24 hr and α-globin mRNA can be detected by 8–12 hr (Nudel *et al.*, 1977). The temporal difference in the appearance of α- and β-globin RNA accounts for the observation that early in the induction with dimethylsulfoxide (at 30 hr) α mRNA is present in a fourfold excess over β mRNA; both species, however, are present in equal amounts by 96 hr (Orkin *et al.*, 1975).

Different inducers of MEL cell differentiation cause differential expression of the β_{major} and β_{minor} genes but the molecular basis for this is not known. MEL cells cultured with dimethylsulfoxide or hexamethylene-bisacetamide for 4 days accumulate two to three times more Hb_{major} than Hb_{minor} whereas hemin induces the appearance of almost exclusively β_{minor} globin (Rovera *et al.*, 1977; Nudel *et al.*, 1977).

The accumulation of globin mRNA in MEL cells appears to be controlled almost entirely at the transcriptional level when dimethylsulfoxide is used as an inducer, leading to a 15- to 25-fold increase in the transcription of α and β major globin genes by day 4 (Ganguly and Skoultchi, 1985). The rate of synthesis of β-globin mRNA was estimated to be 5 molecules/min per cell (Curtis *et al.*, 1977). Butyric acid and hexamethylene-bisacetamide also appear to act by increasing the rate of transcription of globin genes (Profous-Juchelska *et al.*, 1983). However, posttranscriptional but pretranslational events, e.g., stabilization of nuclear transcripts and increased export to the cytoplasm, appear to contribute to the hemin-induced differentiation (Profous-Juchelka *et al.*, 1983), which can increase the level of globin mRNA per cell from 1000 to 12,000 molecules (Lowenhaupt and Lingrel, 1978). In contrast, in K562 cells induced to differentiate with hemin the accumulation of globin mRNA is increased three- to fivefold primarily as a result of increased globin gene transcription (Charnay and Maniatis, 1983). Glucocorticoids inhibit both the transcription and translation of globin mRNA (Lo *et al.*, 1978; Kaneda *et al.*, 1985), whereas certain sex steroids facilitate the expression of globin genes in dimethylsulfoxide-treated MEL cells.

2.2.3. Globin mRNA Degradation

Globin mRNA is very stable (Aviv *et al.*, 1976; Lowenhaupt and Lingrel, 1979) and its half-life does not change during differentiation. In MEL cells the $t_{1/2}$ of globin mRNA has now been revised upwards to a value greater than 60 hr (Volloch and Housman, 1981), which is equal to that observed for ribosomal RNA, and greater than the $t_{1/2}$ of all other mRNAs present (Krowczynska *et al.*, 1985). Likewise, in rabbit bone marrow erythroid cells globin mRNA is much more stable than other bone marrow messages (Langstaff and Arnstein, 1985).

This increased stability of globin mRNA is largely independent of the intracellular environment (Bastos *et al.*, 1977) and is probably inherent in its structure, as suggested by the fourfold shorter half-life of human δ-globin mRNA, compared with β-globin mRNA, which may be related to differences in the 3′ untranslated regions of the messengers (Ross and Pizarro, 1983). The importance of 3′-specific AU-rich sequences in the degradation of other messengers is well documented (Shaw and Kamen, 1986), but recently 5′ sequences were also found to be important for globin mRNA stability (Kabnick and Housman, 1988).

In spite of earlier demonstrations (Huez *et al.*, 1981; Marbaix *et al.*, 1975) that deadenylated globin mRNA is less stable than its native counterpart when injected into *Xenopus* oocytes or HeLa cells, the extent of polyadenylation does not appear to be a critical factor in the changes in globin mRNA primary structure that affect stability. Thus, the original idea (Brawerman, 1981) of a relationship between mRNA translation, poly(A) length, and mRNA decay has not been validated experimentally (Krowczynska *et al.*, 1985). The exact determinants of mRNA stability remain to be elucidated (Brawerman, 1987; Ross, 1988).

There is little information about the mechanism of globin mRNA degradation. Experiments with a lysosomal inhibitor, chloroquin, suggest that in bone marrow erythroid cells there appear to be at least two degradative pathways, involving a lysosomal and a nonlysosomal system with different specificity for different messages (Langstaff and Arnstein, 1985).

In conclusion, it appears that the increased transcription of globin genes and the very long half-life of globin mRNA, combined with the decreased transcription and destabilization of most other mRNAs (described in Section 2.3), account for the observed selective accumulation of globin mRNA during terminal erythroid cell maturation. In this respect the effect of erythropoietin and other inducers of erythroid cell differentiation resembles the action of developmental hormones in other cells. For example, in *Xenopus* hepatocytes, estrogen not only induces the transcription and stabilization of vitellogenin mRNA (Brock and Shapiro, 1983) but also reduces the transcription and stability of albumin mRNA (Wolfe *et al.*, 1985).

2.3. Changes in Nonglobin mRNAs

2.3.1. General Considerations

In MEL cells treated with dimethylsulfoxide, induction of differentiation is accompanied by an overall reduction in RNA synthesis and by a decrease in the appearance of poly(A)$^+$ RNA in the cytoplasm (Shaul *et al.*, 1982). The concentration of most nonglobin mRNAs decreases but that of certain mRNAs is selectively increased. Volloch *et al.* (1987) showed that both nonglobin mRNA synthesis and degradation change during differentiation and that these changes are instrumental in establishing cellular concentrations of specific mRNAs. They found that structural changes observed in some mRNA species during transition from the undifferentiated to the differentiated state occur before the export of mature mRNAs to the cytoplasm and probably result in changes in secondary structure. The stabilities of nonglobin mRNAs fall within two classes, M_1 and M_2, having half lives of approximately 3 and 37 hr, respectively (Aviv *et al.*, 1976). About 90% of the nonglobin mRNAs belong to the M_1 class, while messengers for certain "housekeeping" proteins belong to the M_2 class. It is the decay of these short-lived nonglobin mRNAs, after the final cell division, that accounts for the increase in the relative concentration of globin mRNA. The mRNA degradation process is not clearly linked to the loss of the poly(A) sequences (Krowczynska *et al.*, 1985), but appears to depend on as yet ill-characterized structural features. As may have been expected from the loss of proliferative capacity, the levels of the major (replication-dependent) class of histone mRNAs decrease during MEL differentiation (Brown *et al.*, 1985), whereas the mRNA levels of minor histone variants are either unaffected or are differentiation specific (Brown *et al.*, 1988). Histone mRNA concentrations are regulated at the level of both transcription and mRNA degradation (Sittman *et al.*, 1983), and the rapidly degraded cell-cycle-regulated histone mRNAs lack a poly(A) tail (Ross, 1988).

2.3.2. Erythroid Cell-Specific Nonglobin mRNAs

The expression of the erythroid cell differentiation program requires that, besides the overall decrease in nonglobin mRNAs and selective increase in globin mRNAs, certain messengers are either found exclusively in erythroid cells, or accumulate during specific stages in development.

Attempts at identifying such mRNAs have been made by constructing a cDNA library from 14-day mouse fetal liver cell mRNAs and isolating recombinant cDNAs containing transcripts of nonglobin mRNAs expressed at relatively high levels in erythroblasts compared to other cell types (Affara *et al.*, 1983). Although two such mRNAs were found, 12 S mRNA coding for a 16- to 19-kDa protein and 16 S mRNA coding for a 19-kDa protein (Harrison, 1984), the identity of the proteins is still unknown. The latter polypeptide survives the reticulocyte stage and its mRNA is present at high concentrations (1% of the total mRNA population) in erythroblasts, and also in liver and kidney cells, whereas its level in other tissues tested is 10- to 100-fold lower. Similarly, glutathione peroxidase mRNA accumulates early in erythropoiesis, but is also expressed in liver and kidney cells in contrast to other tissues. The mRNA for lipoxygenase, which represents about 1% of all poly(A)$^+$ RNAs in peripheral blood reticulocytes, is also found in bone marrow erythroblasts, but not in any other tissue tested (Affara *et al.*, 1985). Synthesis of red cell lipoxygenase is regulated by the unmasking of translationally inactive mRNP particles at the reticulocyte stage (Thiele *et al.*, 1982).

The expression of carbonic anhydrase mRNAs has also been studied since there is a parallel increase in carbonic anhydrase and globin both in induced MEL cells (Conscience and Meier, 1980) and in differentiating rabbit erythroblasts (Denton *et al.*, 1975). Using the rabbit cDNA probe isolated by Konialis *et al.* (1985), we found (Ong, 1987) that the concentration of the erythroid cell-specific carbonic anhydrase I mRNA rises during rabbit erythroid cell development, reaches a peak at the orthochromatic cell stage, and then declines in the reticulocyte. The contribution of transcriptional activation and mRNA stability to these qualitative changes has not been determined.

Studies on the expression of genes coding for cytoskeletal membrane proteins have demonstrated some interesting coordinate changes during development. The α- and β-spectrin mRNAs, which are barely detectable in uninduced MEL cells, increase during the first 4 days of induction and thereafter remain constant (Lehnert and Lodish, 1988). In contrast, band 3 mRNA is first detectable in more mature erythroblasts on day 4 of differentiation, whereas actin mRNA levels remain relatively constant. The expression of the chicken erythroid cytoskeletal membrane protein 4.1 variants is also developmentally regulated, and different protein variants are synthesized and accumulate in "early" and "late" cells. This is the result of both differential mRNA splicing and translational control mechanisms (Ngai *et al.*, 1987).

With regard to enzymes of the heme pathway, an early increase in dimethylsulfoxide-treated MEL cells in the activities of δ-aminolevulinic acid synthase and dehydratase was interpreted as being due to transcriptional activation (Sassa, 1976; Chang and Sassa, 1988). A coordinate accumulation of prophobilinogen deaminase and uroporphyrinogen decarboxylase mRNAs was reported to start 6–12 hr after induction and to continue until 48 hr, preceding the accumulation of globin mRNAs, which occurred dramatically between 48 and 96 hr (Grandchamp *et al.*, 1985). Similarly, in bone marrow erythroblasts the erythropoietin-induced increase in porphobilinogen deaminase activity, which parallels the increase in heme synthesis, was also interpreted as due to new gene transcription (Beru and Goldwasser, 1985). The erythroid-specific expression of δ-aminolevulinic acid synthase and porphobilinogen deaminase was due to different genes or alternative transcription (Riddle *et al.*, 1989; Chretien *et al.*, 1988), respectively. Iron is delivered to erythroid cells by transferrin, via receptor-mediated endocytosis, and is stored in ferritin. The level of intracellular iron and heme negatively regulates the level of transcription of transferrin receptor mRNA, which has a $t_{1/2}$ of less than 35 min, thereby determining the level of expression of the transferrin receptor (Rao *et al.*, 1985). During MEL cell differentiation a biphasic increase in the levels of both H and L ferritin mRNAs was observed, but this was not reflected in increased protein synthesis (Beaumont *et al.*, 1987). In contrast to the transferrin receptor, the biosynthetic rates of both ferritin chains are regulated by iron at the translational level, although the same iron-responsive element is present in the 3′ and 5′ untranslated region of the transferrin receptor mRNA and ferritin mRNA, respectively (Casey *et al.*, 1988).

2.3.3. Oncogene mRNAs

It is now recognized that cell proliferation and differentiation is associated with changes in the expression of certain cellular oncogenes. Induction of differentiation of MEL cells by dimethylsulfoxide produces a sequential transient expression of c-fos, c-myb, c-myc, and K-ras protooncogenes (Todokoro and Ikawa, 1986). Expression of c-myc is already high in uninduced cells, decreases 15- to 20-fold 2–4 hr after the addition of dimethylsulfoxide, returns to normal levels by 10–20 hr, and declines again during the next 4–5 days, as the cells terminally differentiate (Lachman and Skoultchi, 1984; Kirsch *et al.*, 1986). Both c-myb and c-myc are expressed at high levels in immature bone marrow erythroblasts from humans (Emilia *et al.*, 1986) and chickens (Coll *et al.*, 1983). Expression of the c-myc oncogene decreases during the differentiation of human BFU-e cells (Umemura *et al.*, 1986), whereas continuous suppression of c-myb expression is required for the commitment of MEL cells to terminal cell division (Ramsay *et al.*, 1986).

The effect of MEL cell differentiation on c-myc and c-myb gene expression is due to premature termination of transcription and increased turnover of mRNA (Watson, 1988). c-myc mRNA has a short half-life (~ 30 min) even before induction (Mechti *et al.*, 1986). Paradoxically, however, the decrease in c-myc mRNA transcripts was not reflected in a corresponding fall in the level of c-myc protein, the $t_{1/2}$ of which was increased from less than 30 to over 75 min (Wingrove *et al.*, 1988). Determinants that contribute to the cytoplasmic stability of c-onc mRNAs reside both at the 3′ end (Shaw and Kamen, 1986) and at the 5′ end (Kabnick and Housman, 1988). There is little information as yet about the nature of the erythroid differentiation "signals" and factors and the protooncogene sequences with which they interact, but since c-myc, c-myb, and c-fos code for nuclear DNA-binding proteins (Bishop, 1985), changes in their levels of expression may be expected to affect transcription of other genes. The mRNA for the transformation-associated protein p53 also declines rapidly during MEL cell differentiation, and its regulation is posttranscriptional (Khochbin *et al.*, 1988). In contrast, the c-erbA protooncogene, which encodes the receptor for thyroid hormone (Sap *et al.*, 1986), is preferentially transcribed in chicken erythrocytic cells during late stages of differentiation (Hentzen *et al.*, 1987), and its expression is necessary for blockage of erythroid differentiation in avian erythroblastosis virus-induced chicken erythroleukemias.

3. CHANGES IN PROTEIN SYNTHESIS AND DEGRADATION

3.1. Nuclear Proteins

In eukaryotes, most histone synthesis is coordinated with DNA synthesis (Brown *et al.*, 1985) and both decline in concert at the time of nuclear condensation in the polychromatic/orthochromatic erythroblast. Besides the ubiquitous histones H_1, H_{2A}, H_{2B}, H_3, and H_4, another histone, H_5, has been found in the nucleated erythrocytes of birds, amphibians, and fish. This histone (Smith *et al.*, 1980) is known to be very similar to H_{1^o}, a variant of histone H_1, which is involved in gene repression (Weintraub, 1984), and its synthesis, which is not coupled to DNA synthesis, increases during avian erythroid cell maturation (Trainor *et al.*, 1987). The differential accumulation of histone H_5 reflected an increase in the transcriptional activity of the H_5 gene, a longer half-life of H_5 mRNA, and a higher stability of the H_5 protein (Affolter *et al.*, 1987). Posttranslational modifications of histones, especially hyperacetylation and ubiquitination, have been correlated with the presence of transcriptionally active chromatin (Gross and Garrard, 1987; Alonso *et al.*, 1987). During rabbit erythroid cell development, histone acetylation decreased in parallel with histone and DNA synthesis, but some acetylation persisted in the late orthochromatic cells (Threadgill and Arnstein, 1985).

Greater interest has centered on the role of nonhistone proteins in differentiation-specific transcription. There is a striking absence of qualitative changes in nonhistone chromatin proteins of induced MEL cells (Peterson and McConkey, 1976) except for a specific basic nonhistone protein termed IP_{25} (Keppel *et al.*, 1977), which may be related to the histone variant H_{1°. Synthesis of a similar protein, of M_r 20,000, also changed during rabbit erythroid cell development, increasing markedly in polychromatic cells, i.e., just before condensation of the nucleus (Threadgill and Arnstein, 1984). The precise involvement of this or other proteins such as hemoglobin, which has also been detected in the nucleus (Granick and Levere, 1965; Stohlman, 1970), and the molecular mechanisms operating in the transcription of erythroid-specific genes and in nuclear condensation are, however, still obscure. The existence of erythroid-specific nuclear factors (Mantovani *et al.*, 1988; Wall *et al.*, 1988; Evans *et al.*, 1988; Kim *et al.*, 1988; Galson and Housman, 1988; Tsai *et al.*, 1989) has been demonstrated recently. It will be of great interest to determine whether these proteins are newly synthesized or modified upon induction and whether they exert positive or negative regulation during erythroid cell differentiation. Elucidation of the role of the nuclear proteins coded by different protooncogenes in development is also wanting, but it is worth noting that different c-myc isoforms are expressed during MEL cell differentiation (Wingrove *et al.*, 1988).

3.2. Membrane Proteins

3.2.1. Membrane-Skeletal Proteins

The terminal differentiation of immature erythroblasts entails the restructuring of the plasma membrane and the formation of the membrane skeleton, a network of actin-associated proteins interacting with the cytoplasmic side of the membrane, which determines the shape and deformability of the erythrocyte (Marchesi, 1985; Lazarides, 1987). This is a complex process, both structurally and temporally, and the most dramatic changes occur during the maturation of orthochromatic erythroblasts to reticulocytes, with spectrin, band 3, and ankyrin segregating away from the extruded cell nucleus (Geiduschek and Singer, 1979).

The major component of the membrane skeleton is spectrin, which accounts for 30% of its protein. It is composed of two nonidentical polypeptides, α- and β-spectrin, which associate to form $(\alpha\beta)_2$ tetramers. Spectrin is linked to two transbilayer plasma membrane proteins, band 3 (the major anion exchange protein) and glycophorin, via two intermediate peripheral proteins, ankyrin and band 4.1, respectively. Protein 4.1 also interacts with actin. Until recently, spectrin, ankyrin, and band 4.1 were thought to be erythroid cell-specific proteins, but antigenically and structurally related polypeptides have now been detected in a variety of nonerythroid cells. Only the glycophorins still appear to be confined to terminally differentiating erythroblasts.

Biosynthesis of plasma membrane and skeletal proteins is asynchronous and controlled in a variety of ways (Pfeffer *et al.*, 1986). In MEL cells, synthesis of spectrin is barely detectable in uninduced cells, increases during induction to a peak by day 4, and remains at that level until after enucleation (Lehnert and Lodish, 1988). In both induced and uninduced cells, the relative synthesis of α-/β-spectrin is 1 : 3, but whereas in uninduced cells they are both degraded with a similar half-life of 10 hr, in differentiated cells α-spectrin is stabilized and the half lives of α- and β-spectrin are 22 and 8 hr, respectively. Thus, the end accumulation of equimolar amounts of α- and β-spectrin is caused by their unequal synthesis and degradation. In contrast, accumulation of band 3 mRNA and synthesis of the anion transporter protein begins on day 4 and continues to day 7, increasing the amount of band 3, relative to that in mature erythrocytes, from 3% in uninduced cells to 50% by day 4 and to 90% by day 7. Both band 3 mRNA and protein are very stable. In FVA-transformed splenic erythroblasts there is some ongoing synthesis of α- and β-spectrin before culturing with erythropoietin, but this increases soon after exposure to the hormone (Koury *et al.*, 1986). Synthesis of band 3 and accumulation of band 4.1 can only be detected after the addition of erythropoietin and reach a peak 48 hr later, paralleling heme and globin synthesis. Glycophorins are also maximally synthesized at the miderythroblast stage and their synthesis declines in the reticulocyte (Kasturi and Harrison, 1985). Minor differences exist between birds and mammals in the assembly of proteins in the membrane skeleton, which in birds is regulated posttranslationally by the availability of band 3. In virally transformed avian erythroid cell lines, the anion transporter is expressed in proliferating cells and after the final cell division any anion transporter molecules that have not been incorporated into the plasma membrane are degraded (Woods *et al.*, 1986). Unassembled spectrin subunits are also rapidly catabolized, β-spectrin ($t_{1/2} \sim 15$ min) by cytoplasmic and α-spectrin ($t_{1/2} \sim 2$ hr) by lysosomal proteases (Woods and Lazarides, 1985). Any excess unincorporated ankyrin and 4.1 proteins are degraded in mature erythrocytes. In avian erythroblasts, synthesis of spectrin, ankyrin, and band 4.1 is present at the CFU-e stage and continues throughout development. The concentration of actin is highest at the CFU-e stage, drops drastically by the polychromatic cell stage, and declines further thereafter. Interestingly, gelsolin, an actin modulator, is down-regulated by a factor of 10^3 during chicken erythroid development (Hinssen *et al.*, 1987).

3.2.2. Membrane Receptors

Erythroid cells express a variety of receptors on their plasma membrane. The recent availability of [^{125}I]erythropoietin has allowed the identification of its receptor by binding studies. The receptor is present in immature erythroblasts (CFU-e) obtained from mouse (Sawyer *et al.*, 1987; Fukamachi *et al.*, 1987), rat (Mayeux *et al.*, 1987), or human (Sawada *et al.*, 1987; Frazer *et al.*, 1988) hematopoietic organs and in a variety of mouse and human erythroleukemic cells (Broudy *et al.*, 1988, and references therein). The receptor binds erythropoietin with high affinity, appears to consist of two polypeptides, and is found at approximately 1000 copies/cell. The murine erythropoietin receptor has recently been cloned (D'Andrea *et al.*, 1989).

The number of erythropoietin receptors per cell decreases during terminal erythroid cell differentiation (Sawada *et al.*, 1987; Fukamachi *et al.*, 1987; Frazer *et al.*, 1988) and no receptors can be detected in reticulocytes or erythrocytes (Mayeux *et al.*, 1987), a finding that remains to be reconciled with the specific effects of the hormone on certain ionic and transport properties (Lawrence *et al.*, 1987; Ghosal *et al.*, 1987) of these mature cells. There is virtually no information as yet about the synthesis and stability of either the erythropoietin receptor or its mRNA, or whether the liganded receptor, which is rapidly endocytosed (Sawyer *et al.*, 1987), is degraded or recycled, but we have evidence for the former since erythropoietin binding is rapidly lost from cycloheximide-treated erythroblasts (S. A. Bonanou-Tzedaki, unpublished observations). Down-regulation of the erythropoietin receptor was also observed when responsive cells were cultured with erythropoietin, IL-3, or phorbol esters (McCaffery *et al.*, 1987; Broudy *et al.*, 1988), whereas up-regulation occurred when MEL cells were induced with dimethylsulfoxide.

The transferrin receptor is already expressed at a high level in human BFU-e cells (Umemura *et al.*, 1986) and in FVA-transformed mouse CFU-e cells at approximately 200,000 sites/cell (Sawyer and Krantz, 1986). Upon culture of these erythroblasts with erythropoietin, synthesis of transferrin receptors increased rapidly and the number of binding sites doubled after 24 hr but decreased to 100,000 sites/cell after 48 hr, as the cells differentiated. Transferrin receptors were still present in reticulocytes but were totally lost from the cell by the mature erythrocyte stage. Cellular uptake of iron from transferrin was limited by the number of transferrin receptors as well as by the level of endocytosis of receptors. Both processes, as well as the release of iron from endocytosed transferrin, were inhibited by heme (Ponka *et al.*, 1988). Evidence for the existence of a separate receptor for heme has been reported (Galbraith *et al.*, 1985). In MEL cells, there were about 10,000 heme binding sites/cell and this number increased during dimethylsulfoxide-induced differentiation. In these cells, iron incorporated into hemoglobin and ferritin could be provided by both transferrin and extracellular hemin (Fibach *et al.*, 1987).

Erythroid cells respond to a variety of hormones and growth factors (see Section 1.4) and therefore presumably express the corresponding receptors. In rabbit erythroblasts, there are about 1000 and 500 β_2-adrenergic receptors per dividing and nondividing cell, respectively (Setchenska and Arnstein, 1983a). After the final cell division, the receptors persist at the same density but become uncoupled from the adenylate cyclase complex. Interestingly, induction of differentiation of Rauscher erythroleukemia cells by erythropoietin resulted in a specific increase in the number of β-adrenergic receptors, which were also not functionally linked to adenylate cyclase (Sytkowsky and Kessler, 1984). Insulin receptors, present at 20,000 and 8000 sites/cell on immature and mature rabbit bone marrow erythroid cells, respectively (Thomopoulos *et al.*, 1980), also decline during reticulocyte maturation but are still present in erythrocytes (Dons *et al.*, 1981). In contrast, the non-plasma-membrane-associated glucocorticoid receptors are present in erythroid cell progenitors, increase in number in basophilic cells, and are absent in erythrocytes (Billat *et al.*, 1981). It can be concluded, therefore, that although, with the exception of the transferrin receptor, little is known about the kinetics of synthesis and degradation of erythroid cell receptors, many receptors appear to be lost or uncoupled from their effector systems at or after the nuclear extrusion stage.

3.2.3. Membrane-Associated Enzymes

Developmentally regulated changes in various enzymatic activities present on erythroid cell membranes have been observed in many instances but there is very little information on the synthesis or degradation of the relevant proteins. Thus, induction of MEL cell differentiation produced a sustained increase in the activity of acetylcholine esterase (Conscience and Meier, 1980), and addition of erythropoietin to FVA-infected spleen erythroblasts resulted in an increase in glucose transport (Koury *et al.*, 1987). The activity of adenylate cyclase, on the other hand, declined continuously during differentiation (Setchenska and Arnstein, 1983b) and became uncoupled from β-adrenergic receptors in rabbits but not in rats, mice, or humans, in which hormone responsiveness persisted in erythrocytes. Addition of erythropoietin to immature rabbit erythroblasts produced a transient activation of adenylate cyclase, which was hormone concentration-dependent and subject to desensitization (Bonanou-Tzedaki *et al.*, 1986). It is unlikely, however, that the hormonal effects are mediated solely by increased cAMP levels (Bonanou-Tzedaki *et al.*, 1987).

A variety of protein kinases are present on the plasma membrane of erythrocytes and have been extensively studied (Boivin, 1988), but much less is known about their changes in activity during erythroid cell differentiation. In rabbit erythroblasts, endogenous phosphorylation of membrane proteins was stimulated by cAMP in dividing, but not in nondividing cells, whereas cAMP-independent phosphorylation declined continuously during maturation up to the late orthochromatic stage and remained constant thereafter (Setchenska *et al.*, 1980). The activity of protein kinase C declined during the differentiation of rabbit erythroid (Setchenska *et al.*, 1988) or MEL cells (Balazovich *et al.*, 1987). In contrast, the activity of tyrosine kinase was found to increase after treatment of MEL cells with dimethylsulfoxide (Earp *et al.*, 1983), but to decrease during the hemin-induced differentiation of K562 cells (Richardson *et al.*, 1987).

3.3. Cytosolic Proteins

3.3.1. Hemoglobin

Synthesis of hemoglobin is observed as soon as its mRNA is transcribed and continues for as long as the cell retains its protein-synthesizing apparatus (Stewart *et al.*, 1976). In rabbit erythroblasts, there is a 50-fold increase in the content of hemoglobin per cell during maturation from the proerythroblast to the reticulocyte stage (Arnstein, 1976). In uninduced MEL cells, hemoglobin accounts for less than 1% of all protein synthesis and rises to greater than 25% after 4–5 days' culturing with dimethylsulfoxide (Marks and Rifkind, 1978). In spite of the noncoordinate transcription of globin genes (see Section 2.2.2), which results in approximately 50% more α- than β-globin mRNA, and the faster rate of translation of α-globin mRNA, which in rabbit reticulocytes incubated at 37°C is 14 sec, as opposed to 21 sec required for the translation of β-globin mRNA (Hunt *et al.*, 1969), almost equal amounts of α- and β-globin chains are produced because of preferential initiation of translation on β mRNAs (Lodish, 1976). The relative overproduction of α-globin gives rise to a small pool of free α chains that is kept constant by continuous degradation (Hanash and Rucknagel, 1978). When greatly unbalanced synthesis of globin chains occurs, as in various thalassemias, or whenever there is an imbalance of heme and globin synthesis, the excess polypeptide chains aggregate and precipitate, forming insoluble inclusion (Heinz) bodies. Lack of heme promotes the formation of a translational inhibitor of protein synthesis, thereby leading to reduced formation of globin (Ochoa and De Haro, 1979), and to a change in the ratio of α/β-globin synthesized. Besides HbA

($\alpha_2\beta_2$), reticulocytes synthesize small amounts of HbF ($\alpha_2\gamma_2$) and HbA_2 ($\alpha_2\delta_2$). The decrease in fetal relative to adult hemoglobin synthesis in reticulocytes is due to the asynchronous transcription of γ- and β-globin genes (Ross and Sullivan, 1985), whereas the decreased stability of the δ- compared to the β-globin mRNA accounts for the small percentage of $\alpha_2\delta_2$ hemoglobin found in erythrocytes (Ross and Pizarro, 1983). Hemoglobin is a very stable protein ($t_{1/2}$ > life span of the erythrocyte) but some molecular variants undergo spontaneous oxidation and precipitation (unstable hemoglobins).

3.3.2. Cytosolic Enzymes

Although dramatic changes in the activity of various cytosolic enzymes have been observed during erythroid cell maturation, in most cases we know little about the quantitative changes of these proteins. Denton *et al.* (1975) showed that during rabbit erythroid cell development, the activity of four enzymes implicated in the nucleotide "salvage pathway," glucose 6-phosphate dehydrogenase, 6-phosphogluconate dehydrogenase, adenosine deaminase, and nucleoside phosphorylase, declined steeply after the final cell division from a high to a very low level. Lactate dehydrogenase activity also declined, but to a lesser extent, as the cell crossed from the dividing to the nondividing stage, with a dramatic change in the isoenzyme pattern involving the complete disappearance of the M_4 and hybrid isoenzymes, present in immature cells, and the persistence of the H_4 isoenzyme to the reticulocyte stage (Setchenska and Arnstein, 1978). This is due to the greater stability of the H compared to the M polypeptide, since both are synthesized in orthochromatic cells while neither polypeptide is synthesized in reticulocytes (Ong, 1987). Likewise, the activity of malate dehydrogenase declined after the final cell division with loss of the mitochondrial isoenzyme and one of the two cytosolic isoenzymes present in immature erythroblasts (Setchenska and Arnstein, 1979). The activity of glutathione peroxidase and reductase, enzymes protecting cells from lipid peroxidation, decreased 2- to 3-fold during erythroid cell maturation, whereas the activities of catalase and superoxide dismutase, enzymes involved in the metabolism of superoxide and H_2O_2, produced largely by hemoglobin, did not change much in dividing erythroblasts but increased 2- to 3-fold between the orthochromatic and erythrocyte stage (Russanov *et al.*, 1981). Adenylate kinase activity also increased after the final cell division, in keeping with the importance of ATP regeneration during glycolysis in mature cells (Denton *et al.*, 1975). The activity of carbonic anhydrase increased in parallel with hemoglobin up to 20-fold in going from the basophilic erythroblast to the reticulocyte (Arnstein, 1976). Since the increase in activity is largely a reflection of the quantity of enzyme present (Spencer and Peller, 1976), the continuing accumulation of this enzyme up to the reticulocyte stage must be due to preferential translation of carbonic anhydrase mRNA or increased stability of its protein product (or both) late in differentiation, given that the actual concentration of carbonic anhydrase I mRNA was found to decrease between the orthochromatic and reticulocyte stage (see Section 2.3.2).

3.3.3. Enzymes of Heme Biosynthesis

With the exception of four enzymes (δ-aminolevulinic acid synthase, coproporphyrinogen and protoporphyrinogen oxidase, and ferrochelatase) that occur in mitochondria,

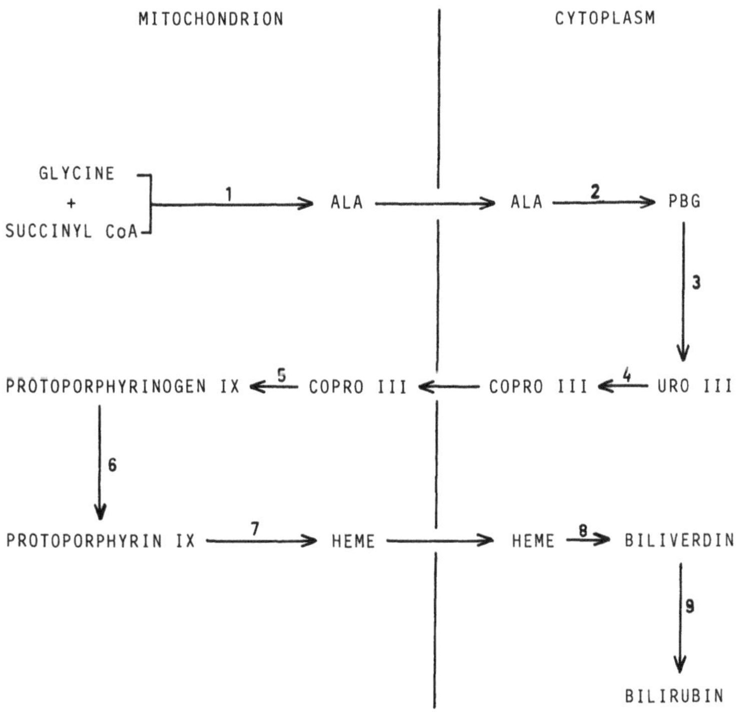

FIGURE 3. Intermediates and enzymes of the heme biosynthetic and degradative pathways. Intermediates: ALA, δ-aminolevulinic acid; PBG, porphobilinogen; URO III, uroporphyrinogen III; COPRO III, coproporphyrinogen III. Enzymes: 1, ALA synthase; 2, ALA dehydratase; 3, PBG deaminase; 4, urodecarboxylase; 5, copro-oxidase; 6, proto-oxidase; 7, ferrochelatase; 8, heme oxygenase; 9, biliverdin reductase.

the remaining enzymes involved in heme synthesis are cytosolic (Figure 3). An asynchrony exists between the synthesis of heme and globin. Heme biosynthesis is already well developed in immature erythroblasts, increases to a peak in basophilic and polychromatic cells, and then declines with further cell maturation, in contrast to the synthesis of hemoglobin, which continues to rise (Glass *et al.*, 1975). The rate-limiting step of heme synthesis is δ-aminolevulinic acid synthase, but in erythroblasts other controlling steps at the level of porphobilinogen deaminase (Beru and Goldwasser, 1985) or ferrochelatase (Rutherford *et al.*, 1979b; Fadigan and Dailey, 1987) have been reported.

In cultures of mouse bone marrow erythroid cell progenitors, the activity of δ-aminolevulinic acid synthase and dehydratase rose progressively, reaching a plateau by about 60 hr and then started to decline (Ibraham *et al.*, 1982), whereas heme oxygenase activity declined up to 60 hr of growth and was elevated at later culture periods, as the cells matured. Maximum heme synthesis was reached at 72 hr, and paralleled the rise in activity of porphobilinogen deaminase and ferrochelatase. The observed increase in the activity of heme enzymes was due to new enzyme synthesis (see also Section 2.3.2) and was suppressed by cycloheximide. A high level of heme inhibits δ-aminolevulinic acid synthase in nonerythroid tissues and in differentiated erythroid cells, but induces its synthesis in immature erythroid cells (Ibraham *et al.*, 1983), as well as stimulating erythroid colony growth (Porter *et al.*, 1979; Mayeux *et al.*, 1986b). Heme also induces the activity of δ-aminolevulinic acid dehydratase and uroporphyrinogen synthase (Granick and Sassa, 1978) and controls the transcription of heme oxygenase (Yoshida *et al.*, 1988), an observation that explains the rise of this enzyme in the later phases of erythroid cell differentiation.

3.4. Other Proteins and Enzymes

Detailed information about the developmental changes in protein levels and enzyme activities associated with different subcellular organelles is lacking. The regulation of enzymes involved in heme biosynthesis and the loss of mitochondrial malate dehydrogenase after the final cell divisions have already been discussed (see Sections 3.3.2 and 3.3.3). The activity of various acid and alkaline ribonucleases declines markedly after cessation of RNA synthesis, with reticulocytes having less than 10% of the activity present in basophilic erythroblasts (Hulea *et al.*, 1975), but the precise location of these enzymes (lysosomal/plasma membrane-bound/cytosolic) has not been established. Similarly, although protein breakdown (McKay *et al.*, 1980; Daniels *et al.*, 1983; Rapoport and Schewe, 1986) and ubiquitin carrier proteins (Pickart and Vella, 1988) are known to decrease during reticulocyte maturation and a soluble ATP-dependent protein degradation system has been reported in MEL cells (Waxman *et al.*, 1985), the activity of a variety of proteolytic systems and their subcellular localization during erythroid cell differentiation have not been systematically investigated. During MEL cell differentiation, there was a decrease in the phosphotyrosine and an increase in the phosphoserine and phosphothreonine acid phosphatases present (Hammond *et al.*, 1985), and a large and selective increase in the synthesis of one of the two species of regulatory subunit of the cAMP-dependent protein kinase was observed (Schwartz and Rubin, 1985). Erythroid cell differentiation was accompanied by a decrease in the activity of the cAMP phosphodiesterase (Setchenska *et al.*, 1981), an enzyme also found in Friend cells (Mason *et al.*, 1985).

There is increasing evidence that the expression of "stress" or "heat-shock" proteins in erythroid cells is inducible and cell stage-specific. Induction of differentiation of K562 cells with hemin led to the accumulation of hsp 70 (Singh and Yu, 1984) and erythroid lineage-specific expression of the inducible hsp 70 was observed during avian embryogenesis (Banerji *et al.*, 1987). Similarly, incubation at 43–44°C of differentiating rabbit erythroblasts resulted in the rapid induction of the major hsp 70 (Bei-Paraskevopoulou *et al.*, 1986), but stress did not induce the major hsp 68 in differentiating mouse erythroleukemic cells, which was constitutively expressed (Aujame, 1988). During erythropoiesis of *Xenopus laevis,* hsp 70 gene transcription was both stage-specific and heat shock-dependent and synthesis of hsp's was controlled at both the transcriptional and the posttranscriptional level (Winning and Browder, 1988).

4. CONCLUDING REMARKS

In this chapter we have attempted to give an account of the major changes in nucleic acids and proteins that are observed during terminal erythroid cell maturation, paying particular emphasis to recent developments in our understanding of how these processes are regulated. Changes in other macromolecules, e.g., carbohydrates, have not been considered due to the scant information available. Likewise, developmental changes in phospholipids or polyamines have not been discussed as they are not macromolecular.

It can be concluded that the greatest changes in macromolecular synthesis and degradation occur at two "critical" stages: at the CFU-e/proerythroblast stage, when the sequential expression of erythroid cell-specific genes is induced, and at the orthochromatic/reticulocyte stage, when nucleic acid synthesis ceases, the plasma membrane is remodeled, and accumulation of erythroid cell-specific proteins occurs. It is also clear that a great deal of data exist on the coordinate changes in the concentration of the various macromolecules that ultimately account for these cellular and biochemical changes. However, in many cases it has not been established whether a change in the amount of a specific nucleic acid or protein is due to an alteration in its rate of synthesis or degradation or, indeed, whether a change in a certain enzyme activity is due to a change in enzyme concentration. More importantly, and perhaps surprisingly for such an intensely studied field, our understanding of the precise molecular mechanisms responsible for the observed changes is far from complete. Thus, in spite of the elucidation of several regulatory processes, which have already been discussed, we do not really know which structural or cellular features determine the stability of different mRNAs and proteins, how various factors control the temporally coordinate expression of erythroid cell-specific genes, what leads to the ultimate repression of DNA and RNA synthesis, or how the same inducer (e.g., erythropoietin or hemin) can produce different effects depending on the developmental stage of the cell. It is in these areas, which are the province of molecular developmental biology in general rather than erythroid cell differentiation in particular, that research is eagerly awaited.

5. REFERENCES

Affara, N., Goldfarb, P. S., Yang, Q.-S., and Harrison, P. R., 1983, Patterns of expression of erythroblast non-globin mRNAs, *Nucleic Acids Res.* **11**:931–945.

Affara, N., Fleming, J., Black, E., Goldfarb, P. S., Thiele, B., and Harrison, P. R., 1985, Analysis of chromatin changes associated with the expression of globin and non-globin genes in cell hybrids between erythroid and other cells, *Nucleic Acids Res.* **13**:5629–5645.

Affolter, M., Cote, J., Renaud, J., and Ruiz-Carrillo, A., 1987, Regulation of histone and beta A-globin gene expression during differentiation of chicken erythroid cells, *Mol. Cell. Biol.* **7**:3663–3672.

Alonso, W. R., Ferris, R. C., Zhang, D. E., and Nelson, D. A., 1987, Chicken erythrocyte α-globin chromatin: Enhanced solubility is a direct consequence of induced histone hyperacetylation, *Nucleic Acids Res.* **15**:9325–9337.

Anagnou, N. P., Yuan, T. Y., Lim, E., Helder, J., Wieder, S., Glaister, D., Marks, B., Wang, A., Colbert, D., and Deisseroth, A., 1985, Regulatory factors specific for adult and embryonic globin genes may govern their expression in erythroleukemia cells, *Blood* **65**:705–712.

Arnstein, H. R. V., 1976, Changes in protein biosynthesis and enzyme activities during erythroid-cell differentiation, *Biochem. Soc. Trans.* **4**:965–968.

Arnstein, H. R. V., Langstaff, J. M., Ong, G., Threadgill, G. J., and Bonanou-Tzedaki, S. A., 1987, Control of macromolecular synthesis and degradation during terminal erythroid cell development, *Biomed. Biochim. Acta* **46**:S115–S119.

Aujame, L., 1988, The major heat-shock protein hsp 68 is not induced by stress in mouse erythroleukemia cell lines, *Biochem. Cell Biol.* **66**:691–701.

Aviv, H., Volloch, Z., Bastos, R., and Levy, S., 1976, Biosynthesis and stability of globin mRNA in cultured erythroleukemic Friend cells, *Cell* **8**:495–503.

Balazovich, K. J., Portnow, P., Boxer, L. A., and Prochownik, E. V., 1987, Changes in protein kinase C activity are associated with the differentiation of Friend erythroleukemia cells, *Biochim. Biophys. Acta* **927**:247–253.

Banerji, S. S., Laing, K., and Morimoto, R. I., 1987, Erythroid lineage-specific expression and inducibility of the major heat shock protein HSP 70 during avian embryogenesis, *Genes Dev.* **1**:946–953.

Bastos, R. N., and Aviv, H., 1977, Globin mRNA precursor molecules: Biosynthesis and processing in erythroid cells, *Cell* **11**:641–650.

Bastos, R. N., Volloch, Z., and Aviv, H., 1977, Messenger RNA population analysis during erythroid differentiation: A kinetic approach, *J. Mol. Biol.* **110**:191–203.

Beaumont, C., Jain, S., Bogard, M., Nordmann, Y., and Drysdale, J., 1987, Ferritin synthesis in differentiating Friend erythroleukemic cells, *J. Biol. Chem.* **262**:10619–10623.

Bei-Paraskevopoulou, T., Bonanou-Tzedaki, S. A. and Arnstein, H. R. V., 1986, The effect of heat shock on differentiating rabbit erythroid cells, *Biochem. Soc. Trans.* **14**:973–974.

Benegra, R., Cantor, C. R., and Axel, R., 1986, Nucleosomes are phased along the mouse β-major globin gene in erythroid and non-erythroid cells, *Cell* **44**:697–704.

Beru, N., and Goldwasser, E., 1985, The regulation of heme biosynthesis during erythropoietin-induced erythroid differentiation, *J. Biol. Chem.* **260**:9251–9257.

Bessis, M., 1973, The erythrocytic series, in: *Living Blood Cells and Their Ultrastructure* (M. Bessis, ed.), pp. 85–87, 110–129, Springer, Berlin.

Beutler, E., 1988, The relationship of red cell enzymes to red cell life-span, *Blood Cells* **14**:69–75.

Billat, C., Felix, J. M., Mayeux, P., and Jacquot, R., 1981, Binding of glucocorticosteroids to hepatic erythropoietic cells of the rat fetus, *J. Endocrinol.* **80**:307–315.

Bishop, J. M., 1985, Viral oncogenes, *Cell* **42**:23–38.

Boivin, P., 1988, Role of the phosphorylation of red blood cell membrane proteins, *Biochem. J.* **256**:689–695.

Bonanou-Tzedaki, S. A., Sohi, M., and Arnstein, H. R. V., 1981, Regulation of erythroid cell differentiation by hemin, *Cell Differ.* **10**:267–279.

Bonanou-Tzedaki, S. A., Sohi, M. K., and Arnstein, H. R. V., 1984, The effect of hemin on RNA synthesis and stability in differentiating rabbit erythroblasts, *Eur. J. Biochem.* **144**:589–596.

Bonanou-Tzedaki, S. A., Setchenska, M. S., and Arnstein, H. R. V., 1986, Stimulation of adenylate cyclase activity of rabbit bone marrow immature erythroblasts by erythropoietin and hemin, *Eur. J. Biochem.* **155**:363–370.

Bonanou-Tzedaki, S. A., Sohi, M. K., and Arnstein, H. R. V., 1987, The role of cAMP and calcium in the stimulation of proliferation of immature erythroblasts by erythropoietin, *Exp. Cell Res.* **170**:276–289.

Bondurant, M. C., Lind, R. N., Koury, M. K., and Ferguson, M. E., 1985, Control of globin gene transcription by erythropoietin in erythroblasts from Friend virus-infected mice, *Mol. Cell. Biol.* **5**:675–683.

Borsook, H., Ratner, K., and Tattrie, B., 1969, Studies on erythropoiesis. II: A method of segregating immature from mature adult rabbit erythroblasts, *Blood* **34**:32–41.

Brawerman, G., 1981, The role of poly (A) sequences in mammalian messenger RNA, *Crit. Rev. Biochem.* **10**:1–38.

Brawerman, G., 1987, Determinants of messenger RNA stability, *Cell* **48**:5–6.

Brock, M. L., and Shapiro, D. J., 1983, Estrogen stabilizes vitellogenin mRNA against cytoplasmic degradation, *Cell* **34**:207–214.

Broudy, V. C., Lin, N., Egrie, J., de Haen, C., Weiss, T., Papayannopoulou, T., and Adamson, J. W., 1988, Identification of the receptor for erythropoietin on human and murine erythroleukemia cells and modulation by phorbol ester and dimethylsulfoxide, *Proc. Natl. Acad. Sci. USA* **85**:6513–6517.

Brown, D. T., Wellman, S. E., and Sittman, D. B., 1985, Changes in the levels of three different classes of histone mRNA during murine erythroleukemia cell differentiation, *Mol. Cell. Biol.* **5**:2879–2886.

Brown, D. T., Yang, Y. S., and Sittman, D. B., 1988, Histone gene switching in murine erythroleukemia cells is differentiation specific and occurs without loss of cell cycle regulation, *Mol. Cell. Biol.* **8**:4406–4415.

Brown, J. E., and Adamson, J. W., 1977, Modulation of *in vitro* erythropoiesis. The influence of β-adrenergic agonists on erythroid colony formation, *J. Clin. Invest.* **60**:70–77.

Burka, E. R., 1968, Hemin: An inhibitor of erythroid cell ribonuclease, *Science* **162**:1287.

Casey, J. L., Hentze, M. W., Koeller, D. M., Caughman, S. W., Rouault, T. A., Klausner, R. D., and Hardford, J. B., 1988, Iron-responsive elements: Regulatory RNA sequences that control mRNA levels and translation, *Science* **240**:924–928.

Chang, C. S., and Sassa, S., 1988, Induction of delta levulinic acid dehydratase in mouse Friend virus transformed erythroleukemia cells during erythroid differentiation, *Blood* **64:**64–70.

Charnay, P., and Maniatis, T., 1983, Transcriptional regulation of globin gene expression in the human erythroid cell line K562, *Science* **220:**1281–1283.

Chretien, S. A., Dubart, A., Beaupain, D., Raich, N., Grandchamp, B., Rosa, J., Goossens, M., and Romeo, P., 1988, Alternative transcription and splicing of the human porphobilinogen deaminase gene result either in tissue-specific or in housekeeping expression, *Proc. Natl. Acad. Sci. USA* **85:**6–10.

Clissold, P. M., Arnstein, H. R. V., and Chesterton, C. J., 1977, Quantitation of globin mRNA levels during erythroid development in the rabbit and discovery of a new β-related species in immature erythroblasts, *Cell* **11:**353–361.

Coll, J., Saula, S., Martin, P., Raes, M. B., Lagrou, C., Graf, T., Beug, H., Simon, I. E., and Stehelin, D., 1983, The cellular oncogenes c-myc, c-myb and c-erb are transcribed in defined types of avian hematopoietic cells, *Exp. Cell Res.* **149:**151–162.

Conkie, D., Kleiman, L., Harrison, P. R., and Paul, J., 1975, Increase in the accumulation of globin mRNA in immature erythroblasts in response to erythropoietin *in vivo* and *in vitro*, *Exp. Cell Res.* **93:**315–324.

Conscience, J. F., and Meier, W., 1980, Coordinate expression of erythroid marker enzymes during dimethylsulfoxide-induced differentiation of Friend erythroleukemia cells, *Exp. Cell Res.* **125:**111–119.

Curtis, P. J., Mantei, N., and Weissmann, C., 1977, Characterization and kinetics of synthesis of 15S β-globin RNA, a putative precursor of β-globin mRNA, *Cold Spring Harbor Symp. Quant. Biol.* **42:**971–984.

D'Andrea, A. D., Lodish, H. F., and Wong, G. G., 1989, Expression cloning of the murine erythropoietin receptor, *Cell* **57:**277–285.

Daniels, R. S., McKay, M. J., Atkinson, E. M., and Hipkiss, A. R., 1983, Subcellular distribution of abnormal proteins in rabbit reticulocytes, *FEBS Lett.* **156:**145–150.

Darnell, J. E., Jr., 1982, Variety in the level of gene control in eukaryotic cells, *Nature* **297:**365–371.

Denton, M. J., and Arnstein, H. R. V., 1973, Characterization of developing adult mammalian erythroid cells separated by velocity sedimentation, *Br. J. Haematol.* **24:**7–17.

Denton, M. J., Spencer, N., and Arnstein, H. R. V., 1975, Biochemical and enzymic changes during erythrocyte differentiation—the significance of the final cell division, *Biochem. J.* **146:**205–211.

Dexter, M. T., 1987, Growth factors involved in haemopoiesis, *J. Cell Sci.* **88:**1–6.

Dons, R. F., Corash, L. M., and Gorden, P., 1981, The insulin receptor is an age-dependent integral component of the human erythrocyte membrane, *J. Biol. Chem.* **256:**2982–2987.

Earp, H. S., Rubin, R. A., Austin, K. S., and Dy, R. C., 1983, DMSO increases tyrosine residue phosphorylation in membranes from murine erythroleukemia cells, *Biochem. Biophys. Res. Commun.* **112:**413–418.

Emilia, G., Donelli, A., Ferrari, S., Torelli, U., Selleri, L., Zucchini, P., Moretti, L., Venturei, D., Cercherelli, G., and Torelli, G., 1986, Cellular levels of mRNA for c-myc, c-myb and c-fes oncogenes in normal myeloid and erythroid precursors of human bone marrow: An *in situ* hybridization study, *Br. J. Haematol.* **62:**287–292.

Evans, T., Reitman, M., and Felsenfeld, G., 1988, An erythrocyte-specific DNA-binding factor recognizes a regulatory sequence common to all chicken globin genes, *Proc. Natl. Acad. Sci. USA* **85:**5976–5980.

Fadigan, A., and Dailey, H. A., 1987, Inhibition of ferrochelatase during differentiation of murine erythroleukaemia cells, *Biochem. J.* **243:**419–424.

Fibach, E., Gambari, R., Shaw, P., Maniatis, G., Reuben, R. C., Sassa, S., Rifkind, R. A., and Marks, P. A., 1979, Tumor promoter-mediated inhibition of cell differentiation: Suppression of the expression of erythroid functions in murine erythroleukemia cells, *Proc. Natl. Acad. Sci. USA* **76:**1906–1910.

Fibach, E., Konijn, A. M., Rauminger, R. E., Ofer, S., and Rachmilewitz, E. A., 1987, Effect of extracellular hemin on hemoglobin and ferritin content of erythroleukemia cells, *J. Cell. Physiol.* **130:**460–465.

Frazer, J. K., Lin, F. K., and Berridge, M., 1988, Expression of high affinity receptors for erythropoietin on human bone marrow cells and on the human erythroleukemic cell line HEL, *Exp. Hematol.* **16:**836–842.

Friend, C., Scher, W., Holland, J. G., and Sato, T., 1971, Hemoglobin synthesis in murine virus-induced leukemic cells *in vitro:* Stimulation of erythroid differentiation by dimethyl sulfoxide, *Proc. Natl. Acad. Sci. USA* **68:**378–382.

Fucci, L., Vitale, E., Cirotto, C., and Gerasi, G., 1987, Evidence that hemoglobin switch in the chick embryo depends on erythroid cell line substitution, *Cell Differ.* **20:**55–63.

Fukamachi, H., Saito, T., Tojo, A., Kitamura, T., Urabe, A., and Takaku, F., 1987, Binding of erythropoietin to CFU-E derived from fetal mouse liver cells, *Exp. Hematol.* **15:**833–837.

Galbraith, R. A., Sassa, S., and Kappas, A., 1985, Heme binding to murine erythroleukemic cells, *J. Biol. Chem.* **260:**12198–12202.

Galson, D. L., and Housman, D. E., 1988, Detection of two tissue-specific DNA-binding proteins with affinity for sites in the mouse β-globin intervening sequence 2, *Mol. Cell. Biol.* **8:**381–392.

Ganguly, S., and Skoultchi, A. I., 1985, Absolute rates of globin gene transcription and mRNA formation during differentiation of cultured mouse erythroleukemic cells, *J. Biol. Chem.* **260:**12167–12173.

Geiduschek, J. B., and Singer, S. J., 1979, Molecular changes in the membranes of mouse erythroid cells accompanying differentiation, *Cell* **16:**149–163.

Ghosal, J., Chakraborty, M., Biswas, T., Ganguly, C. K., and Datta, A. G., 1987, Effect of erythropoietin on the glucose transport of rat erythrocytes and bone marrow cells, *Biochem. Med. Metab. Biol.* **38:**134–141.

Gidari, A. S., 1981, Mechanism of glucocorticoid-mediated inhibition of murine erythroid colony formation *in vitro, J. Cell. Physiol.* **109:**419–427.

Glass, J., Lavidor, L. M., and Robinson, S. H., 1975, Studies on murine erythroid cell development: Synthesis of heme and hemoglobin, *J. Cell Biol.* **65:**298–308.

Golde, D. W., Bersch, N., and Li, C. H., 1977, Growth hormone: Species-specific stimulation of erythropoiesis *in vitro, Science* **196:**1112–1113.

Goodman, J. W., Hall, E. A., Miller, K. L., and Shinpock, S. G., 1985, Interleukin 3 promotes erythroid burst formation in "serum free" cultures without detectable erythropoietin, *Proc. Natl. Acad. Sci. USA* **82:**3291–3295.

Graf, T., and Beug, H., 1983, Role of v-erbA and v-erbB oncogenes of avian erythroblastosis virus in erythroid cell transformation, *Cell* **34:**7–9.

Grandchamp, B., Beaumont, C., de Verneuil, H., and Nordmann, Y., 1985, Accumulation of porphobilinogen deaminase, uroporphyrinogen decarboxylase and α- and β-globin mRNAs during differentiation of mouse erythroleukemic cells, *J. Biol. Chem.* **260:**9630–9635.

Granick, J. L., and Sassa, S., 1978, Hemin control of heme biosynthesis in mouse Friend virus-transformed erythroleukemia cells in culture, *J. Biol. Chem.* **253:**5402–5406.

Granick, S., and Levere, M. D., 1965, The intracellular localization of heme by a fluorescence technique, *J. Cell Biol.* **26:**167–176.

Gregory, C. J., 1976, Erythropoietin sensitivity as a differentiation marker in the haemopoietic system: Studies of three erythropoietic colony responses in culture, *J. Cell. Physiol.* **89:**289–302.

Gross, D. S., and Garrard, W. T., 1987, Poising chromatin for transcription, *Trends Biochem. Sci.* **12:**293–297.

Gross, M., and Goldwasser, E., 1969, On the mechanism of erythropoietin-induced differentiation. V: Characterization of the ribonucleic acid formed as a result of erythropoietin action, *Biochemistry* **8:**1795–1805.

Gross, M., and Goldwasser, E., 1970, On the mechanism of erythropoietin-induced differentiation. VII: The relationship between stimulated DNA and RNA synthesis, *J. Biol. Chem.* **245:**1632–1636.

Groudine, M., Peretz, M., Nakamoto, B., Papayannopoulou, T., and Stamatoyanopoulos, G., 1986, The modulation of HbF synthesis in adult erythroid progenitor (burst-forming unit) cultures reflects changes in γ-globin gene transcription and chromatin structure, *Proc. Natl. Acad. Sci. USA* **83:**6887–6890.

Haas, A. L., and Rose, I. A., 1981, Hemin inhibits ATP-dependent ubiquitin-dependent proteolysis: Role of hemin in regulating ubiquitin conjugate degradation, *Proc. Natl. Acad. Sci. USA* **78:**6845–6848.

Hammond, K. D., Wollbrandt, R. K., and Gilbert, D. A., 1985, Acid phosphatase and phosphoaminoacid phosphatases in murine erythroleukemic cells, *Int. J. Biochem.* **17:**259–264.

Hanash, S. M., and Rucknagel, D. L., 1978, Proteolytic activity in erythrocyte precursors, *Proc. Natl. Acad. Sci. USA* **75:**3427–3431.

Harel, L., Blat, C., Lacour, F., and Friend, C., 1981, Altered RNA/protein ratio associated with the induction of differentiation of Friend erythroleukemia cells, *Proc. Natl. Acad. Sci. USA* **78:**3882–3886.

Harrison, F. L., Beswick, T. M., and Chesterton, C. J., 1981, Separation of haemopoietic cells for biochemical investigation, *Biochem. J.* **194:**789–791.

Harrison, P. R., 1984, Molecular analysis of erythropoiesis, *Exp. Cell. Res.* **155:**321–344.

Hentzen, D., Renucci, A., le Guellec, D., Benchaibi, M., Jurdic, P., Gandrillon, O., and Samarut, J., 1987, The chicken c-erbA proto-oncogene is preferentially expressed in erythrocytic cells during late stages of differentiation, *Mol. Cell. Biol.* **7:**2416–2424.

Hinssen, H., Vandekerckhove, J., and Lazarides, E., 1987, Gelsolin is expressed in early erythroid progenitor cells and negatively regulated during erythropoiesis, *J. Cell Biol.* **105:**1425–1434.

Huez, G., Bruck, C., and Cleuter, Y., 1981, Translational stability of native and deadenylated rabbit globin mRNA injected into HeLa cells, *Proc. Natl. Acad. Sci. USA* **78:**908–911.

Hulea, S. A., Denton, M. J., and Arnstein, H. R. V., 1975, Ribonuclease activity during erythroid cell maturation, *FEBS Lett.* **51:**346–350.

Hunt, T., 1976, Control of globin synthesis, *Br. Med. Bull.* **32:**257–261.

Hunt, T., Hunter, T., and Munro, A., 1969, Control of hemoglobin synthesis: Rate of translation of the messenger RNA for the α and β chains, *J. Mol. Biol.* **43:**123–133.

Ibraham, N. G., Lutton, J. D., and Levere, R. D., 1982, The role of heme biosynthetic and degradative enzymes in erythroid colony development: The effect of hemin, *Br. J. Haematol.* **50:**17–28.

Ibraham, N. G., Friedland, M. L., and Levere, R. D., 1983, Heme metabolism in erythroid and hepatic cells, *Prog. Hematol.* **13:**75–130.

Kabnick, K. S., and Housman, D. E., 1988, Determinants that contribute to cytoplasmic stability of human c-fos and β-globin mRNAs are located at several sites in each mRNA, *Mol. Cell Biol.* **8:**3244–3250.

Kaneda, T., Murate, T., Sheffery, M., Brown, K., Rifkind, R. A., and Marks, P. A., 1985, Gene expression during terminal differentiation: Dexamethasone suppression of inducer-mediated α_1 and β_{maj}-globin gene expression, *Proc. Natl. Acad. Sci. USA* **82:**5020–5024.

Karlsson, S., and Nienhuis, A. W., 1985, Developmental regulation of human globin genes, *Annu. Rev. Biochem.* **54:**1071–1108.

Kasturi, K., and Harrison, P., 1985, The cell specificity and biosynthesis of mouse glycophorins studied with monoclonal antibodies, *Exp. Cell Res.* **157:**253–264.

Kazazian, H. A., Snyder, P., and Cheng, T., 1974, Separation of α- and β-globin messenger RNAs by formamide gel electrophoresis, *Biochem. Biophys. Res. Commun.* **59:**1053–1060.

Keppel, F., Allet, B., and Eisen, H., 1977, Appearance of a chromatin protein during the erythroid differentiation of Friend virus-transformed cells, *Proc. Natl. Acad. Sci. USA* **74:**653–656.

Khochbin, S., Principand, E., Chabanas, A., and Lawrence, J. J., 1988, Early events in murine erythroleukemia cells induced to differentiate: Accumulation and gene expression of the transformation-associated cellular protein p53, *J. Mol. Biol.* **200:**55–64.

Kim, C. G., Barnhart, K. M., and Sheffery, M., 1988, Purification of multiple erythroid cell proteins that bind the promoter of the α-globin gene, *Mol. Cell. Biol.* **8:**4270–4281.

Kirsch, I. R., Bertness, V., Silver, J., and Hollis, G. F., 1986, Regulated expression of c-myb and c-myc oncogenes during erythroid differentiation, *J. Cell Biochem.* **32:**11–21.

Kleiman, L., Peters, S., Woodward-Jack, J., and Myers, J., 1980, Alterations in the metabolism of transfer RNA during erythroid differentiation of the Friend erythroleukemia cells, *Exp. Cell Res.* **129:**415–424.

Konialis, C. P., Barlow, J. H., and Butterworth, P. H. W., 1985, Cloned cDNA for rabbit erythrocyte carbonic anhydrase I: A novel erythrocyte-specific probe to study development in erythroid tissues, *Proc. Natl. Acad. Sci. USA* **82:**663–667.

Koury, M. J., Sawyer, S. T., and Bondurant, M. C., 1984, Splenic erythroblasts in anemia-inducing Friend disease: A source of cells for studies of erythropoietin-mediated differentiation, *J. Cell. Physiol.* **121:**526–532.

Koury, M. J., Bondurant, M. C. and Mueller, T. J., 1986, The role of erythropoietin in the production of principal erythrocyte proteins other than hemoglobin during terminal erythroid differentiation, *J. Cell. Physiol.* **126:**259–265.

Koury, M. J., Bondurant, M. C., and Atkinson, J. B., 1987, Erythropoietin control of terminal erythroid differentiation: Maintenance of cell viability, production of hemoglobin and development of the erythrocyte membrane, *Blood Cells* **13:**217–223.

Krowczynska, A., Yenofsky, R., and Brawerman, G., 1985, Regulation of messenger RNA stability in mouse erythroleukemia cells, *J. Mol. Biol.* **181:**231–239.

Kunzio, G. S., and Goldstein, L., 1981, Small nuclear RNAs in cellular growth and differentiation. I: Metabolic alterations seen in Friend erythroleukemic cells, *J. Cell. Physiol.* **109:**234–241.

Kurtz, A., Jelkmann, W., and Bauer, C., 1983, Insulin stimulates erythroid colony formation independently of erythropoietin, *Br. J. Haematol.* **53:**311–316.

Lachman, H. M., and Skoultchi, A. I., 1984, Expression of c-myc changes during differentiation of mouse erythroleukemia cells, *Nature* **310:**592–594.

Langstaff, J. M., and Arnstein, H. R. V., 1985, Messenger RNA turnover during bone marrow erythroid cell differentiation, *Biochim. Biophys. Acta* **825:**316–325.

Lawrence, W. D., Davis, P. J., and Blas, S. D., 1987, Action of erythropoietin *in vitro* on rabbit reticulocyte membrane Ca^{2+}-ATPase activity, *J. Clin. Invest.* **80:**586–589.

Lazarides, E., 1987, From genes to structural morphogenesis: The genesis and epigenesis of a red blood cell, *Cell* **51:**345–356.

Lehnert, M. E., and Lodish, H. F., 1988, Unequal synthesis and differential degradation of α and β spectrin during murine erythroid differentiation, *J. Cell Biol.* **107:**413–420.

Linch, D. C., and Nathan, D. G., 1984, T cell and monocyte-derived burst-promoting activity directly act on erythroid progenitor cells, *Nature* **312:**775–776.

Lo, S. C., Aft, R., Ross, J., and Mueller, G. C., 1978, Control of globin gene expression by steroid hormones in differentiating Friend leukemia cells, *Cell* **15:**447–453.

Lodish, H. F., 1976, Translational control of protein synthesis, *Annu. Rev. Biochem.* **45:**39–72.

Lowenhaupt, K., and Lingrel, J. B., 1978, A change in the stability of globin mRNA during the induction of murine erythroleukemia cells, *Cell* **14:**337–344.

Lowenhaupt, K., and Lingrel, J. B., 1979, Synthesis and turnover of globin mRNA in murine erythroleukemia cells induced with hemin, *Proc. Natl. Acad. Sci. USA* **76:**5173–5177.

McCaffery, P., Frazer, J. K., Liu, F.-K., and Berridge, M. V., 1987, Erythropoietin receptor identification and modulation of expression on erythroid precursor cells, *Exp. Hematol.* (Suppl.) **15:**436 (abstract).

McKay, M. J., Daniels, R. S., and Hipkiss, A. R., 1980, Breakdown of aberrant proteins in rabbit reticulocytes decreases with cell age, *Biochem. J.* **188:**279–283.

Maniatis, G. M., Rifkind, R. A., Bank, A., and Marks, P. A., 1973, Early stimulation of RNA synthesis by erythropoietin in cultures of erythroid precursor cells, *Proc. Natl. Acad. Sci. USA* **70:**3189–3194.

Mantovani, R., Malgaretti, N., Nicolis, S., Giglioni, B., Comi, P., Cappellini, N., Bertero, M. T., Calligaris-Cappio, F., and Ottolenghi, S., 1988, An erythroid specific nuclear factor binding to the proximal CACCC box of the β-globin gene promoter, *Nucleic Acids Res.* **16:**4299–4313.

Marbaix, G., Huez, G., Burny, A., Cluter, Y., Hubert, E., Leclercq, M., Chantrenne, H., Soreq, H., Nudel, U., and Littauer, U. Z., 1975, Absence of polyadenylate segment in globin messenger RNA accelerates its degradation in *Xenopus* oocytes, *Proc. Natl. Acad. Sci. USA* **72:**3065–3067.

Marchesi, V. T., 1985, Stabilizing infrastructures of membranes, *Annu. Rev. Cell Biol.* **1:**531–561.

Marks, P. A., and Rifkind R. A., 1978, Erythroleukemic differentiation, *Annu. Rev. Biochem.* **47:**419–448.

Martin, P., and Papayannopoulou, T., 1982, HEL cells: A new human erythroleukemia cell line with spontaneous and induced globin expression, *Science* **216:**1233–1235.

Mason, M., Narindrasorasak, S., and Sanwal, B. D., 1985, Regulation of cyclic 3′,5′AMP phosphodiesterase in Friend erythroleukemia cells, *J. Cycl. Nucl. Prot. Phosph. Res.* **10:**129–142.

Mayeux, P., Billat, C., Felix, J. M., and Jacquot, R., 1986a, Mode of action of erythropoietin and glucocorticoids on the hepatic erythroid precursor cells: Role of prostaglandins, *Cell Differ.* **18:**17–26.

Mayeux, P., Felix, J. M., Billat, C., and Jacquot, R., 1986b, Induction by hemin of proliferation and differentiation of progenitor erythroid cells responsible for erythropoietin, *Exp. Hematol.* **14:**801–808.

Mayeux, P., Billat, C., and Jacquot, R., 1987, The erythropoietin receptor of rat erythroid progenitor cells. Characterization and affinity cross-linkage, *J. Biol. Chem.* **262:**13985–13990.

Mechti, N., Piechaczyk, M., Blanchard, J. M., Marty, L., Bonnieu, A., Jeanteur, P., and Lebleu, B., 1986, Transcriptional and post-transcriptional regulation of c-myc expression during the differentiation of murine erythroleukemia Friend cells, *Nucleic Acids Res.* **14:**9653–9666.

Mezl, V. A., Kawasaki, E. S., and Hunt, J. A., 1979, Analysis of the ratio of α- to β-globin and globin messenger RNA content of fractionated rabbit erythroid bone-marrow cells, *Biochem. J.* **179:**525–535.

Ngai, J., Stack, J. H., Moon, R. T., and Lazarides, E., 1987, Regulated expression of multiple chicken erythroid membrane skeletal protein 4.1 variants is governed by differential RNA processing and translational control, *Proc. Natl. Acad. Sci. USA* **84:**4432–4436.

Nijhof, W., and Wierenga, P. K., 1983, Isolation and characterization of the erythroid progenitor cell: CFU-e, *J. Cell Biol.* **96:**386–392.

Nijhof, W., Wierenga, P. K., Sahr, K., Beru, N., and Goldwasser, E., 1987, Induction of globin mRNA transcription by erythropoietin in differentiating erythroid precursor cells, *Exp. Hematol.* **15:**779–784.

Nudel, U., Salmon, J., Fibach, E., Terada, M., Rifkind, R., Marks, P. A., and Bank, A., 1977, Accumulation of α- and β-globin messenger RNAs in mouse erythroleukemia cells, *Cell* **12:**463–469.

Ochoa, S., and De Haro, C., 1979, Regulation of protein synthesis in eukaryotes, *Annu. Rev. Biochem.* **48:**549–580.

Ong, G. C., 1987, Enzyme synthesis and mRNA changes during erythroid differentiation, Ph.D. thesis, University of London.

Orkin, S. T., Swan, D., and Leder, P., 1975, Differential expression of α- and β-globin genes during differentiation of cultured erythroleukemic cells, *J. Biol. Chem.* **250:**8753–8760.

Paul, J., 1976, Haemoglobin synthesis and cell differentiation, *Br. Med. Bull.* **32:**277–281.

Peterson, J. L., and McConkey, E. H., 1976, Proteins of Friend leukemia cells—Comparison of hemoglobin synthesizing and non-induced populations, *J. Biol. Chem.* **251:**555–558.

Pfeffer, S. R., Huima, T., and Redman, C. M., 1986, Biosynthesis of spectrin and its assembly into the cytoskeletal system of Friend erythroleukemia cells, *J. Cell Biol.* **103:**103–113.

Pickart, C. M., and Vella, A. T., 1988, Levels of active ubiquitin carrier proteins decline during erythroid maturation, *J. Biol. Chem.* **263:**12028–12034.

Ponka, P., Schulman, H. M., and Martinez-Medellin, I., 1988, Haem inhibits iron uptake subsequent to endocytosis of transferrin in reticulocytes, *Biochem. J.* **251:**105–109.

Popovic, W. J., Brown, J. E., and Adamson, J. W., 1977, The influence of thyroid hormones on *in vitro* erythropoiesis, *J. Clin. Invest.* **60:**907–913.

Porter, P. N., Meints, R. H., and Mesner, K., 1979, Enhancement of erythroid colony growth in culture by hemin, *Exp. Hematol.* **7:**11–16.

Profous-Juchelka, H. L., Reuben, R. C., Marks, P. A., and Rifkind, R. A., 1983, Transcriptional and posttranscriptional regulation of globin gene accumulation in murine erythroleukemia cells, *Mol. Cell. Biol.* **3:**229–232.

Ramsay, R. G., Ikada, K., Rifkind, R. A., and Marks, P. A., 1986, Changes in gene expression associated with induced differentiation of erythroleukemia: Proto-oncogenes, globin genes and cell division, *Proc. Natl. Acad. Sci. USA* **83:**6849–6853.

Rao, K., Harford, J. B., Rouault, T., McClelland, A., Ruddle, F. H., and Klausner, R. D., 1985, Transcriptional regulation by iron of the gene for the transferrin receptor, *Mol. Cell. Biol.* **6:**236–240.

Rapoport, S. M., and Schewe, T., 1986, The maturational breakdown of mitochondria in reticulocytes, *Biochim. Biophys. Acta* **864:**471–495.

Reuben, R., Rifkind, R., and Marks, P. A., 1980, Chemically induced murine erythroleukemic differentiation, *Biochim. Biophys. Acta* **605:**325–346.

Richardson, J. M., Morla, A. O., and Wang, J. Y. J., 1987, Reduction in protein tyrosine phosphorylation during differentiation of human leukemia cell line K-562, *Cancer Res.* **47:**4066–4070.

Riddle, R. D., Yamamoto, M., and Engel, J. D., 1989, Expression of δ-aminolevulinate synthase in avian cells: Separate genes encode erythroid-specific and nonspecific isozymes, *Proc. Natl. Acad. Sci. USA* **86:**792–796.

Ross, J., 1988, Messenger RNA turnover in eukaryotic cells, *Mol. Biol. Med.* **5:**1–14.

Ross, J., and Pizarro, A., 1983, Human beta and delta globin messenger RNAs turn over at different rates, *J. Mol. Biol.* **167:**607–617.

Ross, J., and Sautner, D., 1976, Induction of globin mRNA accumulation by hemin in cultured erythroleukemic cells, *Cell* **8:**513–520.

Ross, J., and Sullivan, T. D., 1985, Half lives of beta and gamma globin messenger RNAs and of protein synthetic capacity in cultured human reticulocytes, *Blood* **66:**1149–1154.

Rovera, G., Abramczuk, J., and Surrey, S., 1977, The effect of hemin on the expression of β-globin genes in Friend cells, *FEBS Lett.* **81:**366–370.

Russanov, E. M., Kirkova, M. D., Setchenska, M. S., and Arnstein, H. R. V., 1981, Enzymes of oxygen metabolism during erythrocyte differentiation, *Biosci. Rep.* **1:**927–931.

Rutherford, T. R., Clegg, J. B., and Weatherall, D. J., 1979a, K562 human leukemic cells synthesize embryonic hemoglobin in response to hemin, *Nature* **280:**164–165.

Rutherford, T., Thompson, C. G., and Moore, M. R., 1979b, Heme biosynthesis in Friend erythroleukemia cells: Control by ferrochelatase, *Proc. Natl. Acad. Sci. USA* **76:**833–836.

Sap, J., Munoz, A., Damm, K., Goldberg, Y., Ghysdael, J., Leutz, A., Beug, H., and Vennstrom, B., 1986, The c-erb-A protein is a high-affinity receptor for thyroid hormone, *Nature* **324:**635–640.

Sassa, S., 1976, Sequential induction of heme pathway enzymes during erythroid differentiation in mouse Friend leukemia virus-infected cells, *J. Exp. Med.* **143:**305–315.

Sawada, K., Krantz, S. B., Kans, J. S., Dessypris, E. N., Sawyer, S., Glick, A. D., and Civin, C. I., 1987, Purification of human erythroid colony-forming units and demonstration of specific binding of erythropoietin, *J. Clin. Invest.* **80:**357–366.

Sawyer, S. T., and Krantz, S. B., 1986, Transferrin receptor number, synthesis and endocytosis during erythropoietin-induced maturation of Friend virus-infected erythroid cells, *J. Biol. Chem.* **261**:9187–9195.
Sawyer, S., Krantz, S. B., and Goldwasser, E., 1987, Binding and receptor-mediated endocytosis of erythropoietin in Friend virus-infected erythroid cells, *J. Biol. Chem.* **262**:5554–5562.
Schwartz, D. A., and Rubin, C. S., 1985, Identification and differential expression of 2 forms of regulatory subunits R-II of cyclic AMP-dependent protein kinase II in Friend erythroleukemic cells differentiation, *J. Biol. Chem.* **260**:6296–6303.
Setchenska, M. S., and Arnstein, H. R. V., 1978, Changes in lactate dehydrogenase isoenzyme pattern during differentiation of rabbit bone-marrow erythroid cells, *Biochem. J.* **170**:193–201.
Setchenska, M. S., and Arnstein, H. R. V., 1979, Changes in malate dehydrogenase isoenzymes during differentiation of rabbit bone marrow erythroid cells, *Int. J. Biochem.* **10**:817–821.
Setchenska, M. S., and Arnstein, H. R. V., 1983a, Characteristics of the β-adrenergic adenylate cyclase system of developing rabbit bone marrow erythroblasts, *Biochem. J.* **210**:559–566.
Setchenska, M. S., and Arnstein, H. R. V., 1983b, Characteristics of the adenylate cyclase system of differentiating rabbit bone marrow erythroblasts, *Biomed. Biochim. Acta* **42**:1111–1122.
Setchenska, M. S., Vassileva-Popova, J. G., and Arnstein, H. R. V., 1980, Plasma membrane-associated protein kinase activity of differentiating rabbit bone marrow erythroid cells, *Int. J. Biochem.* **11**:393–399.
Setchenska, M. S., Arnstein, H. R. V., and Vassileva-Popova, J. G., 1981, Cyclic AMP phosphodiesterase activity during differentiation of rabbit erythroid bone marrow cells, *Biochem. J.* **196**:887–892.
Setchenska, M. S., Bonanou-Tzedaki, S. A., and Arnstein, H. R. V., 1988, Changes in protein kinase activities and protein phosphorylation during rabbit erythroid cell development, 14th ICB Prague, MO: 263, p. 124 (abstract).
Shaul, Y., Ginzburg, I., and Aviv, H., 1982, Preferential transcription and nuclear transport of globin gene sequences, as control steps leading to final differentiation of murine erythroleukemic cells, *Eur. J. Biochem.* **128**:637–642.
Shaw, G., and Kamen, R., 1986, A conserved AU sequence from the 3′ untranslated region of GM-CSF mRNA mediates selective mRNA degradation, *Cell* **46**:659–667.
Sherton, C., and Kabat, D., 1976, Changes in RNA and protein metabolism preceding onset of hemoglobin synthesis in cultured Friend leukemia cells, *Dev. Biol.* **48**:118–131.
Sieff, C. A., Emerson, S. G., Donahue, R. E., Nathan, D. G., Wang, E. A., Wong, G. G., and Clark, S. C., 1985, Human recombinant granulocyte—macrophage colony-stimulating factor: A multilineage hematopoietin, *Science* **230**:1171–1173.
Singh, M. K., and Yu, J., 1984, Accumulation of a heat shock-like protein during differentiation of human erythroid cell line K562, *Nature* **309**:631–633.
Sittman, D. B., Graves, R. A., and Marzluff, W. F., 1983, Histone mRNA concentrations are regulated at the level of transcription and mRNA degradation, *Proc. Natl. Acad. Sci. USA* **80**:1849–1853.
Smith, B. J., Walker, J. M., and Johns, E. W., 1980, Structural homology between a mammalian H_{1o} subfraction and avian erythrocyte-specific histone H5, *FEBS Lett.* **112**:42–44.
Spencer, N., and Peller, S., 1976, Carbonic anhydrase isoenzymes I and II in rabbit erythroid cells, *Biochem. Soc. Trans.* **4**:1153–1155.
Spivak, J. L., 1986, The mechanism of action of erythropoietin, *Int. J. Cell Cloning* **4**:139–166.
Stewart, A. G., Clissold, P. M., and Arnstein, H. R. V., 1976, The initiation of globin synthesis in differentiating rabbit bone marrow erythroid cells, *Eur. J. Biochem.* **65**:349–355.
Stohlman, F., Jr., 1970, Kinetics of erythropoiesis, in: *Regulation of Hematopoiesis*, Volume I (A. S. Gordon, ed.), pp. 317–326, Educational Division, Meredith Corporation, New York.
Sytkowsky, A. J., and Kessler, C. J., 1984, The β-adrenergic receptor adenylate cyclase complex of Rauscher murine erythroleukemia cells and its response to erythropoietin-induced differentiation, *Blood* **64**:84–90.
Terada, M., Epner, E., Nudel, U., Salmon, J., Fibach, E., Rifkind, R. A., and Marks, P. A., 1978, Induction of murine erythroleukemia differentiation by actinomycin D, *Proc. Natl. Acad. Sci. USA* **75**:2795–2799.
Thiele, B., Andree, H., Hohne, M., and Rapoport, S. M., 1982, Lipoxygenase mRNA in rabbit reticulocytes. Its isolation, characterization and translational repression, *Eur. J. Biochem.* **129**:133–141.
Thomopoulos, P., Testa, U., Flamier, A., and Berthelier, M., 1980, Insulin receptors and protein synthesis in bone marrow cells and reticulocytes, *Diabetes* **29**:820–824.
Threadgill, G. J., and Arnstein, H. R. V., 1984, The nonhistone proteins of developing mammalian erythroid cells, *Cell Differ*, **14**:7–17.

Threadgill, G. J., and Arnstein, H. R. V., 1985, Changes in histone acetylation during the development of rabbit bone marrow erythroid cells, *Biochim. Biophys. Acta* **847**:228–234.

Todokoro, K., and Ikawa, Y., 1986, Sequential expression of proto-oncogenes during a mouse erythroleukemia cell differentiation, *Biochem. Biophys. Res. Commun.* **135**:1112–1118.

Trainor, C. D., Stamler, S. J., and Engel, J. D., 1987, Erythroid-specific transcription of the chicken histone H5 gene is directed by a 3′ enhancer, *Nature* **328**:827–830.

Tsai, S. F., Martin, D. I. K., Zou, L. I., D'Andrea, A. D., Wong, G. G., and Orkin, S. H., 1989, Cloning of cDNA for the major DNA-binding protein of the erythroid lineage through expression in mammalian cells, *Nature* **339**:446–451.

Umemura, T., Umene, K., Nishibura, J., Fukumaki, Y., Sakaki, Y., and Ibayashi, H., 1986, Expression of c-myc oncogene during differentiation of human-burst-forming unit, erythroid (BFU-e), *Biochem. Biophys. Res. Commun.* **135**:521–526.

Urabe, A., Sassa, S., and Kappas, A., 1979, The influence of steroid hormone metabolites on the *in vitro* development of erythroid colonies from human bone marrow, *J. Exp. Med.* **149**:1314–1320.

Volloch, V., and Housman, D., 1981, Stability of globin mRNA in terminally differentiating murine erythroleukemia cells, *Cell* **23**:509–514.

Volloch, V., Schweitzer, B., and Ritz, S., 1987, Messenger RNA changes during differentiation of murine erythroleukemia cells, *Exp. Cell Res.* **173**:38–48.

Wall, L., de Boer, E., and Grosveld, F., 1988, The human β-globin gene 3′ enhancer contains multiple binding sites for an erythroid-specific protein, *Gene Dev.* **2**:1089–1100.

Watson, R. J., 1988, Expression of the c-myb and c-myc genes is regulated independently in differentiating mouse erythroleukemia cells by common processes of premature transcription arrest and increased mRNA turnover, *Mol. Cell Biol.* **8**:3938–3942.

Waxman, L., Fagan, J. M., Tanaka, K., and Goldberg, A. L., 1985, A soluble ATP-dependent system for protein degradation from murine erythroleukaemia cells, *J. Biol. Chem.* **260**:11994–12000.

Weatherall, D. J., and Clegg, J. B., 1981, *The Thalassemia Syndrome,* 3rd ed., Blackwell, Oxford.

Weintraub, H., 1984, Histone H1-dependent chromatin superstructures and the suppression of gene activity, *Cell* **38**:17–27.

Wingrove, T. G., Watt, R., Keng, P., and Macara, I. G., 1988, Stabilization of myc proto-oncogene proteins during Friend murine erythroleukemia cell differentiation, *J. Biol. Chem.* **263**:8918–8924.

Winning, R. W., and Browder, L. W., 1988, Changes in heat shock protein synthesis and hsp70 gene transcription during erythropoiesis of *Xenopus laevis, Dev. Biol.* **128**:111–120.

Wolffe, A. P., Glover, J. F., Martin, S. C., Tenniswood, M. P. R., Williams, J. L., and Tata, J. R., 1985, Deinduction of transcription of Xenopus 74-kDa albumin genes and destabilization of mRNA by estrogen *in vitro* and in hepatocyte cultures, *Eur. J. Biochem.* **171**:457–461.

Woods, C. M., and Lazarides, E., 1985, Degradation of unassembled α- and β-spectrin by distinct intracellular pathways: Regulation of spectrin topogenesis by α-spectrin degradation, *Cell* **40**:959–969.

Woods, C. M., Boyer, B., Vogt, P. K., and Lazarides, E., 1986, Control of erythroid cell differentiation: Asynchronous expression of the anion transporter and the peripheral components of the membrane skeleton in AEV- and S13-transformed cells, *J. Cell Biol.* **103**:1789–1798.

Yoshida, T., Biro, P., Cohen, T., Müller, R. M., and Shibahara, S., 1988, Human heme oxygenase cDNA and induction of its mRNA by hemin, *Eur. J. Biochem.* **171**:457–461.

Yu, J., and Smith, R. C., 1985, Sequential alterations in globin gene chromatin structure during erythroleukemia cell differentiation, *J. Biol. Chem.* **260**:3035–3040.

Chapter 6

Maturational Breakdown of Mitochondria and Other Organelles in Reticulocytes

Samuel M. Rapoport, Tankred Schewe, and Bernd-Joachim Thiele

1. INTRODUCTION

One of the characteristics of the differentiation of erythroid cells is the decay or elimination of organelles, including the nucleus, mitochondria, ribosomes, lysosomes, endoplasmic reticulum, and Golgi apparatus. Many of the changes occur in the nucleated precursors of the erythrocyte. Some organelles, however, primarily mitochondria and ribosomes, but also vestiges of others, remain in the reticulocyte. The mechanisms involved in the degradation of organelles are largely unexplored. The process best understood is the maturational breakdown of mitochondria in reticulocytes, which will therefore be the focus of the present review. The various changes appear to constitute a fixed program of maturation that once started takes it course with little or no outside effectors. The interplay of the various events and their causal relationships are open questions.

1.1. Characteristics of Reticulocytes

1.1.1. Definition, Origin, and General Characteristics of Reticulocytes

The reticulocyte represents the penultimate stage of the differentiation of the erythroid cell (for detailed review see Rapoport, 1986). It derives its name from the characteristic network that appears on staining with basic dyes. This network represents a precipitate of the ribosomal nucleic acid, still containing some tightly bound proteins, with the dye. Various dyes may be used for staining, the most common of which are

Samuel M. Rapoport, Tankred Schewe, and Bernd-Joachim Thiele Institute of Biochemistry, Humboldt University of Berlin, School of Medicine (Charité), DDR-1040 Berlin, German Democratic Republic.

brilliant cresyl blue and acridine orange. The appearance of the precipitate is variable. Very mature reticulocytes may exhibit only one or two specks and may escape detection. The precipitation is a threshold phenomenon so that small amounts of ribonucleic acid cannot be made visible. The staining is postvital.

The unstained living reticulocyte appears polylobulated under the phase-contrast microscope. One may observe slow movements of the protoplasm, which contracts in some regions, while it throws out projections in others. The multishaped and variable form of the reticulocyte is clearly seen with the scanning electron microscope (Bessis, 1973).

Reticulocytes arise in mammals from normoblasts through enucleation. This process represents an abortive cell division with active movements both of the nucleus and of the cell membrane. The nuclei migrate with an assembly of mitochondria and vesicles to one region of the cell membrane that at the same time undergoes remodeling. Concanavalin A and other lectin receptors accumulate in the region destined for the extrusion of the cell nucleus (Schlegel *et al.*, 1982; Geiduschek and Singer, 1979). This region also becomes depleted of spectrin, which aggregates in the remaining portion of the cell membrane to become part of the membrane skeleton of the reticulocyte-to-be. The loss of the nucleus, together with the spectrin-devoid region of the cell membrane (see Chapter 8), results in a decreased cell size and a drastic increase of the packing of spectrin. One of the consequences is the progressive restriction of the movement of membrane proteins. There still remains the capability to form domains depleted of spectrin, which can exhibit pseudopodia, invaginations, and endocytosis (Zweig *et al.*, 1981). These properties decrease during the course of maturation of the reticulocyte to become abolished in the mature erythrocyte. It is remarkable that the neonatal human erythrocyte exhibits domains of membrane mobility and endocytosis of membrane receptors similar to the reticulocyte, properties that are not found in the erythrocytes of adults (Schekman and Singer, 1976). At the same time there are also changes in the order of membrane lipids. The more loosely packed lipid molecules aggregate to a single domain of the cell surface, which surrounds the nucleus as it is being extruded.

In nonmammalian species there appear during anemia red cells with staining properties and other features such as a high respiration of reticulocytes. Thus, it would appear that it is not so much the absence of the nucleus but its inactivation and pyknosis, which is the criterion of the transition to the reticulocyte stage. During differentiation of the erythroid precursors, nuclear activities, such as transcription and export of nucleic acids progressively decrease. At the same time the nucleus becomes pyknotic, a process that involves increase of highly condensed chromatin, as well as loss of nucleoli and of a variety of proteins.

The cell organelles undergo massive changes. The number of both ribosomes and mitochondria per cell decline with each division, even though there is some production of organelles as long as the cells divide and their nuclei are not completely pyknotic. After the last mitosis there is a great decrease of the endoplasmic reticulum, the Golgi apparatus, and the number of lysosomes (Heynen and Verwilghen, 1982). By morphologic criteria there are few if any lysosomes left in the reticulocyte. However, some autophagosomes are found. Enzymatic studies support this conclusion. Enzymes such as β-*N*-acetylglucosaminidase, β-galactosidase, and β-glucuronidase, which are indicative of lysosomes, have very low activities and are entirely located in the cell membrane in both reticulocytes and mature erythrocytes (Yatziv *et al.*, 1979). One may assume that during the maturation process, fusion or trapping of the lysosomes on the inner side of the cell membrane may have occurred. Thus, one may conclude that neither changes of the lysosomes nor their function play a significant role in the maturational changes of the reticulocyte. On the other hand, specific acid endosomes, located near the cell membrane, appear to be functional in the reticulocyte and play an important role in the uptake of iron from iron-transferrin by endocytosis (Van Renswoude *et al.*, 1982).

1.1.2. Different Populations and Maturational Stages of Reticulocytes

As our knowledge of the reticulocyte's biology has progressed it has become clear that the presence of ribosomes is just one of its characteristics. Other features are the presence of mitochondria, various properties of its cell membrane, and the specific metabolic profile of the cell. These characteristics change more or less synchronously but not completely so, during maturation and may differ both between species and within a single species, depending on the state of erythropoiesis, and even according to the manner of its stimulation. Further factors influencing these features may be the nutritional state and the presence of infection.

Under normal conditions the reticulocytes found in the peripheral blood are only slightly larger than the erythrocytes and contain about 90% of their hemoglobin concentration. They are highly mature as judged by their low content of RNA and the small number of mitochondria, which are to a large extent highly degraded and functionally incompetent, as indicated by a low respiration, which is uncoupled, and a small or absent Pasteur effect (Krause *et al.*, 1972; Richter-Rapoport *et al.*, 1977). Their lifetime in the peripheral blood is about 1 day, after having spent about 2 days in the bone marrow.

Severe blood loss or hemolytic agents such as phenylhydrazine produce stimulation of erythropoiesis with the appearance of large numbers of reticulocytes in the peripheral blood. By standardization of the protocol a synchronization of the erythropoietic response can be achieved. The reticulocytes thus produced differ qualitatively from those found under normal circumstances. Their cell volume may be twice as large as that of the normocytes, whereas their hemoglobin concentration may be lower by one-third. The RNA concentration is higher than that of normal reticulocytes and the ATP concentration may exceed twice the normal value (Rapoport *et al.*, 1944; Rosenthal *et al.*, 1964a). A qualitative marker is the presence of large amounts of both lipoxygenase RNA and enzyme protein (Minich *et al.*, 1989).

How do such macro- or stress reticulocytes arise? At present, a definitive answer cannot be given. Several authors have assumed that their emergence is caused by the skipping of a cell division during differentiation in the bone marrow. Thus, a cell of double size would be released prematurely. Within the framework of this hypothesis, it has been further assumed that an early attainment of a critical level of hemoglobin may be responsible for a block of cell division at the polychromatic erythroblast stage rather than at the time of emergence of the orthochromatic erythroblast. Other explanations, such as a premature shift of the reticulocyte from the bone marrow to the peripheral blood, have been put forward. Another possibility may be changes in the differentiation program at the level of transcription. This explanation is best suited to account for the large differences in the content of lipoxygenase and other constituents between normal and stress reticulocytes. In such a case it is conceivable that during the course of an anemia several types of cell populations may arise, depending on the state of stimulation of the bone marrow. Evidence for the appearance of several types of cohorts of cells, differing in several properties among themselves, has been obtained (Rosenthal *et al.*, 1972; Minich *et al.*, 1989).

The maturational changes within one cell type should be clearly distinguished from those of other cell types. For the stress reticulocyte a clear-cut classification according to maturity could be established based on three criteria: (1) the presence of lipoxygenase, (2) ATP-dependent proteolysis, and (3) the presence of mitochondrial susceptibility factor (MSF), the protein rendering mitochondria susceptible to the triggering action of lipoxygenase. Accordingly, four types of reticulocyte populations may be distinguished. The first one, comprising the most immature cells, contains masked lipoxygenase mRNA as well as masked MSF mRNA. They do not exhibit proteolysis of their mitochondria. Such reticulocytes are found in the bone marrow and in the topmost 1–5% fraction of density-separated cells of the peripheral blood. The next stage of maturity is represented by reticulocytes with unmasked lipoxygenase mRNA, which therefore contain sizable amounts of lipoxygenase protein; the MSF mRNA is, however, still masked, so that their mitochondria do not yet undergo proteolysis. They make up the upper 15% of the cell population of the peripheral blood of rabbits on the 7th day of anemia. The third population consists of cells with ongoing activity of lipoxygenase and of the proteolytic system, which may be inhibited by inhibitors of lipoxygenase. The fourth and most mature fraction consists of reticulocytes that have undergone the triggering action of lipoxygenase, but still have on-going proteolysis that can no longer be arrested by inhibitors of lipoxygenase.

These properties show qualitative differences and thus appear adequate for a nonarbitrary classification. Other properties, such as cell density *per se,* which corresponds to the hemoglobin concentration, or the content of RNA, appear to undergo continuous gradual changes. Recently, the respiration and energy status of the various reticulocyte populations of anemic rabbits were studied (Kostić and Rapoport, 1989). It was found that oxygen uptake, particularly coupled respiration, declined progressively, whereas aerobic lactate formation increased, so that a lower Pasteur effect resulted. Total ATP production decreased by about one-half, with the share of aerobic glycolysis increasing from less than 2% to nearly 8%. At the same time the ATP concentration decreased from more than 3 mM to 1.5 mM, close to the value found in erythrocytes. A dramatic 60-fold decrease of the concentration of cAMP was found, the greatest change taking place during the transition from the most immature to the subsequent stage.

1.1.3. Metabolic Characteristics of Reticulocytes

The metabolic profile of the reticulocyte differs from that of the mature erythrocyte in three main respects: (1) respiration, (2) the diversity of its energy-yielding substrates, and (3) its turnover of amino acids with a highly efficient nitrogen economy. All these features are connected with the function of mitochondria.

As to the first point, it is found that whereas mature erythrocytes, if reticulocytes and white blood cells are scrupulously excluded, have an oxygen consumption of less than 1 μmol/ml cells per hr, the oxygen consumption of reticulocytes may reach values as high as 50 μmol/ml cells per hr (Siems *et al.*, 1982, 1984). Correspondingly, the ATP production of reticulocytes amounts to about 150 μmol/ml cells per hr versus only 2 μmol ATP in mature erythrocytes. Oxygen consumption, degree of coupling, and ATP production decline during *in vitro* incubation (Thilo *et al.*, 1979).

As to the substrates utilized, erythrocytes use practically exclusively glucose, whereas reticulocytes utilize also fatty and amino acids both for energy production and for syntheses. The pathway for the activation and β-oxidation of fatty acids as well as those for the utilization of amino acids are found in the mitochondria. The latter include the conversion of serine to glycine mediated by the serine hydroxymethyl transferase and some of the steps of porphyrin synthesis and the Shemin cycle (Rapoport *et al.*, 1980). The reactions are part of a special characteristic feature of the reticulocyte, its nitrogen economy.

The breakdown of organelles, in particular mitochondria, is closely geared to the utilization of the amino acids liberated for the synthesis of hemoglobin. The large amounts of serine and glycine required for the synthesis of heme and globin are formed in two pathways. In both, glucose furnishes the carbon skeleton as 2- or 3-phosphoglycerate, which are converted to hydroxy- and phosphohydroxypyruvate, respectively. These compounds serve as acceptors of amino groups originating from glutamate or alanine, which after deamination are utilized in the citrate cycle. These pathways amount to about 5% of the total glucose utilized and account for about one-half of the serine and glycine required for the synthesis of hemoglobin. In this manner, there results a remarkably efficient nitrogen economy in the reticulocyte, so that only a small portion of the ammonia potentially available from the oxidation of amino acids is actually liberated, whereas the bulk enters in form of serine and glycine for the synthesis of hemoglobin.

1.2. Characteristics of Reticulocyte Mitochondria

In the cells of most tissues, mitochondria exhibit a highly heterogeneous pattern reflecting their asynchronous turnover. An even greater heterogeneity is found in unfractionated reticulocyte populations. In density (age)-fractionated reticulocytes a progression of changes of mitochondrial morphology may be observed (Krause *et al.*, 1972; Heynen and Verwilghen, 1982). Highly immature reticulocytes contain mostly orthodox and condensed forms of mitochondria, many of which exhibit few cristae and a loose matrix structure. In the more mature cells there is an increasing predominance of degraded forms, associated with the appearance of swelling. Also there is a strict correlation between reticulocyte maturity, number of mitochondria, and activity of cytochrome *c* oxidase.

In a recent study, the age dependency of the properties of mitochondria was characterized. A mechanical method of cell disintegration under isotonic conditions followed by fractionation in a Percoll density gradient proved to reflect physiological conditions. The density of mitochondria was found to decrease with progressive maturity, in agreement with the electron micrographic appearance. The mitochondria of mature reticulocytes also exhibit lower specific activities of cytochrome *c* oxidase, perhaps reflecting their proteolytic degradation (Belkner and Rapoport, 1989). In older work it had been found that there also exists a correlation between the structure of mitochondria and the specific activity of succinate dehydrogenase, with the highest activity found in the condensed type, whereas the lowest activity was observed in the degraded forms (Schulz *et al.*, 1971).

The chemical composition of reticulocyte mitochondria differs in some respects from that of liver mitochondria. They contain about double the amounts of cytochromes *a, b, c,* and c_1 and about two and a half times as much phospholipid (Schewe *et al.*, 1972). Phosphatidylcholine (PC) and phosphatidylethanolamine (PE) account for about 40 and 30%, respectively, with smaller amounts of sphingomyelin, cardiolipin, and phosphatidylinositol (PI) each fraction amounting to about 10%. Phosphatidylserine is practically absent (Lutze *et al.*, 1977). It is remarkable that lysophosphatides are present only in small amounts. The sphingomyelin content is nearly an order of magnitude higher than found in liver mitochondria.

The fatty acid composition of the phospholipids does not differ greatly from that of other types of mitochondria, with a high percentage of hexadecanoic and linoleic acids in PC and PE. Linoleic acid accounts for 60% of the fatty acids of cardiolipin, which contains practically only unsaturated C_{18}-fatty acids. The C_{18}-fatty acids make up about 50% of the total, with unsaturated acids predominating. Arachidonic acid accounts for about 4% of PC, 8% of PE, and 12% of PI. Recently, the occurrence of the reduction products of the action of lipoxygenase, particularly 13-hydroxy-9*Z*,11*E*-octadecadienoic acid and other hydroxypolyenoic fatty acids and of some oxodienoic acids in the phospholipids has been established (Kühn *et al.*, 1989). Their amounts appear to be variable but may reach a sizable percentage of the parent compounds. A comparatively high amount of free fatty acids appears to be correlated with the proportion of degraded forms of mitochondria.

The enzymatic machinery and respiratory activity of reticulocyte mitochondria present special features. Isolated mitochondria incorporate actively fatty acids into phospholipids, with acylation of *n*-glycerol-3-phosphate and of lysophospholipids (Augustin *et al.*, 1977). The enzymes of the citrate cycle, as well as glutamate dehydrogenase and aspartate aminotransferase are present, whereas NAD^+-dependent isocitrate dehydrogenase is absent (Greksch *et al.*, 1973). The amounts of these enzymes resemble those found in heart mitochondria rather than those in liver.

1.3. Enzyme Changes and the Degradation of Organelles

Differences between reticulocytes and erythrocytes with respect to their enzymatic machinery reflect to a large degree the degradation of organelles, with which the enzymes are associated. For example, the components of the respiratory chain disappear during maturation, corresponding to their localization in the mitochondrial inner membrane; the same holds true for enzyme activities of the mitochondrial outer membrane, and for most of the enzymes of the mitochondrial matrix. An interesting case is hexokinase. In the reticulocyte, about half of the enzyme activity is found in the form of an isoenzyme bound to the mitochondrial outer membrane. It is preferentially degraded by ATP-dependent proteolysis during maturation (Magnani *et al.*, 1984).

A variety of changes with respect to enzyme activities and related properties occur in the cell membrane. These include losses of the Na^+,K^+-ATPase, nucleoside transport, various amino acid transport systems (Blostein and Grafova, 1987; Antonioli and Christensen, 1969), glucose transport (Augustin *et al.*, 1964; Zeidler and Kim, 1982), the adenylate cyclase system, and activities of phosphoinositide metabolism (Maretzki *et al.*, 1986, 1987). The mechanism responsible for changes of the transferrin receptor is dealt with in Chapter 13 of this volume. Some of the changes observed appear to be ATP-dependent indicating the function of specific processes.

2. THE BREAKDOWN OF MITOCHONDRIA

In this section are discussed three components involved in the breakdown of mitochondria of reticulocytes during their maturation: (1) the erythroid cell-specific lipoxygenase; (2) the mitochondrial susceptibility factor; and (3) the ATP-dependent proteolytic system. Finally, their interplay is outlined in a synopsis.

2.1. The Lipoxygenase System

2.1.1. General Remarks

Lipoxygenases constitute a multigene family of closely related enzymes that catalyze the dioxygenation of polyenoic fatty acids forming hydroperoxy derivatives as primary products. The basic lipoxygenase reaction can be expressed by the general scheme shown in Figure 1. Lipoxygenases are widely distributed in both the animal and the plant kingdom. In many cases their action is preceded by lipid-splitting enzymes (e.g., phospholipase A_2 that liberate their substrates, the free polyenoic fatty acids, which is the case, e.g., in the biosynthesis of lipoxygenase-derived eicosanoids from arachidonic acid (Borgeat *et al.*, 1985; Samuelsson *et al.*, 1987). In great contrast, the erythroid cell-specific lipoxygenase acts directly on membrane phospholipids, i.e., even independently of a phospholipase. Its biological function is closely related to the breakdown of mitochondrial membranes during the maturation process of the reticulocytes.

Before discussing this role of the erythroid-specific lipoxygenase, it is appropriate first to introduce the molecular characteristics, the molecular biology, and the enzymology of this enzyme. Most studies have been performed with the lipoxygenase of rabbit reticulocytes, which was the first mammalian lipoxygenase to be purified and extensively characterized (Rapoport *et al.*, 1979; Schewe *et al.*, 1986). Its discovery was connected with the successful search for a factor in the cytosol of reticulocytes that is responsible for the inactivation of the mitochondrial respiratory chain during the maturation process. A specific inhibitory protein was found (Rapoport and Gerischer-Mothes, 1955; Rapoport and Nieradt-Hiebsch, 1955) that later was identified as lipoxygenase (Rapoport *et al.*,

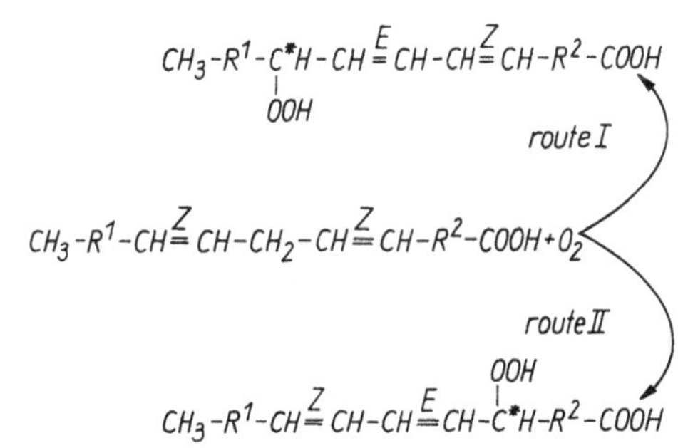

FIGURE 1. The basic lipoxygenase reaction. R^1, R^2 are saturated or *cis*-unsaturated alkyls. Routes I and II are alternative depending on the specificity of the enzyme. *E, Z* specify the geometry of the double bonds. C* marks an asymmetric carbon at which the oxygen is introduced stereospecifically after the initial hydrogen abstraction (for details see Kühn *et al.*, 1986a). The carboxylic group may be esterified in some cases.

1979). The lipoxygenase of rabbit reticulocytes and its capability to attack phospholipids and mitochondrial membranes were first described by Schewe et al. (1975).

2.1.2. Molecular Properties of Erythroid Lipoxygenase

The erythroid lipoxygenase consists of a single polypeptide chain with a molecular mass of 75.3 kDa. Its amino acid sequence was recently established for both the enzyme of rabbits (Fleming *et al.*, 1989) and that of man (Sigal *et al.*, 1988a) by determination of the nucleotide sequences of full-length cDNAs. The rabbit enzyme consists of 662 amino acids and the human enzyme of 661 amino acids. A comparison of the recently available sequence data for the erythroid 15-lipoxygenases as well as those for the lipoxygenases of human leukocytes (5-lipoxygenase, Dixon *et al.*, 1988; Matsumoto *et al.*, 1988), of rat basophilic leukocytes (5-lipoxygenase, Balcarek *et al.*, 1988), and of soybeans (isoenzyme 1, Shibata *et al.*, 1987; isoenzyme 2, Shibata *et al.*, 1988; isoenzyme 3, Yenofsky *et al.*, 1988) shows clearly that the various lipoxygenases constitute a multigene family. An alignment for maximal homology between rabbit reticulocyte lipoxygenase, human leukocyte 5-lipoxygenase, and soybean lipoxygenase-3 is shown in Figure 2. It is evident that a variety of conserved regions occur in all lipoxygenases studied. As expected, the erythroid cell lipoxygenases of rabbit and man exhibit the highest degree of homology (81% identical amino acids) as do the 5-lipoxygenases of rat and man (92%). The degree of homology is considerably lower if the erythroid 15-lipoxygenases are compared with the leukocyte 5-lipoxygenases (38% identical amino acids for the rabbit reticulocyte lipoxygenase and the human leukocyte 5-lipoxygenase), which may reflect structural divergences necessary for the different positional specificities and other properties that are unique to the 5-lipoxygenases such as membrane binding (Rouzer and Samuelsson, 1987) and dependence on Ca^{2+} and ATP (Fuarakawa *et al.*, 1984; Ueda *et al.*, 1986) or those characteristic for the erythroid 15-lipoxygenase such as overexpression in anemia or masking of the mRNA (see Sections 2.1.3 and 2.1.5). Although the degree of homology and analogy between animal and plant lipoxygenases is considerably lower (e.g., 14% identical and 45% similar amino acid residues between the rabbit reticulocyte 15-lipoxygenase and the soybean lipoxygenase-3), the alignment also shows a clear relationship between the plant and the animal enzymes. Immunological and hybridization data indicate a much closer relationship among the erythroid lipoxygenases than between them and lipoxygenases of other cell types (Kroschwald *et al.*, 1989a). These distinctions reflect the evolutionary conservation of the erythroid enzymes as well as their divergence from the other types of lipoxygenase. The high degree of homology found at the level of amino acids between erythroid and leukocyte lipoxygenases is not reflected at the level of nucleotide sequences with frequent diversity of the third bases of the triplets. This circumstance explains the poor cross-hybridization between erythroid lipoxygenase probe and other lipoxygenase mRNAs.

The conserved parts of the primary structure may be involved in the fatty acid binding region and in the catalytic center of the lipoxygenases. Like other lipoxygenases, the erythroid enzyme contains one nonheme iron per mol that is believed to be involved in the mechanism of catalysis (Schewe *et al.*, 1986). The kind of ligandation by the polypeptide chain has not been clarified; however, a cluster of five histidines in a conserved region (Figure 2) and a recently published X-ray absorption fine structure analysis of soybean

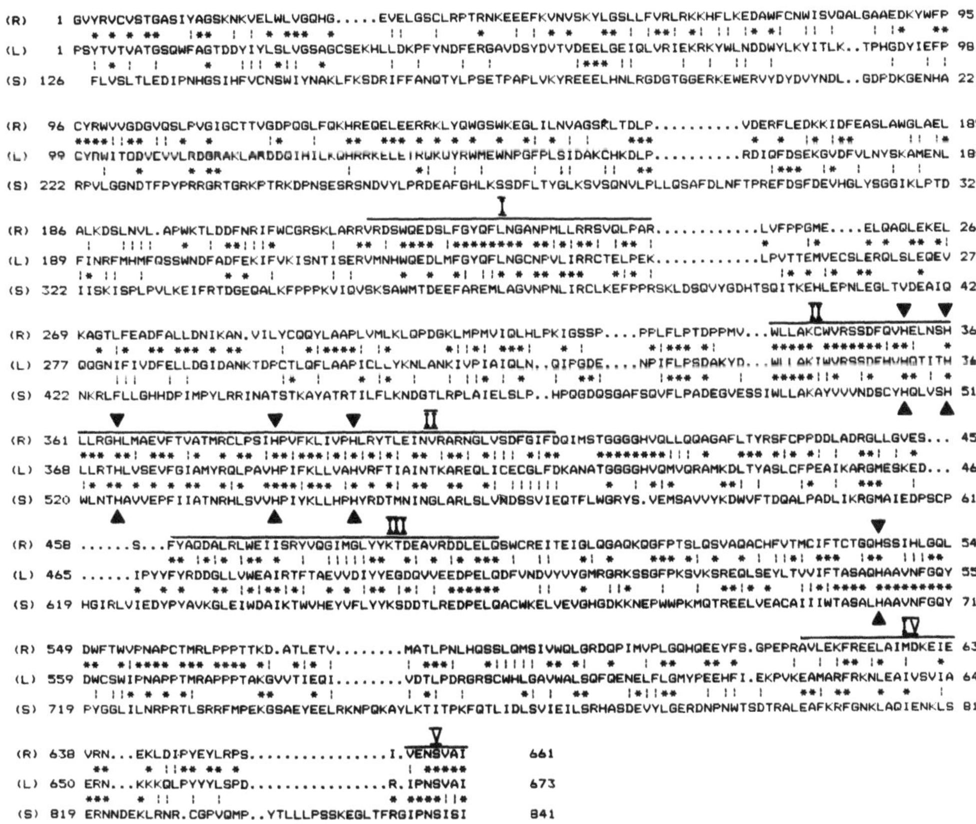

FIGURE 2. Comparison of the amino acid sequence of the rabbit erythroid 15-lipoxygenase with those of other lipoxygenases. The amino acid sequence predicted for the rabbit erythroid 15-lipoxygenase (R) is compared with those encoding the human leukocyte 5-lipoxygenase (L) (Matsumoto *et al.*, 1988; Dixon *et al.*, 1988) and the gene encoding the soybean lipoxygenase-3 (S) (Yenofsky *et al.*, 1988) using the BEST FIT program of the software package of the Genetics Computer Group of the University of Wisconsin, Biotechnology Center. Perfect matches are marked with asterisks and possible conservative changes by vertical dashed lines. The six conserved regions are indicated by solid triangles and the five regions of close homology are overlined and numbered I–V.

lipoxygenase-1 (Navaratnam *et al.*, 1988) point to a role of imidazole residues in chelating the iron moiety in the active site. In the global amino acid composition of rabbit reticulocyte lipoxygenase versus that of soybean lipoxygenase-1, a higher content of cysteine and tryptophan as well as a lower content of charged amino acids is conspicuous; disulfide bridges cannot be detected. The structural peculiarities of the erythroid lipoxygenase impart to it a high hydrophobicity and flexibility that may be important for its interaction with membranes. The isoelectric point is 5.5, which is comparable to other lipoxygenases. The possible presence of pyrroloquinoline quinone (PQQ) in erythroid lipoxygenase is now being studied. PQQ was recently claimed to be present in soybean lipoxygenase-1 (Van der Meer and Duine, 1988). Rabbit reticulocyte lipoxygenase contains about 5% carbohydrate, the functional importance of which remains to be clarified; among the monosaccharides, galactose and galactosamine but not neuraminic acid were detected. A search for the consensus sequence of the type Arg-X-Ser or Arg-X-Thr, which is known as the target for *O*-glycosylation (Wathelet *et al.*, 1986), revealed one element that meets this requirement: it is the sequence Arg-Val-Ser (amino acids 51–53 from the N-terminus). The presence of carbohydrates in erythroid lipoxygenase may be due to cytosolic glycosylation (Holt *et al.*, 1987). Owing to the carbohydrate content the real molecular mass of erythroid cell lipoxygenase is about 78 kDa as determined by three independent methods (Rapoport *et al.*, 1979).

2.1.3. Cell Specificity and Molecular Biology of Erythroid Lipoxygenase

The erythroid lipoxygenase was detected in rabbits (Schewe *et al.*, 1975; Rapoport *et al.*, 1979) and man (Kroschwald *et al.*, 1989a; Sigal *et al.*, 1988a) as well as rhesus monkeys, rats, and mice (Kroschwald *et al.*, 1986); indirect evidence for its presence in chicken reticulocytes was also obtained (Augustin and Rapoport, 1959). Thus, one may assume a general occurrence of this lipoxygenase in reticulocytes of all species. By means of Northern and genomic Southern blot experiments using partial cDNA recombinants derived from the rabbit reticulocyte lipoxygenase, it was shown that the lipoxygenase mRNA is erythroid cell-specific. Erythroid lipoxygenase mRNA is completely absent from spleen, skeletal and heart muscle, liver, kidney, brain, lung, and lens of the rabbit (Thiele *et al.*, 1987a) as well as from human leukocytes (Kroschwald *et al.*, 1989a), which are known to contain both a 5- and a 15-lipoxygenase (Narumiya *et al.*, 1981). It is also not expressed in early erythroid progenitor cells such as the Dexter stem cell line or uninduced Friend murine erythroleukemia cells, but it is inducible by dimethylsulfoxide in Friend cells together with the globin mRNAs (Thiele *et al.*, 1987b). Moreover, human reticulocyte lipoxygenase mRNA obtained from red cells of anemic patients exhibited in Northern blot experiments marked hybridization with a cloned rabbit reticulocyte lipoxygenase cDNA probe under stringent conditions, indicating a close relationship of the two erythroid cell lipoxygenases at the mRNA level. This close relationship also exists at the protein level, as evident by the immunological cross-reactivity of the human enzyme with antiserum against the rabbit enzyme (Kroschwald *et al.*, 1989a), whereas this antiserum does not give positive reactions with homogenates from any other tissue of the rabbit (Rapoport *et al.*, 1979) nor with those from human leukocytes (Kroschwald *et al.*, 1989a). Finally, the relationship is clearly established by comparison of the amino acid sequences (see Section 2.1.2).

The complete base sequence of rabbit erythroid cell lipoxygenase mRNA is now established (Figure 3). Its translated sequence corresponds to 662 nucleotide triplets. As shown by S-1 protection and primer extension experiments with the cDNA and the corresponding portion of a genomic LOX recombinant, it contains a very short untranslated region of 27 bp at the 5′ end, but a long untranslated 3′ region of 581 bp without the poly(A). This portion is characterized by a conspicuous repeated sequence consisting of ten copies with the consensus sequence $C_4PuC_3TCTTC_4AAG$ just after the translational termination codon (Thiele *et al.*, 1987c). The same structural motif is present in slightly modified form in the 5′ untranslated region of the mRNA coding for human reticulocyte 15-lipoxygenase (four copies) and in that for rat leukocyte 5-lipoxygenase (one copy in more distorted form); however, it is missing from human leukocyte 5-lipoxygenase mRNA. Preliminary footprinting experiments with a genomic fragment bearing this re-

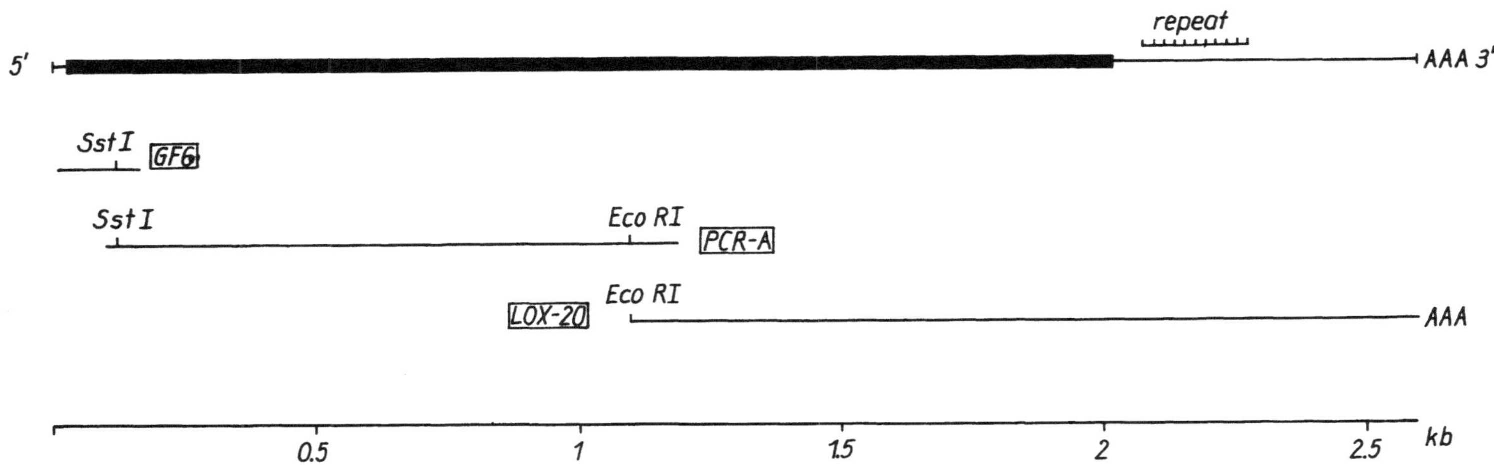

FIGURE 3. Structure of the rabbit erythroid 15-lipoxygenase mRNA. The complete structure was obtained by sequencing the lipoxygenase cDNA clones GF6, LOX-20, and the polymerase chain reaction product PCR-A. The structure of the 5′ end was determined by S1-protection studies and primer extension as described (Fleming *et al.*, 1989).

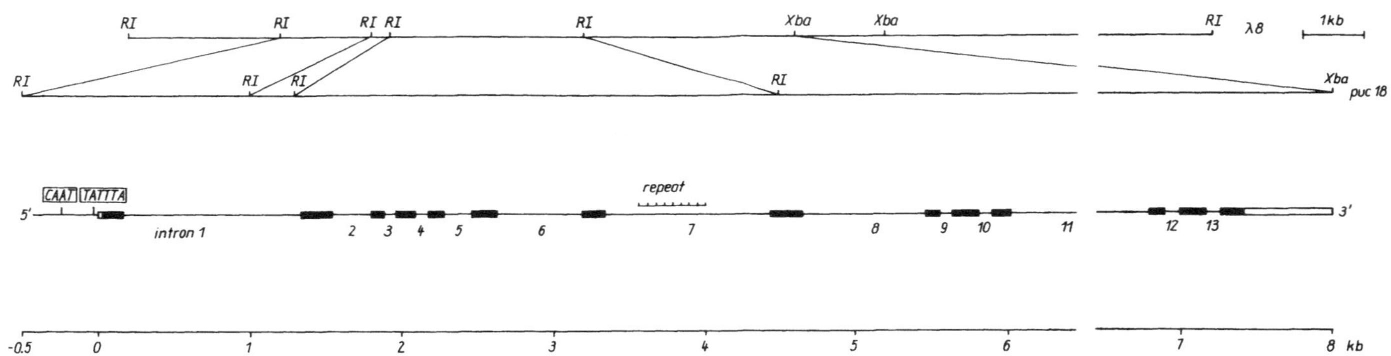

FIGURE 4. Organization of the rabbit erythroid 15-lipoxygenase gene. The intron/exon organization was obtained by comparing the rabbit erythroid 15-lipoxygenase cDNA sequence (Fleming *et al.*, 1989) with the sequences of *Eco*RI or *Eco*RI/*Xba*I fragments subcloned in pUC18 of the rabbit genomic recombinant lambda 8 (Thiele *et al.*, 1989). Exons are represented as solid boxes. The untranslated parts of exon 1 and 14 are shown as open boxes.

petitive element indicate that it could be involved in red-cell-specific transcription: nuclear protein extracts from erythroid cells (erythroleukemia cells) bind specifically to all ten elements of the repeat, but not those from nonerythroid cells (fibroblasts) (P. R. Harrison, personal communication).

The expression of erythroid lipoxygenase mRNA occurs synchronously with the mRNAs for the globin chains, glutathione peroxidase, and other erythroid-specific proteins (Harrison *et al.*, 1987). Therefore, common factors that activate their genes in a coordinated manner may be assumed. Some insight in this regulation may be expected upon elucidation of the structure of the erythroid lipoxygenase gene. The complete nucleotide sequence and the intron/exon organization have now been established (O'Prey *et al.*, 1989). Moreover, about 500 nucleotides of the upstream promoter region have been sequenced. From the cap site to the poly(A) site the transcription unit comprises 8.0 kb. It contains 13 introns (Figure 4). In the 5′ promoter region, typical promoter elements have been found such as a modified TATA box (TATTTA) at −32, a CAAT box at −239, and a CACC element at −214. Their functionality could be demonstrated using constructs with the bacterial chloramphenicol transferase gene in cell transfection experiments (CAT assays) Intron 7 contains another remarkable repetitive element. It consists of eight copies of a highly conserved 54-bp unit. Its possible involvement in transcriptional regulation is currently under investigation. The multiple regulation by promoters and enhancers may be the reason for the overexpression of the erythroid lipoxygenase caused by stress erythropoiesis. Such an overexpression is induced by forced bleeding or hemolytic anemia as judged from Northern blot experiments (Thiele *et al.*, 1987b). This overexpression appears to be the main reason for the high cellular level of lipoxygenase protein in reticulocytes under anemic conditions.

Erythroid lipoxygenase is only synthesized in the reticulocyte, whereas in the preceding stages of the lipoxygenase mRNA is present as free translationally inactive ribonucleoprotein particles (mRNP particles) as detected by Northern blotting with the corresponding lipoxygenase cDNA probe (Thiele *et al.*, 1982; Höhne *et al.*, 1988). There are some indications for the presence of masking proteins in young reticulocytes (Höhne *et al.*, 1988). *In vitro* experiments demonstrated that an unmasking of the lipoxygenase mRNA proceeds in the intact cell and that extracellular factors are not needed. Moreover, it was shown that the cytosol of more mature reticulocytes but not that of erythrocytes causes an unmasking of the lipoxygenase mRNP particles. These findings suggest that the translation of lipoxygenase mRNA in erythroid cells may be regulated by maturation-dependent masking and unmasking proteins that permit a fine-tuning of this specific protein synthesis, which in this way is delayed up to the reticulocyte stage, in which the erythroid lipoxygenase becomes necessary for the breakdown of mitochondria.

2.1.4. Enzymology of Erythroid Lipoxygenase

A characteristic feature of the erythroid lipoxygenase is its ability to react with phospholipids and even with biological membranes without the preceding action of a phospholipase (Schewe *et al.*, 1975; Fiskum *et al.*, 1985; Murray and Brash, 1988). Obviously, the reaction of erythroid lipoxygenase with phospholipids also occurs in intact cells, since oxygenated fatty acids were recently detected in the phospholipids of both mitochondria and plasma membranes of rabbit reticulocytes, whereas these products are practically absent in mature erythrocytes lacking lipoxygenase (Kühn *et al.*, 1990a, 1990b). As in other cells, the primary products of the lipoxygenase reaction, the hydroperoxy compounds, are rapidly reduced to the corresponding hydroxy derivatives. Part of the phospholipid-bound hydroperoxy polyenoic fatty acid is, however, converted to oxodienoic acid (Kühn *et al.*, 1990c), a product of the lipohydroperoxidase activity of lipoxygenases (see below). There is apparently no selectivity of the erythroid lipoxygenase between various phospholipid classes (Kühn *et al.*, 1990). The enzyme also reacts with cholesteryl esters of polyenoic fatty acids (Kühn and Brash, unpublished results).

Reticulocytes of various species metabolize exogenous arachidonic acid to 15*S*-hydroxy-5*Z*,8*Z*,11*Z*,13*E*-eicosatetraenoic acid and linoleic aid to 13*S*-hydroxy-9*Z*,11*E*-octadecadienoic acid (Bryant *et al.*, 1982; Salzmann-Reinhardt *et al.*, 1985; Kroschwald *et al.*, 1986, 1989). The same products are found if pure erythroid lipoxygenase is allowed to react with these polyenoic fatty acids following reduction of the primarily formed hydroperoxy derivatives (Bryant *et al.*, 1982; Kühn *et al.*, 1983a). 13*S*-hydroxy-9*Z*,11*E*-octadecadienoic acid is also the main component among the esterified oxygenated polyenoic fatty acids found in reticulocytes (Kühn *et al.*, 1990b). These observations show that the reaction specificity of the pure enzyme does not differ from that in intact cells and second, that the reaction specificity with free and esterified polyenoic fatty acids is in principle identical. Since the animal lipoxygenases are conventionally classified according to their reaction specificity toward arachidonic acid, the erythroid cell lipoxygenase is a 15-lipoxygenase (Bryant *et al.*, 1982; Kühn *et al.*, 1983a). A report on the supposed presence of 12-lipoxygenase activity in red cells of various species (Kobayashi and Levine, 1983) could not be verified in our laboratory (Ludwig *et al.*, 1988). The erythroid 15-lipoxygenase appears to differ from the 15-lipoxygenase of neutrophils (Narumiya *et al.*, 1981) since a probe for rabbit erythroid lipoxygenase mRNA did not hybridize with mRNA from leukocytes (Kroschwald *et al.*, 1989). The 15-lipoxygenase of human eosinophils seems to be closely related to erythroid lipoxygenase, since the sequence of 14 amino acid residues at the NH_2-terminus proved to be identical except for the N-terminal glycine of erythroid lipoxygenase, which is lacking in the eosinophil lipoxygenase (Sigal *et al.*, 1988a,b).

In vitro, free polyenoic fatty acids are better substrates for the erythroid cell lipoxygenase than the esterified ones, which may be mainly due to their higher solubility in water. For this reason most studies on the enzymology of erythroid cell lipoxygenase were performed with unesterified substrates. These studies are the subject of other reviews (Schewe *et al.*, 1986; Kühn *et al.*, 1986a). In addition to single oxygenation of polyenoic fatty acids, erythroid lipoxygenase is also able to catalyze a variety of other reactions forming a fairly complex product pattern from arachidonic acid, among them the hydrolysis products of an epoxyleukotriene (Bryant *et al.*, 1985; Kühn *et al.*, 1990d). It may be mentioned that the pure reticulocyte lipoxygenase also forms lipoxin B via sequential multiple oxygenation of arachidonic acid (Kühn *et al.*, 1987). Under anaerobic conditions or at low concentrations of oxygen, the enzyme acts as a lipohydroperoxidase. It forms linoleic acid, oxodienic acids, linoleic acid dimers, epoxyhydroxy derivatives, and pentane (Härtel *et al.*, 1982; Kühn *et al.*, 1986b; Salzmann *et al.*, 1984).

The kinetics of rabbit erythroid lipoxygenase was studied in detail using linoleic acid as substrate in the presence of 0.2% sodium cholate (Ludwig *et al.*, 1987). A new kinetic model was developed for lipoxygenases that describes correctly the experimental data. In contrast to earlier models for the aerobic lipoxygenase reaction, this model postulates the presence of only one fatty acid binding site. Little information is available as to the kinetic characteristics of erythroid lipoxygenases from species other than the rabbit. Such investigations are hampered by the lability of these enzymes. The rat erythroid lipoxygenase differs from the rabbit enzyme in two ways: first, it is inhibited rather than activated by sodium cholate, and second, it shows a much higher degree of substrate inhibition (Schewe *et al.*, 1990).

A typical property of all lipoxygenases is their tendency to self-inactivation, i.e., the irreversible loss of activity during their reaction with a substrate or directly by product. For the rabbit reticulocyte lipoxygenase it was shown that the self-inactivation is accompanied by a mono-oxygenation of a single methionine residue forming methionine sulfoxide (Rapoport *et al.*, 1984). The self-inactivation of erythroid lipoxygenase also occurs during the reaction with membranes and may be the cause for the loss of lipoxygenase activity during the maturation process of reticulocytes. The erythroid lipoxygenase is inhibited by a variety of lipoxygenase inhibitors such as 5,8,11,14-eicosatetraenoic acid and other acetylenic fatty acids, nordihydroguaiaretic acid, 4-nitrocatechol, 3-*t*-butyl-4-hydroxyanisol, 3-amino-1-(3-trifluoromethylphenyl)-pyrazoline (BW 755C), and certain hydroxamic acids (Schewe *et al.*, 1986). For studies with red cells and mitochondria, salicylhydroxamic acid (1 mM) proved to be valuable, since this compound does not interfere with the mitochondrial respiratory chain. Owing to the close relationship of the erythroid lipoxygenase with the leukocyte 5-lipoxygenase (see Section 2.1.2), many "specific" 5-lipoxygenase inhibitors also act on the erythroid lipoxygenase.

2.1.5. Biological Dynamics and Cellular Activity of the Lipoxygenase in Rabbit Red Cells

In earlier studies it was shown that lipoxygenase is absent from mature erythrocytes and appears on the fourth to fifth day of forced bleeding, which induces a reticulocytosis (Rapoport *et al.*, 1979). Separation of the cells by buoyant density centrifugation according to their age revealed that the youngest cells may completely lack lipoxygenase. The highest activities were observed in the cells of medium density, whereas the oldest cells with the highest density contained low lipoxygenase activity (Wiesner *et al.*, 1973). These observations indicated that the erythroid lipoxygenase exhibits marked biological dynamics both in the course of an experimental bleeding anemia and during the maturation of reticulocytes. In later studies the biological dynamics of the lipoxygenase during the course of an experimental anemia induced by forced bleeding of rabbits was investigated in more detail (Ludwig *et al.*, 1988). Lipoxygenase activity was assayed both in the supernatant of the hemolysate with linoleic acid as substrate and in intact cells with either [^{14}C]linoleic acid or [^{14}C]arachidonic acid. The concentration of lipoxygenase in the hemolysates was determined by immunoprecipitation. The data of one representative experiment are shown in Figure 5. As can be seen, the three parameters behaved in an approximately parallel manner. Before bleeding, no lipoxygenase could be detected by any of the methods. The lipoxygenase appeared after the third day and increased continuously as long as forced bleeding was carried out. In general, the appearance of large amounts of lipoxygenase goes hand in hand with the outpouring of so-called megaloreticulocytes from the bone marrow (see Section 1.2). The concentration of lipoxy-

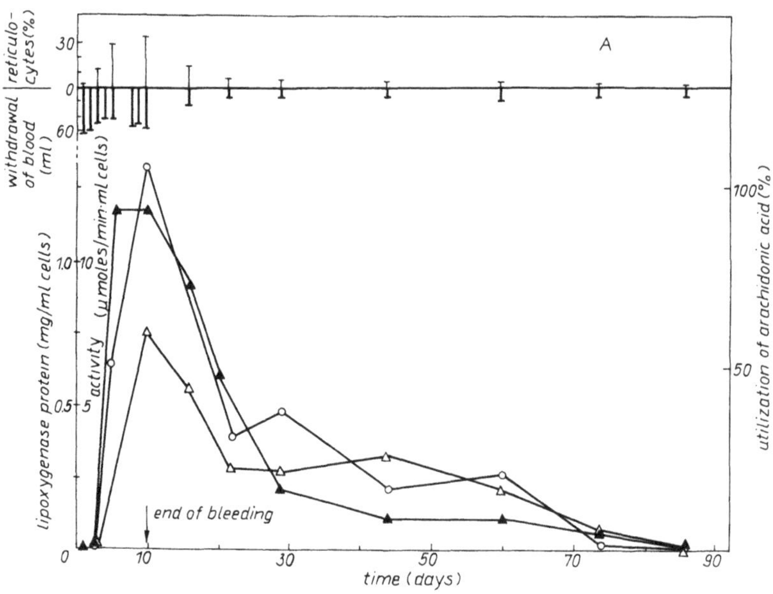

FIGURE 5. Lipoxygenase in rabbit red cells during bleeding anemia and recovery (Ludwig *et al.*, 1988). The bleeding schedule and the reticulocyte percentage are indicated by the bars in the upper part of the figure. Immunoreactive lipoxygenase protein (○) and lipoxygenase activity with linoleic acid as substrate (△) were determined in an ammonium sulfate precipitate from the red cell cytosol. The utilization of [^{14}C]arachidonic acid (▲) of intact cells was estimated by radioscanning thin-layer chromatography with 0.5 ml of packed cells.

genase protein can reach levels up to 4 mg/ml cells (Höhne *et al.*, 1983); under these conditions, it accounts for 3–4% of the newly synthesized proteins including hemoglobin (Thiele *et al.*, 1979). Since such high levels do not appear under nonanemic conditions either in the reticulocytes of circulating blood or in bone marrow cells, the lipoxygenase synthesis in megaloreticulocytes may represent a luxurious overproduction that is mainly due to an overexpression of lipoxygenase mRNA in erythroid bone marrow cells in response to the anemia (see Section 2.1.3). A second reason for the high levels of lipoxygenase in the reticulocytes in severe anemia may be an inhibition of the breakdown of lipoxygenase mRNA. Indirect evidence for such an assumption comes from a recent study, in which the synthesis of lipoxygenase and the content of lipoxygenase mRNA in addition to the cellular amount of lipoxygenase protein were determined during an anemia provoked by administration of phenylhydrazine (Minich *et al.*, 1989). In this study a divergence of lipoxygenase content and capacity for lipoxygenase synthesis was observed in the early recovery period after the anemia. The synthesis of lipoxygenase rapidly dropped to zero, which was shown to be due to a loss of lipoxygenase mRNA, whereas the apparatus for protein synthesis was still functional as judged from the synthesis of globin chains in these cells. Nevertheless, these cells contained a fairly high amount of lipoxygenase. It remains to be clarified whether these recovery reticulocytes lacking lipoxygenase mRNA represent a special cell population unique for the recovery period or an obligatory stage of reticulocyte maturation. In any case, the demonstration of this type of reticulocyte reveals that there occurs a situation in which the lipoxygenase mRNA is selectively broken down even after lipoxygenase synthesis. This selective breakdown of lipoxygenase mRNA may be suppressed in the megaloreticulocytes occurring at the height of an anemia.

From Figure 5 it may be seen that in the recovery period following the bleeding anemia the percentage of reticulocytes dropped rapidly to normal values after about 10 days whereas relatively high enzyme levels persisted for several weeks. Lipoxygenase was still detectable up to 2–3 months after discontinuation of the bleeding, which corresponds to the presumed life span of rabbit red cells. This behavior may be interpreted as follows: The reticulocytes that poured out during the severe anemia possess a strongly elevated content of lipoxygenase, which is not completely inactivated and degraded during the maturation process so that part of it is retained in the erythrocyte stage of this cohort of cells. In this manner the lipoxygenase of mature cells becomes a label for the anemic origin of the cells. Preliminary results suggest that the situation in man may be similar to the rabbit (Kroschwald *et al.*, 1989a). For this reason, the lipoxygenase of human red cells may perhaps serve as a retrospective criterion for a preceding anemia. A study in this direction is planned.

The absence of lipoxygenase from erythrocytes under nonanemic conditions is probably the result of the inactivation and degradation of the enzyme during the maturation of the reticulocytes following its suicidal inactivation during the reaction with the mitochondrial membrane. This self-inactivation may facilitate the proteolytic breakdown of the lipoxygenase. The role of proteinases in the maturational decline of lipoxygenase in reticulocytes is suggested by the observation that leupeptin, an inhibitor of thiol proteinases, prevented the decrease of lipoxygenase during incubation of reticulocyte hemolysates (Rapoport *et al.*, 1979).

A comparison of the lipoxygenase activities assayed in the hemolysates and in the corresponding intact cells with exogenous free polyenoic acids under optimized conditions revealed that the erythroid lipoxygenase is fully active in the cell (Ludwig *et al.*, 1988). In this regard the erythroid lipoxygenase differs from the 15-lipoxygenase of leukocytes, which is in a masked state and is only unmasked if the cells are damaged.

Convincing evidence for the endogenous action of erythroid lipoxygenase is the demonstration of the presence of oxygenated fatty acids in membrane lipids of rabbit reticulocytes (Kühn and Brash, 1989). About nine-tenths of the total amount of the oxygenated polyenoic fatty acids were found to be esterified. Oxygenated fatty acids occur both in mitochondria and in plasma membranes of reticulocytes, their content in mitochondria being three times higher corresponding to one-fifth of the total linoleic acid content (Kühn *et al.*, 1990b). The clear evidence that lipoxygenase-derived products are also fund in the plasma membrane implies that the erythroid lipoxygenase is not only involved in the maturational breakdown of mitochondria but possibly also in restructuring processes of the plasma membrane during maturation. Among the esterified hydroxypolyenoic fatty acids, the enantiomers expected from the reaction specificity of erythroid lipoxygenase predominate, which excludes nonenzymatic lipid peroxidations, such as those catalyzed by heme compounds which would yield racemic mixtures of the hydroxypolyenoic acids. The simultaneous formation of oxopolyenoic fatty acids both in intact reticulocytes and in rat liver mitochondria after incubation with erythroid lipoxygenase indicates that lipohydroperoxidase activity is also operative in addition to that of oxygenase of erythroid lipoxygenase. Low local oxygen concentrations may prevail at the mitochondrial inner membrane owing to the action of the cytochrome *c* oxidase, which has a high affinity for oxygen. Under such conditions the formation of nonchiral products in this membrane would be favored owing to the peculiarities of lipoxygenase kinetics (Ludwig *et al.*, 1987). Indeed, differential analyses of the oxygenated polyenoic fatty acids revealed a high share of racemic products in the mitochondria but not in the plasma membranes of reticulocytes (H. Kühn, unpublished results, 1989).

The endogenous action of the erythroid lipoxygenase in intact cells seems to account for 5–7% of their total oxygen uptake. This percentage was estimated from the inhibition by various lipoxygenase inhibitors in the presence of the respiratory chain inhibitor antimycin A (Salzmann *et al.*, 1985); it corresponds to 1.2 μmol oxygen/ml cells per hr. This value seems to be an overestimate, since antimycin A may facilitate the lipoxygenase pathway by blocking β-oxidation and the incorporation of free polyenoic fatty acids in membrane lipids. In line with such an effect is the augmentation by antimycin A of the conversion of [^{14}C]arachidonic acid to 15*S*-hydroxy-5*Z*,8*Z*,11*Z*,13*E*-eicosatetraenoic acid by rabbit reticulocytes (Salzmann-Reinhardt *et al.*, 1985).

There is an obvious inconsistency between the amount of lipoxygenase products formed from endogenous or exogenous substrates on the one hand and the lipoxygenase-mediated oxygen uptake on the other, the latter being an order of magnitude higher (Kühn *et al.*, 1990a). One has to conclude that in intact cells there is a large excessive oxygen consumption that is initiated by the lipoxygenase reaction but involves secondary oxygen-consuming degradative processes not yet identified. A similar situation holds for the action of purified lipoxygenase on isolated mitochondria (see Section 2.1.6).

In the presence of calcium plus ionophore A23187, rabbit reticulocytes produce pentane. This pentane formation is suppressed by lipoxygenase inhibitors (Salzmann *et al.*, 1984). Two possibilities may be regarded as the cause of the lipoxygenase-mediated pentane formation:

1. Via a lipohydroperoxidase activity of erythroid lipoxygenase, which was demonstrated in a system containing purified lipoxygenase, linoleic acid, and 13*S*-hydroperoxy-9, 11*E*-octadecadienoic acid (Härtel *et al.*, 1982)
2. Lipoxygenase-induced nonenzymatic lipid peroxidation, since lipid peroxidations involving *n*-6 polyenoic fatty acids are generally accompanied by pentane formation

The calcium influx into the cell may facilitate the cellular lipoxygenase reaction by mediating liberation of free fatty acids via activation of phospholipase A_2; free fatty acids are better substrates for the lipoxygenase than esterified ones being oxygenated at an approximately fivefold higher rate.

2.1.6. Interactions of Erythroid Lipoxygenase with Membranes

In early work we demonstrated that the lipoxygenase of rabbit reticulocytes severely attacks isolated rat liver mitochondria (Schewe *et al.*, 1975). The action was evident from the formation of a high amount of thiobarbituric acid-reactive material (hydroperoxylipids, malonyl dialdehyde), high oxygen uptake and release of mitochondrial matrix enzymes.

Moreover, inhibition of the respiratory chain occurred. These effects were thought to be solely due to pure enzymatic reactions with the mitochondrial lipids. However, it was now clearly shown that the total oxygen consumption measured after addition of lipoxygenase to a definite amount of mitochondria greatly exceeded the total amount of polyenoic fatty acids in the membrane lipids assuming a stoichiometric dioxygenase reaction (Belkner *et al.*, 1990). Similar results were obtained with crude phospholipids as substrate for the lipoxygenase. In this respect these reactions differ principally from the reaction of erythroid lipoxygenase with linoleic acid which is stoichiometric (Kühn *et al.*, 1986b). In a comparative study it was also shown that erythroid lipoxygenase attacks rat liver mitochondria, endoplasmic membranes and erythrocyte ghosts to a comparable extent if the percentage of oxygenated polyenoic fatty acids in the membranes is used as criterion, whereas great differences were found for the oxygen uptake (Belkner *et al.*, 1990). These observations lead us to the conclusion that the primary attack of the lipoxygenase on membranes is not selective and, hence, the selective action on mitochondria encompasses two sequential steps:

I. Dioxygenation of the membrane lipids as the lipoxygenase reaction *per se*
II. Secondary oxygen-consuming processes induced by the products formed in step I

Step I seems to occur whenever erythroid lipoxygenase can interact with membranes. Apparently only a small portion of the polyenoic acids of the membranes are accessible to the lipoxygenase, since the share of lipoxygenase products does not exceed a certain percentage of the total polyenoic acids. This action of the lipoxygenase does not show selectivity with respect to membrane types with little difference between the mitochondria and the plasma membrane of reticulocytes.

Step II differs in many respects from step I. It shows a marked selectivity for mitochondrial membranes as revealed by the large differences in oxygen uptake under comparable conditions. Moreover, it depends on the functional state of the mitochondria; addition of ATP or metabolic conditions maintaining a high ATP level protect the mitochondria from this type of lipoxygenase attack (Schewe *et al.*, 1975). In contrast to the primary lipoxygenase action, step II is inhibited by EDTA. EDTA also completely prevents the lipoxygenase-mediated inhibition of some parts of the respiratory chain. The effect of this chelating agent, which is not a lipoxygenase inhibitor, suggests the involvement of iron as catalyst of this step. Mitochondria are rich in iron–sulfur proteins, some of them being destroyed upon lipoxygenase attack (Schewe *et al.*, 1981). Micromolar amounts of ferrous iron are known to be an effective catalyst of lipid peroxidation in mitochondria (Vladimirov *et al.*, 1980). Iron impurities are also present in usual phospholipid preparations, which may account for similar affects during the action of erythroid lipoxygenase on phospholipids without special measures to remove traces of iron.

The action of the erythroid lipoxygenase on beef heart submitochondrial particles leads to an irreversible inhibition of both succinate and NADH oxidase activities. The sites of action are fairly specific; they are located at the interaction sites of uniquinone with the iron–sulfur clusters of Complex I and Complex II of the electron transfer system (Schewe *et al.*, 1981). The iron–sulfur centers of the electron transfer system are not damaged under these conditions, whereas the iron–sulfur centers present in the mitochondrial outer membrane are destroyed (Schewe *et al.*, 1981). The reaction of erythroid lipoxygenase with submitochondrial particles is also accompanied by drastic changes of the passive electric membrane properties (Kühn *et al.*, 1983b). It is remarkable that both respiratory inhibition and changes in the passive electric properties continue with time even if the oxygen in the sample is exhausted; obviously, anaerobic secondary processes or the lipohydroperoxidase of the lipoxygenase are involved in the alterations of the mitochondrial membranes. The incubation of rat liver mitochondria with the pure erythroid lipoxygenase also causes inactivation of the monoamine oxidase activities A and B. The inactivation proceeds in a biphasic manner; a short-term reversible inhibition is followed by a long-term irreversible inactivation. Moreover a conversion of the monoamine oxidase to a diamine oxidase was observed (Wiesner *et al.*, 1989, 1990). It is proposed that the reversible inhibition is a consequence of the alteration of the lipid environment owing to oxygenation of the membrane lipids, whereas the irreversible inactivation may be due to secondary modification of the enzyme protein.

2.2. The Mitochondrial Susceptibility Factor (MSF)

2.2.1. Demonstration of the Occurrence of MSF in Rabbit Reticulocytes

The assumption of a further factor involved in the breakdown cascade of mitochondria in the maturation process of reticulocytes in addition to erythroid lipoxygenase and ATP- and ubiquitin-dependent proteolysis was based on the following considerations: The participation of the lipoxygenase has been clearly shown in experiments on *in vitro* maturation in which an inhibition of the decline of cytochrome c oxidase as a mitochondrial marker enzyme by lipoxygenase inhibitors was observed (Schmidt *et al.*, 1985). The lipoxygenase inhibitor salicylhydroxamic acid also suppressed the ATP-dependent proteolysis in reticulocytes (Dubiel *et al.*, 1981a; Rapoport *et al.*, 1985b). A difficulty arises, however, from the coexistence of active lipoxygenase with functional mitochondria, which has to be presumed for the following reasons:

1. Mitochondria are needed for the synthesis of proteins, primarily hemoglobin, but also lipoxygenase, since the ATP requirement for it can only be met sufficiently by oxidative phosphorylation.
2. The last step in the breakdown cascade of mitochondria is the ATP-dependent proteolysis, which accounts for about 13% of the cellular ATP consumption in rabbit reticulocytes (Siems *et al.*, 1984; Rapoport *et al.*, 1985b; Rapoport, 1986).
3. Mitochondria are protected by ATP from the deleterious lipoxygenase-associated burst of oxygen uptake (see Section 2.1.6). In reticulocytes a sufficiently high ATP level should be maintained that renders the mitochondria nonsusceptible unless the protection is broken by a separate process.

Indeed, electron micrographs of intact reticulocytes revealed the simultaneous presence of both normal condensed mitochondria and swollen, damaged mitochondria in the same cell (Krause *et al.*, 1972). Moreover, there is ample evidence for the occurrence of lipoxygenase-associated oxygen uptake in reticulocytes despite the high ATP level (see Section 2.1.6). All these considerations led to the prediction of a factor that renders the mitochondria of reticulocytes susceptible to deleterious lipoxygenase attack, which in turn enables the ATP- and ubiquitin-dependent proteolytic system to degrade the mitochondrial proteins. This factor—MSF—was actually found to exist. Its discovery resulted from experi-

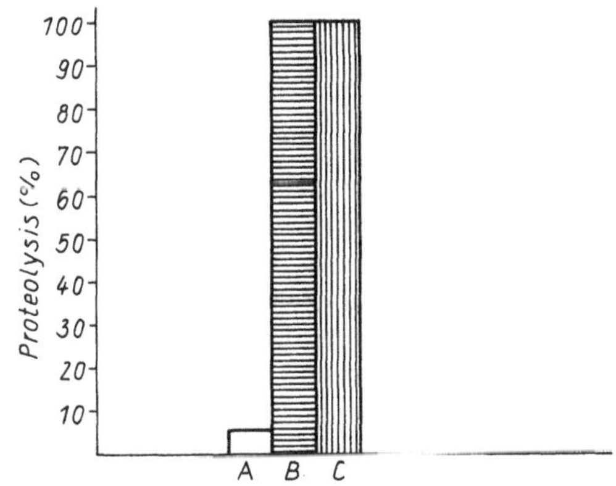

FIGURE 6. Effect of iron ATP-dependent proteolysis on intact immature reticulocytes of rabbits. The cells were fractionated according to maturity as described (Rapoport *et al.*, 1985a); the most immature cell fraction was used for measuring proteolysis after 20 hr incubation at 37°C in a Borsook–Eagle medium without iron (A) or with ferrous sulfate (B) or with ferric–transferrin complex (C).

ments on the *in vitro* maturation of reticulocytes (Rapoport *et al.*, 1986). Reticulocytes that were separated according to their density and, hence, to their maturational stage behave in a different manner. In immature reticulocytes (fraction I) the mitochondria failed to be degraded even though they already contained lipoxygenase. These cells showed, however, extensive proteolysis of the mitochondria, if they were incubated in an iron-containing medium (either as ferrous salt or as ferric–transferrin complex) (Figure 6). This proteolysis was strongly suppressed by cycloheximide, an inhibitor of cytoplasmic protein synthesis, whereas this inhibitor did not affect the proteolysis in more mature reticulocytes (fraction III) (Figure 7). The effect of cycloheximide cannot be explained by inhibition of the synthesis of lipoxygenase, since the cells in fraction I already contained a high level of lipoxygenase prior to *in vitro* maturation. The iron-triggered proteolysis in fraction I was inhibited by salicylhydroxamic acid, a lipoxygenase inhibitor. These observations were interpreted to indicate that in young reticulocytes there occurs an iron-dependent synthesis of a specific protein, the MSF. It was directly detected by incubation of the mitochondrion-containing stroma of immature reticulocytes with the cytosol of cells of the same fraction incubated with the iron-supplemented medium. This system was used as a bioassay for MSF activity. In control experiments it was shown that pure lipoxygenase does not induce proteolysis under these conditions. Quantification revealed that

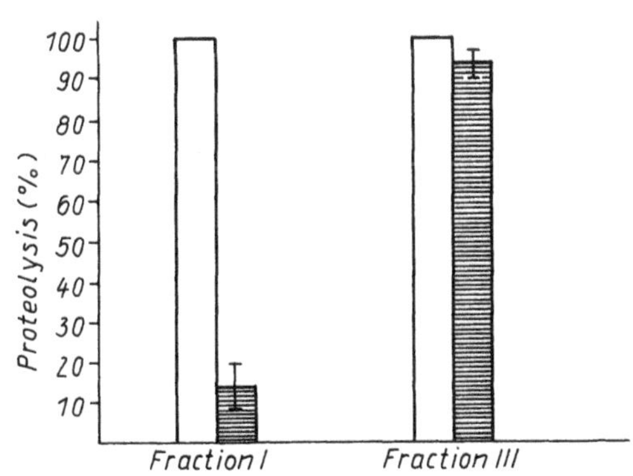

FIGURE 7. Effect of cycloheximide on ATP-dependent proteolysis in reticulocytes of different stages of maturity. Immature cells (fraction I) and cells of medium maturity (fraction III) were incubated in the presence of ferric–transferrin complex without (open columns) or with 20 μM cycloheximide (hatched columns). Other conditions as in Figure 6.

the amount of MSF in the cytosol of more mature reticulocytes exceeds that required for maximal rate of breakdown of mitochondria. Obviously the synthesis of MSF begins only after the synthesis of lipoxygenase. As for the latter, the occurrence of a masked mRNA coding for MSF must be presumed in immature reticulocytes. The role of iron in the synthesis of MSF remains to be clarified. An involvement in the unmasking of the MSF mRNA appears to be more likely than the presence of iron in the MSF molecule.

The MSF was partially purified from stroma-free hemolysates of reticulocyte-rich rabbit red cells by ammonium sulfate fractionation followed by gel chromatography on Bio-Gel P-100 (Rapoport *et al.*, unpublished). A purity of at least 50% was achieved. A molecular mass of about 12 kDa was estimated by both SDS PAGE and sedimentation in a sucrose density gradient.

2.2.2. Action of MSF on Rat Liver Mitochondria

Since the original bioassay described in Section 2.2.1 is fairly time-consuming, the interaction of MSF with rat liver mitochondria was studied as a model for the biological actions of MSF in order to gain insight into its mode of action (Rapoport *et al.*, 1989). Figure 8 shows the results of experiments with freshly prepared mitochondria on the lipoxygenase-associated O_2 uptake. In the absence of any respiratory substrate, MSF alone did not elicit oxygen consumption. Erythroid lipoxygenase when added alone caused a very low O_2 uptake. The simultaneous presence of MSF and lipoxygenase, however, produced a marked burst of O_2 consumption that was inhibited by salicylhydroxamic acid. Addition of sodium cholate (0.2%) further enhanced the lipoxygenase-mediated oxygen uptake irrespective of the presence of MSF. Apparently, sodium cholate renders the mitochondrial phospholipids completely susceptible to the lipoxygenase-associated burst of O_2 consumption, the possible nature of which was discussed in Section 2.1.6. The permissive action of MSF is time-dependent; the full effect is reached after 3–5 min at 37°C.

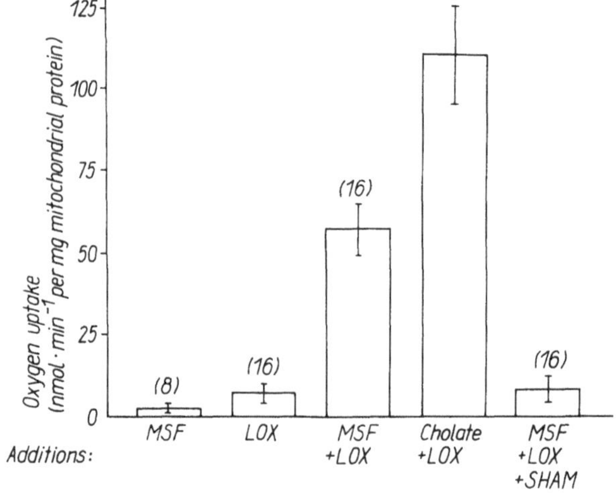

FIGURE 8. Triggering of lipoxygenase-associated oxygen consumption of rat liver mitochondria by the mitochondrial susceptibility factor (MSF). The values (mean ± SEM) were taken from oxygraphic traces with fresh mitochondria; the number of experiments is given in parentheses. LOX, lipoxygenase; SHAM, salicylhydroxamic acid.

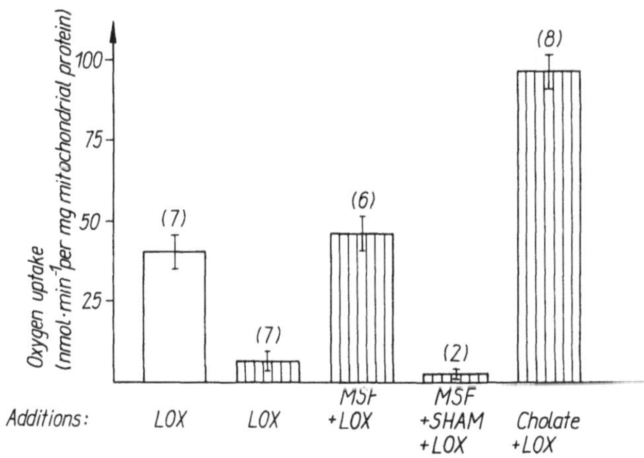

FIGURE 9. Lipoxygenase-associated oxygen consumption of rat liver mitochondria; suppression by ATP is counteracted by MSF. The mitochondria were preincubated for 15 min at 37°C in the presence (hatched columns) or absence (open column) of ATP. See Figure 8 for abbreviations.

Preincubation of mitochondria at 37°C for 15 min rendered them susceptible to lipoxygenase-associated O_2 consumption even in the absence of MSF. The extent of O_2 uptake was about the same as with unincubated mitochondria in the presence of MSF. During incubation a transition of the mitochondria from the condensed to the orthodox state also occurs. Addition of ATP prior to preincubation prevented the appearance of lipoxygenase-associated O_2 uptake; this protective action was, however, abolished by addition of MSF (Figure 9). Under the conditions of the experiment, the percentage of esterified oxygenated fatty acids in the membrane lipids was determined after treatment with lipoxygenase in the presence of ATP both with and without MSF. MSF enhanced the content of lipoxygenase-derived hydroxypolyenoic fatty acids from about 1% to only 3%. This increase did not correlate with the much higher increase in oxygen uptake, which would have accounted for the oxygenation of 20–30% of the total amount of mitochondrial polyenoic fatty acids after a minute, as calculated under the assumption that the oxygen uptake would arise from a pure lipoxygenase reaction. These experiments demonstrate clearly that MSF renders rat liver mitochondria susceptible to lipoxygenase-associated O_2 uptake and also counteracts the protective action of ATP. The reversal of the protection by ATP upon subsequent addition of either MSF or sodium cholate shows that ATP does not protect the mitochondria by inactivation of lipoxygenase but rather by affecting the susceptibility of the mitochondrial lipids to lipoxygenase-associated O_2 uptake.

In further experiments it was observed that MSF renders the mitochondria permeable to NADH as shown by strong stimulation of the NADH oxidase activity. Here again the action of MSF was time-dependent. The full effect is achieved after some minutes. Longer periods of preincubation, however, lead to an inhibition of the respiratory chain (see Section 2.2.3). MSF also lowers the respiratory control index by increasing the state 4 respiration. This uncoupling like action may be caused by the induction of a proton leakage. After longer periods of interaction there is also a decrease in the state 3 respiration, which again appears to be due to the inhibition of the respiratory chain as judged from the fact that uncouplers no longer are effective in increasing the oxygen uptake.

At high MSF concentrations there is profound morphological damage to the mitochondria, as revealed by electron microscopy.

2.2.3. Action of MSF on Submitochondrial Particles

The inhibition of the respiratory chain by MSF was studied in detail with beef heart submitochondrial particles. There occurs a time-dependent inhibition of both succinate oxidase and NADH oxidase activities. The inhibition requires longer preincubations than the permissive action on the lipoxygenase-associated oxygen uptake, the induction of permeability for NADH, or the uncouplerlike action with rat liver mitochondria. Both measurements of partial systems of the mitochondrial electron transfer system and difference spectroscopy revealed that the site of the inhibitory actions is located between cytochromes b and c_1, i.e., identical to the site of action of antimycin A. It is remarkable that the succinate oxidase activity is only partially inhibited even under conditions in which electron transfer between cytochromes b and c_1 is completely blocked. It appears that MSF induces a bypass of the succinate oxidase system, perhaps by rendering cytochrome b autoxidizable. The potency of MSF as respiratory inhibitor is very high; the concentration of half-inhibition of the NADH oxidase activity was estimated to be 17 nM, which is comparable to the efficacies of antimycin A or rotenone indicating a stoichiometric interaction with the inhibitory site. However, the dose–response curve for MSF exhibits a hyperbolic shape in contrast to antimycin A, which produces a sigmoidal curve.

In further experiments it was shown that MSF is bound to submitochondrial particles in the cold without affecting the respiratory chain. The amount of MSF to be bound shows saturation, indicating a receptorlike interaction. The binding was demonstrated after centrifugation of the particles both by subsequent incubation of the MSF-loaded particles at 37°C leading to respiratory inhibition and by addition of the supernatant fluid to fresh particles and subsequent incubation showing that the supernatant fluid is devoid of MSF. The velocity of the MSF binding is markedly higher than that of the production of respiratory inhibition but appears to be in the same range as those for the other actions of MSF. From this observation it follows that the time demand for the actions of MSF is determined by the rate of binding to the mitochondrial membranes, whereas the inhibition of the respiratory chain is a secondary process.

2.2.4. The Possible Nature of MSF

Despite the low molecular mass of MSF it is precipitated by ammonium sulfate at 70% saturation and is inactivated by heat. The UV spectrum of the partially purified compound indicates the possible presence of RNA, which is, however, not functional, since ribonuclease treatment did not affect the activity whereas trypsin treatment completely destroyed the activity of MSF.

The various actions of MSF appear to be initiated by a fairly specific binding to mitochondrial membranes. An enzymatic nature for the MSF action is not probable owing to the stoichoimetric character of the binding. MSF may be a highly membrane-active compound that causes subtle disturbances in the lipid–protein interactions in the membranes, which in turn give rise to changes in permeability, susceptibility to lipoxygenase-mediated oxygen uptake as well as inhibition of the respiratory chain. Among these actions the triggering of the lipoxygenase action seems to be the biologically most important one.

2.3. ATP-Dependent Proteolysis of Reticulocyte Mitochondria

2.3.1. General Remarks

ATP-dependent proteolysis, first observed more than 30 years ago in reticulocytes (Schweiger *et al.*, 1956; Schweiger and Rapoport, 1958), has in the last decade become an area of intensive interest. The reticulocyte has been one of the most convenient objects for study owing to the practical lack of lysosomes and the high degree of ATP dependency of its proteolysis (Müller *et al.*, 1980). Even in the first communication in this area in 1956, it was stated that the ATP-dependent proteolysis is a characteristic feature of erythroid maturation. This view has been confirmed repeatedly (McKay *et al.*, 1980; Speiser and Etlinger, 1982; Raviv *et al.*, 1987; Pickart *et al.*, 1988).

Studies on intact cell populations established the following main points. (1) More than 90% of free lysine liberated, as a measure of overall proteolysis, is ATP-dependent. (2) There exists a close correlation between the percentage of reticulocytes and the rate of proteolysis. (3) Proteolysis is a time-limited process, coming to an end within 4 hr, suggesting the exhaustion of a specific substrate. (4) Proteolysis has a high temperature coefficient, indicating considerable complexity of the proteolytic system. (5) Proteolysis depends on a preceding, triggering action of lipoxygenase. (6) There is a ubiquitin dependency. Long-time incubation experiments on suspensions showed a parallelism between proteolysis and the decline of cytochrome *c* oxidase activity. Both were suppressed by inhibitors of lipoxygenase (Schmidt *et al.*, 1985).

2.3.2. Mitochondria Are the Main Physiological Substrate of Proteolysis during Maturation of Reticulocytes

The first point to be made is the large extent of proteolysis, which requires a substrate in ample amounts. Mitochondria constitute about 10% of the total protein mass of reticulocytes (18–30 mg/ml). No evidence was found for an externalization of enzyme activities characteristic of mitochondria (Orr *et al.*, 1987; Müller *et al.*, 1980). Since lysosomes are practically absent and ATP-independent proteolytic activities amount to only 10% of the total activity in the presence of ATP, the ATP-dependent proteolytic system is the only possible system for the breakdown of mitochondrial proteins in reticulocytes.

Direct evidence for the claim that mitochondria are the main substrate of the ATP-dependent proteolysis is furnished by experiments on fractionated reticulocyte stroma. The extent of proteolysis was found to be proportional to the amount of mitochondria and occurred in the total absence of cell membranes (Dubiel *et al.*, 1981b; Rapoport *et al*, 1981). A further argument against the assumption that the cell membrane may be a major substrate of proteolysis is the fact that the stroma of mature erythrocytes is resistant to the cytosolic ATP-dependent proteolytic system of reticulocytes (unpublished results). These facts allowed the mitochondria-containing stroma of reticulocytes to be used as substrate, which can be obtained and handled easier than mitochondria. Proteolytic activities of cell membranes are so low that they can be neglected.

Earlier suggestions that the biological function of the ATP-dependent proteolysis in reticulocytes is the breakdown of aberrant or denatured proteins (Goldberg and St John, 1976) may apply only to pathological conditions.

2.3.3. The Energy Cost for the Degradation of Mitochondria

Originally it was assumed that the role of ATP in the ATP- and ubiquitin-dependent protein breakdown is exclusively related to the activation of ubiquitin by the ubiquitin-activating enzyme (Hershko and Ciechanover, 1982). An ATP consumption of only a few moles per mole of protein substrate would be required depending on the number of lysine residues in the protein substrate. Correlations between the proteolytic rates and the simultaneously measured liberation of inorganic phosphate were not consistent with such an assumption (Rapoport *et al.*, 1985a; Siems *et al.*, 1984). In the framework of studies on the balance of ATP production and consumption of rabbit reticulocytes the concentration of the ATP-dependent proteolysis has been assessed. For this purpose, several selective inhibitors of ATP-dependent proteolysis in reticulocytes were employed: (1) salicylhydroxamic acid, which acts indirectly by way of inhibition of lipoxygenase; (2) emetine, which inhibits both protein synthesis and ATP-dependent proteolysis (Siems *et al.*, 1984); and (3) hemin, a selective inhibitor of ATP-dependent proteolysis (Etlinger and Goldberg, 1980), the action site of which is beyond the ubiquitin activating step (Haas and Rose, 1981).

From the extent of diminution of ATP hydrolysis and proteolysis produced by salicylhydroxamic acid in hemolysates, a ratio between phosphate and lysine liberation of 48 was estimated. A similar value was obtained from the difference between the coupled oxygen consumption in the presence of cycloheximide, which only inhibits protein synthesis, and that in the presence of emetine, which inhibits both protein synthesis and proteolysis. Ratios between phosphate and lysine liberation of 51 with untreated and of 43 for mitochondria-containing stroma treated with *N*-ethylmaleimide to reduce unspecific ATP hydrolysis, were obtained. Based on a lysine content of 3% for stroma proteins and assuming a nondiscriminating hydrolysis of the proteins, one may estimate from the inorganic phosphate/lysine values that at least one ATP is hydrolyzed per peptide bond cleaved. Since the site of action of hemin was shown to be beyond the ubiquitin-activating reaction, the diminution of the liberation of inorganic phosphate by the addition of hemin corresponds to the amount of ATP required for the reaction steps beyond the conjugation of ubiquitin. It was concluded that the proteolytic breakdown of ubiquitin conjugates requires the bulk of ATP consumed in the ATP- and ubiquitin-dependent proteolytic pathway (Rapoport *et al.*, 1985a).

A relatively large ATP hydrolysis with stroma pretreated with *N*-ethylmaleimide for reduction of unspecific ATPase activity remained in the presence of hemin. Under these conditions, only ubiquitin conjugation reactions and the cleavage of ubiquitin conjugates by isopeptidases should be active. It is possible that there exists a futile cycle consisting of ubiquitin conjugation and isopeptidase reactions (Rapoport *et al.*, 1985a). In parallel with our work, the requirement for ATP in the proteolytic breakdown of ubiquitin conjugates was directly demonstrated on ubiquitin–lysozyme conjugates (Hershko *et al.*, 1984a).

A number of ATP-dependent proteinases fram various mammalian cells degrading both ubiquitin-conjugated (Waxman *et al.*, 1986, 1987; Hough *et al.*, 1986; Ganoth *et al.*,

1988) and nonubiquitinated proteins were partially purified and characterized. They share a high molecular mass (about 1000 kDa), the requirement for hydrolyzable MgATP, and the sensitivity to hemin. Proteinases of a somewhat smaller molecular size that do not require ATP have been purified from a variety of tissues including those of the erythroid system (Ishiura and Sugita, 1986; McGuire and DeMartino, 1986; Hough *et al.*, 1987). They represent a multicatalytic proteinase (for review see Rivett, 1989), which constitutes a large complex, the shape of which corresponds to a hollow cylinder in the electron microscope. It appears to be identical to a particle earlier called the *prosome* for which the name *proteasome* has been proposed (Arrigo *et al.*, 1988). It exhibits various proteolytic activities with both proteins and peptides and is composed of many polypeptides. Whereas in crude preparations ATP enhancement of the activity has been found, this feature has not been consistent in purified preparations (however, see Driscoll and Goldberg, 1989). It is possible that this type of proteinase has lost ATP dependency during purification (Tanaka and Ichihara, 1988).

It is also uncertain whether ATP stabilizes or activates the enzyme complex. There are suggestions that the proteasome, which can be separated from the even larger complex instrumental in the breakdown of ubiquitin-protein conjugates (Waxman *et al.*, 1987; Hough *et al.*, 1987), is in some way involved in their proteolysis, even if it itself does not degrade the ubiquitin conjugates. The mechanism of hemin inhibition of high-molecular-mass ATP-dependent proteinases is completely unclear. The agreement of the degree of inhibition with different substrates makes competition of hemin with substrate unlikely (Dubiel *et al.*, 1987b).

What could be the basis of the need for ATP beyond the activation of ubiquitin? A processive mechanism in which a protein substrate once bound is no longer released but is translocated across a hydrolytically active site of the enzyme complex in an ATP-dependent manner, appears most attractive.

2.3.4. Ubiquitin and tRNA in the ATP-Dependent Degradation of Mitochondria

Both ubiquitin and tRNA are involved in the ATP-dependent degradation of mitochondrial proteins. At present, their exact role, however, is not clear. It has been shown that ubiquitin stimulates the proteolytic breakdown of mitochondria-containing stroma in the presence of the ubiquitin-free fraction of reticulocyte cytosol, which contains the proteolytic system and MgATP, about five- to sixfold and restores proteolytic rates measured in the complete cytosol (Rapoport *et al.*, 1986). The ubiquitin-stimulated proteolysis of mitochondria-containing stroma is inhibited by hemin (Dubiel *et al.*, 1987b). Therefore, it was concluded that the bulk of proteolysis in reticulocytes and the breakdown of mitochondria is ubiquitin-mediated. Moreover, a direct conjugation of [^{125}I]ubiquitin to mitochondria-containing stroma was measured (Dubiel *et al.*, 1986).

The kinetics of [^{125}I]ubiquitin conjugation with native mitochondria-containing stroma is in agreement with that found with endogenous soluble proteins of reticulocyte cytosol (Ciechanover *et al.*, 1980). Taking into account the 1 nmol ubiquitin/ml cells, the residual amount of 0.3 nmol ubiquitin, which is always bound to stromal proteins, and a reticulocytosis of 40%, a total amount of 3.2 nmol ubiquitin/ml reticulocytes can be calculated. By an immunochemical method a value of 4.7 ± 0.2 was found (Haas and Bright, 1985) for the concentration of ubiquitin–protein conjugates of the stromal fraction. These conjugates disappear during maturation. Compared with native mitochondria-containing stroma, only one-fifth of [^{125}I]ubiquitin was conjugated to heat-denatured stroma. It was concluded that the decrease of ubiquitin-conjugate formation in heat-denatured stroma may be the main reason for its lower degradation (Dubiel *et al.*, 1987a,b). The comparably small portion of lysine residues of mitochondrial proteins conjugated with ubiquitin may be explained as follows:

1. The bulk of ϵ-NH_2 groups are masked in mitochondrial membrane proteins and only a small number of amino groups, e.g., the α-NH_2 group (Hershko *et al.*, 1984b), may be sufficient for the degradation of one protein molecule.
2. The ubiquitination of a kind of "receptor" of mitochondria is sufficient to signal the degradation of the whole organelle.

Perhaps ubiquitin has essential functions for proteolysis in addition to direct labeling of the substrate (Speiser and Etlinger, 1983; Rechsteiner, 1987).

tRNAs are involved in the ubiquitin-dependent degradation of albumin and some other proteins with acidic amino-terminal amino acids (Ciechanover *et al.*, 1985). It was found that the arginine tRNA is involved in the posttranslational transfer of its amino acid to the acidic amino-termini of protein. This transfer is catalyzed by an aminoacyl-tRNA-protein transferase. The conjugation of arginine to the acidic amino-terminal acids is essential for the degradation of the protein (Ferber and Ciechanover, 1986, 1987; Ciechanover *et al.*, 1988). Proteins with acidic amino-termini are uncommon, representing perhaps less than 5% of the total found in cells.

We tested the effect of tRNA on the ATP- and ubiquitin-dependent degradation of mitochondria-containing stroma. In a system consisting of the reticulocyte proteolytic system, ubiquitin, and ATP, the complete degradation of RNA in the reaction mixture by micrococcal nuclease reduced both the hemin-sensitive proteolytic rate and the liberation of phosphate with native mitochondria-containing stroma as substrate by about 70%. tRNA, isolated from rabbit reticulocytes, restored about 90% of the original ATP- and ubiquitin-dependent proteolytic activity compared with controls (Dubiel *et al.*, 1988). Such a large extent of tRNA requirement in the mitochondrial breakdown cannot be explained by the mechanism described by Ferber and Ciechanover (1986, 1987) and Ciechanover *et al.* (1988), since the number of proteins with acidic amino acid termini in the mitochondria should be much lower than 50%. The role of tRNA in the degradation of mitochondria is unclear. Perhaps the mitochondrial membrane proteins are cleaved in a first step by an endoproteinase that exposes a great number of acidic amino-termini, or tRNA(s) are essential components of an enzyme involved in the degradation of mitochondria.

2.3.5. ATP-Dependent and Ubiquitin-Independent Release of Peptides from Mitochondria-Containing Stroma

MgATP stimulates the release of trinitrobenzoylsulfonic acid-reactive material, presumably peptides, which are insoluble in trichloroacetic acid, from mitochondria-containing stroma. Some data characterizing this process are summarized in Table I. It may be seen that the release of peptides does not require either cytosol or the addition of ubiquitin. The involvement of tRNA is indicated by the fact that its degradation by micrococcal

Table I
The Effects of ATP, Ubiquitin, Hemin, and Preincubation with Micrococcal Nuclease and Subsequent Complementation with Charged or Uncharged tRNA on the Peptide Release from Mitochondria-Containing Stroma[a]

	3 hr								8 hr	
Additions										
ATP (5 mM)	+	+	−	−	+	+	+	+	+	+
Ubiquitin (50 μg/ml)	+	−	+	−	+	+	+	+	−	−
Others	−	−	−	−	Hemin (100 μM)	Nuclease	tRNA charged	tRNA uncharged	−	−
Mean[b]	33.2	32.8	8.7	10.8	7.1	14.4	37.0	20.6	79.9	31.1
SEM	±1.6	±2.9	±4.4	±6.2	±7.0	±3.3	±5.4	±2.7	±5.2	±12.3
(*n*)	(7)	(5)	(4)	(5)	(3)	(4)	(4)	(3)	(3)	(3)

[a]Measurements were carried out by the trinitrobenzene sulfonic acid method.
[b]Values in nmoles.

nuclease reduced the peptide release by about 50%, whereas addition of tRNA isolated from reticulocytes after inhibition of the nuclease by EGTA restored the original ATP-dependent activity. It is further noteworthy that uncharged tRNA was ineffective. The process is nearly linear over a period of 8 hr. Moreover, neither chloramphenicol, an inhibitor of mitochondrial protein synthesis, nor EGTA is inhibitory (data not shown). Practically negligible amounts of peptides were released in an ATP-dependent manner either from heat-denatured mitochondria-containing stroma or from erythrocyte stroma. Aliquots of the supernatants obtained from preincubation of mitochondria-containing stroma subsequently incubated in the presence of the proteolytic system of reticulocytes produced trichloroacetic acid-soluble trinitrobenzoylsulfonic acid-reactive compounds without a requirement for ubiquitin.

The characteristics of the novel ATP-dependent but ubiquitin-independent release of trichloroacetic acid-insoluble peptides from mitochondria-containing stroma, its sensitivity to hemin, the large dependence on charged tRNA, and its inhibition by heat treatment suggest that the process is catalyzed by a proteinase that is tightly bound to the membranes. It may catalyze the first proteolytic step in the degradation of mitochondria. It possesses characteristics common to the whole proteolytic system of reticulocytes: (1) the ATP-dependence, (2) the hemin-sensitivity, and (3) the large dependence on charged tRNA. The enzyme seems not to be influenced by either Ca^{2+} or *N*-ethylmaleimide.

The question arises as to the ubiquitin-requiring step in the whole proteolytic system. Apparently the ubiquitin exerts its effect when incorporated during the incubation of mitochondria-containing stroma in the presence of the proteolytic system of reticulocytes, presumably by ubiquitination of amino groups, thus favoring proteolysis. The peptide release from the mitochondria could expose amino acids of membrane protein fragments for a subsequent ubiquitination. This assumption is supported by the fact that native mitochondria-containing stroma is ubiquitinated to a much higher extent than the heat-treated one.

One may speculate that both ubiquitin-dependent and -independent pathways are operative in the degradation of mitochondria. The further characterization of the membrane-bound ATP-dependent proteolytic activity is in progress.

In summary, the sequence of the ATP-dependent and ubiquitin-mediated degradation of mitochondrial protein may be imagined as follows. First, there takes place the ATP-dependent and ubiquitin-independent liberation of peptides from membrane proteins by the membrane-bound proteinase which is connected with the exposure of free N-terminal amino acids; second, there occurs the ubiquitination of exposed amino acids, perhaps in a tRNA-mediated manner, favoring further proteolysis; and third, there follows a ubiquitin-independent degradation of the released peptides to amino acids by cytosolic proteinases.

2.4. The Interplay of the Components of the Systems Degrading Mitochondria

The interactions among components of the systems degrading the mitochondria during the maturation of reticulocytes may be outlined on the basis of Figure 10. Their interplay constitutes a cascade of actions, in which the focal event is the lipoxygenase-associated excessive oxygen uptake, which is the immediate trigger for the ATP-dependent proteolysis of the mitochondria. This event is preceded by the unmasking of the

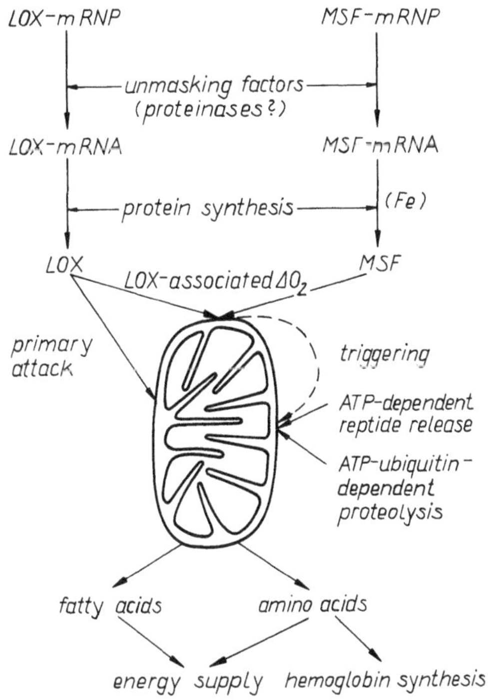

FIGURE 10. Scheme of interplay of systems degrading mitochondria during maturation of reticulocytes. LOX, lipoxygenase; MSF, mitochondrial susceptibility factor; mRNP, translationally inactive mRNA-protein particles.

lipoxygenase mRNA, with subsequent synthesis of the lipoxygenase protein. Then the enzyme attacks the phospholipids of both the mitochondria and the plasma membrane, an action that by itself has no apparent consequences. Only following the presumed unmasking of the MSF mRNA, formation of this protein factor, and its attack on the mitochondrial membrane, there ensues a second reaction by lipoxygenase, which now is accompanied by the excessive oxygen uptake. Also shown in the scheme (Figure 10) is the twofold system of ATP-dependent proteolysis, consisting of ubiquitin-independent peptide release on the one hand, and on the other the further breakdown, which requires both ATP and ubiquitin. The temporal relationships among the systems are interesting. The ATP-dependent proteolytic systems are present even in the most immature reticulocytes and may be functional even in erythroid precursor cells, with possible domination of the ubiquitin-independent system (Waxman *et al.*, 1985). The unmasking of the lipoxygenase mRNA precedes that of the mRNA for the MSF, which indicates some specificity in the mechanisms of the unmasking, the nature of which remains to be elucidated. On the other hand, there are some common features in the decay of the mechanisms instrumental in the degradation of the mitochondria. All of them disappear during maturation and are practically absent in the mature erythrocyte. Therefore, the question arises as to whether there exist common mechanisms for their degradation, in addition to specific ones such as the suicidal inactivation of the lipoxygenase.

3. OTHER ORGANELLES

In this section we will discuss the degradation of ribosomes, endoplasmic reticulum, Golgi apparatus, and components of the plasma membrane. Owing to the scant information available, the accounts will be brief.

3.1. Ribosomes and Protein Synthesis

The total number of ribosomes in rabbit reticulocytes drops continuously during the maturational stage as indicated by a decreasing RNA content with increasing cell density (Rifkind *et al.*, 1964; Knopf and Lamfrom, 1965; Marbaix *et al.*, 1970; Coutelle *et al.*, 1973). The reticulocyte contains up to 100,000 ribosomes/cell, more than half of which take the form of polysomes. Contradictory results have been reported concerning the distribution of ribosomes during maturation. While Glowacki and Millette (1965) and DeBellis (1969) found that during *in vivo* maturation of reticulocytes all classes of ribosomes are lost at rates proportional to their initial concentration, other workers observed that during maturation both *in vivo* and *in vitro* the share of polysomes declines and the share of free ribosomes is correspondingly enhanced (Rifkind *et al.*, 1964; Marbaix *et al.*, 1970; Gasko and Danon, 1972; Knopf and Lamfrom, 1965; Danon *et al.*, 1965; Andree *et al.*, 1980). In parallel, free cytoplasmic α-globin mRNA appears during the maturation of rabbit reticulocytes (Marbaix *et al.*, 1976). From this behavior one may conclude that the decrease in protein synthesis capacity of reticulocytes is not solely due to a diminishing number of ribosomes but rather either to an injury of structural and functional integrity of the ribosomes or to a limitation imposed by other components of the machinery of protein synthesis. The biological kinetics of the degradation processes of ribosomes and of mitochondria appear to differ from each other, since the activity of cytochrome *c* oxidase as a mitochondrial marker enzyme plotted versus the cellular RNA content does not show a straight-line dependence (Wiesner *et al.*, 1973), which suggests that different mechanisms of the breakdown of these organelles may be instrumental.

Nevertheless, there is a common feature of the breakdown of the two kinds of organelles. Like the degradation of the mitochondrial proteins, the breakdown of the ribosomes appears to be energy-dependent (Schweiger and Rapoport, 1958; Park and Morgan, 1984). It was shown that the uncoupler 2,4-dinitrophenol strongly inhibits both the maturational release of free nucleotides from the ribosomes and the loss of the *substantia reticulofilamentosa* (Schweiger and Rapoport, 1958). The reason for the energy demand is not known; it is conceivable that an ATP-dependent proteolytic system of reticulocytes (see Section 2.3) may attack the ribosomal proteins and render in this manner the rRNA species susceptible to ribonucleases. Ribonucleases are found in reticulocytes (Rosenthal *et al.*, 1964b, 1966; Adachi *et al.*, 1964; Farkas and Marks, 1968; Goto and Mizuno, 1971; Prehn *et al.*, 1972; Wreschner *et al.*, 1982), but most of their activity is masked by an inhibitory protein of the cytosol (Rost, 1959; Priess and Zilig, 1967). In apparent contradiction to the energy requirement of the breakdown of ribosomes is the observation that oligomycin, an inhibitor of mitochondrial ATP production, strongly enhances the degradation of rRNA (Kostić, personal communication, 1989). The RNA breakdown during *in vitro* maturation of reticulocytes is also markedly accelerated if unphysiologically high concentrations of inorganic phosphate are present in the incubation medium (Mai *et al.*, 1980). The possible explanation for the conflicting results is not readily apparent. It is conceivable that a certain ATP level, which may be furnished by processes in the cytosol, is provided for the degradation even if the mitochondrial ATP production is suppressed by the respiratory inhibitor. On the other hand, a high phosphorylation potential under normal circumstances may protect the ribosomes, possibly by phosphorylation of some of the ribosomal proteins. Rabbit reticulocytes contain hidden breaks in their rRNAs such as are found in many other types of cells (Andree *et al.*, 1980). The breaks are already detectable in bone marrow cells and may have occurred shortly after the formation of the ribosomes. These breaks in the rRNA molecules are not noticeable in sucrose gradients under nondenaturing conditions and appear not to affect the structural and functional integrity of the ribosomes. Both the secondary structure of the rRNA and its interaction with ribosomal proteins seem to afford sufficient stability to compensate for the hidden breaks. During *in vitro* incubations the number of breaks in the RNA increases. In the 28 S rRNA of the large ribosomal subunit, but not in the 18 S rRNA of the small subunit, there are distinct sites for the primary cleavage that are probably located on the surface of the ribosomes and therefore exposed to the attack by nucleases. The nuclease attack by itself seems not to be the crucial step in the breakdown of ribosomes. It cannot be excluded, however, that a certain number of hidden breaks may confer a loosened structure of the ribosomes more susceptible to proteolytic attack. Further insight into the interaction of ribosomal proteins and rRNA molecules and its consequences for the susceptibility of both proteins and RNA to hydrolytic attack appears necessary to elucidate in detail the breakdown cascade of ribosomes.

Among factors other than ribosomes that may contribute to the decline of protein synthesis, the mRNAs have been of greatest interest. Their stability appears to be determined by its primary structure. The half-life of globin mRNA amounts to 10–20 hr (Burka, 1969; Lodish and Small, 1976). One factor for its stability is the length of the poly(A) tail of the 5′ noncoding region (Nudel *et al.*, 1976; Maniatis *et al.*, 1976). Recently, however, it was shown that the differential stability of mRNA species also seems to reciprocally correlate with the length of $(U)_nA$ sequences in the 3′ noncoding region (Wreschner and Rechavi, 1988). Owing to this differential stability of the various mRNA species, the pattern of protein synthesis should vary with the maturational stage of the reticulocyte.

Protein synthesis in rabbit reticulocytes is strongly inhibited by lack of ATP or by the presence of AMP (Villa-Trevino *et al.*, 1963; Freudenberg *et al.*, 1971; Mosca *et al.*, 1983). This inhibition encompasses two components; (1) a reversible inhibition produced by the increased concentration of AMP, and (2) an irreversible inhibition by a 70-kDa polypeptide that is activated by binding of AMP. It remains to be established whether these inhibitory effects are involved in the maturational changes of reticulocytes and whether they interfere with translation factors.

In reticulocytes there occur the known initiation and elongation factors (Brown-Luedi *et al.*, 1982; Carvalho *et al.*, 1984; Merrick *et al.*, 1975), which are strongly maturation-dependent and therefore may contribute to the decay of protein synthesis. In contrast, the tRNAs, must be excluded as the limiting factor of protein synthesis in maturing reticulocytes since they are much more stable during maturation than the other components of the protein synthesis system; their half-life amounts to 50–60 hr (Smith and Weinberg, 1981).

3.2. Endoplasmic Reticulum and Golgi Apparatus

The amounts of endoplasmic reticulum and Golgi apparatus left in reticulocytes are small. As judged by the activity of the NADPH-cytochrome *c* reductase, which may be considered as a marker enzyme of the endoplasmic reticulum, its amount is only 1% of that in liver cells. The same holds for the activity of CDP-choline-1,2-diglyceride phosphatidylcholine transferase (Augustin *et al.*, 1977). Still there must be some activity of both endoplasmic reticulum and Golgi apparatus left as indicated by various kinds of evidence. For one, incorporation of galactosamine in membrane proteins of reticulocytes has been demonstrated (Harris and Johnson, 1969), which has been confirmed for galactose, mannose, and fucose (Schulze *et al.*, unpublished). Also, glycosylation of endogenous proteins of plasma membranes by dolichol derivatives and the synthesis of mannosyl-phosphorylpolyisoprenes have been found (Parodi and Martin-Barrientos, 1977). According to present knowledge, translocation and glycosylation of integral membrane proteins requires functionality of both the endoplasmic reticulum and the Golgi apparatus. Of particular significance is the demonstration of factors known to be involved in the translocation of peptide chains destined for transport to the cell membrane. Such peptides carry a specific sequence of amino acids, usually N-terminally, which constitutes a signal for translocation. The signal recognition particle, a complex consisting of six proteins and a 7 S RNA, serves to recognize the signal sequence on its emergence from the ribosome during translation and to establish contact with the endoplasmic reticulum. Its presence has been demonstrated in reticulocytes (Meyer *et al.*, 1982). Another factor recently found is the so-called signal recognition receptor (Wiedmann *et al.*, 1987). It is a constituent of the endoplasmic reticulum, thought to serve in the actual translocation process. It also has been found in rabbit reticulocytes (Schulze and Prehn, unpublished). Up to fivefold higher incorporation of amino acids and of carbohydrates have been found in the least mature fraction of rabbit reticulocytes compared to that of highest maturity (Schulze *et al.*, unpublished).

What about the possible mechanism for the decline of the endoplasmic reticulum, or for that matter of other organelles? Clear-cut evidence is not available. As mentioned before, the bulk of the degradation occurs at stages preceding the reticulocyte. The case of methemoglobin reductase exemplifies the difficulties. The system consists of two components, one a form of cytochrome b_5, the other a flavin enzyme, the cytochrome b_5-reductase. In Friend erythroleukemia cells, both are found to be membrane-bound; however, in a soluble form in the cytosol of both reticulocytes and erythrocytes. In the liver cell the components are present in the endoplasmic reticulum. One may assume that the soluble components of the methemoglobin reductase system arise by selective proteolytic processing, and thereby lose their hydrophobic anchors to the endoplasmic reticulum. Model experiments have given controversial answers as to whether a membrane-bound acid proteinase (Schafer and Hultquist, 1981) or the ubiquitin–ATP-dependent system (Raw and DiFini, 1983) are operative. It is even conceivable that both proteolytic systems are needed. Another conceivable mechanism may be exocytosis (see Section 3.3).

3.3. Plasma Membrane

As mentioned in the introduction, manifold changes occur in the plasma membrane during reticulocyte maturation. A decline of a variety of components, including receptors, transporters, enzymes, antigens, and other proteins, is observed (for review see Rapoport, 1986). These changes go hand in hand with a restructuring of the plasma membrane.

A general mechanism for the elimination of components of the plasma membrane of reticulocytes is their externalization in the form of vesicles, so-called exosomes (Johnstone *et al.*, 1987). This process has been studied most thoroughly on the transferrin receptor (Harding *et al.*, 1983, 1984; Pan *et al.*, 1985). It appears to proceed in the following manner: The first step is an endocytosis of surface components, involving clathrin and its associated proteins (for review on clathrin see Rothman and Schmidt, 1986; Brodsky, 1988). The endocytic vesicles combine with the acid endosomes to form multivesicular bodies 0.5–1.5 μm in diameter, in which internal budding occurs. After fusion with the plasma membrane, the buds, which are 50 nm in diameter, undergo exocytosis and are found in the medium (Pan *et al.*, 1985). Thus, the externalized components undergo a double inversion and present their right side to the exterior. In some manner the clathrin-uncoating ATPase is involved and is found in the exosomes (Davis *et al.*, 1986). The clathrin-uncoating ATPase belongs to a group of proteins that are formed in most cells in response to heat shock. It is remarkable that erythroid cells should accumulate copious amounts of a heat-shock protein without external stress (Morimoto and Fodor, 1984; Singh and Yu, 1984). It is conceivable that the decaying components of the plasma membrane are recognized as altered by a specific mechanism related to the heat-shock response to be eliminated from the cell.

Externalization is not limited to the transferrin receptor. The exosomes contain transporters for nucleosides, glucose and Na^+-independent transport of amino acids as well as acetylcholinesterase, all of which diminish during maturation of reticulocytes. Possibly the same type of mechanism is responsible for the elimination of the endoplasmic reticulum and the Golgi apparatus.

The fate of Na^+,K^+-ATPase is unclear. Whereas there is some evidence for the presence of this enzyme, based on ouabain binding in the exosomes (Johnstone *et al.*, 1987), other work on dog reticulocytes suggests the action of ATP-dependent proteolysis (Inaba and Maede, 1986).

A number of questions are unresolved, among them whether the exosomes represent a heterogeneous population according to the variety of specific membrane components they contain. Also, it would be of great interest to find out whether there exists a specific triggering mechanism for the initiation of the externalization.

4. FINAL REMARKS

The present account reflects the fragmentary nature of our knowledge concerning the mechanisms for the degradation of many constituents of the reticulocyte. On the other hand, it justifies the quest for mechanisms underlying the phenomena observed. The discovery of a specific maturational program for the breakdown of mitochondria, constituting a cascade of interlocking steps, may be a paradigm and should stimulate the search for other principles governing maturational changes.

Two main mechanisms for the elimination of organelles and their components may be proposed. One is ATP-dependent proteolysis. It appears to be instrumental in the breakdown not only of mitochondria but also of ribosomes (Schweiger *et al.*, 1956; Park and

Morgan, 1984) of reticulocytes. It is likely to be functional also in preceding stages of maturation. The other principle is the shedding of components, as demonstrated for a variety of receptors and transporters as well as the acetylcholinesterase of the cell membrane. These processes are part of the characteristic restructuring of the plasma membrane and the solidification of the membrane skeleton. As mentioned before, one may speculate whether the loss of the vestiges of endoplasmic reticulum and the Golgi apparatus may occur in an analogous manner.

How are the various maturational changes connected with each other? At present this question cannot be answered. One connecting link appears to be the provision of cellular ATP. It is needed not only for the synthesis of hemoglobin and other proteins, but also for ATP-dependent proteolysis and also in a manner not elucidated as yet for the exocytosis of membrane components, perhaps involving the clathrin-uncoating ATPase. If ATP production plays such a central role, the function of the mitochondria, which are the main source of ATP production, is of key importance. Given such an aspect, the system operative in the reticulocytes for their breakdown appears to be particularly adapted to keep them functioning during the later stages of maturation, despite the presence of the proteolytic system. This is achieved by the masking of the mRNA's essential components of the degradation cascade.

As argued earlier (Rapoport, 1986), one should be cautious in assuming that all differences between reticulocytes and erythrocytes are part of a general maturational program. This view is contradicted by the great biological variation among the mature erythrocytes of various species, with respect to both their enzymatic equipment and their composition including low-molecular-mass compounds and inorganic constituents. Thus, one must assume that both essential and biologically unessential changes occur. The delimitation and determination of the essential general program of maturation appears to be an imminent task.

5. REFERENCES

Adachi, K., Nagano, K., Nakao, T., and Nakao, M., 1964, Purification and characterization of ribonuclease from rabbit reticulocytes, *Biochim. Biophys. Acta* **92:**59–70.

Andree, H. H., Bretschneider, K., Thiele, B. J., and Rapoport, S. M., 1980, Breakdown of ribosomal RNA in rabbit reticulocytes, *Acta Biol. Med. Ger.* **39:**995–1006.

Antonioli, J. A., and Christensen, H. N., 1969, Differences in schedules of regression of transport systems during reticulocyte maturation, *J. Biol. Chem.* **244:**1505–1509.

Arrigo, A. P., Tanaka, K., Goldberg, A. L., and Welch, W. J., 1988, Identity of the 19S "prosome" particle with the large multifunctional protease complex of mammalian cells (the proteasome), *Nature* **331:**192–194.

Augustin, H. W., and Rapoport, S. M., 1959, Über Atmung und Succinatoxydasesystem bei reifen und jugendlichen Hühner-erythrocyten, *Acta Biol. Med. Ger.* **3:**433–449.

Augustin, H. W., Häcker, M. R., and Hofmann, E., 1964, Aufnahme und Phosphorylierung von 2-Desoxy-D-glucose in Kaninchen-erythrocyten und -reticulocyten *Hoppe-Seyler's Z. Physiol. Chem.* **339:**42–58.

Augustin, H. W., Zborowski, J., Baranska, J., Wiswedel, I., and Wojtczak, L., 1977, Synthesis of phospholipids in mitochondria and other membrane fractions of rabbit reticulocytes, *Biochim. Biophys. Acta* **489:**298–306.

Balcarek, J. M., Theisen, T. W., Cook, M. N., Varrichio, A., Hwang, S. M., Strohsacker, M. W., and Crooke, S. T., 1988, Isolation and characterization of a cDNA clone encoding rat 5-lipoxygenase, *J. Biol. Chem.* **263:**13937–13941.

Belkner, J., and Rapoport, S. M., 1989, Approaches to characterize by density-fractionation age-dependent properties of reticulocyte mitochondria, *Biomed. Biochim. Acta* **48**:3–11.

Belkner, J., Kühn, H., and Wiesner, R., 1990, Oxygenation of biological membranes by the reticulocyte lipoxygenase. Lack of stoichiometry between oxygen uptake and product formation, *Biomed. Biochim. Acta* **49**:S31–S34.

Bessis, M., 1973, *Living Blood Cells and Their Ultrastructure*, Springer, Berlin.

Borgeat, P., Nadeau, M., Salari, H., Poubelle, P., and Fruteau de Laclos, B., 1985, Leukotrienes: Biosynthesis, metabolism, and analysis, *Adv. Lipid Res.* **21**:47–77.

Brodsky, F. M., 1988, Living with clathrin. Its role in intracellular membrane traffic, *Science* **24**:1396–1402.

Brown-Luedi, M. L., Meyer, L. J., Milburn, S. C., Mo-Ping Yau, P., Corbett, S., and Hershey, J. W. B., 1982, Protein synthesis initiation factors from human HeLa cells and rabbit reticulocytes are similar: Comparison of protein structure, activities, and immunochemical properties, *Biochemistry* **21**:4202–4206.

Bryant, R. W., Bailey, J. M., Schewe, T., and Rapoport, S. M., 1982, Positional specificity of a reticulocyte lipoxygenase. Conversion of arachidonic acid to 15S-hydroperoxyeicosatetraenoic acid, *J. Biol. Chem.* **257**:6050–6055.

Bryant, R. W., Schewe, T., Rapoport, S. M., and Bailey, J. M., 1985, Leukotriene formation by a purified reticulocyte lipoxygenase enzyme. Conversion of arachidonic acid and 15-hydroperoxyeicosatetraenoic acid to 14,15-leukotriene A_4, *J. Biol. Chem.* **260**:3548–3555.

Burka, E. R., 1969, Characteristics of RNA degradation in the erythroid cell, *J. Clin. Invest.* **48**:1266–1272.

Carvalho, M., Carvalho, J. F., and Merrick, W. C., 1984, Biological characterization of various forms of elongation factor 1 from rabbit reticulocytes, *Arch. Biochem. Biophys.* **234**:603–611.

Ciechanover, A., Heller, H., Elias, S., Haas, A. L., and Hershko, A., 1980, ATP-dependent conjugation of reticulocyte proteins with the polypeptide required for protein degradation, *Proc. Natl. Acad. Sci. USA* **77**:1365–1368.

Ciechanover, A., Wolin, S. L., Steitz, J. A., and Lodish, H. F., 1985, Transfer RNA is an essential component of the ubiquitin- and ATP-dependent proteolytic system, *Proc. Natl. Acad. Sci. USA* **82**:1341–1345.

Ciechanover, A., Ferber, S., Dvorah, G., Elias, S., Hershko, A., and Arfin, S., 1988, Purification and characterization of arginyl-tRNA-protein transferase from rabbit reticulocytes, *J. Biol. Chem.* **263**:11155–11167.

Coutelle, C., Rosenthal, S., Gross, J., David, H., and Uerlings, I., 1973, Leitkriterien der Retikulozytenreifung, VI. Verhalten der RNS und Retikulozytenwerte sowie des Ribosomengehaltes in peripheren erythroiden Zellpopulationen verschiedener Dichte im Verlaufe einer Entblutungsanämie beim Kaninchen, *Acta Biol. Med. Ger.* **31**:781–794.

Danon, D., Zehavi-Willner, T., and Berman, C. R., 1965, Alterations in polyribosomes of reticulocytes maturing in vivo, *Proc. Natl. Acad. Sci. USA* **54**:873–879.

Davis, J. Q., Danserau, D., Johnstone, R. M., and Bennet, V., 1986, Selective externalization of an ATP-binding protein structurally related to the clathrin-uncoating ATPase/heat shock protein in vesicles containing terminal transferrin receptors during reticulocyte maturation, *J. Biol. Chem.* **261**:15368–15371.

DeBellis, R. H., 1969, Fate of reticulocyte ribosomes during *in vivo* maturation, *Biochemistry* **8**:3451–3454.

Dixon, R. A. F., Jones, R. E., Diehl, R. E., Bennett, C. D., Karoman, S,. and Rouzer, C. A., 1988, Cloning of the cDNA for human 5-lipoxygenase, *Proc. Natl. Acad. Sci. USA* **85**:416–420.

Driscoll, J., and Goldberg, A. L., 1989, Skeletal muscle proteasome can degrade proteins in an ATP-dependent process that does not require ubiquitin, *Proc. Natl. Acad. Sci. USA* **86**:787–791.

Dubiel, W., Müller, M., and Rapoport, S. M., 1981a, ATP-dependent proteolysis of reticulocyte mitochondria is preceded by the attack of lipoxygenase, *Biochem. Int.* **3**:165–171.

Dubiel, W., Müller, M., Rathmann, J., Hiebsch, C., and Rapoport, S. M., 1981b, Determination and characteristics of energy-dependent proteolysis in rabbit reticulocytes, *Acta Biol. Med. Ger.* **40**:625–628.

Dubiel, W., Müller, M., and Rapoport, S. M., 1986, Kinetics of ^{125}I-ubiquitin conjugation with and liberation from rabbit reticulocyte stroma, *FEBS Lett.* **194**:50–55.

Dubiel, W., Drung, I., Müller, M., and Rapoport, S. M., 1987a, Effect of substrate heat-denaturation on ATP- and ubiquitin-dependent proteolysis, *Biomed. Biochim. Acta* **46**:159–164.

Dubiel, W., Drung, I., Müller, M., and Rapoport, S. M., 1987b, Kinetic studies on the ATP- and ubiquitin-dependent proteolytic system of reticulocytes, *Biomed. Biochim. Acta* **46**:565–570.

Dubiel, W., Drung, I., Müller, M., and Rapoport, S. M., 1988, The effects of substrate denaturation and tRNA on the ATP- and ubiquitin-dependent proteolytic system of reticulocytes, *Biol. Zentralblatt* **107**:93–96.

Etlinger, J. D., and Goldberg, A. L., 1980, Control of protein degradation in reticulocytes and reticulocyte extracts by hemin, *J. Biol. Chem.* **255**:4563–4568.

Farkas, W., and Marks, P. A., 1968, Partial purification and properties of a ribonuclease from rabbit reticulocytes, *J. Biol. Chem.* **243**:6464–6473.

Ferber, S., and Ciechanover, A., 1986, Transfer RNA is required for conjugation of ubiquitin to selective substrates of the ubiquitin- and ATP-dependent proteolytic system, *J. Biol. Chem.* **261**:3128–3134.

Ferber, S., and Ciechanover, A., 1987, Role of arginine-tRNA in protein degradation by the ubiquitin pathway, *Nature* **326**:808–810.

Fiskum, G., Bryant, R. W., Low, C.-E., Pease, A., and Bailey, J. M., 1985, Lipoxygenation of mitochondrial membranes by reticulocyte lipoxygenase, in: *Prostaglandins, Leukotrienes, and Lipoxins: Biochemistry, Mechanism of Action, and Clinical Applications* (J. M. Bailey, ed.), pp. 87–95, Plenum Press, New York.

Fleming, J., Thiele, B. J., Chester, J., O'Prey, J., Janetzki, S., Aitken, A., Anton, I. A., Rapoport, S. M., and Harrison, P. R., 1989, The complete sequence of the rabbit erythroid cell-specific 15-lipoxygenase mRNA: Comparison of the predicted amino acid sequence of the RBC lipoxygenase with other lipoxygenases, *Gene* **79**:181–188.

Freudenberg, H., and Mager, J., 1971, Studies on the mechanism of the inhibition of protein synthesis induced by intracellular ATP depletion, *Biochim. Biophys. Acta* **232**:537–555.

Furakawa, M., Yoshimoto, T., Ochi, K., and Yamamoto, S., 1984, Studies on arachidonate 5-lipoxygenase of rat basophilic leukemia cells, *Biochim. Biophys. Acta* **795**:458–465.

Ganoth, D., Leshinsky, E., Eytan, E., and Hershko, A., 1988, A multicomponent system that degrades proteins conjugated to ubiquitin: Resolution of factors and evidence for ATP-dependent complex formation, *J. Biol. Chem.* **263**:12412–12419.

Gasko, O., and Danon, D., 1972, Deterioration and disappearance of mitochondria during reticulocyte maturation, *Exp. Cell Res.* **75**:159–169.

Geiduschek, J. B., and Singer, S. J., 1979, Molecular changes in the membranes of mouse erythroid cells accompanying differentiation, *Cell* **16**:149–163.

Goldberg, A. L., and St. John, A. C., 1976, Intracellular protein degradation in mammalian and bacterial cells, *Annu. Rev. Biochem.* **45**:747–803.

Glowacki, E. R., and Millette, R. L., 1965, Polyribosomes and the loss of hemoglobin synthesis in the maturing reticulocytes, *J. Mol. Biol.* **11**:116–127.

Goto, S., and Mizuno, D., 1971, Degradation of RNA in rat reticulocytes. Purification and properties of rat reticulocyte RNase, *Arch. Biochem. Biophys.* **145**:64–70.

Greksch, G., Wiswedel, I., and Augustin, W., 1973, Enzymatic characterization of rabbit reticulocyte mitochondria in: *Abhandlungen der Akademie der Wissenschaften der DDR*, pp. 587–599, Akademie-Verlag, Berlin.

Haas, A. L., and Bright, P. M., 1985, The immunochemical detection and quantitation of intracellular ubiquitin–protein conjugates, *J. Biol. Chem.* **260**:12464–12473.

Haas, A. L., and Rose, I. A., 1981, Hemin inhibits ATP-dependent ubiquitin-dependent proteolysis: Role of hemin in regulating ubiquitin conjugate degradation, *Proc. Natl. Acad. Sci. USA* **78**:6845–6848.

Harding, C., Heuser, J., and Stahl, P., 1983, Receptor-mediated endocytosis and recycling of the transferrin receptor in rat reticulocytes, *J. Cell Biol.* **97**:324–339.

Harding, C., Heuser, J., and Stahl, P., 1984, Endocytosis and intracellular processing of transferrin and colloidal gold transferrin in rat reticulocytes, *Eur. J. Cell Biol.* **35**:256–263.

Harris, E. D., and Johnson, C. A., 1969, Incorporation of glucosamine-^{14}C into membrane proteins of reticulocytes, *Biochemistry* **8**:512–515.

Harrison, P. R., Frampton, J., Chambers, I., Kasturi, K., Thiele, B., Conkie, D., Fleming, J., Chester, J., O'Prey, J., and McBain, W., 1987, Analysis of erythroid cell-specific gene expression, in: *Molecular and Cellular Aspects of Erythropoietin and Erythropoiesis* (I. N. Rich, ed.), pp. 37–50, Springer-Verlag, Berlin.

Härtel, B., Ludwig, P., Schewe, T., and Rapoport, S. M., 1982, Self-inactivation by 13-hydroperoxylinoleic acid and lipohydroperoxidase activity of the reticulocyte lipoxygenase, *Eur. J. Biochem.* **126**:353–357.

Hershko, A., and Ciechanover, A., 1982, Mechanisms of intracellular protein breakdown, *Annu. Rev. Biochem.* **51**:335–364.

Hershko, A., Leshinsky, A., Ganoth, D., and Heller, H., 1984a, ATP-dependent degradation of ubiquitin–protein conjugates, *Proc. Natl. Acad. Sci. USA* **81**:1619–1623.

Hershko, A., Heller, H., Eytan, E., Kaklij, G., and Rose, I. A., 1984b, Role of the α-amino group of protein in ubiquitin-mediated protein breakdown, *Proc. Natl. Acad. Sci. USA* **81**:7021–7025.
Heynen, M. J., and Verwilghen, R. L., 1982, A quantitative ultrastructural study of normal rat erythroblasts and reticulocytes, *Cell Tissue Res.* **224**:397–408.
Höhne, M., Bayer, D., Prehn, S., Schewe, T., and Rapoport, S. M., 1983, In vitro maturation of rabbit reticulocytes. III. Response of lipoxygenase, *Biomed. Biochim. Acta* **42**:1129–1134.
Höhne, M., Thiele, B. J., Prehn, S., Giessmann, E., Nack, B., and Rapoport, S. M., 1988, Activation of translationally inactive lipoxygenase mRNP particles from rabbit reticulocytes, *Biomed. Biochim. Acta* **47**:75–78.
Holt, G. D., Haltiwanger, R. S., Torres, C. R., and Hart, G. W., 1987, Erythrocytes contain cytoplasmic glycoproteins. O-linked GlcNAc on band 4.1, *J. Biol. Chem.* **262**:14847–14850.
Hough, R., Pratt, G., and Rechsteiner, M., 1986, Ubiquitin–lysozyme conjugates: Identification and characterization of an ATP-dependent protease from rabbit reticulocyte lysates, *J. Biol. Chem.* **261**:2400–2408.
Hough, R., Pratt, G., and Rechsteiner, M., 1987, Purification of two high molecular weight proteases from rabbit reticulocyte lysate, *J. Biol. Chem.* **262**:8303–8313.
Inaba, M., and Maede, Y., 1986, Na,K-ATPase in dog red cells. Immunological identification and maturation-associated degradation by the proteolytic system, *J. Biol. Chem.* **261**:16099–16105.
Ishiura, S., and Sugita, H., 1986, Ingensin, a high-molecular mass alkaline protease from rabbit reticulocyte, *J. Biochem.* **100**:753–763.
Johnstone, R. M., Adam, M., Hammond, J. R., Orr, L., and Turbide, C., 1987, Vesicle formation during reticulocyte maturation. Association of plasma membrane activities with released vesicles (exosomes), *J. Biol. Chem.* **262**:9412–9420.
Knopf, P. M., and Lamfrom, H., 1965, Changes in the ribosome distribution during incubation of rabbit reticulocytes *in vitro, Biochim. Biophys. Acta* **95**:398–407.
Kobayashi, T., and Levine, L., 1983, Arachidonic acid metabolism by erythrocytes, *J. Biol. Chem.* **258**:9116–9121.
Kostić, M. M., and Rapoport, S. M., 1989, Maturation-dependent changes of the rabbit reticulocyte energy metabolism, *FEBS Lett.* **250**:400–441.
Krause, W., David, H., Uerlings, I., and Rosenthal, S., 1972, Veränderungen der Mitochondrien-Ultrastruktur von Kaninchenretikulozyten im Reifungsprozeß, *Acta Biol. Med. Ger.* **28**:779–786.
Kroschwald, P., Kroschwald, A., Wiesner, R., Schewe, T., and Kuhn, H., 1986, The occurrence of a lipoxygenase pathway in reticulocytes of various species, *Biomed. Biochim. Acta* **45**:1237–1247.
Kroschwald, P., Kroschwald, A., Kuhn, H., Ludwig, P., Thiele, B. J., Höhne, M., Schewe, T., and Rapoport, S. M., 1989a, Occurrence of the erythroid cell-specific arachidonate 15-lipoxygenase in human reticulocytes, *Biochem. Biophys. Res. Commun.* **160**:954–960.
Kühn, H., and Brash, A. R., 1990, Occurrence of lipoxygenase products in membranes of rabbit reticulocytes. Evidence for a role of the reticulocyte lipoxygenase in the maturation of red cells, *J. Biol. Chem.* **265**: 1454–1458.
Kühn, H., Belkner, J., and Wiesner, R., 1990a, Metabolism of polyenoic fatty acids by rabbit reticulocytes. Intracellular action of the erythroid lipoxygenase on membrane lipids, *Biomed. Biochim. Acta* **49**:S25–30.
Kühn, H., Belkner, J., and Wiesner, R., 1990b, Subcellular distribution of lipoxygenase products in rabbit reticulocyte membranes, *Eur. J. Biochem.*, in press.
Kühn, H., Belkner, J., Wiesner, R., and Alder, L., 1990c, Occurrence of 9- and 13-keto octadecadienoic acid in biological membranes oxygenated by the reticulocyte lipoxygenase, *Arch. Biochem. Biophys.* in press.
Kühn, H., Wiesner, R., and Schewe, T., 1990d, Formation of oxygenase and hydroperoxidase products by the pure reticulocyte lipoxygenase, *Biomed. Biochim. Acta* **49**:S39–S41.
Kühn, H., Wiesner, R., Schewe, T., and Rapoport, S. M., 1983a, Reticulocyte lipoxygenase exhibits both n-6 and n-9 activities, *FEBS Lett.* **153**:353–356.
Kühn, H., Pliquett, F., Wunderlich, S., Schewe, T., and Krause, W., 1983b, Reticulocyte lipoxygenase changes the passive electric properties of bovine heart submitochondrial particles, *Biochim. Biophys. Acta* **735**:283–290.
Kühn, H., Schewe, T., and Rapoport, S. M., 1986a, The stereo chemistry of the reactions of lipoxygenases and their metabolites. Proposed nomenclature of lipoxygenases and related enzymes, *Enzymol. Relat. Areas Mol. Biol.* **58**:273–311.
Kühn, H., Salzmann-Reinhardt, U., Ludwig, P., Pönicke, K., Schewe, T., and Rapoport, S. M., 1986b, The

stoichiometry of oxygen uptake and conjugated diene formation during the dioxygenation of linoleic acid by the pure reticulocyte lipoxygenase. Evidence for aerobic hydroperoxidase activity, *Biochim. Biophys. Acta* **876:**187–193.

Kühn, H., Wiesner, R., Alder, L., Fitzsimmons, B. S., Rokach, J., and Brash, A. R., 1987, Formation of lipoxin B by the pure reticulocyte lipoxygenase via sequential oxygenation of the substrate, *Eur. J. Biochem.* **169:**593–601.

Kühn, H., Wiesner, R., Belkner, J., and Alder, L., 1989, Occurrence of 9- and 13-ketooctadecadienoic acid in biological membranes oxygenated by the reticulocyte lipoxygenase, *Arch. Biochem. Biophys.*, in press.

Lodish, H. F., and Small, B., 1976, Different lifetimes of reticulocyte messenger RNA, *Cell* **7:**59–65.

Ludwig, P., Holzhütter, H.-G., Colosimo, A., Silvestrini, M. C., Schewe, T., and Rapoport, S. M., 1987, A kinetic model for lipoxygenases based on experimental data with the lipoxygenase of reticulocytes, *Eur. J. Biochem.* **168:**325–337.

Ludwig, P., Höhne, M., Kühn, H., Schewe, T., and Rapoport, S. M., 1988, The biological dynamics of lipoxygenase in rabbit red cells in the course of an experimental bleeding anaemia. Unexpected effects of the calcium ionophore A 23187, *Biomed. Biochim. Acta* **47:**593–608.

Lutze, G., Kunze, D., Reichmann, G., Wiswedel, I., and Augustin, H. W., 1977, Phospholipidzusammensetzung und Fettsäuremuster der isolierten Phospholipide von Mitochondrien aus Kaninchenretickulozyten, *Acta Biol. Med. Ger.* **36:**1403–1411.

McGuire, M. J., and DeMartino, C. S., 1986, Purification and characterization of a high molecular weight proteinase (macropain) from human erythrocytes, *Biochim. Biophys. Acta* **873:**279–289.

McKay, M. J., Daniels, R. S., and Hipkiss, A. R., 1980, Breakdown of aberrant protein in rabbit reticulocytes decreases with cell age, *Biochem. J.* **188:**279–283.

Magnani, M., Stocchi, V., Dacha, M., and Fornaini, G., 1984, Rabbit red blood cell hexokinase. Evidence for an ATP-dependent decay during cell maturation, *Mol. Cell. Biochem.* **61:**83–92.

Mai, A., Sandring, D., Belkner, J., Prehn, S., and Rapoport, S. M., 1980, In vitro-Reifung von Retikulozyten. Verhalten von RNS und anorganischer Pyrophosphatase, *Acta Biol. Med. Ger.* **39:**217–222.

Maniatis, G. M., Ramirez, F., Cann, A., Marks, P. A., and Bank, A., 1976, Translation and stability of human globin mRNA in *Xenopus* oocytes, *J. Clin. Invest.* **58:**1419–1427.

Marbaix, G., Burny, A., Huez, G., Lebleu, B., and Temmermann, J., 1970, Evolution of the polyribosome distribution during *in vivo* reticulocyte maturation, *Eur. J. Biochem.* **13:**322–325.

Marbaix, G., Huez, G., Nokin, P., and Cleuter, Y., 1976, Free cytoplasmic α-globin messenger RNA appears during the maturation of reticulocytes, *FEBS Lett.* **66:**269–273.

Maretzki, D., Kostić, M. M., Reimann, B., Schwarzer, E., and Rapoport, S. M., 1986, Maturation of rabbit reticulocytes: Strong decline of the turnover of polyphosphoinositides, *Biomed. Biochim. Acta* **45:**1227–1236.

Maretzki, D., Ueta, N., Reimann, B., Schwarzer, E., Kostić, M., and Rapoport, S. M., 1987, Maturation dependence of the turnover of phosphatidylinositides in rabbit red blood cells, *Biomed. Biochim. Acta* **46:**167–171.

Matsumoto, T., Funk, C. D., Rådmark, O., Höög, J.-O., Jörnvall, H., and Samuelsson, B., 1988, Molecular cloning and amino acid sequence of human 5-lipoxygenase, *Proc. Natl. Acad. Sci. USA* **85:**26–30.

Merrick, W. C., Kemper, W. M., Kantor, J. A., and Anderson, W. F., 1975, Purification and properties of rabbit reticulocyte protein synthesis elongation factor 2, *J. Biol. Chem.* **250:**2620–2625.

Meyer, D. I., Krause, E., and Dobberstein, B., 1982, Secretory protein translocation across membranes—the role of the "docking protein," *Nature* **297:**647–650.

Minich, W. B., Evdokimova, V. M., Oleinikov, A. V., Höhne, M., Thiele, B. J., and Rapoport, S. M., 1989, Evidence for the appearance of a reticulocyte population low in lipoxygenase mRNA during the recovery from a phenylhydrazine-induced anemia in rabbits, *FEBS Lett.* in press.

Morimoto, R., and Fodor, E., 1984, Cell-specific expression of heat shock proteins in chicken reticulocytes and lymphocytes, *J. Cell Biol.* **99:**1316–1323.

Mosca, J. D., Wu, J. M., and Suhadolnik, R. J., 1983, Restoration of protein synthesis in lysed rabbit reticulocytes by the enzymatic removal of adenosine 5′-monophosphate with either AMP deaminase or AMP nucleosidase, *Biochemistry* **22:**346–354.

Müller, M., Dubiel, W., Rathmann, J., and Rapoport, S. M., 1980, Determination and characteristics of energy-dependent proteolysis in rabbit reticulocytes, *Eur. J. Biochem.* **109:**405–410.

Murray, J. J., and Brash, A. R., 1988, Rabbit reticulocyte lipoxygenase catalyzes specific 12(S) and 15(S) oxygenation of arachidonylphosphatidylcholine, *Arch. Biochem. Biophys.* **265:**514–523.

Narumiya, S., Salmon, J. A., Cottee, F. H., Weatherley, B. C., and Flower, R. J., 1981, Arachidonic acid 15-lipoxygenase from rabbit peritoneal polymorphonuclear leukocytes. Partial purification and properties, *J. Biol. Chem.* **256**:9583–9592.

Navaratnam, S., Feiters, M. C., Al-Hakim, M., Allen, J. C., Veldink, G. A., and Vliegenthart, J. F. G., 1988, Iron environment in soybean lipoxygenase-1, *Biochim. Biophys. Acta* **956**:70–76.

Nudel, U., Soreq, H., Littauer, U. Z., Marbaix, G., Huez, G., Leclercq, M., Hubert, E., and Chantrenne, H., 1976, Globin mRNA species containing poly(A) segments of different lengths. Their functional stability in *Xenopus* oocytes, *Eur. J. Biochem.* **64**:115–121.

O'Prey, J., Chester, J., Thiele, B. J., Janetzki, S., Prehn, S., Fleming, J., and Harrison, P. R., 1989, The promoter structure and complete sequence of the rabbit erythroid cell-specific 15-lipoxygenase gene, *Gene* **84**:493–499.

Orr, L., Adam, M., and Johnstone, R. M., 1987, Externalization of membrane-bound activities during sheep reticulocyte maturation is temperature and ATP-dependent, *Biochem. Cell. Biol.* **65**:1080–1090.

Pan, B. T., and Johnstone, R. M., 1983, Fate of the transferrin receptor during maturation of sheep reticulocytes in vitro; selective externalization of the receptor, *Cell* **33**:967–977.

Pan, B. T., Teng, K., Wu, C., Adam, M., and Johnstone, R. M., 1985, Electron microscopic evidence for externalization of the transferrin receptor in vesicular form in sheep reticulocyte, *J. Cell Biol.* **101**:942–948.

Park, E. A., and Morgan, H. E., 1984, Energy dependence of RNA degradation in rabbit reticulocytes, *Am. J. Physiol.* **247**:C390–C395.

Parodi, A. J., and Martin-Barrientes, J., 1977, Glycosylation of endogenous proteins through dolichol derivates in reticulocyte plasma membranes, *Biochim. Biophys. Acta* **500**:80–88.

Pickart, C. M., and Anthony, T. V., 1988, Levels of active ubiquitin carrier proteins decline during erythroid maturation, *J. Biol. Chem.* **263**:12028–12035.

Prehn, S., Rosenthal, S., and Rapoport, S. M., 1972, The temperature-dependent enzymatic breakdown of rRNA of reticulocytes, *Eur. J. Biochem.* **24**:456–460.

Priess, H., and Zillig, W., 1967, Inhibitor für pankreatische Ribonuclease aus roten Blutzellen, *Hoppe-Seyler's Z. Physiol. Chem.* **348**:817–822.

Rapoport, S. M., 1986, *The Retikulocyte,* CRC Press, Boca Raton, Florida.

Rapoport, S. M., and Gerischer-Mothes, W., 1955, Biochemische Vorgänge bei der Erythrocytenreifung: Über einen Hemmstoff des Succinatoxydase-Systems in Reticulocyten, *Hoppe-Seyler's Z. Physiol. Chem.* **302**:167–178.

Rapoport, S. M., and Nieradt-Hiebsch, C., 1955, Biochemische Vorgänge bei der Retikulocytenreifung: Über den Angreiffspunkt des Reticulocytenhemmstoffes in der Atmungskette, *Hoppe-Seyler's Z. Physiol. Chem.* **302**:179–185.

Rapoport, S., Guest, G. M., and Wing, M., 1944, Size, hemoglobin content and acid-soluble phosphorus of erythrocytes of rabbits with phenylhydrazine-induced reticulocytosis, *Proc. Soc. Exp. Biol. Med.* **57**:334–347.

Rapoport, S. M., Schewe, T., Wiesner, R., Halangk, W., Ludwig, P., Janicke-Höhne, M., Tannert, C., Hiebsch, C., and Klatt, D., 1979, The lipoxygenase of reticulocytes. Purification, characterization and biological dynamics of the lipoxygenase, its identity with the respiratory inhibitors of the reticulocyte, *Eur. J. Biochem.* **96**:545–561.

Rapoport, S., Müller, M., Dumdey, R., and Rathmann, J., 1980, Nitrogen economy and the metabolism of serine and glycine in reticulocytes of rabbits, *Eur. J. Biochem.* **108**:449–455.

Rapoport, S. M., Dubiel, W., and Müller, M., 1981, The mechanism of maturation-dependent breakdown of mitochondria in reticulocytes, *Acta Biol. Med. Ger.* **40**:1277–1283.

Rapoport, S., Härtel, B., and Hausdorf, G., 1984, Methionine sulfoxide formation is the cause of self-inactivation of reticulocyte lipoxygenase, *Eur. J. Biochem.* **139**:573–576.

Rapoport, S. M., Dubiel, W., and Müller, M., 1985a, Proteolysis of mitochondria in reticulocytes during maturation is ubiquitin-dependent and is accompanied by a high rate of ATP hydrolysis, *FEBS Lett.* **180**:249–252.

Rapoport, S. M., Schmidt, J., and Prehn, S., 1985b, Maturation of rabbit reticulocytes: Susceptibility of mitochondria to ATP-dependent proteolysis is determined by the maturational state of reticulocytes, *FEBS Lett.* **183**:370–374.

Rapoport, S., Schmidt, J., and Prehn, S., 1986, Fe-dependent formation of a protein that makes mitochondria lipoxygenase-susceptible during maturation of reticulocytes, *FEBS Lett.* **198**:109–111.

Raviv, O., Heller, H., and Hershko, A., 1987, Alterations in components of the ubiquitin-protein ligase system following maturation of reticulocytes to erythrocytes, *Biochem. Biophys. Res. Comm.* **145:**658–665.

Raw, I., and DiFini, F., 1983, The possible role of ATP-dependent proteolysis on the solubilization of methemoglobin reductase during reticulocyte maturation, *Biochem. Biophys. Res. Commun.* **116:**357–359.

Rechsteiner, M., 1987, Ubiquitin-mediated pathways for intracellular proteolysis, *Annu. Rev. Cell Biol.* **3:**1–30.

Richter-Rapoport, S. K. N., Dumdey, R., Hiebsch, C., Thamm, R., Uerlings, I., and Rapoport, S., 1977, Charakterisierung von Retikulozyten des Menschen: Atmung, Pasteur-Effekt und elektronenmikroskopische Befunde an Mitochondrien, *Acta Biol. Med. Ger.* **36:**53–64.

Rifkind, R. A., Danon, D., and Marks, P. A., 1964, Alterations in polysomes during erythroid cell maturation, *J. Cell Biol.* **22:**599–611.

Rivett, A. J., 1989, The multicatalytic proteinase from mammalian cells, *Arch. Biochem. Biophys.* **268:**1–8.

Rosenthal, S., Künzel, W., and Wagenknecht, C., 1964a, Biochemische Charakterisierung der Ribosomen von Kaninchenretikulozyten und ihre Reifungsänderung, *Acta Biol. Med. Ger.* **13:**281–290.

Rosenthal, S., Rapoport, S. M., and Heinemann, G., 1964b, Über eine ribosomale RNase aus Kaninchenretikulozyten, *Acta Biol. Med. Ger.* **13:**946–948.

Rosenthal, S., Prehn, S., and Rapoport, S., 1966, Uridinfreisetzung aus ribosomaler RNS von Kaninchenretikulozyten durch alkalische RNase aus Retikulozytenribosomen und Pankreas, *Acta Biol. Med. Ger.* **17:**667–670.

Rosenthal, S., Gross, J., Grauel, E. L., Papies, B., Schulz, W., Belkner, J., Botscharowa, L., Coutelle, C., Hawemann, M., Nieradt-Hiebsch, C., Müller, M., Opitz, M., Prehn, S., Schultze, M., Staak, R., and Wiesner, R., 1972, Leitkriterien der Retikulozytenreifung, in: *6th Internationales Symposium über Struktur und Funktion der Erythrocyten* (S. Rapoport and F. Jung, eds.), pp. 513–522, Abhandlungen der Deutschen Akademie der Wissenschaften zu Berlin, Akademie-Verlag, Berlin.

Rost, G., 1959, Eigenschaften und Vorkommen eines Ribonuklease-Hemmstoffes im stromafreien Hämolysat roter Blutkörperchen, *Acta Biol. Med. Ger.* **3:**276–283.

Rothman, J. E., and Schmidt, S. L., 1986, Enzymatic recycling of clathrin from coated vesicles, *Cell* **46:**5.

Rouzer, C. A., and Samuelsson, B., 1987, Reversible calcium-dependent membrane association of human leukocyte 5-lipoxygenase, *Proc. Natl. Acad. Sci. USA* **84:**7393–7397.

Salzmann, U., Kühn, H., Schewe, T., and Rapoport, S. M., 1984, Pentane formation during the anaerobic reaction of reticulocyte lipoxygenase. Comparison with lipoxygenases from soybeans and green pea seeds, *Biochim. Biophys. Acta* **795:**535–542.

Salzmann, U., Ludwig, P., Schewe, T., and Rapoport, S. M., 1985, The share of lipoxygenase in the antimycin-resistant oxygen uptake of intact reticulocytes, *Biomed. Biochim. Acta* **44:**211–219.

Salzmann-Reinhardt, U., Kühn, H., Wiesner, R., and Rapoport, S., 1985, Metabolism of polyunsaturated fatty acids by rabbit reticulocytes, *Eur. J. Biochem.* **153:**189–194.

Samuelsson, B., Dahlén, S. E., Lindgren, J. A., and Rouzer, C. A., 1987, Leukotrienes and lipoxins. Structures, biosynthesis and biological effects, *Science* **237:**1171–1175.

Schafer, D., and Hultquist, D. E., 1981, Isolation of an acid protease from rabbit reticulocytes and evidence for its role in processing redox proteins during erythroid maturation, *Biochem. Biophys. Res. Commun.* **100:**1555–1561.

Schekman, R., and Singer, S. J., 1976, Clustering and endocytosis of membrane receptors can be induced in mature erythrocytes of neonatal but not adult humans, *Proc. Natl. Acad. Sci. USA* **73:**4075–4079.

Schewe, T., Wiesner, R., and Schulz, W., 1972, Mitochondrien aus Kaninchen-retikulozyten. III. Zytochrom- und Phospholipid-Gehalt, *Acta Biol. Med. Ger.* **28:**1–6.

Schewe, T., Halangk, W., Hiebsch, C., and Rapoport, S. M., 1975, A lipoxygenase in rabbit reticulocytes which attacks phospholipids and intact mitochondria, *FEBS Lett.* **60:**149–152.

Schewe, T., Albracht, S. P. J., and Ludwig, P., 1981, On the site of action of the inhibition of the mitochondrial respiratory chain by lipoxygenase, *Biochim. Biophys. Acta* **636:**210–217.

Schewe, T., Rapoport, S. M., and Kühn, H., 1986, Enzymology and physiology of reticulocyte lipoxygenase: Comparison with other lipoxygenases, *Adv. Enzymol. Relat. Areas Mol. Biol.* **58:**191–272.

Schewe, T., Kroschwald, P., Kroschwald, A., Ludwig, P., and Kühn, H., 1990, The erythroid arachidonate 15-lipoxygenase in rat reticulocytes, *Biomed. Biochim. Acta* **49:**S42–S46.

Schlegel, R. A., Phelps, B. M., Cofer, G. P., and Williamson, P., 1982, Enucleation eliminates a differentiation-specific marker from normal and leukemic murine cells, *Exp. Cell Res.* **139:**321–328.

Schmidt, J., Prehn, S., and Rapoport, S. M., 1985, Proteolysis during in vitro-maturation of rabbit reticulocytes, *Biomed. Biochim. Acta* **44:**1429–1434.

Schulz, W., Neymeyer, H. G., and Rosenthal, S., 1971, Mitochondrien aus Kaninchenretikulozyten. I. Präparation und zellphysiologische Reinheitskriterien, *Acta Biol. Med. Ger.* **26:**439–456.

Schweiger, H. G., and Rapoport, S. M., 1958, Der N-Stoffwechsel bei Erythrocytenreifung: Die N-Bilanz unter endogenen Bedingungen, *Hoppe-Seyler's Z. Physiol. Chem.* **313:**97–108.

Schweiger, H. G., Rapoport, S. M., and Schölzel, E., 1956, Role of nonprotein nitrogen in the synthesis of haemoglobin in the reticulocyte in vitro, *Nature* **178:**141–142.

Shibata, D., Steczko, J., Dixon, J. E., Hermodson, M., Yazdanparast, R., and Axelrod, B., 1987, Primary structure of soybean lipoxygenase-1, *J. Biol. Chem.* **262:**10080–10085.

Shibata, D., Steczko, J., Dixon, J. E., Andrews, P. C., Hermodson, M., and Axelrod, B., 1988, Primary structure of soybean lipoxygenase L-2, *J. Biol. Chem.* **263:**6816–6821.

Siems, W., Müller, M., Dumdey, R., Holzhütter, H.-G., Rathmann, J., and Rapoport, S. M., 1982, Quantification of pathways of glucose utilization and balance of energy metabolism of rabbit reticulocytes, *Eur. J. Biochem.* **124:**567–576.

Siems, W., Dubiel, W., Dumdey, R., Müller, M., and Rapoport, S. M., 1984, Accounting for the ATP-consuming processes in rabbit reticulocytes, *Eur. J. Biochem.* **139:**101–107.

Sigal, E., Craik, C. S., Highland, E., Grunberger, D., Costello, L. L., Dixon, R. A. F., and Nadel, J. A., 1988a, Molecular cloning and primary structure of human 15-lipoxygenase, *Biochem. Biophys. Res. Commun.* **157:**457–464.

Sigal, E., Grunberger, D., Craik, C. S., Caughey, G. H., and Nadel, J. A., 1988b, Arachidonate 15-lipoxygenase (ω-6 lipoxygenase) from human leukocytes. Purification and structural homology to other mammalian lipoxygenases, *J. Biol. Chem.* **263:**5328–5332.

Singh, M. K., and Yu, J., 1984, Accumulation of a heat-shock-like protein during differentiation of human erythroid cell line, *Nature* **309:**631–633.

Smith, D. W., and Weinberg, W. C., 1981, Transfer RNA in reticulocyte maturation, *Biochim. Biophys. Acta* **655:**195–198.

Speiser, S., and Etlinger, J. D., 1982, Loss of ATP-dependent proteolysis with maturation of reticulocytes and erythrocytes, *J. Biol. Chem.* **257:**14122–14127.

Speiser, S., and Etlinger, J. D., 1983, ATP stimulates proteolysis in reticulocyte extracts by repressing an endogenous protease inhibitor, *Proc. Natl. Acad. Sci. USA* **80:**3577–3580.

Tanaka, K., and Ichihara, A., 1988, Involvement of proteasomes (multicatalytic proteinase) in ATP-dependent proteolysis in rat reticulocyte extracts, *FEBS Lett.* **236:**159–162.

Thiele, B. J., Belkner, J., Andree, H., Rapoport, T. A., and Rapoport, S. M., 1979, Synthesis of non-globin proteins in rabbit erythroid cells. Synthesis of a lipoxygenase in reticulocytes, *Eur. J. Biochem.* **96:**563–569.

Thiele, B. J., Andree, H., Höhne, M., and Rapoport, S. M., 1982, Lipoxygenase mRNA in rabbit reticulocytes. Its isolation, characterization and translational repression, *Eur. J. Biochem.* **129:**133–141.

Thiele, B. J., Black, E., Fleming, J., Nack, B., Rapoport, S. M., and Harrison, P. R., 1987a, Cloning of reticulocyte lipoxygenase mRNA, *Biomed. Biochim. Acta* **46:**S120–S123.

Thiele, B. J., Höhne, M., Nack, B., Harrison, P. R., and Rapoport, S. M., 1987b, Lipoxygenase mRNA during development of red blood cells studied with a cloned probe, *Biomed. Biochim. Acta* **46:**S124–S125.

Thiele, B. J., Fleming, J., Kasturi, K., O'Prey, J., Black, E., Chester, J., Rapoport, S. M., and Harrison, P. R., 1987c, Cloning of a rabbit erythroid cell-specific lipoxygenase mRNA, *Gene* **57:**111–119.

Thilo, C., Schewe, T., Belkner, J., and Rapoport, S., 1979, In Vitro-Reifung von Kaninchenretikulozyten: Verhalten des Sauerstoffverbrauchs, *Acta Biol. Med. Ger.* **39:**1431–1440.

Ueda, N., Kaneko, S., Yoshimoto, T., and Yamamoto, S., 1986, Purification of arachidonate 5-lipoxygenase from porcine leukocytes and its reactivity with hydroperoxyeicosatetraenoic acids, *J. Biol. Chem.* **261:**7982–7988.

Van der Meer, R. A., and Duine, J. A., 1988, Pyrroloquinoline quinone (PQQ) is the organic cofactor in soybean lipoxygenase-1, *FEBS Lett.* **235:**194–200.

Van Renswoude, J., Bridges, K., Harford, J. B., and Klausner, R. D., 1982, Receptor-mediated endocytosis of transferrin and the uptake of Fe in K562 cells: Identification of a nonlysosomal acidic compartment, *Proc. Natl. Acad. Sci. USA* **79:**6186–6190.

Villa-Trevino, S., Shull, K. H., and Farber, E., 1963, The role of adenosine triphosphate deficiency in ethionine-induced inhibition of protein synthesis, *J. Biol. Chem.* **238:**1757–176.

Vladimirov, Y. A., Olenov, V. I., Suslova, T. B., and Cheremisina, Z. P., 1980, Lipid peroxidation in mitochondrial membrane, *Adv. Lipid Res.* **17:**173–249.

Wathelet, M., Moutschen, S., Defilippi, P., Cravador, A., Collet, M., Huez, G., and Content, J., 1986, Molecular cloning, full-length sequence and preliminary characterization of a 56-kDa protein induced by human interferons, *Eur. J. Biochem.* **155:**11–17.

Waxman, L., Fagan, J. M., Tanaka, K., and Goldberg, A. L., 1985, A soluble ATP-dependent system for protein degradation from murine erythroleukemia cells: Evidence for a protease which requires ATP hydrolysis but not ubiquitin, *J. Biol. Chem.* **260:**11994–12000.

Waxman, L., Fagan, J. M., and Goldberg, A. L., 1987, Demonstration of two distinct high-molecular weight proteases in rabbit reticulocytes, one of which degrades ubiquitin conjugates, *J. Biol. Chem.* **262:**2451–2457.

Wiedmann, M., Kurzchelia, T., Hartmann, E., and Rapoport, T. A., 1987, A signal sequence receptor in the endoplasmic reticulum membrane, *Nature* **328:**830–833.

Wiesner, R., Rosenthal, S., and Hiebsch, C., 1973, Leitkriterien der Retikulozytenreifung, II. Das Verhalten von Zytochromoxydase und Hemmstoff F der Atmungskette bei der Retikulozytenreifung, *Acta Biol. Med. Ger.* **30:**631–646.

Wiesner, R., Kasüschke, A., Kühn, H., Anton, M., and Schewe, T., 1989, Oxygenation of mitochondrial membranes by the reticulocyte lipoxygenase. Action on monoamine oxidase activities A and B, *Biochim. Biophys. Acta* **986:**11–17.

Wiesner, R., Kühn, H., Anton, M., and Schewe, T., 1990, Oxygenation of mitochondrial membranes by the erythroid lipoxygenase. Consequences for membrane properties, *Biomed. Biochim. Acta* **49:**S35–S38.

Wreschner, D. H., and Rechavi, G., 1988, Differential mRNA stability to reticulocyte ribonucleases correlates with 3′ non coding $(U)_nA$ sequences, *Eur. J. Biochem.* **172:**333–340.

Wreschner, D. H., Silverman, R. H., James, T. C., Gilbert, C. S., and Kerr, I. M., 1982, Affinity labelling and characterization of the $ppp(A2'P)_nA$-dependent endoribonuclease from different mammalian sources, *Eur. J. Biochem.* **124:**261–268.

Yatziv, S., Kahane, I., Abeliuk, P., Cividally, G., and Rachmilewitz, E. A., 1979, "Lysosomal" enzyme activities in red blood cells of normal individuals and patients with homozygous β-thalassaemia, *Clin. Chim. Acta* **96:**67–72.

Yenofsky, R. L., Fine, M., and Liu, C., 1988, Isolation and characterization of a soybean (Glycine max) lipoxygenase-3 gene, *Mol. Gen. Genet.* **211:**215–222.

Zeidler, R. B., and Kim, M. D., 1982, Pig reticulocytes. IV. In vitro maturation of naturally occurring reticulocytes with permeability loss to glucose, *J. Cell. Physiol.* **112:**360–366.

Zweig, S., Tokuysu, K., and Singer, S., 1981, Membrane-associated changes during erythropoiesis. On the mechanism of maturation of reticulocytes to erythrocytes, *J. Supramol. Struct.* **17:**163–182.

Chapter 7

Function and Structure of the Red Blood Cell Cytoskeleton

Makoto Nakao

1. INTRODUCTION

Mature red blood cells of mammals consist only of cytoplasm and a cell membrane: they lack nucleus, mitochondria, and all other organelles. Cytoplasm contains all enzymes of glycolysis and the hexose monophosphate shunt, and some of those for nucleoside and nucleotide metabolism. These systems provide energy-rich phosphate as ATP and reducing power as NADH and NADPH, as in any other cell. Since most of the synthetic systems for biochemical compounds are lacking in red cells, ATP is mostly consumed on the cell membrane, except for priming to pump glycolysis. The erythrocyte membrane usually obtained by osmotic hemolysis is composed of a lipid bilayer and a fibrous undercoat structure; the latter is generally classified as a part of the cytoskeleton.

According to Singer's theory that membrane proteins are generally divided into integral proteins and peripheral proteins by their solubility, cytoskeletal proteins beneath the lipid bilayer are classified as peripheral proteins. Most proteins in this undercoat structure can be extracted with low- and high-ionic-strength salt solutions, successively. At present, the function of this undercoat cytoskeleton cannot be separated from that of the cell membrane and almost all membrane functions, except transport and receptor phenomena, may be related to the undermembrane structure. Therefore, some functions of the red cell membrane will be included in the following discussion.

Great progress in the field of biological membranes has been made since SDS–PAGE

Abbreviations used in this chapter: ACD blood, blood preserved with acid–citrate–dextrose (NIH formula A); PCD blood, blood preserved with phosphate–citrate–dextrose; Hb, hemoglobin; PC, phosphatidylcholine; PE, phosphatidylethanolamine; PS, phosphatidylserine; PI, phosphatidylinositol; WGA, wheat germ agglutinin.

Makoto Nakao Department of Biochemistry, Tokyo Medical and Dental University School of Medicine, Tokyo, Japan.

(Shapiro, 1967) was applied to red cell membranes and the fluid mosaic model (Singer and Nicolson, 1972) was proposed. As the red cell membrane is the most popular and indeed the most convenient material for studying the cell membrane, it has played an important role in the two developments. All of the red cell membrane proteins are solubilized in SDS solution and most of the membrane proteins could be separated in SDS–PAGE as monomer peptides. The proteins (monomer peptides) or protein bands are termed band 1 (spectrin α, 240 kDa), 2 (spectrin β, 220 kDa), 2.1 (ankyrin, 180 kDa), 3 (100 kDa), 4.1, 4.2, 4.5_s, 4.9, 5 (actin, 43 kDa), 6, 7, Hb (Fairbanks *et al.*, 1971), in the order of their electrophoretic mobility in SDS–PAGE (Fairbanks *et al.*, 1971). In addition, some protein bands appeared as glycoproteins containing a large amount of carbohydrate, after PAS staining instead of Coomassie brilliant blue staining for protein. When low-ionic-strength solutions are applied to red cell membranes, spectrin $\alpha\beta$, actin complexed with band 4.1 are solubilized and when the residue is treated with high concentrations of salts, ankyrin and band 4.9 are extracted. These proteins are components of the undercoat cytoskeleton. On the other hand, when ghosts are treated with detergents such as nonionic Triton X-100, the lipid bilayer containing various phospholipids and cholesterol is extracted. The residue loses most of its lipids when a high concentration of Triton is used for a relatively long time (e.g., 1 hr). This residue contains spectrins, 4.1, 4.9, and actin and is still a biconcave disk or cup, both by dark-field and by phase-contrast microscopy (Yu *et al.*, 1973); it is sometimes called the Triton shell and is believed to contain most of the erythrocyte cytoskeleton. The erythrocyte cytoskeleton has been discussed extensively (Heast, 1982; Palek and Lux, 1983; Cohen, 1983; Knowles *et al.*, 1983; Sheetz, 1983; Shoet and Mohandas, 1983; Bennett, 1985, 1989). However, this is still a very active field and a large amount of new data is added every year. In this chapter, recent progress in the structure and function of the erythrocyte membrane undercoat cytoskeleton will be discussed, with some emphasis on function, including recent findings by our group and others.

2. COMPONENTS OF THE ERYTHROCYTE CYTOSKELETON

The components of the erythrocyte cytoskeleton will be described only briefly. Since SDS–polyacrylamide electrophoretograms provide essentially stable components in the ratio 1 mol spectrin dimer ($\alpha\beta$) : 2–3 mol actin : 1 mol band 4.1 ; 0.1–0.5 mol protein 4.9, many research groups have proposed very similar models (Lux, 1979; Cohen 1983). The model shown in Figure 1 seems to be supported by high-resolution electron microscopy. Tsukita *et al.* (1980) showed a spiderlike structure with a small F-actin fiber in the center using tannic acid–glutaraldehyde fixation and the myosin S-1 arrowhead technique. Polygonal lattice made of spectrin tetramer $(\alpha\beta)_2$ with a short F-actin fiber plus protein 4.1-like particle and an ankyrinlike particle on Triton shells at pH 8.0 was seen after negative staining with uranyl acetate on a 6-Å-thick carbon film by Liu *et al.* (1987). Essentially the same pattern of negative staining was seen by Byers and Branton (1985) using intact cells that were spread on a polylysine-coated 6-Å carbon film, followed by osmotic hemolysis. However, photographs in which units of spectrin tetramer in the lattice are in some places lacking have been published by Palek *et al.* (1987) and Branton *et al.* (1985). According to these authors, they could not distinguish whether the defect is real or an artifact.

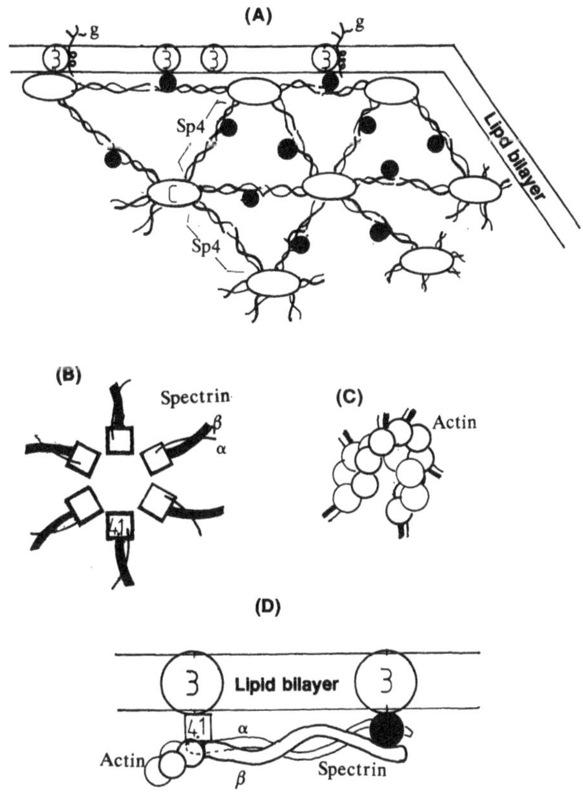

FIGURE 1. A model of erythrocyte undermembrane structure. (A) Overview. The spectrin–actin meshwork lies just beneath the lipid bilayer. (B–D) Lipid bilayer side, cytoplasmic side, and vertical section of actin–spectrin–protein 4.1 complex, which is indicated as C in (A). C also contains band 4.9, tropomyosin, and adducin, in all probability. Spectrin binds to the lipid bilayer at ankyrin, protein 4.1, and directly to phosphatidylserine in the lipid bilayer. These connections are continually releasing and binding. Protein 4.1 does not bind to actin to spectrin. Sp4, spectrin tetramer; 3, band 3; ●, ankyrin; C, protein 4.1–actin–spectrin complex; g, glycophorin.

In vitro binding experiments have shown that these components bind to each other with a K_d of 10^{-7}–10^{-9} M as shown in Table I. The specificity of binding is very high and the activity of binding is easily lost after denaturation or chemical modifications.

This interaction affords a huge, two- or three-dimensional lattice, which warrants various functions as discussed below. Components of this huge lattice are now listed according to the SDS–PAGE pattern. The main components are spectrin, actin, and protein 4.1, which are described in detail in the chapter by Ralston. Therefore, only short comments will be made on the principal molecules and several other accessory proteins, the roles of which have not been elucidated.

2.1. Spectrin

Spectrin is the largest, both in size and in quantity, of the main components of the red cell cytoskeleton. It consists of two subunits, α (240 kDa) and β (220 kDa) (bands 1 and 2 in SDS–PAGE–Coomassie blue staining). The subunits are rod-shaped molecules about 100 nm long, each present in 200,000 copies per erythrocyte, and are arranged side-by-side and antiparallel to form heterodimers. They have many 106-amino-acid repeating units and are homologous (Speicher and Marchesi, 1984). Isolated spectrin shows a varying crevice or gap between the two monomers, judging from negatively stained electron micrographs (Tyler *et al.*, 1979; Shotton *et al.*, 1979), both in dimers and in

Table I
Dissociation Constants for Components of the Erythrocyte Cytoskeleton

	K_d (M)	Reference
Spectrin dimer–dimer	2×10^{-6}	Liu and Palek (1981)
Oligomer formation	$\approx 10^{-5}$	Morrow *et al.* (1984)
Spectrin–PS · PE	7×10^{-7}	Liu *et al.* (1984)
Spectrin dimer–actin	2×10^{-4}	Ohanian *et al.* (1984)
Spectrin–band 4.1	$\approx 10^{-7}$	Tyler *et al.* (1980)
Spectrin–actin–band 4.1	10^{-12} (M^2)	Ohanian *et al.* (1984)
Spectrin–ankyrin	5×10^{-8}	Bennett and Stenbuck (1980)
	$2\text{–}10 \times 10^{-8}$	Litman *et al.* (1980)
Band 3–ankyrin	5×10^{-8}	Hargreaves *et al.* (1980)
	5×10^{-9}, 1.4×10^{-7}	Tyler *et al.* (1980)
		Bennett and Stenbuck (1980)
Band 4.1–glycophorin	1×10^{-7}	Anderson and Lovrien (1984)
Tropomyosin–actin filaments	4×10^{-7} (Mg 2 mM)	Fowler and Bennett (1984)
	4×10^{-8} (Mg 10 mM)	
Tropomyosin–tropomyosin-binding protein	5×10^{-7}	Fowler (1987)
Calmodulin–β-adducin subunit	2×10^{-7}	Gardner and Bennett (1986)
Calmodulin–spectrin (native)	3×10^{-6}	Sobue *et al.* (1982)
Calmodulin–Ca^{2+}-ATPase	3×10^{-9}	Baess and Vincenzi (1980)
Calmodulin–band 4.1	1×10^{-6}	Husain *et al.* (1985a,b)

tetramers. Spectrin heterodimers (A and B) each consisting of an α and β chain are arranged in head-to-head fashion to form tetramers (Shotton *et al.*, 1979; Tyler *et al.*, 1980). Probably the N-terminal end of α in one heterodimer (A) binds to the C-terminal end of β in another heterodimer (B) and the C-terminal end of β in one heterodimer (A) binds to the N-terminal end of α in another heterodimer (B), although no direct evidence for this has been found. Interconversion of dimers and tetramers is a reaction that involves a high activation energy, since at low temperature, i.e., 0°C, little assembly or disassembly of tetramers occurs (Morrow and Marchesi, 1981; Liu and Palek, 1980). The association constant for formation of the hexamer from the dimer and tetramer is lower than the association constant of the two dimers. The association constant for formation of the octamer from the hexamer and dimer is larger and reaches a constant level for formation of higher oligomers (Shahbakhti and Gratzer, 1986). Tetramers are believed to be the major form of spectrin in erythrocyte membranes. Higher-order oligomers, especially long head-to-head oligomers, are very rare in nondenatured polyacrylamide electrophoretograms and in electron micrographs. Spectrin has many sulfhydryl groups, which are reactive to sulfhydryl-reactive reagents (Smith and Palek, 1983), spontaneous oxidative cross-linking in acidic pH (Liu *et al.*, 1977), or when the cell ATP is depleted (Liu and Palek, 1979). Recently, Triton shells consisting of spectrin without any other proteins or lipids were obtained in my laboratory as will be described below. The spectrin shell is very similar in shape to complete Triton shells or normal ghosts obtained by osmotic hemolysis. My collaborators and I propose a hypothesis wherein the free end of a tetramer C-terminal α chain and the N-terminal end of a β chain can bind to each other, as shown in

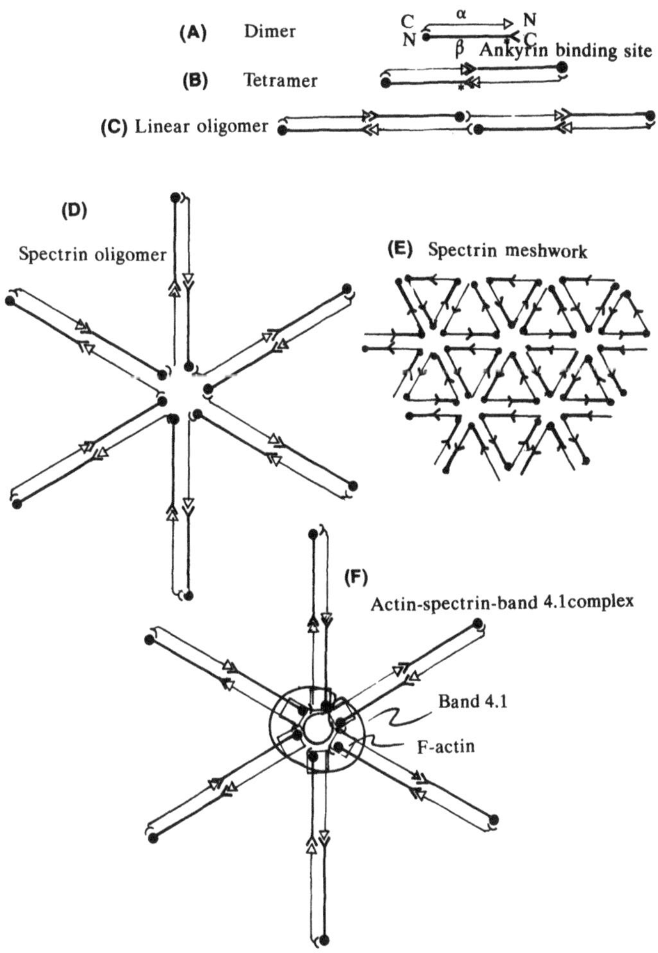

FIGURE 2. Hypothetical molecular structure of erythrocyte spectrin meshwork. (A) Spectrin dimer. α and β bind to each other side-by-side in an antiparallel conformation, but part of the two-peptide attachment is sometimes split. N-terminal ends of α and β are shown as △ and ● and C-terminal ends of α and β as (and <, respectively. (B) Spectrin tetramer. Two heterodimers are connected in a head-to-head manner. The binding between the N-terminus of α of one heterodimer and the C-terminus of β of the other heterodimer is relatively weak and the binding is easily split to make a bridge between the two heterodimers. Thus, a linear polymer as in (C) (very rare) and a branching polymer as in (D) are formed. A macromolecule made of only spectrins can be formed as in (E). (F) A proposed actin–spectrin–protein 4.1 unit that might form a huge oligomeric complex.

Figure 2. Then, the $(\alpha\beta)_2$ heterotetramer, namely two homodimers A, B, each consisting of α and β, forms a cycle, e.g., [αA = βB/βA = αB] and the binding from αA to βB and from αB to βA is tight, forming dimer–dimer association sites and the αβ side-by-side association is also rather tight, but βB–αB and βA–αA associations are loose. The last binding site is easily split to open the site for a new binding to another α and β, each of which can come from different tetramers; spectrin can then make a branching meshwork. This loose end can also bind to protein 4.1 or make a ternary complex, actin–spectrin–

4.1. This may tentatively explain the rare occurrence of higher-order linear oligomerization of tetramers. Thus, many tetramers can bind to each other at these ends and form a branching meshwork. Usually actin filaments or actin–4.1 complexes associate at these ends, instead of the closed ring structure of a tetramer.

2.2. Actin

The presence of actin in erythrocytes was first suggested by T. Ohnishi (1962). Actin isoform β (Pinder *et al.*, 1978) or isoforms β + τ (Nakashima and Beutler 1979) are present in about 400,000–500,000 copies per cell. Most of these molecules are associated with the membrane skeleton and the concentration of free actin in the cytoplasm is 6% of the total, about 15 μg/ml actin (Pinder and Gratzer, 1983), which is nearly the critical association concentration. F-actin filaments comprised of 12–14 monomers cannot be demonstrated directly by decoration with S1 of skeletal myosin, but are clearly indicated after elongation of additional actin (Tsukita *et al.*, 1980). The number of actin filaments is assumed to be about 30,000 by estimation of binding of cytochalasin D, which caps the fast-growing end, and DNase I, which caps the slow-growing end of actin filaments (Brenner and Korn, 1980; Lin and Lin, 1978). Triton-treated ghosts show both calcium-activated and -inhibited ATPases. The calcium-inhibited ATPase is attributed to actin ATPase. This activity may be activated further by crude spectrin–protein 4.1, judging from the activity difference between purified actin and actin–spectrin–4.1 complex (Sato and Nakao, 1986).

2.3. Protein 4.1

This peptide of 78 kDa is present in about 200,000 copies per cell. Its primary structure was completely determined by cDNA cloning (Conboy *et al.*, 1986). After the membrane is extracted at low ionic strengths, 80% of protein 4.1 remains in the membrane, but when the membrane is treated with Triton X-100, almost all of protein 4.1 remains in the shell. It binds to spectrin, accelerating the binding of spectrin to actin (Ungewickell *et al.*, 1979; Ohanian *et al.*, 1984). On SDS–PAGE it consists of a doublet 4.1a and 4.1b, both containing O-linked acetylglucosamine (Holt *et al.*, 1987). The binding site of spectrin to 4.1 is at the free end of the spectrin tetramer, Protein 4.1 does not bind to actin directly, but promotes association of human spectrin with actin (Tyler *et al.*, 1980; Cohen and Foley, 1982; Elbaum *et al.*, 1984). A single molecule of protein 4.1 is sufficient to promote the binding of a spectrin dimer to F-actin, although the molar ratio of protein 4.1/spectrin in the complex is not fixed (Cohen and Foley, 1984). The ternary complex is not affected by magnesium or calcium concentrations, but is markedly affected by KCl above 100 mM. Spectrin binding to F-actin in the presence of protein 4.1 requires the participation of both of spectrin's polypeptide chains (Cohen and Langley, 1984), but the spectrin–actin interaction with protein 4.1 is a property uniquely bestowed on the heterodimer by the β subunit of spectrin (Cohen *et al.*, 1986). Protein 4.1 is phosphorylated in solution by several protein kinases including protein kinase C and cAMP-dependent protein kinase, which catalyze the incorporation of 1–4 mol phosphate per protein 4.1. Phosphorylation of protein 4.1 by each kinase results in 50–80% reduction in the ability of protein 4.1 to promote spectrin binding to F-actin. The phosphorylation might regulate red cell cytoskeletal organization and membrane mechanical properties (Ling *et al.*, 1988), which will be described below. Besides these three main components—spectrin, actin, and protein 4.1—several others will now be discussed. Their precise location and function have not been determined, but their presence may be important with regard to their function.

2.4. Band 4.9

This 48-kDa peptide remains in Triton-treated ghosts even after treatment with high concentrations of salts, but it is extracted from the Triton-treated ghosts or shells with the low-ionic-strength buffers. It exists in solution as trimers with an apparent molecular mass of 145 kDa and a Stokes radius of 50 Å. Electron microscopy shows that the protein is a three-lobed structure. When gel-filtered rabbit muscle actin is polymerized in the presence of band 4.9, actin bundles are generated. The bundles appear brittle and when they are centrifuged, small pieces of filaments break off and remain in the supernatant (Siegel and Branton, 1985). At a low band 4.9 : actin molar ratio such as 1 : 30, band 4.9 reduces the apparent steady-state viscosity by sequestering filaments into thin bundles; at higher ratios the bundles become thicker and an apparent increase in steady-state viscosity is observed. No report on specific binding of band 4.9 to actin–spectrin–4.1 has been published, but band 4.9 cross-links with band 3 when a chemical cross-linker is used on ghosts and intact cells (Sato and Nakao, 1981). Although this protein is a substrate for cAMP-dependent kinase, calcium-stimulated kinase, and protein kinase C, the real function *in situ* is not known and the amount of 4.9 is small, about 1% of the membrane protein (Bennett, 1989).

2.5. Tropomyosin

Erythrocyte tropomyosin was first identified by Fowler and Bennett (1984) who reported on a number of properties shared with other nonmuscle tropomyosins, including common antigenicity, physical properties of an asymmetric dimer, characteristic amino acid composition, isoelectric point (pH 4.5), and heat stability. It consists of two polypeptides of 29 and 27 kDa that are present as dimers, with about 70,000–80,000 copies per cell. Erythrocyte tropomyosin associates with actin filaments with a stoichiometry of 1 dimer per 6–7 actin monomers. The association requires magnesium (> 1 mM). It is assumed that erythrocyte tropomyosin associates with the actin filaments in the membrane skeleton in a manner analogous to tropomyosin and actin in skeletal muscles with a tropomyosin dimer attached to each of the two grooves of the actin helix. Actin filaments identified by electron microscopy have a narrow length distribution of 33–38 nm, approximately the length of a single tropomyosin molecule (Byers and Branton, 1985; Shen *et al.*, 1986). The function of the protein 4.1 may be to stabilize and determine the length of actin filaments or calcium sensitivity, cooperating with other proteins. However, direct evidence is still lacking.

2.6. Tropomyosin-Binding Protein

A tropomyosin-binding protein, presumably different from troponin T from muscle, was reported by Fowler (1987). This 43-kDa protein binds to tropomysoin labeled by the ^{125}I Bolton Hunter reagent, after being separated by SDS–PAGE and nitrocellulose blots.

The affinity of this binding is 5×10^{-7} M at a ratio of about one molecule for every two tropomyosin molecules after extraction of membranes with Triton X-100. Although the mobility of this protein on SDS–PAGE is similar to that of protein 4.9, it is not a substrate for cAMP-dependent protein kinase (Bennett, 1989). Possible functions for the tropomyosin-binding protein are to stabilize and regulate the association of tropomyosin with actin filaments, but this remains to be seen. The tropomyosin-binding protein remains with Triton X-100-treated ghosts even under conditions where tropomyosin is extracted.

2.7. Adducin

Adducin, initially identified on the basis of calmodulin binding activity, is a heterodimer consisting of a 103-kDa α subunit and a 97-kDa β subunit. Adducin is a somewhat irregular disk-shaped molecule with a height of 5.4 nm and diameter of 12.4 nm (Gardner and Bennett, 1987).

Calmodulin binds to the β subunit of adducin in a calcium-dependent manner with a dissociation constant of 200 nM. Calmodulin is present in erythrocytes at a concentration of about 2–3 μM. On the other hand, native spectrin binds calmodulin only weakly and does not associate with calmodulin columns in sharp contrast to adducin. Adducin associates with spectrin–actin complex and promotes binding of spectrin to actin (Mische *et al.*, 1987).

Adducin, at about 30,000 copies per cell, and protein 4.1, at about 100,000 copies per cell, form ternary complexes with spectrin and actin, and promote binding of spectrin to actin. The ternary complex-forming activity of protein 4.1 is stronger than that of adducin, and protein 4.1 interacts with spectrin alone, but not with actin alone, while adducin associates with actin (Mische *et al.*, 1987). Protein kinase C and phosphatidylserine bind to adducin (Wolf and Sahyoun, 1986).

2.8. Myosin

The presence of myosin has been discussed since 1960 when ATP-dependent shape changes were observed in human erythrocytes (M. Nakao *et al.*, 1959, 1960a,b, 1961). Later, the Ca^{2+}/Mg^{2+}-activated ATPase was separated by us from Na^{+},K^{+}-ATPase, using human erythrocyte membranes (T. Nakao *et al.*, 1963). The Ca^{2+}/Mg^{2+}-ATPase is inhibited by *N*-ethylmaleimide (Palek *et al.*, 1971b). Whether all of the calcium-activated ATPase activity in the red cell membrane can be attributed to myosin has not been decided. Myosin, having similar characteristics to smooth muscle myosin, is purified from erythrocyte cytoplasm (Fowler *et al.*, 1985). However, phosphorylation of erythrocyte myosin light chain induces some actin-activated ATPase (Wong *et al.*, 1985). This protein is presumably present in about 6000 copies per cell, existing both in the cytoplasm and as a membrane-associated form. The precise location of myosin and its role in ATP-dependent shape change are unknown.

2.9. Calmodulin

Calmodulin is present in erythrocytes at a concentration of about 2–3 μM (Jarrett and Penniston, 1978), most of which is in the cytoplasm. It inhibits spectrin–actin–protein 4.1

interaction in *in vitro* experiments with a low affinity, about 50 μM, but modulates (Bennett, 1989) adducin–spectrin–actin interactions with a half-maximal effect at 1 μM (Bennett *et al.*, 1988). The dissociation constant between adducin and calmodulin, 0.2 μM, is the lowest among the calmodulin-binding proteins in erythrocytes; e.g., with spectrin it is much larger, and with protein 4.1 it is 2 μM (Anderson and Morrow, 1987). However, the physiological significance of calmodulin in erythrocytes remains to be clarified.

3. FUNCTION OF THE RED CELL MEMBRANE

The red cell membrane creates a specific barrier to various substances and has the ability to transport others. It also has various channels or pumps. Hydrophobic substances such as oxygen and carbon dioxide can easily cross the lipid bilayer, but hydrophilic substances cannot penetrate the bilayer rapidly; they must cross the membrane through channels or some transit hole in the lipid bilayer, usually very slowly. The most abundant channels are anion channels, which are represented by the 100-kDa protein, band 3. The number of cation pumps such as Na^+,K^+-ATPase and Ca^{2+}-ATPase is less than 1000 per cell. There might exist other species of channels, but their number is far smaller than that of anion channels, and they have not as yet been isolated. Channels and pumps move laterally in an artificial lipid bilayer and the physiological location of these molecules may be determined by the cytoskeleton in red cells, as well as in other cells. The main function of the red cell is oxygen and carbon dioxide transport. For this purpose and for maintaining cell life, these channels and pumps are likely to be distributed rather evenly throughout the red cell membrane area.

Besides active transport of cations, several functions of the cell membrane require ATP. This requirement might serve as a key for the understanding of biochemical mechanisms. In 1959 my collaborators and I found that depleted blood cells can be replenished with ATP, when these cells are incubated with adenine plus purine nucleoside, instead of ATP or ADP, which cannot penetrate the membrane (M. Nakao *et al.*, 1959, 1960a, 1961). The mechanism is explained as a sort of positive feedback process, where nucleoside is split into ribose 1-P and purine at first and then ribose 1-P is converted into phosphoribosylpyrophosphate with the aid of a very small amount of ATP remaining inside the cell, and phosphoribosylpyrophosphate reacts with adenine to form ATP (M. Nakao *et al.*, 1960b). This system can work cyclically. The further the cyclic reaction proceeds, the more the reactant ATP increases. As the level of ATP is decreased by the elimination of substrates, or the addition of reversible inhibitors of glycolysis, we can easily control the level of ATP inside the cell (M. Nakao *et al.*, 1961, 1962). Thus, the addition of adenine and nucleoside is useful to clarify the relationship between ATP and cell membrane function.

3.1. Erythrocyte Shape

The shape of a red cell represents a complex phenomenon. The biconcave disk of the human erythrocyte can change reversibly into a crenated shape, a sphere, a cup, and a number of other shapes depending on pH, osmotic pressure, ionic strength, and presence of various reagents. The choice of these shapes depends on the molecular structure of the cell membrane, and the physical and chemical conditions both in the external medium and in the cytoplasm. By using the technique mentioned above, my collaborators and I discovered a close correlation between ATP content and red cell shape.

3.1.1. Intracellular ATP and Shape

The summary of the findings is as follows (M. Nakao *et al.*, 1959, 1960b, 1961):

1. During the preservation of ACD blood at 4°C, the ATP content in erythrocytes decreases gradually accompanied by a change from disk to crenated echinocyte, and echinocyte to sphere. The addition of both adenine and inosine to long-preserved ACD blood replenishes ATP by way of a regeneration process with the aid of the small amount of ATP remaining (see above), but adenine or inosine alone is ineffective and the shape does not change.
2. Both the increase in ATP and the change in shape are inhibited strongly by arsenate, iodoacetate, fluoride, or deoxyglucose. Both phenomena parallel each other.
3. To achieve the increase in ATP content, adenine cannot be replaced by any other purine or pyrimidine base and inosine can be replaced by guanosine but not pyrimidine nucleosides. The increase in ATP and the shape change parallel one another.

   There seem to be two significant levels of ATP. When the ATP content of the erythrocyte is more than one-half that of the fresh erythrocyte, about 0.5 μmol/ml cells, most erythrocytes are biconcave diskocytes, and when the ATP content is one-tenth that of the fresh erythrocyte, most cells are spherical. These erythrocytes are crenated, when the ATP content is between one-half and one-tenth of that in fresh cells. In contrast, the ADP content shows no correlation with shape.
5. Depletion of ATP is achieved by storage for several weeks at 4°C, or several hours at 37°C, after elimination of glucose in the extracellular medium. This is also attained by incubation in 0.154 M sodium fluoride at 37°C for several hours.

On the other hand, the lost ATP is easily replenished when the cells are incubated with adenine plus inosine or guanosine. Adenosine, which is not a substrate for nucleoside phosphorylase, is apparently effective because it is rapidly deaminated by adenosine deaminase. Synthesis of ATP inside the cell is rapid even when the cells had been preserved for 15 weeks at 4°C. Various buffers with or without magnesium or calcium ions can be used if adenine and inosine are added. However, addition of inorganic phosphate to the medium is preferable, because when the depletion medium does not contain chloride ions, phosphate ions moves easily across an anion channel and flows out. This cyclic depletion and repletion of ATP can be repeated several times, with the accompanying shape change.

Some investigators argued that there was some discrepancy between the ATP level and shape when sulfhydryl-reactive reagents were used. However, these reagents also react with spectrin directly and affect dimer–tetramer interconversion (Smith and Palek, 1983), and *N*-ethylmaleimide used in the experiment inhibits the Ca^{2+}ATPase in the erythrocyte membrane and calcium accumulates in the cell. These circumstances complicate interpretation of this discrepancy.

3.1.2. Intracellular Calcium and Shape

When the intracellular ATP decreases during *in vitro* storage at 4°C or prolonged incubation at 37°C, intracellular calcium increases severalfold. The low intracellular calcium ion content is normally maintained by the low permeability of divalent cations across the lipid bilayer and the relatively efficient Ca^{2+}-ATPase pump. Since the ATPase activity is relatively powerful and the K_m for ATP is significantly low, the increased intracellular calcium ion content is rapidly outfluxed when the suspension is incubated at 37°C. Therefore, the ATP-dependent shape change is contrary to the change in intracellular calcium ion content, as shown below (Weed *et al.* 1969). However, when a long-preserved erythrocyte suspension is incubated with adenine plus inosine in a buffer containing EGTA without calcium, the intracellular ATP increases rapidly and the cell changes from a sphere to a disk; whereas the ATP level in control cells without adenine plus inosine does not increase and the cell remains a sphere although the K_m of the calcium pump for ATP is low. It is therefore a little unclear whether the shape change induced by ATP is always preceded by an increase of cell Ca^{2+}. When A23187, one of the most potent ionophores for calcium, is added to the medium, the erythrocyte changes from a disk to a sphere. When the ionophore is eliminated from the cell by the addition of plasma albumin to the medium, the spherical cell becomes a diskocyte again, since the calcium pump is active.

Calcium-activated transglutaminase catalyzes cross-linking of cytoskeletal proteins in the presence of high concentrations of Ca^{2+} (Anderson *et al.*, 1977). This cross-linking is completely irreversible (Lorand *et al.*, 1975), but requires a high concentration of Ca^{2+} inside the cell (> 0.5 mM), which is normally rapidly outfluxed by the Ca^{2+}-ATPase with relatively low concentrations of ATP inside the cell. The diskocyte–spheroechinocyte transformation of erythrocytes undergoing ATP depletion and calcium accumulation is reversible after restoration of ATP content by the addition of adenine and inosine, but becomes irreversible after prolonged incubation without glucose and nucleoside at 37°C or after introduction of a high calcium concentration into the cells (Palek *et al.*, 1978b; White, 1974). This may be due to formation of reducible membrane protein polymers through disulfide coupling initially (Smith and Palek, 1983) and then irreducible polymers formed by transglutaminase (Anderson *et al.*, 1977; Palek *et al.*, 1978a,b).

3.1.3. Intracellular ATP and Change in Surface Area

Since the red cell shape change apparently resembles mitochondria swelling, some change in the cell volume is expected. However, this is not the case. The cell volume has been carefully compared by four methods: (1) comparison of hematocrit after centrifugation (packed cell volume); (2) by Coulter counter model S; (3) estimation of intercellular space using ^{131}I-labeled serum albumin; and (4) calculation of the cell volume from the observed diameter. The materials are ACD or PCD blood preserved for 5 to 10 weeks at 4°C. Each erythrocyte sample is divided between four test tubes and incubated with adenine alone, inosine alone, adenine plus inosine, or no addition. After a 3-hr incubation at 37°C, the ATP content increases about tenfold with adenine plus inosine, two- to fivefold with inosine alone (depending on the preservation period of the specimens), almost no change with adenine alone, and no change or a slight decrease with no addition. The shape of the adenine plus inosine samples changes from sphere to disk, but that of the other samples essentially does not change. Nevertheless, no change in the volume of these erythrocytes is observed in any of the four test tubes (M. Nakao *et al.*, 1981). This strongly suggests that the surface area change depends on the cell ATP level. The calculated increase in surface area is 60–70%. This swelling and shrinking is reversible depending on the cell ATP content. We conclude that the expansion and shrinking of both the lipid bilayer and spectrin–actin network depends on the cell ATP content.

3.1.4. ATP Content and Deformability

Deformability of the erythrocyte can be measured by various techniques each of which is believed to be an indicator of cell viscoelasticity. Teitel (1965) estimated viscosity by the rate of filtration of an erythrocyte suspension through a filter paper and recognized that the rate depended on the ATP content and shape. Later, the cone-plate viscometer, capillary viscometer, ektacytometer, and micropipette-type techniques were used (Mohandas *et al.*, 1983).

The viscosity of an erythrocyte suspension increases with a decrease in ATP content and increase in intracellular calcium; at the same time, the shape changes from disk to sphere. When adenosine is added to the depleted erythrocytes, the ATP content increases and Ca^{2+} content decreases, and at the same time the shape changes from sphere to disk with a reduction in viscosity (Weed *et al.*, 1969), although the effect of adenosine is less than when both adenine and purine nucleosides are added simultaneously. These experiments have also been done using ghosts (see below).

3.1.5. Vesiculation

Human erythrocytes incubated without addition of any energy source at 37°C release a number of small vesicles after 12 or more hours (Lutz *et al.*, 1977). The release is prevented if the intracellular ATP level is maintained completely, but hemolysis takes place whether the cells maintain the ATP level or not. The diameter of the vesicles is 185 ± 23 nm on average, and they contain varying amounts of ATP. The membrane proteins of these vesicles are band 3 and the glycophorins, without spectrin and actin. The lipid composition is similar to that of intact cells except for an increase in phosphatidic acid (tenfold) and a slight decrease in PE (Lutz *et al.*, 1977). Therefore, the released vesicles without cytoskeleton may be due to the destabilization of the cytoskeleton network in the absence of ATP (Lange *et al.*, 1982a). This observation is very indicative of the effect of ATP on the cytoskeleton.

3.1.6. Posttransfusion Survival

Posttransfusion survival may be a complex phenomenon in which shape, surface area, and viscoelasticity of the surface are intimately related. In experiments, ACD blood was divided into three portions; one-tenth volume of physiological saline containing 12 μmol/ml adenine and 25 μmol/ml inosine was then added to one of them (B) and nothing to the other two portions (A, C). After storage for 8 weeks at 4°C, specimen A was incubated with the same final concentrations of adenine and inosine as in B at 37°C for 3 hr. The ATP content and posttransfusion viability measured by a ^{51}Cr method were almost in the range expected for fresh cells in B and A, whereas both decreased in C. However, the osmotic fragility did not decrease, but rather increased in A. This shows that osmotic fragility is not related to posttransfusion viability, but the ATP content is. The erythrocytes are rejuvenated with the regeneration of intracellular ATP. as far as *in vitro* aging is concerned (M. Nakao *et al.*, 1960c,e, 1962, 1983; K. Nakao *et al.*, 1962a,b; T. Nakao *et al.*, 1982; Wada *et al.*, 1960).

The effect of erythrocyte ATP content has been further examined by *in vitro* experiments. Isolated macrophages can phagocytize depleted erythrocytes, but not erythrocytes containing a sufficient amount of ATP (Maruta and Mizuno, 1971).

3.1.7. Distribution of Ion Channels and Pumps

Surface proteins are mobile and can be capped by antibodies or lectins. However, band 3, which is believed to be an anion channel and the most abundant intrinsic protein, is scattered rather evenly. When spectrin in ghosts is depleted or band 3 is incorporated into proteoliposomes, band 3 moves extensively (Elgsaeter and Branton, 1974). To the contrary, oxidative cross-linking (Smith and Palek, 1972) or application of a lectin (Anderson and Lovrien, 1984) limits the lateral mobility of band 3. Recently, Na^+,K^+-ATPase purified from kidney (Nelson and Veshnock, 1987) and voltage-dependent sodium channels from brain (Srinivasan *et al.*, 1988) have been shown to specifically bind to ankyrin from erythrocytes and from brain. These results suggest strongly that ion channels and pumps are evenly distributed in the membrane, via ankyrin and the cytoskeleton. The distribution of some receptors, e.g., glycophorins, may be similar.

3.2. Shape of Ghosts

Data indicative of the shape change of erythrocyte ghosts upon addition of ATP were first obtained by M. Nakao *et al.* (1960d). Sheeetz and Singer (1977) reported that erythrocyte ghosts are spheres in hypotonic solutions, crenated in isotonic KCl and/or NaCl solutions, and after incubation of these ghosts for 30 min with 2 mM MgATP (added to the suspension before raising the ionic strength), these ghosts become biconcave disks. For this change in shape, Mg^{2+} is necessary and MgAMP–PNP is ineffective. When [^{32}P]-ATP is used, the label is incorporated into spectrin β. When antispectrin antibody is added to hemolysis buffers, it accelerates the shape change, but the papain fragment Fab of antispectrin does not. Taking into consideration the time course of ^{32}P incorporation and shape change, the shape change is due to the phosphorylation of spectrin β, but not ATPase, and this transformation is prevented by phosphatases (Sheetz and Singer, 1977; Birchmyer and Singer, 1977). However, there are many contradictions relating phosphorylation to the shape change (Sheetz, 1983; Cohen and Foley, 1986).

Patel and Fairbanks (1983) reported that greater than 95% of the spectrin kinase was removed from the membrane by salt extraction, but no change in the rate of shape transformation was induced by incubation of salt-extracted ghosts with MgATP. Three years before their 1977 report, Sheetz and Singer (1974) had proposed the bilayer couple theory based mainly on two findings: a lipid asymmetry in red cell membranes with a preferential distribution of PE and PS in the cytoplasmic leaflet of the bilayer, and the existence of two sorts of amphipathic agents—stomatocytogenic and echnocytogenic (see

Deuticke *et al.*, this volume). In this theory, lipid phosphorylation could cause the ghost shape change (Lange *et al.*, 1982a). Indeed, diacylglycerol, phosphatidylinositide, and diphosphoinositide can all be phosphorylated by MgATP. However, Patel and Fairbanks (1986) found no correlation between the rate of the MgATP-dependent shape change and the occurrence of lipid phosphorylation. Salt-extracted membranes incubated with adenosine, an inhibitor of spectrin and phosphatidylinositol kinases, underwent normal shape change despite reductions of over 90% in spectrin and phospholipid labeling by [τ-^{32}P]-ATP. This ATP-dependent shape change was blocked by vanadate at micromolar concentrations, but vanadate did not inhibit these phosphorylation reactions or the turnover of spectrin- or lipid-bound phosphate. Thus, vanadate inhibited the part of ATP hydrolysis that accompanied shape change. The vanadate-sensitive MgATPase activity was about 3 nmol P_i/min per mg protein, and the results implied an ATP-dependent shape change (Patel and Fairbanks, 1986). However, it is rather difficult to explain the fact that shape transformation of ghosts by MgATP requires a lag time. Recently, several groups found a protein (or proteins) that accelerates the translation of aminophospholipid from outside the cell to the cytoplasmic side, i.e., from the outer leaflet to the inner leaflet. Seigneuret and Devaux (1984) used spin-labeled phospholipids and ascorbate, an impermeable reducing reagent, to dissipate the spin signal in order to distinguish the inner and outer leaflets. For this translocation, MgATP inside the cells or resealed ghosts was required and incubation was effective even at 0°C. PS is rapidly translocated, followed by PE, but almost no PC and absolutely no sphingomyelin (Heast, 1982; Daleke and Huestis, 1985; Middelkoop *et al.*, 1988). Moreover, even if the erythrocyte suspension was warmed to 50°C, the translocation of PS reached a plateau at 87%, with 1.25 mM MgATP, although the initial velocity was considerably slower (Calvez *et al.*, 1988). This heat treatment denatures spectrin. Therefore, they concluded that the cytoskeleton was not necessary to attain the preferred distribution of aminophospholipid between the inner and outer leaflets, as long as MgATP was present in the cytoplasm.

Recently, Connor and Schroit (1988) synthesized ^{125}I-labeled pyridyldithioethylamine. This compound specifically inhibited the translocation of PS at thiol groups and preferentially labeled a 31-kDa erythrocyte membrane peptide. This compound, which might represent the PS transporter, comigrated with band 7 in SDS–PAGE and might be the same protein that Seigneuret and Deveaux (1984) termed aminophospholipid translocase. However, stoichiometry and other characteristics have not been published. These new findings seem to encourage the bilayer couple hypothesis, but if the following result after warming at 49°C cannot be explained, it would face a major problem. Intact erythrocytes are split to vesicles on warming at 50°C for 30 min, but after incubation at 49°C for 30 min they remain unsplit and become spherical. Under the latter condition, the translocation activity must be sufficiently high and the preferential distribution should be maintained well.

3.2.1. Effect of Calcium on Shape and Size of Ghosts

The presence of contractile protein in the erythrocyte membrane has been strongly suggested, as mentioned before (M. Nakao *et al.*, 1959; Ohnishi, 1962). As muscle contraction is induced by calcium ions, the intracellular calcium has attracted attention. The depletion of ATP induces the intracellular accumulation of calcium, followed by disk–sphere transformation (Weed *et al.*, 1969). Besides this effect of intracellular calcium, a rather high concentration of extracellular calcium (millimolar) affects the permeability of cations and more than 10 mM calcium accelerates the fusion of cells. The effects of calcium ions on the volume of ghosts were precisely described by Palek *et al.* (1971, 1974) and Church *et al.* (1975). The permeability of ghosts to divalent cations and macromolecules is high during hemolysis and then gradually decreases, especially at higher temperatures such as 37°C. Therefore, the preparation of resealed ghosts, with calcium and/or ATP, should be carefully performed. Ghost volume was determined by (1) the Hb concentration in the supernatant, (2) hematocrit, (3) measurement of diameter, and (4) checked by a multichannel particle analyzer and light-scattering. When the osmolarity of the resealed ghosts obtained was rapidly adjusted to 300 mOsm/liter by the addition of inorganic salts, shrinkage of the ghosts immediately occurred although a retarded but smaller shrinkage occurred during further incubation for 20 min, which might be due to the permeability of the lipid membrane barrier. The mean corpuscle volume decreased to less than 40% at 2 mM Ca^{2+}. The lowest Ca^{2+} concentration required for ghost shrinkage both before and after restoration of isotonicity was 2×10^{-6} M. The concentration requirements for other cations increased tenfold in the order $Ca^{2+} < Sr^{2+} < Ba^{2+}$, but saturated at 2×10^{-3} M; the effects of the three cations were the same. These shape changes induced by Ca^{2+} and MgATP were completely inhibited by 2×10^{-3} M *N*-ethylmaleimide (Palek *et al.*, 1974). ATP seemed to inhibit the shrinkage induced by Ca^{2+}. The presence of ATP increased the midpoint of the sigmoidal reaction curve of shrinkage from 2×10^{-6} to 4×10^{-5} M at 4 mM. ADP and AMP were ineffective, but ITP and GTP were effective. This shrinkage occurred throughout the pH range 6 to 9 almost equally, but the volume of control ghosts was largest at pH 7.75. There were two size populations in the ghosts, judging from results obtained with the multichannel particle analyzer. Ghosts from fresh blood showed the smaller size only. After incubation in the buffered isotonic saline at 20°C for 10–40 min, the size increased. For ghosts from long-stored blood, the size population showed two peaks after incubation (Palek *et al.*, 1972). Palek *et al.* (1974) observed from ghost configurations after osmium tetroxide fixation that: (1) flat discoidal electron-penetrable ghosts were prepared without any special additives, or with ATP alone, Mg alone, or EGTA; (2) biconcave and cup-shaped less-electron-penetrable ghosts were prepared with MgATP (0.5–2 mM) during hemolysis; (3) spherical spiculated or markedly deformed electron-dense ghosts of smaller volume were prepared with Ca^{2+} alone during hemolysis; and (4) small, smooth electron-dense spheroids were prepared in the presence of Ca^{2+} and MgATP (2 mM) during hemolysis (Palek *et al.*, 1978a,b). The percentage of biconcave ghost after incubation depended on the concentration of MgATP and reached a plateau of about 90% at about 1 mM MgATP. The calcium concentration required for crenation was about 5×6^{-6} M, giving 90% at 1×10^{-4} M. In the presence of 2×10^{-3} M MgATP, the calcium concentration required for 50% crenation increased to about 5×10^{-5} M. The shape response to MgATP or Ca^{2+} was completely abolished by preincubation of erythrocytes for 30 min at 37°C.

3.2.2. Viscosity of Ghost Suspensions

The fluid viscosity of erythrocyte suspensions is the basis of one of the techniques for determination of erythrocyte deformability, because it is known to be a function of cell shape, surface/volume ratio, fluidity of cell contents, and membrane viscoelasticity.

However, exactly what the viscosity of ghost suspensions implies is not clear. It may be more or less related to viscoelasticity of the cell membrane than in the case of the cell suspension.

Nevertheless, viscosity is closely related to ghost shape (Quist, 1980; Quist and Reece, 1980). In the presence of MgATP, the viscosity of ghost suspensions decreases in 20 min at 22°C, in parallel with the accompanying shape change of ghosts from echinocyte to diskocyte. Micromolar Ca^{2+} inhibits both changes. Ca^{2+} and neomycin can transform diskocytic ghosts, formed by preincubation with 1 mM MgATP, back to echinocytes, by an increase in viscosity. The incorporation of label from [τ^{32}P]-ATP (Quist and Reece, 1980) indicates that the shape transformation is due to the phosphorylation of diphosphatidylinositide.

4. ASSOCIATION OF THE CYTOSKELETON WITH THE LIPID BILAYER

There are many candidates for the cytoskeleton–lipid bilayer connector. Ankyrin–band 3, protein 4.1–band 3, protein 4.1–glycophorin A–band 3, and others have been proposed from *in vitro* experiments. Spectrin can bind to lipid-bilayer PS at multiple sites, including the tail end region (Cohen *et al.*, 1986). Specific interactions are listed in Table I. Some of them will be explained briefly below.

4.1. Association of Band 3–Ankyrin–Spectrin

4.1.1. Association of Ankyrin with Spectrin

The association of ankyrin with isolated spectrin dimer and tetramer is clearly observed by electron microscopy following freezing and shadowing, which shows that ankyrin binds to spectrin at a site about 20 nm from the head-to-head dimer–dimer association contacts (Tyler *et al.*, 1979). Association experiments with isolated peptides indicate that it is the β chain that binds to ankyrin. From the number of spectrin tetramers to ankyrin and assumed ankyrin hydrolysis products 2.2, 2.3, 2.6, etc. in an erythrocyte (Yu and Goodman, 1979; Luna *et al.*, 1979), only one ankyrin binding site per tetramer may be occupied in erythrocytes. Electron micrographs of negatively stained Triton-treated ghosts seem to support the assumption (Shen *et al.*, 1984; Byers and Branton, 1985). Based on binding experiments of [^{125}I]ankyrin to spectrin, the dissociation constant is 5×10^{-8} M, which shows that a very specific binding occurs (Bennett and Stenbuck, 1979). Protein 2.2 has a higher affinity for spectrin than ankyrin (Hall and Bennett, 1987).

4.1.2. Association of Ankyrin with Band 3

Ankyrin and its related peptides (Bennett and Stenbuck, 1979; Jinbu *et al.*, 1984b) bind to band 3, which is a typical intrinsic protein of the erythrocyte membrane. Ankyrin is obtained by treatment with high concentrations of KCl after extraction of ghosts with low concentrations of buffer or EDTA or ATP. Maximum binding of this protein to band 3 is 1 : 1 (Bennett, 1982), but the number of ankyrin molecules per cell is only one-tenth or one-eighth that of band 3 in the cell. The dissociation constant for the binding was very small (Bennett and Stenbuck, 1979, 1980); for the high-affinity site $K_d = 5 \times 10^{-9}$ M, for the low-affinity site $K_d = 1.4 \times 10^{-7}$ M, and for the 43-kDa fragment of the cytoplasmic domain of band 3 that associated with ankyrin $K_d = 10^{-9}$ M. This clearly shows that ankyrin and the cytoplasmic domain of band 3 associate directly and specifically, without requiring intermediary protein or proteins. Proteoliposomes containing purified band 3 without ankyrin, spectrin, actin, 4.1, or 4.2 can bind ankyrin at a 1 : 1 ratio (Bennett and Stenbuck, 1980). This may lead us to assume a dynamic state for this binding site.

There may be some interchange between free and ankyrin-linked band 3. When ^{125}I-labeled Triton supernatants, which are labeled from outside the cell, are incubated with Triton-treated shells at 37°C, the incorporation of label into the shells occurs rapidly and reaches a plateau within 10 min; conversely, when Triton supernatants are incubated with ^{125}I-labeled shells, which are also labeled from outside, the cell radioactivity is released from the ^{125}I-labeled shells. This exchange occurs only in the presence of Triton of more than 0.04% (v/v) because band 3 cannot be solubilized without detergents, but the 43-kDa fragment, which is well established to be the cytoplasmic part of band 3 and water soluble, inhibits ^{125}I-labeled band 3 binding to the shells and also inhibits both the rate and the extent of the ATP-dependent shape change of ghosts. These facts suggest that the exchange between ankyrin and band 3 occurs rather rapidly in cells (Ueno *et al.*, 1987). The binding of cytoskeleton components to each other and the connections between the spectrin–actin network and the lipid bilayer are in dynamic states *in situ* as suggested by *in vitro* binding tests and dissociation constants. The ankyrin–Na^+,K^+-ATPase binding, as described above, may also be in a dynamic state.

4.2. Protein 4.1–Lipid Bilayer Associations

After extraction of spectrin–actin with low-ionic strength solutions, a small fraction is obtained as the protein 4.1–spectrin–actin ternary complex (Beaven *et al.*, 1985; Matsuzaki *et al.*, 1985) and 80% of protein 4.1 remains associated with the lipid bilayer. Several candidates (Steiner and Bennett, 1988) for a membrane-binding site for protein 4.1 have been proposed, including glycophorin A (Anderson and Lovrien, 1984), glycophorin A–polyphosphoinositide complex (Anderson and Marchesi, 1985), band 3 (Pasternack *et al.*, 1985), glycophorin C, and PS in the bilayer (Sato and Ohnishi, 1983). Anderson and Lovrien (1981) reported that only 1.8×10^6 molecules of WGA per cell fixes the erythrocyte morphology. When this amount of WGA is added to the suspension of diskocytes, the change from diskocyte to echinocyte induced by Ca^{2+} and A23187 is completely prevented. Addition of *N*-acetylglucosamine, which has high affinity for the lectin, restores the effect of Ca^{2+} inside and the cell changes to echinocyte both before and after elimination of Ca^{2+} with EGTA. In addition to this, the lectin added to echinocytic cells that are induced by Ca^{2+} plus A23187 fixes the shape of the echinocyte. After elimination of extracellular Ca^{2+} with EGTA and of A23187 with bovine serum albumin, the echinocytic shape is maintained even after an energy source is supplied, but once the ATP level is restored by the addition of inosine, the echinocytic cell changes into a diskocyte on addition of *N*-acetylglucosamine, which releases WGA from the surface. In all cases, the shape is fixed and the shape change due to the content of ATP or of Ca^{2+} does not occur and a large amount of exogenous *N*-acetylglucosamine is required to regenerate the ability to transform shape. This finding has been reconfirmed in our laboratory using ghosts instead of intact cells, and also with shape changes due to acid pH, hyperosmolarity, and the application of amphipathic reagents (unpublished). This indicates that glycophorin A is involved in the connection between the lipid bilayer and cytoskeleton.

However, inherited glycophorin A-deficient cells are normal and glycophorin C-deficient cells exhibit a reduction in membrane stability. The affinity of protein 4.1 for band 3 is less than that for the glycophorin site, but the amount of band 3 is 20 times greater than that of glycophorin C. Therefore, a slight difference of affinity may be overcome by the number of sites (Bennett, 1989).

4.3. Participation of Band 3 in Lipid Bilayer–Cytoskeleton Binding

Intrinsic protein band 3 is a chloride ion channel that spans the lipid bilayer. Various photobleaching techniques have been adopted to determine the movement of this protein in the lipid bilayer. S. Ohnishi and his associates (Tsuji *et al.*, 1988) observed rotational diffusion of band 3 by time-resolved fluosphorescence anisotropy decays of eosinyl-5-maleimide covalently attached to the protein, and compared the results with those on translational diffusion, obtained by themselves and others. Their results, including the effect of temperature, salt concentration, and polyamine, correlate well with translational diffusion data. An increase in the dimer fraction of spectrin, compared with that of the tetramer which greatly increases the lateral diffusion constant of the translationally mobile component of band 3 without changing the fraction of the translationally mobile component (40%), has little influence either on the mobile/immobile fraction of band 3 in terms of rotational mobility or on the rotational correlation time of the mobile fraction of band 3 in ghost membranes. These results are consistent with a cytoskeletal fence model in which: (1) band 3 oligomers, aggregates, and monomers that are bound to the cytoskeletal network via ankyrin are immobile and in temperature-dependent equilibrium in ghost membranes, (2) oligomers and aggregates that are not anchored are rotationally and translationally mobile, and (3) long-range translational diffusion of mobile oligomers and aggregates is restricted by nonspecific barriers imposed by the cytoskeletal network and the rate of translational diffusion is regulated by the fraction of spectrin dimers (open gate) and tetramers (closed gate). The major diffusion barrier is the spectrin tetramer. These data show the participation of band 3 in the connection between the lipid bilayer and the cytoskeletal network.

Our results (Ueno *et al.*, 1987) suggest strongly that the connection via band 3 is also important in the transformation of shape. A water-soluble 43-kDa fragment of band 3 inhibits the incorporation of free ^{125}I-labeled band 3 into the cytoskeleton, which is obtained as Triton shells, and also inhibits the MgATP-dependent shape change of ghosts that are prepared in the presence of the 43-kDa fragment. This inhibition is exhibited on both the rate and the extent of the shape change. The former has also been observed by others (Carter and Fairbanks, 1984). The participation of band 3 does not eliminate the possibility that a quaternary complex, spectrin–protein 4.1–glycophorin–band 3, plays a role in the connection.

4.4. Direct Evidence of the Connection from a Functional Point of View

Spectrin itself has a hydrophobic region and may have some affinity to the lipid bilayer. This is one of the reasons for a fence theory explaining the limited movement of band 3. However, the affinity is not so high that the direct binding of spectrin can maintain the shape of the cell and transmit the power for changing morphology from the cytoskeleton to the lipid bilayer or vice versa, independent of the primary deformation that occurs at the cytoskeleton or at the lipid bilayer. From this and other observations, only two connection systems may be important from a functional point of view at present. The two systems are (1) the lipid bilayer–band 3–ankyrin–spectrin and (2) various components in the bilayer such as the glycophorins–band 3–PS or PI–protein 4.1–spectrin. The association points between protein 4.1 and the bilayer may be multiple. However, the association point between protein 4.1 and the cytoskeleton is clearly spectrin. Both ends of the spectrin tetramer bind to protein 4.1 directly with a dissociation constant of 1×10^{-7} M, although the dissociation constant of ternary complex protein 4.1–spectrin–actin is 10^{-12} M^2 (Coleman *et al.*, 1987). Moreover, the dissociation constants between protein 4.1 and the proposed components in the lipid bilayer are almost 3×10^{-7} M (Cohen *et al.*, 1988). Therefore, the dissociation constants of spectrin–ankyrin (5×10^{-8} M) and ankyrin–band 3 (5×10^{-9} M) are smaller than those related to protein 4.1 binding to the lipid bilayer and spectrin.

Direct evidence of these connections from a functional viewpoint is as follows. (1) WGA, a lectin added from outside the cell, inhibits the shape change (Anderson and Lovrien, 1981), as previously mentioned. The lectin binds to glycophorin A, which is bound to protein 4.1 in inside-out vesicles or when incorporated into artificial lipid vesicles (Anderson and Lovrien, 1984). The binding constant is 1.1×10^7 M. (2) Mohandas *et al.* (1983) reported that protein 4.1-deficient erythrocytes from a family of patients with hemolytic anemia are hemolyzed in the presence and absence of purified protein 4.1 and the effect of shear stress on the stability of ghosts was observed by ektacytometry. The ghosts replenished with protein 4.1 showed a progressive restoration of membrane stability with increasing amounts of added protein 4.1. Near-normal membrane stability can be restored (Takakuwa *et al.*, 1986). (3) Spectrin labeled with $^{32}P_i$ binds to inverted vesicles from erythrocytes with at least two different dissociation constants, showing the presence of at least two different binding sites. Ankyrin binds to the membrane very tightly and the 72-kDa fragment obtained from ankyrin by digestion with trypsin inhibits the binding of ^{32}P-labeled spectrin to inside-out vesicles (Bennett and Branton, 1977). Moreover, antisera raised against this 72-kDa fragment also inhibit the binding of spectrin to inside-out vesicles (Bennett and Stenbuck, 1979) and this water-soluble proteolytic fragment forms a complex with spectrin in solution and increases the rate of lateral mobility of integral membrane proteins (Fowler and Branton, 1978). (4) More direct evidence of the participation of ankyrin in function has been obtained in our laboratory (Jinbu *et al.*, 1982, 1984b). To ghosts obtained by osmotic hemolysis, 1/100th the weight of trypsin is added immediately and brief digestion is performed in 10 mM Tris–HCl buffer (pH 7.4) at 0°C. After the addition of excess soybean inhibitor to stop the reaction, the digestion of individual proteins was examined by SDS–PAGE. Under these digestion conditions, spectrin α, β, actin, protein 4.1a and b, band 4.2, and others are the same for at least 40 sec except for a very small decrease in band 3 at 40 sec and nearly complete loss of ankyrin at 20 sec. The decrease of ankyrin to this point is nearly linear. The sensitivity of ankyrin to proteolytic enzymes is widely accepted and band 3 is more easily digested than spectrin under these hypotonic conditions (Steck *et al.*, 1976). After addition of chlorpromazine to control ghosts without trypsin treatment or resealing with trypsin plus trypsin inhibitor at zero time, the ghosts changed into biconcave disks. Trypsin-treated ghosts are insensitive to chlorpromazine without incubation, in 2 mM MgATP plus incubation at 37°C, and in HEPES plus incubation at 37°C. The number of ghosts that are changed into disks under these three different conditions decreases with the decrease in ankyrin from 90% to only a few percent. This shape change is almost the same in isotonic media as in hypotonic media. After treatment with trypsin for 30–40 sec, ankyrin disappears almost completely and the shape of the ghosts remains echinocytic for at least several minutes at room temperature. Then, small vesicles are pinched off from the surface of the membranes. This also suggests that ankyrin is necessary for stabilizing the cell membrane (Jinbu *et al.*, 1982). The liberation of the spectrin–actin network from the lipid bilayer is confirmed by electron microscopic observation of sections of these ghosts with the aid of tannic acid fixation. After digestion with trypsin for 30 sec at 0°C, the layer of horizontally disposed filamentous components is not destroyed but is distant from the lipid bilayer by 20–100 nm and yet there remain many binding sites scattered at irregular intervals (Figure 2B). As the SDS–PAGE pattern shows that only ankyrin is destroyed (almost completely), the remaining binding sites may be protein 4.1 or any other cytoskeletal protein.

There seems to be two steps in the ATP-dependent shape change of erythrocyte ghosts. One is probably phosphorylation, as incubation with ATP for 10–30 min is necessary, and the other is a rapid reaction that occurs after the first step and immediately after addition of MgATP even at 6°C without incubation at 37°C. The effect of mild trypsin digestion on each step has been examined. The effect on the first step is the same as described above. After ghosts are incubated with MgATP for 30 min at 37°C and washed with 1 mM $MgCl_2$, they immediately become spheroechinocytes. These ghosts are digested with trypsin under hypotonic conditions at 0°C and then trypsin inhibitor is added to stop the reaction. On addition of 2 mM MgATP, the digested ghosts are unchanged in contrast to the control ghosts without digestion. The time course of reducing the ability to change is again paralleled by the disappearance of ankyrin. The effect of chlorpromazine is also the same. MgATP and chlorpromazine act not only on the shape of ghosts but also on the viscosity of ghost suspensions measured by Ostwald-type viscosimeters of microscale. The specific viscosity of ghost suspensions decreases with increasing concentrations of added chlorpromazine and parallels the shape change from echinocyte to diskocyte. The minimum viscosity is observed at 0.25 mM chlorpromazine, where almost all ghosts show the disk form, but with further increase in chlorpromazine, the ghosts become stomatocytic and the viscosity increases gradually again. At 1.0 mM chlorpromazine, the ghosts are completely stomatocytic and the viscosity increases to 0.6 for example, if the control ghost suspension without chlorpromazine is 1.0 and the minimum value at 0.25 mM chlorpromazine is 0.4. When trypsin-digested ghosts are used, the specific viscosity in the presence of 0.5 mM chlorpromazine increases from 0.45 without trypsin digestion to 0.9 within 20 sec digestion. The decrease in the specific viscosity with MgATP both at the first step and at the second step is also prevented by mild trypsin treatment. The prevention of the decrease depends on the duration of trypsin treatment, i.e., from 5 to 30 sec (Jinbu *et al.*, 1983). Taken together, it is concluded that ankyrin is necessary not only for shape change, but also for maintaining erythrocyte deformability.

These facts, from various laboratories, show that ankyrin or protein 4.1 is necessary to support shape change and deformability of erythrocyte membranes (Chasis and Mohandas, 1986). The question is, which is more important from a mechanical point of view? The role mediated by ankyrin seems to be more indispensable, as mentioned before, and the dissociation constants for components of ankyrin–band 3 systems are smaller than those included in the protein 4.1 system. However, more decisive evidence has to be found. At present, a logical conclusion is that both are necessary to maintain the working physiological system. There is a possibility that the band 3–ankyrin binding is constant, but the protein 4.1 binding sites in the lipid bilayer are many and protein 4.1 may move from one site to another, depending on conditions inside the cell and the conformation of the spectrin–actin meshwork, which is, for example, affected by the state of phosphorylation of a component or components. Furthermore, both connections may be in dynamic states. Even ankyrin–band 3 are continually released and bound (Ueno *et al.*, 1987). This may be the mechanism by which the stability of cell membranes is assured.

5. GHOSTS WITHOUT A CYTOSKELETON

Ghosts without a cytoskeleton are obtained by three methods:

1. Extraction with low-ionic-strength media, with or without EDTA, EGTA, or ATP
2. Resealed ghost with proteolytic enzyme
3. Mouse erythrocytes from spectrin-deficient hereditary spherocytosis strains.

The amount of mouse blood is so meager for biochemical studies that ghosts from mouse cells have not been studied extensively. When human erythrocytes are extracted with low-ionic-strength media overnight at 0°C, most of the spectrin and actin and about 20% of the protein 4.1 are extracted; some of these extracted proteins form a spectrin–actin–4.1 complex. Detailed characteristics of this complex are described by Ralston in this volume. When ghosts are incubated in more than 10 volumes of a low-ionic-strength buffer, such as 0.2 mM NaEDTA (pH 7.5) for 30 min at 37°C, spectrin dissociates and the ghosts endovesiculate to form small vesicles that are 80–90% inside-out (Steck *et al.*, 1970). If this extraction of spectrin is performed even at 4°C, these spectrin-depleted ghosts fragment. Ghosts that are made in the presence of trypsin lose almost all spectrin within 10 min at 0°C, judging from the SDS–PAGE pattern. Intact ghosts with an intact spectrin–actin network shrink at acidic pH and in high concentrations of NaCl such as 0.3 or 0.6 M, but spectrin-depleted ghosts do not react with acid media or hyperosmolarity. The diameter of trypsin-treated ghosts (M. Nakao *et al.*, 1987) changes very little at pH 5.5, 7.4, and 9.5 both in 0.14 M NaCl and at the low ionic strength, maintaining their completely spherical shape. The diameter of these ghosts shrinks a little, almost equally at all pH tested (M. Nakao *et al.*, 1987), and these completely spherical ghosts do not change shape on addition of chlorpromazine or MgATP plus incubation. Therefore, the bilayer couple hypothesis does not apply to these cytoskeleton-depleted cells.

6. TRITON SHELLS AND THEIR FUNCTION

Yu *et al.* (1973) treated ghosts with the nonionic detergent Triton X-100. After centrifugation, there remains a reticulum containing most of the spectrin, termed the Triton shell. This reticulum consists of 76% spectrin, 5% actin, and 5% protein 4.1, with band 4.2, 4.9, adducin, and other proteins. The lipid content of this residue depends on the Triton concentration used for extraction, the temperature and duration of action of the detergent, and the ionic strength of the solution. Triton shells retain the shape of the original cell from which they were made—discoid, sickle-shaped, or ellipsoid—to a remarkable degree. Abnormally shaped red cells in several inherited disorders have been associated with defective or absent cytoskeletal proteins (Palek and Lux, 1983). The shape of Triton-treated ghosts is fairly stable. It shrinks when salt is added to the suspension (Heubush *et al.*, 1985). The dissociation is accelerated by hemin (Avissar *et al.*, 1984; Liu *et al.*, 1985) and the association is enhanced by Hb (Liu and Palek, 1984). These intracellular substances may play some role in the physiology of the cell, and the following calcium effect is also very suggestive of this.

Ghosts obtained by hemolysis in 10^{-6} M Ca^{2+} buffer are treated with 0.5% Triton X-100 at low ionic strength (1 mM $MgCl_2$) for 5 min at 0°C. These Triton-treated ghosts or Triton shells lose most of their band 3, 4.2, and 4.5 and do not shrink in hypertonic sucrose solution. Even in 0.15 M NaCl plus 0.6 M sucrose solution they do not shrink any more than the control in 0.15 M NaCl. Electron microscopy of negatively stained specimens shows a number of small vesicles, having diameters from 0.08 to 0.22 μm and a filamentous network. Triton can be used at 1% instead of 0.5%, although 1% Triton-treated ghosts are less readily visible. Triton-treated ghosts are larger when prepared with 10 mM Tris–HCl buffer instead of 10^{-6} M Ca^{2+} buffer, but shrink for a period of several minutes when suspended in Ca^{2+} buffer (10^{-7})–10^{-6} M) (Jinbu *et al.*, 1984c). These shrunken Triton-treated ghosts seem to be spherical and swollen ghosts are diskocytic, although the diameter of shrunken ghosts is approximately 3 μm, which is too small to describe the shape clearly. They swell on addition of 1 mM MgATP at both 6°C and room temperature within several minutes and shrink again with 1 mM $MgCl_2$ (pH 7.4) containing 10^{-6} M Ca^{2+}. This shrinking and swelling is seen at least in the pH range 6.8–7.8. It is also observed when the incubated Triton shells are fixed for 30 min with 1% glutaraldehyde buffered with 10 mM imidazole–HCl at 0°C or room temperature and then postfixed with 2% osmium tetroxide. $MgPP_i$, MgUTP, MgGTP, MgITP, MgAMP-PNP, and MgAMP-PCP are not effective in reversing Ca^{2+}-mediated shrinkage. The $K_{0.5}$ for Ca^{2+} is about 10^{-7} M, in contrast to the relatively high value of 3×10^{-6} M in the presence of MgATP. Triton-treated ghosts, like untreated ghosts, can change shape reversibly depending on MgATP in the presence of 10^{-6} M Ca^{2+} (Jinbu *et al.*, 1984c). Both Triton-treated ghosts from fresh cell ghosts without preincubation and from preserved cell ghosts preincubated with MgATP are highly sensitive to Ca^{2+} in the absence of MgATP. On addition of MgATP, both become relatively insensitive to Ca^{2+}. This action of MgATP in the presence of 10^{-6} M Ca^{2+} is highly specific. ADP, GTP, UTP, ITP, AMP-PNP, AMP-PCP, or PP_i cannot replace ATP at 1 mM. However, Triton-treated ghosts from preserved cells without preincubation are not highly sensitive to Ca^{2+} with or without the addition of MgATP. These results suggest strongly that phosphorylation of Triton-treated ghosts seems to give high sensitivity to the Ca^{2+}-induced shrinking. On the other hand, alkaline phosphatase treatment of fresh cell ghosts seems to make Triton ghosts less sensitive to Ca^{2+} ions, and 10 mM P_i prevented the shift. These Triton ghosts recovered Ca^{2+} sensitivity by reincubation with 2 mM MgATP for 20 min at 37°C (Jinbu *et al.*, 1984a,b). The phosphorylation site is not known at present, but there are many candidates for protein kinases and their substrates (Cohen and Foley, 1986; Janmey and Stossel, 1987; Liu *et al.*, 1988; Wolf and Sahyoun, 1986).

These phenomena occur in low-ionic-strength buffer, without salt, but when the Triton shells are treated with Biobeads to remove excess Triton or hydrophobic substances completely, the shrinking–swelling cycle occurs in 0.1 M KCl media (M. Nakao *et al.*, 1987). When the ghosts are treated with Triton in the presence of salt such as 0.15 or 2 M KCl, the obtained shells lose Ca^{2+} sensitivity completely. These shells also lose most of their phospholipid and many proteins such as ankyrin, adducin, band 4.2, and others, with the exception of spectrin, actin, protein 4.1, and a part of band 4.9.

7. THE MOLECULAR STRUCTURE MAINTAINING (OR CHANGING) CELL SHAPE

It is well known that the shape of intact erythrocytes is cuplike (stomatocytic) at acidic pH (5.5), discoidal at neutral pH (7.4), and echinocytic at alkaline pH (9.5) in 0.154 M NaCl or KCl solution and also that the shape of ghosts obtained by osmotic hemolysis in hypoosmotic solution is, on the contrary, echinocytic at pH 5.5, discoidal or stomatocytic in NaCl solution at pH 9.5, as shown in Table II. How is this discrepancy

Table II
Volume and Apparent Diameter and Shape of Human Red Cells[a]

	pH		
	5.5	7.4	9.5
Intact cell, 0.14 M NaCl + buffer			
Hematocrit	17.4%	15.2%	13.4%
Diameter	6.38 ± 0.14μm	8.43 ± 0.41μm	5.26 ± 0.20μm
Shape	Cup	Disk	Crenate
Ghost, 0.14 M KCl + buffer	5.78 ± 0.34μm Crenate	6.15 ± 0.17μm Crenated disk	8.02 ± 0.83μm Disk, cup
Trypsin-treated ghosts			
−KCl	5.4 ± 0.25μm Sphere	5.70 ± 0.15μm Sphere	5.70 ± 0.13μm Sphere
+KCl	4.7 ± 0.13μm Sphere	5.05 ± 0.22μm Sphere	5.11 ± 0.13μm Sphere
Triton shell			
−KCl	3.7 ± 0.2μm	6.4 ± 0.3μm	10.3 ± 1.0μm
+KCl + MetHb[b]	5.6 ± 0.2μm Cup	6.2 ± 0.2μm Cup	6.0 ± 0.2μm[c] Cup

[a]Two- to four-day PCD blood.
[b]MetHb, methemoglobin induced by $NaNO_2$.
[c]pH 9.0.

explained? The assumed basic structure, the cytoskeleton, shrinks at acidic pH and swells at alkaline pH as do the ghosts, as shown in Table II. Triton shells obtained with 6% Triton and then treated with 2 M K Cl retain only spectrin, actin, protein 4.1, and a small amount of band 4.9. Furthermore, these KCl-treated shells lose almost all protein 4.1, band 4.9, actin, and phospholipids, after further treatment with 1 M KBr at pH 5.5. Both the KCl- and the KCl/KBr-treated shells have discoid or echinocytic shapes very similar to untreated Triton shells. Both shells shrink at acidic pH and swell at alkaline pH, without gross contour changes, although the size is smaller in the KCl-treated shells and much more smaller in the KCl/KBr-treated shells (Kojima *et al.*, 1988). Even the diameter of nontreated Triton shells in 0.15 M KCl is smaller than that of ghosts, but in hemolysate that was treated with $NaNO_2$ to visualize shells in the presence of Hb, it was close to that of ghosts, as indicated in Table II.

Shape, Ca^{2+}-induced, ATP-induced shape change, deformability, and stability should finally be explained by reconstruction of the cytoskeleton from each component. This has not been attained. However, the cup-shape meshwork consisting exclusively of spectrin has been obtained in my laboratory (Kojima *et al.*, 1988). Addition of both protein 4.1 and actin stabilizes the shells during incubation at 37°C in 0.14 M KCl, but not the addition of protein 4.1 alone or actin alone. These shells are deprived of the lipid bilayer, largely or completely. On the other hand, the lipid bilayer lacking cytoskeleton is obtained with trypsin treatment of ghosts that are resealed with trypsin and kept at 0°C with 0.14 M KCl for 10 min. These ghosts are completely smooth spheres without spicules both under hypotonicity and in 0.14 M KCl. These ghosts are no longer reactive to MgATP or chlorpromazine and their diameter is a little smaller at acidic than at neutral pH, but not so different at pH 5.5, 7.4, and 9.5, as shown in Table II. This size (5 μm) roughly corresponds to the size of Triton shells in hemolysate but it is smaller at acidic pH. They are almost the same size as the lipid bilayer, trypsin-treated ghosts. Why is the diameter of the intact erythrocyte at acidic pH larger than that of the cytoskeleton and the lipid bilayer and why is their volume larger than that of intact cells at neutral and at alkaline pH? It is because the cytoplasm contains a large amount of Hb and its amino acid residues dissociate greatly at acidic pH. Thus, hematocrit values express the relative volume of cytoplasm that contains much Hb, the amino acid residues of which dissociate very much at acidic pH. Expansion of the cytoplasm makes both the lipid bilayer and the cytoskeleton expand. Taking all of these findings into account, the present hypothesis suggests that the shape and mechanical properties of erythrocytes are primarily determined by the cytoskeleton and complemented by the tension of the lipid bilayer. Thus, the shape of ghosts is also determined by both of these factors. Finally, they are modified by the relative volume of the cytoplasm compared to the volume of the lipid envelope and that of the cytoskeleton cage.

8. CONCLUDING REMARKS

The erythrocyte inframembrane cytoskeleton plays a basic role in various functions relating to the surface characteristics of the membrane, especially from a mechanical point of view. The cytoskeletal proteins, consisting mainly of spectrin, actin, 4.1 with adducin, tropomyosin, calmodulin, and others, bind to each other with high specific affinity to form a meshwork or cage that is a flexible and elastic, yet stable structure. Each binding is reversible and probably in a dynamic state (Ueno *et al.*, 1987).

Reconstruction of functional shells is expected to lead to understanding of the molecular mechanism of structure and function of the inframembrane cytoskeleton. This may provide a model system to elucidate general characteristics of cell morphology, motility, and mechanical properties.

9. REFERENCES

Anderson, D. R., Davis, J. L., and Larraway, K. L., 1977, Calcium-promoted changes of the human erythrocyte membrane. Involvement of spectrin, transglutaminase and a membrane bound proteinase, *J. Biol. Chem.* **252:**6617–6623.

Anderson, J. P., and Morrow, J. S., 1987, The interaction of calmodulin with human erythrocyte spectrin. Inhibition of protein 4.1-stimulated actin binding, *J. Biol. Chem.* **262:**6365–6372.

Anderson, R. A., and Lovrien, R. E., 1981, Erythrocyte sidedness in lectin control of the Ca-A23187 mediated discocyte–echinocyte conversion, *Nature* **292:**158–161.

Anderson, R. A., and Lovrien, R. E., 1984, Glycophorin is linked by band 4.1 protein to the human erythrocyte membrane skeleton, *Nature* **307:**655–658.

Anderson, R. A., and Marchesi, V. T., 1985, Regulation of the association of membrane skeletal protein 4.1 with glycophorin by a polyphosphoinositide, *Nature* **318:**295–298.

Avissar, N., Inbal, A., Rabizadeh, E., Shaklai, M., and Shaklai, N., 1984, Interaction of spectrin with hemin disaggregates spectrin associations, *Biochem. Int.* **8:**113–120.

Baess, B. U., and Vincenzi, F. F., 1980, Calmodulin activation of red blood cell (Ca^{2+} + Mg^{2+})-ATPase and its antagonism by phenothiazine, *Mol. Pharmacol.* **18:**253–258.

Beaven, G. H., Jean-Baptiste, L., Ungewickell, E., Baines, A. J., Shahbakhti, F., Pinder, J. C., Lux, S. E., and Gratzer, W. B., 1985, An examination of the soluble oligomeric complexes extracted from the red cell membrane and their relation to the membrane cytoskeleton, *Eur. J. Cell Biol.* **36:**299–306.

Bennett, V., 1978, Purification of an active proteolytic fragment of the membrane attachment site for human erythrocyte spectrin, *J. Biol. Chem.* **253:**2292–2299.

Bennett, V., 1982, Isolation of an ankyrin–band 3 oligomer from human erythrocyte membranes, *Biochim. Biophys. Acta* **689:**475–484.

Bennett, V., 1985, The membrane skeleton of human erythrocytes and its implications for more complex cells, *Annu. Rev. Biochem.* **54:**273–304.

Bennett, V., 1989, The spectrin–actin junction of erythrocyte membrane skeleton, *Biochim. Biophys. Acta* **988:**107–121.

Bennett, V., and Branton, D., 1977, Selective association of spectrin with the cytoplasmic surface of human erythrocyte plasma membranes, *J. Biol. Chem.* **252:**2753–2763.

Bennett, V., and Stenbuck, P. J., 1979, Identification and partial purification of ankyrin, the high affinity membrane attachment site for human erythrocyte spectrin, *J. Biol. Chem.* **254:**2533–2541.

Bennett, V., and Stenbuck, P. J., 1980, Association between ankyrin and the cytoplasmic domain of band 3 isolated from the human erythrocyte membrane, *J. Biol. Chem.* **255:**6424–6432.

Bennett, V., Gardner, K., and Steiner, J. P., 1988, Brain adducin: A protein kinase C substrate that may mediate site-directed assembly at the spectrin–actin junction, *J. Biol. Chem.* **263:**5860–5869.

Birchmyer, W., and Singer, S. J., 1977, On the mechanism of ATP-induced shape changes in human erythrocyte membranes. II. The role of ATP, *J. Cell Biol.* **73:**647–659.

Brenner, S., and Korn, E., 1980, Spectrin/actin complex isolated from sheep erythrocytes: Actin polymerization by simple nucleation, *J. Biol. Chem.* **255:**1670–1676.

Burns, N. R., and Gratzer, W. B., 1985, Interaction of calmodulin with the red cell and its membrane skeleton and with spectrin, *Biochemistry* **24:**3070–3074.

Byers, J., and Branton, D., 1985, Visualization of the protein associations in the erythrocyte membrane skeleton, *Proc. Natl. Acad. Sci. USA* **82:**6153–6157.

Calvez, J. Y., Zachowski, A., Herrmann, A., Morrot, G., and Devaux, P. F., 1988, Asymmetric distribution of phospholipids in spectrin-poor erythrocyte vesicles, *Biochemistry* **27:**5666–5670.

Carter, D. P., and Fairbanks, G., 1984, Inhibition of erythrocyte membrane shape change by band 3 cytoplasmic fragment, *J. Cellular Biochem.* **24:**385–393.

Chasis, J. A., and Mohandas, N., 1986, Erythrocyte membrane deformability and stability: Two distinct membrane properties that are independently regulated by skeletal protein associations, *J. Cell Biol.* **103:**343–350.

Church, A., Fairbanks, G., and Palek, J., 1975, Role of calcium in spectrin retention by ghosts of fresh and ATP depleted erythrocytes, *Blood* **46:**1004.

Cohen, A. M., Liu, S.-C., Derick, L. H., and Palek, J., 1986, Ultrastructural studies of the interaction of spectrin with phosphatidylserine liposomes, *Blood* **68:**920–926.

Cohen, A. M., Liu, S.-C., Lawler, J., Derick, L., and Palek, J., 1988, Identification of the protein 4.1 binding site to phosphatidylserine vesicles, *Biochemistry* **27:**614–619.

Cohen, C. M., 1983, The molecular organization of the red cell membrane skeleton, *Semin. Hematol.* **20:**141–158.

Cohen, C. M., and Foley, S. F., 1982, The role of band 4.1 in the association of actin with erythrocyte membranes, *Biochim. Biophys. Acta* **688:**691–701.

Cohen, C. M., and Foley, S. F., 1984, Biochemical characterization of complex formation by human erythrocyte spectrin, protein 4.1, and actin, *Biochemistry* **23:**6091–6098.

Cohen, C. M., and Foley, S. F., 1986, Phorbol ester- and Ca^{2+}-dependent phosphorylation of human red cell membrane skeletal proteins, *J. Biol. Chem.* **261:**7701–7709.

Cohen, C. M., and Langley, R. C., Jr., 1984, Functional characterization of human erythrocyte spectrin α and β chains: Association with actin and erythrocyte protein 4.1, *Biochemistry* **23:**4488–4495.

Choen, C. M., 1983, The molecular organization of the red cell membrane skeleton, *Semin. in Hematol.* **20:**141–158.

Coleman, T. R., Harris, A. S., Mische, S. M., Mooseker, M. S., and Morrow, J. S., 1987, Beta spectrin bestows protein 4.1 sensitivity on spectrin–actin interactions, *J. Cell Biol.* **104:**519–526.

Conboy, J., Kan, Y., Mohandas, N., and Shohet, S., 1986, Molecular cloning of protein 4.1, a major structural element of the human erythrocyte membrane skeleton, *Proc. Natl. Acad. Sci. USA* **83:**9512–9516.

Connor, J. C., and Schroit, A. J., 1988, Transbilayer movement of phosphatidylserine in erythrocytes: Inhibition of transport and preferential labeling of a 31,000-dalton protein by sulfhydryl reactive reagents, *Biochemistry* **27:**848–851.

Daleke, D. L., and Huestis, W. H., 1985, Incorporation and translocation of aminophospholipids in human erythrocytes, *Biochemistry* **24:**5406–5416.

Elbaum, D., Mimms, L. T., and Branton, D., 1984, Modulation of actin polymerization by the spectrin–band 4.1 complex, *Biochemistry* **23:**4813–4816.

Elgsaeter, A., and Branton, D., 1974, Intramembrane particle aggregation in erythrocyte ghosts. 1. The effects of protein removal, *J. Cell Biol.* **63:**1018–1030.

Fairbanks, B., Steck, T. L., and Wallach, D. F. H., 1971, Electrophoretic analysis of the major polypeptides of the human erythrocyte membrane, *Biochemistry* **10:**2607–2617.

Fowler, V. M., 1987, Identification and purification of a tropomyosin binding protein from human erythrocytes, *J. Biol. Chem.* **262:**12792–12800.

Fowler, V., and Bennett, V., 1984, Erythrocyte membrane tropomyosin purification and properties, *J. Biol. Chem.* **259:**5978–5989.

Fowler, V. M., and Bennett, V., 1978, Association of spectrin with its membrane attachment site restricts lateral mobility of human erythrocyte integral proteins, *J. Supramol. Struct.* **8:**215–221.

Fowler, V. M., Davis, V., and Benett, V., 1985, Human erythrocyte myosin: Identification and purification, *J. Cell Biol* **100:**47–55.

Gardner, K., and Bennett, V., 1986, A new erythrocyte membrane-associated protein with calmodulin binding activity. Identification and purification, *J. Biol. Chem.* **261:**1339–1348.

Gardner, K., and Bennett, V., 1987, Modulation of spectrin–actin assembly by erythrocyte adducin, *Nature* **328:**359–362.

Hall, T. G., and Bennett, V., 1987, Regulatory domains of erythrocyte ankyrin, *J. Biol. Chem.* **262:**10537–10545.

Hargreaves, W., Giedd, K., Verkleij, A., and Branton, D., 1980, Reassociation of ankyrin with band 3 in erythrocyte membranes and in lipid vesicles, *J. Biol. Chem.* **255:**11965–11972.

Heast, C. W. M., 1982, Interactions between membrane skeleton proteins and the intrinsic domain of the erythrocyte membrane, *Biochim. Biophys. Acta* **694:**331–352.

Heubusch, P., Jung, C. Y., and Green, F. A., 1985, The osmotic response of human erythrocyte and the membrane cytoskeleton, *J. Cell. Physiol.* **122:**266–272.

Holt, G. D., Haltiwanger, R. S., Torres, C.-R., and Hart, G. W., 1987, Erythrocytes contain cytoplasmic glycoprotin, *J. Biol. Chem.* **262:**14847–14850.

Husain, A., Howlett, G., and Sawyer, W. H., 1985a, The interaction of calmodulin with erythrocyte membrane proteins, *Biochem. Int.* **10:**1 12.

Husain, A., Howlett, G. J., and Sawyer, W. H., 1985b, Calmodulin activation of red blood cell (Ca^{2+} + Mg^{2+})-ATPase and its antagonism by phenothiazine, *Mol. Pharmacol.* **18:**253–258.

Janmey, P. A., and Stossel, T. P., 1987, Modulation of gelsolin function by phosphatidylinositol 4,5-bisphosphate, *Nature* **325:**362–364.

Jarret, H. W., and Penniston, J. T., 1978, Purification of the Ca^{2+}-stimulated ATPase activator from human erythrocytes, *J. Biol. Chem.* **253:**4676–4682.

Jinbu, Y., Sato, S., Nakao, T., and Nakao, M., 1982, Ankyrin is necessary for both drug-induced and ATP-induced shape change of human erythrocyte ghosts, *Biochem. Biophys. Res. Commun.* **104:**1087–1092.

Jinbu, Y., Nakao, M., Otsuka, M., and Sato, S., 1983, Two steps in ATP-dependent shape change of human erythrocyte ghosts, *Biochem. Biophys. Res. Commun.* **112:**384–390.

Jinbu, Y., Sato, S., Nakao, M., and Tsukita, S., 1984a, Ca^{2+}- and Mg-ATP-dependent shape change of human erythrocyte ghosts and Triton shells, *Exp. Cell Res.* **151:**160–170.

Jinbu, Y., Sato, S., Nakao, T., Nakao, M., Tsukita, S., Tsukita, S., and Ishikawa, H., 1984b, The role of ankyrin in shape and deformability change of human erythrocyte ghosts, *Biochim. Biophys. Acta* **773:**237–245.

Jinbu, Y., Sato, S., and Nakao, M., 1984c, Reversible shape change of Triton treated erythrocyte ghosts induced by Ca^{2+} and Mg-ATP, *Nature* **307:**376–378.

Knowles, W., Marchesi, S. L., and Marchesi, V. T., 1983, Spectrin: Structure, function, and abnormalities, *Semin. Hematol.* **20:**159–174.

Kojima, Y., Nakao, M., Sato, S., and Hara, Y., 1988, Role of spectrin in maintaining shape of red cells, *Seikagaku* **60:**789 (in Japanese).

Lange, Y., Gouch, A., and Steck, T. L., 1982a, Role of the bilayer in the shape of the isolated erythrocyte membrane, *J. Membr. Biol.* **69:**113–124.

Lange, Y., Hadesman, R. A., and Steck, T. L., 1982b, Role of the reticulum in the stability and shape of the isolated human erythrocyte membrane, *J. Cell Biol.* **92:**714–721.

Lin, D. C., and Lin, S., 1978, Actin polymerization induced by a motility-related high-affinity cytochalasin binding complex from erythrocyte membrane, *Proc. Natl. Acad. Sci. USA* **75:**2345–2349.

Ling, E., Danilov, Y. N., and Cohen, C. M., 1988, Modulation of red cell band 4.1 function by cAMP-dependent kinase and protein kinase C phosphorylation, *J. Biol. Chem.* **263:**2209–2216.

Litman, D., Hsu, D. J., and Marchesi, V. T., 1980, Evidence that spectrin binds to macromolecular complexes on the linear surface of the red cell membrane, *J. Cell Sci.* **42:**1–22.

Liu, S.-C., and Palek, J., 1979, Metabolic dependence of protein arrangement in human erythrocyte membranes. 2. Cross linking of major proteins in ghosts from fresh and ATP depleted red cells, *Blood* **54:**1117–1130.

Liu, S.-C., and Palek, J., 1980, Spectrin tetramer dimer equilibrium and the stability of erythrocyte membrane skeletons, *Nature* **285:**586–588.

Liu, S.-C., and Palek, J., 1984, Hemoglobin enhances the self-association of spectrin heterodimers in human erythrocytes, *J. Biol. Chem.* **259:**11556–11562.

Liu, S.-C., Fairbanks, G., and Palek, J., 1977, Spontaneous reversible protein cross linking in the human erythrocyte membrane: Temperature and pH dependence, *Biochemistry* **16:**4066–4074.

Liu, S.-C., Palek, J., Prchal, J., and Castleberry, R. P., 1981, Altered spectrin dimer–dimer association and instability of erythrocyte membrane skeletons in hereditary pyropoikilocytosis, *J. Clin. Invest.* **68:**597–605.

Liu, S.-C., Windisch, P., Kim, S., and Palek, J., 1984, Oligomeric states of spectrin in normal erythrocyte membranes: Biochemical and electron microscopic studies, *Cell* **37:**587–594.

Liu, S.-C., Zhai, S., Lawler, J., and Palek, J., 1985, Hemin-mediated dissociation of erythrocyte membrane skeletal proteins, *J. Biol. Chem.* **260:**12234–12239.

Liu, S.-C., Derick, L. H., and Palek, J., 1987, Visualization of the hexagonal lattice in the erythrocyte membrane skeleton, *J. Cell Biol.* **104:**527–536.

Lorand, L., Shishido, R., Parameswaran, K. N., and Steck, T. L., 1975, Modification of human erythrocyte ghosts with transglutaminase, *Biochem. Biophys. Res. Commun.* **67:**1158–1166.

Luna, E. J., Kidd, G. H., and Branton, D., 1979, Identification by peptide analysis of the spectrin-binding protein in human erythrocytes, *J. Biol. Chem.* **254**:2526–2532.

Lutz, H. U., Liu, S.-C., and Palek, J., 1977, Release of spectrin-free vesicles from human erythrocytes during ATP depletion. Part 1. Characterization of spectrin-free vesicles, *J. Cell Biol.* **73**:548–560.

Lux, S. E., 1979, Dissecting the red cell membrane skeleton, *Nature* **281**:426–429.

Maruta, H., and Mizuno, D., 1971, Selective recognition of various erythrocytes in endocytosis by mouse peritoneal macrophages, *Nature* **234**:246–248.

Matsuzaki, F., Sutoh, K., and Ikai, A., 1985, Structural unit of the erythrocyte cytoskeleton. Isolation and electron microscopic examination, *Eur. J. Cell Biol.* **39**:153–160.

Middelkoop, E., Lubin, B. H., Bevers, E. M., Op den Kamp, J. A. F., Comfurius, P., Chiu, D. T.-Y., Zwaal, R. F. A., van Deenen, L. L. M., and Roelofsen, B., 1988, Studies on sickled erythrocytes provide evidence that the asymmetric distribution of phosphatidylserine in the red cell membrane is maintained by both ATP-dependent translocation and interaction with membrane skeletal proteins, *Biochim. Biophys. Acta* **937**:281–288.

Mische, S. M., Mooseker, M. S., and Morrow, J. S., 1987, Erythrocyte adducin: A calmodulin-regulated actin-bundling protein that stimulates spectrin–actin binding, *J. Cell Biol.* **105**:2837–2845.

Mohandas, N., Chasis, J. A., and Shohet, S. T., 1983, The influence of membrane skeleton on red cell deformability, membrane material properties, and shape, *Semin. Hematol.* **20**:225–242.

Morrow, J. S., and Marchesi, V. T., 1981, Self assembly of spectrin oligomers in vitro: A basis for a dynamic cytoskeleton, *J. Cell Biol.* **88**:463–468.

Morrow, J., Haigh, W., and Marchesi, V. T., 1984, Spectrin oligomers in vitro: A structural feature of the erythrocyte cytoskeleton, *J. Supramol. Cell Biochem.* **1**:275–287.

Nakao, K., Wada, T., Kamiyama, T., and Nakao, M., 1962a, Clinical and experimental studies on the post-transfusion viability of the long-term stored erythrocytes, *J. Jpn. Int. Soc.* **51**:211–219 (in Japanese).

Nakao, K., Wada, T., Kamiyama, T., Nakao, M., and Nagano, K., 1962b, A direct relationship between adenosine triphosphatase-level and in vivo viability of erythrocytes, *Nature* **194**:877–878.

Nako, M., Nakao, T., Tatibana, M., Yoshikawa, H., and Abe, T., 1959, Effect of inosine and adenine on adenosine triphosphate regeneration and shape transformation in long-stored erythrocyte, *Biochim. Biophys. Acta* **32**:564–565.

Nakao, M., Nakao, T., and Yamazoe, S., 1960a, Adenosine triphosphate and maintenance of shape of the human red cells, *Nature* **187**:945–946.

Nakao, M., Nakao, T., Tatibana, M., and Yoshikawa, H., 1960b, Phosphorus metabolism in human erythrocyte. III. Regeneration of adenosine triphosphate in long-stored erythrocyte by incubation with inosine and adenine, *J. Biochem.* **47**:661–671.

Nakao, M., Nakao, T., Arimatsu, Y., and Yoshikawa, H., 1960c, A new preservative medium maintaining the level of adenosine triphosphate and the osmotic resistance of erythrocyte, *Proc. Jpn. Acad.* **36**:43–47.

Nakao, M., Nakao, T., Tatibana, M., and Yoshikawa, H., 1960d, Shape transformation of erythrocyte ghosts on addition of adenosine triphosphate to the medium, *J. Biochem.* **47**:694–695.

Nakao, M., Nakao, T., Tatibana, M., and Yoshikawa, H., 1960e, Phosphorus metabolism in human erythrocyte. IV. Destruction of adenine nucleotides in stored blood, *J. Biochem.* **48**:672–684.

Nakao, M., Nakao, T., Yamazoe, S., and Yoshikawa, H., 1961, Adenosine triphosphate and shape of erythrocyte, *J. Biochem.* **49**:487–492.

Nakao, M., Motegi, T., Nakao, T., Yamazoe, S., and Yoshikawa, H., 1962a, A positive feedback mechanism of adenosine triphosphate synthesis in erythrocytes, *Nature* **191**:283–284.

Nakao, M., Nakao, T., Yoshikawa, H., Wada, T., Takaku, H., and Nakao, K., 1962b, A new preservative medium containing adenine and inosine, Proc. 8th Congr. Int. Soc. Blood Transf. Tokyo pp. 455–461.

Nakao, M., Hoshino, K., and Nakao, T., 1981, Constancy of cell volume during shape change of erythrocytes induced by the increasing ATP content, *J. Bioenerg. Biomembr.* **13**:307–316.

Nakao, M., Nakao, T., Komatsu, Y., Sano, K., and Sasakawa, S., 1983, Isoosmotic sucrose, adenine inosine media for preservation of blood, *Biomed. Biochim. Acta* **42**:527–535.

Nakao, M., Jinbu, Y., Sato, S., Ishigami, Y., Nakao, T., Ito-Ueno, E., and Wake, K., 1987, Structure and function of red cell cytoskeleton, *Biomed. Biochim. Acta* **46**:5–9.

Nakao, T., Nagano, K., Adachi, K., and Nakao, M., 1964, Separation of two adenosine triphosphatase from erythrocyte membranes, *Biochem. Biophys. Res. Commun.* **13**:444–448.

Nakao, T., Nagai, F., and Nakao, M., 1982, Posttransfusion viability of rabbit erythrocytes preserved in a medium containing inosine, adenine, and isoosmotic sucrose, *Vox Sang.* **42**:217–222.

Nakashima, K., and Beutler, E., 1979, Comparison of structure and function of human erythrocyte and human actin, *Proc. Natl. Acad. Sci. USA* **76:**935–938.

Nelson, W. J., and Veshnock, P. J., 1987, Ankyrin binding to (Na^+ + K^+)ATPase and implications for the organization of membrane domains in polarized cells, *Nature* **328:**533–536.

Ohanian, V., Wolfe, L. C., John, K. M., Pinder, J. C., Lux, S. E., and Gratzer, W. B., 1984, Analysis of the ternary interaction of the red cell membrane skeletal proteins spectrin, actin, and 4.1, *Biochemistry* **23:**4416–4420.

Ohnishi, T., 1962, Extraction of actin- and myosin-like proteins from erythrocyte membrane, *J. Biochem.* **52:**307–308.

Palek, J., and Lux, S. E., 1983, Red cell membrane skeletal defects on the red cell deformability, membrane material properties and shape, *Semin. Haematol.* **189:**225–240.

Palek, J., Curby, W. A., and Lionetti, F. J., 1971a, Effects of calcium and ATP on volume of human red cell ghosts, *Am. J. Physiol.* **220:**19–26.

Palek, J., Curby, W. A., and Lionetti, F. J., 1971b, Relation of calcium ion activated ATPase to calcium ion linked shrinkage of human red cell ghosts, *Am. J. Physiol.* **220:**1028–1032.

Palek, J., Curby, W. A., and Lionetti, F. J., 1972, Size dependence of ghosts from stored erythrocytes on calcium and ATP, *Blood* **40:**261–275.

Palek, J., Stewart, G., and Lionetti, F. J., 1974, The dependence of shape of human erythrocyte ghosts on calcium, magnesium and ATP, *Blood* **44:**583–597.

Palek, J., Liu, P. A., and Liu, S.-C., 1978a, Polymerization of red cell membrane protein contributes to spheroechinocyte shape irreversibility, *Nature* **274:**505–507.

Palek, J., Liu, S.-C., and Snyder, L. M., 1978b, Metabolic dependence of protein arrangement in human erythrocyte membranes. Part 1. Analysis of spectrin-rich complexes in ATP depleted red cells, *Blood* **51:**385–396.

Pasternack, F. R., Anderson, R. A., Leto, T. L., and Marchesi, V. T., 1985, Interactions between protein 4.1 and band 3, *J. Biol. Chem.* **260:**3676–3683.

Patel, V. P., and Fairbanks, G., 1981, Spectrin phosphorylation and shape changes of human erythrocyte ghosts, *J. Cell Biol.* **88:**430–440.

Patel, V. P., and Fairbanks, G., 1986, Relationship of major phosphorylation reactions and Mg-ATPase activities to ATP-dependent shape change of human erythrocyte membranes, *J. Biol. Chem.* **261:**3170–3177.

Pinder, J. C., and Gratzer, W. B., 1983, Structural and dynamic states of actin in the erythrocyte, *J. Cell Biol.* **96:**768–775.

Pinder, J., Ungewickell, E., Bray, D., and Gratzer, W. B., 1978, The spectrin actin complex and erythrocyte shape, *J. Supramol. Struct.* **8:**435–445.

Quist, E. E., 1980, Regulation of erythrocyte membrane shape by Ca^{2+}, *Biochem. Biophys. Res. Commun.* **92:**631–637.

Quist, E. E., and Reece, K. L., 1980, The role of diphosphatidylinositol in erythrocyte membrane shape regulation, *Biochem. Biophys. Res. Commun.* **95:**1023–1030.

Sato, S., and Nakao, M., 1981, Cross-linking of intact erythrocyte membrane with a newly synthesized cleavable bifunctional reagent, *J. Biochem.* **90:**1177–1185.

Sato, S., and Nakao, M., 1986, Characterization of human erythrocytes cytoskeletal ATPase, *J. Biochem.* **100:**643–649.

Sato, S., and Ohnishi, S., 1983, Interaction of a peripheral protein of the erythrocyte membrane, band 4.1 with phosphatidylserine-containing liposome and erythrocyte inside-out vesicles, *Eur. J. Biochem.* **130:**19–25.

Seigneuret, M., and Devaux, P. F., 1984, ATP-dependent asymmetric distribution of spin-labeled phospholipids in the erythrocyte membrane: Relation to shape changes, *J. Cell Biol.* **81:**3751–3755.

Shahbakhti, F., and Gratzer, W. B., 1986, Analysis of the self-association of human red cell spectrin, *Biochemistry* **25:**5969–5975.

Shapiro, A. L., Vinuera, E., and Maizel, J. V., 1967, Molecular weight estimation of polypeptide chains by electrophoresis in SDS-polycarbonate gels, *Biochem. Biophys. Res. Comm.* **28:**815–821.

Sheetz, M. P. 1983, Membrane skeletal dynamics: Role in modulation of red cell deformability, mobility of transmembrane proteins, and shape, *Semin. Hematol.* **20:**175–188.

Sheetz, M. P., and Singer, S. J., 1974, Biological membranes as bilayer couples. A molecular mechanism of drug-induced interactions, *Proc. Natl. Acad. Sci. USA* **71:**4457–4461.

Sheetz, M. P., and Singer, S. J., 1977, On the mechanism of ATP-induced shape changes in human erythrocyte membranes. I. The role of the spectrin complex, *J. Cell Biol.* **73:**638–696.

Shen, B. W., Josephs, R., and Steck, T. L., 1984, Ultrastructure of unit fragments of the skeleton of the human erythrocyte membrane, *J. Cell Biol.* **99**:810–821.

Shen, B. W., Josephs, R., and Steck, T. L., 1986, Ultrastructure of the intact skeleton of the human erythrocyte membrane, *J. Cell Biol.* **102**:997–1006.

Shotton, D. M., Burke, B. E., and Branton, D., 1979, The molecular structure of human erythrocyte spectrin. Biophysical and electron microscopic studies, *J. Mol. Biol.* **131**:303–320.

Siegel, D. L., and Branton, D., 1985, Partial purification and characterization of an actin-bundling protein, band 4.9, from human erythrocytes, *J. Cell Biol.* **100**:775–785.

Singer, S. J., and Nicolson, G. L., 1972, The fluid mosaic model of the structure of cell membrane, *Science* **175**:720–731.

Smith, D. K., and Palek, J., 1982, Modulation of lateral mobility of band 3 in the red cell membrane by oxidative cross linking of spectrin, *Nature* **297**:424–425.

Smith, D. K., and Palek, J., 1983, Sulfhydryl reagents induce altered spectrin self association skeletal instability and increased thermal sensitivity of red cells, *Blood* **62**:1190–1196.

Sobue, K., Muramoto, Y., Fujita, M., and Kakiuchi, S., 1982, Calmodulin binding protein of erythrocyte cytoskeleton, *Biochem. Biophys. Res. Commun.* **100**:1063–1068.

Speicher, D. W., and Marchesi, V. T., 1984, Erythrocyte spectrin is composed of many homologous triple helical segments, *Nature* **311**:177–180.

Srinivasan, Y., Elmer, L., Davis, J., Bennett, V., and Angelides, K., 1988, Ankyrin and spectrin associate with voltage-dependent sodium channels in brain, *Nature* **333**:177–180.

Steck, T. L., Weinstein, R. S., Straus, J. H., and Wallach, D. F. H., 1970, Inside-out red cell membrane vesicles preparation and purification, *Science* **168**:255–257.

Steck, T. L., Rams, B., and Strapazon, E., 1976, Proteolytic dissection of band 3, predominant polypeptide of the human erythrocyte membrane, *Biochemistry* **15**:1154–1161.

Steiner, J. P., and Bennett, V., 1988, Ankyrin-independent membrane protein-binding sites for brain and erythrocyte spectrin, *J. Biol. Chem.* **263**:11417–11425.

Stromqvist, M., Berglund, A., Shanbhag, V. P., and Backman, L., 1988, Influence of calmodulin on the human red cell membrane skeleton, *Biochemistry* **27**:1104–1110.

Takakuwa, Y., Tchemia, G., Rossi, M., Benabadji, M., and Mohandas, N., 1986, Restoration of normal membrane stability of unstable protein 4.1-deficient erythrocyte membranes by incorporation of purified protein 4.1, *J. Clin. Invest.* **78**:80–85.

Teitel, P., 1965, Disk–sphere transformation and plasticity alteration of red blood cells, *Nature* **26**:409–410.

Tsuji, A., Kawasaki, K., and Ohnishi, S., 1988, Regulation of band 3 mobilities in erythrocyte ghost membranes by protein association and cytoskeletal meshwork, *Biochemistry* **27**:7447–7452.

Tsukita, S., Tsukita, S., and Ishikawa, H., 1980, Cytoskeletal network underlying the human erythrocyte membrane: Thin-section electron microscopy, *J. Cell Biol.* **85**:567–576.

Tsukita, S., Tsukita, S., Ishikawa, H., Sato, S., and Nakao, M., 1981, Electron microscopic study of reassociation of spectrin and actin with the human erythrocyte membrane, *J. Cell Biol.* **90**:70–77.

Tyler, J. M., Hargreaves, W. R., and Branton, D., 1979, Purification of two spectrin-binding proteins: Biochemical and electron microscopic evidence for site-specific reassociation between spectrin and bands 2.1 and 4.1, *Proc. Natl. Acad. Sci. USA* **76**:5192–5196.

Tyler, J., Reinhardt, B., and Branton, D., 1980, Associations of erythrocyte membrane proteins binding of purified bands 2.1 and 4.1 to spectrin, *J. Biol. Chem.* **255**:7034–7039.

Ueno, E., Sato, S., Jinbu, Y., and Nakao, M., 1987, Dynamic association of band 3 with Triton shells in human erythrocyte ghosts, *Biochim. Biophys. Acta* **915**:77–86.

Ungewickell, E., Bennett, P., Calvert, R., Ohanian, V., and Gratzer, W. B., 1979, In vitro formation of complex between cytoskeletal proteins of human erythrocytes, *Nature* **280**:811–814.

Wada, T., Takaku, F., Nakao, K., Nakao, M., Nakao, T., and Yoshikawa, H., 1960, Posttransfusion survival of the red blood cells stored in a medium containing adenine and inosine, *Proc. Jpn. Acad.* **36**:618–623.

Weed, R. I., LaCelle, P. L., and Merrill, E. W., 1969, Metabolic dependence of red cell deformability, *J. Clin. Invest.* **48**:795–809.

White, J. G., 1974, Effects of an ionophore A23187 on the surface morphology of normal erythrocytes, *Am. J. Pathol.* **77**:507–514.

Wolf, M., and Sahyoun, N., 1986, Protein kinase C and phosphatidylserine bind to M_r110,000/115,000 polypeptides enriched in cytoskeletal and postsynaptic density preparations, *J. Biol. Chem.* **261**:13327–13332.

Wong, A. J., Kiehart, D. P., and Pollard, T. D., 1985, Myosin from human erythrocytes, *J. Biol. Chem.* **260**:46–49.

Yu, J., and Goodman, S. R., 1979, Syndeins: The spectrin-binding(s) of the human erythrocyte membrane, *J. Proc. Natl. Acad. Sci. USA* **76**:2340–2344.

Yu, J., Fishman, D. A., and Steck, T. L., 1973, Selective solubilization of proteins and phospholipids from red cell membranes by nonionic detergents, *J. Supramol. Struct.* **1**:233–248.

Chapter 8

Spectrin–Actin Interactions

Greg B. Ralston

1. INTRODUCTION

The protein spectrin was first reported in 1968 in water-soluble extracts from erythrocyte membranes (Marchesi and Steers, 1968). Although for some time it was believed that spectrin represented a specific erythrocyte adaptation, in recent years spectrin and spectrinlike proteins have been found in a wide range of other cell types (e.g., see Goodman *et al.*, 1981). These nonerythroid forms of spectrin all share with erythrocyte spectrin a common set of properties: they are proteins of high molecular mass, comprising two distinct subunits of approximately 250,000 and 230,000 Da, respectively; they bind calmodulin in a calcium-dependent manner, though with greatly differing affinities; they bind to actin filaments; and they are associated with the cell membranes through interaction with other proteins, particularly ankyrin (Bennett, 1985). Certainly in the red blood cell, and probably in other cell types, the major functional role of spectrin is to stabilize the membrane, and to provide a linkage for actin filaments to the membrane. While the red cell may be an inadequate model for other cell types, the organization of the erythrocyte cytoskeleton and its role in maintaining erythrocyte shape and deformability are now reasonably well understood, at least in broad terms, and the findings from the red cell give at least an indication of the organization of such molecules in other cells. The present discussion will be limited to erythrocyte spectrin.

Red cells account for approximately 40% of the volume of blood in humans. Their major function is to provide a controlled environment for the high concentration of hemoglobin required for the efficient distribution of oxygen to the body tissues. The erythrocyte, with its package of 4×10^8 hemoglobin molecules, is organized so as to fulfill effectively its oxygen-transporting role, having lost its nucleus, mitochondria, and other intracellular organelles during maturation. During the average life span of 120 days,

Greg B. Ralston Department of Biochemistry, University of Sydney, Sydney, New South Wales 2006, Australia.

each cell endures more than 500,000 turbulent passages through the heart and travels a distance of about 300 miles (Lux, 1979).

The membrane of the red cell must be deformable in order to maintain a low viscosity of blood for efficient circulation. The viscosity of normal blood under rapid flow can be as little as two to four times greater than that of pure water, and packed cells remain fluid even at hematocrits near 100% (Dintenfass, 1971). If red cells were rigid, blood would have the "consistency of brick or concrete" (Dintenfass, 1971). Even if red cells had the flow properties of white cells, the increase in viscosity of the blood would be so great that efficient pumping of the blood by the heart and, therefore, the efficient distribution of oxygen would be impossible (Lichtman, 1973).

2. THE RED CELL MEMBRANE CYTOSKELETON

2.1. Red Cell Shape and Deformability

The shape of the human red cell at rest in isotonic solution is a smooth biconcave disk with a diameter of ~ 8 μm. This shape is one of minimum curvature, and has been claimed to be one for which bending energy is minimized (Canham, 1970). *In vivo,* however, red cells are subjected to flow and shear deformation in the circulation, and each passage through the vascular system involves large distortions from the biconcave shape; in the microcirculation, erythrocytes are forced through very narrow capillaries, some of which have a diameter only half that of the cell itself. Under these conditions, normal red cells are deformed readily and reversibly and do not present a major obstacle to blood flow.

The morphological behavior is determined by the properties of the membrane, which itself must be highly deformable; a cell with fluid contents and a rigid membrane would behave as a rigid particle. The surface area of the red cell is constant, and the membrane can be ruptured if extended in area by only a few percent. In order to allow the cell to undergo deformation without rupture, the membrane must have a surface area larger than that of a sphere of the same volume.

2.2 Morphology and Composition of the Cytoskeleton

The shape and deformability of red cell membranes, and the distribution of intramembrane particles and glycoproteins within the membrane are controlled by peripheral proteins arranged in a two-dimensional network on the cytoplasmic surface of the membrane. This network has variously been referred to as the erythrocyte membrane cytoskeleton or membrane skeleton. (For recent reviews see Bennett, 1985; Marchesi, 1985; Goodman *et al.*, 1988; Nakao, this volume.) The presence of the network prevents fusion of cells (Mombers *et al.*, 1977; Portis *et al.*, 1979; Shohet, 1979) and restricts the lateral mobility of the integral proteins (Elgsaeter and Branton, 1974; Elgsaeter *et al.*, 1976; Fowler and Branton, 1977; Fowler and Bennett, 1978). The mature erythrocyte differs from other eukaryotic cells in that it lacks intermediate filaments, microfilaments, and microtubules, and the membrane cytoskeleton does not extend into the cytoplasm.

Extraction of erythrocyte membranes (or intact erythrocytes) with the nonionic de-

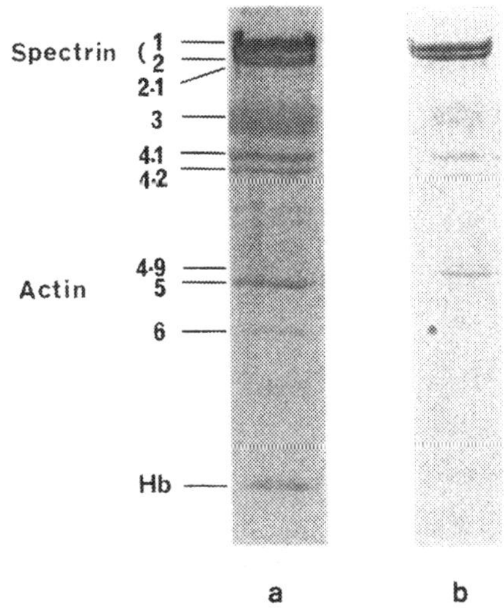

FIGURE 1. Proteins of the human erythrocyte membrane and its cytoskeleton as resolved by acrylamide gel electrophoresis in the presence of sodium dodecyl sulfate and stained with Coomassie blue. (a) The pattern of the entire membrane; (b) the pattern of the Triton shell prepared from 5 mM phosphate buffer, pH 8.0. The proteins are numbered sequentially according to increasing mobility. Note that the glycophorins do not stain with Coomassie blue.

tergents Triton X-100 or Triton X-114 removes most of the lipids and integral proteins, leaving behind an insoluble residue (Yu *et al.*, 1973) termed the "Triton shell" (Sheetz and Sawyer, 1978). This serves as an operational definition of the erythrocyte membrane cytoskeleton, although the detailed composition of such preparations depends on the conditions of extraction. When prepared at low ionic strength (e.g., 5 mM sodium phosphate), the Triton shell carries ankyrin, together with some band 3 protein and associated lipid and accessory proteins (Sheetz, 1979; Shen *et al.*, 1986). Preparations made with isotonic saline solutions of Triton X-100 comprise spectrin, actin, and proteins 4.1 and 4.9; the ankyrin and its attached band 3 protein are dissociated from spectrin under these conditions (Sheetz, 1979). The major proteins of the erythrocyte membrane are numbered sequentially (see Figure 1) in order of increasing mobility on acrylamide gels containing dodecyl sulfate (Fairbanks *et al.*, 1971; Steck, 1974).

The "Triton shell" prepared in this way appears in the electron microscope as a filamentous reticulum, and retains the shape of the cell from which it was derived (Yu *et al.*, 1973; Sheetz and Sawyer, 1978). These Triton shells seem to represent the true cytoskeleton, and do not appear to be artifacts of preparation, since similar structures can be seen in scanning electron micrographs of "isolated membranes torn open to reveal their interior space" (Hainfeld and Steck, 1977).

Further supporting evidence is provided by electron micrographs of sectioned membranes that have been fixed with tannic acid and glutaraldehyde (Tsukita *et al.*, 1980). In these preparations, filamentous structures similar to Triton shell material can be seen lying parallel to the membrane surface and apparently attached to the membrane through particles protruding from the cytoplasmic face.

More recently, clear images of the erythrocyte membrane cytoskeleton have been obtained from negatively stained preparations of erythrocyte membranes lysed *in situ* on polylysine-coated grids (Byers and Branton, 1985). By enhancing the spreading of the attached membrane skeleton, extremely clear images of the morphology and organization of the network were obtained (Byers and Branton, 1985). Very similar structures have

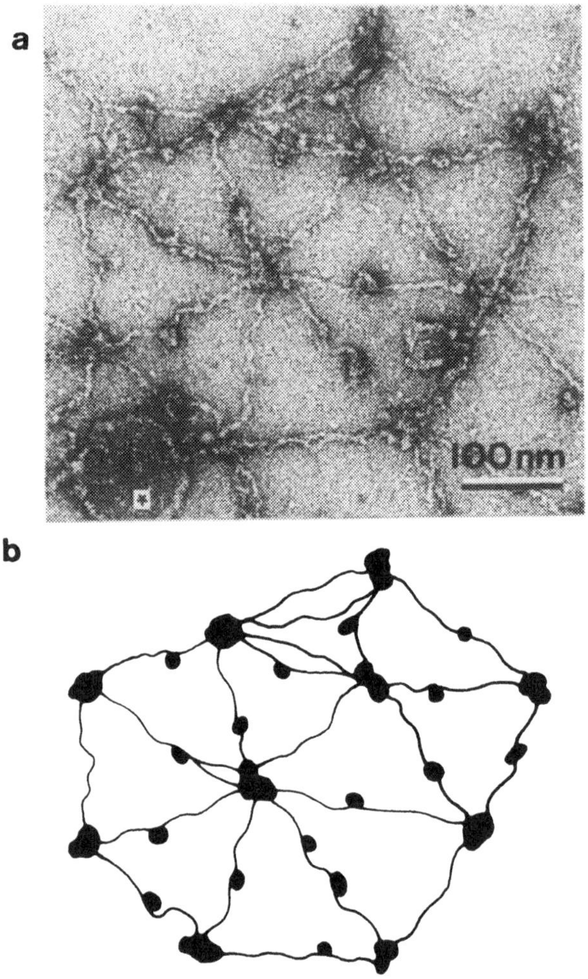

FIGURE 2. The erythrocyte membrane cytoskeleton. (a) Electron micrograph of a spread section of the cytoskeleton. Spectrin tetramers (radial arms) link actin protofilaments (at the vertices). Globular material bound near the centers of the spectrin tetramers may represent ankyrin molecules. (b) Diagrammatic representation of the field shown in panel a. Reproduced from the *Journal of Cell Biology*, 1987, **104:**527–536, by copyright permission of the Rockefeller University Press.

been reported for isolated membrane skeletons, freed of band 3 protein, ankyrin, and band 4.9 by washing with 1.5 M NaCl (Shen *et al.*, 1986; Liu *et al.*, 1987).

The basic structure of the skeleton thus appears to comprise spectrin, actin, and band 4.1. Electron microscopic images show short F-actin filaments each linked to five or six spectrin molecules, and forming a more or less hexagonal array, particularly when the cytoskeleton is highly expanded (Figure 2). The spectrin molecules in turn appear to be largely tetrameric although a significant proportion of them appear to be three- or four-armed structures, suggestive of higher oligomers (Liu *et al.*, 1987). The spectrin molecules thus appear to cross-link the actin filaments, resulting in a highly branched and cross-linked two-dimensional network. The site of spectrin attachment to the actin filaments often shows a thickening, which may represent the attachment of band 4.1 (Shen *et al.*, 1986).

Other proteins may participate in the functioning cytoskeleton. Although band 4.9 can bundle actin filaments *in vitro* (Siegel and Branton, 1985), phosphorylation of band 4.9 protein by a cAMP-dependent kinase abolishes the bundling activity (Husain-Chishti *et al.*, 1988) and its role *in vivo* remains uncertain. Erythrocyte tropomyosin has been isolated from red cell membranes prepared in the presence of magnesium salts (Fowler and Bennett, 1984), and may play a role in regulating the length of the actin filaments. The proteins calmodulin and adducin may play important roles in controlling the interactions within the cytoskeleton (see Sections 5.3 and 5.4).

2.3. Membrane Attachment Sites for the Cytoskeleton

The best characterized membrane attachment site for the skeletal network involves spectrin, ankyrin, and the major integral membrane protein band 3 (Bennett and Stenbuck, 1979, 1980a,b; Haest, 1982). Ankyrin binds to the spectrin heterodimer 20 nm from the self-association interface (Tyler *et al.*, 1980). Only the β chain of spectrin possesses an ankyrin binding site (Calvert *et al.*, 1980a; Morrow *et al.*, 1980). The interaction between ankyrin and spectrin *in vitro* is moderately strong ($K \approx 10^7$ M^{-1}) (Tyler *et al.*, 1980), while that between ankyrin and band 3 is significantly stronger ($K \approx 2 \times 10^8$ M^{-1}) (Bennett and Stenbuck, 1980b). Thus, the cytoskeleton is attached firmly to the membrane through the bifunctional binding of ankyrin.

Band 4.1 has also been implicated as a second membrane attachment site for the cytoskeleton. While low-ionic-strength incubation removes almost all the spectrin and actin from the membrane, more than 80% of the band 4.1 protein remains associated with the membrane under these conditions. The band 4.1 protein, however, can be removed by incubation with 0.8–1.0 M KCl (Tyler *et al.*, 1979; Ungewickell *et al.*, 1979), high concentrations of Tris-HCl (Ohanian and Gratzer, 1984), or 1 mM *p*-mercuribenzene sulfonate (Ralston and Crisp, 1981), indicating that it is not an integral protein.

Extracted band 4.1 protein rebinds to the membrane with high affinity, and with a maximal binding capacity close to the amount of band 4.1 protein found on the intact membrane (Shiffer and Goodman, 1984). The binding is reduced by 60% after proteolysis of the membrane with papain (but not with chymotrypsin), indicating that the binding site is a protein; glycophorin A was suggested as a possible candidate as it is digested by papain, but not by chymotrypsin. Pasternack *et al.* (1985) also reported a high-affinity binding site on the membrane for band 4.1, but in addition they reported a weaker affinity site with a much greater capacity, which may represent band 3 protein.

Extraction of erythrocyte membranes with Triton X-100 solubilizes most of the integral proteins, but leaves glycophorin C associated with the cytoskeletal pellet (Mueller and Morrison, 1981; Owens *et al.*, 1980). Solubilization of glycophorin C could only be achieved after prior removal of band 4.1. Although these results might imply that glycophorin C is a high-affinity site for band 4.1, there are insufficient copies of glycophorin C (~ 35,000 per cell; Furthmayr, 1978) for complete binding of band 4.1 (~ 200,000 per cell). Selective extraction of band 4.1 from spectrin-depleted membranes by the nonionic detergent Tween 20 leads to the coextraction of glycophorins A and C with the band 4.1 protein (Elliott and Ralston, 1984). These proteins were associated with band 4.1 protein throughout all attempts at purification. Band 4.1 is capable of binding to phosphatidylcholine vesicles containing glycophorin A, and the binding can be abolished by inclusion of anti-glycophorin A antibodies (Anderson and Lovrien, 1984). Inclusion of anti-glycophorin A antibodies in binding studies using red cell membrane vesicles stripped of peripheral proteins led to abolition of only 60% of the band 4.1 binding.

Studies with a heterofunctional cross-linking reagent have shown that close interaction may occur between band 4.1 and glycophorins B and C, as well as less specific interactions with glycophorin A (Whitfield *et al.*, 1986). Other evidence also implicates glycophorin B: while some patients completely lacking glycophorin A seem to possess erythrocytes of normal morphology, some cases of elliptocytosis have been correlated with deficiencies of glycophorins B and C (Anstee *et al.*, 1984), as well as with band 4.1 deficiency (Sondag *et al.*, 1987). The tentative conclusion may be drawn that high-affinity binding sites on the membrane are located on glycophorins B and C, with a larger number of possibly weaker sites on glycophorin A; an even larger number of low-affinity sites may exist on band 3 protein.

It has been suggested (Anderson and Marchesi, 1985) that interaction between band 4.1 and glycophorin A may be regulated by polyphosphoinositides; band 4.1 protein cosediments with glycophorin micelles in the presence of phosphatidylinositol 4-phosphate and phosphatidylinositol 4,5-bisphosphate. Furthermore, in the presence of the polyphosphoinositides, exogenous glycophorin competitively inhibits the binding of band 4.1 protein to stripped red cell membrane vesicles. In the absence of phosphatidylinositol 4,5-bisphosphate, the linkage between glycophorin and band 4.1 protein may be displaced to other sites, such as that proposed to exist on band 3 (Pasternack *et al.*, 1985).

Both spectrin and band 4.1 protein interact with phospholipids (Marinetti and Crain, 1978) and both appear to bind preferentially to negatively charged phospholipids such as phosphatidylserine (Cohen *et al.*, 1988), which, in normal red cells, are restricted largely to the inner leaflet (Mombers *et al.*, 1977; Shiffer *et al.*, 1988). Furthermore, spectrin protects phospholipid bilayers against the action of Ca^{2+} and Mg^{2+}, thereby preventing fusion of lipid vesicles (Mombers *et al.*, 1977). Thus, the integrity of the red cell may depend not only on the interactions of the cytoskeleton with the integral membrane proteins, but also on the interactions of spectrin and band 4.1 protein with the inner leaflet of the lipid bilayer.

3. SPECTRIN

The currently accepted model of the erythrocyte membrane cytoskeleton is dominated by the huge spectrin heterodimer. Low-ionic-strength treatment of erythrocyte membranes results in the extraction of spectrin. If this treatment is performed at low temperatures the tetramer is extracted, whereas at 37°C the dimer is the principal form of extracted spectrin (Ralston *et al.*, 1977; Kam *et al.*, 1977). Solubilization of spectrin from the membrane leads to fragmentation of the membrane into small vesicles, predominantly with an inside-out orientation (Marchesi *et al.*, 1970). The erythrocytes of some laboratory-bred strains of mice with a severe form of hereditary spherocytosis have abnormally low levels of spectrin (Greenquist *et al.*, 1978). These erythrocytes are extremely fragile; the membrane fuses easily and lacks elasticity. The degree of severity of hemolytic anemia in these mice correlates well with the extent of spectrin deficiency. Reconstitution with normal spectrin of red cell ghosts from these mice leads to a reduction in fragility and the tendency to fusion (Shohet, 1979).

3.1. The Structure of Spectrin

Human erythrocyte spectrin is comprised of two distinct polypeptide chains of very high molecular mass: the α chain (band 1 in the nomenclature of Steck, 1974) of ~ 250 kDa and the β chain (band 2) of ~ 230 kDa. Although the two polypeptides have distinct amino acid sequences and appear to be coded by separate genes, their overall amino acid compositions are very similar and they show clear homology (Speicher, 1986).

Both polypeptides exhibit a repeating homologous motif of 106 residues, indicating the evolution of both chains by duplication of an ancestral gene. A segment in the middle of the α chain, and one at the carboxyl-terminal portion of both chains appear to be nonhomologous with the 106-residue repeat (Speicher, 1986). The carboxyl-terminal region of the β chain is further distinguished by being phosphorylated at four potential sites (Harris and Lux, 1980).

The 106-residue repeating units appear to be folded into triple-α-helical segments (Speicher and Marchesi, 1984); within the α-helical regions of each 106-residue repeat there also appear to be two reverse turns each of four contiguous residues, possibly allowing the helices to reverse direction, and a region that is nonhelical and that may act as a flexible connecting segment between triple-α-helical segments (Speicher, 1986).

This conformational model for spectrin is consistent with the high degree of α-helical conformation (> 75%) estimated from optical rotatory dispersion and circular dichroism (Ralston, 1978; Calvert *et al.*, 1980b). In addition, a structure of this type is consistent with the images of a long, flexible structure, seen with platinum-shadowed preparations of spectrin in electron microcopy (Shotton *et al.*, 1979). Given the number of residues (~ 2000 per chain), the high degree of α-helix content, and the distance between residues of the α helix (0.15 nm), a triple-α-helical conformation would allow each chain to be folded into a single flexible rod of contour length ~ 100 nm and diameter ~ 5 nm, consistent with electron microscopy (Shotton *et al.*, 1979) and with physical measurements in solution (Stokke and Elgsaeter 1981).

The two subunits of spectrin appear to associate noncovalently in an antiparallel manner, with the two chains loosely coiled about each other to form a heterodimer (Gratzer and Beaven, 1975; Ralston, 1975, 1978; Shotton *et al.*, 1979; Stokke and Elgsaeter 1981). The spectrin dimers can undergo further self-association (see Section 3.3) to form tetramers consisting of two α and two β chains (Ralston, 1975, 1978; Ralston *et al.*, 1977) and higher oligomers.

3.2. Domain Structure of Spectrin

While each of the chains of spectrin appears to be folded into repeating 106-residue units, operationally the protein can be cleaved by trypsin into well-defined fragments that have been termed "chemical domains" (Speicher *et al.*, 1980). The individual chemical domains of spectrin are capable of refolding into apparently native conformation after urea- or SDS-induced denaturation, following the removal of the denaturant (Knowles *et al.*, 1979). The availability of well-defined fragments has allowed the otherwise unmanageable size of spectrin to yield to sequence studies.

The amino-terminal segment of the α chain has been isolated as an 80-kDa fragment (Speicher *et al.*, 1980; Hanspal and Ralston 1981). Sequencing of this fragment provided the first evidence of the repeating structure of spectrin (Speicher *et al.*, 1980). This domain was subsequently shown to carry the binding site on the α chain for the carboxyl-terminus of the β chain (Morrow *et al.*, 1980; Hanspal and Ralston, 1982). Another well-characterized domain to which an important function has been assigned is a 50-kDa segment generated from the β chain by cleavage at cysteine residues with 2-nitro-5-

thiocyanobenzoic acid; this fragment is thought to carry the binding site for ankyrin, which attaches spectrin to the membrane (Morrow *et al.*, 1980).

Binding sites on each of the chains for the lateral association of the other chain have also been identified and mapped onto the domain structure (Morrow *et al.*, 1980). The sites of phosphorylation of the β chain have been mapped to the carboxyl end of this polypeptide (Harris and Lux, 1980; Speicher *et al.*, 1982). There are four sites of phosphorylation: three phosphoserine and one phosphothreonine, all clustered within 100 residues of the carboxyl-terminus (Harris and Lux, 1980).

3.3. Spectrin Self-Association

In electron micrographs, the α and β chains of the heterodimer appear to be associated through multiple weak interactions along the length of the chains; the two chains appear loosely twisted about each other with varying degrees of contact, and with both left- and right-hand twist (Shotton *et al.*, 1979). Nevertheless, no unambiguous evidence of dissociation of the two chains in the absence of denaturant has been presented. At the two ends of the heterodimer the two chains appear strongly bound together.

The tetramer appears as an extended flexible molecule having a contour length approximately twice that of the heterodimers; i.e., the tetramer appeared to be formed by a "head-to-head" form of association, without measurable overlap (Shotton *et al.*, 1979). Shotton and his colleagues postulated a mechanism of reciprocal binding at two identical pairs of binding sites in which the α subunit of one heterodimer binds to the β subunit in the second heterodimer (Figure 3). The formation of the spectrin tetramer from dimers is reversible and depends on the temperature and on the ionic strength (Ungewickell and Gratzer, 1978). The dimer is the more stable form at low ionic strength; higher ionic strength favors the tetramer and higher oligomers. The equilibrium is rapid at temperatures above 30°C whereas at low temperatures spectrin is kinetically trapped in its existing oligomeric state.

Although the association of spectrin to form tetramers appeared to be of a closed type, with all valences satisfied, occasional three-armed structures consistent with a hexamer (Figure 3) were seen subsequently in electron micrographs of spectrin (Tyler *et al.*, 1980). Additional bands, presumably representing spectrin oligomers greater than the tetramer, were also seen on nondenaturing gel electrophoresis of spectrin, incubated in 0.1 M NaCl at 30°C to promote association and subsequently chilled to prevent reequilibrium during electrophoresis (Morrow and Marchesi, 1981; Hanspal and Ralston, 1982; Liu *et al.*, 1984).

Sedimentation equilibrium studies of spectrin at 30°C showed that the self-association reaction could not be described adequately by a dimer–tetramer equilibrium, but that higher oligomers participated in the reaction (Morris and Ralston, 1984). Analysis of gel electrophoresis data indicated that the tetramer is the predominant species formed, i.e., that the equilibrium constant for the dimer/tetramer reaction is substantially greater than that for subsequent steps (Morrow and Marchesi, 1981; Shahbakhti and Gratzer, 1986). However, experiments of this type carry the risk of perturbation of the original equilibrium distribution during electrophoretic separation of the oligomers. Sedimentation equilibrium analysis, on the other hand, suggests that while the dimer/tetramer step appears to be slightly favored, the equilibrium constant for subsequent steps may be up to half the value of that of the dimer/tetramer step (Morris and Ralston, 1985).

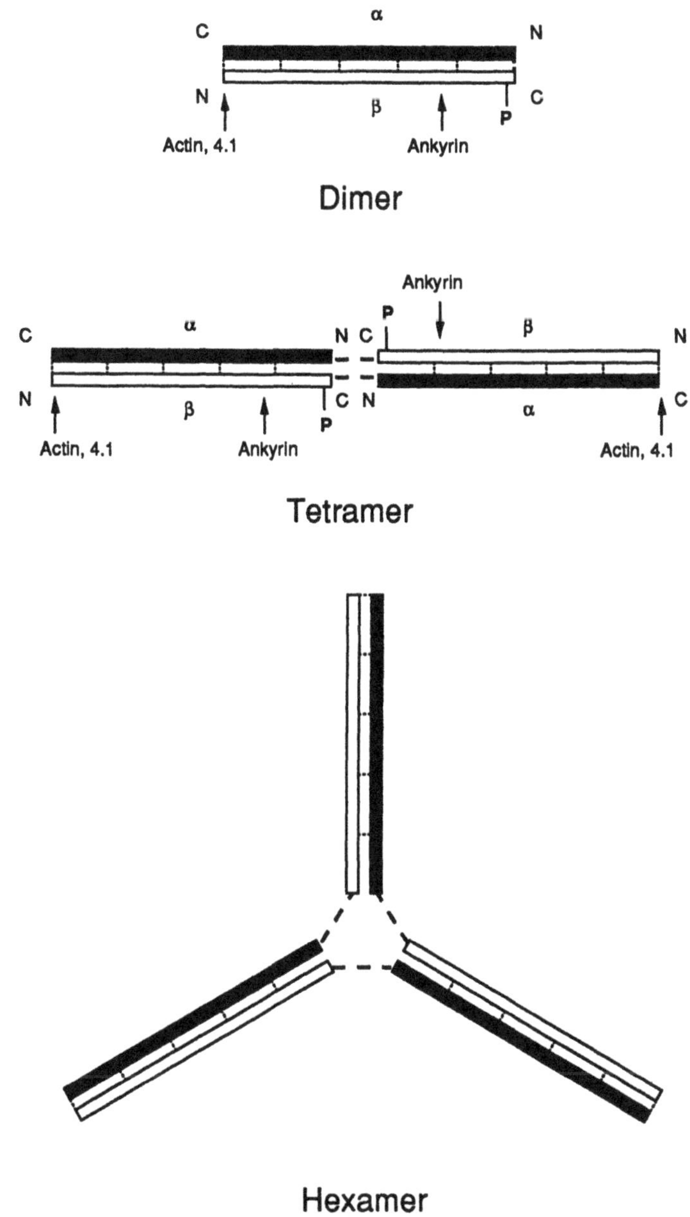

FIGURE 3. Diagrammatic representation of the dimer, tetramer, and hexamer of spectrin. The α and β chains are represented by filled and open rectangles, respectively, with the N- and C-termini indicated on the dimer and tetramer. Phosphorylation sites on the β chains are shown by "P" and approximate locations of interactions with ankyrin, actin, and band 4.1 protein are indicated with arrows. Note that the heterodimer is univalent toward actin, while the tetramer and hexamer are di- and trivalent, respectively.

Analysis of the distribution of spectrin oligomers extracted from erythrocyte membranes at low ionic strength and low temperature (conditions that minimize dissociation of spectrin oligomers) indicated that tetramers may comprise 45–55% of the total spectrin on the membrane, dimers 5–10%, and higher oligomers 25–35% (Liu *et al.*, 1984).

Whatever the detailed distribution of spectrin oligomers on the membrane, it is clear that conditions that result in dissociation to the heterodimer result in markedly more fragile membranes. Manipulation of temperature and ionic strength in order to promote dissociation of spectrin to the heterodimer results in membranes that are mechanically unstable (Liu and Palek, 1980; Palek and Liu, 1981; Liu *et al.*, 1981, 1982). Furthermore, in some cases of hereditary pyropoikilocytosis (Liu *et al.*, 1981) and hereditary elliptocytosis (Liu *et al.*, 1982), membranes of erythrocytes show mechanical instability, and the oligomeric distribution of spectrin shows a marked shift toward the heterodimer.

While studies of spectrin self-association at moderate concentration (up to 10 g/liter) have given an indication of the pattern of spectrin association, the behavior *in vivo* may be quite different. Within a narrow zone adjacent to the membrane, the local concentration of spectrin is likely to be very high. Nonideality contributions arising from these high concentrations, and from the very high hemoglobin concentration within the cell, are likely to have enormous effects on the detailed distribution of oligomers on the membrane (Minton, 1983). Some data already show that the presence of "bystander" proteins, such as hemoglobin and albumin, enhances spectrin self-association (Liu and Palek, 1984).

4. ACTIN

Actin is a major component of the cytoskeleton in eukaryotic cells, and is responsible for many types of cell motility, as well as for muscle contraction. In spite of the relatively small size of the actin molecule (43 kDa), this protein is capable of recognizing and binding to a wide range of other molecules, including other actin molecules in the formation of actin filaments. The functions of actin are controlled by a plethora of other molecules including a wide range of proteins that are able to bind specifically to either the filamentous (F-actin) form or the monomeric globular (G-actin) form; these actin-binding proteins may cap the ends of filaments, cross-link filaments, stabilize nucleation sites for polymerization, bind to and sever filaments, or sequester monomeric actin. The literature of actin is too vast to be reviewed adequately here. Pollard and Cooper (1986) have reviewed actin and actin-binding proteins, and also refer back to earlier reviews on actin.

Higher vertebrates express at least six different actins. There are four closely related muscle actins (cardiac muscle actin, skeletal muscle actin, and two forms of smooth muscle actin) and in the cytoplasm of most eukaryotic cells, two actins, referred to as the β and γ isoforms, are normally expressed together. Pinder *et al.* (1978a) demonstrated, using two-dimensional isoelectric focusing, that erythrocyte actin is exclusively of the β type.

Differences have been observed in the properties of muscle and nonmuscle actins. Nonmuscle actins activate the heavy meromyosin ATPase from skeletal muscle with only 75% the efficiency of muscle actin (Gordon *et al.*, 1977). The critical concentrations for a few nonmuscle actins have been determined under optimal and suboptimal conditions (Gordon *et al.*, 1977). Under suboptimal ionic conditions, the critical concentrations have been shown to be significantly higher than for muscle actin, presumably reflecting differences in one or more of the rate constants for the polymerization reaction.

4.1. Erythrocyte Actin

Tilney and Detmers (1975) were the first to characterize the 43-kDa protein, band 5, which is invariably extracted with spectrin on low-ionic-strength treatment of erythrocyte ghosts, and to identify it as actin. Erythrocyte actin constitutes about 4–5% by weight of the membrane proteins (Fairbanks *et al.*, 1971; Steck, 1974), and exists in about 400,000 copies per cell. Purified erythrocyte actin is capable of polymerizing into helical filaments 5–7 nm wide, which can be decorated with heavy meromyosin (Tilney and Detmers, 1975) and which are effective in stimulating muscle myosin ATPase (Sheetz *et al.*, 1976; Nakashima and Beutler, 1979).

Tilney and Detmers (1975) were also the first to recognize that spectrin binds to actin and postulated that, in the erythrocyte, these proteins "associate to form an anastomosing network beneath the erythrocyte membrane."

4.2. Polymerization of Erythrocyte Actin

The polymerization behavior of erythrocyte actin shows some differences from that of muscle actin. Nakashima and Beutler (1979) found similar viscosity increases in solutions of erythrocyte and muscle actin after addition of 50 mM KCl and 2 mM $MgCl_2$. They also examined the effects of certain modulators of actin polymerization: phalloidin, cytochalasin B, and DNase I were all found to affect the time courses for erythrocyte actin polymerization in a manner similar to their effect on muscle actin. However, under suboptimal conditions for polymerization, erythrocyte actin polymerizes less rapidly and to a lesser extent than does rabbit muscle actin under identical conditions (Tilley and Ralston, 1984). In particular, the polymerization of erythrocyte actin in the presence of calcium ions is even more severely retarded.

Since the initial discovery of erythrocyte actin, there has been considerable controversy over the association state of actin in the red cell membrane. No characteristic long actin filaments have been seen in electron micrographs of the membrane or isolated cytoskeletons. Consequently, some investigators originally suggested that actin exists in its monomeric form stabilized by spectrin on the erythrocyte membrane (Tilney and Detmers, 1975; Pinder *et al.*, 1978a; Ralston, 1978). It has now been established conclusively that, in the erythrocyte membrane cytoskeleton, actin exists as short "protofilaments" of between 12 and 20 monomers in length (Brenner and Korn, 1980; Atkinson *et al.*, 1982; Pinder *et al.*, 1981; Pinder and Gratzer, 1983). This length is close to that of a single turn of the actin long-pitch (or "double") helix.

However, in spite of differences in the polymerization behavior of muscle and erythrocyte actins, the organization of actin in the erythrocyte membrane cytoskeleton into protofilaments does not seem to derive from a particular property of erythrocyte actin. Skeletal muscle actin, in the presence of either spectrin tetramers alone or spectrin tetramers plus band 4.1, can be induced to form high-molecular-weight complexes (Brenner and Korn, 1979; Fowler and Taylor, 1980) in which the actin exists as short filaments cross-linked with spectrin.

5. SPECTRIN–ACTIN INTERACTIONS

5.1. Binding of Spectrin to Actin Filaments

The functional heart of the erythrocyte membrane cytoskeleton is the interaction between spectrin and short filaments of actin, resulting in a closed, two-dimensional anastomosing network, imparting elasticity and extensibility to the membrane and yet not impeding the fluid motion of the membrane around the cell contents.

Early experiments (Tilney and Detmers, 1975) showed that water-soluble extracts from erythrocyte membranes (predominantly spectrin, actin, and band 4.1 protein) increased the final viscosity of preformed actin filaments. However, when a solution of G-actin was allowed to polymerize in the presence of the extract, the resulting final viscosity was less than that when the extract was absent.

Those results were taken as evidence that spectrin was capable of binding to F-actin, but that spectrin reduced the polymerization of actin by inhibiting the elongation of the filaments. Although this suggestion was remarkably close to current ideas concerning the spectrin–actin interaction, resolution of some of the complexities proceeded slowly over a number of years. Subsequently, it has been shown by Brenner and Korn (1979, 1980) that the viscosity increase with preformed actin is due to cross-linking of actin filaments by spectrin tetramers or higher oligomers. Copolymerization of actin with spectrin is now known to lead to a slight enhancement of the degree of actin polymerization, reflected in the fluorescence of pyrene-labeled actin (Stromqvist *et al.*, 1985; Tilley and Ralston, 1987), but the filaments are shorter (Cohen *et al.*, 1980), leading to a lower final viscosity than when preformed F-actin is used.

Pinder *et al.* (1975) showed that crude water-soluble extracts were capable of inducing the polymerization of G-actin. This polymerization-inducing activity was subsequently assigned to the high-molecular-weight complexes of spectrin, actin, and band 4.1 found in such crude extracts (Lin and Lin, 1979; Pinder *et al.*, 1979; Brenner and Korn, 1980), and now referred to as "junctional complexes" (Shen *et al.*, 1984). These junctional complexes presumably acted as nuclei for actin polymerization.

Careful studies with purified spectrin dimers and tetramers labeled with ^{32}P showed unequivocally that spectrin could bind only to F-actin and not to G-actin (Brenner and Korn, 1979). The binding of spectrin tetramers to F-actin occurs at both ends of the spectrin tetramer molecule (see Figure 3), potentially allowing tetramers to cross-link actin filaments and leading to marked increases in the viscosity of actin solutions (Brenner and Korn, 1979). Binding of the univalent spectrin dimers to F-actin resulted in only a small increase in viscosity. In spite of earlier indication that phosphorylation of spectrin was required for its binding to actin (Pinder *et al.*, 1978b), Brenner and Korn (1979) showed that such binding was independent of phosphorylation, and the site of phosphorylation is now known to be at the opposite end of the spectrin β chain from the actin-binding site (Speicher *et al.*, 1982).

The binding of spectrin to preformed F-actin can be visualized in rotary-shadowed preparations by means of electron microscopy (Cohen *et al.*, 1980). Electron microscopy confirmed the bivalent binding of tetramers and the monovalent binding of dimers. An intriguing and not yet fully explained observation is that while spectrin tetramers appear simply to decorate and cross-link preformed F-actin filaments, the copolymerization of spectrin tetramer with actin results in small, tightly aggregated clusters in which the actin exists predominantly as short protofilaments (Cohen *et al.*, 1980).

Both α and β chains of spectrin seem to be required for the binding to F-actin. Isolated α and β chains, prepared from spectrin dissociated by urea or SDS, are not able to bind F-actin (Calvert *et al.*, 1980a; Cohen and Langley, 1984), while reconstituted $\alpha\beta$ heterodimers regain binding ability (Calvert *et al.*, 1980a; Cohen and Langley, 1984).

Electron microscopic images seem to show that spectrin is capable of binding along the length of actin filaments (Ungewickell *et al.*, 1979; Cohen *et al.*, 1980). No clear-cut evidence of preferential binding to one or other end compared to the length of the filaments has been presented, although there have been indications that binding may be enhanced at the "pointed," or slow-growing end of the filament. Ca^{2+}-actin displays a lower critical concentration in the presence of spectrin dimer than does Mg^{2+}-actin, suggesting that in the presence of Ca^{2+}, spectrin may be preferentially bound at the pointed end of actin filaments (Stromqvist *et al.*, 1985). The rates of actin polymerization are also enhanced by spectrin, particularly in the presence of Ca^{2+}, and although this behavior may be a reflection of the stabilization of actin nuclei by bound spectrin, it is also consistent with spectrin acting as a filament-severing protein (Stromqvist *et al.*, 1985).

Pinder *et al.* (1984) showed that cytochalasin E prevents the stimulation of actin polymerization by the high-molecular-weight complexes from water-soluble extracts of red cell membranes. Since cytochalasin E preferentially binds to the "barbed" (or fast-growing) end of actin filaments, the conclusion was drawn that the pointed end must be capped or blocked by the binding of spectrin.

However, if the concentration of G-actin used in these experiments is below critical at the pointed end of the actin filaments, no elongation would be possible. Even with G-actin concentrations slightly above critical, growth at the pointed end may be slow, due to the small rate constant for association to this end. Other workers have shown with the aid of electron microscopy that exogenous actin added to spectrin–actin–band 4.1 complexes is capable of extending filament growth in both directions (Shen *et al.*, 1984; Tsukita *et al.*, 1984). In an elegant series of experiments, Podolski and Steck (1988) demonstrated convincingly that the pointed end of actin in the erythrocyte membrane cytoskeleton protofilaments is free for the accretion of additional monomeric actin units. DNAse I was found to bind to the pointed end of the actin protofilaments, blocking addition of actin units at this end, while cytochalasin D blocked addition to the barbed end. The ends of the actin filament thus do not appear to be blocked, but quantitative data are lacking concerning the *rates* and *equilibria* for binding of actin to and the dissociation of actin from filaments containing bound spectrin.

While the binding of spectrin to actin *in vitro* has been demonstrated unequivocally, and the pattern of binding resembles that in fragments of the cytoskeleton (Shen *et al.*, 1986), the binding is quite weak. Estimates for the equilibrium constant for binding of spectrin dimer to F-actin range from $5 \times 10^3\ M^{-1}$ (Ohanian *et al.*, 1984) to $10^4\ M^{-1}$ (Brenner and Korn, 1979), although with binding constants of this low magnitude, the precision of the estimate is low. Measurements of the affinity of spectrin dimers for F-actin from quantitative binding assays are in concordance with the observation by Cohen *et al.* (1980) that relatively few molecules of spectrin could be seen associated with actin filaments in shadowed preparations of actin filaments incubated with spectrin.

Electron micrographs of the cytoskeletal network show that each actin protofilament

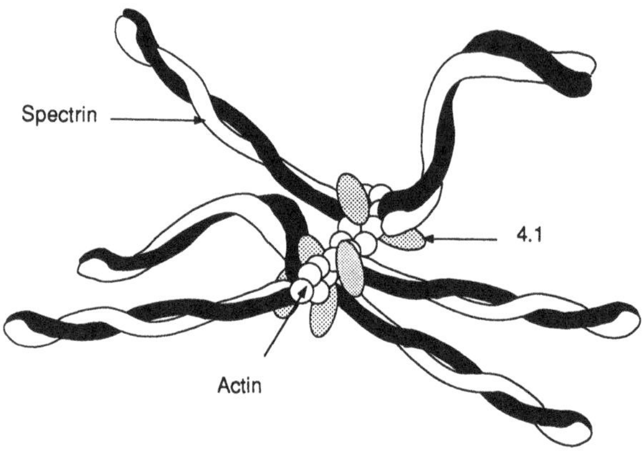

FIGURE 4. Diagrammatic representation of a junctional complex from the erythrocyte membrane cytoskeleton. The actin protomers are indicated by open circles, and the spectrin chains by the wormlike bodies. Band 4.1 is also shown on the complexes as stippled ellipses.

is associated with five to eight spectrin molecules (Shen *et al.*, 1986) located at both ends, as well as along the length of the protofilaments (Figure 4). Junctional complexes have been isolated that contain an actin protofilament core to which are attached five to eight spectrin heterodimers, and associated band 4.1 molecules. These complexes are extraordinarily stable to further dissociation, and appear to have a relatively well-defined stoichiometry of 1 spectrin dimer : 2 actin protomers : 1 band 4.1 molecule (Shen *et al.*, 1984; Beaven *et al.*, 1985; Matsuzaki *et al.*, 1985). These junctional complexes appear to correspond to the vertices of the cytoskeletal network, and thus may be thought of as structural units of the cytoskeleton.

The length of the actin protofilaments both in junctional complexes and in the intact cytoskeleton, 33–38 nm (Byers and Branton, 1985; Shen *et al.*, 1984, 1986), corresponds to 12–14 actin protomers, i.e., a single turn of the actin long-pitch or "double" helix. This length is also close to that of a single tropomyosin heterodimer, and it is likely that the binding of erythrocyte tropomyosin to the actin protofilaments in the cytoskeleton, in the presence of physiological concentrations of Mg^{2+} (Fowler and Bennett, 1984), is a means of regulating their length, preventing both their dissociation to smaller, and their growth to larger units.

While the weight of evidence seems to suggest that actin can be added to each end of the filaments within the complex, dissociation of actin from filaments seems to halt at the spectrin junctions. The binding of spectrin to F-actin frequently shows clustering of spectrin molecules, often at the ends (Cohen *et al.*, 1980; Shen *et al.*, 1984), and there is some suggestion of cooperativity. Some selectivity of binding to the ends of actin filaments is evident from studies of the lowering of the critical concentration of pyrene-labeled actin in the presence of spectrin (Tilley and Ralston, 1987). On the assumption that spectrin bound equally to all actin protomers within the filament, an equilibrium constant of 8×10^5 M^{-1} was determined for the binding of spectrin to actin protomers within the filament. However, subsequent direct measurement of binding in the analytical ultracentrifuge, under identical solution conditions, failed to detect binding at all (G. Nisbet and G. B. Ralston, unpublished data). This finding puts an upper limit on the association constant of 10^4 M^{-1}, a value consistent with that obtained in other laboratories. This discrepancy can be most simply explained if spectrin is capable of binding preferentially to the pointed (or slow-growing) end of the actin filament. If the bound spectrin were then to inhibit dissociation of actin from this end, the measured critical concentration would approach more closely the lower value of the barbed end.

In the presence of ATP, the critical concentration at the barbed end of actin filaments is below that at the pointed end; net polymerization occurs at the barbed end, and net dissociation occurs from the pointed end. This treadmilling mechanism may result in erosion of actin promoters from the pointed end, down to a spectrin–actin junction. If bound spectrin were to inhibit further dissociation, the pointed end would appear to be preferentially decorated with spectrin, and the measured critical concentration would fall toward that of the barbed end.

5.2. Enhancement of Spectrin–Actin Binding by Band 4.1

Although the binding of spectrin to actin filaments is rather weak, in the presence of band 4.1 protein the binding of spectrin to actin is enhanced dramatically, as detected both by electron microscopy (Cohen *et al.*, 1980) and by direct binding assays (Ungewickell *et al.*, 1979; Ohanian *et al.*, 1984). The equilibrium constant for formation of the ternary actin–spectrin–band 4.1 complex has been estimated from binding studies to be $1.2 \times 10^{12}\ M^{-2}$ (Ohanian *et al.*, 1984), an enhancement by eight orders of magnitude. This large enhancement is probably the origin of the extraordinary stability of the junctional complexes.

High-resolution electrophoresis on acrylamide gels containing SDS reveals that band 4.1 is comprised of two closely spaced subbands (4.1 a and b). These have been shown to be sequence-related polypeptides differing by ~ 2 kDa (Goodman *et al.*, 1982; Leto and Marchesi, 1984); both bind spectrin equally well.

Spectrin has been shown to form a binary complex with band 4.1 with a binding constant near $10^7\ M^{-1}$ (Tyler *et al.*, 1980; Wolfe *et al.*, 1982), although more recent measurements of fluorescence energy transfer indicate a value of $8.7 \times 10^7\ M^{-1}$ and a stoichiometry of 1 : 1 (Podgorski and Elbaum, 1984). The stoichiometry of 1 molecule of band 4.1 per spectrin molecule in the complexes seems to be at variance with the finding that band 4.1 binding to spectrin saturates at a stoichiometry of 2 (Tyler *et al.*, 1980). This can perhaps be resolved by the finding that each of the spectrin chains is capable of binding a band 4.1 molecule (Cohen and Langley, 1984; Cohen *et al.*, 1984). However, it appears that only a single band 4.1 molecule participates in the ternary complex. Given the number of copies of spectrin heterodimers and band 4.1 protein per erythrocyte (~ 200,000 each), there appears to be just sufficient band 4.1 to saturate the spectrin.

Gel electrophoresis of the ternary complex formed when band 4.1 protein was limiting indicated a stoichiometry of 1 : 1 for spectrin : band 4.1 protein (Ohanian *et al.*, 1984), corroborated by quantitative binding assays (Cohen *et al.*, 1984; Podgorski and Elbaum, 1985), and suggests that only a single molecule of band 4.1 protein is required to promote the binding of a spectrin molecule to actin. Saturation of actin with spectrin and band 4.1 protein reveals that each actin protomer within the filament is capable of binding a spectrin molecule in the presence of band 4.1 protein (Cohen *et al.*, 1984).

Recently, evidence has been presented that band 4.1 is capable of binding directly to actin to form a binary complex (Cohen *et al.*, 1984; Podgorski and Elbaum, 1985; M. Morris, personal communication).

The site of band 4.1 binding to spectrin has been located at the end of the dimer distal to the tetramerization interface (i.e., the "tail" of spectrin) (Tyler *et al.*, 1979, 1980) at a site indistinguishable from that for actin binding within the resolution of electron microscopy (Tyler *et al.*, 1980; Cohen *et al.*, 1980).

The formation of the spectrin–actin–band 4.1 complex is calcium sensitive, and the viscosity of mixtures of these proteins shows a maximum near 10^{-7} M Ca^{2+} (Fowler and Taylor, 1980; Cohen and Korsgren, 1980). The binding between actin and band 4.1 also shows calcium sensitivity, but with a maximum near 10^{-5} M (Cohen *et al.*, 1984); both above and below this concentration, the binding of band 4.1 to actin is reduced. However, caution is needed in interpreting these results, since they pertain to pH 6.5; at more closely physiological pH values, the calcium sensitivity is not seen (Cohen *et al.*, 1984). It is unclear whether calmodulin is involved in these calcium-dependent effects.

Band 4.1 protein may also provide an additional anchoring site on the membrane for the cytoskeleton (see Section 2.3), leading to the attachment of spectrin to the membrane at a site near the "head" region, via ankyrin, and at a site near the "tail" via band 4.1 protein. Thus, with multiple attachment sites, spectrin may be more closely adherent to the membrane and may interact more directly with it.

5.3. Erythrocyte Adducin

Recently, an additional protein has been isolated that serves to enhance the linkage between spectrin and actin (Ling *et al.*, 1986; Gardner and Bennett, 1987). This protein, termed adducin, interacts only weakly with spectrin alone, or with actin alone, but cosediments with actin filaments in the presence of spectrin (Gardner and Bennett, 1987), binding to spectrin–actin complexes with an affinity approximately ten times greater than that for spectrin alone.

Adducin is comprised of two partially homologous polypeptides of 103 and 97 kDa, respectively, and associated as a heterodimer (Gardner and Bennett, 1986). It is a substrate for Ca^{2+}- and phospholipid-dependent protein kinase C, accepting up to 3 mol phosphate per polypeptide chain, and thus is subject to regulation by calcium by two separate pathways (Ling *et al.*, 1986).

In adducin–spectrin–actin complexes, the stoichiometry approaches 2 molecules of spectrin per adducin molecule. It has been suggested (Gardner and Bennett, 1987) that adducin may be involved in the biogenesis of the erythrocyte membrane cytoskeleton, possibly recruiting spectrin onto actin filaments, prior to the synthesis of band 4.1 protein.

The behavior of adducin is modulated by calmodulin and calcium (Gardner and Bennett, 1987; Mische *et al.*, 1987). It has been shown that calmodulin binds to adducin in the presence of micromolar calcium with an association constant of $\sim 5 \times 10^6$ M^{-1} (Gardner and Bennett, 1986) and specifically inhibits the adducin-dependent binding of spectrin to actin filaments (Mische *et al.*, 1987).

The concentration of adducin in the mature erythrocyte is relatively low: 50 pmol/mg membrane protein, corresponding to ~ 30,000 copies/ghost (Gardner and Bennett, 1986). This is an order of magnitude less than the number of copies of spectrin (~ 200,000 heterodimer copies/ghost), and thus the role of adducin in the mature erythrocyte is somewhat open to question. However, this is a sufficient number of copies for a single adducin molecule to be associated with each actin protofilament (Gardner and Bennett, 1986) and the possibility remains that if adducin were to have an actin filament-severing activity, this protein could be important for regulating the number and the size of the protofilaments. However, no information exists on the distribution of adducin over the population of actin protofilaments; is there one adducin on each protofilament, or is the distribution random?

5.4. Calmodulin Binding to the Cytoskeleton

Calmodulin is present in erythrocytes at micromolar concentrations (Jarrett and Penniston, 1977). Calmodulin is a calcium-binding protein, of 17 kDa, that mediates calcium-dependent effects in the proteins to which it binds. On binding calcium, calmodulin undergoes a conformational change, leading to its binding to the target protein, and thereby bringing about metabolic change (Cheung, 1980). The main target within erythrocytes is the Ca^{2+}/Mg^{2+}-ATPase (the calcium pump), which maintains a low level of cytosolic free Ca^{2+} ($< 10^{-6}$ M; Schatzman, 1975). A rise in intracellular calcium leads to changes in red cell shape and deformability, but the molecular nature of those changes is far from clear (see Section 5.5).

The amount of calmodulin in the red cell is orders of magnitude greater than that required to effectively saturate the Ca^{2+}/Mg^{2+}-ATPase, and the reason for this extra calmodulin is also a mystery. Calmodulin has been reported to bind to erythrocyte spectrin (Sobue *et al.*, 1981), but the binding is relatively weak to human spectrin in the native state, although potentially, the number of binding sites is large. Of more likely physiological relevance is the binding of calmodulin to band 4.1 protein (Husain *et al.*, 1985) and adducin (Gardner and Bennett, 1987; Mische *et al.*, 1987).

While the binding of calmodulin to native human spectrin is relatively weak, with an association constant of $0.5–1.0 \times 10^5$ M^{-1} (Husain *et al.*, 1984; Burns and Gratzer, 1985), in the presence of urea, the binding is an order of magnitude greater (Husain *et al.*, 1984) and is restricted to the β subunit (Sears *et al.*, 1986). It is interesting in this context to note that other forms of spectrin, notably brain spectrin and avian erythrocyte spectrin (Glenney *et al.*, 1982; Husain *et al.*, 1984), have a higher affinity for calmodulin, even in the native state, and the α subunit is responsible for binding. In the presence of the higher concentrations of free calcium found in neurons than in erythrocytes, the binding of calmodulin to brain spectrin may play a more direct role.

The photolytic coupling of spectrin to an azido derivative of calmodulin results in inhibition of the binding of spectrin to actin, even in the presence of band 4.1 protein (Anderson and Morrow, 1987). Native calmodulin similarly reduced the binding of spectrin to actin, but in a calcium-dependent manner. Thus, the possibility exists that calmodulin may provide a mechanism for calcium-dependent alterations in the assembly of the cytoskeleton. While calmodulin may allow a calcium-dependent weakening of spectrin–actin–band 4.1 interaction (Anderson and Morrow, 1987), only about 5% of the spectrin molecules should contain bound calmodulin (Burns and Gratzer, 1985), However, in the presence of raised free calcium levels, this degree of occupancy would be expected to rise markedly.

5.5. Metabolic Regulation of Spectrin–Actin Interactions

Red cell shape is intimately linked with metabolism in the red cell; ATP depletion of red cells results in shape changes and a rise in intracellular calcium (Weed *et al.*, 1969), and isolated red cell membranes undergo similar shape changes modulated by ATP and Ca^{2+} (Sheetz and Singer, 1977; Quist, 1980).

The role of phosphorylation of erythrocyte cytoskeletal proteins has long been an enigma; early reports that the phosphorylation of spectrin by ATP may be an important regulator of cytoskeleton assembly have not been supported by experiment: the phosphorylation of spectrin has no apparent effect on its self-association (Ungewickell and Gratzer, 1978), its interaction with actin (Brenner and Korn, 1979), its interaction with the membrane (Anderson and Tyler, 1980), or the shape of the cell (Patel and Fairbanks, 1981, 1986). There is some suggestion that phosphorylation of spectrin may reduce its affinity for band 4.1 protein (Eder *et al.*, 1986).

The recent isolation of adducin (see Section 5.3) raises the possibility that calcium-mediated phosphorylation of this protein could modulate the assembly of the cytoskeleton (Ling *et al.*, 1986). Furthermore, band 4.1 protein is also phosphorylated by several different kinases (Cohen and Foley, 1986; Eder *et al.*, 1986; Ling *et al.*, 1988). Phosphorylation of band 4.1 has been shown to reduce both its affinity for spectrin (Eder *et al.*, 1986; Ling *et al.*, 1988) and its ability to induce the binding of spectrin to actin (Ling *et al.*, 1988). The degree of phosphorylation of band 4.1 brought about by protein kinase C is greater than that due to the cAMP-dependent kinase, and the degree of phosphorylation correlates with the reduction in promoting the binding of spectrin to actin (Ling *et al.*, 1988).

Any alteration in the spectrin–actin–band 4.1 network may be expected to have profound effects on the shape of the cell and its deformability. It has been suggested (Ling *et al.*, 1988) that control of the dynamic state of association of the cytoskeleton through phosphorylation of band 4.1 protein may allow the membrane to yield reversibly under extended stress. Dephosphorylation brought about by ATP depletion may then result in a more rigid network with reduced deformability.

Elevated calcium concentrations can have a number of effects: an increase in diacylglycerol content can stimulate protein kinase C (Nishizuka, 1986), and elevated calcium may directly stimulate Ca^{2+}-dependent protein kinases (Cohen and Foley, 1986). Phosphorylation of band 4.1 and adducin leads to a weakening of the spectrin–actin interaction. In the presence of calmodulin, calcium concentrations above 1 μM lead to a *destabilization* of the cytoskeleton (Takakuwa and Mohandas, 1988), probably through a calmodulin-dependent weakening of the spectrin–actin–band 4.1 interaction, although the role of adducin in this phenomenon is unknown. The spectrin–actin–band 4.1 protein interaction may be itself directly calcium sensitive (Cohen and Foley, 1986); at calcium concentrations above 10 μM, the *deformability* of the cytoskeleton decreases in a calmodulin-independent manner (Takakuwa and Mohandas, 1988). Alterations in the levels of polyphosphoinositides may affect the shape and deformability of the cell through modulation of the attachment site for band 4.1 protein (see Section 2.3).

Polyphosphates such as ATP and 2,3-DPG can bring about dissociation of the cytoskeleton (Schindler *et al.*, 1980; Sheetz and Casaly, 1981), but the concentrations of these compounds required to bring about changes (> 2 mM) are greater than those normally found *in vivo*. Nevertheless, even at physiological concentrations, these compounds may cause a small reduction in the affinity of protein–protein interactions that could conceivably have measurable consequences for the network as a whole, particularly as the concentrations of these agents fluctuate in response to the oxygenation state of hemoglobin.

The number of components believed to be involved in the organization and control of the erythrocyte cytoskeleton increases annually. How many of these are important functioning components of the mature cytoskeleton and how many are simply vestiges of the earlier stages in erythrocyte maturation is still to be unraveled. Nevertheless, a coherent picture is emerging in which the fundamental structural units of the cytoskeleton, the spectrin oligomers and the actin protofilaments, interact in ways that are sensitive to the metabolic state of the cell in order to maintain the membrane in an intact, highly deformable state, capable of enduring the stresses of the circulation for its 120-day life span. Already, the lessons learned from the simple red cell are being applied to other, more complex cell types.

6. REFERENCES

Anderson, J. M., and Tyler, J. M., 1980, State of spectrin phosphorylation does not affect erythrocyte shape or spectrin binding to erythrocyte membranes, *J. Biol. Chem.* **255:**1259–1265.

Anderson, J. P., and Morrow, J. S., 1987, The interaction of calmodulin with human erythrocyte spectrin, *J. Biol. Chem.* **262:**6365–6372.

Anderson, R. A., and Lovrien, R. E., 1984, Glycophorin is linked by band 4.1 to the human erythrocyte membrane skeleton, *Nature* **307:**655–658.

Anderson, R. A., and Marchesi, V. T., 1985, Regulation of the association of membrane skeletal protein 4.1 with glycophorin by a polyphosphoinositide, *Nature* **318:**295–298.

Anstee, D. J., Parsons, S. E., Ridgwell, K., Tanner, M. J. A., Merry, A. H., Thomson, E. E., Judson, P. R., Johnson, P., Bates, S., and Fraser, I. D., 1984, Two individuals with elliptocytocytic red cells apparently lack three minor erythrocyte membrane sialoglycoproteins, *Biochem. J.* **218:**615–619.

Atkinson, M. A. L., Morrow, J. S., and Marchesi, V. T., 1982, The polymeric state of actin in the human erythrocyte cytoskeleton, *J. Cell Biochem.* **18:**493–505.

Beaven, G. H., Jean-Baptiste, L., Ungewickell, E., Baines, A. J., Shahbakhti, F., Pinder, J. C., Lux, S. E., and Gratzer, W. B., 1985, An examination of the soluble oligomeric complexes extracted from the red cell membrane and their relation to the membrane cytoskeleton, *Eur. J. Cell Biol.* **36:**299–306.

Bennett, V., 1985, The membrane skeleton of human erythrocytes and its implications for more complex cells, *Annu. Rev. Biochem.* **54:**273–304.

Bennett, V., and Stenbuck, P., 1979, The membrane attachment protein for spectrin is associated with band 3 in human erythrocyte membranes, *Nature* **280:**468–473.

Bennett, V., and Stenbuck, P., 1980a, Human erythrocyte ankyrin: Purification and properties, *J. Biol. Chem.* **255:**2540–2548.

Bennett, V., and Stenbuck, P., 1980b, Association between ankyrin and the cytoplasmic domain of band 3 from the human erythrocyte membrane, *J. Biol. Chem.* **255:**6424–6432.

Brenner, S. L., and Korn, E., 1979, Spectrin–actin interaction, *J. Biol. Chem.* **254:**8620–8627.

Brenner, S. L., and Korn, E., 1980, Spectrin–actin complex isolated from sheep erythrocytes accelerates actin polymerization by simple nucleation, *J. Biol. Chem.* **255:**1670–1676.

Burns, N. R., and Gratzer, W. B., 1985, Interaction of calmodulin with the red cell and its membrane skeleton and with spectrin, *Biochemistry* **24:**3070–3074.

Byers, T. J., and Branton, D., 1985, Visualization of the protein associations in the erythrocyte membrane skeleton, *Proc. Natl. Acad. Sci. USA* **82:**6153–6155.

Calvert, R., Bennett, P., and Gratzer, W. B., 1980a, Properties and structural role of the subunits of human spectrin, *Eur. J. Biochem.* **107:**355–361.

Calvert, R., Ungewickell, E., and Gratzer, W. B., 1980b, A conformational study of human spectrin, *Eur. J. Biochem.* **107:**363–367.

Canham, P. B., 1970, Minimum energy of bending as a possible explanation of the biconcave shape of the human red blood cell, *J. Theor. Biol.* **26:**61–81.

Cheung, W. Y., 1980, Calmodulin plays a pivotal role in cellular regulation, *Science* **207:**19–27.

Cohen, C. M., and Foley, S. F., 1986, Organization of the spectrin–actin–band 4.1 ternary complex and its regulation by band 4.1 phosphorylation, in: *Membrane Skeletons and Cytoskeletal–Membrane Associations* (V. Bennett, C. M. Cohen, S. E. Lux, and J. Palek, eds.), pp. 211–222, Liss, New York.

Cohen, C. M., and Korsgren, C., 1980, Band 4.1 causes spectrin–actin gels to become thixotropic, *Biochem. Biophys. Res. Commun.* **97:**1429–1435.

Cohen, C. M., and Langley, R. C., Jr., 1984, Functional characterization of human erythrocyte spectrin α and β chains: Association with actin and protein 4.1, *Biochemistry* **23:**4488–4495.

Cohen, C. M., Tyler, J. M., and Branton, D., 1980, Spectrin–actin associations studied by electron microscopy of shadowed preparations, *Cell* **21:**875–883.

Cohen, C. M., Langley, R. C., Foley, S. F., and Korsgren, C., 1984, Functional associations of band 4.1 in the erythrocyte membrane skeleton and their role in inherited membrane skeletal abnormalities, *Prog. Clin. Biol. Res.* **159:**13–29.

Cohen, A. M., Liu, S.-C., Lawler, J., Derick, L., and Palek, J., 1988, Identification of the protein 4.1 binding site to phospholipid vesicles, *Biochemistry* **27:**614–619.

Dintenfass, L., 1971, The rheology of blood in vascular disease, *J. R. Coll. Physicians London* **5:**231–240.

Eder, P. S., Soong, C.-J., and Tao, M., 1986, Phosphorylation reduces the affinity of protein 4.1 for spectrin, *Biochemistry* **25:**1764–1770.

Elgsaeter, A., and Branton, D., 1974, Intramembrane particle aggregation in erythrocyte ghosts. I. The effects of protein removal, *J. Cell Biol.* **63:**1018–1030.

Elgsaeter, A., Shotton, D., and Branton, D., 1976, Intramembrane particle aggregation in erythrocyte ghosts. II. The influence of spectrin aggregation. *Biochim. Biophys. Acta* **426:**101–122.

Elliott, C., and Ralston, G. B., 1984, Solubilization of human erythrocyte band 4.1 protein in the non-ionic detergent Tween 20, *Biochim. Biophys. Acta* **775:**313–319.

Fairbanks, G., Steck, T. L., and Wallach, D. F. H., 1971, Electrophoretic analysis of the major polypeptides of the human erythrocyte membrane, *Biochemistry* **10:**2606–2617.

Fowler, V., and Bennett, V., 1978, Association of spectrin with its membrane attachment site restricts lateral mobility of human erythrocyte integral membrane proteins, *J. Supramol. Struct.* **8:**215–221.

Fowler, V., and Bennett, V., 1984, Erythrocyte membrane tropomyosin, *J. Biol. Chem.* **259:**5978–5989.

Fowler, V., and Branton, D., 1977, Lateral mobility of human erythrocyte integral membrane proteins, *Nature* **268:**23–26.

Fowler, V., and Taylor, D. L., 1980, Spectrin plus band 4.1 cross-link actin. Regulation by micromolar calcium, *J. Cell Biol.* **85:**361–376.

Furthmayr, H., 1978, Glycophorins A, B, and C: A family of sialoglycoproteins. Isolation and preliminary characterisation of trypsin derived peptides, *J. Supramol. Struct.* **9:**79–95.

Gardner, K., and Bennett, V., 1986, A new erythrocyte membrane-associated protein with calmodulin binding activity: identification and purification, *J. Biol. Chem.* **261:**1339–1348.

Gardner, K., and Bennett, V., 1987, Modulation of spectrin–actin assembly by erythrocyte adducin, *Nature* **328:**359–362.

Glenney, J. R., Glenney, P., and Weber, K., 1982, Erythroid spectrin, brain fodrin, and intestinal brush border proteins (TW-260/240) are related molecules containing a common calmodulin-binding subunit bound to a variant cell type-specific subunit, *Proc. Natl. Acad. Sci. USA* **79:**4002–4006.

Goodman, S. R., Zagon, I. S., and Kulikowski, R. R., 1981, Identification of a spectrin-like protein in non-erythroid cells, *Proc. Natl. Acad. Sci. USA* **78:**7570–7574.

Goodman, S. R., Yu, J., Whitfield, C. F., Culp, E. N., and Posnak, E. J., 1982, Erythrocyte membrane skeletal protein bands 4.1a and b are sequence-related phosphoproteins, *J. Biol. Chem.* **257:**4564–4569.

Goodman, S. R., Krebs, K. E., Whitfield, C. F., Riederer, B. M., and Zagon, I. S., 1988, Spectrin and related molecules, *CRC Crit. Rev. Biochem.* **23:**171–234.

Gordon, D. J., Boyer, J. L., and Korn, E. D., 1977, Comparative biochemistry of non-muscle actins, *J. Biol. Chem.* **252:**8300–8309.

Gratzer, W. B., and Beaven, G. H., 1975, Properties of the high molecular weight protein spectrin from human erythrocyte membranes, *Eur. J. Biochem.* **58:**403–409.

Greenquist, A. C., Shohet, S. B., and Bernstein, S. E., 1978, Marked reduction of spectrin in hereditary spherocytosis in the common house mouse, *Blood* **51:**1149–1155.

Haest, C. W. M., 1982, Interactions between membrane skeleton proteins and the intrinsic domain of the erythrocyte membrane, *Biochim. Biophys. Acta* **694:**331–352.

Hainfeld, J. F., and Steck, T. L., 1977, The sub-membrane reticulum of the human erythrocyte: A scanning electron microscope study, *J. Supramol. Struct.* **6:**301–311.

Hanspal, M. K., and Ralston, G. B., 1981, Purification of a trypsin-insensitive fragment of spectrin from human erythrocyte membranes, *Biochim. Biophys. Acta* **669:**133–139.

Hanspal, M. K., and Ralston, G. B., 1982, Binding of an 80,000 dalton fragment of spectrin to intact spectrin, *Biochim. Biophys. Acta* **709:**105–109.

Harris, H. W., and Lux, S. E., 1980, Structural characterization of the phosphorylation sites of human erythrocyte spectrin, *J. Biol. Chem.* **255:**11965–11972.

Husain, A., Howlett, G. J., and Sawyer, W. H., 1984, The interaction of calmodulin with human and avian spectrin, *Biochem. Biophys. Res. Commun.* **122:**1194–1200.

Husain, A., Howlett, G. J., and Sawyer, W. H., 1985, The interaction of calmodulin with erythrocyte membrane proteins, *Biochem. Int.* **10:**1–12.

Husain-Chishti, A., Levin, A., and Branton, D., 1988, Abolition of actin-bundling by phosphorylation of human erythrocyte protein 4.9, *Nature* **334:**718–720.

Jarrett, H. W., and Penniston, J. T., 1977, Partial purification of the Ca^{2+}-Mg^{2+} ATPase activator from human erythrocytes: Its similarity to the activator of 3′,5′-cyclic nucleotide phosphodiesterase, *Biochem. Biophys. Res. Commun.* **77:**1210–1216.

Kam, Z., Josephs, R., Eisenberg, H., and Gratzer, W. B., 1977, Structural study of spectrin from human erythrocyte membranes, *Biochemistry* **16:**5568–5572.

Knowles, W. J., Speicher, D. W., Morrow, J. S., and Marchesi, V. T., 1979, Renaturation of the chemical domains of human erythrocyte spectrin, *J. Cell Biol.* **83**(2, Pt. 2):272a.

Leto, T. L., and Marchesi, V. T., 1984, A structural model of human erythrocyte protein 4.1, *J. Biol. Chem.* **259:**4603–4608.

Lichtman, M. A., 1973, Rheology of leukocytes, leukocyte suspensions and blood in leukaemia. Possible relationships to clinical manifestations, *J. Clin. Invest.* **52:**350–358.

Lin, D. C., and Lin, S., 1979, Actin polymerization induced by a motility-related high-affinity cytochalasin binding complex from human erythrocyte membranes, *Proc. Natl. Acad. Sci. USA* **76:**2345–2349.

Ling, E., Gardner, K., and Bennett, V., 1986, Modulation of red cell band 4.1 function by c-AMP-dependent kinase and protein kinase C phosphorylation, *J. Biol. Chem.* **261:**13875–13878.

Ling, E., Danilov, Y. N., and Cohen, C. M., 1988, Modulation of red cell band 4.1 function by cAMP-dependent kinase and protein kinase C phosphorylation, *J. Biol. Chem.* **263:**2209–2216.

Liu, S.-C., and Palek, J., 1980, Spectrin tetramer–dimer equilibrium and the stability of erythrocyte membrane cytoskeletons, *Nature* **285:**586–588.

Liu, S.-C., and Palek, J., 1984, Hemoglobin enhances the self-association of spectrin heterodimers in human erythrocytes, *J. Biol. Chem.* **259:**11556–11562.

Liu, S.-C., Palek, J., Prchal, J., and Castleberry, R. P., 1981, Altered spectrin dimer–dimer association and instability of erythrocyte membrane skeletons in hereditary pyropoikilocytosis, *J. Clin. Invest.* **68:**597–605.

Liu, S.-C., Palek, J., and Prchal, J. T., 1982, Defective spectrin dimer–dimer association in hereditary elliptocytosis. *Proc. Natl. Acad. Sci. USA* **79:**2072–2076.

Liu, S.-C., Windisch, P., Kim, S., and Palek, J., 1984, Oligomeric states of spectrin in normal erythrocyte membranes: Biochemical and electron microscopic studies, *Cell* **37:**587–594.

Liu, S.-C., Derick, L. H., and Palek, J., 1987, Visualization of the hexagonal lattice in the erythrocyte membrane skeleton, *J. Cell Biol.* **104:**527–536.

Lux, S. E., 1979, Spectrin–actin membrane skeleton of normal and abnormal red blood cells, *Semin. Hematol.* **16:**22–51.

Marchesi, S. L., Steers, E., Marchesi, V. T., and Tillack, T. W., 1970, Physical and chemical properties of a protein isolated from red cell membranes, *Biochemistry* **9:**50–57.

Marchesi, V. T., 1985, Stabilizing infrastructure of cell membranes, *Annu. Rev. Cell Biol.* **1:**531–561.

Marchesi, V. T., and Steers, E., 1968, Selective solubilization of a protein component of the red cell membrane, *Science* **159:**203–204.

Marinetti, G. V., and Crain, R. C., 1978, Topology of amino-phospholipids in the red cell membrane, *J. Supramol. Struct.* **8:**191–213.

Matsuzaki, F., Sutoh, K., and Ikai, A., 1985, Structural unit of the erythrocyte cytoskeleton. Isolation and electron microscopic examination, *Eur. J. Cell Biol.* **39:**153–160.

Minton, A. P., 1983, The effect of volume occupancy upon the thermodynamic activity of proteins: Some biochemical consequences, *Mol. Cell. Biochem.* **55:**119–140.

Mische, S. M., Mooseker, M. S., and Morrow, J., 1987, Erythrocyte adducin: A calmodulin-regulated actin-bundling protein that stimulates spectrin–actin binding, *J. Cell Biol.* **105:**2837–2845.

Mombers, C., Van Dijck, P. W. M., Van Deenen, L. L. M., De Gier, J., and Verkleij, A., 1977, The interaction of spectrin–actin and synthetic phospholipids, *Biochim. Biophys. Acta* **470:**152–160.

Morris, M. B., and Ralston, G. B., 1984, A reappraisal of the self-association of human spectrin, *Biochim. Biophys. Acta* **788:**132–137.

Morris, M. B., and Ralston, G. B., 1985, Determination of the parameters of protein self-association by direct fitting to the omega function, *Biophys. Chem.* **23:**49–61.

Morrow, J. S., and Marchesi, V. T., 1981, Self-assembly of spectrin oligomers *in vitro:* Basis for a dynamic cytoskeleton, *J. Cell Biol.* **88:**463–468.

Morrow, J. S., Speicher, D. W., Knowles, W. J., Hsu, C. J., and Marchesi, V. T., 1980, Identification of functional domains of human erythrocyte spectrin, *Proc. Natl. Acad. Sci. USA* **77:**6592–6596.

Mueller, T. J., and Morrison, M., 1981, Glycoconnectin (PAS 2) a membrane attachment site for the human erythrocyte cytoskeleton, in: *Erythrocyte Membranes 2: Recent Clinical and Experimental Advances* (W. C. Kruckeberg, J. W. Eaton, and G. J. Brewer, eds.), pp. 95–112, Liss, New York.

Nakashima, K., and Beutler, E., 1979, Comparison of structure and function of human erythrocyte and human muscle actin, *Proc. Natl. Acad. Sci. USA* **76:**935–938.

Nishizuka, Y., 1986, Studies and perspectives of protein kinase C, *Science* **233:**305–312.

Ohanian, V., and Gratzer, W. B., 1984, Preparation of red-cell membrane cytoskeletal constituents and characterization of protein 4.1, *Eur. J. Biochem.* **144:**375–379.

Ohanian, V., Wolfe, L. C., John, K. M., Pinder, J. C., Lux, S. E., and Gratzer, W. B., 1984, Analysis of the ternary interaction of the red cell membrane skeletal proteins spectrin, actin, and 4.1, *Biochemistry* **23:**4416–4420.

Owens, W., Mueller, T. J., and Morrison, M., 1980, A minor sialoglycoprotein of the human erythrocyte membrane, *Arch. Biochem. Biophys.* **204:**247–254.

Palek, J., and Liu, S.-C., 1981, Alterations of spectrin assembly in the red cell membrane: Functional consequences. *Scand. J. Clin. Lab. Invest.* **41**(Suppl. 156)**:**131–138.

Pasternack, G. R., Anderson, R. A., Leto, T. L., and Marchesi, V. T., 1985, Interactions between protein 4.1 and band 3. An alternative binding site for an element of membrane skeleton, *J. Biol. Chem.* **260:**3676–3683.

Patel, V. P., and Fairbanks, G., 1981, Spectrin phosphorylation and shape change of human erythrocyte ghosts, *J. Cell Biol.* **88:**430–440.

Patel, V. P., and Fairbanks, G., 1986, Relationship of major phosphorylation reactions and MgATPase activities to ATP-dependent shape change of human erythrocyte membranes, *J. Biol. Chem.* **261:**3170–3177.

Pinder, J. C., and Gratzer, W. B., 1983, Structural and dynamic states of actin in the erythrocyte, *J. Cell Biol.* **96:**768–775.

Pinder, J. C., Bray, D., and Gratzer, W. B., 1975, Actin polymerization induced by spectrin, *Nature* **258:**765–766.

Pinder, J. C., Ungewickell, E., Bray, D., and Gratzer, W. B., 1978a, The spectrin–actin complex and erythrocyte shape, *J. Supramol. Struct.* **8:**439–445.

Pinder, J. C., Bray, D., and Gratzer, W. B., 1978b, Control of interaction of spectrin and actin by phosphorylation, *Nature* **270:**752–754.

Pinder, J. C., Ungewickell, E., Calvert, R., Morris, E., and Gratzer, W. B., 1979, Polymerization of G-actin by spectrin preparations: Identification of the active constituent, *FEBS Lett.* **104:**396–400.

Pinder, J. C., Clerk, S. E., Baines, A. J., Morris, E., and Gratzer, W. B., 1981, The construction of the red cell cytoskeleton, in: *The Red Cell: Fifth Ann Arbor Conference* (G. M. Brewer, ed.). pp. 343–354, Liss, New York.

Pinder, J. C., Ohanian, V., and Gratzer, W. B., 1984, Spectrin and protein 4.1 as an actin filament capping complex, *FEBS Lett.* **169:**161–164.

Podgorski, A., and Elbaum, D., 1985, Properties of red cell membrane proteins: Mechanism of spectrin and band 4.1 interaction, *Biochemistry* **24:**7871–7876.

Podolski, J. L., and Steck, T. L., 1988, Association of deoxyribonuclease I with the pointed ends of actin filaments in human red blood cell membrane skeletons, *J. Biol. Chem.* **263:**638–645.

Pollard, T. D., and Cooper, J. A., 1986, Actin and actin-binding proteins, *Annu. Rev. Biochem.* **55:**987–1035.

Portis, A., Newton, C., Pangborn, W., and Papahadjopoulos, D., 1979, Studies on the mechanism of membrane fusion: Evidence for an intermembrane Ca^{2+}–phospholipid complex, synergism with Mg^{2+}, and inhibition by spectrin, *Biochemistry* **18:**780–790.

Quist, E., 1980, Regulation of erythrocyte membrane shape by Ca^{2+}, *Biochem. Biophys. Res. Commun.* **92**:631–637.

Ralston, G. B., 1975, The isolation of aggregates of spectrin from bovine erythrocyte membranes, *Aust. J. Biol. Sci.* **28**:259–266.

Ralston, G. B., 1978, Physical chemical studies of spectrin, *J. Supramol. Struct.* **8**:361–374.

Ralston, G. B., and Crisp, E. A., 1981, The action of organic mercurials on the erythrocyte membrane, *Biochim. Biophys. Acta* **649**:98–104.

Ralston, G. B., Dunbar, J. C., and White, M. D., 1977, The temperature dependent dissociation of spectrin, *Biochim. Biophys. Acta* **491**:345–348.

Schatzman, H. J., 1975, Active calcium transport and Ca^{2+}-activated ATPase in human red cells, *Curr. Top. Membr. Transp.* **6**:125–168.

Schindler, M., Koppel, D. E, and Sheetz, M. P., 1980, Modulation of membrane protein lateral mobility by polyphosphates and polyamines, *Proc. Natl. Acad. Sci. USA* **77**:1457–1461.

Sears, D. E., Marchesi, V. T., and Morrow, J. S., 1986, A calmodulin and α-subunit binding domain in human erythrocyte spectrin, *Biochim. Biophys. Acta* **870**:432–442.

Shahbakhti, F., and Gratzer, W. B., 1986, Analysis of the self-association of human red cell spectrin, *Biochemistry* **25**:5969–5975.

Sheetz, M. P., 1979, Integral membrane protein interaction with Triton cytoskeletons of erythrocytes, *Biochim. Biophys. Acta* **557**:122–134.

Sheetz, M. P., and Casaly, J., 1981, Phosphate metabolite regulation of spectrin interactions, *Scand. J. Clin. Lab Invest.* **41**(Suppl. 156):117–122.

Sheetz, M. P., and Sawyer, D., 1978, Triton shells of intact erythrocytes, *J. Supramol. Struct.* **8**:399–412.

Sheetz, M. P., and Singer, S. J., 1977, On the mechanism of ATP-induced shape changes in human erythrocyte membranes, *J. Cell Biol.* **73**:638–646.

Sheetz, M. P., Painter, R. G., and Singer, S. J., 1976, Relationships of the spectrin complex of human erythrocyte membranes to the actomyosins of muscle cells, *Biochemistry* **15**:4486–4492.

Shen, B. W., Josephs, R., and Steck, T. L., 1984, Ultrastructure of unit fragments of the skeleton of the human erythrocyte membrane, *J. Cell Biol.* **99**:810–821.

Shen, B. W., Josephs, R., and Steck, T. L., 1986, Ultrastructure of the intact skeleton of the human erythrocyte membrane, *J. Cell Biol.* **102**:997–1006.

Shiffer, K. A., and Goodman, S. R., 1984, Protein 4.1: Its association with the human erythrocyte membrane, *Proc. Natl. Acad. Sci. USA* **81**:4404–4408.

Shiffer, K. A., Goerke, J., Duzgunes, N., Fedor, J., and Shohet, S. B., 1988, Interactions of erythrocyte protein 4.1 with phospholipids. A monolayer and liposome study, *Biochim. Biophys. Acta* **937**:269–280.

Shohet, S. B., 1979, Reconstitution of spectrin-deficient, spherocytic mouse erythrocyte membranes, *J. Clin. Invest.* **64**:483–494.

Shotton, D. M., Burk, B. E., and Branton, D., 1979, The molecular structure of human erythrocyte spectrin. Biophysical and electron microscopic studies, *J. Mol. Biol.* **131**:303–329.

Siegel, D. L., and Branton, D., 1985, Partial purification and characterization of an actin-bundling protein, band 4.9, from human erythrocytes, *J. Cell Biol.* **100**:775–785.

Sobue, K., Muramoto, Y., Fujita, M., and Kakiuchi, S., 1981, Calmodulin-binding protein of erythrocyte cytoskeleton, *Biochem. Biophys. Res. Commun.* **100**:1063–1070.

Sondag, D., Alloisio, N., Blanchard, D., Ducluzeau, M.-T., Colonna, P., Bachir, D., Bloy, C., Cartron, J.-P., and Delaunay, J., 1987, Gerbich reactivity in 4.1 (−) hereditary elliptocytosis and protein 4.1 level in blood group Gerbich deficiency, *Br. J. Haematol.* **65**:43–50.

Speicher, D. W., 1986, The present status of erythrocyte spectrin structure: The 106–residue repetitive structure is a basic feature of an entire class of proteins, *J. Cell Biochem.* **30**:245–258.

Speicher, D. W., and Marchesi, V. T., 1984, Erythrocyte spectrin is comprised of many homologous triple helical segments, *Nature* **311**:177–180.

Speicher, D. W., Morrow, J. S., Knowles, W. J., and Marchesi, V. T., 1980, Identification of proteolytically resistant domains of human erythrocyte spectrin, *Proc. Natl. Acad. Sci. USA* **77**:5673–5677.

Speicher, D. W., Morrow, J. S., Knowles, W. J., and Marchesi, V. T., 1982, A structural model of human erythrocyte spectrin, *J. Biol. Chem.* **257**:9093–9101.

Steck, T. L., 1974, Organization of proteins in the human red blood cell membrane, *J. Cell Biol.* **62**:1–19.

Stokke, B. T., and Elgsaeter, A., 1981, Human spectrin VI. A viscometric study, *Biochim. Biophys. Acta* **640**:640–645.

Stromqvist, M., Backman, L., and Shanbhag, V., 1985, Effect of spectrin dimer on actin polymerization, *FEBS Lett.* **190**:15–20.

Takakuwa, Y., and Mohandas, N., 1988, Modulation of erythrocyte membrane material properties by Ca^{2+} and calmodulin, *J. Clin. Invest.* **82**:394–400.

Tilley, L., and Ralston, G. B., 1984, Purification and kinetic characterization of human erythrocyte actin, *Biochim. Biophys. Acta* **790**:46–52.

Tilley, L., and Ralston, G. B., 1987, Effect of erythrocyte spectrin on actin self-association, *Aust. J. Biol. Sci.* **40**:27–36.

Tilney, L. G., and Detmers, P., 1975, Actin in erythrocyte ghosts and its association with spectrin, *J. Cell Biol.* **66**:508–520.

Tsukita, S., Tsukita, S., and Ishikawa, H., 1980, Cytoskeletal network underlying the human erythrocyte membrane, *J. Cell Biol.* **85**:567–576.

Tsukita, S., Tsukita, S., and Ishikawa, H., 1984, Bidirectional polymerization of g-actin on the human erythrocyte membrane, *J. Cell Biol.* **98**:1102–1110.

Tyler, J., Hargreaves, W., and Branton, D., 1979, Purification of two spectrin binding proteins: Biochemical and electron microscopic evidence for site-specific reassociation between spectrin and band 2.1 and 4.1, *Proc. Natl. Acad. Sci. USA* **76**:5192–5196.

Tyler, J. M., Reinhardt, B. N., and Branton, D. 1980, Associations of erythrocyte membrane proteins: Binding of purified bands 2.1 and 4.1 to spectrin, *J. Biol. Chem.* **255**:7034–7039.

Ungewickell, E., and Gratzer, W. B., 1978, Self-association of human spectrin. A thermodynamic and kinetic study, *Eur. J. Biochem.* **88**:379–385.

Ungewickell, A., Bennett, P. M., Calvert, R., Ohanian, V., and Gratzer, W. B., 1979, *In vitro* formation of a complex between cytoskeletal proteins of the human erythrocyte, *Nature* **280**:811–814.

Weed, R. I., LaCelle, P. L., and Merrill, E. W., 1969, Metabolic dependence of red cell deformability, *J. Clin. Invest.* **48**:795–809.

Whitfield, C. F., Culp, E. N., and Goodman, S. R., 1986, Transfer of label from protein 4.1-crosslinker complex to 4.1 membrane binding sites, *J. Cell Biol.* **103**:542a.

Wolfe, L. C., John, K. M., Falcone, J. C., Byrne, A. M., and Lux, S. E., 1982, A genetic defect in the binding of protein 4.1 to spectrin in a kindred with hereditary spherocytosis, *N. Engl. J. Med.* **307**:1367–1374.

Yu, J., Fischman, D. A., and Steck, T. L., 1973, Selective solubilization of proteins and phospholipids from red blood cell membranes by nonionic detergents, *J. Supramol. Struct.* **1**:233–248.

Chapter 9

Some High-Molecular-Weight Oligomeric Proteins and Enzymes of Reticulocytes and Erythrocytes

J. Robin Harris

1. INTRODUCTION

The mammalian erythrocyte has occupied a position at the forefront of enzymology since the early days of physiological chemistry. Indeed, the erythrocyte continues to occupy a place of prominence, with respect to the enzymology of both its membranous and cytosolic systems (Beutler, 1975, 1986; Brewer, 1984; Friedman and Rapoport, 1974; Schrier, 1977).

Within this review, examples of several high-molecular-weight membrane-bound, membrane-associated, and cytosolic enzymes and proteins will be discussed. A somewhat arbitrary minimum native molecular weight or relative molecular mass (M_r) of 200,000 has been imposed for most of the examples selected. This molecular weight cutoff does, of necessity, mean that many enzymes and proteins of biochemical interest will be excluded. It is, nevertheless, hoped that the material presented will be of interest and relevance to those working on the biochemistry of the erythrocyte. The prime reason for this approach to the subject has been the author's long-standing interest in the application of transmission electron microscopy (TEM) to the study of biological macromolecules and the contribution that this very direct biophysical technique can make when integrated with

Abbreviations used in this chapter: AChE, acetylcholinesterase; eIF, eukaryotic initiation factor; G3PD, glyceraldehyde-3-phosphate dehydrogenase; HS, hereditary spherocytosis; M_r, relative molecular mass (molecular weight); PAGE, polyacrylamide gel electrophoresis; PNH, paroxysmal nocturnal hemoglobinuria; PRPP, phosphoribosylpyrophosphate; ScRNP, small cytoplasmic ribonucleoprotein; SDS-PAGE, sodium dodecyl sulfate polyacrylamide gel electrophoresis; TEM, transmission electron microscopy; TPP II, tripeptidyl peptidase II.

J. Robin Harris North East Thames Regional Transfusion Centre, Brentwood, Essex CM15 8DP, England.

the more conventional biochemical techniques such as polyacrylamide gel electrophoresis (PAGE), density gradient centrifugation, and column chromatography.

This theme will underlie much of what follows. Protein molecules of M_r less than 200,000 are rather difficult to define ultrastructurally, whereas a large number of molecules of M_r greater than this value have been shown in negatively stained specimens to have discrete and readily recognizable quaternary conformations. That TEM has actually contributed to the discovery of a number of erythrocyte proteins indicates further the value of this technique, which should perhaps be given more serious overall consideration by protein biochemists and enzymologists in general.

Early attempts to produce erythrocyte ghosts by low-ionic-strength hemolysis and washing generally yielded a product that was grossly contaminated with hemoglobin; indeed, hemoglobin has even been considered to be an integral part of the membrane. This situation was largely overcome as a result of the work of Dodge *et al.* (1963) in which the ionic parameters and pH conditions were defined (~ 10 mM phosphate buffer, pH 7.4) for the maximal release of hemoglobin, with retention of the biconcave morphology of the erythrocyte ghosts. That hemoglobin and other cytosolic proteins can interact with the internal/cytoplasmic surface of the erythrocyte membrane (i.e., with membrane skeletal and other exposed integral membrane proteins) is clearly implied (Mitchell *et al.*, 1965; Eisinger *et al.*, 1982; Salhany, 1983; Lilley and Fung, 1987). Although the concept that hemoglobin might be an integral part of the erythrocyte membrane has been proposed by Hollan *et al.* (1977), it had earlier been concluded by Weed *et al.* (1963) that hemoglobin was not a structural component of the erythrocyte membrane, a view that has been widely supported by more recent work. In our own studies (Harris, 1968, 1969a,b), it was shown that hemolysis and centrifugal washing of human erythrocyte ghosts with cold (4°C) 10 mM phosphate buffer (pH 7.4) generally produced a slightly pink product, but that further washing in 10 mM Tris-HCl buffer (pH 7.4) yielded a completely hemoglobin-free suspension of erythrocyte ghosts. In hindsight, it is clear that erythrocyte ghosts washed with 10 mM phosphate buffer (pH 7.4) have a tendency to reseal, thereby entrapping hemoglobin and cytoplasmic proteins (Tillmann *et al.*, 1975) and that further washing in Tris-HCl buffer (pH 7.4) reopens the membrane lesions allowing these proteins to be released. Support for this concept comes from the investigation of Brown and Harris (1970) in which phosphate-buffer-washed erythrocyte ghosts were shown to present a permeability barrier to externally added ferritin molecules, whereas ghosts subsequently washed in Tris-HCl buffer allowed the entry of ferritin molecules. Significantly, it was possible to reseal the Tris-HCl-washed ghosts, if they were transferred back to phosphate buffer and Ca^{2+} (1 mM) was added to the suspension, but this did not occur when Ca^{2+} (1–6 mM) was added to the ghosts in Tris-HCl buffer. That osmotic hemolysis does indeed produce a large number of small (20–100 nm) membrane lesions and that they have a tendency to reseal were shown by Seeman (1967, 1974) and Seeman *et al.* (1973). This work has been challenged by more recent studies (Lieber and Steck, 1982a,b) that propose the existence of large static lesions, but it is, nevertheless, clear that erythrocyte ghosts have the property of "molecular sieves" and that they thereby tend to entrap the higher-molecular-weight cytosolic proteins (Harris, 1984; MacGregor and Tobias, 1972). The precise ionic and temperature conditions under which spontaneous or experimentally induced resealing occur are obviously of great importance within this context (Bramley *et al.*, 1971), yet even now are not fully defined.

Another possibility that must be considered is that the transfer from an isotonic to a low-ionic-strength environment that occurs at the time of hemolysis and is maintained throughout the washing scheme may itself promote adsorption of certain cytosolic proteins onto the erythrocyte membrane or the membrane skeleton. This has been shown to be particularly so when Ca^{2+} is added to the cells before hemolysis or to the hemolysis and washing buffers (Allen and Cadman, 1979). This situation is thought to mimic a number of hemolytic anemias in which there is an increased intracellular Ca^{2+} level, although this is apparently not the case with ATP-depleted normal erythrocytes and with pyruvate kinase-deficient reticulocytes and mature cells, which also adsorb increased quantities of cytosolic proteins (Allen *et al.*, 1983).

The relevance of the above discussion of erythrocyte ghost permeability and protein adsorption to a number of the enzymes presented in this review will become self-evident throughout several of the sections below, but for the moment it serves to underline the difficulty of making precise classifications as to the localization of some of these proteins and the interactions they may have with the erythrocyte membrane and/or the membrane skeleton.

Clearly a complete survey of the enzymology of the erythrocyte is not possible in the space available. Thus, in this review I will concentrate on a very limited number of high-molecular-weight oligomeric proteins and enzymes isolated primarily from the human erythrocyte, the study of some of which over the past years I have been personally involved with.

2. PROTEASES, PROSOMES, CYLINDRIN, AND PEPTIDASES

2.1. Proteases

The presence of a number of neutral and acidic proteases within the erythrocyte and associated with its membrane has been known for many years (Ballas and Burka, 1979; Bernacki and Bosmann, 1972; Burkholder and Brecher, 1972; McGuire and DeMartino, 1986; Melloni *et al.*, 1982; Murakami *et al.*, 1979; Pontremoli *et al.*, 1979; Scott and Kee, 1979; Tarone *et al.*, 1979; Wang *et al.*, 1988; Witheiler and Willson, 1972; Yamamoto *et al.*, 1988). That increased adsorption of cytosolic proteases onto the erythrocyte membrane occurs in the presence of calcium ions has also been indicated by the work of Anderson *et al.* (1977) and Allen and Cadman (1979).

Since the late 1970s, studies on reticulocyte proteases have led to an understanding of the energy-dependent and independent proteolytic processes occurring in the non-lysosomal protein degradation pathway (Boches and Goldberg, 1982; Etlinger and Goldberg, 1977; Muller *et al.*, 1980; Tanaka *et al.*, 1983; Waxman *et al.*, 1987) and the involvement of the heat-stable low-molecular-weight protein "ubiquitin." Further details of this ubiquitin-dependent system are presented in Chapter 6 (this volume).

Recently, however, it has been shown in reticulocytes (Hough *et al.*, 1987; Tanaka *et al.*, 1986) that a very large 26 S ($M_r \sim 10^6$) protease is implicated in the ATP-dependent degradation of ubiquitin–protein conjugates; and that a smaller 20 S enzyme complex ($M_r \sim 7 \times 10^5$) is present that may or may not require ATP for its proteolytic activity, depending on the degree of erythrocytic maturation. A high-molecular-weight protease

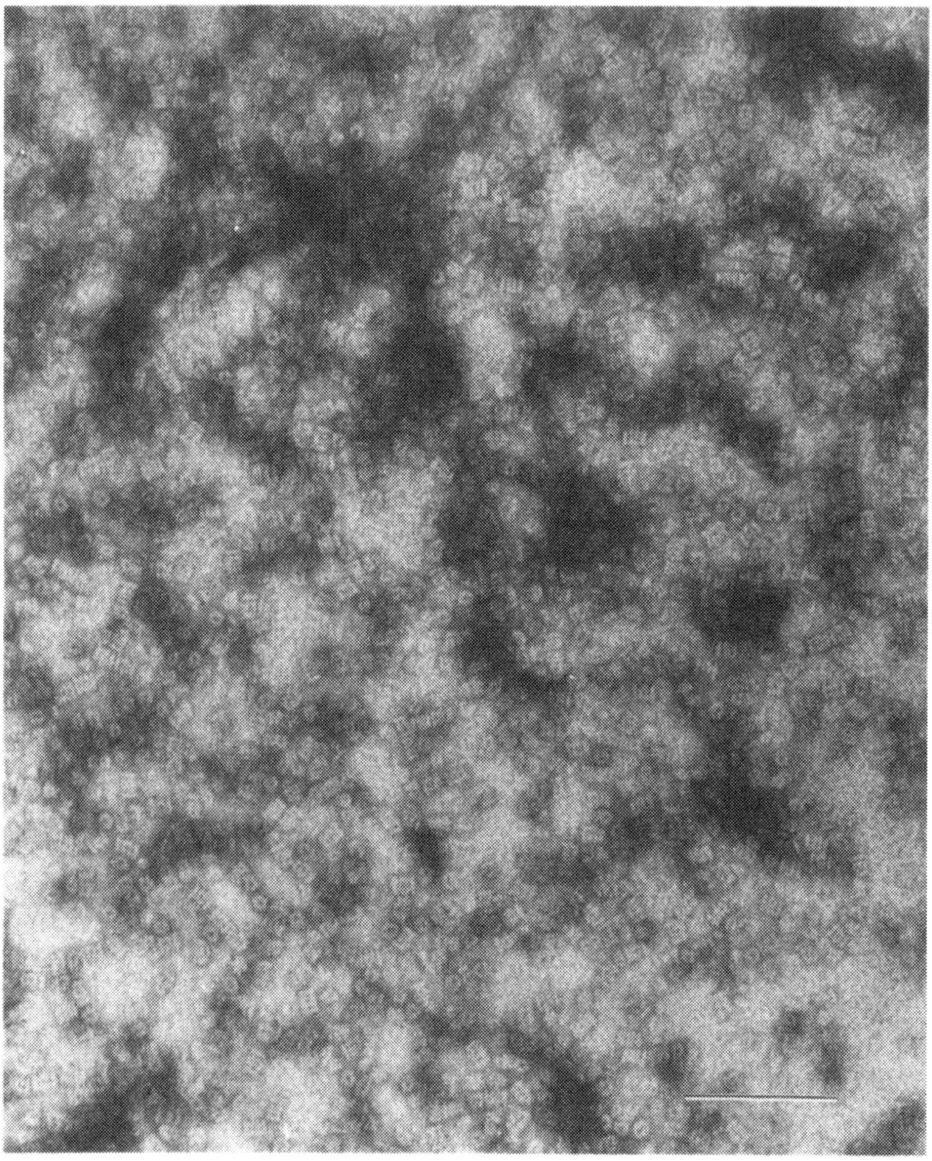

FIGURE 1. A transmission electron micrograph of randomly spread human erythrocyte 20 S multicatalytic protease (purified by Dr. R. Hough, University of Utah). Negatively stained with 2% uranyl acetate. Bar = 100 nm.

complex of identical composition to the 20 S reticulocyte protease has also been isolated from mature erythrocytes by Ohkubo *et al.* (1988) and by Hough *et al.* (previously unpublished data), as shown in Figure 1. In the mature erythrocyte, the 20 S protease apparently does not require ATP. An identical protease, termed "ingensin," has been isolated independently from K562 erythroleukemia cells by Tsukahara *et al.* (1988). The polypeptide composition of the 20 S reticulocyte protease revealed by SDS-PAGE indicates that it is equivalent to the multicatalytic protease complex isolated from muscle and a number of other tissues (Arrigo *et al.*, 1988; Fagan *et al.*, 1987; Falkenburg *et al.*, 1988; Hendil, 1988; Tanaka *et al.*, 1986; 1988a,b; Wilk and Orlowski, 1983). In all cases this 20 S protease complex has been found to contain up to ten subunits, all within the M_r range 21,000 to 32,000, which provides a very characteristic electrophoretic pattern.

2.2. Prosomes

For technical and logistic reasons, the above presentation on the 20 S multicatalytic protease complex must be considered separately from any discussion of the 19 S "prosome" ScRNP particle. "Prosome" is the term given initially by Klaus Scherrer and his colleagues (Schmid *et al.*, 1984) to the duck and mouse erythroid, and HeLa, ScRNP particle shown by TEM to possess a cylindrical quaternary conformation (Figure 2). This ScRNP appears, to all intents and purposes, to be identical to the 20 S multicatalytic protease (Section 2.1) and to cylindrin (Section 2.3). It should, however, be emphasized that the studies on the 19 S prosome differ fundamentally from the other two topics in that the prosome is always isolated as a subcomponent from previously purified 20 S repressed

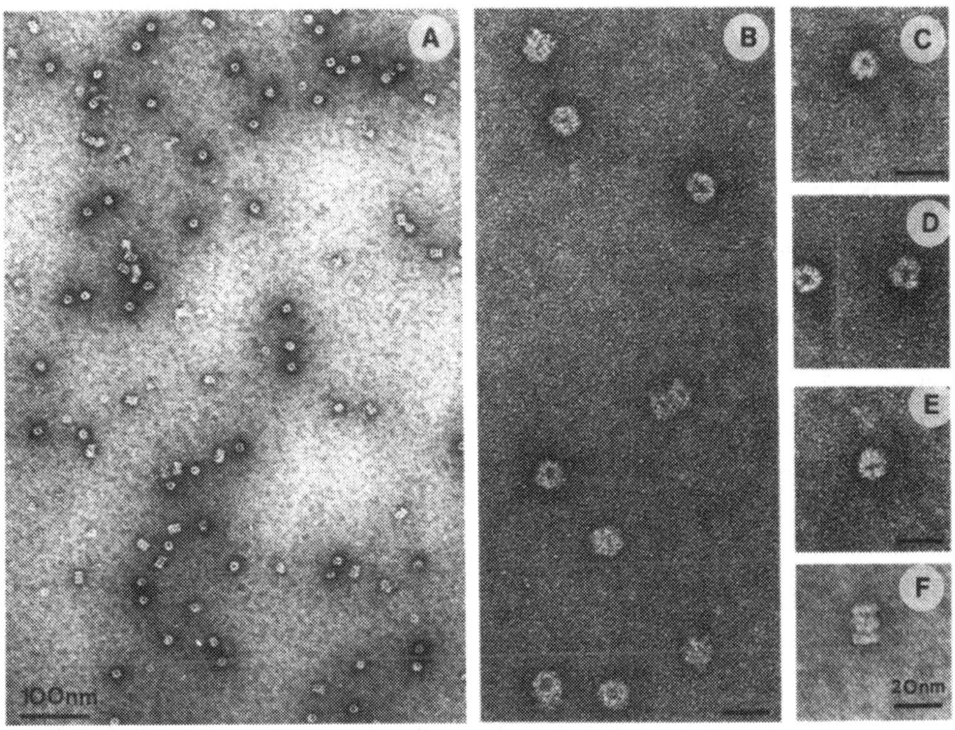

FIGURE 2. Transmission electron micrographs of prosome particles isolated from duck globin mRNP. A survey micrograph (A) shows the two orientations of the cylindrical prosome (viz. circular and rectangular images). At higher magnifications (B–F) the varying morphology of individual particles is more clearly shown. All material is negatively stained with uranyl acetate. Bar = 100 nm (A) and 20 nm (B–F). Micrographs were produced by Dr. F. Puvion, Villejuif, using prosomes prepared by K. Scherrer and colleagues (Schmid *et al.*, 1984; Martins de Sa *et al.*, 1986).

globin mRNP (Granboulan *et al.*, 1970; Schmid *et al.*, 1984), by treatment with 0.5 M KCl or EDTA. The "core" globin mRNP can be separated from the 19 S cylindrical prosome, which has been found to retain two specific small cytoplasmic RNA (ScRNA) molecules containing 70–90 nucleotides. This RNA is apparently internalized to the extent that it is protected from exogenous RNase. It is a remarkable coincidence that apart from possessing a similar quaternary structure to the 20 S protease and cylindrin, the prosome particle also possesses subunits (ca. 10) in the M_r range 20,000 to 30,000, although there may be differences in the minor polypeptide components. Also, the prosome particle has the ability to inhibit protein synthesis *in vitro*, a property used as a major characterizing feature.

Further studies on the cylindrical 19 S prosome have shown its presence in a number of different species and cell types (Akhayat *et al.*, 1987a; Falkenburg *et al.*, 1988; Kloetzel *et al.*, 1987; Kremp *et al.*, 1986; Martins de Sa *et al.*, 1986). Although there appear to be variations in the ScRNA components, the quaternary conformation and polypeptide composition remain extremely consistent. The equivalence of the 20 S protease and the 19 S prosome has been clearly proposed by Arrigo *et al.* (1988), who have introduced the term "proteasome," but this is not currently accepted by Scherrer and his colleagues (personal communication) who consider that their prosome possesses the characteristic features mentioned above as well as proteolytic activity. For the sake of completeness, it should perhaps be mentioned that similar 19 S cylindrical complexes had, previous to the prosome work, been isolated from *Xenopus* oocytes (Kleinschmidt *et al.*, 1983), HeLa and a number of other cultured cell lines (Domae *et al.*, 1982; Harmon *et al.*, 1983; Narayan and Rounds, 1973; Smulson, 1974) as well as several animal and plant cell types (Shelton *et al.*, 1970), a number of which were apparently associated with ribosomes.

Evidence for a new type of duck erythroblast prosome (M_r 600,000) was advanced by Akhayat *et al.* (1987b). In this instance, translationally repressed nonglobin mRNA-containing 35 S mRNP particles were dissociated by 0.5 M KCl or 0.2% SDS to yield a 32 S core complex that contained the mRNA, and an approximately 12-nm spherical RNP particle. This RNP particle contains a single, approximately 21,000 subunit and an ScRNA component with 80 to 100 nucleotides. Again, this new prosome particle has found to inhibit mRNA translation *in vitro* and the proposition was therefore advanced that it may be involved in the regulation of translation.

From the extensive literature on the characterization of ferritin and apoferritin, which have long been known to be predominant proteins within erythroid cells (Bessis and Breton-Gorius, 1959), it seems possible that they may equate with the new spherical prosome. Ultrastructurally, apoferritin is revealed as an approximately 12-nm spherical particle (Harris, 1982a,b) that sediments at 17 S (M_r 445,000). Clearly, if there is any iron hydroxide accumulation within the apoferritin spherical shell, there will be a progressive increase in the sedimentation coefficient and M_r. This may account for the M_r of 600,000 for the spherical prosome, since from its molecular dimensions alone this would seem to be excessive.

2.3. Cylindrin

All of the recent work on the 20 S protease complex correlates very well with the studies of the author on the 22.5 S erythrocyte protein that has been termed "cylindrin"

(Harris, 1980). The structural and biochemical parallel between cylindrin, the 20 S protease, and the prosome particle has been expounded in some detail by Harris (1988b). Thus, only a brief outline will now be presented. Cylindrin was discovered by TEM in 1967, along with the somewhat smaller ringlike protein "torin" (see Section 6.2), in extracts obtained from human and cattle erythrocyte ghosts. Although it was initially thought that these proteins were associated with the erythrocyte membrane, it is now believed that they are primarily cytoplasmic, but that they tend to become entrapped within the hemolysed erythrocyte and retained throughout the washing scheme used to remove the hemoglobin, due to molecular sieving (Harris, 1984). In a series of publications, some of the biochemical characteristics of cylindrin were defined (Harris, 1968, 1969a,b, 1971, 1974, 1980; Harris and Naeem, 1981); these have now received considerable support from the data of others (Howe and Bächi, 1973; Lande *et al.*, 1984; Malech and Marchesi, 1981; White and Ralston, 1979). The characteristic quaternary conformation of cylindrin (Figure 3) is clearly seen to be identical to that of the 20 S protease complex and the prosome particle (see Figures 1 and 2). The dimensions of the cylindrical complex are ~ 12-nm external diameter/width and ~ 17-nm length, with eight-fold rotational symmetry (Harris, 1969b, 1980, 1988b). Although electron optical images of this molecular complex do vary slightly in different negative stains, there is remarkable uniformity of the images in the published electron micrographs from the different groups. Indeed, from the author's experience gained from the study of many different macromolecules by TEM (Harris, 1982b), it would seem very likely that they all represent the same macromolecular complex. A recent electron microscopic investigation by Baumeister *et al.* (1988) has utilized digital image analysis to study the structure of the skeletal muscle 20 S multicatalytic protease. In general, the interpretation of the electron optical images advanced by these workers agrees strongly with that advanced previously by the present author using erythrocyte cylindrin. The sedimentation coefficient of human erythrocyte cylindrin was determined to be 22.5 S by Harris (1971), but White and Ralston (1979) obtained the slightly lower figure of 19.3 S. Estimates of the molecular weight likewise show some discrepancy (810,000 ± 21,000, Harris, 1980; 747,000 ± 38,000, White and Ralston, 1979). PAGE under nondissociating conditions reveals a slowly migrating band of protein, which positions itself between the apoferritin dimer and the thyroglobulin monomer on gradient (exclusion) gels. Following dissociation with SDS, cylindrin is split into at least five subunits, of M_r in the approximate range 22,000 to 32,000 (Harris and Naeem, 1981), as shown in Figure 4. This subunit composition was in general supported and extended by the studies of Malech and Marchesi (1981) and Lande *et al.* (1984). The latter group also performed two-dimensional electrophoretic analysis, from which a more complex subunit composition emerged showing 17 spots. Studies on the 20 S multicatalytic protease, which reveal a larger number of subunits (i.e., ten), appear to have obtained this figure because of a superior electrophoretic resolution, which separates each of the five principal subunits (Figure 4; also Malech and Marchesi, 1981) into two (Arrigo *et al.*, 1988). The isoelectric point of native cylindrin was determined by Harris and Naeem (1981) to be 4.6, a figure very close to that obtained for the 20 S protease by Hough *et al.* (1987).

Clearly, the situation as it stands is not totally satisfactory. Further immunological and enzymological comparisons of cylindrin and the 20 S protease/prosome particle must be performed, together with high-resolution electron microscopic analysis. This latter approach may well be achieved by the formation of two-dimensional crystals, as indicated

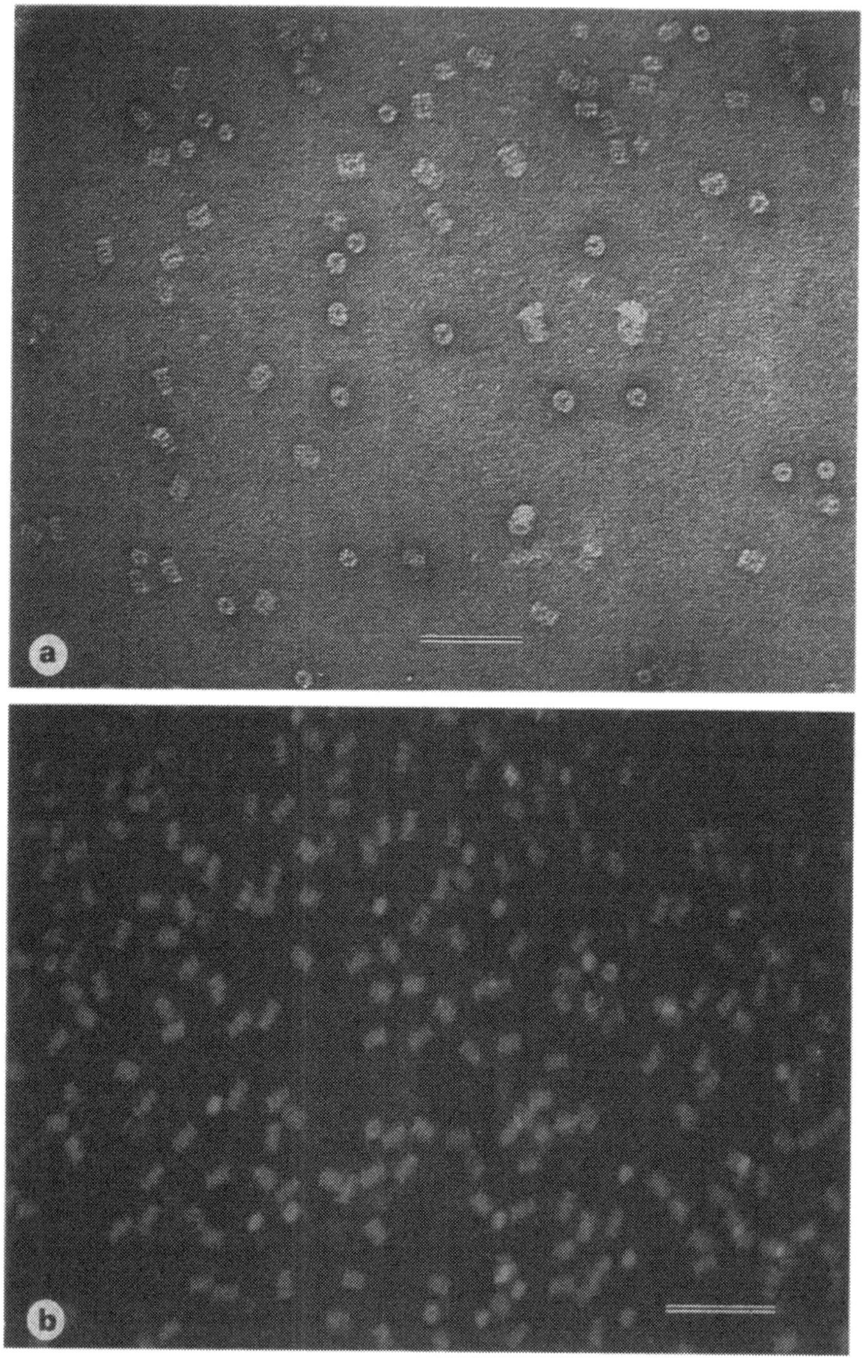

FIGURE 3. A transmission electron micrograph of human erythrocyte cylindrin (a) and rabbit reticulocyte aminoacyl-tRNA synthetase (b). Micrograph b was provided by Mr. R. Oliver and prepared from an enzyme sample purified by Dr. B. Hardesty, both of the University of Texas at Austin. It should be noted that the claim that this cylindrical protein complex is an oligomeric aminoacyl-tRNA synthetase complex is not supported by recent studies. The sample in micrograph a is negatively stained with 2% uranyl acetate and in b with 2% methylamine tungstate. Bars = 50 nm.

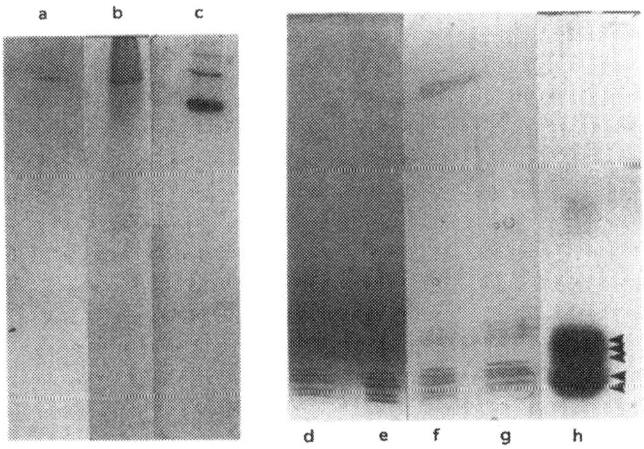

FIGURE 4. PAGE of human erythrocyte cylindrin and human erythrocyte 20 S multicatalytic protease complex. Channels a–c show electrophoresis on a 2.5 to 27% polyacrylamide gradient gel of native human erythrocyte cylindrin, human erythrocyte 20 S multicatalytic protease, and horse apoferritin (monomer, dimer, and trimer), respectively. Channels d–h show SDS-PAGE on a 2.5 to 27% polyacrylamide gradient gel of human erythrocyte cylindrin and human erythrocyte 20 S multicatalytic protease, respectively.

in Figure 5, prepared by the mica-spreading "negative-staining carbon film" technique of Horne and Pasquali-Ronchetti (1974). Nevertheless, it can be stated with some degree of confidence that a remarkable correlation of data obtained from widely diverse tissues by several different research groups is now being achieved. The question as to whether proteolytic activity is the true/major physiological function of cylindrin, and indeed of the 20 S protease and the prosome particle, remains to be fully established. For instance, it has been shown by Castaño *et al.* (1986) that a 16 S cylindrical complex, apparently identical to cylindrin and the 20 S protease complex, possesses pre-tRNA 5′ processing nuclease activity (see also Section 3).

2.4. Peptidases

Although there have been a number of reports of erythrocytic peptidase (EC 3.4.11–3.4.19) (Abramic *et al.*, 1988; Witheiler and Wilson, 1972), within this review I will restrict myself to a discussion of the recent exciting work on human erythrocyte tripeptidyl peptidase II (TPP II) (Bålöw *et al.*, 1986; Macpherson *et al.*, 1987). This enzyme has also been purified to homogeneity from rat liver by the same group of workers.

TPP II is classified as an extralysosomal serine exopeptidase, and is unique in possessing an extremely large native M_r ($>10^6$) as judged by gel filtration chromatography. A single subunit of M_r 135,000 is present within the native homocomplex, as assessed by SDS-PAGE under reducing conditions. It has proved difficult to perform PAGE with the native TPP II complex, due probably to its asymmetric elongated conformation, which results in an oligomeric particle that will not readily migrate through the aqueous channels of the gel (Harris and Tomkinson, 1990). An electron micrograph of the native TPP II complex is shown in Figure 6, negatively stained with ammonium molybdate. The two predominant images present have been described by Macpherson *et al.* (1987) as the "single bow" and the "double bow." Both are some 50 nm long; the single

FIGURE 5. A transmission electron micrograph of human erythrocyte 20 S multicatalytic protease, prepared by spreading the enzyme on mica in the presence of 0.05% PEG (M_r 10,000). The protein complex has formed a quasi-two-dimensional paracrystalline array. It is considered that if further improvement of the regularity of arrays such as this can be achieved, the specimens will be suitable for digital image analysis. Negatively stained with 2% uranyl acetate. Bar = 50 nm.

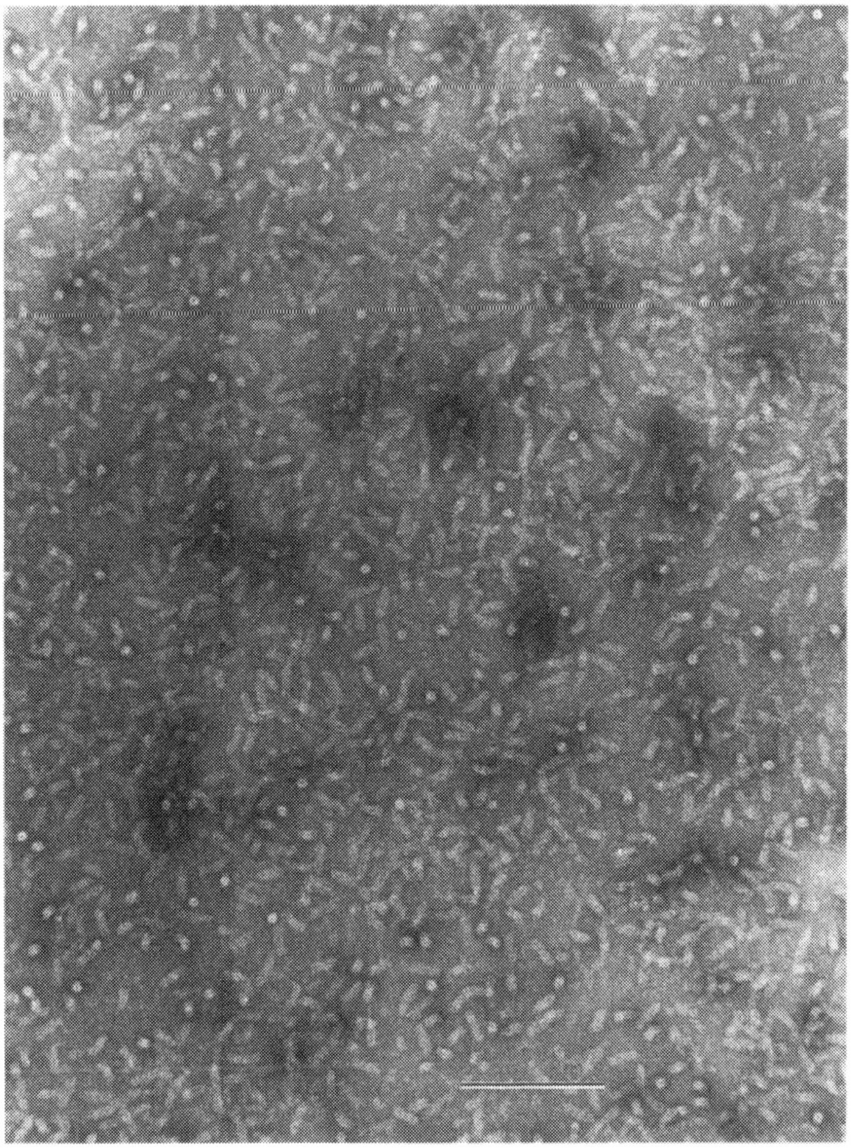

FIGURE 6. A transmission electron micrograph of the native complex of human erythrocyte tripeptidylpeptidase II (TPP II). Negatively stained with 2% ammonium molybdate. This undissociated TPP II contains only the very high-molecular-weight "double bow" oligomeric structures, which are oriented predominantly on their sides, but within the relatively deep negative stain present many "double bows" are supported on-end, resulting in the circular electron-transparent structures containing an electron-dense stain-filled center. Bar = 200 nm. The enzyme sample, in the presence of 30% glycerol, 1 mM dithiothreitol, and 50 mM phosphate buffer (pH 7.5), was provided by Dr. B. Tomkinson, University of Uppsala.

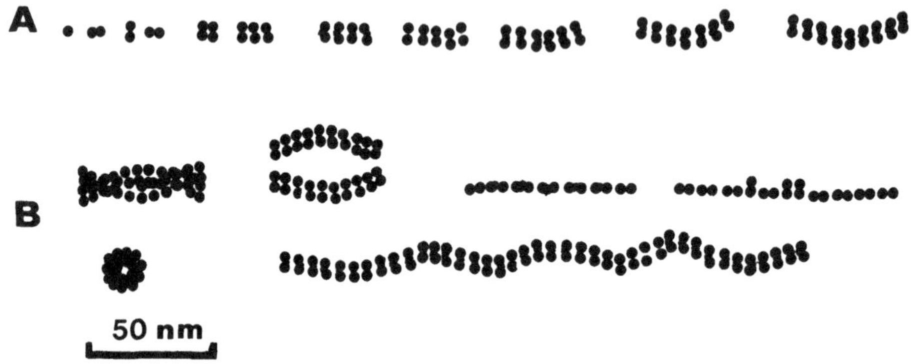

FIGURE 7. Diagrammatic representations of the high-molecular-weight native human erythrocyte TPP II shown electron optically in Figure 6, together with some of the dissociation products produced by dialysis of TPP II against low-ionic-strength buffer with reduced or zero glycerol (A). The elongated single and double bow structures are usually oriented on their sides (B) but occasionally the "double bows" are supported on-end by the negative stain. The limiting unit in the absence of SDS appears to be a dimeric particle (M_r 270,000) and this has been used to construct this diagram, but on SDS-PAGE the TPP II completely dissociates into a M_r 135,000 peptide.

bow is some 10 nm wide, while the double bow, consisting of two intersecting single bows, is 20 to 24 nm wide. A diagrammatic representation of these two structures is given in Figure 7. It is clear that these large polymeric/oligomeric complexes of TPP II consist of a regular number of smaller "limiting units" ~ 3 by 10 nm that are stacked side by side to produce the arclike single bow. The exact location and orientation of the M_r 135,000 subunit within the native complexes have not been precisely defined, but it may be that some form of regular helical arrangement of subunits is present to account for the production of the arclike structures.

Reduction of the ionic strength of the purified enzyme by dialysis against 2 mM Tris-HCl buffer (pH 8.0) produced a loss of enzyme activity, but this also produced a progressive dissociation of the native complex (Figure 8). Dissociation *per se* may not have been the prime reason for the loss of enzyme activity, since the stabilizing glycerol was also removed by dialysis. Application of the mica-spreading technique for the preparation of negatively stained specimens to this partly dissociated TPP II has produced regions containing regular arrays of particles, as shown in Figure 9. The individual particles in Figure 9 are of somewhat greater M_r than 135,000; indeed, they may represent groups of two or four M_r 135,000 subunits.

3. AMINOACYL-tRNA SYNTHETASE COMPLEX AND INITIATION FACTORS

3.1. Aminoacyl-tRNA Synthetase Complex

That erythroblasts have an exceptionally large capacity for globin synthesis is self-evident, and the continued presence of the amino acid activating enzymes (the aminoacyl-

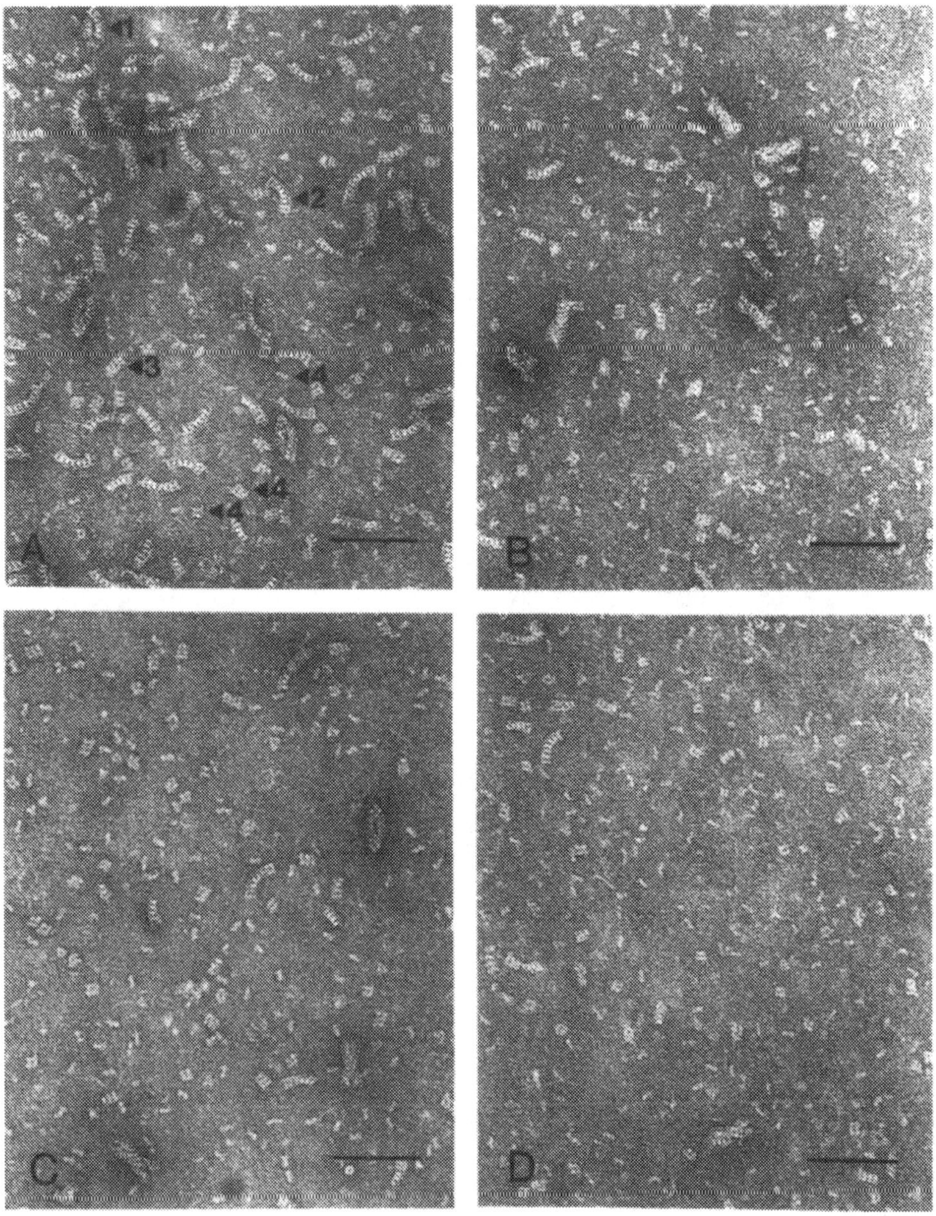

FIGURE 8. Transmission electron micrographs of human erythrocyte TPP II at varying stages of dissociation produced by dialysis against 2 mM Tris-HCl buffer containing 3% (w/w) glycerol and 0.5 mM 2-mercaptoethanol. (A) Before dialysis; (B) after 1 hr dialysis; (C) after 2 hr dialysis; and (D) after 4 hr dialysis. All the specimens were negatively stained with 2% ammonium molybdate (pH 7.5). Bars = 100 nm. In the undissociated sample (A) the particles present can be divided into four different size classes: structures containing 14 to 18 limiting units, structures containing 7 to 9 limiting units, structures containing 4 to 6 limiting units, and structures containing 1 to 3 limiting units. From Macpherson *et al.* (1987) with permission.

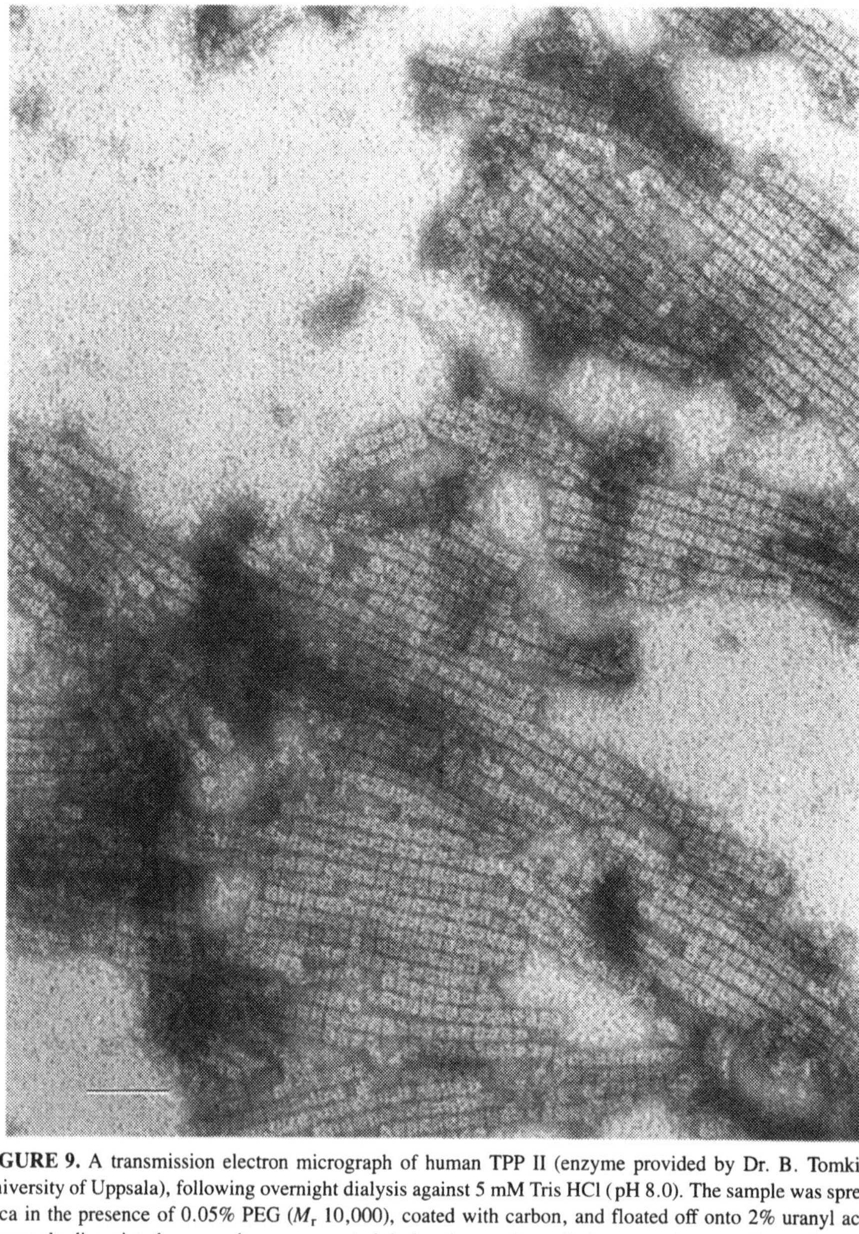

FIGURE 9. A transmission electron micrograph of human TPP II (enzyme provided by Dr. B. Tomkinson, University of Uppsala), following overnight dialysis against 5 mM Tris HCl (pH 8.0). The sample was spread on mica in the presence of 0.05% PEG (M_r 10,000), coated with carbon, and floated off onto 2% uranyl acetate. The partly dissociated enzyme has reaggregated during the specimen drying to produce quasi-two-dimensional arrays. Although there are some small regions of paracrystallinity, in general considerable structural deformity is present within the lattices, which prevents the application of digital image analysis. Bar = 50 nm.

tRNA synthetases, EC 6.1.1.–) in reticulocytes is well established. Indeed, these enzymes are also present in the mature normal erythrocyte, with increased levels in erythrocytes from patients with macrocytic and microcytic anemia (Izak *et al.*, 1960). Furthermore, it is also known that a number of the mammalian aminoacyl-tRNA synthetases have a tendency to aggregate or group together as a high-molecular-weight multifunctional complex. With respect to the rabbit reticulocyte, it has been shown that some nine synthetases tend to associate as an 18 S to 25 S complex (Som and Hardesty, 1975; Tanaka *et al.*, 1976; Ussery *et al.*, 1977). Aminoacyl-tRNA synthetase activity is also present in a "free" form and associated with the ribosomal fraction (Smith *et al.*, 1979). The 18 S to 25 S synthetase complex purified finally by sucrose density gradient centrifugation was found to contain activity for the incorporation of arginine, aspartic acid, glutamine, glutamic acid, isoleucine, leucine, lysine, methionine, proline, and valine.

Significantly, by electron microscopy this 18 S to 25 S synthetase complex (Oliver, 1973) was shown to possess a cylindrinlike quaternary conformation, as expounded by Harris (1983). Unfortunately, it now seems likely that this "purified" synthetase complex was grossly contaminated by the 20 S multicatalytic protease complex (see Section 2.1). The hazards of cosedimentation of molecular complexes in sucrose or glycerol density gradients has already been mentioned; indeed, molecules such as thyroglobulin and α_2-macroglobulin have presented major challenges for this separation procedure (Harris, 1982b). This problem was emphasized earlier by the work of Shelton *et al.* (1970) who also detected a cylindrinlike macromolecule in the 18 S to 20 S fraction from sucrose density gradient centrifugation in isolates from rabbit reticulocytes and human erythrocytes. It was maintained by these workers that this fraction possessed aminoacyl transferase I (EC 2.3.2.6) activity, yet the work of Smulson (1974) using an extract from HeLa cells again suggested the possibility of aminoacyl-tRNA synthetase activity. It is, however, worth bearing in mind that the energy-dependent aminoacyl-tRNA synthetase and transferase activities are essentially the reverse of the polypeptide-splitting vity of the protease and peptidases. This concept may receive support from the recent observation that arginyl-tRNA protein transferase from rabbit reticulocytes is involved in the post-translational modification and degradation of proteins (Ciechanover *et al.*, 1988). That both aminoacyl-tRNA synthetases and the aminoacyl-tRNA protein transferases are recoverable from reticulocyte lysates by their affinity to heparin–Sepharose (Hradec, 1980) is of interest, as is the suggestion that these two enzyme systems may be structurally and functionally coupled within a multienzyme complex.

Despite the above comments, the available evidence from those performing electron microscopic work on the aminoacyl-tRNA synthetase complex is that it does not possess a cylindrical quaternary conformation (Gulik and Orsini, 1984). This is supported by as yet unpublished studies of the author, performed in collaboration with J.-P. Waller. Rat liver aminoacyl-tRNA synthetase complex, which is of identical subunit composition by SDS-PAGE to the reticulocyte complex, has been found to be an irregular, ~ 15-nm particle. A study by Godar *et al.* (1988) has shown that impure preparations of reticulocyte aminoacyl-tRNA synthetase do indeed contain cylindrin as a contaminant and that more highly purified fractions (18 S) contain amorphous particles that tend to agree with our own data and those of Gulik and Orsini (1984). A recent paper by Norcum (1989) on the aminoacyl-tRNA synthetase complex from murine erythroleukemia cells (the DS-19E5 cell line) indicates that this high-molecular-weight complex is relatively stable when subjected to negative staining with uranyl acetate. The unfixed native complex and the glutaraldehyde-fixed complex both yield images of irregularly shaped particles, with approximate dimensions 27 x 27 nm. It appears that this particle may be orientated in several different positions, resulting in squarish, U-shaped, cup-shaped, elongated cup-shaped, rectangular, and triangular profiles. Since no one profile predominates, it may be difficult to assign a precise quaternary structure from electron microscopy, despite the apparent constancy of the subunit stoichiometry.

The subunit molecular weights of the aminoacyl-tRNA synthetase complex, which displays an SDS-PAGE pattern containing ten major components of M_r ranging from 40,000 to 140,000 (Kellerman *et al.*, 1982; Mirande *et al.*, 1982), are significantly higher than the subunit molecular weights of cylindrin and the 20 S multicatalytic protease complex. The possibility that proteolytic cleavage of the native aminoacyl-tRNA synthetase complex may occur during isolation, particularly in extracts from reticulocytes, was also emphasized by Kellerman *et al.* (1982). Nevertheless, the possibility that the 26 S protease complex isolated by Hough *et al.* (1987). Arrigo *et al.* (1988), and Waxman *et al.* (1987) might be equivalent to the native aminoacyl-tRNA synthetase complex remains to be shown, and indeed our understanding of the composition of the aminoacyl-tRNA synthetase "core" complex remains to be defined. Likewise, the physiological significance of the structural grouping of seven to nine of the aminoacyl-tRNA synthetases within an apparently fragile supramolecular assembly is not established. That proteins other than the synthetases are present within the high-molecular-weight complex has already been implied, and Pendergast *et al.* (1987) have shown the presence of a M_r 37,000 protein that in a phosphorylation study was identified as casein kinase I.

3.2. Initiation Factors

Intimately associated with protein synthesis are the eukaryotic initiation factors (eIFs), which along with the aminoacyl-tRNA synthetases and aminoacyl protein transferases have been found in the fraction binding to heparin–Sepharose (Waldman *et al.*, 1975). The five principal rabbit reticulocyte initiation factors, eIF-2, eIF-3, eIF-4A, eIF-4B, and eIF-5, have been studied by Brown-Luedi *et al.* (1982), who found a strong identity with the eIFs of HeLa cells. The molecular weight of eIF-3 was determined by sedimentation equilibrium ultracentrifugation to be ~ 410,000 (14.2 S), with some material of approximately twice this mass also present. This higher M_r would be more equivalent to the figures obtained by Schreier *et al.* (1977) and Trachsel and Staehelin (1979), who suggested that eIF-3 was a complex of some eight to nine different subunits with molecular weights in the range 40,000 to 100,000.

The structure of rabbit reticulocyte eIF-3 has been presented by Boublik *et al.* (1983), as shown in Figure 10. No clearly definable quaternary structure is present, but some particles present a circular image (arrowheads), whereas others present an approximately square or rectangular profile, which might be compatible with the sideways orientation of a compact or short cylindrical structure. These and other workers (Emanuilov *et al.*, 1978) have attempted to identify by TEM the location of eIF-3 when bound to the 40 S ribosomal subunit. This has proved to be technically very difficult, because of the varying orientations and profiles of the overall ribosome–eIF-3 complex. Nevertheless, the 40 S ribosomal subunit–eIF-3 complex, which was characterized by gradient centrifugation as

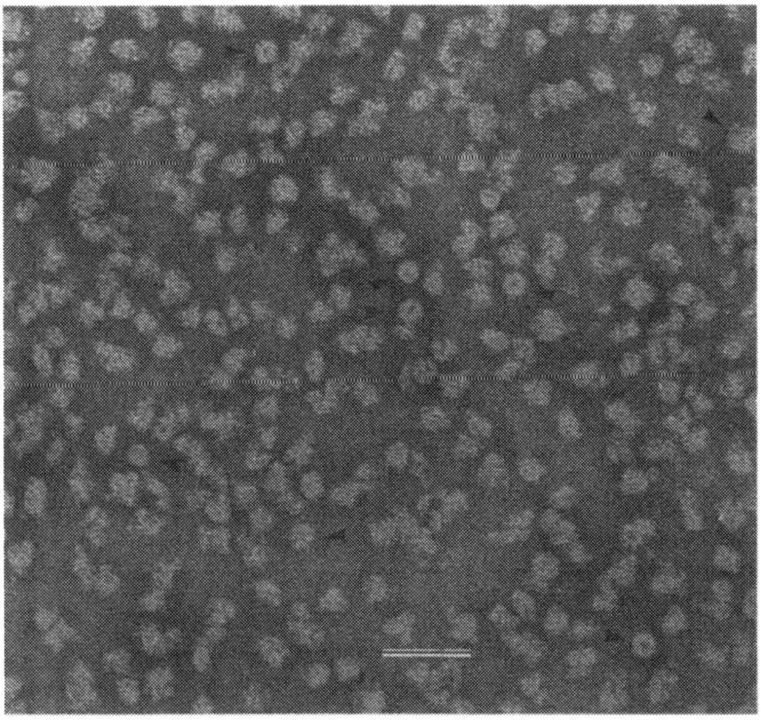

FIGURE 10. Eukaryotic initiation factor-3 (eIF-3) from rabbit reticulocytes. The arrowheads point to molecules that seem to show a definable "circular" quaternary structure. Negatively stained with 0.5% uranyl acetate. Bar = 50 nm. Previously unpublished micrograph, provided by Dr. M. Boublik.

possessing a sedimentation coefficient of 46 S, does have a characteristic protrusion that was thought to represent bound eIF-3. The situation with ribosomally bound eIF-2 is less clear, but attempts have been made (Bommer *et al.*, 1988) to perform immunoelectron microscopy of rat liver 40 S ribosomal subunits containing bound eIF-2.

4. ACETYLCHOLINESTERASE

The presence of acetylcholinesterase (AChE) (EC 3.1.1.7) on the surface of the mammalian erythrocyte has been known for many years, but even now no satisfactory reason has been advanced to account for the location of this enzyme or for its physiological role. It should be remembered that the role of the broader-specificity, predominantly plasmatic, butyrylcholinesterase (EC 3.1.1.8) is equally elusive. A great volume of work has been published on the molecular characterization, immunodetermination, and pathology of AChE from nervous tissues and muscles, from a variety of different species, which cannot be dealt with in this review. The reader is referred to the recent reviews by Brodbeck (1986), Lawson and Barr (1987), and Rakonczay and Brimijoin (1988), which thoroughly survey the topic.

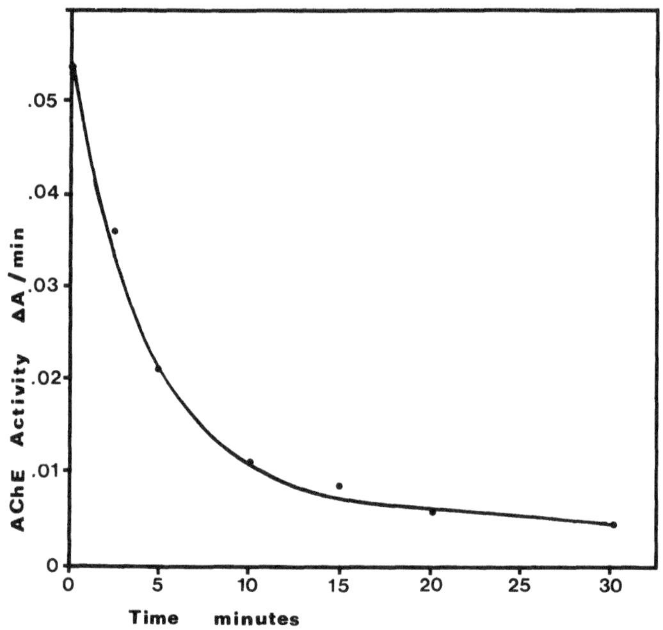

FIGURE 11. A graph showing the loss of membrane-bound AChE activity in a suspension of human erythrocytes when incubated with cysteine-activated papain (1.0 mg/ml). Negligible enzyme activity is released from the erythrocytes into the supernatant because at this high papain concentration inactivation of the AChE proceeds very rapidly.

4.1. Release and Inactivation of AChE by Proteases

Early studies on human erythrocyte AChE linked the reduction of this enzyme activity in the rare anemia, paroxysmal nocturnal hemoglobinuria (PNH), to the increased susceptibility of the erythrocyte to complement-mediated hemolysis in this condition (Burapakulsolsri *et al.*, 1979; Siriwittayakorn and Yuthavong, 1979). It is now known, however, that in PNH there is a mixed population of circulatory erythrocytes, some of which are normal and the others are AChE deficient (Brimijoin *et al.*, 1986; Chow *et al.*, 1985; Dockter and Morrison, 1986). Also, it is apparent that the complement sensitivity is linked not to AChE but to the loss of another membrane protein, so-called "decay-accelerating factor" (Medof *et al.*, 1987; Sugerman *et al.*, 1986), which is also the major Fc receptor (Selvaraj *et al.*, 1988).

This work on PNH is of relevance to AChE isolation, because it was shown that limited protease treatment of erythrocytes would experimentally produce AChE-depleted cells (Carraway *et al.*, 1971; Herz and Kaplan, 1974; Herz *et al.*, 1963), which it was thought at the time might mimic PNH (Firkin *et al.*, 1963). In fact, this is not the case, for it is now clear that AChE is released from the erythrocyte surface and rapidly inactivated, by the action of a number of different proteases. At the same time, there is progressive removal of the external portion of the "band 3" glycoprotein from the erythrocyte surface (Steck, 1978). This early work has been supported by Harris (1988a) who employed the relatively high concentrations of papain (i.e., 1.0 mg/ml) that are used for serological

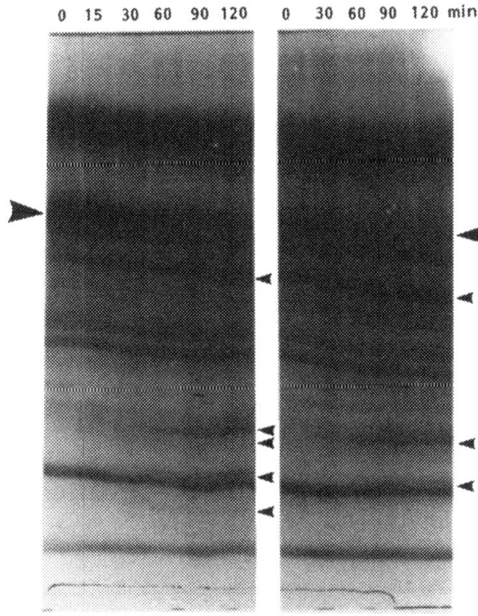

FIGURE 12. SDS-PAGE of human erythrocyte ghosts prepared from erythrocytes treated with papain (1.0 mg/ml). There is a progressive loss of the band 3 region (large arrowheads), with a steady increase in a number of lower M_r bands (small arrowheads).

premodification of human erythrocytes. It was found that there was a rapid loss of AChE from the erythrocyte surface, as shown in Figure 11. Although solubilized under these conditions, AChE is at the same time rapidly inactivated. The loss of "band 3" occurred more slowly (Figure 12). However, when low concentrations of papain are used (i.e., 0.01 to 0.25 mg/ml) there is a progressive release of functional AChE from the erythrocyte surface (Figure 13), with some indication of loss of total activity at the top end of this concentration range (Table I), although almost 25% of the functional AChE was recoverable in a solubilized state.

Partial purification of the AChE released from human erythrocytes by limited papain digestion was achieved by precipitation with 60% ammonium sulfate, followed by gel filtration chromatography and sucrose density gradient centrifugation. This proteolytically cleaved AChE appears to be a dimer of $M_r \sim 150{,}000$. There is only slight indication of oligomer formation, which is markedly different from the neutral surfactant-solubilized AChE when the surfactant is removed (see below).

Table I
Percentage Degradation of Total RBC Acetylcholinesterase[a]

Papain concentration	30 min	60 min	90 min	135 min
0.025 mg/ml	ND	ND	ND	ND
0.05 mg/ml	ND	ND	9%	20%
0.1 mg/ml	ND	ND	18%	20%
0.25 mg/ml	5%	10%	30%	50%

[a]ND, not detectable.

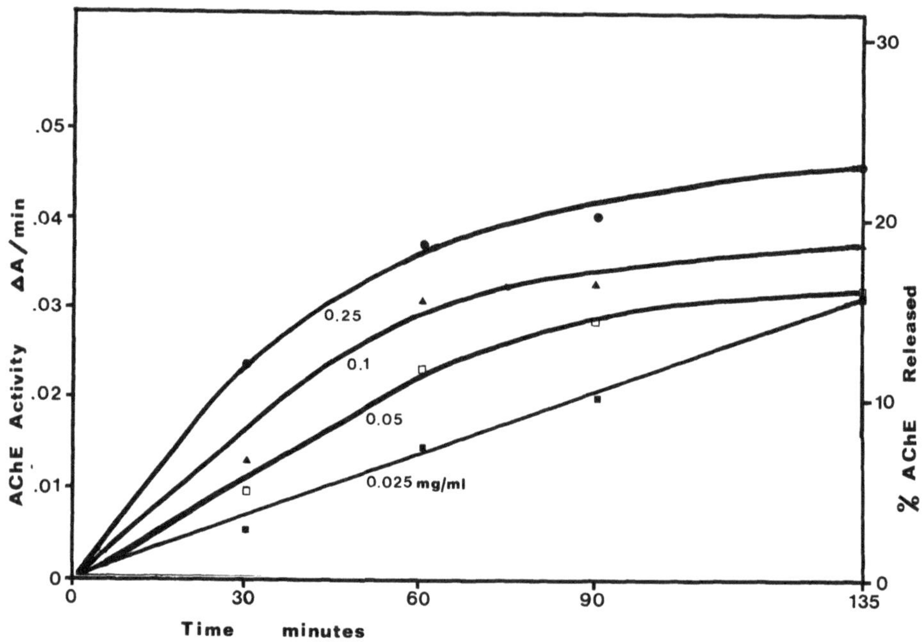

FIGURE 13. Graphs showing the progressive release of active AChE into the cell-free supernatant when human erythrocytes are incubated with cysteine-activated papain at 0.025, 0.05, 0.1, and 0.25mg/ml. Almost 25% of the cellular AChE was recovered in the supernatant following treatment with papain at 0.25 mg/ml, although at this papain concentration considerable inactivation of the solubilized enzyme also occurs.

4.2. Neutral Surfactant Solubilization of AChE

The main approach used for the solubilization of AChE from nervous tissues, muscle, and erythrocytes has been to employ a neutral surfactant such as Triton X-100 (Brockman *et al.*, 1986; Ciliv and Özand, 1972; Ott *et al.*, 1975; Paniker *et al.*, 1973; Rosenberry and Scroggin, 1984; Wright and Plummer, 1972). This initial solubilization of AChE from erythrocyte ghosts has been followed by a variety of purification procedures, viz. DEAE-cellulose and calcium phosphate-gel chromatography (Ciliv and Özand, 1972); gel filtration chromatography on Sephadex G-200 (Paniker *et al.*, 1973); ammonium sulfate precipitation, calcium phosphate-gel and gel filtration chromatography (Wright and Plummer, 1972), affinity chromatography and density gradient centrifugation (Ott *et al.*, 1975), gel filtration chromatography and density gradient centrifugation (Weitz *et al.*, 1984); affinity chromatography on immobilized acridinium resin (Rosenberry and Scroggin, 1984).

A suggestion that emerged from the early purification studies was that purified AChE might exist in solution in varying states of molecular aggregation or oligomerization (Wright and Plummer, 1973) and this was subsequently supported by the work of Brodbeck and his colleagues (Brodbeck *et al.*, 1975; Ott and Brodbeck, 1978; Ott *et al.*, 1975, 1982), together with the work of Biagioni *et al.* (1982) and Rosenberry and Scroggin (1984). Essentially, these workers have shown that following Triton X-100 depletion of the purified enzyme by dialysis, oligomers were produced that were multiples of the G_2

FIGURE 14. A transmission electron micrograph of human erythrocyte AChE extracted from erythrocyte ghosts by Triton X-100 and purified by sucrose density gradient centrifugation. Negatively stained with 2% sodium phosphotungstate (pH 7.0). A single, possible dimeric, particle predominates. Bar = 100 nm.

dimeric form of the enzyme. Nondissociating PAGE revealed seven enzymatically active bands, six of which could also be detected by staining for carbohydrate (Ott *et al.*, 1975). It was shown by zonal centrifugation that up to eight molecular forms of human erythrocyte AChE could be detected, with the four most abundant oligomers ranging in sedimentation coefficient from 12.5 S to 19.0 S (M_r 500,000 to 1,131,000, composed of ca. 6 to 14 of the M_r 80,000 subunit). Readdition of Triton X-100 disaggregated the oligomers into

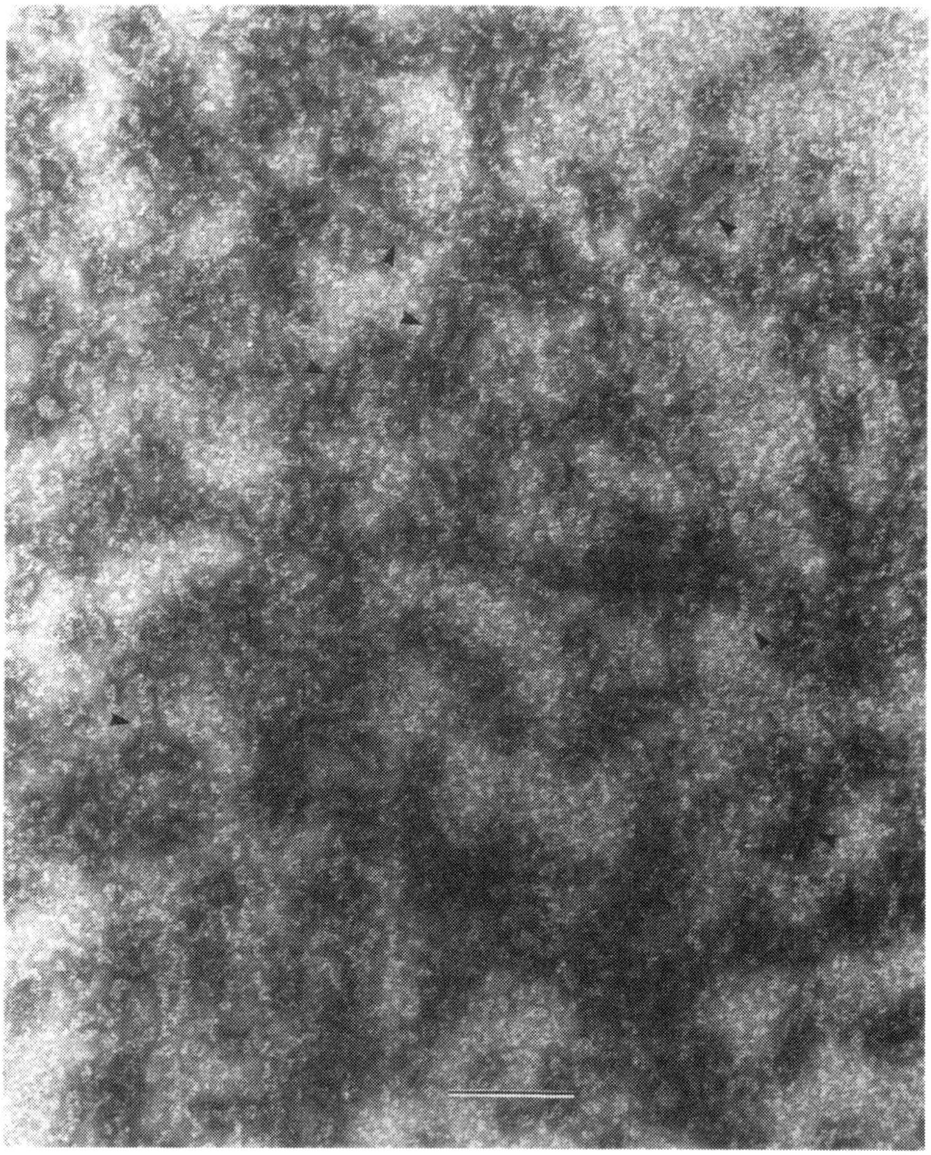

FIGURE 15. A transmission electron micrograph of Triton X-100-depleted (oligomerized) human erythrocyte AChE (enzyme sample prepared by Professor U. Brodbeck, University of Bern). Negatively stained with uranyl acetate, after spreading on mica. Numerous molecular aggregates are present (arrowheads). Bar = 100 nm.

the ~ 7 S G_2 dimer, which would again aggregate on further surfactant removal. Electron micrographs of purified Triton X-100-solubilized human erythrocyte AChE and the mixture surfactant-depleted oligomers are shown in Figures 14 and 15. In the former case, the enzyme was applied directly to the sucrose density gradient and this resulted in a single narrow zone of sedimenting AChE activity, whereas after detergent depletion by dialysis overnight against 20 mM Tris-HCl (pH 7.4), 1% methyl-D-mannoside, and 2% sucrose, a

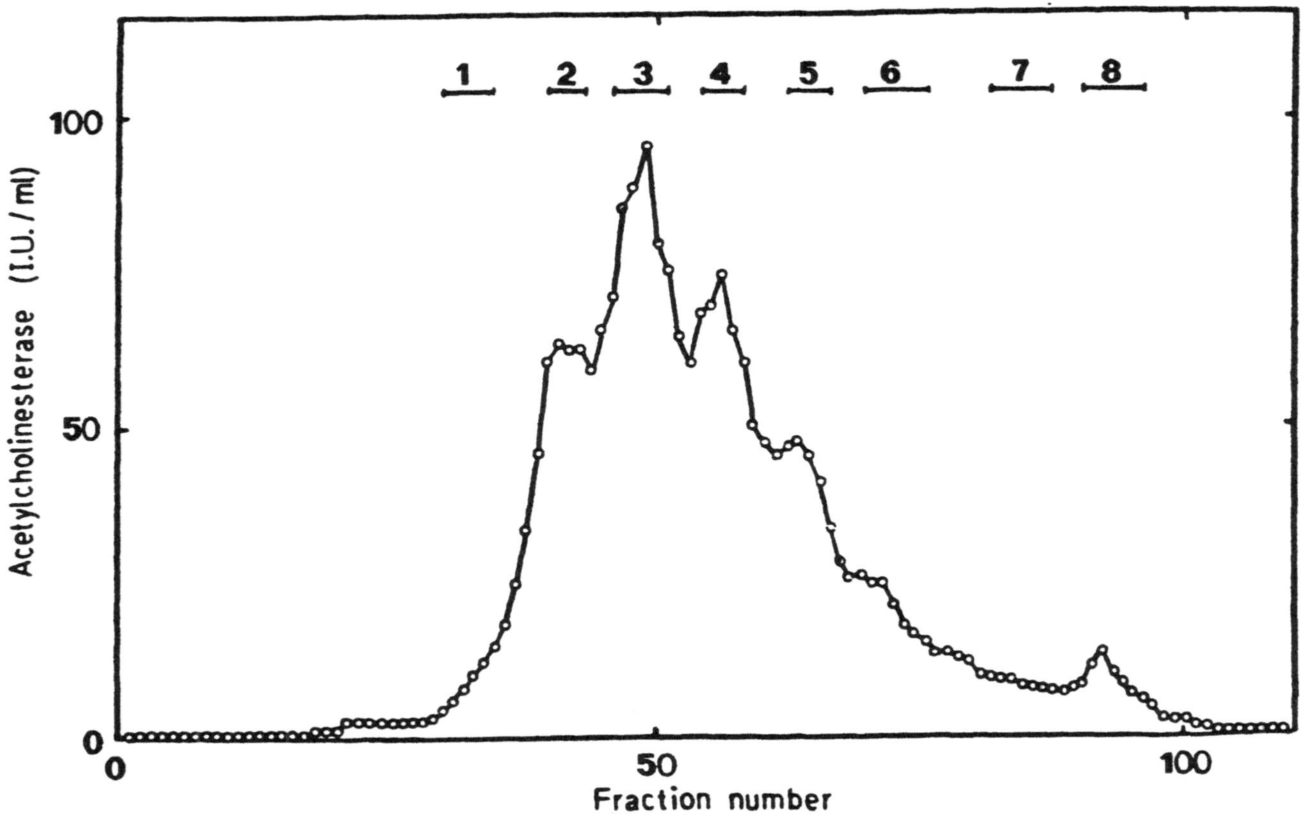

FIGURE 16. The sedimentation of human erythrocyte AChE on a sucrose density gradient, following Triton X-100 depletion. The AChE oligomeric mass increases from 1 to 8. From Ott *et al.* (1975).

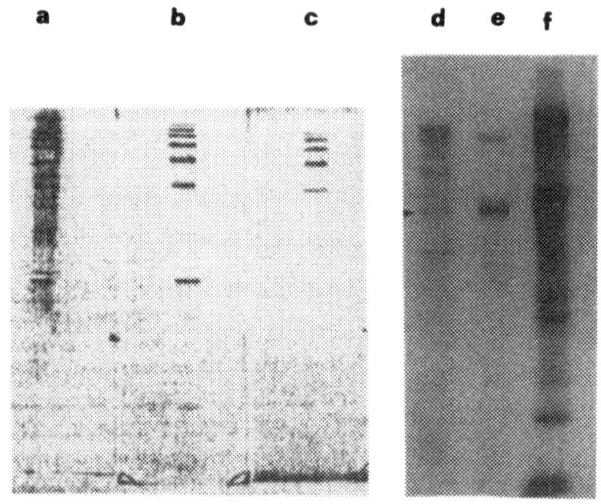

FIGURE 17. Native PAGE of oligomeric human erythrocyte AChE, Channels a–c stained for enzyme activity, protein, and carbohydrate, respectively (from Ott *et al.*, 1975). Channel d shows undissociated oligomeric human erythrocyte AChE and channel e, 2% SDS–β-mercaptoethanol-dissociated human erythrocyte AChE (enzyme provided by Professor U. Brodbeck), electrophoresed by SDS-PAGE. Channel f shows human erythrocyte ghosts. Note the location of the leading $M_r \sim 80{,}000$ AChE peptide (arrowhead), which bands slightly ahead of the erythrocyte membrane band 3 zone.

very broad distribution of AChE activity was found, ranging in sedimentation coefficient from ~ 7 S to 19 S (Brodbeck *et al.*, 1975; Ott and Brodbeck, 1978), as shown in Figure 16. PAGE of this oligomeric mixture is shown in Figure 17.

Molecular cross-linking studies using bifunctional diimidates (Römer-Lüthi *et al.*, 1979) indicated that significant cross-linking (i.e., beyond the dimer) was only obtained in the absence of Triton X-100. Similarly, glutaraldehyde (20 mM) treatment of the detergent-depleted AChE (Römer-Lüthi *et al.*, 1980) produced stabilization of the oligomeric aggregates and prevented subsequent SDS dissociation. The spontaneous oligomerization of human erythrocyte AChE on removing the Triton X-100 is not thought to be of significance within the context of the membranous environment of the native enzyme. Indeed, the Triton X-100-solubilized dimeric form (7 S; M_r 151,000 ± 8000) is thought to represent the functional unit (Ott *et al.*, 1982). SDS-PAGE of the Triton X-100-solubilized and purified human AChE reveals the M_r 151,000 dimeric form when run under nonreducing conditions and the M_r 75,000 subunit when a reducing agent is present.

Isoelectric focusing of the Triton X-100-solubilized AChE dimer has been shown to produce five molecular forms, ranging from pI 4.55 to 5.18 (Ott *et al.*, 1975), and the suggestion was advanced that this was a reflection of the varying sialic acid content of the carbohydrate moiety since neuraminidase treatment produced increasing amounts of the higher pI species. An even greater range of active AChE bands (13), which possessed a significantly higher pI range (pI 4.97 to 5.85), was detected by Biagioni *et al.* (1982).

4.3. Membrane Attachment of AChE

In common with many amphipathic membrane proteins, such as the glycohydrolases and hydrolases in general (Hauri, 1988), AChE contains a carboxy-terminal membrane-binding domain or "tail." In this instance, however, this domain represents a very small portion of the overall molecular mass and contains only two amino acids (histidine and glycine), which are themselves hydrophilic rather than hydrophobic (Roberts and Rosenberry, 1986). The hydrophobicity of the membrane-binding domain is imparted by the presence of covalently attached glycolipid, and the diacylglycerol moiety serves to anchor the enzyme to the lipid bilayer.

Papain and pronase cleavage of the Triton X-100-extracted human erythrocyte AChE releases the small (M_r 2000) hydrophobic C-terminal domain from the larger enzymatically active hydrophilic glycoprotein domain of AChE (Dutta-Choudhury and Rosenberry, 1984). This papain-cleaved AChE loses its ability to bind nonionic surfactants and cannot be reconstituted into liposomes. It does, however, remain unaggregated on removing the Triton X-100. It is also of significance in relation to the data presented above (Section 4.1) that Dutta-Choudhury and Rosenberry were only able to release up to 10% of the erythrocyte surface AChE by direct proteolysis, whereas the author has achieved a much greater

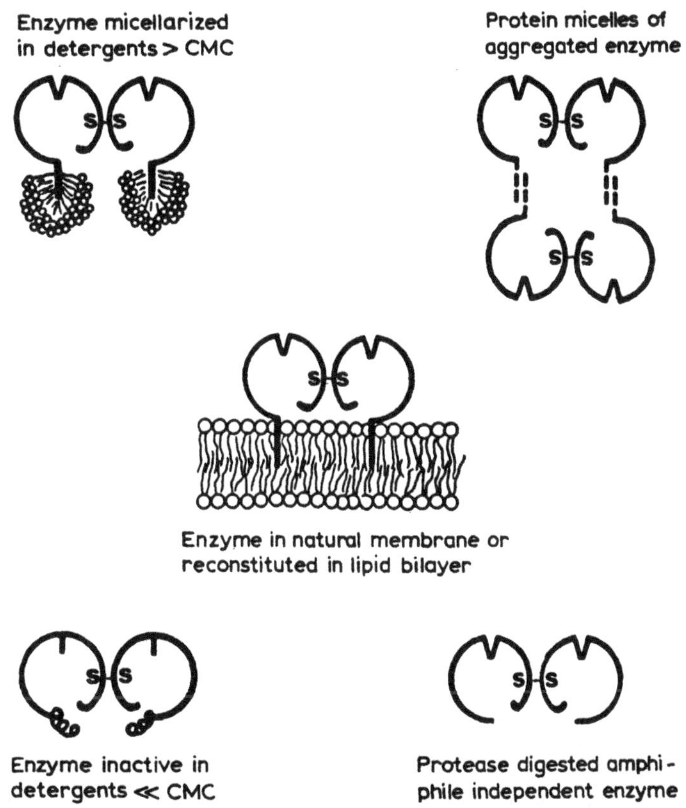

FIGURE 18. Diagrammatic representations of the dimeric D_2-AChE, within a natural or artificial membrane, and the various soluble forms of the enzyme. The native amphiphilic enzyme is active only when it undergoes hydrophobic interactions with lipid and detergents or self-aggregation in the absence of detergents. When the enzyme is deprived of stabilizing amphiphiles, it irreversibly loses its activity. Proteolytic digestion converts the D_2 form into a G_2, globular hydrophilic enzyme, which does not undergo hydrophobic interactions and does not require the presence of a stabilizing amphiphile. Modified from Brodbeck (1986).

level when using low papain concentrations over an extended period of time. A minimal change in the M_r of the papain-digested Triton X-100-solubilized human erythrocytye AChE was also shown by Weitz *et al.* (1984), although the Stokes radius decreased from 8.2 to 5.9 nm and the sedimentation coefficient increased from 6.3 to 7.0 S. Human erythrocyte AChE short hydrophobic membrane-binding domain was defined by Roberts and Rosenberry (1985, 1986) as being of glycolipid composition and attached to the C-terminus of the polypeptide chain. Methanolysis released covalently bound fatty acids present in the approximate ratio of 1 mol saturated fatty acid and 1 mol of unsaturated fatty acid per mol of hydrophobic domain, which has a M_r of approximately 3000 and as mentioned above is surprising because it contains only histidine and glycine.

It is of interest that direct treatment of erythrocytes with phosphatidylinositol-specific phospholipase C (EC 3.1.4.10) is able to release a high proportion (85%) of the AChE from bovine erythrocytes but only a very small proportion (5%) from human erythrocytes (Low *et al.*, 1986; Roberts *et al.*, 1987; Taguchi and Ikezawa, 1987). This species variability may indicate a varying mode of AChE attachment in different species. This topic has been reviewed within the broader context of AChE from a range of different tissues as well as species, by Silman and Futerman (1987), who interpreted the available data to indicate that human erythrocyte AChE may be totally covalently bound to phosphatidylinositol, but that the phosphatidylinositol may itself be inaccessible or resistant to the action of *Staphylococcus aureus* phosphatidylinositol-specific phospholipase C. A diagrammatic model for the attachment of AChE to the erythrocyte membrane is given in Figure 18.

5. GLYCOGEN SYNTHASE AND Mg^{2+}-ATPase

One of the principal ATP-consuming enzymes of the reticulocyte and erythrocyte is widely accepted as being the Na^+/K^+-ATPase of the plasma membrane, but more recent studies have also implicated the cytosolic ATP–ubiquitin-dependent protease complex. Protein synthesis and the aminoacyl-tRNA synthetases also require ATP. In addition, a number of other enzymes require ATP for their function, such as the membrane-bound Ca^{2+}/Mg^{2+}-ATPase, adenylate kinase, soluble and membrane-bound protein kinases, glycogen synthase, and both soluble and membrane-bound Mg^{2+}-ATPases. The difficulty, nevertheless, of accounting for total ATP consumption within the reticulocyte is clearly apparent, as indicated by Siems *et al.* (1984). Within this section, comment will be restricted to a consideration of two ATP-consuming enzymes, the soluble Mg^{2+}-ATPase and glycogen synthase of human erythrocytes, both of which exist as high-molecular-weight complexes.

5.1. Glycogen Synthase

That reticulocytes and erythrocytes possess the enzymes required for glycogen accumulation and breakdown is firmly apparent from the fact that glycogen particles can be detected in the cytoplasm of normal erythroblasts, with greater amounts in erythroblasts from anemic patients, and in erythrocytes from patients with glycogen-storage disease types III and VI. The mature normal erythrocyte retains glycogen synthase activity in a vestigial yet functional form (Moses *et al.*, 1972a,b), possibly because of conversion of the active dephosphorylated (I) form into the less active glucose-6-phosphate phosphorylated (D) form, together with the limiting allosteric activation of the D form of glycogen synthase by cellular glucose-6-phosphate levels and inhibition by physiological concentrations of ATP, ADP, and P_i. Although, to my knowledge, glycogen synthase has not been isolated in a purified form from human erythrocytes, a considerable volume of data is available on the enzyme from skeletal muscle (Nimmo *et al.*, 1976), heart muscle (Dickey-Dunkirk and Killilea, 1985), and kidney (Issa and Mendicino, 1973). In general, the emphasis in these studies has been placed on the kinetic properties of glycogen synthase, but electrophoretic data have indicated a subunit M_r of 90,000, with a native M_r of slightly less than 400,000 (13.3 S), although Nimmo *et al.* (1976) also detected an enzymatically active higher M_r species (19.0 S). It was suggested that the 13.3 S species might be a tetramer, but within the bounds of experimental error it might possibly be a trimer, with the larger 19.0 S species then being a hexamer, of the M_r 90,000 subunit.

A confusing paper on the electron microscopy of muscle glycogen synthase (Rebhun *et al.*, 1973) presented images of negatively stained molecular complexes, some of which are clearly identical to cylindrin and the 20 S protease complex (see Section 2) and others that possess a more compact hexagonal structure. The undoubtedly erroneous suggestion was advanced that these two quaternary conformations might represent the glucose-6-phosphate-dependent and -independent forms of glycogen synthase. Gel filtration and gradient centrifugation clearly did not in this instance provide a satisfactory procedure for separating these two high-molecular-weight proteins. It is thus possible that the hexagonal protein shown by Rebhun *et al.* (1973) represents the 19 S glycogen synthase, with cylindrin present as a major contaminant.

Our own studies on skeletal muscle glycogen synthase (Sigma) provide some support for this interpretation. This commercial glycogen synthase has been found to contain a single predominant high-molecular-weight protein by nondissociating PAGE and a single subunit of M_r ~ 90,000 by SDS-PAGE. There is no trace of the cylindrinlike contaminant, either by electrophoresis or by electron microscopy. The native glycogen synthase stains for glycogen synthase activity (Krisman and Blumenfeld, 1986) and Mg^{2+}-ATPase activity, and comigrates with human erythrocyte Mg^{2+}-ATPase (Figure 19). TEM of

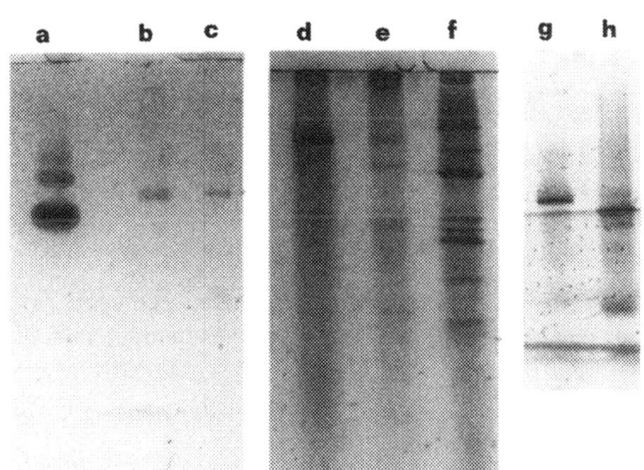

FIGURE 19. Native PAGE of muscle glycogen synthase (channels b and e) and human erythrocyte Mg^{2+}-ATPase (channels c, d, and h) electrophoresed on a 2.5 to 27% polyacrylamide gradient gel. A predominant M_r 600,000 protein is present in both enzymes. Channels a, f, and g show horse apoferritin, *E. coli* phosphoribosylpyrophosphate synthetase, and human erythrocyte cylindrin, respectively.

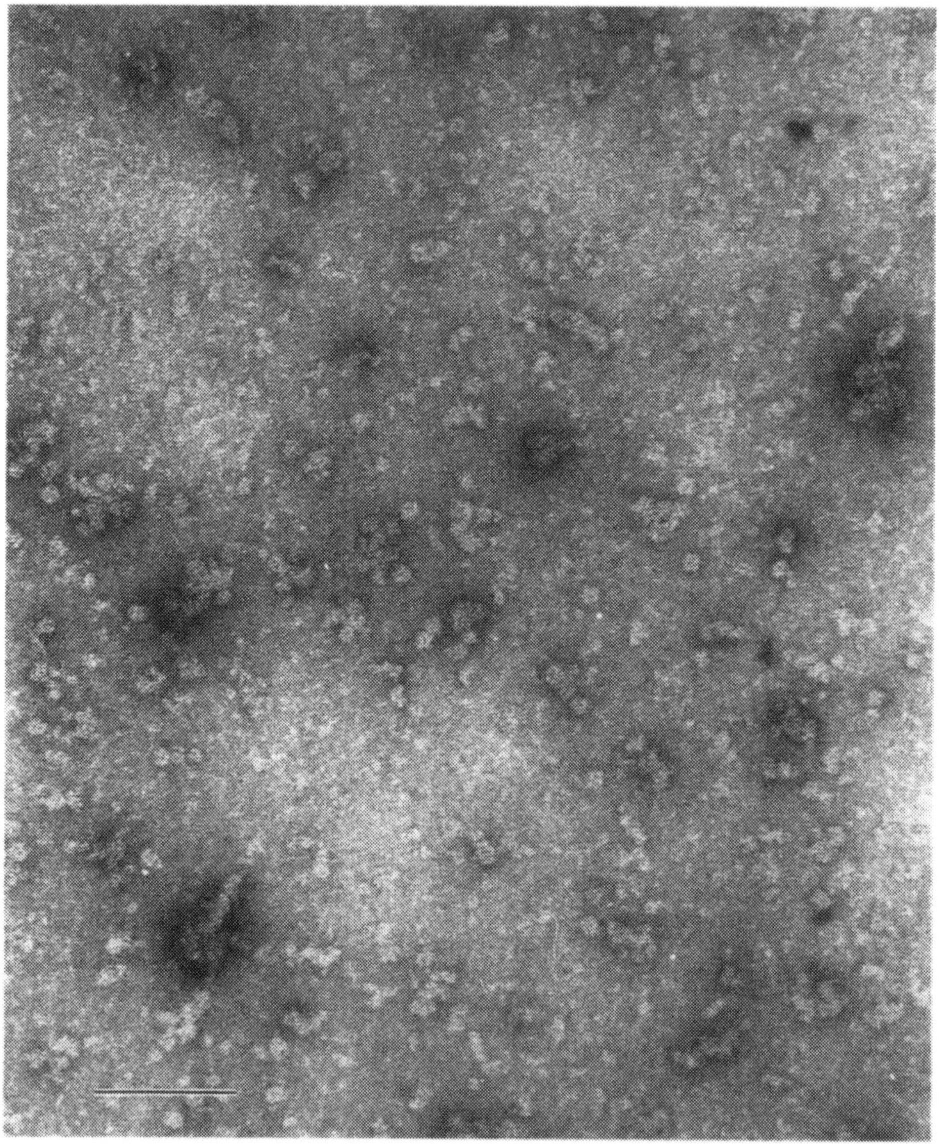

FIGURE 20. A transmission electron micrograph of muscle glycogen synthase (Sigma). Individual hexagonal particles (M_r 600,000) and small aggregates are present. Negatively stained with 2% uranyl acetate. Bar = 100 nm.

muscle glycogen synthase has revealed a compact hexagonal particle (probably equivalent to the 19.0 S, $M_r \sim 600{,}000$ species mentioned above), as shown in Figure 20. To date, attempts to stain for glycogen synthase activity using the native Mg^{2+}-ATPase from human erythrocytes have not provided conclusive evidence for the identity of these protein complexes. This being the case, the Mg^{2+}-ATPase will now be considered separately.

5.2. Mg^{2+}-ATPase

The presence of Mg^{2+}-dependent ATPase activity in the low-ionic-strength extract prepared from erythrocyte ghosts was indicated some years ago by the work of Hoogeveen *et al.* (1970) and Kirkpatrick *et al.* (1975, 1976). Subsequently, White and Ralston (1976, 1980) purified this enzyme from human erythrocyte ghosts and showed that it migrated electrophoretically ahead of the cylindrin molecule in a nondissociating PAGE system. The M_r of the native Mg^{2+}-ATPase was estimated to be 600,000 and after dissociation with SDS the enzyme was revealed as a single subunit of M_r 100,000. Independent studies by Harris (1984) have shown that human erythrocyte Mg^{2+}-ATPase can also be recovered from the cytosol. This was made possible because the Mg^{2+}-ATPase precipitates isoelectrically in the region of pH 5.0, along with cylindrin and a number of other cytosolic proteins. Purification of Mg^{2+}-ATPase was achieved by sucrose density gradient centrifugation, the fractions containing the enzyme being located by direct assay for Mg^{2+}-ATPase and histochemical staining of Mg^{2+}-ATPase on nondissociating PAGE tube gels, and protein staining, as shown in Figure 19. The Mg^{2+}-ATPase sediments less rapidly than cylindrin, but migrates electrophoretically slightly faster than cylindrin, in agreement with the observations of White and Ralston (1980). SDS-PAGE of purified Mg^{2+}-ATPase likewise indicates a single M_r 90,000 subunit (Figure 21). Kinetic studies have also been performed by Mostafa and Hanahan (1984) on a human erythrocyte Mg^{2+}-ATPase, but these workers did not provide a native M_r for their enzyme. The subunit M_r was, however, reported to be 113,000, which is in reasonable agreement with the figure obtained by White and Ralston (1976, 1980) and Harris (1984).

TEM of negatively stained Mg^{2+}-ATPase has revealed the molecule as a compact pseudohexagonal particle, which has a tendency to form aggregates from a number of

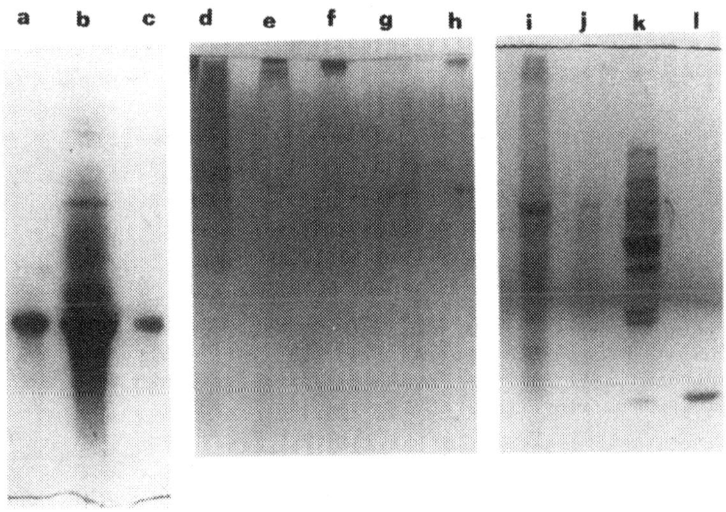

FIGURE 21. SDS-PAGE of human erythrocyte Mg^{2+}-ATPase (channels a–e, i, and j) and muscle glycogen synthase (Sigma; channels f–h). Note that both enzymes possess a predominant M_r 90,000 subunit. Channels k and l show rat liver aminoacyl-tRNA synthetase complex and horse apoferritin, respectively.

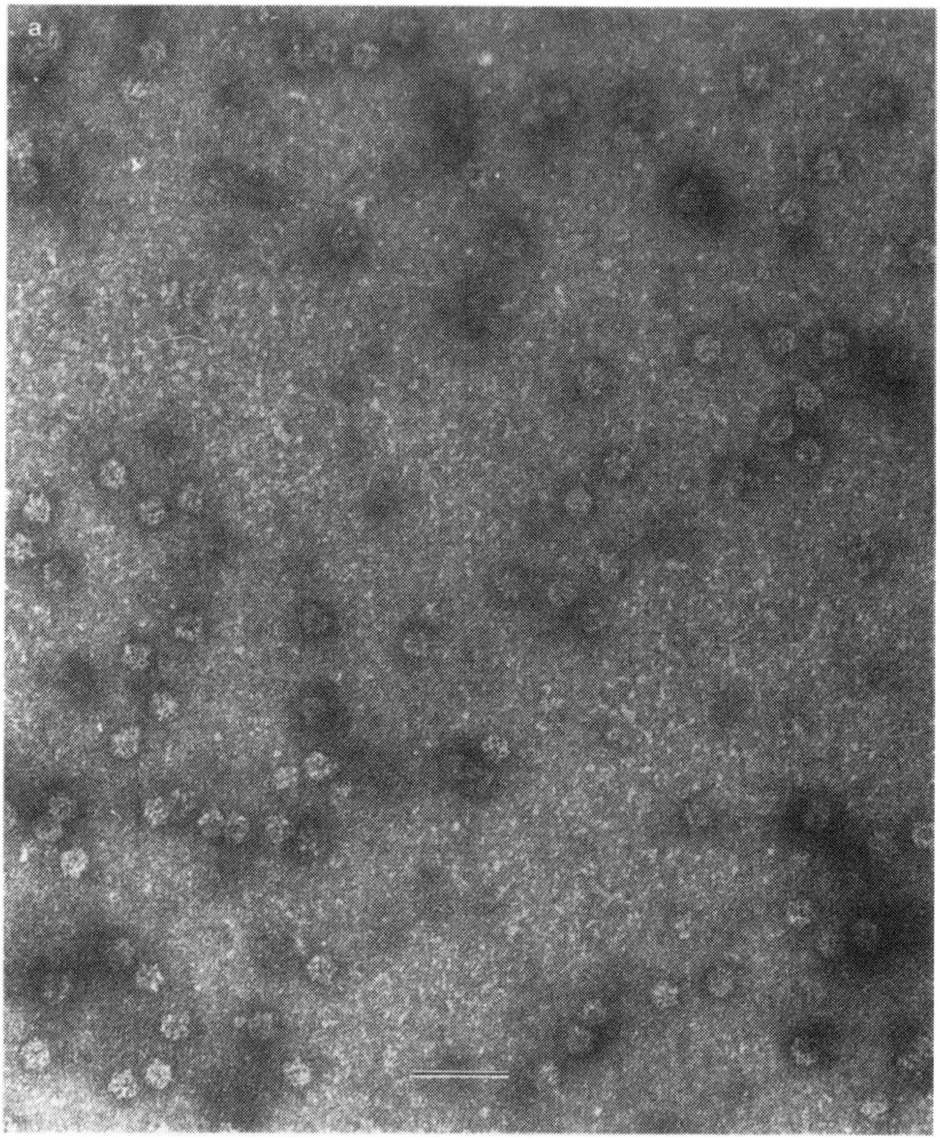

FIGURE 22. Transmission electron micrographs of human erythrocyte Mg^{2+}-ATPase, showing single M_r 600,000 hexagonal molecules (a) and molecular aggregates (b). When oriented on-edge within the aggregates, the molecules are revealed as double-layered structures (arrowheads). Negatively stained with 2% uranyl acetate after spreading on mica. Bars = 50 nm.

stacked "edge-on" molecules when specimens are prepared by the negative-staining carbon film technique (Figure 22). Comparison of the electron optical images of negatively stained Mg^{2+}-ATPase with those generated from muscle glycogen synthase (Figure 20) (see Section 5.1) indicates a considerable degree of morphological similarity, which is also supported by the native M_r of glycogen synthase (600,000) and principal subunit M_r

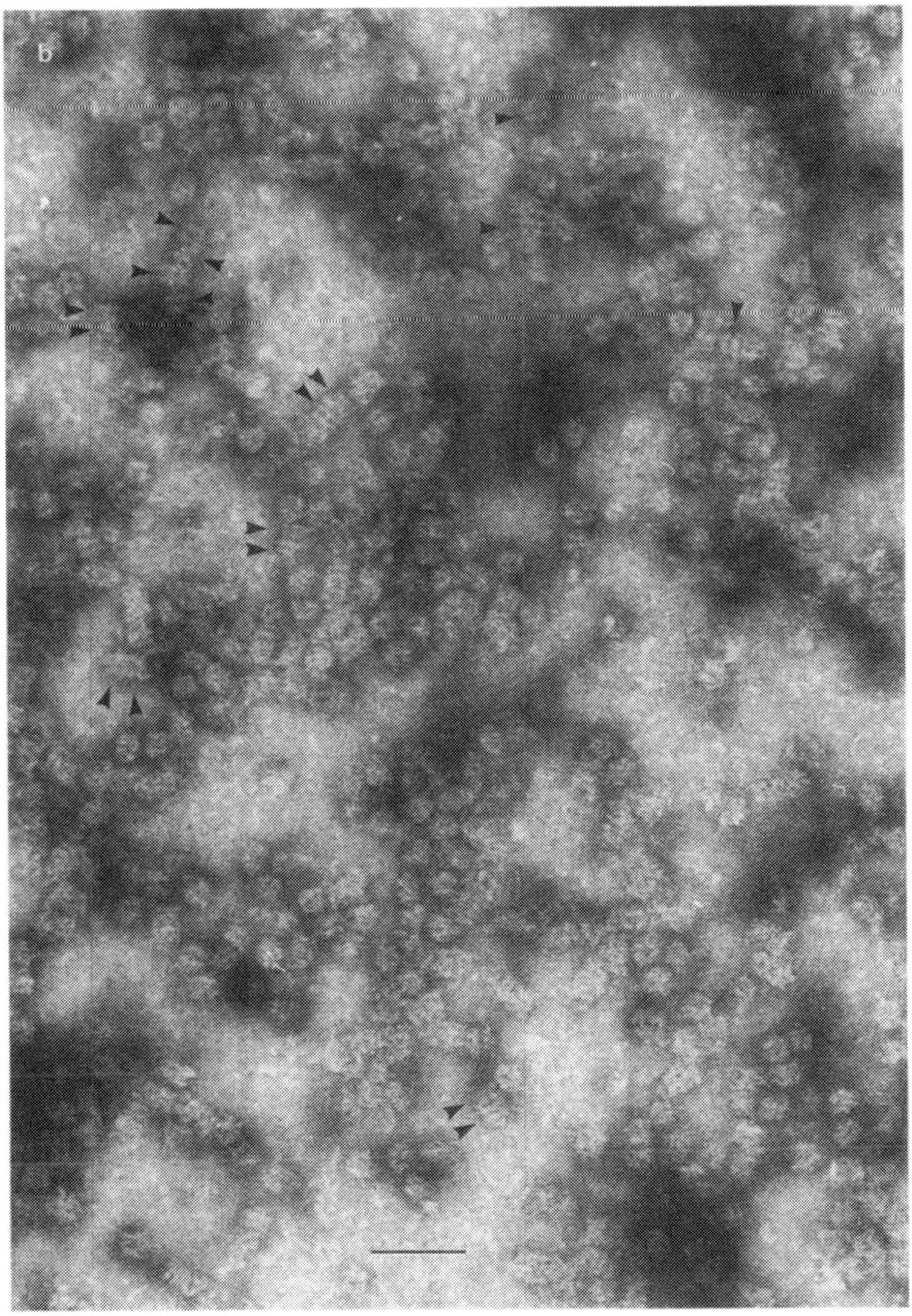

FIGURE 22. *(Continued)*

of 90,000. Furthermore, the fact that both muscle glycogen synthase and the erythrocyte Mg^{2+}-ATPase stain histochemically for ATPase on nondissociating PAGE also indicates their identity, but it must be emphasized that it is possible that the glycogen synthase is contaminated by a copurifying Mg^{2+}-ATPase. Alternatively, it is certainly not inconceivable that the enzyme is bifunctional, but extreme care must be taken and further experimentation performed before pursuing this hypothesis.

6. CATALASE AND TORIN

6.1. Catalase

Catalase (band 4.5) (EC 1.11.1.6) is one of the most abundant enzymes in animal, plant, and bacterial cells. Historically, it was probably the first enzymatic function to be defined and it is one of the predominant enzymes of the erythrocyte (Deisseroth and Dounce, 1970; Rapoport and Müller, 1974). Erythrocyte catalase has a native M_r of 256,000 and contains four identical subunits of M_r 64,000; these figures are somewhat higher than the reported values obtained from the liver enzyme. Multiple molecular forms of catalase have been reported (Aebi *et al.*, 1974; Contz *et al.*, 1968), yet the enzyme activity within reticulocytes and erythrocytes is apparently the same. Considerable variation is, however, shown for the catalase activity within erythrocytes of different species. Catalase has been purified from human and cattle erythrocytes (Bonaventura *et al.*, 1972; Deisseroth and Dounce, 1969; Stansell and Deutsch, 1965) and human erythrocyte catalase is now available commercially (Behring Diagnostics, La Jolla, Calif.). Both human erythrocyte and bovine liver catalase have been studied in the TEM by Kiselev *et al.* (1967) who showed that crystallization from 1.5 M ammonium sulfate produced tubular arrays with monomolecular walls. Tubes of human erythrocyte catalase possessed an external diameter of 42 nm, whereas those formed from bovine liver catalase had an external diameter of 31 nm. It was suggested that this variation in the diameter of the tubular aggregates might be a reflection of minor structural differences between the two molecules. Figure 23 shows individual molecules of human erythrocyte catalase stained with uranyl acetate by the negative-staining carbon film procedure (Horne and Pasquali-Ronchetti, 1974). This procedure, which has the ability to induce the formation of two-dimensional crystals, particularly in the presence of polyethylene glycol (0.05%), has been found to work extremely well with erythrocyte catalase, which readily forms two-dimensional arrays, as shown in Figure 24 (Harris and Holzenburg, 1989). Specimens of this type hold considerable potential for further high-resolution TEM study, digital image analysis and reconstruction, for correlation with data obtained by X-ray diffraction of enzyme crystals. That cytosolic catalase has an affinity for the erythrocyte membrane or membrane skeletal proteins was indicated by its presence in low-ionic-strength extracts prepared from erythrocyte ghosts (Deas *et al.*, 1978). This interaction has been shown by Aviram and Shaklai (1981) to be dependent on the pH and ionic strength of the suspension medium. Furthermore, Allen *et al.* (1977) and Allen and Cadman (1979) have shown that normal erythrocytes incubated or hemolysed in the presence of 1.0 mM calcium yielded ghosts that possessed an increased quantity of the SDS-PAGE band 4.5 ($M_r \sim 60{,}000$), which was identified as the catalase subunit. The band 4.5 protein was also found to be

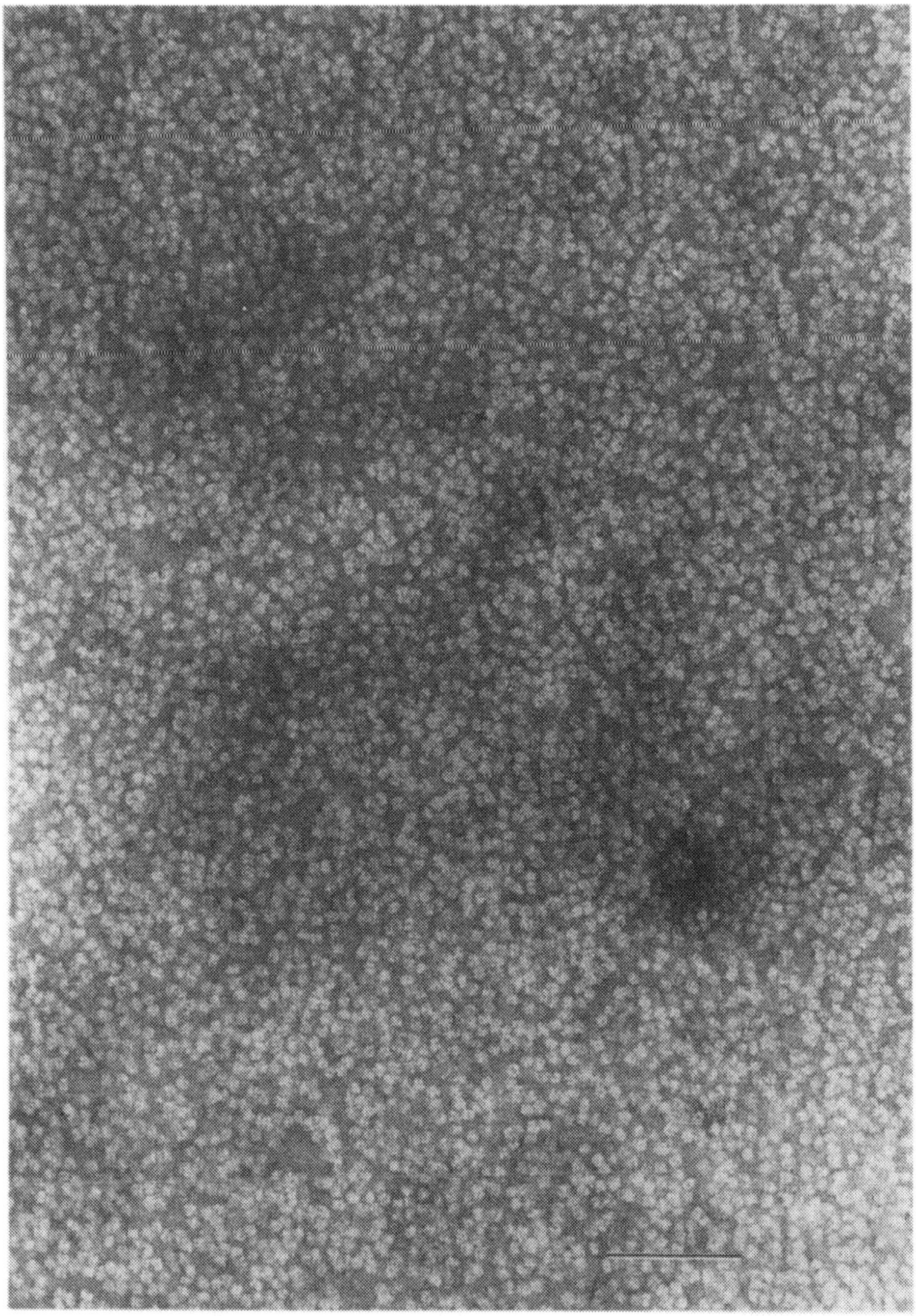

FIGURE 23. A transmission electron micrograph of human erythrocyte catalase, showing single molecules. Negatively stained with 2% uranyl acetate after spreading on mica in the absence of PEG. Bar = 100 nm.

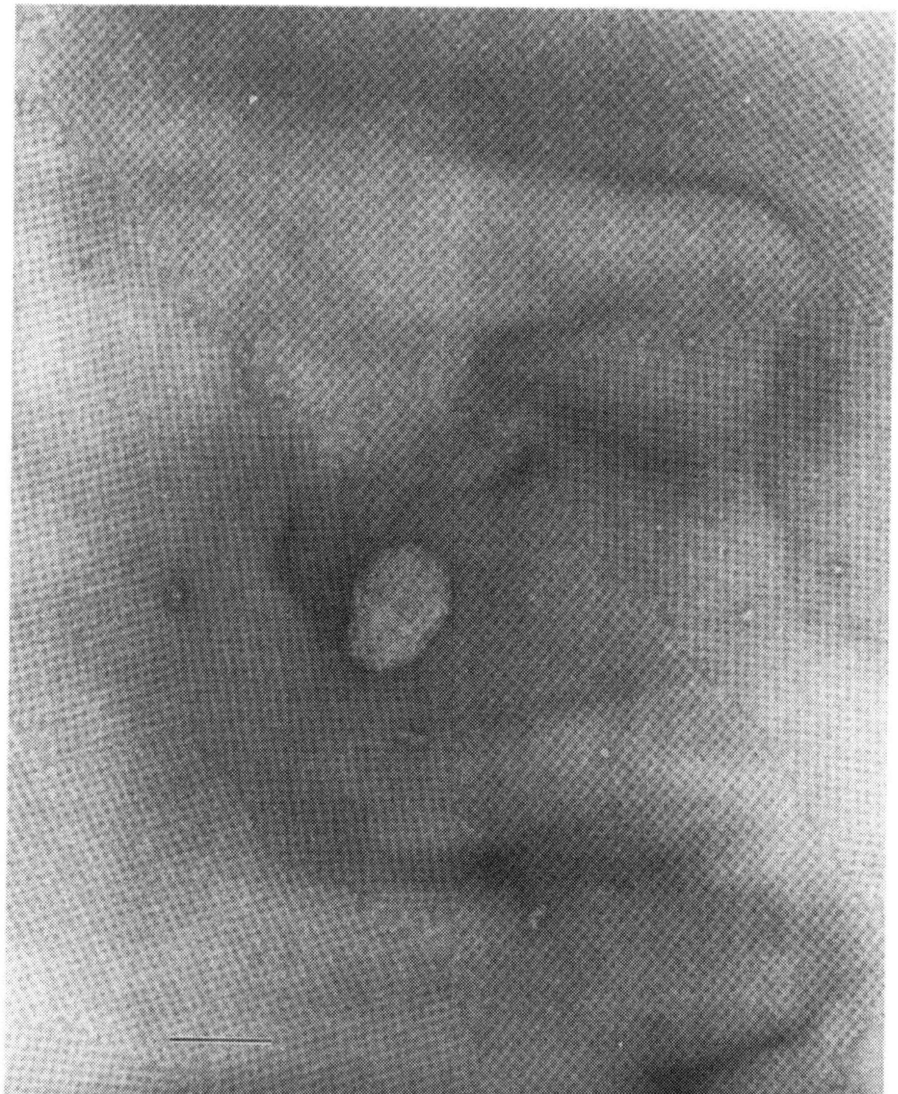

FIGURE 24. A transmission electron micrograph of human erythrocyte catalase showing two-dimensional molecular arrays, with well-defined crystallization discontinuities. Negatively stained with 2% uranyl acetate after spreading on mica in the presence of 0.05% PEG (M_r 10,000). Bar = 100 nm. (See Harris and Holzenburg, 1989.)

slightly increased in erythrocyte ghosts prepared from hereditary spherocytosis (HS) erythrocytes, compared to those from normal erythrocytes. This was thought to be due to the increased cytosolic calcium level within HS erythrocytes, resulting from an increased membrane permeability to plasmatic calcium. Figure 25 shows PAGE and SDS-PAGE of purified human erythrocyte catalase and of catalase present in the low-ionic-strength extract (pH 5.0 supernatant) from erythrocyte ghosts (J.R.H., previously unpublished data) partially purified by DEAE chromatography. It is significant that both catalase and

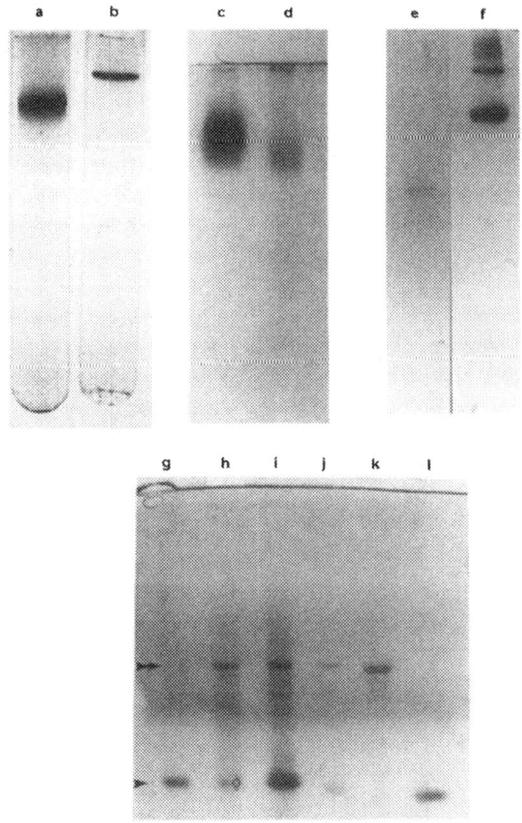

FIGURE 25. Native PAGE and SDS-PAGE of human erythrocyte catalase and human erythrocyte torin. Channels a and e show native torin electrophoresed on a 7% polyacrylamide tube gel and on a 2.5 to 27% polyacrylamide gradient gel, respectively. Channels c and d show human erythrocyte catalase electrophoresed on a 2.5 to 27% polyacrylamide gel. Channels b and f show human erythrocyte cylindrin and horse spleen apoferritin (monomer, dimer, trimer, and higher oligomers). After dissociation with 2% SDS–β-mercaptoethanol and SDS-PAGE, purified torin (channel g) and purified catalase (channel k) are revealed as single peptides (M_r 20,000 and 60,000, respectively; single and double arrowheads). Channels h–j show mixtures of torin and cylindrin obtained during purification, and channel 1 shows the M_r 18,000 subunit of horse spleen apoferritin.

the protein torin (band 8) (see below) have a tendency to copurify, but they are readily distinguished by their differing subunit molecular weights (i.e., M_r 60,000 and 20,000) as shown in Figure 25, whereas the electrophoretic separation of the native proteins is very much less distinct.

6.2. Torin

The human erythrocyte protein torin (band 8) was discovered by the author in 1967, along with the larger protein cylindrin (Section 2.2), when TEM studies were performed on negatively stained specimens made using the low-ionic-strength (water dialysis) extract obtained from human erythrocyte ghosts (Harris, 1968). Initially, the protein was termed the "single torus protein," but this was subsequently changed to the simpler term "torin" (Harris, 1980). It may be of significance that of the limited number of investigations performed using cattle erythrocytes, only the protein cylindrin has been detected. These studies were, however, performed at the early stages of this investigation, when it was thought that torin might be a dissociated form of cylindrin; this is now known to not be the case.

While it is clearly apparent that TEM has made a major contribution to these studies, and indeed to all studies dealing with the 20 S protease complex, which is to all intents

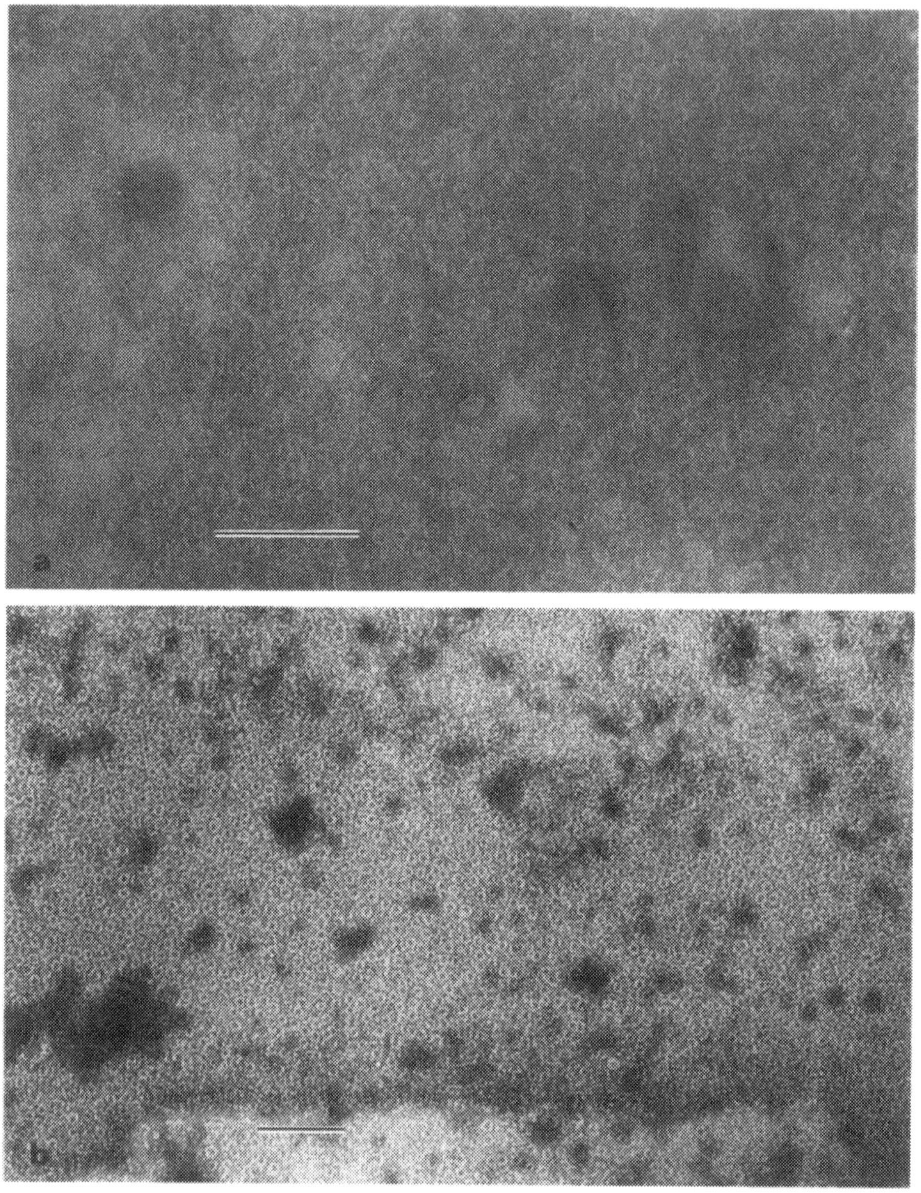

FIGURE 26. Transmission electron micrographs of human erythrocyte torin, showing individual molecules oriented horizontally (ringlike images). Negatively stained with 2% sodium phosphotungstate (a) and uranyl acetate (b). Bars = 100 nm.

purposes identical to cylindrin (Section 2), it does not follow that progress would not have been made in the absence of this extremely direct biophysical technique. The ultrastructural characterization of torin was presented in some detail by Harris (1969a). This ringlike protein has an external diameter of 13.1±0.5 nm and an inner diameter of 6.0±0.4 nm, determined from negatively stained images of the protein oriented horizon-

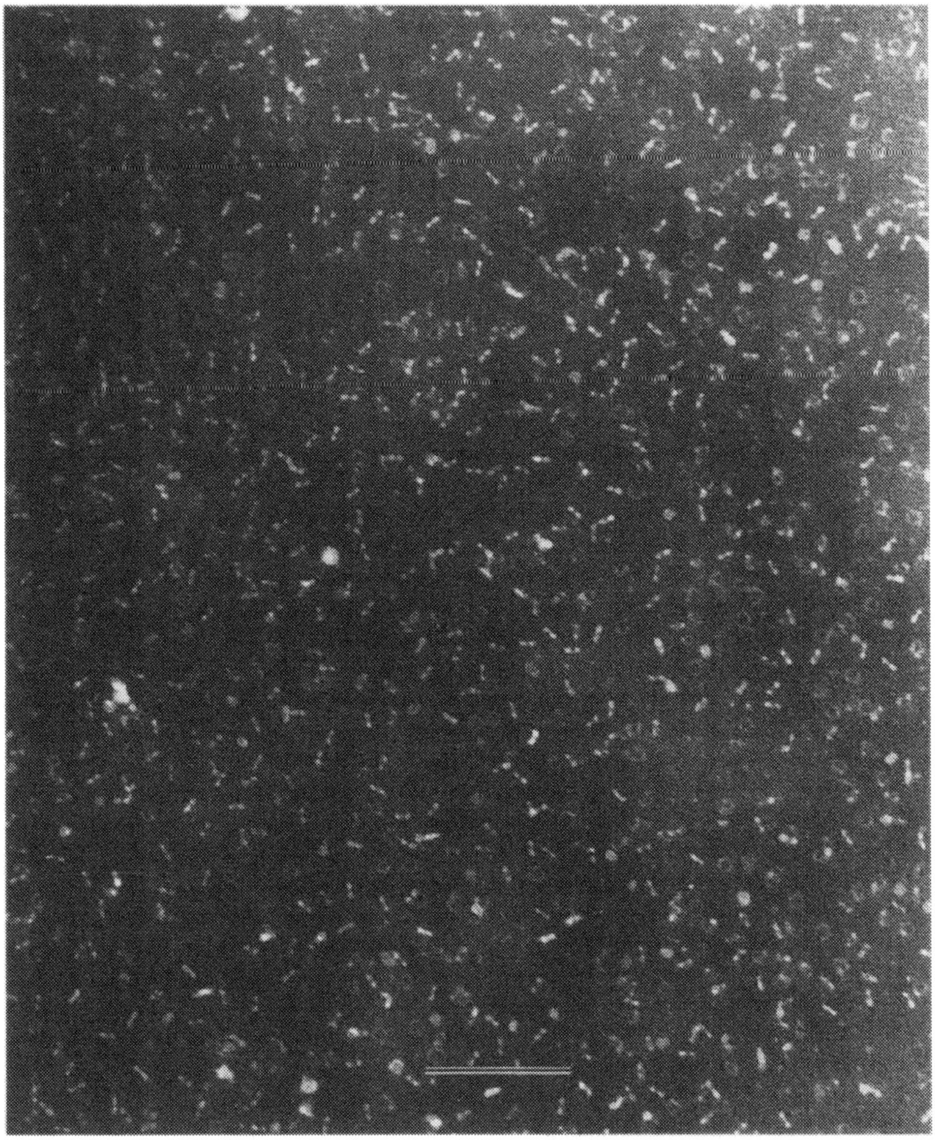

FIGURE 27. A transmission electron micrograph of human erythrocyte torin, showing individual molecules oriented on-edge (double-dot images) within relatively deep negative stain (2% sodium phosphotungstate). Numerous, somewhat less distinct, oval and circular images of torin are also present. Bar = 100 nm.

tally (lying flat, as shown in Figure 26). When oriented vertically (on-edge) in relatively deep negative stain, the protein is revealed as the "double dot" image, which from theoretical considerations is the expected electron optical image derived from a ring of electron-transparent protein embedded within electron-dense stain (Figure 27). Application of the photographic rotation technique for contrast enhancement (Markham *et al.*, 1963) indicates the presence of tenfold rotational symmetry, therefore suggesting the presence of ten subunits within the native molecule. Biochemical characterization of torin has also been performed (Harris, 1969b, 1971, 1974, 1980; Harris and Naeem, 1981). Briefly, the native protein, purified initially by sucrose density gradient centrifugation, was found to migrate electrophoretically as a single band, ahead of cylindrin on non-dissociating PAGE, and to possess a sedimentation coefficient of 9.0 S (Harris, 1971). Purification was also achieved by preparative PAGE (Harris, 1974), but this proved to be a rather tedious and unreliable approach. Further analysis of torin by SDS-PAGE (Harris and Naeem, 1978, 1981) has revealed the presence of a single subunit of M_r 20,000 (Figure 25b). This implies that the native M_r will be 200,000, a figure compatible with the sedimentation coefficient (9.0 S) and with the molecular dimensions revealed by electron microscopy. The isoelectric point (pI) of torin was determined to be 4.8 (Harris and Naeem, 1981), but unlike cylindrin, torin does not precipitate from solution of this pH, a fact now utilized to separate cylindrin and torin. It should be mentioned that the work of Howe and Bächi (1973) independently confirmed the existence of torin and cylindrin, and showed that antibody against purified torin bound specifically to the internal aspect of the erythrocyte ghosts. Despite this evidence and the earlier claims of the author that both torin and cylindrin are "membrane associated," it now appears that this association is due primarily to the steric entrapment of the proteins within the erythrocyte ghost, rather than a firm binding to the cytoplasmic surface of the membrane (see Section 1). This binding may, however, play some as yet poorly defined role.

Even though the author has made considerable efforts to determine a biochemical functional for torin, no positive function can as yet be ascribed to this protein. Since foraminotransferase-cyclodeaminase from porcine liver possesses a similar ringlike structure (eight subunits), this indicates strongly that torin will eventually be shown to possess an enzyme activity.

7. GLYCERALDEHYDE-3-PHOSPHATE DEHYDROGENASE

Although the native form of glyceraldehyde-3-phosphate dehydrogenase (G3PD) (EC 1.2.1.12) has a M_r of 148,000 and is composed of four identical subunits, there are no reported electron microscopic studies on this erythrocyte enzyme. G3PD has been included because of the considerable interest in the adsorption of this cytosolic enzyme to the erythrocyte membrane. The original work of Tanner and Gray (1971) identified a membrane polypeptide of $M_r \sim 33{,}000$ by SDS-PAGE, which by N-terminal sequencing was identified as G3PD. It was concluded from this study that G3PD was unlikely to be an integral part of the erythrocyte membrane. Subsequent work by Tanner and others (Kant and Steck, 1973; McDaniel *et al.*, 1974; Shin and Carraway, 1973) defined the G3PD polypeptide as band 6 in the erythrocyte membrane polypeptide profile and indicated that up to 80% of the erythrocyte enzyme remained within the erythrocyte ghost fraction after hemolysis and washing. Elution of G3PD from the membrane was achieved by increasing salt concentrations, and a reversible ionic association of the enzyme with the cytoplasmic surface of the membrane was shown to occur *in vitro*, which was thought to be of relevance *in vivo* during metabolic variations.

More recently, Allen and his colleagues (Allen *et al.*, 1987) have provided further biochemical and genetic data (i.e., peptide maps and nucleotide sequences) that support the identity of the M_r 37,000 polypeptide with G3PD. Electrofocusing studies (Lin and Allen, 1986) indicated microheterogeneity of G3PD, which probably accounts for the varied binding of this enzyme to the erythrocyte membranes isolated from different species (Ballas, 1987; Yu and Steck, 1975). Furthermore, a recent study by Ryazanov *et al.* (1988) has suggested that rabbit reticulocyte G3PD is associated with mono- and polyribosomes. Using erythrocytes and ghosts in conjunction with a double immunofluorescence technique for the detection of G3PD, Rogalski *et al.* (1989) have shown that in humans the enzyme is located primarily at the cell membrane, whereas for rats it is distributed more evenly throughout the cytoplasm.

8. PHOSPHORIBOSYLPYROPHOSPHATE SYNTHETASE

Phosphoribosylpyrophosphate synthetase (PRPP synthetase) (EC 2.7.6.1) is a key enzyme in the pathway of *de novo* purine nucleotide synthesis. Within the context of this review, PRPP synthetase exhibits several interesting properties that integrate appropriately within the overall theme.

The synthesis of PRPP has been extensively reviewed by Becker *et al.* (1979), but there is a notable lack of more recent research data relating to this enzyme. Purified human erythrocyte PRPP synthetase has been found to have a single subunit of M_r 33,000 (Becker *et al.*, 1977; Fox and Kelley, 1971), whereas the rat liver enzyme has a subunit M_r of 48,000 (Roth *et al.*, 1974). In both cases, however, the enzyme forms a range of high-molecular-weight aggregates, purportedly composed of up to 16 or 32 subunits (Becker *et al.*, 1977), which collate with a range of native M_r ranging from 60,000 (dimer) up to 1,200,000. Fox and Kelley (1971) maintained that the M_r 720,000 and 1,200,000 species predominated. TEM data from rat liver PRPP synthetase (Roth *et al.*, 1974) indicate the presence of stacked aggregates of varying length, but to date no equivalent data have been generated using the erythrocyte enzyme. However, electron micrographs of *E. coli* PRPP synthetase (Sigma) indicate that this enzyme possesses a high molecular weight (J.R.H., unpublished data) and that it tends to aggregate in the presence of inorganic phosphate and Mg-ATP. This is supported by the work of Switzer (1969) using *Salmonella typhimurium* PRPP synthetase, which was shown by sucrose density gradient centrifugation to have a M_r of 540,000, and by gel electrophoresis to form different oligomeric states in the presence or absence of inorganic phosphate.

Self-association of the human erythrocyte enzyme is also promoted by increasing enzyme concentration and by Mg-ATP and Mg alone, in the presence of inorganic phosphate. Conversely, dissociation is promoted by 2,3-diphosphoglycerate and purine nucleotides. While the physiological significance of the aggregation of PRPP synthetase is not known, there is evidence indicating that full activity is only present when the enzyme is highly oligomerized (Mayer and Becker, 1977) (see TPP-II, Section 2.4). Whether or not this is the case in the presence of cellular 2,3-diphosphoglycerate levels is not clear. A number of hereditary hyperactive variants of human PRPP synthetase have been identified and there is some indication that this may be due in certain instances to an increased stabilization of the high-molecular-weight aggregates (Becker, 1976; Becker *et al.*, 1979), whereas in others the increased activity was thought to be due to a diminished inhibition by 2,3-diphosphoglycerate and purine nucleotides.

9. CONCLUDING COMMENTS

Erythrocyte enzymology is a steadily expanding subject, and no attempt has been made in this review to cover it in its entirety. Rather, a very selective and personal approach has been adopted, which provides some insight into the study of a number of high-molecular-weight oligomeric enzymes.

The role of TEM in this area, when placed alongside the more conventional biochemical approaches such as enzymology, gradient centrifugation, gel electrophoresis, and the various forms of chromatography, is shown to be extremely powerful, in particular because directly interpretable molecular information is made available. Nevertheless, as with all biophysical techniques, TEM has its limitations. Because of the subjective nature of the data, great care has often to be taken to avoid personal interpretations made from a limited selection of data. In the future this aspect is likely to be lessened by the more widespread availability of the advanced procedures for digital image analysis of single molecules (Baumeister *et al.*, 1988) and of two-dimensional crystals, which essentially removes the subjective interpretation (see *Journal of Electron Microscopy Technique*, Volume 9, Number 4, 1988).

The impact of immunoelectron microscopy and affinity labeling at the macromolecular level is already beginning to yield interesting data on the location of specific subunits within oligomeric complexes. It can be confidently expected that through the application of monoclonal antibodies, the location of individual epitopes on subunits within oligomeric enzymes will be revealed and their accessibility within the overall native complex defined. TEM studies on partly and completely dissociated and reassociated oligomeric complexes are also likely to yield useful information, as may limited proteolytic cleavage. In the long run, advances are likely to result from the continued application of biochemical, biophysical, and immunological studies performed in parallel with TEM.

10. REFERENCES

Abramic, M., Zubanovic, M., and Vitale, L., 1988, Dipeptidyl peptidase III from human erythrocytes, *Biol. Chem. Hoppe-Seyler* **369:**29–38.

Aebi, H., Wyss, S. R., Scherz, B., and Skvaril, F., 1974, Heterogeneity of erythrocyte catalase. II. Isolation and characterization of normal and variant erythrocyte catalase and their subunits, *Eur. J. Biochem.* **48:**137–145.

Akhayat, O., Grossi de Sa, M.-F., and Infante, A. A., 1987a, Sea urchin prosome: Characterization and changes during development, *Proc. Natl. Acad. Sci. USA* **84:**1595–1599.

Akhayat, O., Infante, A. A., Infante, D., Martins de Sa, C., Grossi de Sa, M.-F., and Scherrer, K., 1987b, A new type of prosome-like particle, composed of small cytoplasmic RNA and multimers of a 21-kDa protein, inhibits protein synthesis *in vitro*, *Eur. J. Biochem.* **170:**23–33.

Allen, D. W., and Cadman, S., 1979, Calcium-induced erythrocyte membrane changes, the role of adsorption of cytosol proteins and proteases, *Biochim. Biophys. Acta* **551:**1–9.

Allen, D. W., Cadman, S., McCann, S. R., and Finkel, B., 1977, Increased membrane binding of erythrocyte catalase in HS and in metabolically stressed normal cells, *Blood* **49:**113–123.

Allen, D. W., Groat, J. D., Finkel, B., Rank, B. H., Wood, P. A., and Eaton, J. W., 1983, Increased adsorption of cytoplasmic proteins to the erythrocyte membrane in ATP-depleted normal and pyruvate kinase-deficient mature cells and reticulocytes, *Am. J. Hematol.* **14:**11–25.

Allen, R. W., Trach, K. A., and Hoch, J. A., 1987, Identification of the 37kDa protein displaying a variable interaction with the erythroid membrane as glyceraldehyde-3-phosphate dehydrogenase, *J. Biol. Chem.* **262:**649–653.

Anderson, D. R., Davis, J. L., and Carraway, K. L., 1977, Ca-promoted changes of the human erythrocyte membrane. Involvement of spectrin, transglutaminase and a membrane-bound protease, *J. Biol. Chem.* **252:**6617–6623.

Arrigo, A.-P., Tanaka, K., Goldberg, A. L., and Welch, W. J., 1988, Identity of the 19 S 'prosome' particle with the large multifunctional protease complex of mammalian cells (the proteasome), *Nature* **331:**192–194.

Aviram, I., and Shaklai, N., 1981, The association of human erythrocyte catalase with the cell membrane, *Arch. Biochem. Biophys.* **212:**329–337.

Ballas, S. K., 1987, Comparative distribution of glyceraldehyde-3-phosphate dehydrogenase activity in human, guinea pig, rabbit and mouse erythrocytes, *Comp. Biochem. Physiol. B* **87:**837–842.

Ballas, S. K., and Burka, E. R., 1979, Protease activity in the human erythrocyte—Localization to the cell membrane, *Blood* **53:**875–882.

Bålöw, R.-M., Tomkinson, B., Ragnarsson, U., and Zetterqvist, Ö., 1986, Purification, substrate specificity and classification of tripeptidylpeptidase II, *J. Biol. Chem.* **261:**2409–2417.

Baumeister, W., Dahlmann, B., Heegerl, R., Kopp, F., Kuehn, L., and Pfeifer, G., 1988, Electron microscopy and image analysis of the multicatalytic proteinase, *FEBS Lett.* **241:**239–245.

Becker, M. A., 1976, Patterns of phosphoribosylpyrophosphate and ribose-5-phosphate concentration and generation in fibroblasts from patients with group and purine overproduction, *J. Clin. Invest.* **56:**308–318.

Becker, M. A., Meyer, L. J., Huisman, W. H., Lazar, C., and Adams, W. B., 1977, Human erythrocyte phosphoribosylpyrophosphate synthetase, *J. Biol. Chem.* **252:**3911–3918.

Becker, M. A., Raivio, K. O., and Seegmiller, J. E., 1979, Synthesis of phosphoribosylpyrophosphate in mammalian cells, *Adv. Enzymol.* **49:**281–306.

Bernacki, R. J., and Bosmann, H. B., 1972, Red cell hydrolases. II. Proteinase activities in human erythrocyte plasma membranes, *J. Membr. Biol.* **7:**1–14.

Bessis, M. C., and Breton-Gorius, J., 1959, Ferritin and ferruginous micelles in normal erythroblasts and hypochromic hyperesideremic anemias, *Blood* **14:**423–432.

Beutler, E., 1975, *Red Cell Metabolism: A Manual of Biochemical Methods,* Grune & Stratton, New York.

Beutler, E., 1986, *Methods in Hematology,* Volume 16, Churchill Livingstone, Edinburgh.

Biagioni, S., Scarsella, G., Settimi, L., and Traina, M. E., 1982, Acetylcholinesterase molecular forms from rat and human erythrocyte membrane, *Mol. Cell. Biochem.* **47:**183–190.

Boches, F. S., and Goldberg, A. L., 1982, Role for the adenosine triphosphate-dependent proteolytic pathway in reticulocyte maturation, *Science* **215:**978–980.

Bommer, U.-A., Lutsch, G., Behlke, J., Stahl, J., Nesytova, N., Henske, A., and Bielka, H., 1988, Shape and location of eukaryotic initiation factor eIF-2 on the 40S ribosomal subunit of rat liver, *Eur. J. Biochem.* **172:**653–662.

Bonventura, J., Schroeder, W. A., and Fang, S., 1972, Human erythrocyte catalase: An improved method of isolation and a reevaluation of reported properties, *Arch. Biochem. Biophys.* **150:**606–617.

Boublik, M., Hellman, W., Staehelin, T., and Trachsel, H., 1983, Electron microscopic study of eukaryotic 40S initiation complex in protein synthesis, *Eur. J. Cell Biol.* **32:**136–142.

Bramley, T. A., Coleman, R., and Finean, J. B., 1971, Chemical, enzymological and permeability properties of human erythrocyte ghosts prepared by hypotonic lysis in media of different osmolarities, *Biochim. Biophys. Acta* **241:**752–769.

Brewer, G. J., 1984, *Red Cell: Progress in Clinical and Biological Research,* Volume 165 (G. J. Brewer, ed.), Liss, New York.

Brimijoin, S., Hammond, P. I., and Petit, R. M., 1986, Paroxysmal nocturnal hemoglobinuria: Erythrocyte deficit analyzed immunoassay and fluorescence-activated sorting, *Mayo Clin. Proc.* **61:**522–529.

Brockman, S. K., Usiak, M. F., and Younbkin, S. G., 1986, Assembly of monomeric acetylcholinesterase into tetrameric and asymmetric forms, *J. Biol. Chem.* **261:**1201–1207.

Brodbeck, U., 1986, Amphiphilic acetylcholinesterase: Properties and interactions with lipids and detergents, in: *Progress in Protein–Lipid Interactions 2* (W. Watts and J. J. H. H. M. de Pont, eds.), pp. 303–338, Elsevier, Amsterdam.

Brodbeck, U., Ott, P., and Wiedmer, T., 1975, Comparative studies on the molecular properties of purified acetylcholinesterase from human erythrocytes and the electric organ of *Electrophorus electricus, Croat. Chem. Acta* **47:**201–210.

Brown, J. N., and Harris, J. R., 1970, The entry of ferritin into hemoglobin-free human erythrocyte ghosts prepared under different conditions, *J. Ultrastruct. Res.* **32**:405–416.

Brown-Luedi, M. L., Meyer, L. J., Milburn, S. C., Yau, P. M.-P., Corbet, S., and Hershey, J. W. B., 1982, Protein synthesis initiation factors from human HeLa cells and rabbit reticulocytes are similar, *Biochemistry* **21**:4202–4206.

Burapakulsolsri, N., Yuthavong, Y., and Wilairat, P., 1979, An examination of complement proteins on membranes of paroxysmal nocturnal haemoglobinuria (PNH) and PNH-like red cells, *Br. J. Haematol.* **41**:393–398.

Burkholder, D. E., and Brecher, A. S., 1972, Interaction between proteases and bovine erythrocyte membranes, *Biochim. Biophys. Acta* **282**:135–145.

Carraway, K. L., Kobylka, D., and Triplett, R. B., 1971, Surface proteins of erythrocyte membrane, *Biochim. Biophys. Acta* **241**:934–940.

Castaño, J. G., Ornberg, R., Koster, J. G., Tobian, J. A., and Zasloff, M., 1986, Eukaryotic pre-tRNA 5′ processing nuclease: Copurification with a complex cylindrical particle, *Cell* **46**:377–387.

Chow, F.-L., Telen, M. J., and Rosse, W. F., 1985, The acetylcholinesterase deficit in paroxysmal nocturnal hemoglobinuria. Evidence that the enzyme is absent from the cell membrane, *Blood* **66**:940–945.

Ciechanover, A., Ferber, S., Ganoth, D., Elias, S., Hershko, A., and Arfin, S., 1988, Purification and characterization of arginyl-tRNA protein transferase from rabbit reticulocytes, *J. Biol. Chem.* **263**:11155–11167.

Ciliv, G., and Özand, P. T., 1972, Human erythrocyte acetylcholinesterase purification, properties and kinetic behavior, *Biochim. Biophys. Acta* **284**:136–156.

Contz, M., Mörkorfer-Zwez, S., Bossi, E., Kaufmann, H., van Wartburg, J. P., and Aebi, H., 1968, Alternative molecular forms of erythrocyte catalase, *Experientia* **24**:119–121.

Dahlmann, B., Rutschmann, M., Kuehn, L., and Reinauer, H., 1985, *Biochem. J.* **228**:171–177.

Deas, J. E., Lea, L. T., and Howe, C., 1978, Peripheral proteins of human erythrocytes, *Biochem. Biophys. Res. Commun.* **82**:296–304.

Deisseroth, A., and Dounce, A. L., 1969, Purification and crystallization of beef erythrocyte catalase, *Arch. Biochem. Biophys.* **131**:18–29.

Deisseroth, A., and Dounce, A. L., 1970, Catalase: Physical and chemical properties, *Physiol. Rev.* **50**:315–375.

Dickey-Dunkirk, S., and Killilea, S. D., 1985, Purification of bovine heart glycogen synthase, *Anal. Biochem.* **146**:199–205.

Dockter, M. E., and Morrison, M., 1986, Paroxysmal nocturnal hemoglobinuria erythrocytes are of two distinct types: Positive or negative for acetylcholinesterase, *Blood* **67**:540–543.

Dodge, J. T., Mitchell, C., and Hanahan, D. J., 1963, The preparation and chemical characteristics of hemoglobin-free ghosts of human erythrocytes, *Arch Biochem. Biophys.* **100**:119–130.

Domae, N., Harmon, F. R., Busch, R. K., Spohn, W., Subrahmanyam, C. S., and Busch, H., 1982, Donut-shaped "miniparticles" in nuclei of human and rat cells, *Life Sci.* **30**:469–477.

Dutta-Choudhury, T. A., and Rosenberry, T. L., 1984, Human erythrocyte acetylcholinesterase is an amphipathic protein whose short membrane-binding domain is removed by papain digestion, *J. Biol. Chem.* **259**:5653–5660.

Eisinger, J., Flores, J., and Salhany, J. M., 1982, Association of cytosol hemoglobin with the membrane in intact erythrocytes, *Proc. Natl. Acad. Sci. USA* **79**:408–412.

Emanuilov, I., Sabatini, D. D., Lake, J. A., and Freienstein, C., 1978, Localization of eukaryotic initiation factor 3 on native small ribosomal subunits, *Proc. Natl. Acad. Sci. USA* **75**:1389–1393.

Etlinger, J. D., and Goldberg, A. L., 1977, A soluble ATP-dependent proteolytic system responsible for the degradation of abnormal proteins in reticulocytes, *Proc. Natl. Acad. Sci. USA* **74**:54–58.

Fagan, J. M., Waxman, L., and Goldberg, A. L., 1987, Skeletal muscle and liver contain a soluble ATP+ubiquitin-dependent proteolytic system, *Biochem. J.* **243**:335–343.

Falkenburg, P.-E., Haass, C., Kloetzel, P.-M., Niedel, B., Kopp, F., Kuehn, L., and Dahlmann, B., 1988, *Drosophila* small cytoplasmic 19S ribonucleoprotein is homologous to the rat multicatalytic proteinase, *Nature* **331**:190–192.

Firkin, R. G., Beal, R. W., and Mitchell, G., 1963, The effects of trypsin and chymotrypsin on the acetylcholinesterase content of human erythrocytes, *Aust. Ann. Med.* **12**:26–29.

Fox, I. H., and Kelley, W. N., 1971, Human phosphoribosylpyrophosphate synthetase, *J. Biol. Chem.* **246**:5739–5748.

Friedman, H., and Rapoport, S. M., 1974, Enzymes of the red cell; A critical catalogue in: *Cellular and Molecular Biology of the Erythrocyte* (H. Yoshikawa and S. M. Rapoport, eds.), pp. 181–259, University Park Press, Baltimore.

Godar, D. E., Godar, D. E., Garcia, V., Jacob, A., Aebi, U., and Yang, D. C. H., 1988, Structural organization of the multienzyme complex of mammalian aminoacyl-tRNA synthetases, *Biochemistry* **27**:6921–6928.

Granboulan, N., Spohr, G., Kayibanda, B., and Scherrer, K., 1970, Examen au microscope electronique de complexes RNA messager-proteins, 7th International Congress on Electron Microscopy, Grenoble, pp. 591–592, Societé Français de Microscopie Electronique, Paris.

Gulik, A., and Orsini, G., 1984, Electron microscopy of the aminoacyl-tRNA synthetase multienzyme complex purified from rabbit reticulocytes, *Mol. Biol. Rep.* **10**:23–30.

Harmon, F. R., Spohn, W. H., Domae, N., Ha, C. S., and Busch, H., 1983, Purification and partial characterization of ring-shaped miniparticles, *Cell Biol. Int. Rep.* **7**:333–343.

Harris, J. R., 1968, Release of a macromolecular protein component from human erythrocyte ghosts, *Biochim. Biophys. Acta* **150**:534–537.

Harris, J. R., 1969a, Some negative contrast staining features of a protein from erythrocyte ghosts, *J. Mol. Biol.* **46**:329–335.

Harris, J. R., 1969b, The isolation and purification of a macromolecular protein component from the human erythrocyte ghost, *Biochim. Biophys. Acta* **188**:31–42.

Harris, J. R., 1971, Further studies on the proteins released from haemoglobin-free erythrocyte ghosts at low ionic strength, *Biochim. Biophys. Acta* **229**:761–770.

Harris, J. R., 1974, The purification of some membrane-associated proteins from erythrocyte ghosts, in: *Methodological Developments in Biochemistry*, Volume 4 (E. Reid, ed.), pp. 395–404, Longman Group, London.

Harris, J. R., 1980, Torin and cylindrin, two extrinsic proteins of the erythrocyte membrane: A review, *Nouv. Rev. Fr. Hematol.* **22**:411–448.

Harris, J. R., 1982a, Some negative staining electron microscopic and biochemical studies on apoferritin and its oligomers, *Micron* **13**:169–184.

Harris, J. R., 1982b, Nonenzymic proteins, in: *Electron Microscopy of Proteins*, Volume 2 (J. R. Harris, ed.), pp. 49–103, Academic Press, New York.

Harris, J. R., 1983, Comparative studies on cylindrin: Identity with aminoacyl-tRNA synthetase, *Microsc. Acta* **14**:193–205.

Harris, J. R., 1984, Biochemical and ultrastructural characterization of a high molecular soluble Mg^{2+}-ATPase from human erythrocytes, *J. Mol. Biol.* **174**:705–721.

Harris, J. R., 1988a, Release of acetylcholinesterase by limited papain digestion of red blood cells, Proc. ISBT/BBTS Meet., London, p. 157.

Harris, J. R., 1988b, Erythrocyte cylindrin: Possible identity with the ubiquitous 20S high molecular weight protease complex and the prosome particle, *Ind. J. Biochem. Biophys.* **25**:459–466.

Harris, J. R., and Holzenburg, A., 1989, Transmission electron microscopic studies on the quaternary structure of human erythrocyte catalase, *Micron Microsc. Acta,* In press.

Harris, J. R., and Naeem, I., 1978, The subunit composition of two high molecular weight extrinsic proteins from human erythrocyte membranes, *Biochim. Biophys. Acta* **537**:495–500.

Harris, J. R., and Naeem, I., 1981, Further studies on the characterization of cylindrin and torin, two extrinsic proteins of the erythrocyte membrane, *Biochim. Biophys. Acta* **670**:285–290.

Harris, J. R., and Tomkinson, B., 1990, Electron microscopical and biochemical studies on the oligomeric states of human erythrocyte tripeptidyl peptidase-II, *Micron Microsc. Acta,* In press.

Hauri, H.-P., 1988, Biogenesis and intracellular transport of intestinal brush border membrane hydrolases: Use of antibody probes and tissue culture, in: *Subcellular Biochemistry,* Volume 12 (J. R. Harris, ed.), pp. 155–219, Plenum Press, New York.

Herz, F., and Kaplan, E., 1974, *In vitro* modifications of red cell acetylcholinesterase activity, *Br. J. Haematol.* **26**:165–178.

Herz, F., Kaplan, E., and Stevenson, J. H., Jr., 1963, Acetylcholinesterase inactivation of enzyme-treated erythrocytes, *Nature* **200**:901–902.

Hollan, S. R., Szelenyi, J. G., Hasitz, M., Szasa, I., and Gardos, G., 1977, Haemoglobin and the red cell membrane, *Physiol. Bohemoslov.* **26**:219–224.

Hoogeveen, J. T., Juliano, R., Coleman, J., and Rothstein, A., 1970, Water-soluble proteins of the human red cell membrane, *J. Membr. Biol.* **3**:156–172.

Horne, R. W., and Pasquali-Ronchetti, I., 1974, A negative staining-carbon film technique for studying viruses. I. Preparative procedure for studying icosahedral and filamentous viruses, *J. Ultrastruct. Res.* **47**:361–383.

Hough, R., Pratt, G., and Rechsteiner, M., 1987, Purification of two high molecular weight proteases from rabbit reticulocyte lysate, *J. Biol. Chem.* **262**:8303–8313.

Howe, C., and Bächi, T., 1973, Localization of erythrocyte membrane antigens by immune electron microscopy, *Exp. Cell Res.* **76**:321–332.

Hradec, J., 1980, Incorporation of labelled amino acids into proteins, from rabbit reticulocytes, retained on heparin–Sepharose, *Biochim. Biophys. Acta* **610**:285–296.

Issa, H. U., and Mendicino, J., 1973, Role of enzyme–enzyme interactions in the regulation of glycolysis and gluconeogenesis, *J. Biol. Chem.* **248**:685–696.

Izak, G., Wilner, T., and Mager, J., 1960, Amino acid activating enzymes in red blood cells of normal, anemic and polycythemic subjects, *J. Clin. Invest.* **39**:1763–1770.

Kant, J. A., and Steck, T. L., 1973, Specificity in the association of glyceraldehyde-3-phosphate dehydrogenase with isolated human erythrocyte membrane, *J. Biol. Chem.* **248**:8457–8464.

Kellerman, O., Tonetti, H., Brevet, A., Mirande, M., Pailliez, J.-P., and Waller, J.-P., 1982, Macromolecular complexes from sheep and rabbit containing seven aminoacyl-tRNA synthetases, *J. Biol. Chem.* **257**:11041–11048.

Kirkpatrick, F. H., Woods, G. M., LaCelle, P. L., and Weed, R. I., 1975, Calcium and magnesium ATPases of the spectrin fraction of human erythrocytes, *J. Supramol. Struct.* **3**:415–425.

Kirkpatrick, F. H., Woods, G. M., LaCelle, P. L., and Weed, R. I., 1976, Calcium and Mg-ATPases of the spectrin fraction of human erythrocytes, *J. Supramol. Struct.* **3**:415–425.

Kiselev, N. A., Sphitzberg, C. L., and Vainshtein, B. K., 1967, Crystallization of catalase in the form of tubes with monomolecular walls, *J. Mol. Biol.* **25**:433–441.

Kleinschmidt, J. A., Hügle, B., Grund, C., and Franke, W. W., 1983, The 22S cylinder particles of *Xenopus laevis*. I. Biochemical and electron microscopic characterization, *Eur. J. Cell Biol.* **32**:143–156.

Kloetzel, P.-M., Falkenburg, P.-E., Hössl, P., and Glätzer, K. H., 1987, The 19S ring-type particles of *Drosophila, Exp. Cell Res.* **170**:204–213.

Kremp, A., Schliephacke, M., Kull, U., and Schmid, H.-P., 1986, Prosomes exist in plant cells too, *Exp. Cell Res.* **166**:553–557.

Krisman, C. R., and Blumenfeld, M. L., 1986, A method for the direct measurement of glycogen synthase activity in gels after polyacrylamide gel electrophoresis, *Anal. Biochem.* **154**:409–413.

Lande, W. M., Thjiemann, P. W., Fiosher, K. A., and Mentzer, W. C., 1984, Two-dimensional electrophoretic analysis of human erythrocyte cylindrin, *Biochim. Biophys. Acta* **778**:105–111.

Lawson, A. A., and Barr, R. D., 1987, Acetylcholinesterase in red blood cells, *Am. J. Hematol.* **26**:101–112.

Lieber, M. R., and Steck, T. L., 1982a, A description of the holes in human erythrocyte membrane ghosts, *J. Biol. Chem.* **257**:11651–11659.

Lieber, M. R., and Steck, T. L., 1982b, Dynamics of the holes in human erythrocyte membrane ghosts, *J. Biol. Chem.* **257**:11660–11666.

Lilley, G. E., and Fung, C. W.-M., 1987, Hemoglobin–membrane interaction at physiological ionic strength and temperature, *Life Sci.* **41**:2429–2444.

Lin, T., and Allen, R. W., 1986, Isolation and characterization of a 37,000-dalton protein associated with the erythrocyte membrane, *J. Biol. Chem.* **261**:4594–4599.

Low, M. G., Ferguson, M. A. J., Futerman, A. H., and Silman, I., 1986, Covalently attached phosphatidylinositol as a hydrophobic anchor for membrane proteins, *Trends Biochem. Sci.* **11**:212–215.

McDaniel, C. F., Kirtley, M. E., and Tanner, M. J. A., 1974, The Interaction of glyceraldehyde-3-phosphate dehydrogenase with human erythrocyte membranes, *J. Biol. Chem.* **249**:6478–6485.

MacGregor, R. D., and Tobias, C. A., 1972, Molecular sieving of red cell membranes during gradual osmotic hemolysis, *J. Membr. Biol.* **10**:345–356.

McGuire, M. J., and DeMartino, G. N., 1986, Purification and characterization of a high molecular weight proteinase (macropain) from human erythrocytes, *Biochim. Biophys. Acta* **873**:279–289.

Macpherson, E., Tomkinson, B., Bålöw, Höglund, S., and Zetterqvist, Ö., 1987, Supramolecular structure of tripeptidyl peptidase II from human erythrocytes as studied by electron microscopy, and its correlation to enzyme activity, *Biochem. J.* **248**:259–263.

Malech, H. L., and Marchesi, V.T., 1981, Hollow cylinder protein in the cytoplasm of human erythrocytes, *Biochim. Biophys. Acta* **554**:469–478.

Markham, R., Frey, S., and Hills, G. J., 1963, Method for enhancement of image detail and accentuation of structure in electron microscopy, *Virology* **22**:88–102.

Martins de Sa, C., Grossi, de Sa, M. -F., Akhayat, O., Broders, F., Scherrer, K., Horsch, A., and Schmid, H. -P., 1986, Prosomes ubiquity and inter-species structural variation, *J. Mol. Biol.* **187**:479–493.

Mayer, L. J., and Becker, M. A., 1977, Human erythrocyte phosphoribosylpyrophosphate synthetase, *J. Mol. Biol.* **252**:3919–3925.

Medof, M. E., Gottlieb, A., Kinoshita, T., Hall, S., Silber, R., Nussenzweig, V., and Rosse, W. F., 1987, Relationship between decay accelerating factor deficiency, diminished acetylcholinesterase activity, and defective terminal complement pathway restriction in paroxysmal nocturnal hemoglobinuria erythrocytes, *J. Clin. Invest.* **80**:165–174.

Melloni, E., Sparatore, B., Salamino, F., Michetti, M., and Pontremoli, S., 1982, Proteolysis on human reticulocyte membrane proteins: Evidence for a physiological role of the acid endopeptidase, *Arch. Biochem. Biophys.* **218**:579–584.

Mirande, M., Gache, Y., Le Corre, D., and Waller, J.-P., 1982, Seven mammalian aminoacyl-tRNA synthetases co-purified as high molecular weight entities are associated within the same complex, *EMBO J.* **1**:733–736.

Mitchell, C. D., Mitchell, W. B., and Hanahan, D. J., 1965, Enzyme and hemoglobin retention in human erythrocyte stroma, *Biochim. Biophys. Acta* **104**:348–358.

Moses, S. W., Bashan, N., and Gutman, A., 1972a, Glycogen metabolism in the normal red blood cell, *Blood* **40**:836–843.

Moses, S. W., Bashan, N., and Gutman, A., 1972b, Properties of glycogen synthetase in erythrocytes, *Eur. J. Biochem.* **30**:205–210.

Mostafa, M. A., and Hanahan, D. J., 1984, Partial purification of a novel Mg^{2+}-ATPase from human erythrocytes, *Biochim. Biophys. Acta* **802**:490–500.

Muller, M., Dubiel, W., Rathmann, J., and Rapoport, S., 1980, Determination and characteristics of energy-dependent proteolysis in rabbit reticulocytes, *Eur. J. Biochem.* **109**:405–410.

Murakami, T., Suzuki, Y., and Murachi, T., 1979, An acid protease in human erythrocytes and its localization in the inner membrane, *Eur. J. Biochem.* **96**:221–227.

Narayan, K. S., and Rounds, D. E., 1973, Minute ring-shaped particles in cultured cells of malignant origin, *Nature New Biol.* **243**:146–150.

Nimmo, H. G., Proud, C. G., and Cohen, P., 1986, The purification and properties of rabbit skeletal muscle glycogen synthase, *Eur. J. Biochem.* **68**:21–30.

Norcum, M. T., 1989, Isolation and electron microscopic characterization of the high molecular mass aminoacyl tRNA synthetase complex from murine erythroleukemia cells, *J. Biol. Chem.* **264**:15043–15051.

Ohkubo, I., Namikawa, C., and Sasaki, M., 1988, Purification and characterization of HMW proteinase from human erythrocytes, Proc. Int. Symp. Intracellular Protein Catabolism, Shimoda, Japan, P-5.

Oliver, R., 1973, Negative stain electron microscopy, *Methods Enzymol.* **27**:616–672.

Ott, P., and Brodbeck, U., 1978, Multiple forms of acetylcholinesterase from human erythrocyte membranes, *Eur. J. Biochem.* **88**:119–125.

Ott, P., Jenny, B., and Brodbeck, U., 1975, Multiple molecular forms of purified human erythrocyte acetylcholinesterase, *Eur. J. Biochem.* **57**:469–480.

Ott, P., Lustig, A., Brodbeck, U., and Rosenbusch, J. P., 1982, Acetylcholinesterase from human erythrocyte membranes: Dimers as functional units, *FEBS Lett.* **138**:187–189.

Paniker, N. V., Arnold, A. B., and Hartmann, R. C., 1973, Solubilization and purification of human erythrocyte membrane acetylcholinesterase, *Proc. Soc. Exp. Biol. Med.* **144**:492–497.

Pendergast, A. M., Venema, R. C., and Traugh, J. A., 1987, Regulation of phosphorylation of aminoacyl-tRNA synthetases in the high molecular weight core complex in reticulocytes, *J. Biol. Chem.* **262**:5939–5942.

Pontremoli, S., Sparatore, B., Melloni, E., Morelli, A., Benatti, U., and De Flora, A., 1979, Isolation and partial characterization of three acidic proteinases in erythrocyte membranes, *Biochem. J.* **181**:559–568.

Rakomczay, Z. and Brimijain, S., 1988, Biochemistry and pathophysiology of the molecular forms of cholinesterases in *Subcellular Biochemistry, Volume 12* (Harris, J. R., ed.) pp. 335–378, Plenum, New York.

Rapoport, S. M., and Müller, M., 1974, Catalase and glutathione peroxidase, in: *Cellular and Molecular Biology of Erythrocytes* (H. Yoshikawa and S. M. Rapoport, eds.), pp. 167–179, University Park Press, Baltimore.

Rebhun, L. I., Smith, C., and Larner, J., 1973, Electron microscope studies on glycogen synthase, *Mol. Cell. Biochem.* **1**:55–61.

Roberts, W. L., and Rosenberry, T. L., 1985, Identification of covalently attached fatty acids in the hydrophobic membrane-binding domain of human erythrocyte acetylchoinesterase. *Biochem. Biophys. Res. Commun.* **133**:621–627.

Roberts, W. L., and Rosenberry, T. L., 1986, Selective radiolabeling and isolation of the hydrophobic membrane-binding domain of human erythrocyte acetylcholinesterase, *Biochemistry* **25**:3091–3098.

Roberts, W. L., Kim, B. H., and Rosenberry, T. L., 1987, Differences in the glycolipid membrane anchors of bovine and human erythrocyte acetylcholinesterase, *Proc. Natl. Acad. Sci. USA* **84**:7817–7821.

Rogalski, A. A., Steck, T. L., and Waseem, A., 1979. Association of glyceraldehyde-3-phosphate dehydrogenase with the plasma membrane of the intact human red blood cell, *J. Biol. Chem.* **264**:6438–6446.

Römer-Lüthi, C. R., Hajdu, J., and Brodbeck, U., 1979, Molecular forms of purified human erythrocyte acetylcholinesterase investigated by crosslinking with diimidates, *Hoppe-Seyler's Z. Physiol. Chem.* **360**:929–934.

Römer-Lüthi, C. R., Ott, P., and Brodbeck, U., 1980, Reconstitution of human erythrocyte membrane acetylcholinesterase in phospholipid vesicles, *Biochim. Biophys. Acta* **601**:123–133.

Rosenberry, T. C., and Scroggin, D. M., 1984, Structure of human erythrocyte acetylcholinesterase, *J. Biol. Chem.* **259**:5643–5652.

Roth, D. G., Shelton, E., and Deuel, T. F., 1974, Purification and properties of phosphoribosyl pyrophosphate synthetase from rat liver, *J. Biol. Chem.* **249**:291–296.

Ryazanov, A. G., Ashmarina, L. I., and Muronetz, V. I., 1988, Association of glyceraldehyde-3-phosphate dehydrogenase with mono- and polyribosomes of rabbit reticulocytes, *Eur. J. Biochem.* **171**:301–305.

Salhany, J. M., 1983, Binding of cytosolic proteins to the erythrocyte membrane, *J. Cell. Biochem.* **23**:211–222.

Schmid, H. P., Akhayat, O., Martins de Sa, C., Puvion, F., Koehler, K., and Scherrer, K., 1984, The prosome: An ubiquitous morphologically distinct RNP particle associated with repressed mRNPs and containing specific ScRNA and a characteristic set of proteins, *EMBO J.* **3**:29–34.

Schreier, M. H., Erni, B., and Staehelin, T., 1977, Initiation of mammalian protein synthesis: The importance of ribosome and initiation factor quality for the efficiency of in vitro systems, *J. Mol. Biol.* **73**:329–349.

Schrier, S. L., 1977, Human erythrocyte membrane enzymes: Current status and clinical correlates, *Blood* **50**:227–237.

Scott, G. K., and Kee, T. B., 1979, Neutral protease from human and ovine erythrocyte membranes, *Int. J. Biochem.* **10**:1039–1044.

Seeman, P., 1967, Transient holes in the erythrocyte membrane during hypotonic hemolysis and stable holes in the membrane after lysis by saponin and lysolecithin, *J. Cell Biol.* **32**:55–70.

Seeman, P., 1974, Ultrastructure of membrane lesions in immune lysis, osmotic lysis and drug-induced lysis, *Fed. Proc.* **33**:2116–2124.

Seeman, P., Cheng, D., and Iles, G. H., 1973, Structure of membrane holes in osmotic and saponin hemolysis, *J. Cell Biol.* **56**:519–527.

Selvaraj, P., Tosse, W. F., Silber, R., and Springer, T. A., 1988, The major Fc receptor in blood has a phosphatidylinositol anchor and is deficient in paroxysmal nocturnal haemoglobinuria, *Nature* **333**:565–567.

Shelton, E., Kuff, E. L., Maxwell, E. S., and Harrington, J. T., 1970, Cytoplasmic particles and aminoacyl transferase I activity, *J. Cell Biol.* **45**:1–8.

Shin, B. C., and Carraway, K. L., 1973, Association of glyceraldehyde-3-phosphate dehydrogenase with the human erythrocyte membrane, *J. Biol. Chem.* **248**:1436–1444.

Siems, W., Dubiel, W., Dumdey, R., Muller, M., and Rapoport, S. M., 1984, Accounting for the ATP-consuming processes in rabbit reticulocytes, *Eur. J. Biochem.* **139**:101–107.

Silman, I., and Futerman, A. H., 1987, Modes of attachment of acetylcholinesterase to the surface membrane, *Eur. J. Biochem.* **170**:11–22.

Siriwittayakorn, J., and Yuthavong, Y., 1979, Relation between low erythrocyte acetylcholinesterase activity and membrane lipids in paroxysmal nocturnal haemoglobinuria, *Br. J. Haematol.* **41**:383–391.

Smith, D. W. E., Silbert, P. E., and McNamara, A. L., 1979, The association of histidyl-tRNA synthetase with reticulocyte ribosomes, *Biochim. Biophys. Acta* **562**:453–461.

Smulson, M., 1974, Subribosomal particles of HeLa cells, *Exp. Cell Res.* **87**:253–258.

Som, K., and Hardesty, B., 1975, Isolation and partial characterization of an aminoacyl-tRNA synthetase complex from rabbit reticulocytes, *Arch. Biochem. Biophys.* **166**:507–517.

Stansell, M. J., and Deutsch, H. F., 1965, Preparation of crystalline erythrocuprein and catalase from human erythrocytes, *J. Biol. Chem.* **240**:4299–4305.

Steck, T. L., 1978, The band 3 protein of the human red cell membrane: A review, *J. Supramol. Struct.* **8**:311–324.

Sugarman, J., Devine, D. V., and Rosse, W. F., 1986, Structural and functional differences between decay-accelerating factor and red cell acetylcholinesterase, *Blood* **68**:680–684.

Switzer, R. L., 1969, Regulation and mechanism of phosphoribosylpyrophosphate synthetase, *J. Biol. Chem.* **244**:2854–2863.

Taguchi, R., and Ikezawa, H., 1987, Properties of bovine erythrocyte acetylcholinesterase solubilized by phosphatidylinositol-specific phospholipase C, *J. Biochem.* **102**:803–811.

Tanaka, K., Waxman, L., and Goldberg, A. L., 1983, ATP serves two distinct roles in protein degradation in reticulocytes, one requiring and one independent of ubiquitin, *J. Cell Biol.* **96**:1580–1585.

Tanaka, K., Ii, K., Ichihara, A., Waxman, L., and Goldberg, A. L., 1986, A high molecular weight protease in the cytosol of rat liver, *J. Biol. Chem.* **261**:15197–15203.

Tanaka, K., Yoshimura, T., Ichihara, A., Ikai, A., Nishigai, M., Morimoto, Y., Sato, M., Tanaka, N., Katsube, Y., Kameyama, K., and Takagi, T., 1988a, Molecular organization of a high molecular weight multi-protease complex from rat liver, *J. Mol. Biol.* **203**:985–996.

Tanaka, K., Yoshimura, T., Kumatori, A., Ichihara, A., Ikai, A., Nishigai, M., Kameyama, K., and Takagi, T., 1988b, Proteasomes (multi-protease complexes) as 20S ring-shaped particles in a variety of eukaryotic cells, *J. Biol. Chem.* **263**:16209–16217.

Tanaka, W. K., Som, K., and Hardesty, B., 1976, Comparison of free and ribosome-bound phenylalanine-tRNA synthetase from rabbit reticulocytes, *Arch. Biochem. Biophys.* **172**:252–260.

Tanner, M. J. A., and Gray, W. R., 1971, The isolation and functional identification of a protein from the human erythrocyte 'ghost,' *Biochem. J.* **125**:1109–1117.

Tarone, G., Hamasaki, N., Fukudoa, M., and Marchesi, V. T., 1979, Proteolytic degradation of human erythrocyte band 3 by membrane-associated protease activity, *J. Membr. Biol.* **45**:1–12.

Tillmann, W., Cordua, A., and Schroter, W., 1975, Organization of enzymes of glycolysis and of glutathione metabolism in human red cell membranes, *Biochim. Biophys. Acta* **382**:157–171.

Trachsel, H., and Staehelin, T., 1979, Initiation of mammalian protein synthesis the multiple functions of the initiation factor eIF-3, *Biochim. Biophys. Acta* **565**:305–331.

Tsukahara, T., Ishiura, S., and Sugita, H., 1988, An ATP-dependent protease and ingensin, the multicatalytic proteinase, in K562 cells, *Eur. J. Biochem.* **177**:261–266.

Ussery, M. A., Tanaka, W. K., and Hardesty, B., 1977, Subcellular distribution of aminoacyl-tRNA synthetases in various eukaryotic cells, *Eur. J. Biochem.* **72**:491–500.

Waldman, A. L., Marx, G., and Goldstein, J., 1975, Isolation of rabbit reticulocyte initiation factors by means of heparin bound to Sepharose, *Proc. Natl. Acad. Sci. USA* **72**:2352–2356.

Wang, K. K. W., Villalobo, A., and Roufogalis, B. D., 1988, Activation of the Ca^{2+}-ATPase of human erythrocyte membrane by an endogenous Ca^{2+}-dependent neutral protease, *Arch. Biochem. Biophys.* **260**:696–704.

Waxman, L., Fagan, J. M., and Goldberg, A. L., 1987, Demonstration of two distinct high molecular weight proteases in rabbit reticulocytes, one of which degrades ubiquitin conjugates, *J. Biol. Chem.* **262**:2451–2457.

Weed, R. I., Reed, C. F., and Berg, G., 1963, Is hemoglobin an essential structural component of human erythrocyte membranes? *J. Clin. Invest.* **42**:581–588.

Weitz, M., Bjerrum, O. J., and Brodbeck, U., 1984, Characterization of an active hydrophilic erythrocyte membrane acetylcholinesterase obtained by limited proteolysis of the purified enzyme, *Biochim. Biophys. Acta* **776**:65–74.

White, M. D., and Ralston, G. B., 1976, A water-soluble Mg^{2+}-ATPase from erythrocyte membranes, *Biochim. Biophys. Acta* **436**:567–576.

White, M. D., and Ralston, G. B., 1979, The 'hollow cylinder' protein of erythrocyte membranes, *Biochim. Biophys. Acta* **554**:469–478.

White, M. D., and Ralston, G. B., 1980, Purification of a water-soluble Mg^{2+}-ATPase from human erythrocyte membranes, *Biochim. Biophys. Acta* **599**:569–579.

Wilk, S., and Orlowski, M., 1983, Evidence that pituitary cation-sensitive neutral endopeptidase is a multicatalytic protease complex, *J. Neurochem.* **40**:842–849.

Witheiler, J., and Wilson, D. B., 1972, The purification and characterization of a novel peptidase from sheep red cells, *J. Biol. Chem.* **247**:2217–2221.

Wright, D. L., and Plummer, D. T., 1973, Multiple forms of acetylcholinesterase from human erythrocytes, *Biochem. J.* **133**:521–527.

Yamamoto, K., Yamamoto, H., Takeda, G., and Kato, Y., 1988, An aspartic proteinase of erythrocyte membranes. Proposed mechanism of activation and further molecular properties, *J. Biol. Chem.* **369**:315–322.

Yu, J., and Steck, T. L., 1975, Association of band 3, the predominant polypeptide of the erythrocyte membrane, *J. Biol. Chem.* **250**:9176–9184.

Chapter 10

Molecular Genetics of Human Erythrocyte Sialoglycoproteins
Glycophorins A, B, C, and D

Jean-Pierre Cartron, Yves Colin, Shinichi Kudo, and Minoru Fukuda

1. INTRODUCTION

The membrane of human erythrocytes has been a model system for studying the structure and function of the plasma membrane. By analyzing the components of the human erythrocyte membrane, the most important concepts in membrane biochemistry and biology have been borne out. Isolation and characterization of a major erythrocyte sialoglycoprotein, now called glycophorin A, provided the essential characteristics of transmembrane glycoprotein (Tomita and Marchesi, 1975). Membrane proteins associated less strongly with the lipid bilayer—peripheral proteins—were recognized at the same time (Singer and Nicolson, 1972). The erythrocyte cytoskeleton was identified by investigation of the interaction of these proteins with the filamentous network beneath the lipid bilayer (Yu *et al.*, 1973). Thus, this system provides an interesting model for studying the structure and organization of membrane glycoproteins, how those glycoproteins interact with other proteins, and cytoskeletal protein interactions. Furthermore, the structural change of cell-surface carbohydrates during development was elucidated for the first time in this system (Fukuda *et al.*, 1979), which corresponds to fetal (i) and to adult (I) erythrocyte antigens.

The human erythrocyte membrane may be the first membrane for which the major sialoglycoproteins were isolated and characterized (Tomita and Marchesi, 1975; Furth-

Jean-Pierre Cartron and Yves Colin Unité INSERM U76, Institut National de Transfusion Sanguine, 75015 Paris, France. **Shinichi Kudo and Minoru Fukuda** La Jolla Cancer Research Foundation, La Jolla, California 92037, USA.

mayr, 1978a; Blanchard *et al.*, 1987a). The amino acid sequences, the nucleotide sequences of cDNA, and the genomic organization for these glycoproteins have been elucidated. Recent advances in these fields are summarized here with the objectives of determining whether the red cell membrane glycoproteins are the products of the same gene family and of understanding how these molecules are synthesized during erythroid differentiation and expressed at the cell surface. The structure and antigenic properties of the erythrocyte membrane glycoproteins will only be briefly discussed as they have been reviewed extensively (Anstee, 1980, 1981, Dahr, 1986; Gahmberg *et al.*, 1988; Lisowska, 1989).

2. NOMENCLATURE AND PROPERTIES OF GLYCOPHORINS

The human erythrocyte carries at least four main sialic acid-rich glycoproteins that can be specifically detected by periodic acid–Schiff (PAS) staining (Fairbanks *et al.*, 1971) after separation of membranes by SDS–polyacrylamide gel electrophoresis in the discontinuous buffer system of Laemmli (1970). These glycoproteins represent approximately 2% of the total erythrocyte membrane proteins and have been given various names (Table I). Here, the designation of glycophorins A, B, C (Furthmayr, 1978a,b), and D (Dahr, 1986) will be used.

Glycophorin A (GPA) and B (GPB) are the major components representing approximately 85% and 10% of the PAS-positive material, respectively, whereas glycophorin C (GPC) and D (GPD) are minor species contributing only 4 and 1% to the PAS-positive material. The monomeric form of these glycoproteins have an apparent molecular mass of 36, 20, 32, and 23 kDa, respectively, as estimated on 10% polyacrylamide gels in the presence of SDS (Blanchard *et al.*, 1982), but differences in these values have been recorded between laboratories. Although the glycophorins are responsible for most (60%) of the negative charge of red cells, they can be distinguished according to several other properties such as the presence of blood group determinants or of N- and/or O-glycosidically linked oligosaccharide chains (Table I).

The primary structure of GPA, GPB, and GPC has been determined both by protein sequence analysis (Tomita and Marchesi, 1975; Furthmayr, 1978b; Tomita *et al.*, 1978; Dahr *et al.*, 1980, 1982; Blanchard *et al.*, 1987a) and by nucleotide sequencing of cDNAs representing their RNA (Siebert and Fukuda, 1986a, 1987; Colin *et al.*, 1986). The primary structure of GPD, however, is still not known completely but recent evidence suggests that this glycoprotein is immunologically and structurally related to GPC. Indeed, both glycoproteins carry blood group Gerbich determinants (Reid *et al.*, 1987a; Dahr *et al.*, 1987) and polyclonal antibodies raised against purified GPC or GPD react with antigenic determinants common to the C-terminal or intramembranous regions of both glycoproteins (Reid *et al.*, 1987a; El-Maliki *et al.*, 1989). The structural homology between GPC and GPD is even more obvious when the peptides generated by enzymatic and acid cleavage of the purified glycoproteins are compared in one-dimensional analysis by high-performance thin-layer chromatography (El-Maliki *et al.*, 1989). Indeed, most iodopeptides migrated identically, but some are present exclusively in the GPC digest (Figure 1). Since murine monoclonal antibodies reacting with the N-terminal region of GPC, but not with GPD, have been characterized, these findings suggest that GPD might

Table I
Nomenclature and Properties of the Major Red Cell Membrane Glycoproteins

Properties	Red cell membrane glycophorins[a]			
	GPA	GPB	GPC	GPD
Synonyms				
Fairbanks *et al.* (1971)	PAS-1	PAS-3	PAS-2	
Dahr *et al.* (1975)	MN SGP	Ss SGP	Component D	Component E
Anstee *et al.* (1979)	GPα	GPδ	GPβ	GPγ
Blood group antigens	MN	Ss	Ge:3	Ge:2,3
PAS-positive material (%)	85	10	4	1
Apparent molecular mass (kDa)[b]	36	20	32	23
Copies/cell (× 10^{-3})[c]	500–900	80–300	50–100	20
Polypeptide chain (residues)	131	72	128	107[d]
O-linked sugar chains	15	11	12	6[d]
N-linked sugar chains	1	0	1	0
Percent solubilized by Triton X-100[e]				
0.1%	25	70	15	
0.5%	35	82	20	
1.0%	82	95	19	4
5.0%	89	92	20	30

[a]Nomenclature according to Furthmayr (1978a) and extended by Dahr (1986).
[b]In SDS–PAGE according to Blanchard *et al.* (1982).
[c]Values derived from estimations based on PAS-staining and Scatchard plot analysis according to Fairbanks *et al.* (1971), Furthmayr (1978b), Gahmberg *et al.* (1979), Anstee *et al.* (1984a), Merry *et al.* (1984, 1986), and Dahr (1986).
[d]From the suggested protein sequence (El-Maliki *et al.*, 1989).
[e]Glycoproteins from red cell membranes prepared by hypotonic lysis were extracted at 4°C with 0.1 to 5% (w/v) Triton X-100 in phosphate-buffered saline. The glycoprotein composition of the solubilized fractions was analyzed by SDS–PAGE followed by PAS-staining. Adapted from Bloy *et al.* (1987).

be an abridged version of GPC in its N-terminal region (Blanchard *et al.*, 1987b; Dahr *et al.*, 1989). Preliminary protein sequence analysis is consistent with this assumption, since GPD was found to be identical to residues 21 to 128 of GPC (El-Maliki *et al.*, 1989). Additional recent evidence indicates in addition that both proteins are encoded by the same gene (Le Van Kim *et al.*, 1987).

The biological functions of glycophorins have not been clarified. These glycoproteins have been reported as ligands for viruses, bacteria, and parasites. The viruses bind first to the sialic acid residues attached to glycophorin molecules and subsequently the sialic acids are cleaved by a viral neuraminidase, a process that may facilitate dissemination of the viruses (Burness, 1981). Some strains of *E. coli* causing pyelonephritis (Jokinen *et al.*, 1985) or others responsible for sepsis and meningitis in newborns (Parkkinen *et al.*, 1986) bind similarly to a blood group M glycopeptide of GPA or to the alkali-labile tetrasaccharide carried by glycophorins, respectively. The sialic acid residues of glycophorins have similarly been involved as potential ligands for certain strains of *Plasmodium falciparum* malarial merozoites (reviews by Hadley *et al.*, 1986; Mitchell and Bannister, 1988) but it is possible that a peptide region of GPA located close to the double layer also participates in this binding function (Tounkara *et al.*, 1986). Although

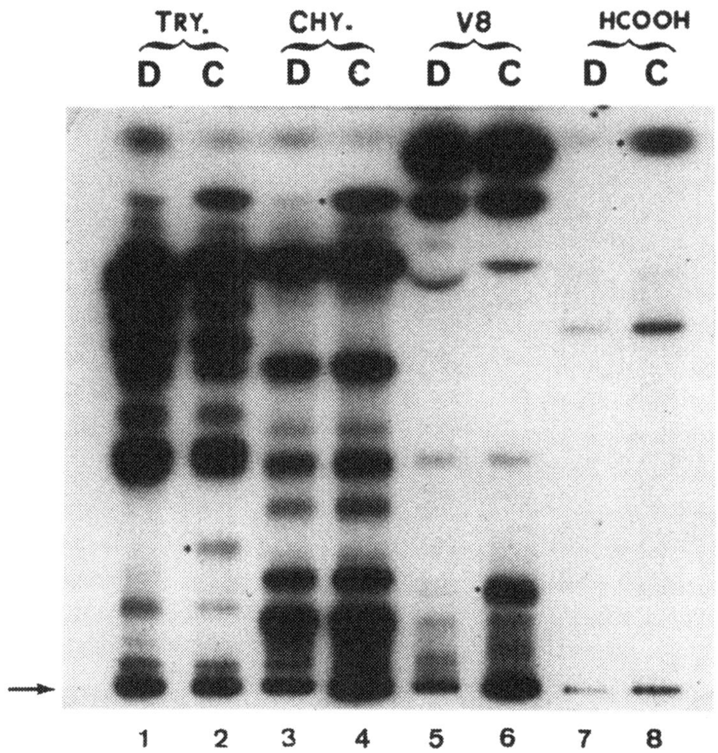

FIGURE 1. One-dimensional peptide map analysis of glycophorins C and D. GPC and GPD fractions labeled with [^{125}I]iodine were purified by polyacrylamide gel electrophoresis in the presence of SDS. Gel slices containing GPC and GPD were excised and subjected to proteolytic digestion (trypsin, chymotrypsin, and V8 protease at 50 μg/ml) or to formic acid cleavage for 24 and 48 hr at 37°C, respectively. Soluble iodopeptides (80% of the input radioactivity) were resolved by high-performance TLC in chloroform : methanol 25% (v/v) : ammonium hydroxide (2 : 2 : 1, by volume) and revealed by autoradiography. Lanes 1, 3, 5, and 7 correspond to GPD whereas lanes 2, 4, 6, and 8 correspond to GPC. Arrow marks the origin of migration; asterisks indicate peptides present in the GPC profile but absent from the GPD pattern. From El-Maliki *et al.* (1989).

these studies give interesting information on the properties of glycophorins, they do not yield any knowledge of the physiological function of these molecules in the red cell membrane.

Whether glycophorins are structurally important constituents of red cell membranes is still a matter of debate. One way of understanding the physiological function of these molecules was to analyze the membrane composition and abnormalities associated with glycophorin-deficient erythrocytes. Accordingly, it is likely that the major glycophorins do not play a significant structural role, at least on mature erythrocytes, since the red cells from variants lacking GPA, GPB, or even both GPA and GPB do not exhibit any abnormal morphology, survival, or transport function (Furthmayr, 1978a; Anstee, 1980; Reid *et al.*, 1987b). However, other abnormalities such as increased glycosylation of band 3 (Gahmberg *et al.*, 1976; Tanner and Anstee, 1976) or a greater lipid fluidity (Jansson *et al.*, 1981) have been described in such cells. In contrast, variant red cells lacking GPC and GPD ("Leach phenotype") are markedly elliptocytic (Anstee *et al.*, 1984a,b; Daniels *et al.*, 1986), suggesting that these glycoproteins contribute in some way as factors that may influence erythrocyte morphology. Further, those GPC- and GPD-deficient erythrocytes have altered mechanical stability and membrane deformability (Reid *et al.*, 1987b). It is well established that the general red cell shape is determined by peripheral membrane proteins (spectrin, actin, ankyrin, protein 4.1, and some others), which are organized in a highly ordered three-dimensional network known as the membrane skeleton (for review see Bennett, 1985, 1989). Integral membrane proteins like band 3 and glycophorins that interact via their cytoplasmic domain with the skeletal proteins ankyrin (Bennett, 1982) and protein 4.1 (Mueller and Morrison, 1981), respectively, serve as attachment sites of the membrane skeleton on the internal face of the phospholipid bilayer. Such protein–protein interactions are important for assembly of the peripheral membrane proteins during biogenesis of red cells (Lazarides, 1987) and for regulating the deformability and mechanical stability of these cells (Mohandas *et al.*, 1983; Chasis and Mohandas, 1986). The association of glycophorins with the membrane skeleton can be simply demonstrated by analyzing the protein composition of the soluble and insoluble fractions obtained following extraction of erythrocyte membrane preparations by varying concentrations of Triton X-100 (Yu *et al.*, 1973; Sheetz and Sawyer, 1978). Comparative analysis indicates that GPC and GPD bind strongly to the membrane skeleton and are difficult to solubilize (Table I) whereas GPB, and GPA to a lesser extent, are mostly recovered in the soluble fraction (Bloy *et al.*, 1987).

The interaction of GPA and GPC with the skeletal proteins is mediated through binding to protein 4.1 (Mueller and Morrison, 1981; Andersson and Lovrien, 1984), possibly by a polyphosphoinositide-dependent mechanism (Anderson and Marchesi, 1985), but the above considerations on red cells deficient in glycophorins suggest that only the interaction involving GPC might be of functional importance.

It is certainly significant that red cells from homozygous patients with protein 4.1 deficiency are also elliptocytic and have in addition a severe decrease of GPC and GPD, but a normal level of GPA and GPB (Alloisio *et al.*, 1985; Sondag *et al.*, 1987). All these findings, therefore, suggest that protein 4.1 might be critical for the correct insertion or transport of GPC and GPD to the red cell surface. In this context, it is also interesting to recall that a selective depression of blood group antigens, among which are the blood group Gerbich determinants, has been found associated with hereditary ovalocytosis among Melanesians (Booth *et al.*, 1977), but it is not known whether these abnormalities are directly related to some abnormality of GPC and GPD.

It is likely that greater knowledge of the biochemistry and biosynthesis of these glycophorins will help to clarify these questions.

3. CLONING OF cDNAs FOR GLYCOPHORINS

3.1. Isolation and Nucleotide Sequence of cDNA Encoding GPA

GPA was the first membrane protein for which the amino acid sequence was elucidated (Tomita and Marchesi, 1975; Tomita *et al.*, 1978). GPA is made up of 131 amino acids as a mature protein, and is synthesized as a precursor with a cleavable signal peptide

GPA/GPB

```
                    A   ATG TAT GGA AAA ATA ATC TTT GTA TTA CTA TTG TCA GCA ATT GTC AGC ATA TCA GCA
                    B   --- --- --- --- --- --- --- --- --- --- --- --- -A- --- --- --- --- --- ---
                    a   Met Tyr Gly Lys Ile Ile Phe Val Leu Leu Leu Ser Ala Ile Val Ser Ile Ser Ala   -1
                    b   --- --- --- --- --- --- --- --- --- --- --- --- Glu --- --- --- --- --- ---   -1

A   TCA AGT ACC ACT GGT GTG GCA ATG CAC ACT TCA ACC TCT TCT TCA GTC ACA AAG AGT TAC ATC TCA TCA
B   -T- --- --- --- -AG --- --- --- --- --- --- --- --- --- --- --- --- --- --- --- --- --- ---
a   Ser Ser Thr Thr Gly Val Ala Met His Thr Ser Thr Ser Ser Ser Val Thr Lys Ser Tyr Ile Ser Ser   23
b   Leu --- --- --- Glu --- --- --- --- --- --- --- --- --- --- --- --- --- --- --- --- --- ---   23

A   CAG ACA AAT GAT ACG CAC AAA CGG GAC ACA TAT GCA GCC ACT CCT AGA GCT CAT GAA GTT TCA GAA ATT
B   --- --- ---
a   Gln Thr Asn Asp Thr His Lys Arg Asp Thr Tyr Ala Ala Thr Pro Arg Ala His Glu Val Ser Glu Ile   46
b   --- --- ---                                                                                   26

A   TCT GTT AGA ACT GTT TAC CCT CCA GAA GAG GAA ACC GGA GAA AGG GTA CAA CTT GCC CAT CAT TTC TCT
B                                                   --- --- -C- -G- --- --- -T- --- -G- --- A--
a   Ser Val Arg Thr Val Tyr Pro Pro Glu Glu Glu Thr Gly Glu Arg Val Gln Leu Ala His His Phe Ser   69
b                                                   --- --- Thr Gly --- --- Val --- Arg --- Thr   37

A   GAA CCA             GAG ATA ACA CTC ATT ATT TTT GGG GTG ATG GCT GGT GTT ATT GGA ACG ATC CTC
B   -T-     GCT CCT GTA -T- --- -T- --- --- --- --G T-T --- --- --- --- A-- --- --- --- --- ---
a   Glu Pro             Glu Ile Thr Leu Ile Ile Phe Gly Val Met Ala Gly Val Ile Gly Thr Ile Leu   89
b   Val --- Ala Pro Val Val --- Ile --- --- --- Leu Cys --- --- --- --- Ile --- --- --- --- ---   60

A   TTA ATT TCT TAC GGT ATT CGC CGA CTG ATA AAG AAA AGC CCA TCT GAT GTA AAA CCT CTC CCC TCA CCT
B   --- --- --- --- AC- --- --- --- --- --- --- GC- TGA
a   Leu Ile Ser Tyr Gly Ile Arg Arg Leu Ile Lys Lys Ser Pro Ser Asp Val Lys Pro Leu Pro Ser Pro  112
b   --- --- --- --- Thr --- --- --- --- --- --- Ala end                                           72

A   GAC ACA GAC GTG CCT TTA AGT TCT GTT GAA ATA GAA AAT CCA GAG ACA AGT GAT CAA TGA
a   Asp Thr Asp Val Pro Leu Ser Ser Val Glu Ile Glu Asn Pro Glu Thr Ser Asp Gln end              131
```

GPC

```
ATG TGG TCG ACG AGA AGC CCC AAC AGC ACG GCG TGG CCT CTC AGC CTC GAG CCT GAT CCG GGG ATG GCC
Met Trp Ser Thr Arg Ser Pro Asn Ser Thr Ala Trp Pro Leu Ser Leu Glu Pro Asp Pro Gly Met Ala   23

TCT GCC TCC ACC ACA ATG CAT ACT ACC ACC ATT GCA GAG CCT GAT CCA GGG ATG TCT GGA TGG CCG GAT
Ser Ala Ser Thr Thr Met His Thr Thr Thr Ile Ala Glu Pro Asp Pro Gly Met Ser Gly Trp Pro Asp   46

GGC AGA ATG GAG ACC TCC ACC CCC ACC ATA ATG GAC ATT GTC GTC ATT GCA GGT GTG ATT GCT GCT GTG
Gly Arg Met Glu Thr Ser Thr Pro Thr Ile Met Asp Ile Val Val Ile Ala Gly Val Ile Ala Ala Val   69

GCC ATC GTC CTA GTC TCC CTC CTC TTC GTC ATG CTG CGC TAC ATG TAC CGG CAC AAG GGC ACG TAC CAC
Ala Ile Val Leu Val Ser Leu Leu Phe Val Met Leu Arg Tyr Met Tyr Arg His Lys Gly Thr Tyr His   92

                                                                   A
ACC AAT GAG GCC AAG GGC ACG GAG TTT GCT GAG AGT GCA GAT GCA GCC CTG CAG GG GAC CCT GCC CTC
                                                                   G
Thr Asn Glu Ala Lys Gly Thr Glu Phe Ala Glu Ser Ala Asp Ala Ala Leu Gln Gly Asp Pro Ala Leu  115

CAA GAT GCT GGT GAT AGC AGC AGA AAG GAG TAC TTT ATT TGA
Gln Asp Ala Gly Asp Ser Ser Arg Lys Glu Tyr Phe Ile end                                      128
```

FIGURE 2. Nucleotide sequences of the coding regions of glycophorin A, B, and C cDNAs and their deduced amino acid sequences. Comparison of GPA/GPB. Each set of four lines shows (from top to bottom) the cDNA sequences of GPA and GPB followed by the amino acid sequences deduced from cDNA sequences, as indicated by the letters, A, B, a, and b to the left. Regions of identical sequences are shown by dashes. Amino acid residues are numbered at the far right. From Siebert and Fukuda (1986a, 1987). The cDNA sequence of GPC and the deduced amino acid sequence according to Colin *et al.* (1986, 1989).

(Jokinen *et al.*, 1981; Morrow and Rubin, 1987). The cDNA sequence of GPA (Figure 2) shows that 19 amino acids, which all have characteristics of a cleavable leader peptide, are encoded at the GPA NH_2-terminus (Siebert and Fukuda, 1986a). In particular, the presence of a basic amino acid, lysine, close to the initiation methionine (Sabatini *et al.*, 1982), and the amino acid residues at the third and first positions from the COOH-terminus of the leader peptide fits very well with the so-called (−3, −1) rule for signal peptide cleavage (Von Heijne, 1983). The native GPA protein is arranged into three parts: the extracellular domain (residues 1 to 70) carrying the blood group and ligand properties, the transmembrane portion (residues 71 to 92), and the cytoplasmic segment (residues 93 to 131), which may interact with cytosolic and peripheral membrane proteins. The MN blood group specificity is determined by amino acid polymorphism at positions 1 and 5 (M = Ser^1/Gly^5; N = Leu^1/Glu^5) of GPA (Furthmayr, 1978a; Dahr *et al.*, 1977; Wasniowska *et al.*, 1977). GPB also carries an unusual form of blood group N antigen (noted "N"), which, unlike the N antigen carried by GPA, is resistant to trypsin hydrolysis on native red cells (Dahr *et al.*, 1975).

Each cDNA clone encoding blood group M or N was isolated from the K562 cDNA library, and its nucleotide sequence indicates that three out of six nucleotides specifying amino acids 1 and 5 of mature GPA are different between the two blood groups (Figure 2). Although the amino acid sequence was known when the attempt to isolate cDNA was made, it was not entirely simple. Initial efforts using mixed oligonucleotide probes failed. The first isolation of cDNA was made only after the cDNA library was enriched by priming cDNA with mixed oligonucleotides based on the amino acid sequence close to the COOH-terminus (Siebert and Fukuda, 1986a). This was probably because the abundance of mRNA for GPA is low, even in erythroleukemic K562 cells. The latter observation is consistent with previous results showing that GPA is a minor component in K562 cells (Fukuda, 1980). The mRNA abundance of GPA is low also in human reticulocyte and human fetal cDNA libraries, which were, however, successfully screened with synthetic oligonucleotides (Rahuel *et al.*, 1988a; Tate and Tanner, 1988). The nucleotide sequences from these clones are completely consistent with those from the K562 library indicating that the erythroleukemic cells synthesize identical GPA polypeptide chain as erythrocytes, in contrast to previous conclusions deduced from studies with monoclonal antibodies (Rearden *et al.*, 1985). Together, these data indicate a further difference between M and N types of GPA in the cleavable leader peptide (Figure 2), since position −7 was alanine for M (Siebert and Fukuda, 1987) and glutamic acid for N (Tate and Tanner, 1988).

3.2. Isolation and Nucleotide Sequence of cDNA Encoding GPB

The establishment of the nucleotide sequence for GPA was useful to isolate cDNAs for GPB because a portion of the GPA and GPB amino acid sequences are almost identical. As shown by Furthmayr (1978b) and Dahr *et al.* (1980), and later by Blanchard *et al.* (1987a), the amino acid sequence of GPA and GPB are almost identical from residues 1 to 26. However, the amino acid sequence of residues 27 to 55 is unique to GPA. Therefore, the oligonucleotide probe GC encoding amino acid residues 18–26, which are identical for GPA and GPB, and the oligonucleotide GE encoding residues 28–36, which are unique to GPA, were constructed. Among cDNA clones isolated from a K562 library screened with these oligonucleotides, one clone was reactive with the GC probe but was unreactive with the GE probe, and this latter clone was found to encode GPB (Siebert and Fukuda, 1987).

Comparison of the nucleotide sequence of GPA and GPB cDNAs revealed the following (Figure 2). First, GPB has a leader peptide sequence almost identical to that of GPA, with glutamic acid at position −7 as found in GPA from N individuals. Second, GPB has leucine and glutamic acid at residues 1 and 5, which define blood group N, and the nucleotide sequences in these residues are identical to those found in GPA of blood type N. Third, the segment of residues 27–55 of GPA is absent from GPB. Because of this deletion, GPB lacks the Asn-Asp-Thr sequence, which is the N-glycosylation site for GPA at residues 26–28. This result is consistent with the fact that GPB contains only O-linked oligosaccharides (Furthmayr, 1978b). This region of GPB carries for the blood group Ss polymorphism, which is determined by amino acid exchange at position 29, a methionine and a threonine defining the S and s antigens, respectively (Dahr *et al.*, 1980).

The next region encoding residues 60–71 of GPA and 27–39 of GPB have some homologous sequences, whereas the sequences encoding residues 72–100 of GPA and residues 34–71 of GPB are strikingly similar in both the amino acid and nucleotide sequences. It is noteworthy that GPB carries a cysteine at position 50 and alanine at the C-terminus, which are absent in GPA and were not detected by protein sequencing (Blanchard *et al.*, 1987a). Finally, residues 101 to 131 in GPA are unique and absent from GPB. This latter segment is actually the antigenic site that is recognized by polyclonal antibodies specific to GPA (Cotmore *et al.*, 1977).

A similar cDNA encoding for GPB has been isolated from a human fetal liver library, but this clone differs slightly in the 3′UT region (Tate and Tanner, 1988). In addition, the clone includes an insert of 84 unrelated nucleotides within the codon for Glu-^{7} (in the signal peptide) at the exact position where a large intron (> 30 kb) is normally present (see below and Figure 4). Whether this message is a partially processed RNA or a cloning artifact is not clear. Although several examples of intron-encoding cDNAs have been reported (Yamanashi *et al.*, 1987; Katamine *et al.*, 1988), the cloning artifact cannot be ruled out since this clone was isolated as a head-to-head concatenate with 418 nucleotides of α_1-globin cDNA. Alternatively, this unusual mRNA might represent the transcript of an altered form of the GPB gene from the genome of an individual whose red cells lack GPB (Tate and Tanner, 1988). Although other evidence indicates that GPB-deficient individuals exhibit a large deletion of the GPB gene (Huang *et al.*, 1987), it is still possible that the GPB protein defect is a more heterogeneous condition than previously thought.

It is interesting to note that GPA and GPB, like other class I membrane proteins, contain a stretch of approximately 20 hydrophobic residues followed by four or three positively charged residues on its carboxy-terminal side that acts as a typical stop-transfer signal (Von Heijne, 1988). These residues are ideally located to interact with the polar head groups of the phospholipids inserted into the cytoplasmic side of the lipid bilayer (Segrest *et al.*, 1972; Anderson and Marchesi, 1985), an orientation that might contribute to lock the proteins into the membrane. Another interesting point is that the transmembrane portion of GPA and GPB are highly homologous and only a few differences can be noted. It has been shown that GPA and GPB behave as dimer, trimer, and tetramer, and the hydrophobic transmembrane domain is responsible for such multimer function

(Silverberg *et al.*, 1976). It is, therefore, likely that the positions where amino acid residues differ are not directly involved in such interaction, although most of the changes are conservative, with the exception of glycine at residue 79, which is replaced by cysteine at the counterpart position of GPB. The absence or presence of the cytoplasmic segment may be critical when the glycoproteins need to interact with membrane skeletal proteins. Although it is possible that GPA interacts with actin filaments through band 4.1 protein (Anderson and Lovrien, 1984), this hypothetical interaction should not play any significant biological role since GPA-deficient cells have normal functions (Furthmayr, 1978b). Recently, however, GPA was found to be a new interesting model for studying transmembrane signaling since the binding of ligands to the extracellular domain of this glycoprotein induces the association of its COOH-terminal domain with skeletal proteins (Chasis *et al.*, 1988). GPB has only six residues in the cytoplasmic tail, and it is unlikely that such a short segment, which even includes three basic amino acids for membrane anchoring, could have specific interaction with other components. The role of GPB is, therefore, expected to be much less significant, compared to GPA, in both membrane architecture and putative signal transduction.

3.3 Isolation and Nucleotide Sequence of cDNA Encoding GPC

Until recently, the information available on the primary structure of GPC was fragmentary, since only an NH_2-terminal tryptic peptide was sequenced (Dahr *et al.*, 1982). The complete protein sequence was deduced from cloning of the mRNA in a human reticulocyte cDNA library using an oligonucleotide approach. Mixed oligonucleotides specific for the amino acid region Asp-Pro-Gly-Met-Ala, which is partially repeated in the protein sequence at positions 19–23 and 38–42, were used. A first clone covering the entire coding sequence of GPC was characterized (Colin *et al.*, 1986) and subsequently other clones covering the entire 3′-untranslated region and almost all the 5′-untranslated region were isolated from a human liver cDNA library (Le Van Kim *et al.*, 1987; High and Tanner, 1987; Colin *et al.*, 1989).

The amino acid sequence deduced from the cDNA analysis indicated that GPC is a single polypeptide chain of 128 amino acid residues (Figure 2). The calculated molecular mass of the polypeptide backbone is about 14 kDa. However, this molecule is glycosylated and carries both N- and O-glycosylation sites (Dahr *et al.*, 1982). There is only one potential site (Asn-X-Ser/Thr) for attachment of an N-linked oligosaccharide to Asn at position 8. A variant of GPC, probably lacking this N-linked sugar chain, has been recently described on red cells from an individual carrying the low-incidence blood group antigen Webb, a trypsin-sensitive determinant postulated to be located in the vicinity of amino acid residue 8 (Reid *et al.*, 1985; Macdonald and Gerns, 1986). Approximately 12 O-glycosylated chains are attached to Ser or Thr residues of GPC (Dahr *et al.*, 1982). From the known structure of these sugar chains, which resemble those carried by GPA (Thomas and Winzler, 1969; Yoshima *et al.*, 1980; Irimura *et al.*, 1981), the calculated molecular mass of GPC is close to 28 kDa, a value that correlates well with the apparent molecular mass found by SDS-PAGE (see Table I).

Examination of the protein sequence indicates that GPC, like GPA, is clearly organized in three distinct domains:

1. The N-terminal domain extending approximately from residues 1 to 57 contains hydrophilic amino acids and the sites for N- and O-glycosylation. This fragment is exposed to the outer surface of the red cell and carries the Gerbich blood group determinants. Recent evidence based on various fractionation, fragmentation, and chemical modifications of the sialoglycoproteins from normal Gerbich-positive red cells used in hemagglutination inhibition tests has shown that the Ge : 3 antigens are located within residues 41 to 50 (Dahr *et al.*, 1987).
2. The N-terminal domain is followed by a stretch of 24 nonpolar amino acids (residues 58 to 81), which defines the intramembranous domain crossing the phospholipid bilayer.
3. The third domain is defined by the COOH-terminal end of the molecule (residues 82 to 128), which contains several basic and acid residues and is located inside the cell, where it is available for interaction with skeletal membrane proteins such as protein 4.1. All the tyrosine residues are present in this domain, which explains the observation that GPC cannot be radiolabeled with [^{125}I]iodine in intact red cells.

Other information deduced from a comparison of protein and nucleotide sequences is that there is no structural homology between GPA/GPB and GPC. A further noticeable difference between GPA and GPB comes from examination of the 5′ region of the cDNA, indicating that GPC, unlike GPA and GPB, is synthesized without any cleavable peptide signal (High and Tanner, 1987; Colin *et al.*, 1989), since the AUG codon for Met at position 1 on the mature GPC is preceded by an in-phase termination codon (see below and Figure 8).

GPC belongs to the type III class of membrane proteins that have an uncleaved signal sequence and the NH_2-terminus exposed outside the cell (Von Heijne, 1988). The internal apolar domain of these glycoproteins, which acts as a "start–stop" transfer sequence, is characterized by a charge distribution different from that of other membrane proteins. Indeed, the topological orientation of the GPC N-terminus satisfies the charge difference rule proposed recently by Hartmann *et al.* (1989).

3.4. What mRNA for GPD?

As discussed earlier, immunochemical and biochemical analyses demonstrate that GPD is an abridged version of GPC, since both glycoproteins have an identical COOH-terminal sequence extending over about 100 amino acid residues, but differ in their NH_2-terminal region (El-Maliki *et al.*, 1989).

Because of this high degree of homology, it is expected that the GPC cDNA should cross-hybridize on Northern blots and Southern blots with a specific GPD mRNA and a specific GPD gene, respectively. Surprisingly, a single mRNA of 1.1 kb and a single copy of the GPC gene per haploid genome could be detected with such a probe, even under low stringency conditions (Cartron *et al.*, 1986; Le Van Kim *et al.*, 1987). Further, the simultaneous lack of GPC and GPD on red cells from certain blood group Gerbich variants is associated with a 3.4-kb deletion within the GPC gene (Le Van Kim *et al.*,

1987; Tanner *et al.*, 1988; Colin *et al.*, 1989). From these findings, it was postulated that GPC and GPD are encoded by the same gene (Le Van Kim *et al.*, 1987). In general, the production of two proteins (or more) from a single gene may arise by the use of different promoters, alternative splicing of a primary transcript, or posttranslational modification (Leiff *et al.*, 1986; Breitbart *et al.*, 1987). There is, however, no evidence for the presence of two promoters or alternative splicing generating two different mRNAs from a unique primary transcript of the GPC gene (Colin, unpublished). That the two proteins might be generated by endoproteolytic cleavage of a unique polypeptide chain as it occurs in many other protein systems like those of the coagulation and complement protein factors (Mann *et al.*, 1988; Müller-Eberhard and Miescher, 1985) cannot be easily ruled out, but the degradation should be incomplete and allow both the substrate (GPC) and the proteolytic cleavage product (GPD) to coexist. There is still another plausible mechanism for generating GPC and GPD from a single message (Blanchard *et al.*, 1987b; Tanner *et al.*, 1988; Colin *et al.*, 1989). Indeed, examination of the GPC cDNA sequence indicates that the nucleotides surrounding the ATG codon for Met^{1} (5′-CA GGA *ATG* TG-3′) do not fit properly with the consensus translation initiation sequence (5′-CC$^{A}_{G}$ CC *ATG* GC-3′) found in viral and eukaryotic mRNAs (Kozak, 1981, 1984). However, it is worth noticing that the ATG codon for Met^{22} is present in the optimal context for binding of ribosomes, with a purine at position -3 from the initiator ATG and a guanine at position $+4$ (Figure 2). Since both AUGs are in the same reading frame, it is proposed that GPC and GPD might be produced by the alternative use of these translation initiation sites from a unique mRNA, according to the leaky scanning mechanism permitting the ribosome to initiate either at the first or at the second internal AUG, as described for some viral mRNAs (Kozak, 1986a,b, 1989). The use of the second AUG would result in the translation of a polypeptide chain of 107 amino acid residues having the expected size for GPD, assuming the presence of only six O-glycosidically linked tetrasaccharide chains.

Interestingly, the first example of this mechanism occurring in eukaryotes has been described for the biosynthesis of two forms of the invariant chain protein that is transiently associated intracellularly, with the α and β subunits of the murine Ia and human HLA-DR histocompatibility antigens during their transport to the cell surface (Strubin *et al.*, 1986). Another example of such a mechanism has been proposed to explain that the single-copy c-*myc* gene encodes for two proteins with distinct NH_2-termini (Hann *et al.*, 1988). This system is, however, most unusual since the two proteins are derived by alternative translational initiations at an AUG and a non-AUG codon of a single mRNA. When it is established on direct experimental evidence, the biosynthesis of GPC and GPD might represent a further example of such a mechanism in eukaryotes.

4. GENES FOR GLYCOPHORINS

4.1. Isolation of GPA and GPB Genomic Clones

Preliminary results obtained by Southern analysis of total genomic DNA with synthetic oligonucleotides as probes clearly indicate that GPA and GPB are encoded by separate and single-copy genes (Siebert and Fukuda, 1986b). The nucleotide sequences of GPA and GPB cDNAs, therefore, suggest that these two genes evolved from a common ancestral gene. By *in situ* hybridization on human chromosomes hybridized with a cDNA probe, the structural gene for GPA was assigned to chromosome 4q28-q31 (Rahuel *et al.*, 1988a), in agreement with studies by linkage analyses that have attributed the MNSs blood group locus to chromosome 4 (Cook *et al.*, 1981).

In order to investigate the genomic structures of GPA and GPB, a first genomic DNA library was constructed in phage vector Lambda fix from the total DNA of human erythroleukemia cell line K562 (Kudo and Fukuda, 1989). It had been shown previously that K562 cells contain both M and N blood group active GPA by isolating both GPA cDNAs encoding the M and N blood group (Siebert and Fukuda, 1987). Independently, a second library was constructed in λEMBL4 phages with genomic DNA prepared from human leukocytes (Vignal *et al.*, 1989).

The genomic DNA libraries were screened by oligonucleotides synthesized according to the cDNA sequences where GPA and GPB contain identical sequences, corresponding to the leader peptide sequences and to the sequences close to the NH_2-termini of mature glycophorins. Isolated genomic clones from each library were divided into two groups, one representing the 5′ extreme of the gene, including exon 1, and the other representing the rest of the genes. The respective restriction maps of these clones are shown in Figure 3A and 3B. Although different nomenclatures have been used for designation of some exons (Figure 3C), both maps are in agreement and show that the GPA and GPB genes are composed of seven and five exons, respectively (Kudo and Fukuda, 1989; Vignal *et al.*, 1989). Exon 1 and exon 2 are probably separated by more than 30 kb, since one cosmid clone encompassing 30 kb from exon 1 was still found not to contain exon 2. The nucleotide sequences upstream from the initiation site of exon 1 are very similar among GPA and GPB genes.

One group of genomic clones in the K562 library, which provide *Bgl*II fragment of 2.4 kb, has the same 5′-untranslated sequence as one of the isolated GPA cDNA, so that this group is judged to code for GPA. The second group of the genomic clones, which provide *Bgl*II fragment of 2.4 kb, contains the same 5′ upstream sequence as that of GPB cDNA isolated by Tate and Tanner (1988), so that this group is judged to be from the GPB gene. Similarly, among the clones sorted out from the leukocyte genomic library, some contain exon 1 within a *Bam*HI restriction fragment of 7 kb and others within a *Bam*HI restriction fragment of 11 kb, and were attributed to the GPA and GPB genes, respectively (Vignal *et al.*, 1989).

During these investigations, another group of clones containing exon 1 within a *Bam*HI fragment of 9.5 kb (Vignal *et al.*, 1989) or within a *Bgl*II fragment of 3.8 kb (Fukuda *et al.*, unpublished) was identified. These clones probably encode the 5′ region of a third glycophorin gene that has previously been reported as "invariant" (Rahuel *et al.*, 1988b). The "invariant" gene is present even when the GPA gene is depleted, as seen in En(a−) individuals and when both the GPA and GPB genes are deleted as seen in homozygous Mk individuals (Rahuel *et al.*, 1988b; Vignal *et al.*, 1989). It is likely to be present also in $S^-s^-U^-$ individuals, where the GPB gene is deleted (Huang *et al.*, 1987). Apparently, this novel gene encompasses the full length of the GPB gene, and further studies are in progress to isolate and characterize this "novel" glycophorin gene (Vignal *et al.*, and Kudo and Fukuda, submitted).

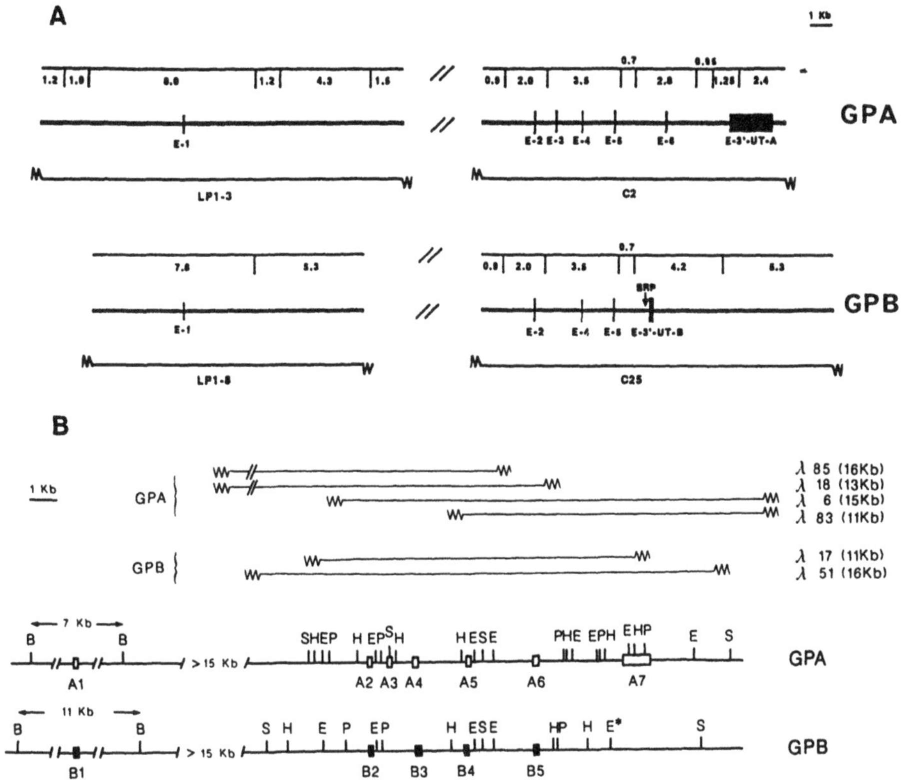

C

Genomic clones	Reference	Nomenclature of exons						
GPA	(a)	E-1	E-2	E-3	E-4	E-5	E-6	E-3'UT-A
	(b)	A-1	A-2	A-3	A-4	A-5	A-6	A-7
GPB	(a)	E-1	E-2		E-4	E-5		E-3'UT-B
	(b)	B-1	B-2		B-3	B-4		B-5

FIGURE 3. Organization of the glycophorin A and B genes. (A) The gene organization was constructed by combining two different groups of genomic clones isolated from a K562 library, which represent the extreme 5′ end and the rest of the genes, respectively. Each set of two lines (from top to bottom) shows *Eco*RI sites with sizes between sites in kilobases, genomic structures, and genomic clones in which flanking sequences are indicated by zigzags. For brevity, the restriction sites were shown only for *Eco*RI. Exons are indicated by vertical lines or black boxes, according to the proportions of the sizes. The recombination site (or breakpoint) is indicated by BRP. Exons 1 and 2 are separated by an intron of more than 30 kb. From Kudo and Fukuda (1989). (B) Organization of the GPA and GPB genes deduced from studies of genomic clones λ85, λ18, λ6, λ83 (for GPA) and λ17 and λ51 (for GPB) isolated from a human leukocyte library. Exons for GPA and GPB are numbered from A1 to A7 and B1 to B5, respectively. The first exon of each gene is at least 15 kb from the second exon, and is included in distinct *Bam*HI restriction fragments of 7 and 11 kb, respectively. Symbols for restriction endonucleases are: S, *Sac*I; H, *Hind*I; E, *Eco*RI; Pv, *Pvu*II; B, *Ban*HI. Adapted from Vignal *et al.* (1989). (C) Nomenclature of exons for the GPA and GPB genes. (a)According to the restriction map shown in panel A; (b) according to the restriction map shown in panel B.

The genomic DNA clones representing the second exon to the last exon for GPA and GPB can be classified as follows. The genomic clones derived from GPA contain *Eco*RI fragments with sizes of 2.0, 3.5, 2.8, and 1.25 kb, which are also detected in total genomic DNA by hybridization with GPA cDNA. The genomic clones containing *Eco*RI fragments with sizes of 2.0, 3.5, and 4.2 kb are judged to be derived from the GPB gene, since the same *Eco*RI fragments can be detected in total genomic DNA by GPB cDNA. These results are summarized in Figure 3A and 3B, respectively.

4.2. Structural Organization of the GPA Gene

The transcription initiation site was determined by S1 nuclease mapping and primer extension analysis. The results of both experiments agree well within a few bases, and adenosine, which is 55 (Kudo and Fukuda, 1989) or 56 nucleotides (Rahuel *et al.*, 1989) upstream from the A in the AUG translation initiation codon, was identified to be the transcription initiation site.

Nucleotide sequences were determined around exon–intron junctions and at 5′ and 3′ flanking regions of the GPA gene whose restriction map is shown in Figure 3A (Kudo and Fukuda, 1989). Each boundary was confirmed by comparison of obtained sequences to cDNA sequences. These results are shown in Figure 4 and reveal the following features. The first exon encodes two-thirds of the leader peptide, and the second exon encodes the rest of the leader peptide plus the NH_2-terminal 26 residues of the native protein. The codons specifying blood groups M and N are also present on exon 2, and serine and glycine are found on the first and fifth codons, indicating group M, as expected. Following these two exons, five more exons are present in the GPA gene. Among them, the fifth exon encodes for the transmembrane domain, the sixth exon for the cytoplasmic segment, and the seventh exon for the 3′ untranslated region. All the 5′ donor and 3′ acceptor sites conform with the exon/gt . . . intron . . . ag/exon consensus sequences (Breathnach and Chambon, 1981). There are also branch points in every intron, which are necessary for lariat formation of precursor mRNA (Ruskin *et al.*, 1984; Padgett *et al.*, 1986) positioned within 37 bp from the 3′-splicing sites (Figure 4).

4.3. Structural Organization of the GPB Gene

From the 5′-untranslated sequence to the region about 1 kb downstream from exon 5, the nucleotide sequences of the GPB gene are very similar to those of the GPA gene. The first and fifth amino acid residues of a mature protein encoded in the second exon are leucine and glutamic acid, respectively. These residues define N blood group activity, as expected from a GPB gene.

Comparative analysis with the GPA gene indicates that GPB mRNA lacks exon 3 found in GPA (Siebert and Fukuda, 1987). However, nucleotide sequence determination of GPB genomic DNA revealed the existence of sequences quite similar to GPA exon 3 and the adjacent introns of the GPA gene. There are two distinct differences from the GPA gene that are likely to affect splicing. The GPA has two possible branch sites in the second intron, and one site perfectly conforms with the consensus sequence (PyNPyTPuAPy) (Ruskin *et al.*, 1984). The GPB gene has substitutions in this site that abolish the branch site in this sequence. The second significant difference in this region is that the 5′-splicing

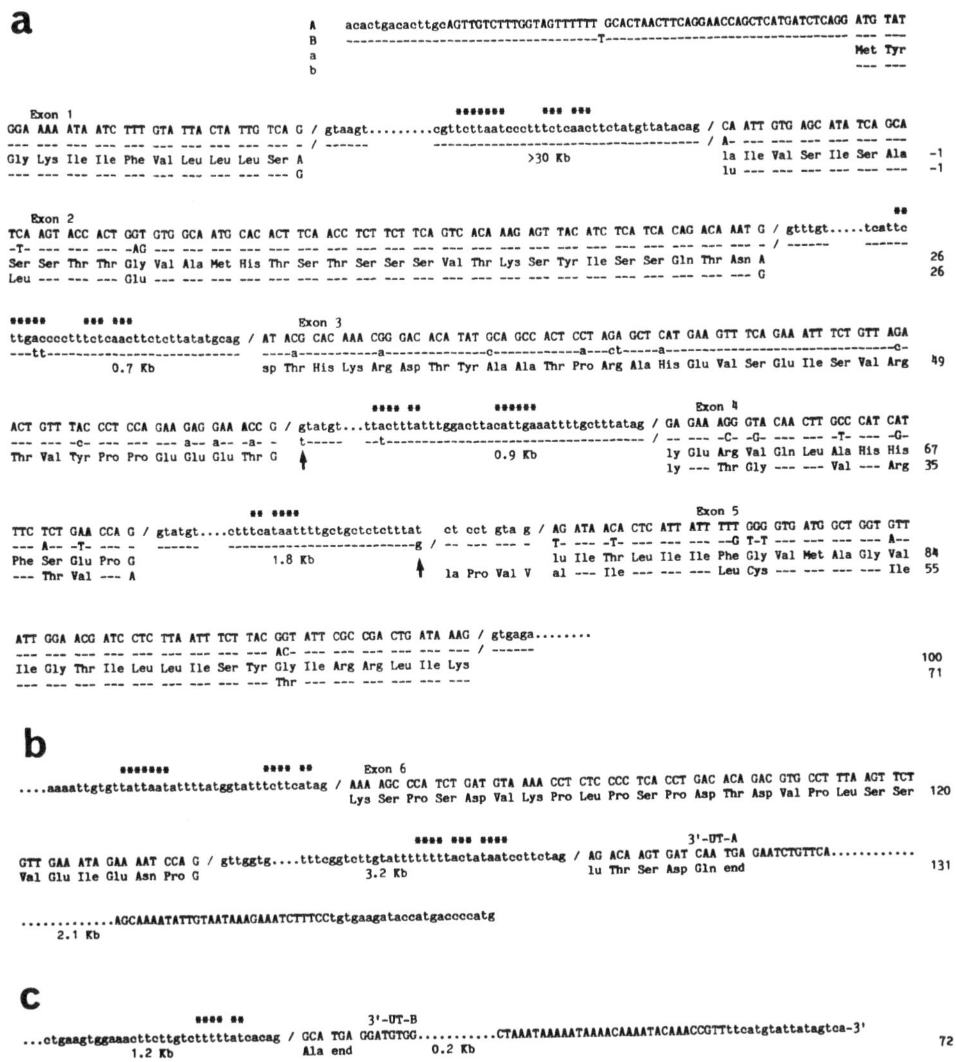

FIGURE 4. Comparison of the nucleotide sequences of glycophorin A and B genes. The sequences of all exons, exon–intron junctions, and 5′ and 3′ flanking regions and the amino acid sequences coded by exons are shown. (a) The sequences where GPA and GPB genes are almost identical. The arrows indicate two point mutations in the GPB gene, resulting in different exons. Each set of four lines (from top to bottom) shows the nucleotide sequences of GPA and GPB (A,B) followed by amino acid sequence (a,b). Identical sequences are shown by dashes. (b,c) The 3′-end sequence, which is unique to the GPA (b) or GPB (c) gene. The sequences in (b) and (c) are those directly following the transition site from homologous to nonhomologous sequences (BRP in Figure 3A). The splicing branch points are indicated by asterisks. Exon sequences are in uppercase intron and flanking sequences in lowercase. Amino acid residues are numbered at the far right. From Kudo and Fukuda (1989).

site of intron 3 is disrupted in the GPB gene. The change from CG/*gt*at to CG/*tt*at resulted in the loss of the 5′-splicing site (arrows in Figure 4a). Because of this disruption in the 5′-splicing site, the third exon of GPB is skipped, and direct ligation between the second and fourth exon apparently takes place.

In order to explain this result, it is proposed that the splicing machinery moves from the 3′ to the 5′ end of precursor RNA so that the 5′ cleavage of intron 3, as well as the 3′ cleavage of intron 2, are ignored. In relation to this hypothesis, it is noteworthy that one type of β-thalassemia was produced by a similar splicing pattern due to a mutation at the 5′-splicing site (Padgett *et al.*, 1986). Since no mutation was detected at the upstream splicing sites, in this case, the results strongly favor the 3′-to-5′ movement of the splicing machinery.

Compared with GPA, GPB has an additional three amino acid residues in the transmembrane domain encoded by exon 5. This difference is due to a change at the 3′-splicing site generated by a single base pair substitution (T→G), resulting in the formation of a tt*ag*/CT sequence from tt*atct* (arrows in Figure 4a). An alternative splicing site exists 9 bp upstream from the corresponding GPA splicing site, and this is the site used in GPB mRNA production.

4.4. Transition Sites in GPA and GPB Genes

Although the nucleotide sequences encompassing exon 1 to exon 5 are extremely homologous between GPA and GPB, the sequences of the 3′-proximal regions are entirely different. The transition sites from homologous to nonhomologous sequences could be located approximately 1–3 kb from the 3′ end of exon 5. The existence of human *Alu* repeat sequences was noted when the last 350 bp of the sequences homologous between GPA and GPB genes was searched with the aid of a computer data base. The *Alu* sequences lie in inverted orientation and are composed of 304 bp in the GPA gene and 294 bp in the GPB gene, and differing in the length of the 5′ poly (T) stretch (Figure 5). In general, *Alu* sequences are flanked on both sides by direct repeats 7–20 bp in size, suggesting that *Alu* could act as mobile genetic elements. In fact, it has been reported that the transposition of the *Alu* element led to inactivation of the target gene when cells were fused with UV-irradiated cells (Lin *et al.*, 1988). The direct repeats were conserved in the GPA gene, whereas those in GPB apparently lost homology. On the other hand, the upstream element is well conserved between GPA and GPB genes. Therefore, it can be judged that the original genomic structure around the *Alu* repeat is repressed by the GPA gene, whereas the GPB gene acquired a different *Alu* sequence on the downstream sequence. The homology between GPA and GPB sequences is lost within the last 23 bp of the 3′-*Alu* repeats having 6 bp mismatches. These results strongly suggest that recombination took place somewhere 40 to 24 bp upstream from the 3′ end of the *Alu* sequences (Figure 5). This recombination site in *Alu* sequences lies between A and B promoter sequences for RNA polymerase II (Paollela *et al.*, 1983), as has been observed in the majority of *Alu* recombination sites such as LDL receptor gene of familial hypercholesterolemia (Lehrman *et al.*, 1987), human α-globin genes having natural mutation (Nicholls *et al.*, 1987), and sex chromosome interchange in XX males (Rouyer *et al.*, 1987).

Based on these results, it can be postulated that the GPB gene was produced by duplication of the original gene, diversion by recombination through *Alu* repeats, and point mutation of the duplicated gene. It is possible that *Alu* repeats were directly involved

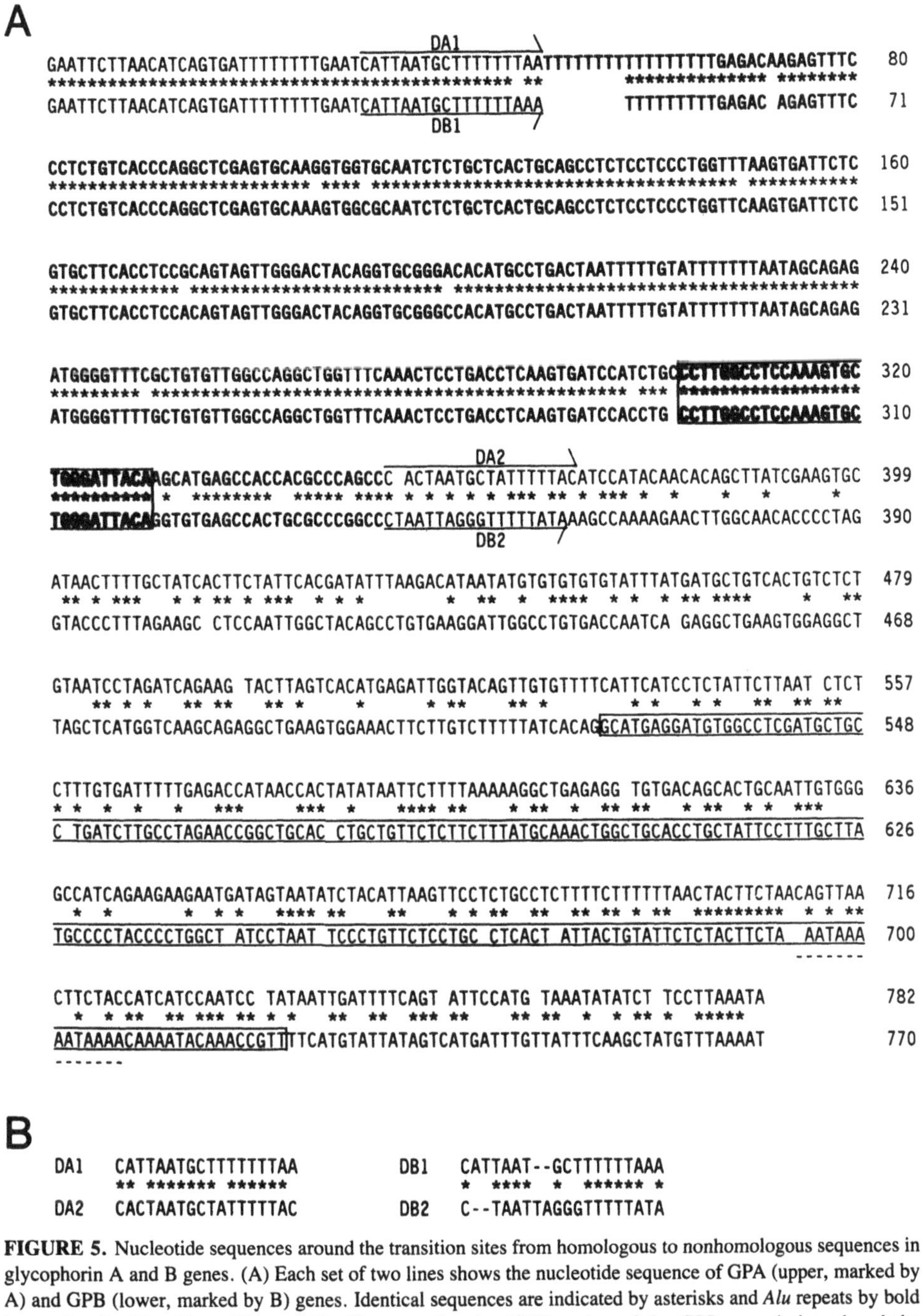

FIGURE 5. Nucleotide sequences around the transition sites from homologous to nonhomologous sequences in glycophorin A and B genes. (A) Each set of two lines shows the nucleotide sequence of GPA (upper, marked by A) and GPB (lower, marked by B) genes. Identical sequences are indicated by asterisks and *Alu* repeats by bold letters (nucleotides 50 to 353 or 343). The exon for the 3′UT region of the GPB gene is boxed and the polyadenylation signal is underlined with dashes. Arrows indicate direct repeats flanking *Alu* sequences. The presumptive recombination site (or breakpoint) is indicated by the shaded box. These sequences represent the 5′-ends of 2.8- or 4.8-kb segments that follow exon 5 (see Figure 3A). (B) Comparison of the direct repeats flanking *Alu* sequences. DA1 and DA2 for GPA direct repeats and DB1 and DB2 for GPB, which are indicated in (A). From Kudo and Fukuda (1989).

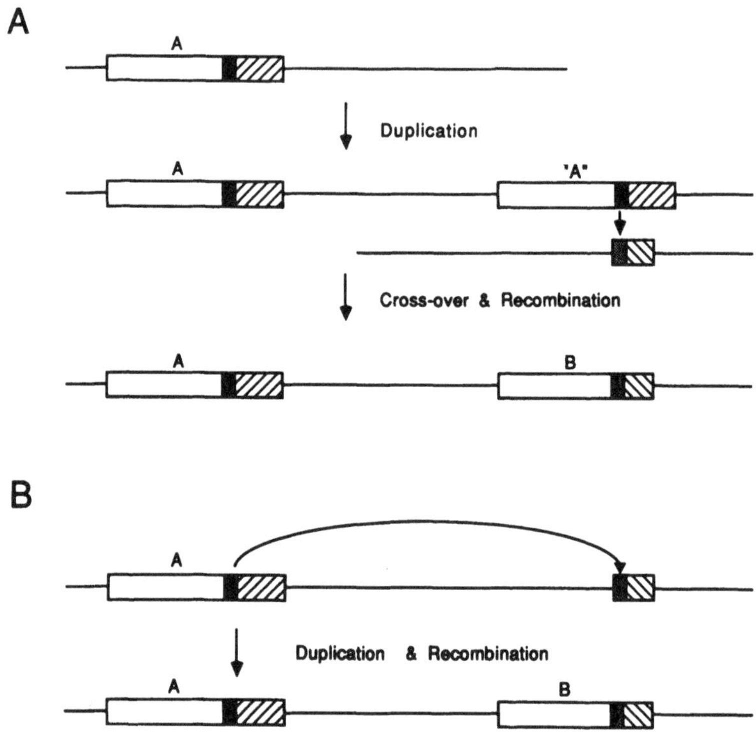

FIGURE 6. Two possible mechanisms for evolution of glycophorin A and B genes. (A) The original gene was first duplicated, and the duplicated gene acquired the GPB-specific 3′ end by crossover at *Alu* repeats. (B) Duplication of the original gene took place at *Alu* sequences, generating the GPB-specific 3′ end in one step. In both cases, the original gene is a direct precursor of the GPA gene. *Alu* sequences in GPA and GPB genes are indicated by a solid box and solid and speckled boxes, respectively. Slashed boxes indicate the 3′-end sequences unique to the GPA or GPB gene, while open boxes indicate the homologous sequences shared by GPA and GPB genes. From Kudo and Fukuda (1989).

during duplication (Figure 6B) or during recombination upon crossover after the gene was duplicated (Figure 6A). In both cases, homologous recombination through *Alu* repeats resulted in a 3′-end sequence unique to the GPB gene and the loss of exon 6. The original gene, on the other hand, directly evolved into the GPA gene. The results strongly suggests that *Alu* repeats might be involved in the generation of a number of genes.

By analyzing the sequences in the homologous region, it is possible to estimate the length of time since the gene duplication. Intron sequence flanking exon 1 to exon 5 between the GPA and GPB genes have 95.5% identity in a total of more than 4 kb sequence determined. It was, therefore, estimated that the duplication of the ancestral gene took place about 9–35 million years ago, according to two different divergence rates: 0.5% per million years for an average mammalian rate (Hayashida and Miyata, 1983) or 0.13% per million years for an average primate rate (Britten, 1986). It will be interesting to analyze the glycophorin genes in primates in order to estimate the time when the actual gene duplication occurred for GPB formation.

4.5. Isolation of Genomic Clones and Organization of the GPC Gene

By *in situ* hybridization on human metaphasic chromosomes with a cDNA probe, the structural gene for GPC has been assigned to chromosome 2ql4-q21 (Mattei *et al.*, 1986). This gene is therefore localized on a different chromosome than the structural gene for GPA (Rahuel *et al.*, 1988a).

Preliminary analysis by Southern blotting of human genomic DNA digested by the *Eco*RI endonuclease and hybridized with cDNA and oligonucleotide probes indicated that most of the GPC gene is carried by a 16-kb restriction fragment except for the 5′ region, which was detected in a 2.2 kb fragment.

A partial genomic library enriched in the 16-kb *Eco*RI fragments by preparative agarose gel electrophoresis was therefore constructed in λCharon 4A and screened with the GPC cDNA probe. On the other hand, oligonucleotide probes specific for the 5′ region of the GPC mRNA were used to identify genomic clones covering the 5′ end of the GPC gene. Positive clones isolated from these two libraries were subjected to restriction mapping and nucleotide sequencing (Colin *et al.*, 1989). The resulting map of the GPC gene, which is shown in Figure 7, indicates that this gene is organized in four exons distributed over a 13.5-kb-long DNA fragment. Southern blots with human genomic DNA and with the cloned GPC gene exhibited the same pattern when hybridized with the GPC cDNA probe, indicating that there is a single gene for GPC in the human genome. From sequence analysis (Figure 8), it was deduced that all exon/intron junctions within the GPC gene conform with the exon/gt . . . polypyrimidines . . . ag/exon consensus sequence (Breathnach and Chambon, 1981). Exons 1, 2, and 3 encode for the extracellular domain of GPC, whereas exon 4 encodes for the hydrophobic membrane-spanning region and for the COOH-terminal domain of the molecule.

The transcription initiation of the GPC gene was determined by S1 mapping and primer extension analysis with RNA preparations from erythroid tissues. Both techniques localized the cap site of the GPC mRNA at a position 156 ± 1 bp upstream from the translation initiation codon (Figure 8).

An interesting feature of the GPC gene structure is the presence of two 3.4-kb-long direct repeated domains in tandem (Figure 8). It is suggested that these two domains result from the duplication of the unique domain of an ancestral gene and that this event should have occurred relatively recently in evolution, since the nucleotide sequence divergence between the two duplicated domains is less than 5% over 1.7 kb of DNA. The significance of this duplication event is not known, since there is presently no indication that the protein sequence encoded by these repeated segments plays any biological function. As the homologous exons 2 and 3 are each carried by one of the two repeated domains, it is speculated that the putative ancestral form of the GPC gene should have been organized in three exons and lacked either exon 2 or 3, which differ essentially by the insertion of 27 nucleotides in exon 3. These 27 nucleotides encode amino acids 42–50 of GPC, where the Ge : 3 antigens have been localized.

The presence of the long direct repeats certainly explains the difficulty of isolating nonrearranged genomic clones after propagation of the recombinant phages in *E. coli*. Indeed, 80% of the GPC clones isolated from the λCharon 4A library shared an insert of 12.6 kb instead of the 16-kb fragment initially cloned in the *Eco*RI phage arms. Restriction mapping and sequence analysis of some of these clones indicate that they were

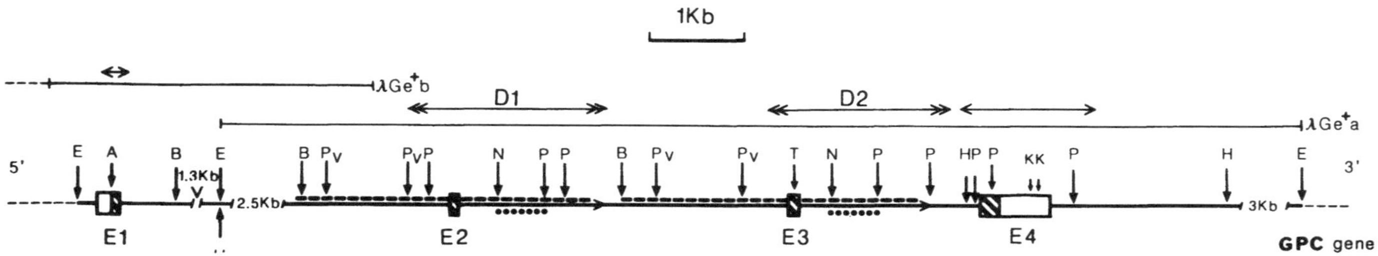

FIGURE 7. Structure of the glycophorin C gene. The gene organization was established from analysis of λGe^+ and λGe^+ genomic clones isolated from a human DNA leukocyte library. DNA regions analyzed by nucleotide sequencing on both strands are indicated between arrows. Double arrows refer to sequencing data of the repeated domains D1 and D2. Exons are numbered from the 5′ to the 3′ end and their coding and noncoding parts are indicated by hatched and open boxes, respectively. Introns and flanking sequences are shown by black lines. Relevant endonuclease sites for the restriction map are *Acc*I (A), *Bam*HI (B), *Eco*RI (E), *Hind*III (H), *Kpn*I (K), *Nco*I (N), *Pst*I (P), and *Pvu*II (P_V). Adapted from Colin *et al.* (1989).

```
             5'-ggaaaccccaccagccaagtctgcagaaaaaaataaatgaagtctgcctatctccgggccAGAGCCCCTCCCCTCGGCCCGCGCGGGAGGAGTGTGACCCAG
     GTGCCGCTTCCTCTCGCCGCCGAGGGTCAGGAGCCCGGGAGCGCGACCCTCCCCGGCCCGGCCTGGCCCGGCCTGGCCAGTCCCCGCGGTCTCTGCCCGGGCTGACGCCCAGGA

         Exon 1
     ATG TGG TCG ACG AGA AGC CCC AAC AGC ACG GCG TGG CCT CTC AGC CTC G/gtgag....(6.2 Kb)....ctctgacctcagattcttgtcctctgtt
1    Met Trp Ser Thr Arg Ser Pro Asn Ser Thr Ala Trp Pro Leu Ser Leu G

                   Exon 2
     cacag/AG CCT GAT CCG GGG ATG GCC TCT GCC TCC ACC ACA ATG CAT ACT ACC ACC ATT GCA G/gtgag.....(3.4 Kb).....ctctgacat
17         lu Pro Asp Pro Gly Met Ala Ser Ala Ser Thr Thr Met His Thr Thr Thr Ile Ala G

                                    Exon 3
     cagattcttgtcctctgttcacag/AG CCT GAT CCA GGG ATG TCT GGA TGG CCG GAT GGC AGA ATG GAG ACC TCC ACC CCC ACC ATA ATG GAC
36                              lu Pro Asp Pro Gly Met Ser Gly Trp Pro Asp Gly Arg Met Glu Thr Ser Thr Pro Thr Ile Met Asp

                                                                                 Exon 4
     ATT GTC GTC ATT GCA G/gtgag.....(1.8 Kb).....gactgaccccttcgacctcttcccacctgcag/GT GTG ATT GCT GCT GTG GCC ATC GTC CTA
59   Ile Val Val Ile Ala G                                                         ly Val Ile Ala Ala Val Ala Ile Val Leu

     GTC TCC CTC CTC TTC GTC ATG CTG CGC TAC ATG TAC CGG CAC AAG GGC ACG TAC CAC ACC AAT GAG GCC AAG GGC ACG GAG TTT GCT
74   Val Ser Leu Leu Phe Val Met Leu Arg Tyr Met Tyr Arg His Lys Gly Thr Tyr His Thr Asn Glu Ala Lys Gly Thr Glu Phe Ala

                                         A
     GAG AGT GCA GAT GCA GCC CTG CAG GG  GAC CCT GCC CTC CAA GAT GCT GGT GAT AGC AGC AGA AAG GAG TAC TTT ATT TGA GGGACAA
                                         G
103  Glu Ser Ala Asp Ala Ala Leu Gln Gly Asp Pro Ala Leu Gln Asp Ala Gly Asp Ser Ser Arg Lys Glu Tyr Phe Ile end

     CAGACTTCACTTCCCTGAATGCCTCCCCCATCTCCATCAGGAAAAATACACCCCATCGCCCAGTATCCCCGTCGATACCACCAGACAGAGAGAGAGAGTACACTTGATTTCTTCCC
     GAGATAGTTACCTAGAAACACTAGGTGCCTGCCCAAGGAGGAACGGAGGAGGACTCGCGCTACAAGAGGCCACTCCCAGGGACCCAGGGAGGCGATGGCCACCCCAGAGGCCACCT
     TTTGCTGCACGGAGGTGGGAGAGAATCTGGGCACATGGGGCCCCCTAGGGCAGTGCAGGACAACATCAGCTCACTGGCAGGAAAGTCCTTGTTGAGGGTGAGGGGGTGCTGGGGTA
     CCCGGGGGCTGGGGAAGCAAGGAAATAAGTCATCTGTATGCTGACTGGGGATAATGGCATCAATGTCAGTCCTTGACTTTGGGGGGAACAGCAGGTGCCAGAGCTAAAAGGTACCT
     TTGTCTGCCATTGATCCAGCTAAGAACGATTGGAAATAAATTGGAAATGTAACCGAGCattccgagtccaacagtattactgtagttatct-3'
```

FIGURE 8. Nucleotide sequence of the glycophorin C gene. The sequence of all exons, exon–intron junctions, and 5′ and 3′ flanking regions and the amino acid sequences coded by exons are given. Exon sequences are in uppercase, intron and flanking sequences in lowercase. The splicing branch points are underlined. The large arrow shows the start of transcription in erythroid tissues (nucleotide +1) whereas the small arrows indicate the predominant start of transcription in the promyelocytic cell line HL-60 (position −57) and in the megakaryotic cell line MEG-01 (position −11), respectively. Compiled from Colin *et al.* (1989) and Le Van Kim *et al.* (1989).

spontaneously deleted of 3.4 kb and that the breakpoints of these deletions were located within two perfectly homologous 500-bp *NcoI–PstI* fragments carried by the D1 and D2 repeated domains (Figure 8). Although repetitive sequences such as *Alu* sequences are involved in a great number of gene rearrangements by unequal crossing over, in the case of the GPC gene the sequences around the deletion breakpoints are unique in the human genome. It is interesting that the same deletion within the GPC gene occurs also in the genome of certain individuals whose red cells lack Gerbich blood group antigens and are deficient in GPC and GPD (Colin *et al.*, 1989).

5. EXPRESSION AND TISSUE SPECIFICITY OF GLYCOPHORINS A, B, AND C

The expression of GPA, GPB and GPC can be measured by Northern analysis using isolated cDNAs as probes with RNA preparations from a variety of cell lines and tissues and by analysis of the cell surface expression with polyclonal and monoclonal antibodies.

5.1. Transcription of Glycophorin mRNAs

cDNA clones encoding the full length of GPA detect four different mRNA species in K562 extracts, with molecular sizes of approximately 2.8, 1.7, 1.0, and 0.6 kb. If a GPA cDNA clone is short and encodes for only two-thirds of cDNA from the 3′ end, it detects only mRNA of 2.8, 1.7, and 1.0 kb. Since the homology between GPA and GPB cDNAs is more significant in one-third of the total cDNA from the 5′ end, it was judged that mRNA of 0.6 kb encodes for GPB, whereas mRNA of 2.8, 1.7, and 1.0 kb encode for GPA (Siebert and Fukuda, 1986a). These results were confirmed recently (Tate and Tanner, 1988), and multiple mRNAs were similarly detected in human adult erythroblasts but not in human adult liver (Rahuel *et al.*, 1988a). Interestingly, the cDNA for GPB detected only the 0.6-kb mRNA, failing to react with the other three mRNA species detected by full-length GPA cDNA (Siebert and Fukuda, 1987). It appears that some secondary structure is involved in interaction between cDNA probes and mRNA on the filter, even when hybridization was made under strong denaturing conditions.

The finding that three mRNAs were recognized by the GPA cDNA probes has now been clarified following isolation of new clones from a human fetal liver cDNA library with long 3′-termini that carry six polyadenylation signals (Rahuel *et al.*, 1989). Using oligonucleotide probes located between these signals, it was demonstrated by Northern blot analysis that two of the polyadenylation signals, which are not adjacent, are used for generating the 1.0- and 1.7-kb mRNA species of GPA (Figure 9). Only a few nucleotides, however, from the 2.2 (or 2.8)-kb species could be characterized, suggesting that a longer GPA mRNA is expected, possibly with even more polyadenylation signals. It was found also by primer extension analysis with two different oligonucleotide probes that all the GPA mRNAs from a spleen erythroblast extract are transcribed from the same initiation site, therefore indicating that the production of multiple GPA mRNAs is governed by poly(A) site choice and not by 5′ size heterogeneity (Rahuel *et al.*, 1989). Examination of the 3′-untranslated sequence shown in Figure 9 did not provide any particular support for consensus sequences responsible for the poly(A) site choice used to generate the GPA mRNAs.

GTT GAA ATA GAA AAT CCA GAG ACA AGT GAT CAA TGAGAATCTGTTCACCAAACCAAATGTGGAAAGAACACAAAGAAGACATAAGACTTCAGTCAAGTGAAAAATTAA
Val Glu Ile Glu Asn Pro Glu Thr Ser Asp Gln end

CATGTGGACTGGACACTCCAATAAATTATATACCTGCCTAAGTTGTACAATTTCAGAATGCAATTTTCATTATAATGAGTTCCAGTGACTCAATGATGGGGAAAAAAATCTCTGCTCAT
1

TAATATTTCAAGATAAAGAACAAATGTTTCCTTGAATGCTTGCTTTTGTGTGTTAGCATAATTTTTAGAATTGTTTGAGAATTCTGATCCAAAACTTTAGTTGAATTCATCTACGTTTG

TTTAATATTAACTTAACCTATTCTATTGTATTATAATGATGATTCTGTCAAATGAAAGGCTTGAAATACCTAGATGAAGTTTAGATTTTCTTCCTATTGTAAACTTTTGAGTCTGGTTT
P28

CATTGTTTTAAATAAATTAAGGGGACACTAAAGTCCTATCATTCATTTCCTTCATTGCTGAACAGGCAAGATATAATATTACATGAATGATTACTATATTTTGTTCACACTAATAAAGC
2 3

TTATGCTCAGAAATGCCATACACACACACAAACACACACATTTATCATTTAATGCATAAATCAACACAAAAGGTTTTCCCATTAATATGAAATATTACATATATATAAGTGCCATATTT

AAAATAATTTGTCTAACAGTAGAACTATGTCGGAGCACTCACTGAAGCTTGCATTCCACTGAAAGAGTTATTTGTGTAAGTAGAGTATCCGGAGAAGGAAAAGAACTTACGACCTTTCT

TTATAACAGAAACTCAACTCTAAATTCAACAAGATGTGCAAACCGGACATGCAGGTGAATATTTTAATAGGTTACTATAAGGTTCTCAATTAAATTCTTTAATCTGTCCAGTCCCAGTT

TCTCTTATTAATAAAACTTTGGAAATTGCTTTAAACCATTTAAAGGAAAGGTTTCTAGATATAGAAACTAAGGACTGTGACTATACAGCTGTCACTCATTTGTAGTAAAACTTAAAAAG
4

CAAAAACAAAAAACAAAAAAGACCTTCCTGTGATACTTTATTTCCGAACTAATAAAAATCTATATGACTTTTTATTATTGTGTGATAACCAAGTAAATGTTTTCTATTTTGCATATTTT
5

CAGGCATGGTAACAGAAATTTACCTTTTAATAAATTAAAAAATCTAAATTTTAACCTACTTGTATGTTCGGAGAGTGTTTTTGTACTATATTGACT-3'
P29 6 P30

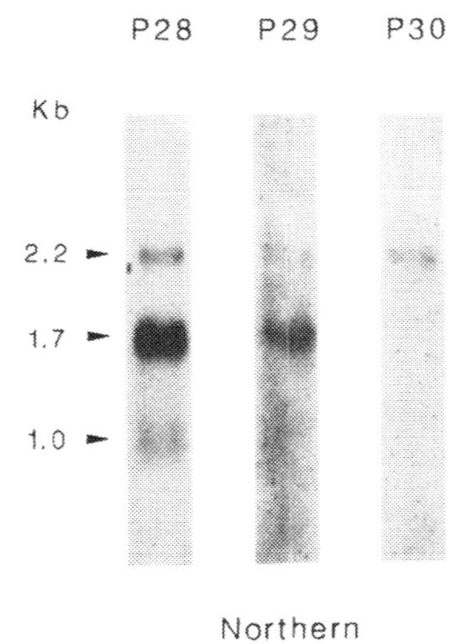

FIGURE 9. 3′-end analysis of GPA mRNA. (Left) Composite nucleotide sequence of the 3′ end of GPA cDNAs isolated from a human fetal liver library. A short segment encoding the COOH-terminal end of GPA is shown, followed by a continuous stretch of 1242 nucleotides containing six potential polyadenylation signals AATAAA (underlined and numbered from 1 to 6). P28, P29, and P30 correspond to oligonucleotides used for Northern blot analysis (on the right). (Right) Northern blot analysis of poly(A+) mRNAs isolated from human spleen erythroblasts hybridized with oligonucleotide probes P28, P29, and P30 deduced from the nucleotide sequence of the 3′-untranslated region of the GPA mRNA shown on the left. P28, P29, and P30 hybridized respectively with three (2.2, 1.7, and 1.0 kb), two (1.7 and 1.0 kb) and one (2.2 kb) mRNA species, indicating that polyadenylation signals 2 and 6 are used to generate the different mRNAs. Adapted from Rahuel *et al.* (1989).

Table II
Tissue Specificity of Human Glycophorins[a]

Blood cells or cell lines	Immunofluorescence[b]				Northern[c]	
	A		C			
	R18	L885	MR4-130	APO3	A,B	C
Peripheral blood						
Red cells	++	+++	+++	++		
B lymphocytes	−	−	−	((+))	−	(+)
T lymphocytes	−	−	−	+	−	+
Granulocytes (N,Eo)	−	−	−	−	−	−
Monocytes	−	−	−	+/−	−	+
Platelets	−	−	−	−	ND	ND
Cell lines						
K562	+++	+++	+++	+++	+++	+++
HEL	++	++	++	+++	+	++
KU-812	++	++	++	+++	+++	+++
KG-1	−	−	(+)	++	−	+++
HL-60	−	−	−	((+))	−	+
U937	−	−	(+)	+	−	+
Jurkat	−	−	(+)	++	−	+
Raji	−	−	−	−	−	−
MEG-01	−	−	+	++	−	++

[a]Expression of glycophorins on peripheral blood cells and cell lines was examined by indirect immumnofluorescence for cell surface expression and by Northern blot analysis for presence of mRNA. Data compiled from Villeval *et al.* (1989) and Le Van Kim *et al.* (1989).

[b]Indirect immunofluorescence was carried out with murine monoclonal antibodies against GPA (R18) and GPC (MR4-130, APO3) as well as with a polyclonal anti-GPA antibody (L885) raised in rabbits. Cell binding was revealed with anti-IgG $F(ab')_2$ fragments conjugated to fluorescein. Fluorescence intensity graded from ((+)) to +++.

[c]Northern blot analysis was carried out with total RNA extracts resolved on formaldehyde agarose gels, transferred to nylon filters, and hybridized with 32p-labeled cDNA probes for GPA, B, and C. The intensity of bands on the X-ray film was scored from (+) to +++. ND, not done.

RNA processing mechanisms are generally used to generate several protein products from a complex transcription unit and are often tissue-specific or developmentally regulated (Leiff *et al.*, 1986; Breitbart *et al.*, 1987). In the glycophorin model, however, the three GPA mRNAs are exclusively present in cell lines (K562, HEL) and tissues (fetal liver) of erythroid origin (Table II), but are absent from continuous cell lines of myeloid (HL-60, U937), lymphoid (Jurkat, Raji), and megakaryoblastic (MEG-01) origin or from other human adult tissues like liver, kidney, breast, or thymus (Le Van Kim *et al.*, 1989; Rahuel *et al.*, 1989; Siebert and Fukuda, 1986b, 1987). Therefore, the biological significance for GPA mRNA size heterogeneity presently remains unclear. Obviously, all these messages should be translated into the same protein product, but it is not known whether they are translated with the same efficiency or whether the 3′-untranslated regions play a role in the stability of the mRNAs.

From similar studies carried out with GPB mRNA (Table II), it was concluded that GPA and GPB are specifically transcribed and translated in erythroid cells or tissues. In addition, since one of the primers used in these experiments hybridized with a nucleotide sequence common to the GPA and GPB mRNAs, it is likely that the GPB mRNA is initiated at the same site as the GPA mRNA (Rahuel *et al.*, 1989). Further, both GPA and GPB mRNAs are down-regulated by treatment of cells with tumor-promoting phorbol esters (Siebert and Fukuda, 1986b, 1987).

The cDNA probe for GPC was also used in Northern blot analysis and shown to detect a single band of 1.1 kb in RNA preparations from erythroid tissues, both under high and low stringency conditions, thus indicating no evidence for cross-hybridization with another related mRNA species of different size (Colin *et al.*, 1986; Cartron *et al.*, 1986). This argues, but does not prove, that GPC and GPD could be encoded by the same mRNA, as discussed earlier. Interestingly, analysis of RNA preparations from erythroid and nonerythroid cell lines and tissues provided new unexpected information (Table II). Indeed, the GPC mRNA was present in both erythroid and nonerythroid cells, although the transcription level was much higher in the former than in the latter (Le Van Kim *et al.*, 1989). The GPC mRNA, however, was not detected in granulocytes or in a Burkitt lymphoma cell line (Raji), although it is present at a very low level in normal B lymphocytes (Table II). The GPC mRNA was present at a low level in all human tissue examined so far including thymus, breast, and liver. As noted earlier (High and Tanner, 1987), it is also detectable in human kidney preparations. These results clearly indicate that, in contrast to the GPA and GPB genes, the expression of the GPC gene is not restricted to erythroid lineage. In addition, they suggest that the level of transcription of the GPC gene is regulated differently in erythroid and nonerythroid cells.

In order to determine whether the GPC mRNAs detected by Northern blot analysis in erythroid and nonerythroid cells are synthesized from the same transcription initiation sites, cap site determinations were carried out by primer extension analysis and the polymerase chain reaction with RNA preparations from nonerythroid cell lines. It was found that in myelomonocytic (HL-60) and lymphoid (Jurkat) cell lines, the transcription initiation site of the GPC mRNA is located predominantly 11 nucleotides upstream from the erythroid cap site (Figure 8), whereas it is 57 nucleotides upstream from this position in the megakaryocytic (MEG-01) cell line (Le Van Kim *et al.*, 1989). It is possible that the use of these different initiation sites regulates the different level of expression of the GPC gene in erythroid and nonerythroid tissues, but further analysis of the 5′ flanking sequence is required to address these questions. The presence of multiple transcription initiation sites in differential gene expression in various tissues has been reported for other genes like c-myc (Battey *et al.*, 1983), IL-2 receptor (Leonard *et al.*, 1985), IL-6 (Yasukawa *et al.*, 1987), c-myb (Watson *et al.*, 1987), human EGF (Ishii *et al.*, 1985), and band 3 (Kopito *et al.*, 1987a), but the mechanism regulating this phenomenon remains unknown.

5.2. Cell Surface Expression

Whether in the different cells the glycophorin mRNAs are translated can be deduced from immunochemical studies with monoclonal and polyclonal antibodies. The expression of GPA has been studied extensively using specific antibodies, reacting either with its extracellular portion (Gahmberg *et al.*, 1978) or with the cytoplasmic segment (Yurchenco and Furthmayr, 1980; Fukuda, 1981). More recently, a number of murine monoclonal antibodies against GPA have been produced and characterized by immunochemical analyses, binding studies, flow cytometry, and MN blood typing of normal erythrocytes. A comparative analysis of these monoclonal antibodies has been performed during an International Workshop (Rouger *et al.*, 1988). Using one of these monoclonal antibodies (R18) and a rabbit anti-GPA (L885), it was confirmed that the cell surface expression of GPA is restricted to erythroid cell lineage (Table II).

Antibodies specific to GPB that are of high enough specificity to perform Western blots of immunofluorescence microscopy have not yet been obtained. This is because the majority of antibodies obtained react with the NH_2-terminal portion of the molecules, a region that is shared by GPA and GPB.

The cell surface expression of GPC has been examined similarly by flow cytometry analysis (Villeval *et al.*, 1989; Le Van Kim *et al.*, 1989) using well-characterized murine monoclonal antibodies (MR4-130 and APO3) that react with the NH_2-terminal domain of GPC, but not with GPD (Dahr *et al.*, 1989). The results are summarized in Table II. The MR4-130 antibody is glycosylation-dependent and reacts strongly with erythroid cell lines but not significantly with nonerythroid cells. In contrast, APO3 is sialic acid-independent and reacts with erythroid and nonerythroid cell lines. A good correlation was found between the cell surface expression of APO3 epitopes and the presence of mRNA in cell extracts (Table II). These findings indicate that GPC is expressed at the surface of many cells with the exception of granulocytes, platelets and the Raji cell line, and that the nonerythroid form of this glycoprotein is glycosylated differently than the erythroid molecular species. Whether these cells produce GPC and also GPD is not known since there is as yet no monoclonal antibodies specific for GPD and the polyclonal antibodies that react with this glycoprotein either also recognized GPC (Dahr *et al.*, 1987; Reid *et al.*, 1987a) or are not strong enough for reliable studies (human anti-Ge : 2 antibodies). Immunoprecipitation analyses should provide an answer but these experiments are difficult since the number of glycophorin molecules is limited on nonerythroid cells.

5.3. Expression during Erythroid Differentiation

Since GPA, GPB, and GPC have been well characterized, they are useful markers to follow the sequential steps of erythroid differentiation.

During normal erythroid development, GPA is expressed rather late on morphologically recognizable precursors (Gahmberg *et al.*, 1978) but is not detectable on the progenitors BFU-E (burst-forming-unit-erythroid) or CFU-E (colony-forming-unit-erythroid), as deduced from indirect immunofluorescence studies by FACS (fluorescence-activated cell sorter) and by biochemical analyses (Robinson *et al.*, 1981; Sieff *et al.*, 1982; Fukuda, 1985). Similar investigations of human bone marrow cells carried out by FACS with murine monoclonal antibodies have shown that GPC is present on the erythroid progenitors (Figure 10), since it is detected on mature BFU-E by a non-glycosylation-dependent antibody (APO3) but only at the CFU-E stage with a glycosylation-dependent antibody (MR4-130). These findings indicate that GPC is expressed earlier than GPA during erythroid differentiation and that further maturation is accompanied by a change in GPC glycosylation (Villeval *et al.*, 1989). This is reminiscent of the increased O-glycosylation of GPA observed during the differentiation of erythroid precursors (Gahmberg *et al.*, 1984). It is worth mentioning that glycosylation changes of several other cell surface glycoconjugates do occur during development and differentiation of erythroid cells (Fukuda and Fukuda, 1984).

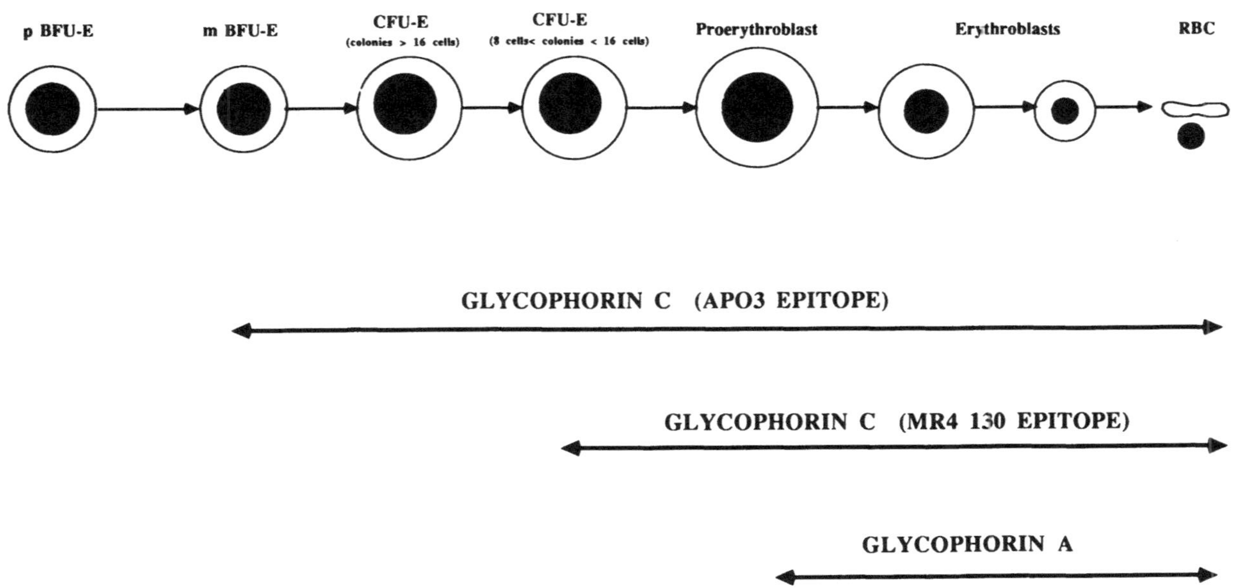

FIGURE 10. Schematic diagram showing expression of glycophorins A and C during normal erythroid differentiation. p BFU-E, primitive BFU-E (day 16); m BFU-E, mature BFU-E (day 12).

The cell surface expression of GPA has been used as a marker for identification of "cryptic" or early erythroleukemia among previously diagnosed cases of acute lymphoblastic or myeloblastic leukemia (Anderson *et al.*, 1979, 1980; Greaves *et al.*, 1983). Indeed, in some cases, a significant proportion of GPA-positive blast cells were detected, but the percentage of these patients is controversial (Liska *et al.*, 1983). In a more recent study, however, it was found that markers expressed earlier than GPA, like the carbonic anhydrase I, blood group A, and Gerbich (glycophorin C) antigens, can be useful in identifying additional examples of such patients whose blast cells are blocked at a "CFU-E-like" stage of differentiation (Villeval *et al.*, 1986). Among these patients, some have an intermediate phenotype with a positive expression of GPA, but a cell surface expression of GPC detectable with the glycosylation-independent (APO3) but not with the glycosylation-dependent (MR4-130) monoclonal antibodies. This is evocative of an altered posttranslational modification of GPC (not GPA) in these blast cells. Additional studies by Northern analysis and Western blotting indicated that the mRNA encoding GPC (and GPD, see above) is normally present but is translated mostly in GPD (Villeval *et al.*, 1989). Further investigations should clarify whether these abnormalities are due to the leukemic process or to a posttranscriptional control of GPC/GPD production in immature cells.

6. SUMMARY AND PROSPECTS

Recent studies have provided important information on the biochemistry and molecular biology of the major integral membrane proteins of the human erythrocyte, glycophorins A, B, C, and D, which are the carriers of MN, Ss, and Gerbich blood group antigens, respectively. These glycoproteins have been purified to homogeneity and their genes cloned. The main conclusions from these investigations can be summarized as follows:

1. GPA and GPB are homologous glycoproteins produced by genes that clearly have been derived from a duplication of the original gene. The GPA and GPB genes have been sequenced and it was found that a recombination between *Alu* sequences identified within intron sequences might have generated the present genes, which are highly homologous in their 5′ region but differ in their 3′-untranslated region. It is possible that *Alu* repeats were directly involved during duplication or during recombination upon crossing-over after duplication of the original gene.

Because of mutation of a 5′ donor splice junction and of a lariat branch site, exon 3, which is constitutively expressed in GPA, is spliced out in the GPB counterpart. These results together with analysis of variants indicate that the GPA gene is localized 5′ upstream of the GPB gene on chromosome 4, but the distance between these genes is not known.

2. Surprisingly, a new gene called *inv*, closely resembling the GPA and GPB genes, has been identified in the human genome. The *inv* gene is present in all human DNA samples investigated to date, including those from individuals carrying large deletions of GPA, GPB, or both the GPA and GPB genes. The chromosomal localization of the *inv* gene has not been established, but it is attractive to speculate that if it is present on chromosome 4, as a further duplication of the GPB gene, it should be located 3′ down-

stream of that gene, so that the order of the three genes on chromosome 4 should be GPA, GPB, and *inv*.

It is not known whether the *inv* gene is functional and whether it is transcribed and translated into a new glycoprotein. If it is, GPA- and/or GPB-deficient erythrocytes might carry an as yet unrecognized glycoprotein that might prevent structural and functional abnormalities in red cells.

3. The gene encoding for GPC has also been cloned and sequenced. It is characterized by the presence of two internal direct repeats in tandem, each extending over 3.4 kb of DNA. These repeats show little divergence and are likely to have been derived from the recent duplication of a unique ancestral domain. In addition, biochemical and DNA analyses of GPC and GPD indicate that the same gene encodes GPC *and* GPD. This was also substantiated by molecular analysis of DNA samples from individuals deficient in GPC and GPD expression on their red cells (Colin *et al.*, 1989). The gene encoding GPC and GPD has been named *Ge* (Gerbich) since both GPC and GPD carry blood group Gerbich antigens.

The mechanism generating GPC and GPD from the same gene is presently under study, but most likely involves the translation of the same mRNA at two in-phase AUGs by leaky translation. If this is true, the unique mRNA transcribed from the GPC gene should be translated into a long (GPC) and a short (GPD) protein form that differ in their NH_2-terminus. Indeed, present sequencing information on GPD is consistent with this model (El-Maliki *et al.*, 1989).

4. Now that the structure of the genes encoding GPA, GPB, and GPC has been defined precisely, a detailed analysis of the genetic alterations accompanying the red cell glycophorin defects identified in a large number of variants can be conducted. Such studies are presently in progress in several laboratories and already indicate that a variety of complex recombinations and deletions within and between the glycophorin genes are involved. The study of such a diversity makes the glycophorins an interesting model for further analysis of human gene polymorphism. GPA appears also as an original model of biological cumulative dosimeter to estimate the level of somatic cell mutations following past radiation exposures or the high degree of mutations that accumulate in Bloom's syndrome (Langlois *et al.*, 1987, 1989).

5. Another interesting piece of information deduced from these studies arises from the finding that GPC and GPD are not specific for the erythroid cell lineage, since the mRNA and the proteins have been detected in a number of nonerythroid tissues and cell lines. However, the level of mRNA transcription is much higher in erythroid than in nonerythroid tissues and in addition the proteins are differently glycosylated.

This situation is clearly distinct from that found for GPA and GPB gene expression, which are strictly specific for erythroid tissues. Of interest, it was found that the different tissues that express GPC transcribe the message from different initiation sites, but the basis of this mechanism has not been determined. The erythroid and nonerythroid forms of the GPC polypeptide are identical, however, as seen by nucleotide sequencing of the mRNA from human spleen erythroblasts and lymphocytes (Le Van Kim *et al.*, 1989). Obviously a fine analysis of the promoter region and of the *cis*-acting sequences and *trans*-activating factors that regulate gene expression should clarify the mechanism of differential tissue specificity of this gene.

The presence of GPC (and GPD?) at the surface of nonerythroid cells raises the question of its biological role and function in nonerythroid tissues and opens a new area of investigation. Tissue analogues of erythrocyte membrane proteins have been recognized for some time by immunological cross-reactivity of antibodies (for review see Bennett, 1985). In mammals, the erythroid and nonerythroid proteins might be the product of the same gene or of related nonidentical genes. For instance, tissue analogues of α-spectrin (brain α-fodrin), protein 4.1 (brain synapsin I), and band 3-like proteins are probably the products of distinct genes (Baines and Bennett, 1985; Demuth *et al.*, 1986; Leto *et al.*, 1988). However, the protein 4.1 isoforms of human red cells and lymphoid tissues are the product of a single gene and arise by alternative splicing of separated exons from a common RNA precursor (Tang *et al.*, 1988). Similarly, the mouse erythrocyte, lymphocyte, and renal transcripts of band 3 are issued from the same gene possibly by alternate use of a tissue-specific promoter element (Kopito *et al.*, 1987a,b; Alper *et al.*, 1988).

The widespread tissue distribution of erythrocyte membrane protein analogues has led to the speculation that these proteins may play a similar cell function by mediating, for instance, association of integral membrane proteins with the cytoskeletal proteins. The finding that the human epithelial lens cell, before its maturation to a fiber cell, contains many antigenically related erythrocyte membrane proteins is in good agreement with this assumption (Allen *et al.*, 1987).

In conclusion, all the information deduced from recent studies indicates that the genes encoding GPA and GPB on the one hand and the GPC and GPD on the other, do not belong to the same gene family since this pair of genes is not homologous in structure and further is not carried by the same chromosome. Accordingly, the nomenclature of these glycoproteins might appear as a misnomer unless the terminology "glycophorin" proposed first by Marchesi *et al.* (1972) is restricted to the general meaning of carbohydrate-rich protein.

Future studies should determine which factors are involved in the tissue specificity of these glycoproteins for a comparison with those regulating other erythroid-specific genes. It will also be of interest to investigate in detail the molecular events responsible for the altered expression of glycophorins in erythroleukemias, since these data may provide additional information useful for the diagnosis of these diseases and clarify the mechanisms involved in the biosynthesis and transport of glycophorins to the cell surface.

ACKNOWLEDGMENTS. We thank Drs. Paul Siebert (La Jolla Cancer Research Institute), Caroline Le Van Kim, Cecile Rahuel, Alain Vignal, Jacqueline London (INSERM U76 Research Unit) and Dr. William Vainchencker (INSERM U91 Research Unit) for their contribution to the data summarized in this chapter. We also thank Claire Brankovic, Henny Bierhuizen, and Kaarin Soma for secretarial assistance. This work was supported by the Institut National de la Santé et de la Recherche Médicale (to J.P.C.) and by R01 CA33000 from the National Cancer Institute (to M.F.).

7. REFERENCES

Allen, D. P., Low, P. S., Dola, A., and Maisel, H., 1987, Band 3 and ankyrin homologues are present in eye lens: Evidence for all major erythrocyte membrane components in same non-erythroid cell, *Biochem. Biophys. Res. Commun.* **149:**266–275.

Alloisio, N., Morlé, L., Bachir, D., Guetarni, D., Colonna, P., and Delaunay, J., 1985, Red cell membrane sialoglycoprotein in homozygous and heterozygous 4.2(−) hereditary elliptocytosis, *Biochim. Biophys. Acta* **816:**57–62.

Alper, S. L., Kopito, R. R., Libresco, S. M., and Lodish, H. F., 1988, Cloning and characterization of a murine band 3-related cDNA from kidney and from a lymphoid cell line, *J. Biol. Chem.* **263:**17092–17099.

Andersson, L. C., Gahmberg, C. G., Teerenhovi, L., and Vuopio, P., 1979, Glycophorin A as a cell surface marker of early erythroid differentiation in acute leukemia, *Int. J. Cancer* **23:**717–723.

Anderson, L. C., Wegelius, R., Borgström, G. H., and Gahmberg, C. G., 1980, Change in cellular phenotype from lymphoid to erythroid in a case of ALL, *Scand. J. Haematol.* **23:**115–121.

Anderson, R. A., and Lovrien, R. E., 1984, Glycophorin is linked by band 4.1 protein to the human erythrocyte membrane skeleton, *Nature* **307:**655–658.

Anderson, R. A., and Marchesi, V. T., 1985, Associations between glycophorin and protein 4.1 require a polyphosphoinositide cofactor: A mechanism for membrane skeleton association, *Nature* **318:**295–298.

Anstee, D. J., 1980, Blood group MNSs-active sialoglycoproteins of the human erythrocyte membrane in: *Immunology of the Erythrocyte* (S. G. Sandler, M.D. Nusbacher, and M. S. Schanfield, eds.), Volume 43, pp. 67–98, Liss, New York.

Anstee, D. J., 1981, The blood group MNSs-active sialoglycoproteins, *Semin. Hematol.* **18:**13–31.

Anstee, D. J., Mawby, W. J., and Tanner, M. J. A., 1979, Abnormal blood-group-Ss-active sialoglycoproteins in the membrane of Miltenberger class III, IV and V human erythrocytes, *Biochem. J.* **183:**193–203.

Anstee, D. J., Parsons, S. F., Ridgwell, K., Tanner, M. J. A., Merry, A. H., Thomson, E. E., Judson, P. A., Johnson, P., Bates, S., and Fraser, I. D., 1984a, Two individuals with elliptocytic red cells lack three minor erythrocyte membrane sialoglycoproteins, *Biochem. J.* **218:**615–619.

Anstee, D. J., Ridgwell, K., Tanner, M. J. A., Daniels, G. L., and Parsons, S. F., 1984b, Individuals lacking the Gerbich blood-group antigen have alterations in the human erythrocyte membrane sialoglycoproteins β and γ, *Biochem. J.* **221:**97–104.

Baines, A. J., and Bennett, V., 1985, Synapsin I is a spectrin-binding protein immunologically related to erythrocyte protein 4.1, *Nature* **315:**410–413.

Battey, J., Moulding, C., Taub, R., Murphy, W., Stewart, T., Potter, H., Lenoir, G., and Leder, P., 1983, The human c-*myc* oncogene: Structural consequences of translocation into the IgH locus in Burkitt lymphoma, *Cell* **34:**779–787.

Bennett, V., 1982, The molecular basis for membrane–cytoskeleton association in human erythrocytes, *J. Cell. Biochem.* **18:**49–65.

Bennett, V., 1985, The membrane skeleton of human erythrocytes and its implications for more complex cells, *Annu. Rev. Biochem.* **54:**273–304.

Bennett, V., 1989, The spectrin–actin junction of erythrocyte membrane skeletons, *Biochim. Biophys. Acta* **988:**107–121.

Blanchard, D., Cartron, J. P., Rouger, P., and Salmon, C., 1982, Pj variant, a new hybrid MNSs glycoprotein of the human red cell membrane, *Biochem. J.* **203:**419–426.

Blanchard, D., Dahr, W., Hummel, M., Latron, F., Beyreuther, K., and Cartron, J. P., 1987a, Glycophorins B and C from human erythrocyte membranes: Purification and sequence analysis, *J. Biol. Chem.* **262:**5808–5811.

Blanchard, D., El-Maliki, B., Hermand, P., Dahr, W., and Cartron, J. P., 1987b, Structural homology between glycophorins C and D, two minor glycoproteins of the human erythrocyte membrane carrying blood group Gerbich antigen, Proc. IXth Int. Symp. Glycoconjugates, p. F54, Lille, France.

Bloy, C., Blanchard, D., Lambin, P., Goossens, D., Rouger, P., Salmon, C., and Cartron, J. P., 1987, Human monoclonal antibody against Rh(D) antigen: Partial characterization of the Rh(D) polypeptide from human erythrocytes, *Blood* **69:**1491–1497.

Booth, P. B., Serjeantson, S., Woodfield, D. G., and Amato, D., 1977, Selective depression of blood group antigens associated with hereditary ovalocytosis among Melanesians, *Vox Sang.* **32:**99–110.

Breathnach, R., and Chambon, P., 1981, Organization and expression of eukaryotic split genes coding for proteins, *Annu. Rev. Biochim.* **50:**349–383.

Breitbart, R. E., Andreadis, A., and Nadal-Ginard, B., 1987, Alternative splicing: A ubiquitous mechanism for the generation of multiple protein isoforms from single genes, *Annu. Rev. Biochem.* **56:**467–495.

Britten, R. J., 1986, Rates of DNA sequence evolution differ between taxonomic groups, *Science* **231:**1393–1398.

Burness, A. T. H., 1981, Glycophorin and sialylated components as receptor for viruses, in: *Virus Receptors*, Part 2 (K. Lonberg-Holm and L. Philipson, eds.), pp. 63–84, Chapman & Hall, London.

Cartron, J. P., Colin, Y., Le Van Kim, C., Rahuel, C., Blanchard, D., Bloy, C., and London, J., 1986, Structure of human erythrocyte glycophorin C deduced from cDNA analysis, *Blood Transf. Immunohaemat.* **29:**267–285.

Chasis, J. A., and Mohandas, N., 1986, Erythrocyte membrane deformability and stability: Two distinct membrane properties that are independently regulated by skeletal protein associations, *J. Cell Biol.* **103:**343–350.

Chasis, J. A., Reid, M. E., Jensen, R. H., and Mohandas, N., 1988, Signal transduction by glycophorin A: Role of extracellular and cytoplasmic domains in a modulatable process, *J. Cell Biol.* **107:**1351–1357.

Colin, Y., Rahuel, C., London, J., Romeo, P. H., d'Auriol, L., Galibert, F., and Cartron, J. P., 1986, Isolation of cDNA clones for human erythrocyte glycophorin C, *J. Biol. Chem.* **261:**229–233.

Colin, Y., Le Van Kim, C., Tsapis, A., Clerget, M., d'Auriol, L., London, J., Galibert, F., and Cartron, J. P., 1989, Human erythrocyte glycophorin C gene structure and rearrangement in genetic variants, *J. Biol. Chem.* **264:**3773–3780.

Cook, P. J. L., Lindenbaum, R. H., Salonen, R., De La Chapelle, A., Daker, M. G., Buckton, K. E., Noades, J. E., and Tippett, P., 1981, The MNSs blood group of families with chromosome 4 rearrangements, *Ann Hum. Genet.* **45:**39–47.

Cotmore, S. F., Furthmayr, H., and Marchesi, V. T., 1977, Immunochemical evidence for the transmembrane orientation of glycophorin A. Localization of ferritin–antibody conjugates in intact cells, *J. Mol. Biol.* **113:**539–553.

Dahr, W., 1986, Immunochemistry of sialoglycoproteins in human red blood cell membranes, in: *Recent Advances in Blood Group Biochemistry* (V. Vengelen-Tyler and W. J. Judd, eds.), pp. 23–65, American Association of Blood Banks, Arlington, Va.

Dahr, W., Uhlenbruck, G., and Knott, H., 1975, Immunochemical aspects of the MNSs blood group system, *J. Immunogenet.* **2:**87–100.

Dahr, W., Uhlenbruck, G., Janssen, E., and Schmalisch, R., 1977, Different N-terminal amino acids in the M,N glycoprotein from MM and NN erythrocytes, *Hum. Genet.* **35:**335–343.

Dahr, W., Beyreuther, K., Steinbach, H., Gielen, W., and Krüger, J., 1980, Structure of the Ss blood group antigens. II. A methionine/threonine polymorphism within the *N*-terminal sequence of Ss glycoprotein, *Hoppe-Seyler's Z. Physiol. Chem.* **361:**895–906.

Dahr, W., Beyreuther, K., Kordowicz, M., and Krüger, J., 1982, N-terminal amino acid sequence of sialoglycoprotein D (glycophorin C) from human erythrocyte membranes, *Eur. J. Biochem.* **125:**57–62.

Dahr, W., Kiedrowski, S., Blanchard, D., Hermand, P., Moulds, J. J., and Cartron, J. P., 1987, High frequency of human erythrocyte membrane sialoglycoproteins. V. Characterization of the Gerbich blood group antigens: Ge2 and Ge3, *Biol. Chem. Hoppe-Seyler* **368:**1375–1383.

Dahr, W., Blanchard, D., Kiedrowski, S., Poschmann, A., Cartron, J. P., and Moulds, J., 1989, High frequency antigens of human erythrocyte membrane sialoglycoproteins. VI. Monoclonal antibodies reacting with the N-terminal domain of glycophorin C, *Biol. Chem. Hoppe-Seyler* **370:**849–854.

Daniels, G. L., Shaw, M. A., Judson, P. A., Reid, M. E., Anstee, D. J., Colpitis, P., Cornwall, J., Moore, B. P. L., and Lee, S., 1986, A family demonstrating inheritance of the Leach phenotype: A Gerbich-negative phenotype associated with elliptocytosis, *Vox Sang.* **50:**117–121.

Demuth, D. R., Showe, L. C., Ballantinne, M., Palumbo, A., Fraser, P. J., Cioe, L., Rovera, G., and Curtis, P. J., 1986, Cloning and structural characterization of a human non-erythroid band 3-like protein, *EMBO J.* **5:**1205–1214.

El-Maliki, B., Blanchard, D., Dahr, W., Beyreuther, K., and Cartron, J. P., 1989, Structural homology between glycophorins C and D of human erythrocytes, *Eur. J. Biochem.* **183:**639–643.

Fairbanks, G., Steck, T. L., and Wallach, D. F. H., 1971, Electrophoretic analysis of the major polypeptides of the human erythrocyte membrane, *Biochemistry* **10:**2606–2617.

Fukuda, M., 1980, K562 human leukaemic cells express fetal type (i) antigen on different glycoproteins from circulating erythrocytes, *Nature* **285:**405–407.

Fukuda, M., 1981, Tumor-promoting phorbol diester-induced specific changes in cell surface glycoprotein profile of K562 human leukemic cells, *Cancer Res.* **41:**4621–4628.

Fukuda, M., 1985, Cell surface glycoconjugates as onco-differentiation markers in hematopoietic cells, *Biochim. Biophys. Acta* **780:**119–150.

Fukuda, M., and Fukuda, M. N., 1984, Cell surface glycoproteins and carbohydrate antigens in development and differentiation of human erythroid cells in: *The Biology of Glycoproteins* (R. J. Ivatt, ed.), pp. 183–234, Plenum Press, New York.

Fukuda, M., Fukuda, M. N., and Hakomori, S., 1979, Developmental change and genetic defect in the carbohydrate structure of band 3 glycoprotein of human erythrocyte membrane, *J. Biol. Chem.* **254**:3700–3703.

Furthmayr, H., 1978a, Glycophorins A, B, C: A family of sialoglycoproteins. Isolation and preliminary characterization of trypsin-derived peptides, *J. Supramol. Struct.* **9**:79–95.

Furthmayr, H., 1978b, Structural comparison of glycophorins and immunochemical analysis of genetic variants, *Nature* **271**:519–524.

Gahmberg, C. G., Myllyla, G., Leikola, J., Pirkola, A., and Nordling, S., 1976, Absence of the major sialoglycoprotein in the membrane of human En(a−) erythrocytes and increased glycosylation of band 3, *J. Biol. Chem.* **251**:6108–6116.

Gahmberg, C. G., Jokinen, M., and Andersson, L. C., 1978, Expression of the major sialoglycoprotein (glycophorin) on erythroid cells in human bone marrow, *Blood* **52**:379–387.

Gahmberg, C. G., Jokinen, M., and Andersson, L. C., 1979, Expression of the major red cell sialoglycoprotein glycophorin A, in the human leukemic cell line K562, *J. Biol. Chem.* **254**:7442–7448.

Gahmberg, C. G., Ekblom, M., and Anderson, L. C., 1984, Differentiation of human erythroid cells is associated with increased O-glycosylation of the major sialoglycoprotein, glycophorin A, *Proc. Natl. Acad. Sci. USA* **81**:6752–6756.

Gahmberg, C. G., Autero, M., and Hermonen, J., 1988, Major O-glycosylated sialoglycoproteins of human hematopoietic cells: Differentiation antigens with poorly understood functions, *J. Cell. Biochem.* **37**:91–105.

Greaves, M. F., Sieff, C., and Edwards, P. A. W., 1983, Monoclonal antiglycophorin as a probe for erythroleukemias, *Blood* **61**:645–651.

Hadley, T. J., Klotz, F. W., and Miller, L. H., 1986, Invasion of erythrocytes by malaria parasites: A cellular and molecular overview, *Annu. Rev. Microbiol.* **40**:451–477.

Hann, S. R., King, M. W., Bentley, D. L., Anderson, C. W., and Eisenman, R. N., 1988, A non-AUG translation initiation in c-myc exon 1 generates an N-terminal distinct protein whose synthesis is disrupted in Burkitt's lymphomas, *Cell* **52**:1985–1995.

Hartmann, E., Rapoport, T. A., and Lodish, A. F., 1989, Predicting the orientation of eukaryote membrane-spanning proteins, *Proc. Natl. Acad. Sci.* **86**:5786–5790.

Hayashida, H., and Miyata, T., 1983, Unusual evolutionary conservation and frequent DNA segment exchange in class I genes of the major histocompatibility complex, *Proc. Natl. Acad. Sci. USA* **80**:2671–2675.

High, S., and Tanner, M. J. A., 1987, Human erythrocyte membrane sialoglycoprotein β. The cDNA sequence suggests the absence of a cleaved N-terminal signal sequence, *Biochem. J.* **243**:277–280.

Huang, C. H., Johe, K., Moulds, J. J., Siebert, P. P., Fukuda, M., and Blumenfeld, O. O., 1987, δ glycoprotein (glycophorin B) gene deletion in two individuals homozygous for the $S^-s^-U^-$ blood group phenotype, *Blood* **70**:1830–1835.

Irimura, T., Tsuji, T., Togami, K., Yamamoto, K., and Osawa, T., 1981, Structure of a complex-type sugar chain of human glycophorin A, *Biochemistry* **20**:560–566.

Ishii, S., Xu, Y.-H., Stratton, R. H., Roe, B. A., Merlino, G. T., and Pastan, I., 1985, Characterization and sequence of the promoter region of the human epidermal growth factor receptor gene, *Proc. Natl. Acad. Sci. USA* **82**:4920–4924.

Jansson, S. E., Gripenburg, J., Hekali, R., and Gahmberg, C. G., 1981, Organization of membrane lipids and proteins in human En(a−) erythrocytes that lack major sialoglycoprotein, glycophorin A, *Biochem. J.* **195**:123–128.

Jokinen, M., Ulmanen, I., Andersson, L. C., Kääriäinen, L., and Gahmberg, C. G., 1981, Cell-free synthesis and glycosylation of the major human-red-cell sialoglycoprotein, glycophorin A, *Eur. J. Biochem.* **114**:393–397.

Jokinen, M., Ehnholm, C., Vaisanen-Rhen, V., Korhonen, T., Pipkorn, R., Kalkkinen, N., and Gahmberg, G., 1985, Identification of the major human sialoglycoprotein from red cells, glycophorin A^M, as the receptor for Escherichia coli IH11165 and characterization of the receptor site, *Eur. J. Biochem.* **147**:47–52.

Katamine, S., Notario, V., Rao, D., Miki, T., Cheah, M. S. C., Tronick, S. R., and Robbins, K. C., 1988, Primary structure of the human *fgr* proto-oncogene product $p55^{c\text{-}fgr}$, *Mol. Cell. Biol.* **8**:259–266.

Kopito, R. R., Andersson, M. A., and Lodish, H. F., 1987a, Multiple tissue-specific sites of transcriptional initiation of the mouse anion antiport gene in erythroid and renal cells, *Proc. Natl. Acad. Sci. USA* **84:**7149–7153.

Kopito, R. R., Andersson, M., and Lodish, H. F., 1987b, Structure and organization of the murine band 3 gene, *J. Biol. Chem.* **262:**8035–8040.

Kozak, M., 1981, Possible role of flanking nucleotides in recognition of the AUG initiator codon by eukaryotic ribosomes, *Nucleic Acids Res.* **9:**5223–5252.

Kozak, M., 1984, Compilation and analysis of sequences upstream from the translational start site in eukaryotic mRNAs, *Nucleic Acids Res.* **12:**857–872.

Kozak, M., 1986a, Point mutations define a sequence flanking the AUG initiator codon that modulates translation by eukaryotic ribosomes, *Cell* **44:**283–292.

Kozak, M., 1986b, Bifunctional messenger RNAs in eukaryotes, *Cell* **47:**481–483.

Kozak, M., 1989, The scanning model for translation: An update, *J. Cell Biol.* **108:**229–241.

Kudo, S., and Fukuda, M., 1989, Structural organization of glycophorin A and B genes: Glycophorin B gene evolved by homologous recombination at *Alu* repeat sequences, *Proc. Natl. Acad. Sci. USA* **86:**4619–4623.

Langlois, R. G., Bigbee, W. L., Kyoizumi, S., Nakamura, N., Bean, M. A., Akiyama, M., and Jensen, R. H., 1987, Evidence for increased somatic cell mutations at the glycophorin A locus in atomic bomb survivors, *Science* **236:**445–448.

Langlois, R. G., Bigbee, W. L., Jensen, R. H., and German, J., 1989, Bloom's syndrome: Use of the glycophorin A (GPA) assay to demonstrate an excessive accumulation of mutations in vivo, *Am. J. Hum. Genet.* **43:**A26 (abstract).

Lazarides, E., 1987, From genes to structural morphogenesis: The genesis and epigenesis of a red blood cell, *Cell* **51:**345–356.

Laemmli, U. K., 1970, Cleavage of structural proteins during the assembly of the head of bacteriophage T4, *Nature* **227:**680–681.

Lehrman, M. A., Golstein, J. L., Russel, D. W., and Brown, M. S., 1987, Duplication of seven exons in LDL receptor gene caused by Alu–Alu recombination in a subject with familial hypercholesterolemia, *Cell* **48:**827–835.

Leiff, S. E., Rosenfeld, M. G., and Evans, R. M., 1986, Complex transcriptional units: Diversity in gene expression by alternative RNA processing, *Annu. Rev. Biochem.* **55:**1091–1097.

Leonard, W. J., Depper, J. M., Kanehisa, M., Krönke, M., Peffer, N. J., Svettik, P. B., Sullivan, M., and Greene, W. C., 1985, Structure of the human interleukin-2 receptor gene, *Science* **230:**633–639.

Leto, T. L., Fortugno-Erikson, D., Barton, D., Yang-Feng, T. L., Francke, U., Harris, A. S., Morrow, J. S., Marchesi, V. T., and Benz, E. J., 1988, Comparison of nonerythroid α-spectrin genes reveals strict homology among diverse species, *Mol. Cell. Biol.* **8:**1–9.

Le Van Kim, C., Colin, Y., Blanchard, D., Dahr, W., London, J., and Cartron, J. P., 1987, Gerbich group deficiency of the Ge:1,-2,-3 and Ge:1,-2,-3 types, *Eur. J. Biochem.* **165:**571–579.

Le Van Kim, C., Colin, Y., Mitjavila, M. T., Clerget, M., Dubart, A., Nakazawa, M., Vainchenker, W., and Cartron, J. P., 1989, Structure of the promoter region and tissue specificity of the human glycophorin C, *J. Biol. Chem.* **264:**20407–20414.

Lin, C. S., Goldthwait, D. A., and Samols, D., 1988, Identification of Alu transposition in human lung carcinoma cells, *Cell* **54:**153–159.

Liska, K., Majdic, O., Bettelheim, P., and Knapp, W., 1983, Glycophorin A expression in malignant hematopoiesis, *Am. J. Hematol.* **15:**219–226.

Lisowska, E., 1989, Antigenic properties of human erythrocyte glycophorins, in: *Molecular Immunology of Complex Carbohydrates* (A. M. Wu, ed.), Plenum Press, New York, in press.

Macdonald, E. B., and Gerns, L. M., 1986, An unusual sialoglycoprotein associated with the Webb-positive phenotype, *Vox Sang.* **50:**112–116.

Mann, K. G., Jenny, R. J., and Krishnaswamy, S., 1988, Cofactor proteins in the assembly and expression of blood clotting enzyme complexes, *Annu. Rev. Biochem.* **57:**915–956.

Marchesi, V. T., Tillack, T. W., Jackson, R. L., Segrest, J. P., and Scott, R. E., 1972, Chemical characterization and surface orientation of the major glycoprotein of the human erythrocyte membrane, *Proc. Natl. Acad. Sci. USA* **69:**1445–1449.

Mattei, M. G., Colin, Y., Le Van Kim, C., Mattei, J. F., and Cartron, J. P., 1986, Localization of the gene for human erythrocyte glycophorin C to chromosome 2q14-q21, *Hum. Genet.* **74:**420–422.

Merry, A. H., Thomson, E. E., Anstee, D. J., and Stratton, F., 1984, The quantification of erythrocyte antigen sites with monoclonal antibodies, *Immunology* **51**:793–800.

Merry, A. H. J., Hodson, C., Thomson, E., Mallinson, G., and Anstee, D. J., 1986, The use of monoclonal antibodies to quantify the levels of sialoglycoproteins α and δ variant sialoglycoproteins in human erythrocyte membranes, *Biochem. J.* **233**:93–98.

Mitchell, G. H., and Bannister, L. H., 1988, Malarial parasites invasion: Interaction with the red cell membranes, in: *CRC Critical Reviews in Oncology/Hematology,* Volume 8, pp. 255–310, CRC Press, Cleveland, Ohio.

Mohandas, N., Chasis, J. A., and Shohet, S. B., 1983, The influence of membrane skeleton on red cell deformability, membrane material properties and shape, *Semin. Hematol.* **20**:225–242.

Morrow, B., and Rubin, C. S., 1987, Biogenesis of glycophorin A in K562 human erythroleukemia cells, *J. Biol. Chem.* **262**:13812–13820.

Mueller, T. J., and Morrison, M., 1981, Glycoconnection (PAS 2), a membrane attachment site for the human erythrocyte cytoskeleton, in: *Erythrocyte Membrane 2: Recent Clinical and Experimental Advances* (W. C. Kruckenberg, J. W. Eaton, and G. J. Brewer, eds.), pp. 95–112, Liss, New York.

Müller-Eberhard, H. J., and Miescher, P., 1985, *Complement,* Springer-Verlag, Berlin.

Nicholls, R. D., Fishel-Ghodsian, N., and Higgs, D. R., 1987, Recombination at the human α-globin gene cluster: Sequence features and topological constraints, *Cell* **49**:369–378.

Padgett, R. A., Grabowski, P. J., Konarska, M. M., Seiler, S., and Sharp, P. A., 1986, Splicing of messenger RNA precursors, *Annu. Rev. Biochem.* **55**:1119–1150.

Paollela, G., Lucero, M. A., Murphy, M. H., and Baralle, F. E., 1983, The *Alu* family repeat promoter has transfer RNA like bipartite structure, *EMBO J.* **2**:691–696.

Parkkinen, J., Rogers, G. N., Korhonen, T., Dahr, W., and Finne, J., 1986, Identification of the O-linked sialyloligosaccharides of glycophorin A as the erythrocyte receptors for S-fimbriated *Escherichia coli, Infect. Immun.* **54**:37–42.

Rahuel, C., London, J., d'Auriol, L., Mattei, M. G., Tournamille, C., Skrzynia, C., Lebouc, Y., Galibert, F., and Cartron, J. P., 1988a, Characterization of cDNA clones for human glycophorin A. Use for gene localization and for analysis of normal and glycophorin A deficient (Finnish type) genomic DNA, *Eur. J. Biochem.* **172**:147–153.

Rahuel, C., London, J., Vignal, A., Cherif-Zahar, B., Colin, Y., Siebert, P., Fukuda, M., and Cartron, J. P., 1988b, Alteration of the genes for glycophorin A and B in glycophorin A deficient individuals, *Eur. J. Biochem.* **177**:605–614.

Rahuel, C., Vignal, A., London, J., Hamel, S., Romeo, P. H., Colin, Y., and Cartron, J. P., 1989, Structure of the 5′ flanking region of the glycophorin A gene and analysis of its multiple transcripts, *Gene* **85**:471–477.

Rearden, A., Taetle, R., Elmajian, D. A., Majda, J. A., and Baird, S. M., 1985, Glycophorin A on normal and leukemia cells detected by monoclonal antibodies, including a new monoclonal antibody reactive with glycophorins A and B, *Mol. Immunol.* **22**:368–378.

Reid, E. M., Shaw, M. A., Rowe, G., Anstee, D. J., and Tanner, M. J. A., 1985, Abnormal minor human erythrocyte membrane sialoglycoprotein β in association with the rare blood-group antigen Webb (Wb), *Biochem. J.* **232**:289–291.

Reid, E. M., Anstee, D. J., Tanner, M. J. A., Ridgwell, K., and Nurse, C. T., 1987a, Structural relationships between human erythrocyte sialoglycoproteins β and γ and abnormal sialoglycoproteins found in certain rare human erythrocyte variants lacking the Gerbich blood-group antigen(s), *Biochem. J.* **244**:123–128.

Reid, E. M., Chasis, J. A., and Mohandas, N., 1987b, Identification of a functional role for human erythrocyte sialoglycoproteins β and γ, *Blood* **69**:1068–1072.

Robinson, J., Sieff, C., Delia, D., Edwards, P. A. W., and Greaves, M., 1981, Expression of cell-surface HLA-DR, HLA (ABC) and glycophorin during erythroid differentiation, *Nature* **289**:68–71.

Rouger, P., Anstee, D. J., and Salmon, C., 1988, First International Workshop on monoclonal antibodies against human red blood cell and related antigens, Paris 1987, *Blood Transf. Immunohaemat.* **31**(2).

Rouyer, F., Simmler, M. C., Page, D. C., and Weissenbach, J., 1987, A sex chromosome rearrangement in a human XX male caused by Alu–Alu recombination, *Cell* **51**:417–425.

Ruskin, B., Kariner, A. R., Maniatis, T., and Green, M. R., 1984, Excision of an intact intron as a novel lariat structure during pre-mRNA splicing in vitro, *Cell* **38**:317–331.

Sabatini, D. D., Kreibich, G., Morimoto, T., and Adesnik, M., 1982, Mechanisms for the incorporation of proteins in membranes and organelles, *J. Cell Biol.* **92**:1–22.

Segrest, J. P., Jackson, R. L., Marchesi, V. T., Guyer, R. B., and Terry, W., 1972, Red cell membrane glycoprotein: Amino acid sequence of an intramembranous region, *Biochem. Biophys. Res. Commun.* **49**:964–969.

Sheetz, M. P., and Sawyer, D., 1978, Triton shells of intact erythrocytes, *J. Supramol. Struct.* **8**:399–412.

Siebert, P. D., and Fukuda, M., 1986a, Isolation and characterization of human glycophorin A cDNA clones by a synthetic oligonucleotide approach: Nucleotide sequence and mRNA structure, *Proc. Natl. Acad. Sci. USA* **83**:1665–1669.

Siebert, P. D., and Fukuda, M., 1986b, Human glycophorin A and B are encoded by separate single copy genes coordinately regulated by a tumor-promoting phorbol ester, *J. Biol. Chem.* **261**:12433–12436.

Siebert, P. D., and Fukuda, M., 1987, Molecular cloning of human glycophorin B cDNA: Nucleotide sequence and genomic relationship to glycoprotein A, *Proc. Natl. Acad. Sci. USA* **84**:6735–6739.

Sieff, C., Bicknell, D., Caine, G., Robinson, J., Lam, G., and Greaves, M., 1982, Changes in cell surface antigen expression during hematopoietic differentiation, *Blood* **60**:703–713.

Silverberg, M., Furthmayr, H., and Marchesi, V. T., 1976, The effect of carboxymethylating a single methionine residue on the subunit interactions of glycophorin A, *Biochemistry* **15**:1448–1454.

Singer, S. J., and Nicholson, G. L., 1972, The fluid mosaic model of the structure of cell membranes, *Science* **175**:720–731.

Sondag, D., Alloisio, N., Blanchard, D., Ducluzeau, M. T., Colonna, P., Bachir, D., Bloy, C., Cartron, J. P., and Delaunay, J., 1987, Gerbich reactivity in 4.1 (−) hereditary elliptocytosis and protein 4.1 level in blood group Gerbich deficiency, *Br. J. Haematol.* **65**:43–50.

Strubin, M., Long, E. O., and Mach, B., 1986, Two forms of the Ia antigen-associated invariant chain result from alternative initiations at two in-phase AUGs, *Cell* **47**:619–625.

Tang, K. T., Leto, T. L., Correas, I., Alonso, M. A., Marchesi, V. T., and Benz, E. J., Jr., 1988, Selective expression of an erythroid specific isoform of protein 4.1, *Proc. Natl. Acad. Sci. USA* **85**:3713–3717.

Tanner, M. J. A., and Anstee, D. J., 1976, The membrane change in En(a−) human erythrocytes. Absence of the major erythrocyte sialoglycoprotein, *Biochem. J.* **153**:271–277.

Tanner, M. J. A., High, S., Martin, P. G., Anstee, D. J., Judson, P. A., and Jones, T. J., 1988, Genetic variants of human red-cell membrane sialoglycoprotein β. Study of the alterations occurring in the sialoglycoprotein-β gene, *Biochem. J.* **250**:407–414.

Tate, C. G., and Tanner, M. J. A., 1988, Isolation of cDNA clones for human erythrocyte membrane sialoglycoproteins α and δ, *Biochem. J.* **254**:743–750.

Thomas, D. B., and Winzler, R. J., 1969, Structural studies on human erythrocyte sialoglycoproteins: Alkali-labile oligosaccharides, *J. Biol. Chem.* **244**:5943–5946.

Tomita, M., and Marchesi, V. T., 1975, Amino-acid sequence and oligosaccharide attachment sites of human erythrocyte glycophorin, *Proc. Natl. Acad. Sci. USA* **72**:2964–2968.

Tomita, M., Furthmayr, H., and Marchesi, V. T., 1978, Primary structure of human erythrocyte glycophorin A. Isolation and characterization of peptides and complete amino acid sequence, *Biochemistry* **17**:4756–4770.

Tounkara, A., Piller, F., Blanchard, D., and Cartron, J. P., 1986, Evaluation of receptor sites for *Plasmodium falciparum* (malaria) on human erythrocytes. Chemical synthesis of peptides relevant to glycophorin A, in *Proceedings, European Peptide Symposium* (D. Theodoropoulos, ed.), de Gruyter, Berlin.

Treisman, R., Proudfoot, N. J., Shandler, M., and Maniatis, T., 1982, A single-base change at a splice site in a β°-thalassemic gene causes abnormal RNA splicing, *Cell* **29**:903–911.

Vignal, A., Rahuel, C., El Maliki, B., Le Van Kim, C., London, J., Blanchard, D., d'Auriol, L., Galibert, F., Blajchman, M. A., and Cartron, J. P., 1989, Molecular analysis of glycophorin A and B gene structure and expression in homozygous Miltenberger class V (Mi^{V}) human erythrocytes, *Eur. J. Biochem.* **184**:337–344.

Villeval, J. L., Cramer, P., Lemoine, F., Henri, A., Bettaieb, A., Bernaudin, F., Beuzard, Y., Berger, R., Flandrin, G., Breton-Gorius, J., and Vainchenker, W., 1986, Phenotype of early erythroblastic leukemia, *Blood* **68**:1167–1174.

Villeval, J. L., Le Van Kim, C., Bettaieb, A., Debili, N., Colin, Y., El Maliki, B., Blanchard, D., Vainchenker, W., and Cartron, J. P., 1989, Early expression of glycophorin C during normal and leukemic human erythroid differentiation, *Cancer Res.* **49**:2626–2629.

Von Heijne, G., 1983, Patterns of amino-acids near signal-sequence cleavage sites, *Eur. J. Biochem.* **133**:17–21.

Von Heijne, G., 1988, Transcending the impenetrable: How proteins come to terms with membranes, *Biochim. Biophys. Acta* **947**:307–333.

Wasniowska, K., Drzeniek, Z., and Lisowoska, E., 1977, The amino acids of M and N blood group glycopeptides are different, *Biochem. Biophys. Res. Commun.* **76**:385–390.

Watson, R. J., Dyson, P. J., and McMahon, J., 1987, Multiple c-myb transcript cap sites are variously utilized in cells of mouse haematopoietic origin, *EMBO J.* **6**:1643–1651.

Yamanashi, Y., Fukushige, S.-H., Semba, K., Sukegawa, J., Miyajima, N., Matsubara, K. I., Yamamoto, T., and Toyoshima, K., 1987, The *yes*-related cellular gene *lyn* encodes a possible tyrosine kinase similar to $p56^{lck}$, *Mol. Cell. Biol.* **7**:237–243.

Yasukawa, K., Hirano, T., Watanabe, Y., Muratini, K., Matsuda, T., Nakai, S. and Kishimoto, T., 1987, Structure and expression of human B cell stimulatory factor-2 (BSF-2/IL-6) gene, *EMBO J.* **6**:2939–2945.

Yoshima, H., Furthmayr, H., and Kobata, A., 1980, Structures of the asparagine-linked sugar chains of glycophorin A, *J. Biol. Chem.* **255**:9713–9718.

Yu, J., Fischman, D. A., and Steck, T. L., 1973, Selective solubilization of proteins and phospholipids from red cell membranes by nonionic detergents, *J. Supramol. Struct.* **1**:232–248.

Yurchenco, P. D., and Furthmayr, H., 1980, Expression of red cell membrane proteins in erythroid precursor cells, *J. Supramol. Struct.* **13**:255–269.

Chapter 11

The Anion Transport System of Red Blood Cell Membranes

Z. Ioav Cabantchik

1. INTRODUCTION: ANION TRANSPORT SYSTEMS OF MAMMALIAN CELL MEMBRANES

The Cl–HCO_3 exchanger (CBE) is the most prominent transport system of red blood cells (RBC). This system has been widely explored in terms of its physiological role (Cabantchik *et al.*, 1978; Knauf, 1979; Gunn, 1979; Jennings, 1985; Frohlich and Gunn, 1986; Passow, 1987), kinetic mechanisms (Cabantchik *et al.*, 1978; Knauf, 1979; Gunn, 1979; Jennings, 1985; Frohlich and Gunn, 1986; Passow, 1987), chemical architecture (Cabantchik *et al.*, 1978; Jay and Cantley, 1986; Passow, 1987), and encoding gene structure (Kopito and Lodish, 1985; Demuth *et al*, 1986; Kopito *et al.*, 1987; Cox and Lazarides, 1988; Alper *et al.*, 1988; Tanner *et al.*, 1988), and it has provided one of the best conceptual and methodological models for understanding protein-mediated transport. In RBC, CBE subserves the systemic transport of CO_2 from tissues to lungs, while in other cells, CBE activity might be involved directly in regulated cellular pH homeostasis (Reinertsen *et al.*, 1988; Ganz *et al.*, 1989) and in conjunction with other transport systems it might also participate in cell volume regulation (Finn, 1985; Cala, 1985; Hoffman, 1986; Grinstein *et al.*, 1985; Spring and Ericson, 1985) and epithelial salt secretory or absorptive processes (Greger, 1988; Muallem *et al.*, 1988; Tago *et al.*, 1986; Schuster *et al.*, 1986; Wagner *et al.*, 1987; Schwartz *et al.*, 1985). The two other major mammalian cell membrane Cl transporters whose functions have been identified are the Na,K,Cl cotransporter (SPCC) and the Cl channel (ClCh). The relative activities of these systems vary among cells, and in the case of epithelial cells, they vary even between luminal and contraluminal membranes (Figure 1). The SPCC has recently been recognized

Z. Ioav Cabantchik Department of Biological Chemistry, Institute of Life Sciences, Hebrew University, Jerusalem, Israel 91904; and Laboratory of Cell Biology and Genetics, National Institute of Diabetes and Kidney Diseases, National Institutes of Health, Bethesda, Maryland 20892, USA.

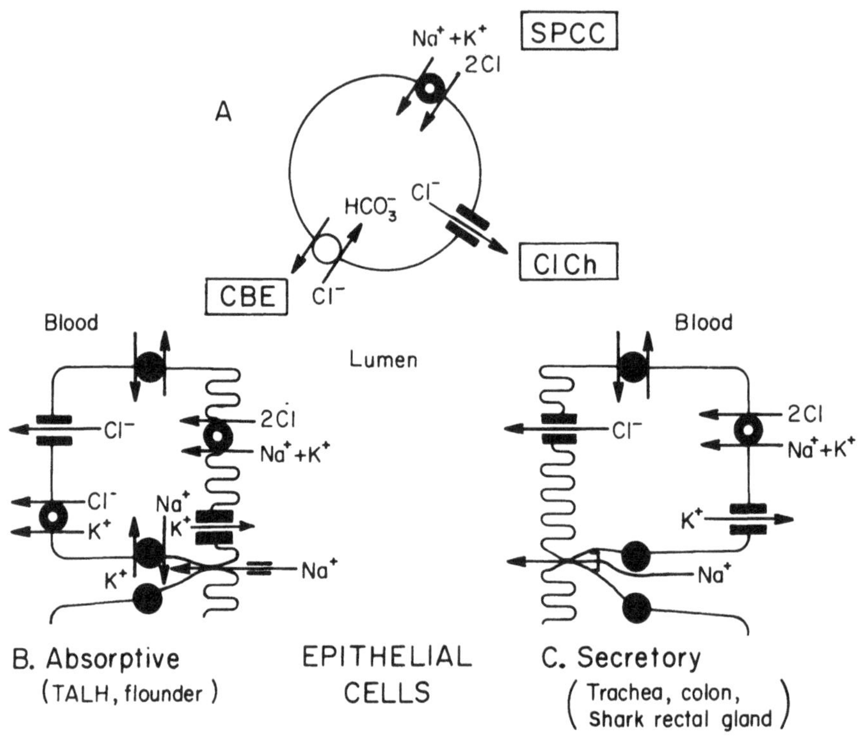

FIGURE 1. The major Cl transport systems. The various systems relevant for Cl transport in single cells (A), and in cells of absorptive (B) and secretory (C) epithelia.

(Hoffman, 1986; O'Grady *et al.*, 1987) as an anion transport system of particular importance in salt absorptive and secretory epithelia (Kinne and Heinz, 1987), but it has also been shown to be involved in down-regulation of cell volume upon hyperosmotic shock and hormonal stimulation (Hoffman, 1986). Various types of Cl channels have been described in electrophysiological terms, primarily in salt secretory and absorptive epithelia (Greger, 1988; Frizzell *et al.*, 1986; Gogelein, 1988), nerve (Hamill *et al.*, 1983), muscle (Blatz and Magleby, 1985), lymphocytes (Chen *et al.*, 1989), and intracellular membranes (Glickman *et al.*, 1983; Xie *et al.*, 1983). One particular type of Cl-channel activity that is normally hormone-regulated (Donowitz and Welsh, 1986) was found to be defective in salt secretory tissues of cystic fibrosis patients, as evidenced by an apparent lack of response to hormonal activation (Frizzell *et al.*, 1986; Welsh and Liedtke, 1986; Schoumaker *et al.*, 1987). However, the present state of knowledge in the field of biochemistry and biophysics of SPCC and Cl-channel systems is rather limited and therefore will not be specifically reviewed here.

In addition to physiological and mechanistic differences, in practice the three major Cl transport systems are identified by their differential susceptibilities to pharmacological agents with demonstrable inhibitory activity (Figure 2):

1. CBE is highly susceptible to disulfonic stilbene derivatives, the most prominent representatives of which are the covalently reactive derivative DIDS (4,4′-di-

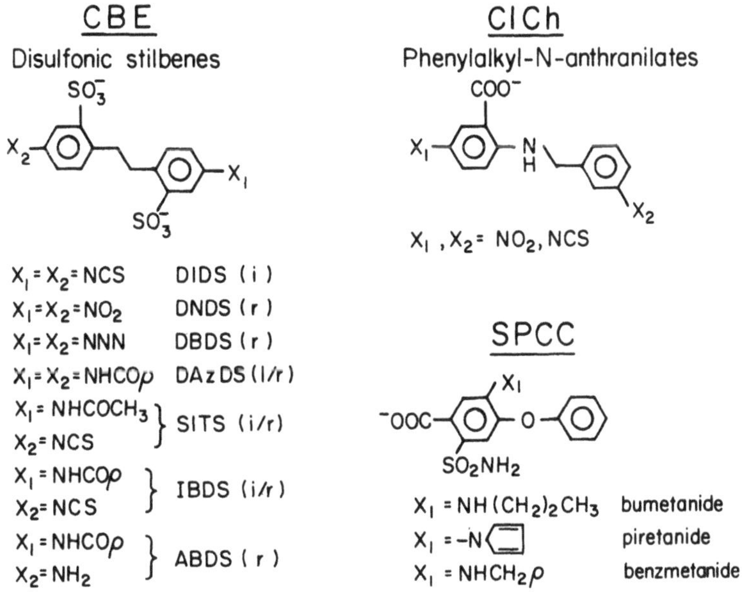

FIGURE 2. Inhibitors of Cl transport systems.

isothiocyano-2,2′-disulfonic stilbene) and the reversibly acting derivatives DNDS and DBDS (the 4,4′-dinitro and the 4,4′-dibenzamido analogues of DIDS, respectively). In RBC, these derivatives inhibit CBE competitively in the lower micromolar range (Cabantchik *et al.*, 1978; Barzilay *et al.*, 1979; Passow, 1987).

2. SPCC is specifically inhibited by nanomolar to micromolar concentrations of loop diuretics such as furosemide, piretanide, benzmetanide, and bumetanide (O'Grady *et al.*, 1987; Schlatter *et al.*, 1983) particularly in the presence of the substrates.
3. The Cl channel is particularly sensitive (in the nanomolar to micromolar range) to phenyl-aminoalkyl-phenylcarboxylates, of which the nitro benzoate derivatives are among the most potent (Wangemann *et al.*, 1986) and to indanyloxyacetic derivatives (Landry *et al.*, 1987).

The use of these inhibitors as biochemical, physiological, and pharmacological tools has contributed enormously to our present understanding of anion transport. In fact, the inhibitory action of various "specific" blockers has been used frequently as the major tool for identification of particular transport activities in intact cells or tissues, and in this respect, anion transport is not exceptional. However, the inhibitory efficacy of the various agents for a particular transport system varies considerably among cell types and organisms, and cross-inhibitory effects are not unusual, indicating that the presumed specificities are only relative in character. Therefore, the spectrum of specificities would need to be determined on an individual basis for each probe, transport system, and cell type, before any conclusion can be drawn as to the operation of a particular system in a given cell. This is particularly the case for CBE, which displays a marked susceptibility to all three classes of agents described above, and despite the fact that differential specificities are observed. Nevertheless, it is also conceivable that these cross-reactivities among high-affinity binding inhibitors are indicative of a common structural domain that is shared by the different anion transport systems, some of which might also represent different isoforms of the same protein which might display different substrate specificities. Minor structural variations in these domains or in access routes to these domains might be the reason for the differences found in the efficacy of the inhibitors toward the various classes of anion transport systems (Barzilay *et al.*, 1979; Schlatter *et al.*, 1983).

With the exception of CBE in RBC, our knowledge of the pathways for Cl transport and the forces involved in Cl translocation in mammalian cell membranes is very limited. The fact that various Cl transporters are copresent in the same cells and that the spectrum of specificities of the above inhibitors is variable, poses some additional problems for understanding the determinants and consequences of intracellular Cl levels as well as of transcellular anion movements. Taking into consideration the electrical properties of Cl translocation and the associated thermodynamic equilibrium conditions for the different Cl transport systems, we obtain:

1. For the electroneutral CBE, at equilibrium, the electrochemical potential $\delta\mu$ difference of Cl^- and of HCO_3^- is independent of membrane potential δV and is given by:

$$\delta\mu_{Cl//HCO_3} = \delta\mu_{Cl} + \delta\mu_{HCO_3}$$

and, since at equilibrium $\delta_{Cl/HCO_3} = 0$, we get

$$\frac{[Cl^-]_i}{[Cl^-]_o} = \frac{[HCO_3^-]_i}{[HCO_3^-]_o}$$

Moreover, in the absence of NA–H antiporter, the pH gradient across the cell membrane will be coupled, through the action of carbonic anhydrase, to bicarbonate, so that

$$\frac{[HCO_3^-]_i}{[HCO_3^-]_o} \equiv \frac{[H^+]_o}{[H^+]_i}$$

2. For the electroneutral SPCC, the combined electrochemical potential gradient for the three ions (Na^+, K^+, and 2 Cl^-) is given by the sum of the respective chemical potentials of each of the transported ions:

$$\delta\mu_{Na/K/Cl} = \delta\mu_{Na} + \delta\mu_K + 2\delta\mu_{Cl}$$

which at equilibrium ($\delta\mu_{Na/K/Cl} = 0$) gives

$$[Na]_i/[K]_i/[Cl]_i^2 = [Na]_o/[K]_o/[Cl]_o^2$$

3. For the Cl channel, the electrochemical potential gradient of Cl is given by:

$$\delta\mu_{c1} = zF \cdot V_m + RT \cdot \ln([Cl]_i/[Cl]_0)$$

which at equilibrium ($\delta\mu_{Cl} = 0$) gives

$$\delta\Omega = E_1^C = \frac{RT}{zF} \ln \frac{[Cl]_1}{[Cl]_o}$$

where E_{c1} is the equilibrium or Nernst potential of the Cl anion.

The RBC is a rather unique example of a cellular system in which Cl^- as well as H^+ are in thermodynamic equilibrium (Cabantchik *et al.*, 1978; Knauf, 1979; Gunn, 1979; Jennings, 1985; Passow, 1987). Due to the Donnan effect, the δpH of RBC determines to a large extent the cellular Cl concentration ($[Cl]_1$). In other cells, $[Cl]_i$ is markedly above the equilibrium level, the deviation determined by the weighed activities of the various Cl transporters as well as (1) by the intracellular pH (for cells with relatively high CBE activity), (2) by the Na gradient generated by the Na pump (for cells with relatively high SPCC activity), or (3) by the membrane potential (for cells with relatively high Cl-channel activity). In epithelial cells, the vectorial–transcellular movement of salt and water, which is dictated by the activity of the various transporters located on different membranes and by the ion gradients across the different membranes, also has a profound effect on the steady-state [Cl].

2. THE Cl–HCO_3 EXCHANGER OF RED BLOOD CELLS

2.1. Physiological Aspects

The transport function for which the RBC is most highly specialized is that of anion exchange (Cabantchik *et al.*, 1978; Knauf, 1979; Gunn, 1979; Jennings, 1985; Passow, 1987). This membrane specialization, which subserves CO_2 removal from tissues to lungs, is comparable to cytosolic hemoglobin (Hb) specialization, which subserves O_2 delivery from lungs to tissues. In fact, these two properties are intimately related by the process of intracellular formation of HCO_3^- and H^+ mediated by cytosolic carbonic anhydrase-catalyzed hydration of CO_2 (e.g., in peripheral blood capillaries), while the released H^+ is buffered by Hb, thereby promoting O_2 release from oxygenated Hb, known as the Bohr effect). Up to three-fourths of the HCO_3^- formed inside the RBC (corresponding to three-fifths of the CO_2 species carried in the venous blood) leaves the cell and is carried in the plasma toward the lungs (Figure 3).

The massive egress of HCO_3^- from RBC passing through tissue capillaries is accompanied by an equivalent ingress of Cl^- with no apparent changes in cell volume and membrane potential. This process, known as the Cl shift or the Hamburger shift, takes place in the reverse direction in RBC passing through lung capillaries. In order to be effective, the combined processes depicted in Figure 3, namely CO_2 diffusion, HCO_3^- formation, and HCO_3^-/Cl^- exchange, must be completed within the mean passage time of RBC through capillaries, which is about 1 sec (Klocke, 1976). This apparently is the case *in vivo,* as deduced from experimental observations showing that transport rates of

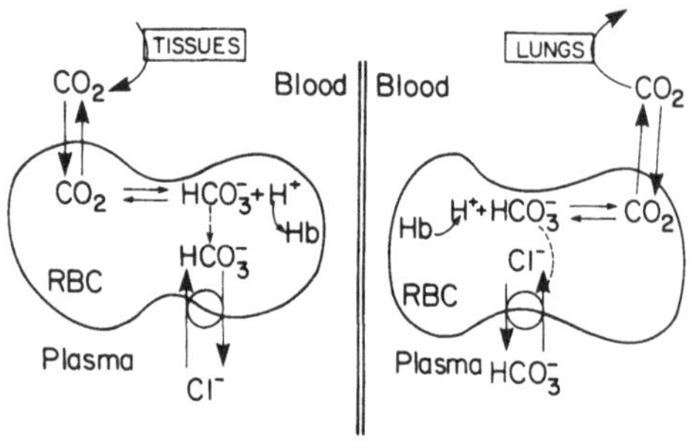

FIGURE 3. The role of Cl^-–HCO_3^- exchange (CBE) in blood transport of CO_2. CO_2 generated in tissue cells diffuses through the plasma into red cells passing through tissue capillaries. The CO_2 is rapidly converted into HCO_3^- and H^+ by cellular carbonic anhydrase and the HCO_3^- is exchanged for extracellular Cl^- in an electroneutral fashion by the CBE system, thereby increasing the CO_2-carrying capacity of the blood. In the lungs, the reverse process takes place, as the removal of CO_2 drives HCO_3^- into the cells in exchange for intracellular Cl^-.

CBE are fast enough to transport 80% of HCO_3^- produced within one-third of a second (Klocke, 1976).

2.2. Kinetics and Mechanism

The anion exchange protein of RBC is a simple carrier system that operates as an electroneutral, obligatory 1 : 1 exchanger of anions, with broad substrate admission properties but with high specialization for fast translocation of halides and HCO_3^- (Cabantchik *et al.*, 1978; Knauf, 1979; Gunn, 1979; Jennings, 1985; Passow, 1987). Its main characteristics are (Frohlich and Gunn, 1986):

1. All halides as well as HCO_3^-, NO_3^-, SO_4^{2-}, and HPO_4^{2-} display saturation kinetics, compete with each other, have similar K_s values (in the 20–50 mM range), but differ considerably in their transport rates (e.g., $Cl^- > Br^- > I^- >> SO_4^{2-}$). The list of anionic compounds, both inorganic and organic, that can be translocated by the exchanger is remarkable (e.g., aliphatic monocarboxylic and dicarboxylic acids, aliphatic, aromatic, and heterocyclic phosphates and sulfonates, cAMP). However, their transport rates span an enormous range of values and are for most of them orders of magnitude slower than those of Cl^- or HCO_3^-, the two physiologically important substrates (e.g., Cl^- is transported about 10^4-fold faster than SO_4^{2-}). The list of anionic compounds that can be admitted, but not necessarily translocated by the system (i.e., impermeant competitive inhibitors), is even larger. The structure–activity relationship of these inhibitors (Barzilay *et al.*, 1978; Motais and Cousin, 1978) provided the basis for the design of affinity labels for the exchanger, which led to the identification of the band 3 protein as the relevant functional component of the RBC membrane anion exchange (Cabantchik *et al.*, 1978).

2. All transportable anions, monovalent and divalent alike, show a similar value for the energy of activation (E_a of 30–40 kcal/mol or 120–160 kJ/mol or Q_{10} of 5–6), which is considerably higher than that of free diffusion (2–4 kcal/mol), diffusion through lipid bilayers (10–12 kcal/mol), and even higher than some active transport mechanisms (20–30 kcal/mol). For the particular case of halides, the temperature profiles of self-exchange show a break at a temperature that differs for the various anions (for halides, below the break point, the E_a is 20 kcal/mol) (Passow, 1987).

3. Whereas halide exchange rates have pH optima in the 7.5–8.5 pH range, the divalent sulfate and phosphate have pH optima at 6.4. This, as well as other properties, were best explained by the mechanism of a titratable carrier that can be converted from a

chloride-transporting form (E^+) into a sulfate-transporting form (E^{2+}) by protonation of a single group (Knauf, 1979; Gunn, 1979; Jennings, 1985; Frohlich and Gunn, 1986; Passow, 1987):

$$E^+ + H^+ \leftrightarrow E^{2+}$$
$$E^+ + Cl^- \leftrightarrow E\text{–}Cl$$
$$E^{2+} + SO^{2}{}_4{}^- \leftrightarrow E\text{–}SO_4$$

so that the translocating unit is electroneutral. In the E^+ conformation, the exchanger can also transport anionic ion pairs (such as $NaCO_3{}^-$ and $LiCO_3{}^-$, both analogues of $HCO_3{}^-$) in an electroneutral fashion. At present, additional effects of pH on the transport mechanism cannot be ruled out, demanding modifications of the titratable carrier model (Knauf, 1986).

4. Halides and sulfate also interact with an additional site on the anion exchange protein (AEP), the so-called modifier site. This interaction, which causes a noncompetitive self-inhibition of anion exchange, is of no obvious physiological relevance, as it is of a relatively low affinity character (K_I of 200–300 mM) (Frohlich and Gunn, 1986).

The high degree of coupling between inward and outward translocation of anions (net Cl permeability is about four orders of magnitude slower than exchange permeability) could be explained by simultaneous as well as by sequential models. However, detailed kinetic analysis demonstrated that the basic mechanism of anion translocation could be best explained by a Ping-Pong model, with a single site alternating between an outward-facing conformation and an inward-facing one (Knauf, 1979; Gunn, 1979; Frohlich and Gunn, 1986) (Figure 4).

According to the Ping-Pong mechanism, the transport site E alternates between two topologically distinct states: one (E_o) accessible to external solute, and the other (E_i) accessible to internal ones. The site can cross the permeability barrier only as an electroneutral binary complex (e.g., $E^+\text{–}X^-$ or $E^{2+}\text{–}X^{2-}$, where X is a permeant anion) but not as a charged entity (i.e., E^+, $E^+\text{–}SO_4^{2-}$, or $E^{2+}\text{–}X^-$. This model was put to test by a variety of strategies, including studies of single turnover rates of the anion-complexed carrier (Jennings, 1985). The RBC also offered the possibility of exposing the two sides of

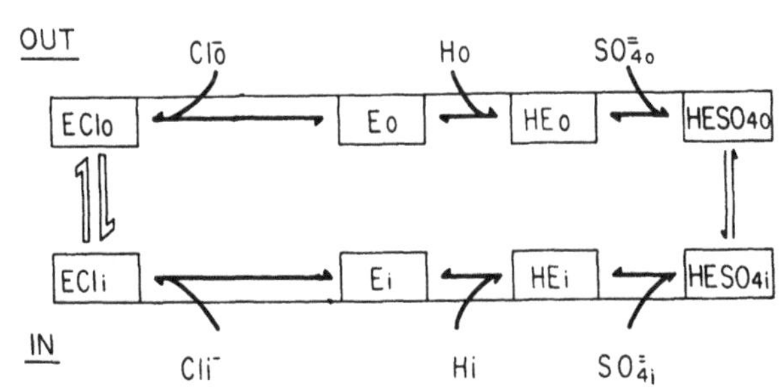

FIGURE 4. Ping-Pong mechanism of the titratable AEP system. The titratable carrier (E) is depicted at both membrane surfaces in terms of the H^+-titratable forms E_o, E_i and HE_o, He_i, where o and i denote the inner- and outer-facing forms of the carrier. The E forms can form the neutral translocatable complexes ECl upon binding of Cl^-, while the HE forms yield the neutral $HESO_4$ translocatable complexes upon binding of SO_4^{2-}. The horizontal arrows indicate the possible interconversions of the carrier forms, while the two vertical arrows indicate the two possible translocation steps, one for Cl^- and the other for SO_4^{2-} (the latter is about 10^4 slower than the former).

the membrane to media of different compositions, one containing a permeant anion and the other lacking such anion or containing a slowly permeant anion. In such situations, all transport sites are "recruited" away from the surface containing the fast permeating anion toward the opposite surface (Jennings, 1985; Frohlich and Gunn, 1986). Interestingly, an analogous asymmetric distribution of transport sites at the membrane surface could be generated by varying the concentration of halides in a given compartment while keeping the concentration at the opposite compartment constant (Knauf, 1986). The means to determine the availability of sites at a given surface were both of kinetic (Jennings, 1985) as well as of chemical nature (Knauf, 1986; Frohlich, 1982; Eidelman and Cabantchik, 1983a,b). These studies indicated that a single transport site can face only one side of the membrane at a time, as predicted by the Ping-Pong model.

5. In addition to the highly coupled ingress and egress of anions, the RBC has an additional mode of Cl translocation, which is an electrically active (i.e., charge conductive) mechanism. It is about four orders of magnitude slower than anion exchange and is partially sensitive to disulfonic stilbenes (Knauf, 1979). This mode of translocation has been associated with the same CBE system, except that the uncoupled anion flux was attributed to either slippage (Knauf, 1979) or tunneling (Frohlich, 1988) through the CBE carrier. However, the possibility that a chemically different entity might be responsible for net fluxes cannot be fully excluded at this stage.

2.3. Biochemical Aspects

The operation in the RBC membrane of a high-capacity system for the exchange of anions is attributed to band 3, the 95-kDa AEP (Cabantchik *et al.*, 1978; Passow, 1987). Although other transport functions have been attributed to the AEP (e.g., water, sugar, and urea transport), no solid evidence supporting those conjectures has been presented. The identification of band 3 as the AEP has been provided by a variety of studies that encompass selective and stoichiometric labeling of the polypeptides with specific inhibitors of the function (Cabantchik *et al.*, 1978; Passow, 1987; Jay and Cantley, 1986), functional isolation (Cabantchik *et al.*, 1980; Cabantchik and Darmon, 1985) and functional reconstitution into artificial membranes (Cabantchik and Darmon, 1985; Kohne *et al.*, 1981; Lukacovic *et al.*, 1981; Scheuring *et al.*, 1986), implantation of the isolated protein into cells lacking AEP (Cabantchik *et al.*, 1980), and expression of a DIDS-sensitive anion transport activity in *Xenopus* eggs microinjected with erythroid cell mRNA (Morgan *et al.*, 1985).

AEP is the predominant polypeptide of the membrane of avian and mammalian RBC (1×10^6 copies per mature human RBC, amounting to about 8% of the membrane dry mass of 20% of the membrane protein; in mouse RBC, about half that number of copies per cell was observed) (Figure 5).

AEP is a glycoprotein whose 3- to 8-kDa carbohydrate branch is of the complex type, carrying in its core the sequence α-Man$_2$-β-Man-β-GlcNAc, the disaccharide Gal-1→4GLcNAc-β1→3 as a variable repeating sequence with branching points at C-6 of some Gal residues (Tsuji *et al.*, 1980), Fuc residues both in the periphery as well as in the core portions, and some sialic acid at the periphery (Fukuda *et al.*, 1984). This variable moiety confers to band 3 an apparent heterogeneity that is manifested as a broad band in SDS-PAGE (Figure 5) but that can be removed by exofacial cleavage of the 95-kDa protein,

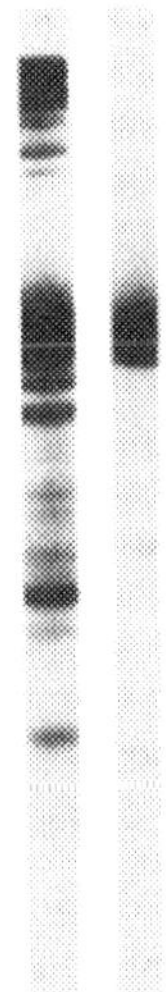

FIGURE 5. SDS–PAGE of (a) human RBC membranes and (b) purified band 3 reconstituted according to Darmon *et al.* (1983).

resulting in a 65-kDa carbohydrate-free fragment that is of normal band-width and a carbohydrate-containing wide-band counterpart of 30–35 kDa (Cabantchik *et al.*, 1978). Other posttranslational modifications of band 3 are phosphorylation at Try-8 located at the N-terminal cytosolic domain (Waxman, 1979) and methylation of Asp residues at the hinge region between the cytoplasmic and the transmembrane domain of the protein (Terwilliger and Clarke, 1981). No functional significance has been attached to the chemical modification of these residues.

AEP is the carrier for Ii antigens, which have been regarded as the precursors of blood group ABH antigens (Childs *et al.*, 1978). In umbilical cord vessels, the antigen activity is of the Oi type, whereas in adult blood it is of the OI type. The antigenic change associated with development has been recognized as branching of a single exofacial oligosaccharide chain N-linked Asp and the AEP (Fukuda *et al.*, 1984). A claim has also been made that the RBC AEP might contribute surface chemical determinants that function as receptors for malaria *Plasmodium falciparum* invasion into RBC (Okoye and Bennet, 1985).

2.3.1. Origin and Life Span

AEP is absent in the early stages of erythroid cell differentiation, but from the polychromic normoblast stage up to just before the reticulocyte stage, increasing amounts of the protein are synthesized and incorporated into the plasma membrane (Lazarides, 1987; Foxwell and Tanner, 1981). In undifferentiated cells that are transformed with various viruses, AEP normally is not expressed; however, upon terminal differentiation of cells treated with various types of inducers, AEP begins to be expressed concomitantly with the onset of globin expression. Similar to other integral membrane proteins, AEP biosynthesis occurs on polysomes by cotranslational insertion into RER components, initial glycosylation (but no proteolytic excision of signal sequence) at the ER, subsequent modification at the Golgi apparatus, and migration to the cell membrane (Lodish *et al.*, 1981; Braell and Lodish, 1981). It appears as though the signal sequence of insertion into the ER is near the N-terminus of the membrane-associated domain (Lodish *et al.*, 1981). In analogy with what has been observed with glycophorin, the insertion of the protein into the plasma membrane is not contingent upon its glycosylation (Sabban *et al.*, 1981, 1982). The expression of band 3 has recently been shown to be modulated by the v-erb gene in an avian model system (Zenke *et al.*, 1988); the significance of such observations is yet to be determined.

The appearance of AEP in the cell membrane is also paralleled by a progressive increase in the organization of the cell membrane cytoskeletal components (Lazarides, 1987). Stabilization of these preassembled components is another major role played by AEP (Low, 1988). Approximately 10-20% of AEP polypeptides are associated with ankyrin, which links AEP with other components of the cytoskeleton (Low, 1986; Marchesi, 1985; Palek and Lux, 1983; Bennet, 1985; Branton *et al.*, 1981).

In the time span of 4 months between the appearance of AEP on the RBC membrane and the removal of the RBC from the circulation, the protein appears to retain the basic structural as well as functional features. Claims have been made that a very small fraction of the polypeptides (0.1%) undergo conformational changes as a result of oxidative and other chemically damaging events (Lutz, 1987; Lutz *et al.*, 1988; Kanna *et al.*, 1988) or proteolytic incision at the exofacial domain (Kay, 1986). These putative products were postulated to induce structural changes that are recognized as senescence (damage) signals by circulating autoimmune senescence antibodies and that trigger the removal of the senescent (damaged) cells from the circulation (Lutz, 1987; Kay, 1986).

2.3.2. Molecular Biology

The band 3 gene has been cloned and sequenced from full-length cDNA clones of murine erythroid cells (Kopito and Lodish, 1985), human reticulocytes (Tanner *et al.*, 1988), human K562 cells (Demuth *et al.*, 1986), human erythroid cells (Tanner *et al.*, 1988), chicken erythroid cells (Cox and Lazarides, 1988), murine kidney (Alper *et al.*, 1988), human fetal liver (Lux *et al.*, 1989), and mouse genomic DNA (Kopito *et al.*, 1987). In the mouse genome, the 17-kb band 3 gene is present in a single copy and has 19 intervening sequences, with intron/exon junctions well correlated with predicted structural and functional domains of the protein and possibly also with evolution (Kopito *et al.*, 1987). The location of the introns that interrupt the protein-coding regions of band 3 and the residues presumed to form loops between membrane-spanning segments were shown to be organized in a pattern similar to that observed with other integral membrane proteins (Kopito *et al.*, 1987). The sequence of the murine band 3 gene shows a single open reading frame that extends for 929 codons (with two ATG codons at the beginning of the frame) flanked by untranslated regions at both ends. The deduced amino acid sequences from the above sources show a greater than 70% homology and a conservation of the overall topography of the alleged transmembrane segments (in hydropathy plots), particularly those assumed to be relevant to the function. The calculated size of 103 kDa is within the size range estimated by SDS–PAGE. The murine kidney band 3 analogue cDNA has a predicted mass of 135 kDa (1237 amino acids), and when compared to the erythroid band 3 sequence has 81% homology in the membrane domain (61% in the entire C-terminal end of the protein), 35% homology in the domain adjacent to the membrane-spanning fraction of the protein, and virtually no homology in the hydrophilic N-terminal domain (Alper *et al.*, 1988). Two cDNA clones derived from human nonerythroid K562 cells were shown to code for a band 3-like polypeptide of 865 amino acids that shows an overall 71% homology with murine erythroid band 3: up to 90% homology in the membrane domain but only 35% homology in the cytoplasmic domain of the polypeptide (Demuth *et al.*, 1986). This polypeptide has a major insert of 29 predominantly hydrophilic amino acids appearing after residue 582 of the mouse or human liver gene sequence, probably representing an additional hydrophilic loop in a region occupied by transmembrane segments (Demuth *et al.*, 1986). This major insert and an additional minor (3 amino acid) insert are of hydrophilic character, and appear to be exofacial in hydropathy plots. Both are likely to result from alternative splicing of the mRNA. At the N-terminus, a stretch of about 100 amino acids is missing from the sequence (compared to murine band 3). The chick band 3 polypeptides are also coded by a single gene, but the presence of alternative primary structures in its membrane-spanning segments would indicate that, in analogy to the K562 band 3, one of the exons described in murine band 3 mRNA is removed in the chicken by alternative splicing. The human erythroid band 3 sequence contains 911 amino acids (Tanner *et al.*, 1988) and has a strong homology to murine erythroid band 3 (929 amino acids), particularly in the membrane-associated N-terminal domain.

In general, three distinct domains can be identified in the hydropathy plots of the various band 3 sequences (Figure 6 shows the one urine erythroid band 3):

1. An N-terminal 46-kDa cytoplasmic domain of hydrophilic character with an approximately 2 : 1 ratio between acidic and basic amino acids. This domain shows poor homology between human erythroid, human nonerythroid, murine, and chicken sequences, except for a single region assumed to be the ankyrin-binding site of AEP. This is also in line with the fact that murine and especially chicken band 3 have poor binding capacities for glycolytic enzymes (Jay and Cantley, 1986).
2. A central region of 50 kDa of mixed polar/apolar character that is in intimate contact with the lipid matrix. This intramembranous domain could form 12–14 hydrophobic segments, half of which represent transmembrane-spanning segments that comply with Eisinger's criteria for producing membrane-spanning α-helices (Eisinger, 1984) and at least half of which are predicted to form amphipathic helices. Most of the transmembrane segments are bounded by Arg residues residing either exofacially or endofacially, possibly in interaction with negatively charged lipidic groups (Demuth *et al.*, 1986) or, more likely, composing charged microdomains for the concentration of anions at the entrance to the transport domain of the protein (Jay and Cantley, 1986).
3. A 3.5-kDa C-terminal hydrophilic region containing 30% acidic amino acids whose functional significance is yet to be determined.

2.3.3. Membrane Disposition of AEP

The character of AEP as an integral protein spanning the membrane several times, has been known for years, especially from studies using chemical probes in conjunction

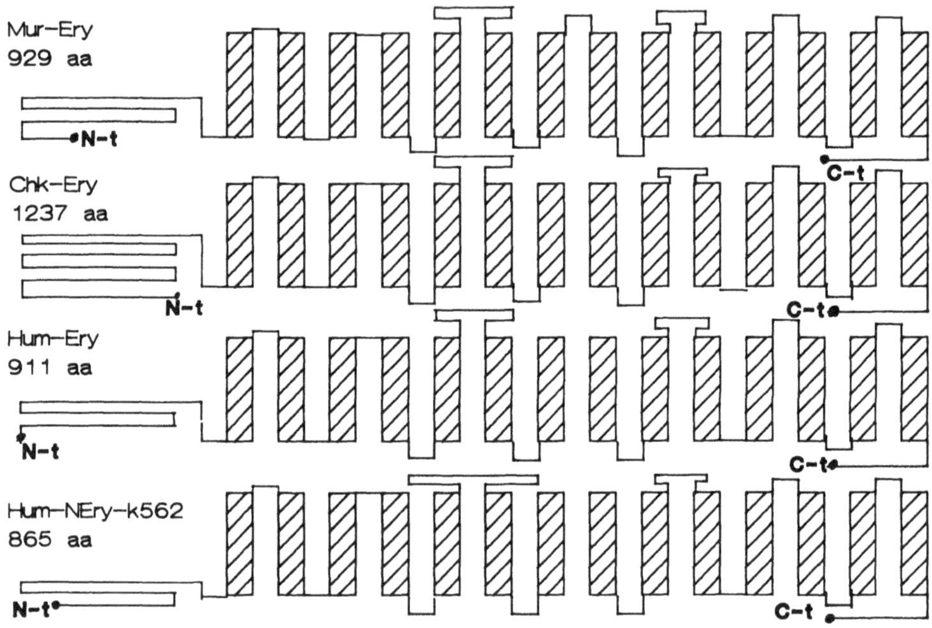

FIGURE 6. Scheme of hydropathy plots based on the amino acid sequences generated from the cDNA of band 3 protein. The various sources of band 3 are: murine erythroid (Mur-Ery, 929 amino acids, Kopito and Lodish, 1985); chicken erythroid (Chk-Ery, 1237 amino acids, Cox and Lazarides, 1988); human erythroid (Hum-Ery, 911 amino acids, Tanner *et al.*, 1988); and human K562 nonerythroid line (Hum-NEry-k562, 865 amino acids, Demuth *et al.*, 1986). The plots were generated using a modified Kyte and Doolittle algorithm with 7-amino-acid window and were aligned so that the homologous transmembranous domains (slashed areas) are interconnected by hydrophilic stretches. The highly variable N-terminal cytoplasmic domains are depicted on the left side. The bottom-facing interconnecting lines protrude into the cytosolic surface while the top ones correspond to the outward-facing fragments.

with specific polypeptide-cleaving procedures. These studies, which were conducted primarily with human AEP, provided much of the insight and present views about the structural and functional organization of the polypeptides. The findings obtained by several groups (Steck, 1974; Steck *et al.*, 1978; Drickamer, 1980; Tanner *et al.*, 1980; Rothstein *et al.*, 1980) were integrated in the scheme depicted in Figure 7. The polypeptides have an N-acetylated Met residue at the amino-terminus (N_t) and a Val residue at the carboxy-terminus (C_t). Similar to the information gained from the above hydropathy plots, two distinct domains could be defined experimentally on the basis of proteolytic cleavages (with trypsin or chymotrypsin) from the inner membrane surface (sizes are based on SDS–PAGE estimates): the cytoplasmic water-soluble domain (42K) carrying the protein N_t and the membrane-associated carbohydrate-containing domain (50K), which remain in the membrane after chymotrypsin cleavage, or even after subsequent cleavages by externally applied proteases, which eventually yield the subfragments of 15–17 and 35 kDa. When applied to intact cells, these proteases generate fragments of 60–65 (42 + 15) and 35 kDa, which also remain tightly associated with the membrane matrix (Drickamer, 1980; Tanner *et al.*, 1980; Rothstein *et al.*, 1980). Excision of the 42 kDa cytoplasmic fragment from

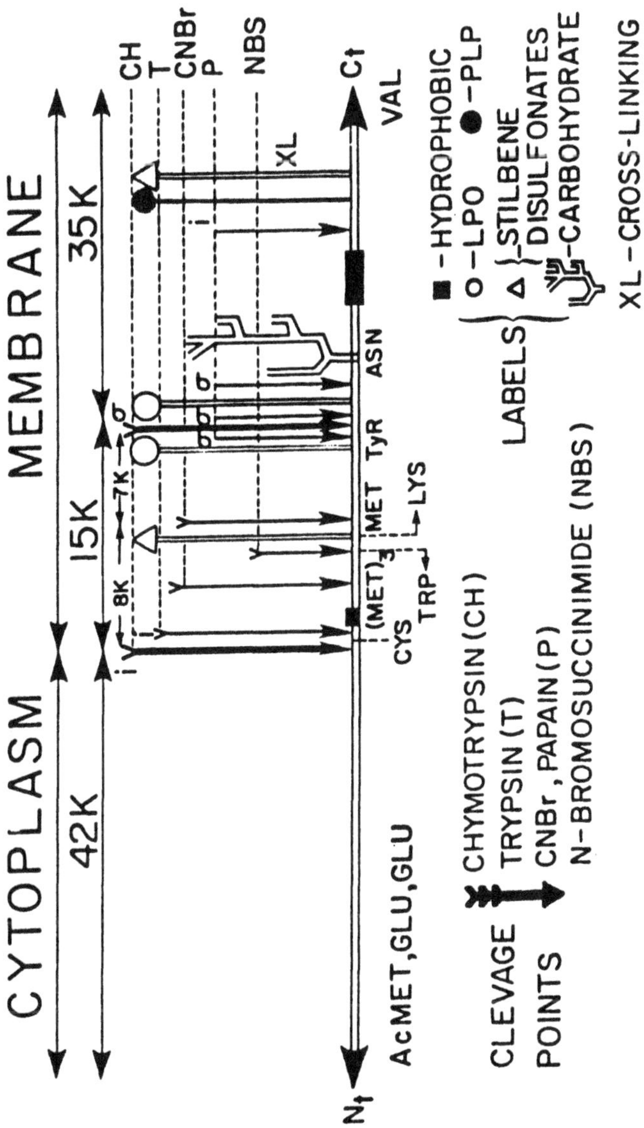

FIGURE 7. Labeling and protease cleavage points of band 3 polypeptides.

band 3 has no demonstrable effect on anion transport function (Rothstein *et al.*, 1980). The alleged role of this fragment in some species is binding of cytosolic enzymes (Low, 1988; Klimann and Steck, 1980) and hemoglobin (Sayare and Fikiet, 1981; Low, 1988) and in most species, binding of ankyrin, the protein link to the RBC membrane cytoskeleton (Bennet, 1985; Low, 1988). Approximately one-fifth of the total number of band 3 monomers in human erythroid band 3 appear to be tightly bound to the membrane cytoskeleton (Low, 1988). Kopito *et al.* (1989) have recently demonstrated that a neuronal homolog of the erythrocyte anion exchanger lacking about half the NH_2 terminal amino acids was sufficient for anion exchange activity.

The anion transport-relevant 95-kDa polypeptides and the proteolytic fragments of 60, 55, 15, and 35 kDa have been studied in some detail for the past 10 years with respect to amino acid sequence (Mawby and Findlay, 1982; Brock and Tanner, 1986) and spatial disposition (Cabantchik *et al.*, 1978; Tanner *et al.*, 1980; Drickamer, 1980; Rothstein *et al.*, 1980; Passow, 1987; Jay and Cantley, 1986). The basic methodology for studying the topology of the membrane-associated fragments employed a combination of labeling and cleaving agents having defined chemical, permeation, and sidedness-of-action properties on cells, open membranes, isolated band 3 polypeptides of fragments derived thereof. Some of the sites of action of these agents are depicted in Figure 7.

The match between the amino acid sequences deduced from nucleic acid analysis and from sequence data of discrete areas of the protein (i.e., proteolytic fragments) is remarkable (Kaul *et al.*, 1983; Mawby and Findlay, 1983; Brock and Tanner, 1984). The same holds also for the location of protease and chemical cleavage sites as well as for the labeling sites for a variety of surface and other chemical labels. Differences obtained in cleavage patterns between human and murine AEP are indicative of differences in primary structure with possible implications for the understanding of structure–function relationships of the protein. Of particular interest are the domains which carry the sites susceptible to membrane-impermeant specific affinity labels such as DIDS, its reduced analogue H_2DIDS, and other structurally related analogues, as these might help in defining the architecture of the transport site. Unfortunately this information is available only for the human and murine erythroid band 3 polypeptides, so that extrapolations based on sequence homologies and segmental conservation (Tanner *et al.*, 1988) should be regarded with caution.

The information gathered from chemical labeling and cleavages of AEP (summarized in Figure 7) in conjunction with the hydropathy plots based on the amino acid sequence of human erythroid band 3 (Tanner *et al.*, 1988; Lux *et al.*, 1989) was used in order to gain some insight about the disposition of polypeptide segments in the membrane matrix (Figure 8).

Predictions based on hydropathy plots show that 12–14 segments can form membrane-spanning hydrophobic α-helices. However, it is clear that structural models for the intramembranous parts of the protein, which are built entirely of such α-helices, do not provide an adequate model for the membrane disposition of AEP or for the functional components lining the surface of the anion passageway. First, the nature of the intramembranous segments is apparently not of a pure hydrophobic character, as indicated by labeling studies with lipophilic reagents (Guidotti, 1980; Brunner *et al.*, 1981; Figure 8), which reveal only two such hydrophobic domains, one in the 15 kDa and one in the 35-kDa fragments. Moreover, calculations of local amphipathic moments (for either α-helical

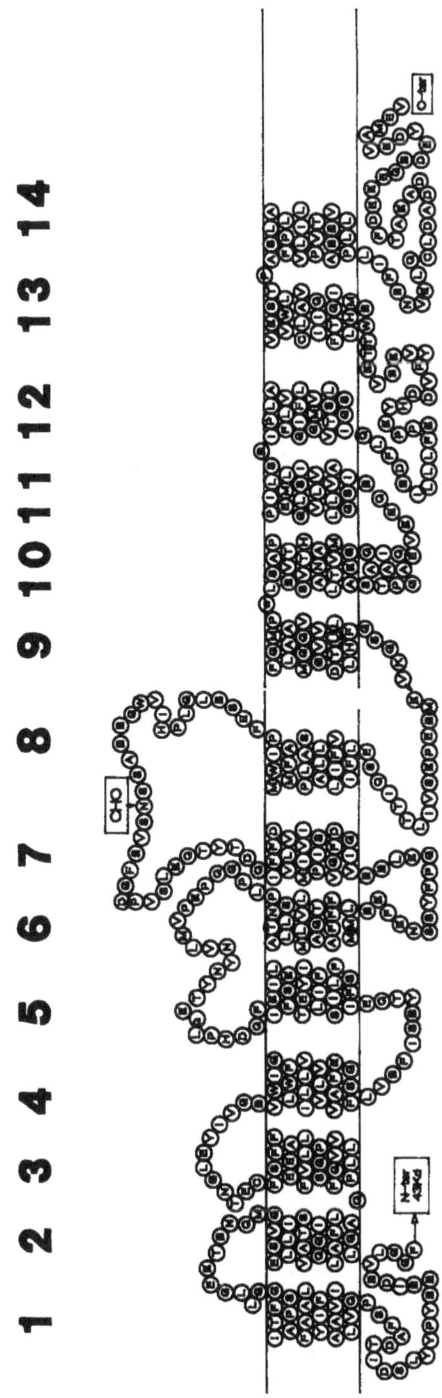

FIGURE 8. Secondary structure model for the transmembrane disposition of AEP. Membrane-associated fragments were identified based on hydropathy plots and other structural considerations. (Adapted from Lux *et al.*, 1989.)

or β-sheet periodicities and a 7-amino-acid-long window) show that indeed some of the proposed α-helical structures are in regions of the sequence that have high amphipathic character (Eidelman and Cabantchik, in press). This seems to be the case particularly for stretches 544–618 and 712–797 of the human erythroid band 3. Second, transmembrane α-helices define cylindrical pores that have constant cross sections along the entire thickness of the membrane. However, all available kinetic data as well as structural (see Section 2.3.4) and functional information obtained with anionic probes (Jay and Cantley, 1986; Barzilay *et al.*, 1978; Cabantchik, 1983) indicate that anion translocation occurs over a small fraction of the anatomical width of the membrane. The access routes to the translocation site are apparently of hydrophilic character, inasmuch as they can admit charged molecules of the size of pyridoxal-5′-phosphate and disulfonic stilbenes. It is therefore evident that the formation of hydrophilic channels leading toward the anion exchange site(s) requires, in addition to full-length α-helices, other structural elements such as short amphipathic helices (Lodish, 1988; McCrea *et al.*, 1988) or "broken" helices (e.g., see Figure 10) as suggested by the occurrence of prolines in several α-helical stretches.

2.3.4. The Anion Transport Site

The identification of the particular chemical groups comprising the functional sites and their tentative location in the two fragments has relied on (1) labeling (Figure 7) and structure–activity relationship studies (Cabantchik *et al.*, 1978; Barzilay *et al.*, 1979; Motais and Cousin, 1978), (2) the effects of chemical modification of one segment on labeling of the other, and (3) site directed mutagenesis of specific amino acid residues (Garcia and Lodish, 1989; Bartel *et al.*, 1989). Of particular relevance for correlating structure with function is the spatial relationship that different transmembrane segments have with respect to each other (Figure 9).

The first indication of such a relationship was provided by cross-linking studies with H_2DIDS, which suggested not only a geographical proximity (1.5–2.0 nm) of the 15-kDa and a segment of the 35-kDa fragment but also the participation of both fragments in anion transport (Jennings and Passow, 1979). Second, these two segments appear to interact rather tightly as judged by the fact that they coextract in nonionic detergent and migrate as a single charged complex (due to the carbohydrate-containing 35-kDa segment) in electrofocusing systems (Jennings, 1984). Third, chemical modification of the transport site on the 15-kDa segment affects labeling of the 35-kDa segment, and vice versa. Fourth, proteolytic degradation of a discrete area of 8 kDa at the N-terminus of the 35-kDa fragment (in the intact membrane) is apparently correlated with inhibition of the function (Jennings and Adams, 1981). These data, in conjunction with the inhibitory effect of the agents on anion transport, are taken to indicate that at least two transmembrane segments of band 3 cooperate in the formation of the putative transport site. Figure 10 is an artistic view of a possible three-dimensional arrangement that highlights the organization of secondary structure elements into two hydrophilic funnels leading anions toward the translocation site from either membrane facing medium.

The location of the transport site in relation to the membrane matrix was probed from the outer surface with disulfonic stilbenes coupled to dextrans via narrow spacers (aliphatic chains or polyethyleneglycol chains of variable length) and from the inner surface

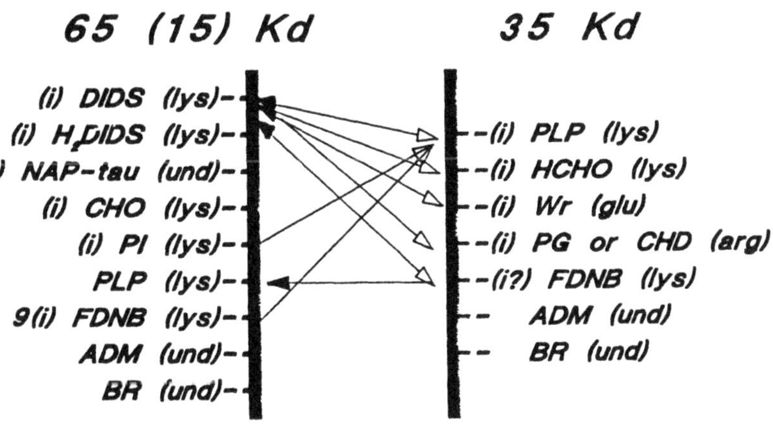

FIGURE 9. Sites of labeling and inhibition of anion transport. Binding sites for the various chemical agents in the 60- to 65-kDa (or 15-kDa subfragment) and 35-kDa fragments of the membrane-associated fragment of AEP. The arrows indicate mutual inhibitions of one labeling agent by the other. Labeling agents (chemical target groups are given in brackets) that are inhibitors of AEP function include:

- DIDS [amino], H_2DIDS [amino]; PLP (pyridoxal-5′-phosphate) [amino]
- NAP-tau (nitroazidophenyl-taurine) [nonspecific]
- HCHO (formaldehyde) [amino]; PI (phenyl-isothiocyanate) [amino]
- IPS (sulfophenyl-isothiocyanate) [amino]
- PG (phenylglyoxal) [guanido]; CHD (1,2-cyclobutanedione) [guanido]
- Wr (Woodward's reagent K: N-ethyl-5-phenylisoxazolium-3′-sulfonate) [carboxyl]
- ETC (1-ethyl-3-(3-trimethylaminopropyl)-carbodiimide) [carboxyl]
- FDNB (fluorodinitrobenzene) [amino]

ADM (adamantylidene) and BR (Brunner's reagent: 3-(trifluoromethyl)-3-(*m*-(iodophenyl)-diazirine) are nonspecific labels for hydrophobic domains while PI is a label for amino groups, located in hydrophobic as well as hydrophilic areas.

with amino-phenyl-methyl-benzene-thiazole disulfonate (APMB) similarly coupled to dextrans (Barzilay *et al.*, 1978). These studies, as well as several more recent ones, indicate that the sites reside away from the membrane boundaries, deep within the matrix, but still close to the outer membrane surface. This interpretation is in agreement with estimation of distances based on fluorescence energy transfer measurements between fluorophores placed at the cytoplasmic domain of band 3 and disulfonic stilbenes attached to band 3 exofacially. These have placed the transport site 2.8–4.2 nm away from the inner membrane surface (Rao *et al.*, 1979).

The transport domain apparently comprises a small area within the protein itself, flanked by two transmembrane segments and separating two aqueous compartments formed by indentations or foldings of the polypeptide into the middle of the membrane matrix (Cabantchik, 1983). Thus, translocation of substrates would not need to occur over the entire anatomical width of the membrane but rather over a small ($<$ 1 nm) distance. This is consistent with the observations that the protein as a whole does not rotate across the membrane (Cherry *et al.*, 1976), that transport occurs at the remarkable speed of 2×10^4 ions/site per sec at 37°C (Knauf, 1979; Gunn, 1979; Jennings, 1985) and with the

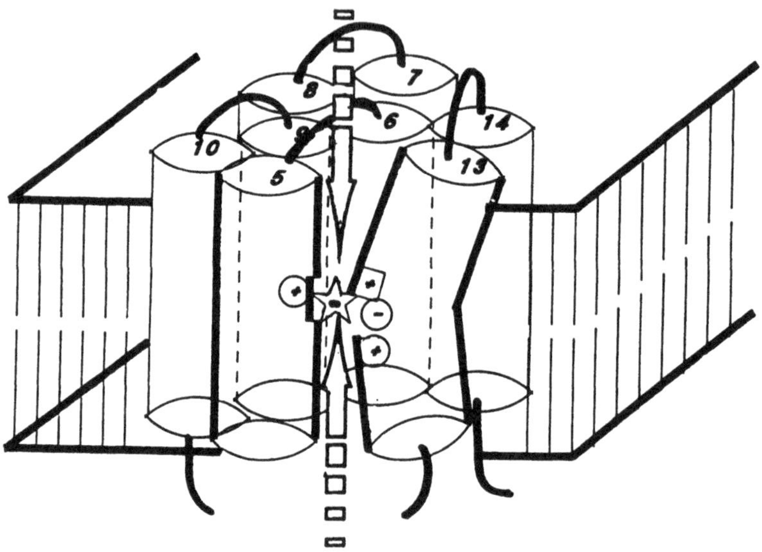

FIGURE 10. Spatial organization of the functional domain of AEP (schematic). Shown is a cross section of the membrane-spanning domains of AEP. The numbers correspond to membrane-spanning segments as depicted in Figures 6 and 8. The transport site is comprised of at least two noncontiguous transmembrane segments of AEP, 5 and 13, both of which can form amphipathic helices. One or both of these structures can form either a broken helix or span only part of the membrane width. The pathways for anions from either surface of the membrane toward the translocation site have a funnellike shape accessible to the respective aqueous environment (shaded area). The exofacial and endofacial funnel domains are clearly different, as seen from the fact that DIDS has access only from the relatively wider exofacial surface, whereas PLP can gain access to the site from either surface. The anion transport domain contains a relatively high charge density with groups such as Lys (⊕), Arg (□), Asp and/or Glu (⊖), and possibly His (not shown) and an adjacent hydrophobic domain. Binding of an anion (star) induces a conformational change that causes the transport site to move (a short distance) across the permeability barrier. The aqueous funnel can accommodate not only halides, but also relatively larger anions, including disulfonic stilbenes and even anions coupled to macromolecules via appropriate spacers (arrows).

kinetic features of Cl binding and translocation, as analyzed by Cl-NMR (Falke *et al.*, 1984a and b). The particular groups found to comprise the transport niche are two Lys (Jennings, 1982; Cabantchik *et al.*, 1975; Ramjeesingh *et al.*, 1981; Passow *et al.*, 1977), one or two carboxyl-containing Glu (Wieth *et al.*, 1982; Bjerrum, 1983; Jennings and Anderson, 1987; Jennings and Al-Rhaiyel, 1988) and possibly an Arg (Wieth *et al.*, 1982; Bjerrum, 1983; Zaki and Julien, 1985) and an electron-donating group (His?) located in an adjacent hydrophobic pocket (Barzilay *et al.*, 1979; Cabantchik, 1983; Chiba *et al.*, 1986). However, the precise location of these amino acid residues, let alone their direct involvement in anion binding and translocation are yet to be firmly established. On the basis of chemical studies and sequence analogies, it was proposed that Lys-558 or -561 (mouse erythroid) or Lys-539 or -542 (human erythroid) and Lys-869 (mouse erythroid) or Lys-852 (human erythroid) are probably the best candidates for comprising the functional domain of AEP. This was suggested by the presumptive labeling of particular residues by DIDS (Ramjeesingh *et al.*, 1981), diiodosulfophenylthiocyanate (Mawby and Findlay, 1983), and PLP (Kawano *et al.*, 1988), respectively and the presence of such sources in various band 3 sources (Tanner, *et al.*, 1988). For example, recent studies of site-directed mutagenesis provided clear demonstration that Lys-539 (human) or Lys-558 (mouse) are neither essential for anion transport (Cl-fluxes) nor for conferring susceptibility to reversibly acting disulfonic stilbenes (Garcia and Lodish, 1989; Bartel *et al.*, 1989). However, these studies have reached different conclusions regarding the possibility that Lys-539 (human) or its analogous Lys-558 in mouse are the actual residues modified irreversibly by disulfonic stilbenes, the first rejecting that possibility while the other supporting it. Further studies using different substituted residues would help define more precisely the involvement of particular amino acids in the transport function. A Glu was recently shown to participate in the transport mechanism as the pH-titratable group that is involved in the conversion of the monovalent (Cl binding) to divalent (sulfate binding) forms of the carrier (Jennings and Al-Rhaiyel, 1988).

2.3.5. Models for the Anion Exchange Mechanism

The essential features of AEP activity that need to be accounted for by any mechanistic model of transport include: (1) on the functional side, the electroneutral character of anion translocation, the alternation of a single transport site between the two membrane surfaces, and the admission and translocation of a wide spectrum of anions, and (2) on the structural side, the involvement of various protein residues located on different transmembrane segments of the band 3 polypeptide in the recognition and translocation of the anions and a conformational flexibility that can account for the transport of all those anions.

Some of these features were incorporated into a model that suggested that the functional unit might be comprised of two interconvertible channel-like domains formed by nine amphiphilic helices and lined by charged hydrophilic residues (Kopito and Lodish, 1985) (Figure 11). Binding of an anion at one membrane face was thought to induce the alteration in the conformational state of the protein. Although visually attractive, this model does not account for the existence of a single transport site per band 3 monomer (Cabantchik *et al.*, 1978; Passow, 1987) and it invokes excessive conformational changes associated with anion translocation.

On the other hand, the multicharged character of the anion transport domain (depicted also in Figure 10) has served previously as the basis for the proposed "zipper" model of anion translocation (Wieth *et al.*, 1982). According to this model, salt bridges formed between protein carboxyl and amino (or guanido) groups are proposed to line up in a channellike structure, and are replaced one at a time with an incoming anion as in an internal transferlike mechanism (zipper action). So as to account for the nonconductive nature of translocation, the opening of a bridge by the permeant anion is followed by regeneration (or closing) of the bridge as the anion proceeds into the following bridge. The "half site reactivity" properties observed in those sites toward some chemical modifications (Wieth *et al.*, 1982; Zaki and Julien, 1985) suggested that the bridges might be contributed cooperatively by adjacent band 3 monomers. This would seem to agree with the idea that the protein resides in the membrane as dimers with an average distance between adjacent transport sites of 2.8–4.8 nm (Macara and Cantley, 1981). However, recent information that implicated in the transport domain a single Glu residue that can

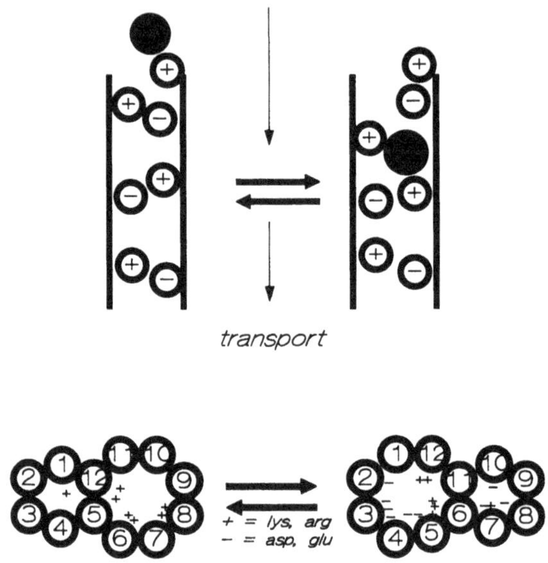

FIGURE 11. *Upper:* Zipper model of anion exchange. The filled circle represents the permeating anion as it is sequentially transferred between functional groups which form internal salt bridges either among themselves or with the translocating anion (after Wieth *et al.*, 1987). *Lower:* Model for the possible nearest-neighbor arrangement of band 3 transmembrane segments as membrane spanning amphipathic helices (after Kopito *et al.*, 1987).

alternate between inwards and outwards transport-related conformations, has raised questions about the validity of the "zipper" model (Jennings and Al-Rhaiyel, 1988).

The proposed mechanistic model of anion transport (Figure 12) is based on the operation of a transport unit comprised of an exofacial domain built as a cluster of positively charged groups (Cabantchik, 1983; Jay and Cantley, 1986; Passow, 1987) that serves as an anion "collector." This structure is provided by Lys and Arg groups on the 15-kDa (of the CHT65) fragment and on the CHT35 fragment. One of these groups is salt bridged to a Glu group, with both of these groups residing on the 35-kDa CHT fragment. Three of these groups, two Lys and one Glu, comprise the anion translocating group (ATG) (depicted as a circle), which is part of a mobile unit that alternates between the [i] (inner) and [o] (outer) domains, provided it is bound to an appropriate anion. While the ATG faces a given surface, it induces a conformational change so that another anion cannot approach the transport domain at the *cis* surface. This would explain the fact that the site can be recruited to either membrane surface by manipulation of the anion composition, as reflected in DNDS binding (which interacts with the 60-kDa fragment) (Frohlich, 1982; Eidelman and Cabantchik, 1983a,b) and PLP binding to the 35-kDa segment (BarNoy *et al.*, 1989) and probably also in binding of Woodward's reagent K to the 35-kDa segment (Jennings and Al-Rhaiyel, 1988).

Inorganic anions, as well as test anions (such as DIDS or PLP) approaching from outside will be admitted into the anion-binding domain provided the ATG is facing that surface. Binding is followed by translocation of anions, but for PLP the $t1/2$ of translocation is approximately 4 hr at 37°C (Cabantchik *et al.*, 1975), whereas for the bulky and highly reactive DIDS, irreversible modification is rather fast and translocation virtually

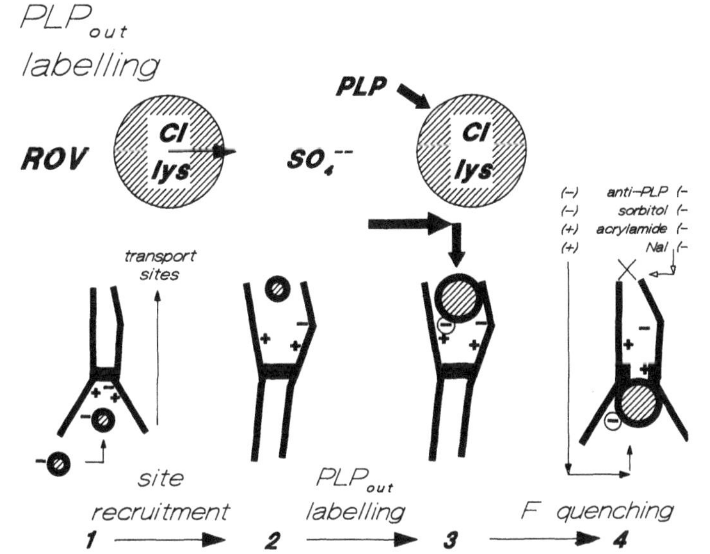

FIGURE 12. Mechanistic model of the AEP as revealed by the transportable affinity label pyridoxal-5′-phosphate (PLP). The model depicts the formation of the transport site by two membrane-crossing polypeptide stretches, one from the 60- to 65-kDa fragment with one Lys group, and the other, containing one Lys and one Glu, from the 35-kDa fragment. These groups (and others) form the anion-translocating domain, which comprises a restricted area of the transmembrane segments, and which can translocate across the transport barrier. The sites can be recruited to either surface by imposing on the system a gradient of anion composition, in this case Lys-HCl inside and either Na-sulfate or Na-gluconate outside. The test anion PLP binds to the site after approaching from the external medium (corresponding to the outer surface in right-side-out vesicles and to the inner surface in inside-out vesicles), which contains the slowly translocating anion (adapted from BarNoy *et al.*, 1990). After an initial interaction, the reversibly bound PLP are fixed with $NaBH_4$, and fluorescence quenching by the indicated agents applied from either surface is then assessed. The quenching of the bound PLP can be achieved only from the surface opposite the one from which binding has occurred, indicating that the probe became inaccessible to the (small) quencher molecules approaching it from the surface of labeling while becoming exposed to the opposite surface.

undetected (although the noncovalent reactive analogue DNDS can permeate through AEP at a relatively slow rate at 37°C; Barzilay and Cabantchik, 1979). For PLP, the covalent reaction to form a Schiff base is reversible unless the initial reaction is stopped or irreversibly fixed with a reducing agent. Interestingly, the binding of PLP to the putative Lys group at the transport site, induces a conformational change in the protein whereby the bound probe becomes accessible to collisional quenchers added only from the surface opposite that from where the PLP was originally added. These results are interpreted to mean that the initial binding of an anion to the transport site at the surface of approach ("o") leads to the occlusion of that anion from "o," and only after its release at the opposite surface ("i") and binding of another anion there will that particular site become exposed again to "o." The binding of either disulfonic stilbenes (Passow, 1987), PLP

(BarNoy *et al.*, 1990), or eosin-isothiocyanate (Jay and Cantley, 1986) on the exofacial domain of AEP has been shown to induce marked conformational changes, which are reflected in the occlusion of the bound probes from the *cis* face while becoming accessible from the *trans* face of the anion transport domain. These studies were conducted with fluorescence collisional quenchers of the bound probes, such as acrylamide, NaI, and CsCl (BarNoy *et al.*, 1990). The model is reminiscent of other "lock-in knock-on"-type models (Knauf, 1979; Passow, 1987; Falke *et al.*, 1984a) and is supported by a variety of biochemical studies and by recent NMR studies of Cl binding to AEP transport sites (Falke *et al.*, 1984a,b). One of the interesting mechanistic features of this model is that it provides for an intrinsically high conformational flexibility that allows the AEP to admit a variety of anionic agents, from simple inorganic anions such as halides to relatively complex organic acids such as PLP, disulfonic stilbenes, and others. The molecular features underlying the recognition of various classes of anions have been previously described (Barzilay *et al.*, 1979; Cabantchik, 1983). For hydrophilic–polar substrates, the conformational change might be substantial, depending on the size and charge density of the molecule. For instance, PLP and *trans*-disulfonic stilbenes such as *trans*-DNDS, both planar and elongated molecules, can be admitted into the cleft, while the more bulky analogues *cis*-DNDS or disulfonic naphthalenes are not admitted. The ensuing translocation event is accompanied by an additional conformational change, which might be sufficient to allow passage of PLP and to some extent even of *trans*-disulfonic stilbenes. It should be stressed that slow but measurable efflux of *trans*-DNDS has been observed in resealed ghosts (Barzilay and Cabantchik, 1979). For anions of a more dual hydrophilic/hydrophobic character, the conformational change might be more subtle, as the translocation of the admitted molecules can also involve "sliding" through a neighboring apolar face of the amphiphilic helix.

The translocation capacity of AEP for different anions might vary according to the tissue or the cell organelle in which the protein operates. Various insoforms of the band 3 protein have indeed been identified in different tissues (Cox *et al.*, 1985; Alper *et al.*, 1988; Kopito *et al.*, 1988, 1989) and in Golgi membranes of mammalian cells (Kellokumpu *et al.*, 1988) by genetic, chemical, and immunochemical probing. A variety of anion transport systems that mediate translocation of organic acids (Aronson, 1989) and CBE (Wagner *et al.*, 1987; Schuster *et al.*, 1987) share a common susceptibility to disulfonic stilbenes, with minor variations that reflect either differential accessibilities to the probe and/or structural differences in the anion transport domain. Hitherto, none of these systems have been studied in detail in terms of structure or function, although it is expected that their amino acid sequences might soon be available. However, many efforts will have to be invested in order to gain the level of understanding that has been reached with the CBE in erythrocytes, as witnessed by the information provided in the present review of the field.

ACKNOWLEDGMENTS. This work was supported in part by NIH grant RO1 HL40685 and by the German–Israel Foundation for Scientific Research and Development. The work was written while on sabbatical leave from the Hebrew University of Jerusalem, sponsored by the U.S. Cystic Fibrosis Foundation and the Laboratory of Cell Biology and Genetics, NIDDK, NIH, Bethesda, Maryland. Dr. O. Eidelman is kindly acknowledged for critical reading of the manuscript and for interesting discussions.

3. REFERENCES

Alper, S. L., Kopito, R., Libresco, S. M., and Lodish, H. F., 1988, Cloning and characterization of murine band 3-related cDNA from kidney and from a lymphoid cell line, *J. Biol. Chem.* **263:**17902–17909.

Aronson, P. S., 1989, The renal proximal tubule: a model for diversity of anion exchangers and stilbene sensitive anion transporters. *Ann. Dev. Physiol.* **51:**419–441.

BarNoy, S. and Cabantchik, Z. I., 1990, The transport domain of the erythrocyte anion exchange protein, *J. Membr. Biol.* in press.

Bartel, D., Lepke, S., Layh-Schmitt, G., Legrum, B., and Passow, H., 1989, Anion transport in oocytes of Xenopus Laevis induced by expressions of mouse erythroid band 3 protein-encoding CDNA and of a cRNA derivative obtained by site-directed mutagenesis at the stilbene disulfonate bonding site *EMBO J.* **12:**3601–3609.

Barzilay, M., and Cabantchik, Z. I., 1979, Anion transport in red blood cells. Sites and sidedness of inhibition by high-affinity reversible binding probes, *Membr. Biochem* **2:** 297–322.

Barzilay, M., Jones, D., and Cabantchik, Z. I., 1978, Sidedness of inhibitory effects as evidence for asymmetric location of the anion transport system of red blood cell membranes, *Fed. Proc.* **37:**1295.

Barzilay, M., Ship, S., and Cabantchik, Z. I., 1979, Anion transport in red blood cells. 1. Chemical properties of anion recognition sites as revealed by structure activity relationships of aromatic sulfonic acids, *Membr. Biochem.* **2:**227–254.

Bennet, V., 1985, The membrane skeleton of human erythrocytes and its implications for more complex cells, *Annu. Rev. Biochem* **54:**273–304.

Bjerrum, P., 1983, Identification and location of amino acid residues essential for anion transport in red cell membranes, in: *Structure and Function of Membrane Proteins* (E. Quagliariello and F. Palmieri, eds.), pp. 107–115, Elsevier, Amsterdam.

Blatz, A. E., and Magleby, K. L., 1985, Single chloride-selective channels active at resting membrane potentials in cultured rat skeletal muscle, *Biophys. J.* **47:**119–123.

Boron, W. F., Hogan, E., and Russell, J. M., 1988, pH-sensitive activation of the intracellular-pH regulation system in squid axons by ATP-γ-S, *Nature* **332:**262–265.

Braell, W. A., and Lodish, H. F., 1981, Biosynthesis of the erythroctye anion transport protein, *J. Biol. Chem.* **256:**11337–11344.

Braell, W. A., and Lodish, H. F., 1982, The erythrocyte anion transport protein is cotranslationally inserted into microsomes, *Cell* **28:**23–31.

Branton, D., Cohen, C. M., and Tyler, J., 1981, Interactions of cytoskeletal proteins on the human erythrocyte membrane, *Cell* **24:**24–32.

Brock, C. J. and Tanner, M. J. A., 1986. The human erythrocyte anion-transport protein. Further amino acid sequence from the integral membrane domain homologous with the murine protein, *Biochem. J.* **235:**899–901.

Brunner, J., and Semenza, G., 1981, Selective labeling of the hydrophobic core of membranes with 3-(trifluoromethyl)-3-(m[^{125}I]iodophenyl) diazirine, a carbene-generating reagent, *Biochemistry* **20:**7174–7182.

Cabantchik, Z. I., 1983, Structure–function relations in band 3 protein, in *Structure and Function of Membrane Proteins* (E. Quagliariello and F. Palmieri, eds.), pp. 271–281, Elsevier, Amsterdam.

Cabantchik, Z. I., and Darmon, A., 1985, Reconstitution of membrane transport systems, in *Structure and Properties of Membranes* (G. Benga, eds.), pp. 123–165, CRC Press, Boca Raton.

Cabantchik, Z. I., Balshin, M., Breuer, V. W., and Rothstein, A., 1975, Pyridoxal phosphate. An anionic probe for protein amino groups exposed on the outer and inner surfaces of intact human red blood cells, *J. Biol. Chem.* **250:**5130–5136.

Cabantchik, Z. I., Knauf, P. A., and Rothstein, A., 1978, The anion transport system of the red blood cell. The role of membrane protein evaluated by use of "probes," *Biochim. Biophys. Acta* **515:**239–303.

Cabantchik, Z. I., Volsky, D. J., Ginsburg, H., and Loyter, A., 1980, Reconstitution of the erythrocyte anion transport system, *Ann. N.Y. Acad. Sci.* **34:**444–454.

Cala, P., 1985, Volume regulation by Amphiuma red blood cells: Strategies for identifying alkali metal/H^+ transport, *Fed Proc.* **44:**2500–2507.

Chen, J. H., Schulman, H., and Gardner, P., 1989, c-AMP regulated chloride channel in lymphocytes that is affected in cystic fibrosis, *Science* **243:**657–660.

Cherry, R. J., Burkli, A., Busslinger, M., Schneider, G., and Parish, R. G., 1976, Rotational diffusion of band 3 proteins in the human erythrocyte membrane, *Nature* **263:**389–393.

Chiba, T., Sato, Y., and Suzuki, Y., 1986. Characterization of eosin 5-isothiocyanate binding site in band 3 protein of the human erythrocyte, *Biochim. Biophys. Acta* **858:**107–117.

Childs, R. A., Feizi, T., Fukuda, M., and Hakomori, S., 1978, Blood group-I activity associated with band 3, the major intrinsic membrane protein of human erythrocyte, *Biochem. J.* **173:**333–336.

Cox, J. K., and Lazarides, E., 1988, Alternative primary structures in the transmembrane domain of the chicken erythroid anion transporter, *Mol. Cell. Biol.* **8:**1327–1335.

Cox, J. V., Moon, R. T., and Lazarides, E., 1985, Anion transporter: High cell specific expression of distinct polypeptides and transcripts in erythroid and non-erythroid cells, *J. Cell Biol.* **100:**1548–1557.

Darmon, A., Zangvill, M., and Cabantchik, Z. I., 1983, New approaches for the reconstitution and functional assay of membrane transport proteins. Application to the anion transporter of human erythrocytes, *Biochim. Biophys. Acta* **727:**77–88.

Demuth, D. R., Showe, L. C., Ballantine, M., Palumbo, A., Fraser, P. J., Cioe, L., Rovera, R., and Curtis, R. J., 1986, Cloning and structural characterization of a human non-erythroid band 3-like protein, *EMBO J.* **5:**1205–1214.

Eidelman, O., and Cabantchik, Z. I., 1983a, The mechanism of anion transport across human red blood cell membranes as revealed with a fluorescent substrate: I. Kinetic properties of NBD-taurine transfer in symmetric conditions, *J. Membr. Biol.* **71:**141–148.

Eidelman, O., and Cabantchik, Z. I., 1983b, The mechanism of anion transport across human red blood cell membranes as revealed with a fluorescent substrate: II. Kinetic properties of NBD-taurine transfer in asymmetric conditions, *J. Membr. Biol.* **71:**149–161.

Eisinger, D., 1984, Three-dimensional structure of membrane and surface proteins, *Annu. Rev. Biochem.* **53:**595–623.

Falke, F., Pace, R. J., and Chan, S. I., 1984a, Direct observation of the transmembrane recruitment of band 3 transport sites by competitive inhibitors. A (^{35}Cl] NMR study, *J. Biol. Chem.* **259:**6481–6491.

Falke, J. F., Pace, R. J., and Chan, I., 1984b, Chloride binding to the anion transport binding sites of band 3. A [^{35}Cl] NMR study, *J. Biol. Chem.* **259:**6472–6480.

Finn, A. L., 1985, Volume-dependent pathways in animal cells, *Fed Proc.* **44:**2599.

Foxwell, B. M. J., and Tanner, M. J. A., 1981, Synthesis of the erythrocyte anion exchange protein. Immunochemical study of its incorporation into the plasma membrane of erythroid cells, *Biochem. J.* **195:**129–137.

Frizzell, R. A., Halm, D. R., Rechkemmer, G., and Shoemaker, R. L., 1986, Chloride channel regulation in secretory epithelia, *Fed. Proc.* **45:**2727–2731.

Frohlich, O., 1982, The external anion binding site of the human erythrocyte anion transport: DNDS binding and competition with chloride, *J. Membr. Biol.* **65:**111–123.

Frohlich, O., 1988, The "tunneling" mode of biological carrier mediated transport, *J. Membr. Biol.* **101:**189–198.

Frohlich, O., and Gunn, R. B., 1986, Erythrocyte anion transport: The kinetics of a single site obligatory exchange system, *Biochim. Biophys. Acta* **864:**169–194.

Fukuda, M., Dell, A., Oates, J. E., and Fukuda, M. N., 1984, Structure of branched lactoseamino-glycan, the carbohydrate moiety of band 3 isolated from adult erythrocytes, *J. Biol. Chem.* **259:**8260–8279.

Ganz, M. B., Boyarsky, G., Sterzel, R. B., and Boron, W. F., 1989, Arginine vasopressin enhances pHi regulation in the presence of HCO_3^- by stimulating three acid–base transport systems, *Nature* **332:**648–651.

Garcia, A. M. and Lodish, H., 1989, Lysine 539 of human band 3 is not essential for ion transport or inhibition by stilbene disulfonates, *J. Biol. Chem.* **264:**19607–19613.

Glickman, J., Croen, K., Kelly, S., and Al-Awqati, Q., 1983, Golgi membranes contain an electrogenic H-pump in parallel with a chloride conductance, *J. Cell Biol.* **97:**1303–1308.

Gogelein, H., 1988, Chloride channels in epithelia, *Biochim. Biophys. Acta* **947:**521–547.

Greger, R., 1987, Ion transport mechanisms in thick ascending limb of Henle's loop of mammalian nephron, *Physiol. Rev.* **65:**755–797.

Greger, R., 1988, Chloride transport in thick ascending limb, distal convolution, and collecting duct, *Annu. Rev. Physiol.* **50:**111–122.

Guidotti, G., 1980, The structure of the band 3 polypeptide, in: *Membrane Transport in Erythrocytes, Alfred Benzon Symposium 14* (U. V. Lassen, H. H. Ussing, and J. O. Wieth, eds.), pp. 300–308, Munksgaard, Copenhagen.

Gunn, R. B., 1979, Anion transport in red cells: An asymmetric, ping pong mechanism, in *Mechanisms of Intestinal Secretion* (H. J. Binder, ed.), pp. 25–43, Liss, New York.

Hamill, O. P., Bormann, J., and Sackmann, 1983, Activation of multiple conductance state chloride channels in spinal neurones by glycine and GABA, *Nature* **305:**805–808.

Hoffman, E. K., 1986, Anion transport systems in the plasma membrane of vertebrate cells, *Biochim. Biophys. Acta* **864:**1–32.

Jay, D., and Cantley, L. C., 1986, Structural aspects of the red cell anion exchange protein, *Annu. Rev. Biochem.* **55:**511–538.

Jennings, M. L., 1982, Reductive methylation of the two H_2DIDS-binding lysine residues of band 3, the human erythrocyte anion transport protein, *J. Biol. Chem.* **257:**7554–7559.

Jennings, M. L., 1984, Oligomeric structure and the anion transport function of human erythrocyte band 3 protein, *J. Membr. Biol.* **80:**105–117.

Jennings, M. L., 1985, Kinetics and mechanism of anion transport in red blood cells, *Annu. Rev. Physiol.* **47:**519–533.

Jennings, M. L., and Adams, M. F., 1981, Modification by papain of the structure and function of band 3, the erythrocyte anion transport protein, *Biochemistry* **20:**7118–7122.

Jennings, M. L., and Al-Rhaiyel, S., 1988, Modification of a carboxyl group that appears to cross the permeability barrier in the red cell anion transporter, *J. Gen. Physiol.* **92:**161–178.

Jennings, M. L., and Anderson, M. P., 1987, Chemical modification of glutamate residues at the stilbene-disulfonate site of human red blood cell band 3 protein, *J. Biol. Chem.* **262:**1691–1697.

Jennings, M. L., and Passow, H., 1979, Anion transport across the erythrocyte membrane: In situ proteolysis of band 3 protein and cross linking of proteolytic fragments by 4,4′-diisothiocyano-dihydrostilbene-2,2′-disulfonate, *Biochim. Biophys. Acta* **554:**498–519.

Kanna, K., Labotka, R., and Low, P. S., 1988, Isolation and characterization of hemichrome-stabilized membrane protein aggregates from sickle erythrocytes, *J. Biol. Chem.* **263:**13766–13773.

Kawano, Y., Okubo, K., Tokunaga, F., Miyata, T., Iwanaga, S., and Hamasaki, N., 1988, Localization of the pyridoxal phosphate binding site at the COOH-terminal region of the erythrocyte band 3 protein, *J. Biol. Chem.* **263:**8232–8238.

Kay, M. B. B., 1986, Red cell clearance, in *Red Cell Antigens and Antibodies* (G. Garratty, eds.), pp. 35–82, American Association of Blood Banks, Arlington, Va.

Kellokumpu, S., Neff, L., Jamsa-Kellokumpu, S., Kopito, R., and Baron, R., 1988, A 115-kD polypeptide immunologically related to erythrocyte band 3 is present in Golgi apparatus, *Science* **242:**1308–1311.

Kinne, R., and Heinz, E., 1987, Role of potassium in cotransport systems, *Curr. Top. Membr. Transp.* **28:**73–85.

Klimann, H. J., and Steck, T. L., 1980, Kinetic analysis of the association of glyceraldehyde-3-phosphate dehydrogenase with the human red cell membrane, in: *Membrane Transport in Erythrocytes, Alfred Benzon Symposium 14* (U. V. Lassen, H. H. Ussing, and J. O. Wieth, eds.), pp. 312–322, Munksgaard, Copenhagen.

Klocke, R. A., 1976, Rate of bicarbonate–chloride exchange in human red cells at 37°C, *J. Appl. Physiol.* **40:**707–714.

Knauf, P. A., 1979, Erythrocyte anion exchange and the band 3 protein; transport kinetics and molecular structure, in: *Current Topics in Membrane Transport* (F. Bronner and A. Kleinzeller, eds.), pp. 249–363, Academic Press, New York.

Knauf, P. A., 1986, Anion transport in erythrocytes, in: *Membrane Transport Disorders* (T. Andreoli, J. F. Hoffman, S. G. Schultz, and D. D. Fanestil, eds.), pp. 191–220, Plenum Press, New York.

Kohne, W., Haest, C. W. M., and Deuticke, B., 1981, Mediated transport of anions in band 3–phospholipid vesicles, *Biochim. Biophys. Acta* **229:**547–556.

Kopito, R. R., and Lodish, H. F., 1985, Structure of the murine anion exchange protein, *J. Cell. Biochem.* **29:**1–17.

Kopito, R. R., Anderson, M., and Lodish, H. V., 1987, Structure and organization of the murine band 3 gene, *J. Biol. Chem.* **262:**8035–8040.

Kopito, R. R., Andersson, M. A., Herzlinger, D. A., Al-Awqati, Q. and Lodish, H. F., 1988, Structure and tissue-specific expressions of the mouse anion exchanger gene in erythroid and renal cells, in: *Cell Physiology of Blood* (R. B. Gunn and J. C. Parker, eds.), pp. 151–161. Rockefeller University Press, New York.

Kopito, R. R., Lee, D. S., Simmons, D. M., Lindsey, A. E., Morgans, C. W., and Schneider, K., 1989, Regulation of intracellular pH by a neuronal homology of the erythrocyte anion exchanger, *Cell* **59**:927–937.

Landry, D. W., Reitman, M., Cragoe, E. J., and Al-Awqati, Q., 1987, Epithelial chloride channel: Development of inhibitory ligand, *J. Gen. Physiol.* **90**:779–798.

Lazarides, E., 1987, From genes to structural morphogenesis: The genesis and epigenesis of a red blood cell, *Cell* **51**:345–356.

Lodish, H., 1988, Multispanning membrane proteins: How accurate are the models? *Trends Biochem. Sci.* **13**:332–335.

Lodish, H. F., Braell, W. A., Schwarz, A. L., Strous, G. J. A. M., and Zilberstein, A., 1981, Synthesis and assembly of membrane and organelle proteins, *Int. Rev. Cytol.* **12**:247–307.

Low, P. S., 1986, Structure and function of the cytoplasmic domain of band 3: Center of erythrocyte–peripheral protein interactions, *Biochim. Biophys. Acta* **864**:145–167.

Lukacovic, M. F., Feinstein, M. B., Sha'afi, R. I., and Perrie, S., 1981, Purification of stabilized band 3 protein of the human erythrocyte membrane and its reconstitution into liposomes, *Biochemistry* **20**:3145–3151.

Lutz, H. U., 1987, Red cell clearance: A review, *Biomed. Biochem. Acta* **46**:65–71.

Lutz, H. U., Fasler, S., Stammler, P., Bussolino, F., and Arese, P., 1988, Naturally occurring anti-band-3 antibodies and complement in phagocytosis of oxidatively-stressed and in clearance of senescent red cells, *Blood Cells* **14**:175–195.

Lux, S. A., John, K. M., Kopito, R. R., and Lodish, H. F., 1989, Cloning and characterization of band 3, the human erythrocyte anion exchange protein, *Proc. Natl. Acad. Sci.* **86**:9089–9093.

Macara, I. G., and Cantley, L. C., 1981, Mechanism of anion exchange across the red cell membrane by band 3: Interactions between stilbene-disulfonate and NAP-taurine binding sites, *Biochemistry* **20**:5695–5701.

McCrea, R., Engelman, D. M., and Popot, J. L., 1988, Topography of membrane proteins: Hydrophobicity analysis versus immunolocalization, *Trends Biochem. Sci.* **13**:289–291.

Marchesi, V. T., 1985, Stabilizing infrastructures of membranes, *Annu. Rev. Cell Biol.* **1**:531–561.

Mawby, W. J., and Findlay, J. B. C., 1982, Characterization and partial sequence of di-iodo-sulphenylisothiocyanate binding protein from human erythroctye anion transport protein, *Biochem. J.* **205**:465–475.

Morgan, M., Hanke, P., Grygorczyk, K. R., Tintschl, A., Fasold, H., and Passow, H., 1985, Mediation of anion transport in oocytes of Xenopus laevis by biosynthetically inserted band 3 protein from mouse spleen erythroid cells, *EMBO J.* **4**:1927–1931.

Motais, R., and Cousin, J. L., 1978, A structure activity study of some drugs acting as reversible inhibitors of chloride permeability in red cell membranes: Influence of ring substituents, in: *Cell Membrane Receptors for Drugs and Hormones: A Multidisciplinary Study* (R. W. Straub and L. Bolis, eds.), pp. 219–225, Raven Press, New York.

Muallem, S., Blissard, D., Cragoe, E. J., and Sachs, G., 1988, Activation of Na/H and Cl/HCO_3 exchange by stimulation of acid secretion in the parietal cell, *J. Biol. Chem.* **263**:14703–14711.

O'Grady, Z. M., Palfrey, H. C., and Field, M., 1987, Characteristics and functions of Na-K-Cl cotransport in epithelial tissues, *Am. J. Physiol.* **253**:C177–C192.

Okoye, V. C. N., and Bennet, V., 1985, Plasmodium falciparum: Band 3 as a possible receptor during invasion of human erythrocytes, *Science* **227**:169–171.

Palek, J., and Lux, S. E., 1983, Red cell membrane skeletal defects in hereditary and acquired hemolytic anemias, *Semin. Hematol.* **20**:189–224.

Passow, H., 1987, Molecular aspects of band 3 protein-mediated anion transport across the red blood cell membrane, *Rev. Physiol. Biochem. Pharmacol.* **103**:62–217.

Passow, H., Fasold, H., Lepke, S., Pring, M., and Schuhmann, B., 1977, Chemical and enzymic modification of membrane proteins and anion transport in human red blood cells, in: *Advances in Experimental Medical Biology* (M. W. Miller and A. E. Shamoo, eds.), pp. 353–379, Plenum Press, New York.

Pimplikar, S. W., and Reithmeier, R. A. F., 1988, Identification, purification and characterization of a stilbene disulfonate binding glycoprotein from canine kidney brush border membranes. A candidate for a renal exchanger, *J. Biol. Chem.* **263**:4485–4493.

Ramjeesingh, M., Gaarn, A., and Rothstein, A., 1981, The amino acid conjugate formed by the interaction of the anion transport inhibitor DIDS with band 3 protein from human red blood cell membranes, *Biochim. Biophys. Acta* **641**:173–182.

Rao, A., Martin, P., Reitheimer, R. A. F., and Cantley, L. C., 1979, Location of the stilbene disulfonate binding site of the human erythrocyte anion-exchange system by resonance energy transfer, *Biochemistry* **18:**4505–4516.

Reinertsen, K. V., Tonnessen, T. I., Jacobsen, J., Sandvig, K., and Olsnes, S., 1988, Role of chloride/bicarbonate antiport in control of cytosolic pH. Cell-line differences in activity and regulation of antiport, *J. Biol. Chem.* **263:**11117–11125.

Rothstein, A., Ramjeesingh, M., Grinstein, S., and Knauf, P. A., 1980, Protein structure in relation to anion transport in red cells, *Ann. N.Y. Acad. Sci.* **34:**433–443.

Sabban, E., Marchesi, V. T., Adesnik, M., and Sabatini, D. D., 1981, Erythrocyte membrane protein band 3: Its biosynthesis and incorporation into membranes, *J. Cell Biol.* **91:**637–646.

Sabban, E., Sabatini, D. D., Marchesi, V. T., and Adesnik, M., 1982, Biosynthesis of erythrocyte membrane protein band 3 in DMSO-induced Friend erythroleukemia cells, *J. Cell Physiol.* **104:**261–268.

Sayare, M., and Fikiet, M., 1981, Cross-linking of hemoglobin to the cytoplasmic surface of human erythrocyte membranes, *J. Biol. Chem.* **256:**13152–13158.

Scheuring, U., Kollewe, K., Haase, W., and Schubert, D., 1986, A new method for the reconstitution of the anion transport system of the human erythrocyte membrane, *J. Membr. Biol.* **90:**123–135.

Schlatter, E., Greger, R., and Weidtke, C., 1983, Effect of high "ceiling" diuretics on active salt transport in the cortical thick ascending limb of Henle's loop of rabbit kidney: Correlations of chemical structure and inhibitory potency, *Pfluegers Arch.* **396:**210–217.

Schoumacher, R. A., Shoemaker, R. L., Halm, D. R., Tallant, E. A., Wallace, R. W., and Frizzell, R. A., 1987, Phosphorylation fails to activate chloride channels from cystic fibrosis airway cells, *Nature* **330:**752–754.

Schuster, V. L., Bonsib, S. M., and Jennings, M. L., 1987, Two types of collecting duct mitochondria-rich (intercalated) cells: Lectin and band 3 cytochemistry, *Am. J. Physiol.* **251:**C347–C355.

Schwartz, G. J., Barasch, J., and Al-Awqati, Q., 1985, Plasticity of functional epithelial polarity, *Nature* **318:**368–371.

Spring, K. R., and Ericson, A., 1983, Epithelial cell volume modulation and regulation, *J. Membr. Biol.* **69:**167–176.

Steck, T. L., 1974, The organization of proteins in the human red cell membrane, *J. Cell Biol.* **62:**1–19.

Steck, T. L., Koziarz, J. J., Singh, M. K., Reddy, R., and Kohler, H., 1978, Preparation and analysis of seven major, topographically defined fragments of band 3, the predominant transmembrane polypeptide of human erythrocyte membranes, *Biochemistry* **17:**1216–1222.

Tago, K., Schuster, V. L., and Stokes, J. B., 1986, Regulation of Cl self exchange by cAMP in cortical collecting tubule, *Am. J. Physiol.* **251:**F40–F48.

Tanner, M. J. A., Williams, D. G., and Jenkins, R. E., 1980, Structure of the erythrocyte anion transport protein, *Ann. N.Y. Acad. Sci.* **34:**455–464.

Tanner, M. J. A., Marti, P. G., and High, S., 1988, The complete amino acid sequence of the human erythrocyte membrane anion-transport protein deduced from the cDNA sequence, *Biochem. J.* **256:**703–712.

Terwilliger, T. L., and Clarke, S., 1981, Methylation of membrane proteins in human erythrocytes, *J. Biol. Chem.* **256:**3067–3076.

Tsuji, T., Irimura, T., and Ozawa, T., 1980, The carbohydrate moiety of band 3 glycoprotein of human erythrocyte membranes, *Biochem. J.* **187:**677–685.

Wagner, S., Vogel, R., Lietzke, R., Koob, R., and Drenckhahn, D., 1987, Immunochemical characterization of band 3-like anion exchanger in collecting duct of human kidney, *Am. J. Physiol.* **253:**F213–F221.

Wangemann, P., Wittner, M., DiStefano, A., Englert, H. C., Lang, H. J., Schlatter, E., and Greger, R., 1986, Cl-channel blockers in the thick ascending limb of the loop of Henle: Structure activity relationship, *Pfluegers Arch.* **407**(Suppl.):S128–S141.

Waxman, L., 1979, The phosphorylation of the major proteins of the human erythrocyte membrane, *Arch. Biochem. Biophys.* **195:**300–314.

Welsh, M. J., and Liedtke, C. M., 1986, Chloride and potassium channels in cystic fibrosis airway epithelia, *Nature* **322:**467–470.

Wieth, J. O., Andersen, O. S., Brahm, J., Bjerrum, P. J., and Borders, C. L., 1982, Chloride–bicarbonate exchange in red blood cells, *Philos. Trans. Ry. Soc.* **299:**383–399.

Xie, X. S., Stone, D. K., and Racker, E., 1983, Determinants of clathrin-coated vesicle acidification, *J. Biol. Chem.* **258:**14834–14838.

Zaki, L., and Julien, T., 1985, Anion transport in red blood cells and arginine specific reagents. Interaction between the substrate binding site and the binding site of arginine specific reagents, *Biochim. Biophys. Acta* **818:**325–332.

Zenke, M., Kahn, P., Disela, C., Vennstrom, B., Leutz, A., Keega, K., Hayman, M. J., and Choi, H. R., 1988, v-erbA specifically suppresses transcription of the avian erythrocyte anion transporter (band 3) gene, *Cell* **52:**107–119.

Chapter 12

Receptors for Erythropoietin
Distribution, Structure, and Role in Receptor-Mediated Endocytosis in Erythroid Cells

Stephen T. Sawyer

1. INTRODUCTION

Erythropoietin (EP) is the glycoprotein hormone that is the primary regulator of red cell development. It is synthesized and released from the kidneys and livers of animals and man in response to a deficit of oxygen. Erythroid precursor cells respond to the hormone through surface receptors by maturing to functional erythrocytes. Increased erythrocytes in turn result in an increased availability of oxygen. The increased oxygen in the tissues leads to a decrease in synthesis and release of EP. These events comprise the elements of the reflex pathway that is the primary if not exclusive control of normal erythropoiesis. Normally the need for new erythrocytes is small and the circulating levels of the hormone are quite low; however, a decreased delivery of oxygen through hemorrhage, high-altitude hypoxia, red cell disorders, and other causes results in up to 1000-fold induction of the mRNA encoding EP in the kidney and a corresponding increase in the circulating levels of the hormone.

Within the past few years, considerable progress has been made in the understanding of both the sites of EP production and the possible mechanisms that directly couple oxygen tension to the regulation of EP synthesis. In addition, the identification of receptors for EP in responsive erythroid cells has increased our knowledge of how EP exerts its effects. This review will concentrate on the interaction of EP with receptors on erythroid progenitor cells and possible mechanisms by which the hormone activates these cells to proceed with the program of erythroid maturation. Also, a brief review of recent advances in the control of EP production will be presented.

Stephen T. Sawyer Division of Hematology, Department of Medicine, Vanderbilt University School of Medicine, Nashville, Tennessee 37232, USA.

1.1. Discovery of Erythropoietin, Purification, and Cloning

The presence of a humoral factor controlling erythropoiesis had been speculated since the beginning of this century (Carnot and Deflandré, 1906). In the previous century, it was recognized that increased erythropoiesis occurred in humans living at high altitude. Therefore, the connection of erythropoiesis to oxygen delivery was thought to be either a direct action of oxygen on erythroid progenitors in the bone marrow or the indirect effect of oxygen at another site and the elaboration of a factor that stimulated or inhibited erythropoiesis (for a complete review of this history see Krantz and Jacobson, 1970). Direct demonstration of a humoral factor that stimulated erythropoiesis was carried out by Reissmann (1950) using parabiotic rats. When one parabiotic rat was placed in a hypoxic atmosphere, erythropoiesis in the normal paired rat was increased. Other workers showed that the serum from anemic or hypoxic animals contained a factor that enhanced the production of red blood cells when injected into normal or plethoric animals. Since this factor did not affect the number of white cells or platelets, it was called erythropoietin.

Purification of EP from the sera of anemic animals was attempted with little success. It is now known that normal levels of EP in the serum are 10 to 30 mU in laboratory animals and man and these levels can increase more than 1000-fold under the most severe anemia or hypoxia. If an animal had 10 U EP/ml (in the range of the highest reported), the EP would be 83 ng/ml or about 0.00008% of total serum protein. Until very recently, EP used in tissue culture was primarily that from the plasma of anemic sheep having 2–4 U EP/mg protein or 0.002% of total protein. Urine from anemic animals or humans proved to be a much better source for the purification of EP due to a lesser number of contaminating proteins. The purification of EP from human urine was reported by Miyake *et al.* (1977). This material was found to be a glycoprotein that migrated with a molecular weight of 34,000 on SDS–PAGE and contained 30% carbohydrate. The specific activity of this material was reported to be 74,000 U/mg.

Cloning of the gene for EP and enhanced expression of the gene in cultured cells has greatly facilitated study of the hormone. Purified urinary EP was only available in small quantities, but the cloned gene product is produced in quantities ($>$ 100 g/year) sufficient to treat patients suffering from certain anemias. The single-copy gene consists of five exons and four introns and include a leader sequence that is cleaved during translation in a similar fashion as other secreted proteins (Jacobs *et al.*, 1985; Lin *et al.*, 1985). Cloning of the gene by groups at both Genetics Institute and AMGen required information on the sequences of amino acids from the purified urinary EP to construct probes to detect the gene in human libraries. The cloned gene was used to show the expression of mRNA encoding EP in human fetal liver. Also, both groups expressed the cloned gene in mammalian cells [COS monkey kidney cells or Chinese hamster ovary (CHO) cells] to give an active glycosylated hormone.

Complete glycosylation of EP is critical for *in vivo* studies and administration to humans. Earlier studies using purified urinary EP demonstrated that the sialic acid residues were required for *in vivo* biological activity but not *in vitro* activity (Goldwasser, 1975). This requirement is due to the rapid clearance of asialylated EP in the liver. However, expression of the gene in bacteria results in a totally unglycosylated EP that has no activity *in vivo* and partial activity *in vitro*. This may be due to incorrect disulfide bridges, poor solubility of unglycosylated EP, or rapid clearance *in vivo*.

The sequence of nucleotides in the cloned gene encodes a peptide of 166 amino acids with a molecular weight of 18,398 (Jacobs *et al.*, 1985; Lin *et al.*, 1985). However, the carboxy-terminal arginine predicted from the gene is missing in both the purified urinary EP and EP produced in CHO cells from the cloned gene (Recny *et al.*, 1987). The molecular weight of recombinant human EP was determined to be 29,000 ± 400 or 30,400 ± 400 by velocity sedimentation (Brown *et al.*, 1986; Davis *et al.*, 1987). This measurement indicated that carbohydrate comprised 38% or 39% of the total mass of EP. By SDS–PAGE recombinant EP migrated at a position corresponding to a molecular weight of 35,000 to 40,000, which is similar to that of the purified urinary EP. Glycosylated proteins characteristically have anomalous migration in SDS–PAGE due to the inability of carbohydrate to bind SDS and the contribution of carbohydrate to the net charge. In this laboratory, iodinated human recombinant EP (AMGEN) migrated on SDS–PAGE at a position corresponding to a molecular weight of 40,000 while in the same gel iodinated urinary EP migrated at a position corresponding to 35,000 molecular weight. Slight differences in glycosylation (Sasaki *et al.*, 1987a) or site of iodination of the recombinant protein compared to the urinary EP may account for this observation.

Purified urinary EP has been reported to have a potency of 70,400 U/mg protein (Miyake *et al.*, 1977; Goldwasser, 1981), which corresponds to 82,800 U/A_{280}. It is not clear if these measurements were based on the total weight of the purified protein or on a protein measurement of EP in solution. A unit is arbitrarily defined as one-tenth the activity in an ampule of the International Reference Preparation and was originally defined as the EP response of a starved rat to 5 μmol cobalt. Purified human EP expressed in CHO cells has a potency of 174,000 U/A_{280} (Browne *et al.*, 1986) which is 2-fold greater than that reported for the purified urinary EP. This discrepancy is unexplained but created a great deal of confusion when recombinant EP became commercially available. AMGen sells EP for research in units but the specific activity was considered a trade secret and was not released until recently. Since the extinction coefficients (absorbance at 280 nm) are different for purified urinary EP, 8.5, and the recombinant EP, 7.4 (Goldwasser, 1981; Davis *et al.*, 1987), the specific activity of urinary EP seems to have been based on the peptide content of EP rather than the weight of the glycoprotein. Therefore, the specific activity of urinary EP reported seems to be 70,400 U/mg protein (peptide) and 50,000 U/mg total weight [used by Krantz and Goldwasser (1984) for the determination of the number of molecules of urinary EP bound to cells]. The specific activity of the recombinant EP is 129,000 U/mg total weight and 210,000 U/mg protein (peptide). These assays for EP were carried out in mice. Partial deglycosylation of the urinary EP before or during purification may explain the 2.5-fold lower specific activity reported. *In vitro* assays of purified urinary EP versus recombinant EP in our laboratory showed that the potency of the urinary EP was 70% of the recombinant EP.

1.2. Sites and Regulation of Erythropoietin Production

Cloning of the mouse gene for EP (McDonald *et al.*, 1986; Shoemaker and Mistock, 1986) has made possible studies in mice on the accumulation of mRNA coding for EP during stimulation of erythropoiesis. Early work had shown in humans and laboratory animals that the kidney was the primary regulatory organ of erythropoiesis (Jacobson *et al.*, 1957). However, very little EP could be extracted from kidneys of normal or anemic animals and two theories evolved to explain this apparent discrepancy. One theory proposed the elaboration by the kidneys of an erythrogenin that acted on a circulating precursor of EP to produce EP (Zanjani *et al.*, 1971). Peschle and Condorelli (1975) proposed that a circulating enzyme cleaved a precursor of EP secreted from the kidney to form active EP. Cloning of the gene from human, monkey, and mouse showed that EP was not synthesized in a larger proerythropoietin molecule. Also, the transfected gene was expressed in cultured cells to yield an active hormone. Therefore, the direct synthesis of EP in the kidney was a logical conclusion.

A direct demonstration of mRNA encoding EP in the kidneys and liver of mice made anemic by bleeding (Bondurant and Koury, 1986) and kidneys of rats injected with cobalt or bled (Beru *et al.*, 1986) has been reported. No mRNA for EP was detected in normal kidneys or liver in mice by Northern blot analysis, and only a very faint signal was detected using an EP mRNA protection assay. Upon severe bleeding of adult mice, the EP mRNA protection assay detected 500- to 1000-fold more mRNA in the anemic kidneys than in normal kidneys (M. J. Koury *et al.*, 1988). However, the liver in anemic mice produced only 7% of the total mRNA coding for EP and mRNA was undetectable in normal liver. Other tissues from anemic animals had no detectable mRNA for EP.

The mRNA for EP was induced within 1 hr of bleeding and was maintained until the hermatocrit rose to normal levels. Recent information has shown that this increase in EP mRNA is at least partially due to an increased rate of transcription of the gene (Schuster *et al.*, 1989).

Two independent groups have shown a rare cell containing EP mRNA in the interstitium between tubules in the inner cortex of kidneys of anemic mice using *in situ* hybridization (S. T. Koury *et al.*, 1988a; Lacombe *et al.*, 1988). This cell may be an endothelial cell, but additional experiments are necessary to confirm this. Only extremely rare cells are found to have mRNA for EP in the liver. These cells do not appear to be hepatocytes but their morphology is uncertain. Some workers have proposed that macrophages produce EP (Rich *et al.*, 1982) but the cells responsible for production of EP in the kidneys and liver do not have the morphology of macrophages.

Reducing the hematocrit to varying degrees by bleeding leads to an increasing number of cells transcribing mRNA encoding EP in the kidney (S. T. Koury *et al.*, 1989). The level of regulation appears to be the number of cells transcribing EP mRNA rather than the control of the amount of EP produced per cell. Careful examination of normal kidneys by *in situ* hybridization shows that very rare cells transcribe the maximum level of mRNA encoding EP (S. T. Koury *et al.*, 1989). This result is consistent with the kidneys being the major source of EP in the circulation for both normal and stressed erythropoiesis. There is a strong inverse correlation between the hematocrit, circulating EP as determined by radioimmuno assay, and the number of cells in the kidney that transcribe EP mRNA when the hematocrit is decreased by bleeding to varying degrees or during the recovery from a severe bleed. It does not appear that EP is synthesized and stored to be released in the event of hypoxia, nor does the circulating level of EP feed back to control EP synthesis. Rather, individual cells in the kidney (and liver) apparently respond to hypoxia by switching on the transcription and translation of EP mRNA to form EP, which is directly secreted into the circulation.

The regulation of EP production at the molecular level is still unknown. The cell line

Hep 3B has been shown to respond to hypoxia by increasing accumulation of EP mRNA and synthesis of secreted EP (Goldberg *et al.*, 1987). This cell line also responds to cobalt and carbon monoxide in a similar fashion as to hypoxia. A heme-containing protein that shifts from an oxy form to a deoxy form has been proposed to be responsible for the regulation of transcription of the EP mRNA (Goldberg *et al.*, 1988). This hypothesis is based on earlier observations in hemoglobin that cobalt and carbon monoxide shift the conformation of hemoglobin to a deoxy conformation. However, others have speculated that the state of oxidation of the electron transport chain is involved in the control of EP production based on the observation that mitochondrial poisons increase the secretion of EP. It is of great interest that the 5′ upstream sequences of human, monkey, and mouse EP genes are more highly conserved than the coding sequence of the gene (McDonald *et al.*, 1986; Shoemaker and Mistock, 1986). This probably indicates that the control of transcription of the gene is through a common mechanism that acts on this segment of DNA. Perhaps even the direct binding of an oxygen-sensing heme protein to this segment of DNA is responsible for the transcription of the EP mRNA. Introduction of the human EP gene into transgenic mice resulted in the expression of mRNA for human EP in a variety of tissues including kidney. However, the gene was only regulated in response to hypoxia or anemia in the liver (Semenza *et al.*, 1989). This result suggests that transcription of the EP gene may be regulated differently in the kidney and liver.

The liver is an organ that produces a small fraction of EP compared to the kidney. Earlier studies with anephric laboratory animals and humans indicated a source of EP other than the kidney (Jacobson *et al.*, 1959). The liver is apparently the only extrarenal site of EP synthesis since circulating levels of EP fall from low to undetectable when the liver of anephric rats is removed (Fried, 1972). EP production in the adult liver may be a vestige of that in the fetal liver. The latter is a source of EP in humans (Lin *et al.*, 1985) and sheep (Zanjani *et al.*, 1977) but a careful study of mice showed only a trace of EP mRNA in fetal liver at midgestation (M. J. Koury *et al.*, 1988). The transplacental transfer of EP from the maternal circulation to the fetus (M. J. Koury *et al.*, 1988) and placental receptors for EP (Sawyer *et al.*, 1989; Figure 2) that may mediate this transfer have been documented in mice. The transfer of EP across the placenta may only occur in certain species such that the fetal liver is most involved in the production of EP where a maternal source of EP is not available.

It is possible that another source of EP other than kidney and liver exists. The discovery of murine erythroleukemia cell lines that secrete EP has increased interest in the possibility that autocrine production of EP by erythroid progenitor cells or synthesis of the hormone by accessory cells in the marrow might contribute to the normal level of erythropoiesis (Tambourin *et al.*, 1983). Manipulation of the erythropoietic state by hypoxia or anemia substantially increases the number of late-stage erythroid progenitor cells, which undoubtedly results from the tremendous increase in EP in the circulation. On the other hand, hyperoxia or hypertransfusion results in a great reduction of late-stage erythroid progenitor cells in the marrow. Since the normal levels of EP mRNA in the kidneys and liver and EP in the circulation are either below the level of detection or at the limit of detection, it is not certain that hyperoxia results in a decrease of the level of circulating EP that correlates with the loss of erythroid progenitors. This leaves open the possibility that EP from a source in the marrow or even another factor that regulates erythropoiesis is suppressed under conditions of hyperoxia.

2. CELLS THAT RESPOND TO ERYTHROPOIETIN

2.1. Erythropoietin-Responsive Erythroid Progenitor Cells

Erythropoiesis occurs at a steady-state level in the marrow of nonanemic mammals such that a full spectrum of erythroid progenitor cells from the earliest committed erythroid cells through reticulocytes is present. Since differentiation occurs through successive divisions of the progenitor cells, the more mature cells greatly exceed the number of earlier progenitor cells. Changing the level of erythropoiesis by inducing anemia through bleeding or reducing oxygen delivery by placing an animal in a hypobaric chamber increases the number of erythroid cells (both early EP-responsive cells and later nonresponsive cells, while hypertransfusion and/or hyperoxia virtually eliminate the recognizable, later stage erythroid cells and substantially reduce the earlier progenitors in the marrow). Introduction of EP into cultures of marrow cells from hypertransfused mice resulted in the appearance of recognizable erythroblasts and reticulocytes within 72 hr (Jacobson *et al.*, 1959). Addition of a neutralizing antibody to EP to these marrow cultures demonstrated that the initial erythroid response was the direct result of EP's effect on erythroid progenitors but that the later stages of erythroid maturation (late erythroblast to reticulocyte to erythrocyte) did not require EP.

In the early erythroid developmental stages, EP-responsive cells are rare and morphologically undefined; therefore, indirect methods were established to define early stage EP-responsive cells. These methods depended on semisolid tissue culture systems (plasma clot and methylcellulose) that allowed single erythroid progenitor cells to develop into colonies of morphologically defined erythroblasts in the presence of EP. Two major subdivisions of EP-responsive erythroid cells were described using this system in mice (Axelrad *et al.*, 1974; Iscove and Sieber, 1975) and humans (Tepperman *et al.*, 1974; Gregory and Eaves, 1977). These are known as colony-forming units–erythroid (CFU-E) and burst-forming units–erythroid (BFU-E) and are defined on the basis of the number of mature cells per colony and the time required for colonies of mature, hemoglobin-containing colonies to develop. CFU-E are present at a frequency of 0.3% of nucleated marrow cells in mouse (0.1% in normal human marrow) and are defined as cells that give rise to small colonies (8 to 50 cells) of erythroblasts in 2 days for mouse cells and 7 days for human cells after initiation of culture in the presence of small quantities of EP (0.01 to 1.0 U/ml). BFU-E are present at 0.02% of marrow cells in mice and humans and are defined as cells that give rise to very large colonies or clusters of colonies of mature, hemoglobin-containing cells numbering from several hundred to several thousand after 8 days in culture with mouse cells and 15 days with human cells. BFU-E require more EP in the culture (1.0 to 5.0 U/ml) than CFU-E but this may be due to the degradation of EP in culture over a longer period of culture. CFU-E and BFU-E can be separated by physical means; CFU-E are larger cells than BFU-E and are easily separated from BFU-E by velocity sedimentation of marrow cells (Heath *et al.*, 1976).

Even earlier erythroid progenitor cells are identifiable using other methods. Injection of marrow cells into lethally irradiated mice leads to development of mixed colonies of erythroid, granulocytic, and megakaryocytic cells in the spleen. It is speculated that these colonies result from the transplantation of a single pluripotent hematopoietic stem cell known as spleen colony-forming units (CFU-S) (Till and McCulloch, 1961). More recently, culture techniques have been perfected to demonstrate a range of primitive cells that give rise to mixed colonies of separate lineages in tissue culture. Mixed colonies of granulocytic, erythrocytic, megakaryocytic, and monocytic cells, CFU-GEMM, can be grown from human marrow (Ogawa *et al.*, 1983) and probably represent a slightly more mature human equivalent of CFU-S in the mouse. These definitions are obviously oversimplified as there are continuous intermediate stages between stem cells that can reconstitute a lethally radiated animal and the terminal cell of each line of hematopoietic differentiation.

It is not known exactly which cells in the erythroid line of differentiation are responsive to EP with the exception of very mature erythroblasts, reticulocytes, and erythrocytes, which are not EP-responsive (M. J. Koury and Bondurant, 1988). Administration of EP to animals or manipulations that result in high levels of endogenous EP result in much higher numbers of CFU-E, erythroblasts, and mature BFU-E (a cell of maturation intermediate between BFU-E and CFU-E such that burst form in culture after 5 days rather than 8 days in mice); however, the CFU-S and BFU-E compartments are not increased. This suggests that the earliest cell that responds to EP by increased proliferation is the BFU-E or a slightly more mature progenitor cell.

Whereas the majority of CFU-E are continuously in cell division as demonstrated by thymidine suicide experiments, most BFU-E are not in the DNA synthesis phase of the cell cycle (only 30% are killed by [^{3}H]thymidine suicide) (Iscove, 1977). Administration of EP to marrow cells results in a dramatic increase of the percentage of BFU-E that enter DNA synthesis (Dessypris and Krantz, 1984), indicating the responsiveness of these cells to EP. The increased concentration of EP needed to culture BFU-E compared to CFU-E has resulted in the hypothesis that BFU-E are less responsive to EP and the proliferating daughter cells of BFU-E become increasingly more sensitive as they reach the CFU-E stage. However, the greater requirement for EP necessary to culture BFU-E is a laboratory artifact due to the degradation of EP during 8 to 15 days in culture. Multiple additions of very small amounts of EP to these cultures result in the same number of BFU-E as one large dose of EP. Therefore, it is unlikely that BFU-E and CFU-E are very different in their requirements for EP. BFU-E can be maintained in culture without EP until erythroid differentiation nears the CFU-E stage of maturation by the addition of burst-promoting activity (BPA) (Iscove and Guilbert, 1978). An erythroid-specific BPA has never been identified but interleukin-3 (IL-3) and granulocyte–macrophage colony-stimulating factor (GM-CSF) have BPA activity. Serum-free culture of human BFU-E requires IL-3, GM-CSF, and granulocyte colony-stimulating factor (G-CSF) as well as EP for maximum development (Sawada *et al.*, 1989), but the *in vivo* significance of these regulatory molecules in erythropoiesis is uncertain. On the other hand, purified CFU-E are not affected by these factors and can be grown without serum in the presence of EP with only transferrin and insulin or insulinlike growth factor I (Sawada *et al.*, 1987, 1988a,b).

Administration of EP to humans suffering from anemia due to renal failure results in increased BFU-E as well as megakaryocytic colony-forming units (CFU-Meg) and granulocytic and macrophage colony-forming units (CFU-GM) in the marrow (Dessypris *et al.*, 1988). Administration of large amounts of EP can increase the number of CFU-Meg in mouse marrow and the number of platelets in mice and rats (McDonald *et al.*, 1987; Berridge *et al.*, 1988). EP also potentiates the proliferation of CFU-Meg in human and mouse marrow cultures (Dessypris *et al.*, 1987). EP may act through an indirect mechanism on these nonerythroid cells such as CFU-GM and CFU-Meg. However, since CFU-Meg are more related to erythroid progenitors, these cells may respond to EP directly (Fraser *et al.*, 1989).

Currently, two theories have been seriously considered to explain the humoral regulation of the differentiation of a stem cell into the terminal cells of hematopoiesis. The most accepted view is that the earliest progenitors are self-renewing stem cells that randomly (often referred to as stochastically) give rise to progenitor cells that are either unipotent or multipotent but not self-renewing (Till *et al.*, 1964; Ogawa *et al.*, 1983). These mostly multipotent cells are stimulated by growth factors like IL-3 to proliferate and undergo differentiation. As the daughter cells give rise to even more mature progenitors, the cells become unipotent and lose receptors for the multipotent growth factors and acquire lineage-specific receptors. During the transition from earlier to later progenitor cells, the lineage-specific factor receptors have completely replaced other receptors for growth multipotent factors. In the case of erythropoiesis, the earliest unipotent erythroid progenitor recognized is the BFU-E, which is apparently dependent on factors other than EP, while the more mature CFU-E is probably dependent on EP and no other hematopoietic factors. In this model, the hematopoietic factors do not direct differentiation directly but maintain the proliferation of preprogrammed unipotent progenitor cells. For the erythroid lineage, EP would not recruit cells into the erythroid arm of differentiation but rather would maintain the viability of cells that randomly have become committed to erythroid differentiation. When the level of EP is high, a larger number of progenitor cells complete the program of differentiation to erythrocytes than when the level of EP is normal or low. The minority view is that the lineage-specific factors such as EP act directly on the stem cell and instruct it to proceed toward a particular arm of hematopoietic differentiation (Goldwasser, 1981). This scheme suggests a competition between factors such that the continual occupancy of receptors for one factor leads to the loss of other receptors for other hematopoietic factors and the cell becomes committed to a particular lineage of differentiation. The basic difference in these two models is the signal that EP or other hematopoietic factors transmit to the cell after binding to surface receptors; in this second scheme the factor transmits a signal that is capable of directing differentiation, whereas in the first scheme only a mitogenic signal is transmitted. This question remains to be answered.

2.2. Model Systems of Erythroid Maturation

2.2.1. *In Vitro* Models of Erythropoiesis Using Erythroid Cells from Bone Marrow, Fetal Liver, and Spleen

In normal bone marrow, EP-responsive cells make up less than 1% of total nucleated cells and a continuum of erythroid precursors are present. Obviously, the ability to study the effect of EP on these cells is very limited; however, the effects of EP on strictly erythroid-specific events such as hemoglobin synthesis and ^{59}Fe incorporation into hemoglobin have been studied in liquid cultures of normal marrow cells from many animals. Hypertransfusion or hyperoxia in laboratory animals results in a dramatic drop in the number of CFU-E and erythroblasts in the marrow, and yet has little effect on the earlier erythroid progenitors, BFU-E. Therefore, hypertransfusion gives a more uniform population of progenitor cells compared to the heterologous mixture of erythroid cells in normal marrow, but EP-responsive cells represent only a small fraction of the total cells. Marrow from hypertransfused rats and mice has been used to study the effect of EP and has suggested an early effect of EP and RNA synthesis (Krantz and Goldwasser, 1965). However, one should remember that all experiments of this era were done with extremely crude EP in cell preparations in which the responsive cells were a tiny minority. Experiments in this system measuring erythroid-specific events such as ^{59}Fe incorporation into hemoglobin (Krantz *et al.*, 1963) are likely to be correct while the measurements of other, more general physiological events may not be due to the interaction of EP and EP-responsive cells.

To obtain more EP-responsive cells, some workers have made animals hypoxic to increase the percentage of CFU-E and erythroblasts in the marrow. In rodents, the anemia results in the proliferation of erythroid cells in the spleen such that up to 80% of the cells there are erythroid progenitors and erythroblasts. While the number of nonerythroid cells is greatly reduced, the erythroid cells generated represent the full spectrum of precursor cells such that mature erythroblasts are the predominating cells in the culture. These late-stage cells have lost their responsiveness to EP and earlier cells have been exposed to high levels of EP such that experiments on the mechanism of action of EP are complicated. This cell system is used extensively for an assay of the biological activity of EP (Krystal, 1983). The effect of EP on [^{3}H] thymidine incorporation is correlated with a known standard; however, the effect is not a direct stimulation of DNA synthesis in the cell population by EP, but the maintenance of cell division in progenitor cells for several days in the presence of EP. These cells would die in the absence of the hormone. Some workers have also investigated the number of receptors for EP in this mixed population of cells (Mufson and Gesner, 1987; Fraser *et al.*, 1988c).

Fetal liver from rodents is another rich source of erythroid precursor cells. The midgestation fetal liver is an active site of erythropoiesis and apparently is also a source of EP in some nonrodent species. While the erythroid cells in fetal liver may not have been exposed to the greatly elevated level of EP experienced when animals are hypoxic, the erythroid cells are a mixture of earlier progenitor cells, more mature precursors, and predominantly mature erythroblasts. Fetal liver cells have also been used to investigate the receptors for EP and the mechanism of action of EP (Fukamachi *et al.*, 1987a; Tojo *et al.*, 1987; Sasaki *et al.*, 1987b; Mayeux *et al.*, 1987b).

2.2.2. Friend Virus-Infected Erythroleukemia Cell Lines

Friend virus was isolated from a mouse that developed erythroleukemia as a result of inoculation of a cell-free extract of Ehrlich ascites cells (Friend, 1957). Plasma from such infected mice transmitted to other mice a disease characterized by rapid enlargements of the spleen due to erythroblastosis. The disease is characterized by (1) the presence of such great numbers of proerythroblasts in the spleen that the size of the organ increases tenfold, (2) the presence of proerythroblasts in the circulation, and (3) a decreased hematocrit after the initial phase of the disease. Because of the lowered hematocrit, the strain of virus initially isolated is known as the anemia-inducing strain of Friend virus (FVA). FVA does not cause anemia in mice because the red cell mass actually increases but not as much as the concomitant expansion of plasma volume. During Friend virus passage in mice, a mutation or selection event occurred resulting in virus that produced a similar but more aggressive disease in which the hematocrit rose to high levels. This strain is known as the polycythemia-inducing strain of Friend virus (FVP) (Mirand *et al.*, 1968; Sassa *et al.*, 1968). Later studies showed that FVA infection resulted in the proliferation of pro-erythroblasts that were responsive to EP both *in vivo* (Tambourin, 1978) and *in vitro* (Hankins and Troxler, 1980) but that the corresponding infection with FVP resulted in the proliferation of erythroid cells that were independent of EP *in vivo* and *in vitro* (Mirand *et al.*, 1968). Analysis of these viruses revealed that both FVA and FVP contained two components: a replication-competent helper virus (Friend murine leukemia virus) and a replication-defective spleen focus-forming virus ($SFFV_A$ or $SFFV_P$). Although the Friend helper virus can induce erythroleukemias in neonates (Troxler and Scolnick, 1978) and in adult mice (Dawson *et al.*, 1979), the anemia-inducing or polycythemia-inducing manifestations of Friend disease can be linked to the genomes of $SFFV_A$ and $SFFV_P$, respectively, which are recombinant viruses between the Friend helper and endogenous ex-otropic viral sequences. In particular, the glycoprotein gp52, which is encoded by the recombinant envelope gene, seems responsible for the action of both $SFFV_A$ and $SFFV_P$. Distinctions in the sequence of amino acids of gp52 account for the different diseases of FVA and FVP (reviewed by Ruscetti and Wolff, 1984). It is interesting that gp52 is derived entirely from viral sequences and its mode of action is unknown, however, preliminary evidence suggests that gp52 interacts directly with the receptor for EP, resulting in activation of the cell in the absence of EP.

There are two phases of Friend disease, an early erythroblastosis in which the erythroid progenitors are not malignant and a later stage in mice that survive the early disease and in which malignant erythroleukemia later develops. Cell lines have been established from mice infected with either FVA or FVP at the later stages of erythroleukemia. These cells are known as murine erythroleukemia (MEL) cells and have been widely used for the study of erythroid differentiation. These cells were initially observed to undergo partial erythroid maturation in the presence of chemical inducers such as dimethylsulfoxide (DMSO) after the cells were frozen in this solvent and then cultured in the presence of DMSO after thawing (Friend *et al.*, 1971). In general, MEL cells are rapidly proliferating cells that stop or are greatly slowed in cell division by exposure to an inducer during a commitment window. Following treatment with an inducer such as DMSO, hexamethylene bisacetamide, or butyrate for a period of time, 12 to 24 hr, the cells are cultured and erythroid differentiation is determined by staining for hemoglobin. Typically, the percentage of cells showing any evidence of staining with benzidine is reported as erythroid maturation.

MEL cells are well suited for some studies such as the changes in gene expression during maturation because the cells are homogeneous in differentiation. Changes such as the transcription rate of erythroid-specific genes and changes in the chromatin structure around these genes can be easily studied after induction to differentiation. Regulation of globin gene expression is an area in which MEL cells have proven useful for studying certain differentiation events (Hofer *et al.*, 1982). These cells have also been studied as a model for the cellular events responsible for differentiation. For example, a specific chromosomal protein, IP-25, was found to accumulate after exposure of MEL to inducers (Keppel *et al.*, 1977). Calcium ions and protein kinase C have been implicated in the mechanism through which the differentiation signal is transmitted to the nucleus after

MEL cells are exposed to inducers (Melloni *et al.*, 1987). In addition, these cells have been used extensively to study transferrin receptors due to their abundance in erythroid cells. The wide use of MEL cells results from the ready availability, malignant nature of the cells (which translates into very easy tissue culture), and simple induction into differentiation (reviewed by Marks and Rifkind, 1978; Marks *et al.*, 1987).

MEL cells are unresponsive to EP (some Friend and Rauscher virus-infected cell lines are weakly responsive to the hormone) such that the differentiation mechanism is not the normal pathway. MEL cells are a poor model for the study of erythroid maturation due to the very incomplete maturation program after induction. Although induced cells synthesize hemoglobin, the amount produced is often trivial compared to normal mature erythroblasts; and often the ratio of α- and β-globins is unbalanced (Orkin *et al.*, 1975). In addition, the accumulation of other erythroid-specific proteins such as the spectrins, glycophorin, and band 3 is very low and α- and β-spectrins are synthesized in unusual ratios (Pfeffer *et al.*, 1986). MEL cells do not enucleate in culture after induction without unusual steps to stabilize the cells (Volloch and Housman, 1982; Patel and Lodish, 1987).

2.2.3. Friend Virus-Infected Primary Erythroid Cells: FVA Cells

In vitro infection of mouse bone marrow cells with either FVA or FVP results in the proliferation of erythroid progenitor cells (mature BFU-E) into very large colonies or bursts of erythroblasts within 5 days of culture (Hankins and Troxler, 1980; Hankins *et al.*, 1978). In the case of infection with FVP, the erythroblasts spontaneously mature in the absence of EP such that the bursts become red. However, the corresponding infection of bone marrow with FVA results in the same size burst at the same time, but the cells are not red. In fact, the cells appear to be proerythroblasts with no accumulation of hemoglobin. Addition of EP to cultures infected with FVA at day 3 results in bursts of mature erythroblasts at day 5 that are identical to the bursts produced by FVP infection. Fortunately, the FVA-infected erythroid cells recognized in culture at day 5 were still responsive to EP and resulted in hemoglobinized mature erythroblasts by culture day 7. This observation paved the way to establish the FVA cell as the first model system that was fully responsive to EP and yet was comprised of a developmentally synchronized population of very pure precursor cells that had not been exposed to EP (M. J. Koury *et al.*, 1982; Bondurant *et al.*, 1983).

The initial system for FVA cells that was developed in the Hematology Research Laboratory at Vanderbilt University required that proerythroblasts generated from the infection of mouse marrow with FVA be harvested from cultures after 5 days by plucking the bursts of FVA cells using drawn-out Pastrue pipets, dissecting microscopes, and skilled technical assistance. FVA cells obtained were greater than 90% proerythroblasts but the number of cells per experiment was limited to 10^6 to 10^7 cells. These cells were used to study the time course of hemoglobin and spectrin synthesis (M. J. Koury *et al.*, 1982) as well as ^{45}Ca flux after the stimulation of maturation by EP (Sawyer and Krantz, 1984).

This preliminary model system using *in vitro* infection of marrow cells with FVA evolved into a related method that depended upon *in vivo* infection of mice. The proerythroblasts that accumulate in the spleens of mice infected with FVA are morphologically and biochemically identical to the FVA cells that accumulate in marrow cultures.

The spleen enlarged 10- to 20-fold after FVA infection, and the enlargement was due exclusively to erythroblasts. Therefore, the nucleated cell population of the enlarged spleen was greater than 90% FVA erythroid cells before any purification. Immature proerythroblasts were separated from splenic erythrocytes and more mature erythroblasts by unit gravity sedimentation through a gradient of bovine serum albumin (M. J. Koury *et al.*, 1984; Sawyer *et al.*, 1987a). A spleen from a mouse infected with FVA has $1-2 \times 10^9$ erythroid cells; using a large 1% to 2% gradient of albumin, 5×10^8 immature erythroid cells can be recovered. These cells are greater than 95% erythroid and 97–98% are negative for benzidine staining for hemoglobin. The contaminating cells, which are 10% or less of the population, are mature erythroblasts, erythrocytes, and lymphocytes in that order. These FVA erythroid cells are at the stage of CFU-E (40% give rise to $\geq$ 8-cell colonies) or slightly more mature when cultured in plasma clot or methycellulose medium and respond to the same concentration of EP as CFU-E from normal mouse marrow. In contrast to MEL cell lines, FVA cells undergo complete enucleation to form reticulocytes in liquid culture in the period from 35 to 72 hr of culture in the presence of EP (S. T. Koury *et al.*, 1988b,c). In the absence of EP, FVA cells lose the ability to respond to EP between 4 and 18 hr and many disintegrate before 48 hr of culture (M. J. Koury and Bondurant, 1988).

The characteristics of an erythroid cell model system critical for the study of the mechanism by which EP acts are: (1) high purity, (2) adequate cell number, (3) synchrony of differentiation, (4) responsiveness to the hormone, and (5) normal erythroid maturation. The FVA cells fulfill each aspect and represent the best available model to study the effects of EP on erythroid precursor cells as well as the events of erythroid maturation from the proerythroblast to the reticulocyte.

Thus far, there is no evidence that FVA infection results in an abnormal proerythroblast other than the shedding of FVA viral particles as these cells mature in culture. These cells are well suited for the study of receptors for EP, the mechanism of action of EP, the molecular events responsible for initiation of transcription and translation of erythroid-specific genes such as hemoglobin and components of the erythrocyte membrane and cytoskeleton, the assembly of the cytoskeleton, the process of enucleation, and remodeling of the reticulocyte into the erythrocyte. A number of these investigations have already begun in this laboratory. Hemoglobin and heme synthesis (M. J. Koury *et al.*, 1982, 1984, 1987a,b; M. J. Koury and Bondurant, 1988; Bondurant *et al.*, 1983), ^{45}Ca flux and intracellular Ca (Sawyer and Krantz, 1984; Thompson *et al.*, 1988), transferrin receptors and metabolism of iron (Sawyer and Krantz, 1986a), globin gene transcription (Bondurant *et al.*, 1985), development of the erythrocyte membrane (M. J. Koury *et al.*, 1986, 1987a,b) and enucleation (M. J. Koury *et al.*, 1987b; S. T. Koury *et al.*, 1988b,c) as well as the receptor for EP (Krantz and Goldwasser, 1984; Sawyer *et al.*, 1987b,c; Sasaki *et al.*, 1987b) have been studied in FVA-infected cells.

FVP-infected mice can also be used as a source of erythroid cells. Spleens from FVP-infected mice become even more enlarged than those from FVA-infected mice; however, FVP-infected spleens contain more mature erythroblasts and less proerythroblasts than FVA-infected spleens. Velocity sedimentation can be used to obtain proerythroblasts from the spleens of FVP-infected mice. However, these erythroid precursor cells are totally unresponsive to EP and spontaneously differentiate into reticulocytes in culture.

2.2.4. Mouse and Human CFU-E and EP-Responsive Cell Lines

A technique has been developed to generate a relatively synchronized population of erythroid cells from mice (Nijhof and Wierenga, 1983). Bleeding of the mice results in elevated EP levels, which drives erythropoiesis in this technique. However, the anemia is induced in mice recovering from thiamphenicol treatment, which greatly reduces the number of erythroid precursors. Therefore, EP acts on the progenitor cells not killed by thiamphenicol such that a wave of erythroid precursor cells develop. By precise timing and the use of centrifugal elutriation to purify a subpopulation of progenitor cells from the spleens of these mice, it is claimed that relatively pure CFU-E are recovered. It appears that cells synchronized in the cell cycle are recovered from these bled, thiamphenicol-treated mice since these cells go through two or three waves of thymidine incorporation. However, these cells are under the influence of very high levels of EP in the mouse such that the hormone is not required in the initial culture of these cells. This is not the case for CFU-E from normal mice and FVA cells. The initial lack of requirement for EP appears to be due to the *in vivo* saturation of receptors for EP with the hormone. An acid wash was required to remove bound EP from the cells so that [^{125}I]-EP binding could be fully expressed (Landschulz *et al.*, 1989). This finding limits the use of this system to study the mechanism of action of EP, and the low number of cells (10^7) (Nijhof *et al.*, 1984) generated per mouse requires careful planning to carry out biochemical experiments on the differentiation events in these cells. Nevertheless, this system is useful for the study of a number of events in erythroid maturation and to verify results found in virus-induced erythroid systems. Thus far, the metabolism of the receptors for EP in these regenerating CFU-E is similar to that found in FVA cells (Landschulz *et al.*, 1989).

A related strategy has been to culture human CFU-E (Sawada *et al.*, 1987). The observation that BFU-E are found in the circulation of normal human blood while the more mature precursors are absent was the basis for the partial purification of BFU-E and the resulting generation of CFU-E by culturing these BFU-E for 8 or 9 days. Twentyfold purification of blood BFU-E was accomplished by removal of adherent cells and cells that expressed antigens recognized by antibodies raised against granulocytes, monocytes, CFU-GM, and natural killer cells. The resulting cells were cultured in methylcellulose-containing medium for 8 or 9 days in the presence of conditioned medium and 2 U EP/ml. The precursor cells were then removed from culture and additionally purified by centrifugation over Ficoll–Hypaque and removal of adherent cells. The resulting cells are from 50 to 90% CFU-E as defined by the growth of colonies consisting of 8–49 hemoglobinized cells after 7 days in culture (Sawada *et al.*, 1987).

Since approximately 10^7 CFU-E are obtained from 450 ml of blood using this method, biochemical experiments are somewhat limited. In addition, since the cells are exposed to 2 U EP/ml in culture, studies on the mechanism of action of EP are complicated. However, this system is the only one available to study normal human CFU-E and offers the potential to study CFU-E from patients with diseases that may affect erythroid precursor cells such as polycythemia vera and Diamond–Blackfan syndrome. Recently, EP receptors were studied on CFU-E from a patient with polycythemia vera (Means *et al.*, 1989).

The ideal erythroid cell for studying the action of EP and erythroid maturation would be a continuous cell line that did not require EP and yet differentiated completely to an erythrocyte when the hormone was added. Such a cell line does not exist. If EP binding to a cell controlled the differentiation mechanism, a proliferating cell line might be discovered that was induced into differentiation by EP. On the other hand, if EP acts on the cells to maintain viability as the erythroid precursor cell proceeds through preprogrammed differentiation, one would expect to find cell lines that depended upon EP for survival. Cells lines have been reported that partially meet both criteria.

One must remember that erythroid maturation is a terminal process and a self-regenerating cell line that responded to EP in a normal mode would cease to exist as a result of terminal differentiation due to EP in trace levels in serum preparations used for tissue culture. It is probable that when erythroid cell lines are established, the cells that respond to EP by differentiation are selected against such that only cells that do not respond to EP at all or are only weakly responsive survive. If EP is added to the medium so as to establish cell lines that require EP for proliferation, cells responding by erythroid maturation will certainly not survive.

Most erythroleukemia cell lines that have been studied are those established from mice in the later stages of Friend disease (both FVA and FVP). These MEL cells are totally unresponsive to EP but differentiate when exposed to chemical inducers such as DMSO (Friend, *et al.*, 1971; Marks and Rifkind, 1978). However, there are reports of cell lines that are weakly responsive to EP or respond to the combination of EP and chemical inducers. DeBoth *et al.* (1978) established a cell line from MEL cells in DBA mice infected with Rauscher murine leukemia virus. These cells synthesized some hemoglobin when induced with either EP and DMSO but responded in a synergistic fashion when treated with both DMSO and EP. Sytkowski *et al.* (1980) reported that a subclone of these cells was more responsive to EP; however, the responsiveness of these cells is questionable as no substantial studies have followed this initial observation. Weiss *et al.* (1989) reported that another subclone of these cells was responsive to EP. This response was not detected in cell proliferation but as a twofold increase in RNA synthesis and twofold increase in total accumulation of iron.

EP-responsive cell lines have been established from erythroleukemias in mice infected with FVA. Mishina and Obinata (1985) have reported that the TSA8 cell line established by Shibuya and Mak (1983) becomes committed to EP-responsive, differentiation-competent CFU-E after DMSO treatment. However, others claim that the proliferation of TSA8 cells is stimulated by EP, but not differentiation (Fukamachi *et al.*, 1987b). Todokoro *et al.* (1987) have recently reported the establishment of an EP-responsive cell line (SKT6). EP apparently does not affect the proliferation of these cells but increases the number of cells synthesizing hemoglobin. It will be interesting to see if additional studies with SKT6 cells confirm these observations and quantitate the magnitude of the erythroid differentiation response to EP.

Hankins *et al.* (1987) have established several erythroleukemia cell lines (including HCD33, HCD57) from mice infected with the Friend helper virus free of SFFV. These cells are dependent upon EP for survival and proliferation but they differentiate only in the presence of a chemical inducer (hemin). Apparently, EP alone of the hematopoietic factors is sufficient for the survival of these cells as GM-CSF, IL-1, IL-2, or IL-3 could not substitute for EP. The morphology of these HCD33 and HCD57 cells resembles that of erythroblasts, which suggests a role for EP in maintenance of the erythroid precursor distinct from the differentiation program, which can be induced by hemin. If differentiation of SKT6 cells indeed can be induced with EP, it is possible that EP might act through a mechanism(s) that generates two distinct signals of cell proliferation and erythroid differentiation. However, this question is far from resolution, and these two possibilities cannot be distinguished in model systems that mimic normal erythropoiesis, i.e., FVA cells, mouse and human CFU-E. In these erythroid systems, cell proliferation and differentiation are inseparable such that the hypothesis that the total effect of EP is on the proliferation and viability of the cell, which in turn proceeds through preprogrammed differentiation, is more reasonable.

In contrast to the cell lines described above, some cell lines that are dependent on multilineage hematopoietic factors such as IL-3, GM-CSF, and G-CSF have been discovered to respond to EP. In the case of DA-1 cells, EP can substitute for the requirement of IL-3 or GM-CSF for the proliferation of these cells (Branch *et al.*, 1987; Tsao *et al.*, 1988; Sakaguchi *et al.*, 1987) but has no ability to induce erythroid differentiation. Other murine cell lines that are similar are IC-2 and FDC-P2 (Tsao *et al.*, 1988). The NSF-60 cell line is claimed to be a murine bipotential progenitor cell that can differentiate into neutrophils and macrophages in the presence of IL-3 and GM-CSF, or erythroblasts in the presence of EP, or remain in the promyelocytic stage when exposed to G-CSF (Hara *et al.*, 1988).

Human erythroleukemia cell lines have been established, and some have been reported to be slightly responsive to EP. K562 cells have been reported to have about ten receptors for EP per cell and respond to EP by a 6% increase in colony growth (Fraser *et al.*, 1988b); however, this result has not been confirmed and no receptors and no effect of EP on K562 cells were found in other laboratories (Todokoro *et al.*, 1988). Hitomi *et al.* (1988) have established a cell line from human erythroleukemia, JK1, in which 30% of the cells spontaneously differentiate. JK1 cells double their rate of proliferation when stimulated by EP but the hormone has no effect on differentiation of these erythroid cells.

3. DEMONSTRATION AND QUANTITATION OF RECEPTORS FOR ERYTHROPOIETIN

Radiolabeled EP, required to demonstrate binding sites for EP, and a population of EP-responsive cells to which EP could be bound have only been available within the last 5 years. Therefore, the quantity and quality of research in the field of EP receptors are somewhat lagging compared to other hormone and growth factor receptors.

Subsequent to the purification of human urinary EP, Goldwasser (1981) reported that iodination of EP either by oxidation and direct iodination of tyrosine residues or indirectly by cross-linking of iodinated compounds through N-terminal or amino side chains (Bolton–Hunter method) resulted in complete loss of biological activity. A fluorescent conjugate of EP was prepared that resulted in the labeling of 1–2% of bone marrow cells from anemic mice and rats (Weiss *et al.*, 1985); however, this result was not a conclusive demonstration of receptors since the fluorescent cells could not be determined to be erythroid.

The existence of surface receptors for EP was proposed from experiments showing that protease digestion of marrow cells and inhibition of protein synthesis obliterated the effect of EP but reversal of the inhibition of protein synthesis allowed the recovery of responsiveness to EP (Chang *et al.*, 1974). Others claimed that EP interacted through surface receptors since EP attached to beads by absorption or through lectins allowed some erythroid growth in culture surrounding these beads (Dukes and Einbinder, 1975; Roodman *et al.*, 1981). However, since the linkage of EP to a lectin or adsorption to a glass bead is not irreversible, the action of EP at the surface of the cell was not proven as the EP could have entered the cells after being displaced from the lectin by cell surface carbohydrate.

EP was tritiated by oxidation of the carbohydrate moiety of EP and subsequent reduction with tritiated borohydride, which resulted in [^{3}H]-EP with five ^{3}H atoms per EP molecule and full biological activity. This material was used to demonstrate specific binding of [^{3}H]-EP to the pure population of EP-responsive, FVA cells developed in the Hematology Research Laboratory at Vanderbilt (Krantz and Goldwasser, 1984). Unfortunately, the specific activity was so low that Krantz and Goldwasser were required to use extremely high cell numbers in very small volumes to detect binding of [^{3}H]-EP. This resulted in the aggregation of cells such that the number of binding sites was underestimated, the binding parameters were incorrectly measured, and the binding affinity measured was 10- to 60-fold less than subsequent experiments demonstrated (Sawyer *et al.*, 1987b).

Iodinated human urinary EP (^{125}I-EP) from Dr. Goldwasser's laboratory, which had been reported to be devoid of biological activity (Goldwasser, 1981), was found to bind to the FVA cells. Upon further investigation, we found that pure recombinant EP (AMGEN) could be iodinated and still retain full biological activity when assayed using the response of FVA cells to EP in an *in vitro* assay. Furthermore, by varying the iodination conditions, the extent of iodination was found to be inversely related to the ability of the hormone to bind to the receptor and the biological activity of ^{125}I-EP (see Figure 1). Under no condition tested was the binding of biologically inactive EP observed. Purified urinary EP (from E. Goldwasser) could also be iodinated and still retain greater than 90% of biological activity when

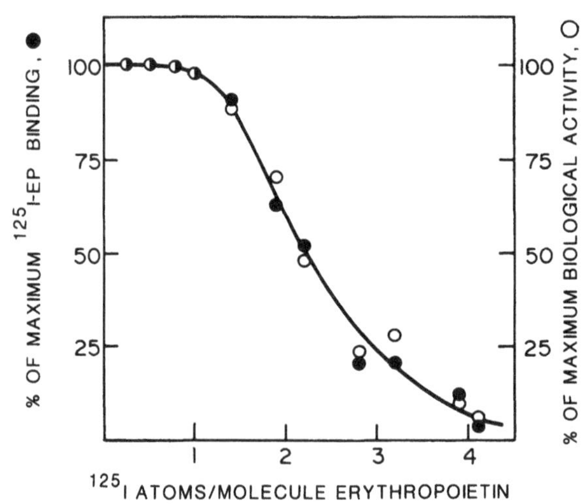

FIGURE 1. Effect of iodination on the biological activity of EP and ability of EP to bind to its receptor. Recombinant human EP was iodinated by oxidation of NaI for 1 to 15 min in the presence of 0.5 to 5.0 μg of Iodo-gen (Pierce Chemical Co.). These conditions resulted in ^{125}I-EP with the indicated ranges in the extent of iodination. ^{125}I-EP binding to FVA cells was quantitated as described previously (Sawyer *et al.*, 1987b). Biological activity of ^{125}I-EP was determined by measuring the incorporation of ^{59}Fe into hemoglobin in FVA cells cultured in the presence of ^{125}I-EP for 48 hr (Sawyer *et al.*, 1987a).

assayed in the *in vitro* FVA cell assay (Sawyer *et al.*, 1987a). ^{3}H-labeled urinary EP obtained from E. Goldwasser, metabolically labeled [^{35}S]-EP obtained from J. Browne at AMGEN, and ^{125}I-EP prepared from AMGEN recombinant human EP gave similar numbers of binding sites and binding affinities when the binding was simultaneously tested on FVA cells. Self-displacement analysis in this laboratory and others (Mayeux *et al.*, 1987b) has shown that ^{125}I-EP binds with the same affinity as native EP. It is not clear why the previous report found no activity in iodinated human urinary EP (Goldwasser, 1981); however, since the specific activity of the recombinant EP is manyfold higher than urinary EP, more atoms of ^{125}I may have been incorporated into urinary EP than calculated resulting in inactivation.

In contrast to the first report of [^{3}H]-EP binding to FVA cells, ^{125}I-EP binding to the surface of FVA cells at 0°C showed a nonlinear biphasic plot when the binding of increasing concentrations of EP was analyzed by the Scatchard method. This nonlinearity was interpreted as the existence of two classes of EP receptors that differ in the affinity for EP. FVA cells were found to have approximately 1000 receptors for EP of which 300 to 400 have a higher affinity for EP (K_d of 0.08 to 0.10 nM) while the remaining receptors have a lower affinity (K_d of 0.6 to 1.0 nM) (Sawyer *et al.*, 1987b,c). This biphasic Scatchard plot can also be interpreted as negative cooperativity in which the interaction of bound receptors with unbound receptors leads to the free receptors binding EP with lesser affinity. Negative cooperativity can be tested by examining the dissociation of ^{125}I-EP from occupied receptors in the presence and absence of an excess unlabeled EP. If there is negative cooperativity, ^{125}I-EP will be released faster in the presence of unlabeled EP. Experiments in this laboratory have shown that bound ^{125}I-EP is released from FVA cells at the same rate in the presence and absence of unlabeled EP, indicating no negative cooperativity.

During these experiments, the binding of ^{125}I-EP to Friend MEL cells, clone 745, was investigated as a control. The binding to these cells was not expected as they are totally unresponsive to EP. However, almost as many EP receptors were found on these cells as the FVA cells. These cells expressed only the lower-affinity EP receptors (K_d 0.5 to 1 nM) (Sawyer *et al.*, 1987b, 1989). Mayeux *et al.* (1987a) also reported that clone 745 MEL cells had 500 surface receptors with an affinity of 0.49 nM while Sasaki *et al.* (1987b) reported that a subline, 745A MEL cells, had 100 receptors with a binding constant of 0.21 nM. Because the MEL cells that have only low-affinity receptors are totally unresponsive to EP and because of the predominance of high-affinity receptors occupied on FVA cells at the EP concentration that gives the maximum biological activity, we have proposed that interaction of EP with the high-affinity receptors is necessary for the full biological effect of the hormone.

Thus far, erythroid cells that respond to EP by full differentiation and that show the two classes of EP receptors include fetal mouse liver cells (Fukamachi *et al.*, 1987a; Tojo *et al.*, 1987), mouse CFU-E induced by thiamphenicol treatment (Landschulz *et al.*, 1989), and human CFU-E derived *in vitro* from circulating BFU-E (Sawada *et al.*, 1988b) as well as the FVA cells. In addition, some cell lines that are partially responsive to EP also have higher- and lower-affinity receptors: TSA8 cells have higher-affinity receptors of K_d = 0.25nM and lower-affinity receptors of K_d = 6 nM (Fukamachi *et al.*, 1987b), and JK1 human erythroleukemia cells have higher-affinity receptors of K_d = 0.06 nM and lower-affinity receptors of K_d = 0.40 nM. These JK1 receptors have almost exactly the same affinity as the receptors in human CFU-E and FVA cells (Hitomi *et al.*, 1988). In addition, transplantable erythroleukemia cells (Hitomi *et al.*, 1989) and COS cells expressing the cloned receptor have both higher and lower affinity receptors (D'Andrea *et al.*, 1989). It is interesting that both the human and regenerating mouse CFU-E cells were initially reported to express only receptors for EP of the lower-affinity type (Krantz *et al.*, 1988; Landschulz and Boyer, 1988). This was due to the occupancy of the higher-affinity receptors with EP when the cells were recovered from a mouse with elevated EP or from tissue cultures in the presence of high concentrations of EP. Endogenous EP occupying higher-affinity receptors may account for the conflicting observations of only a single class of EP receptors in fetal mouse liver cells (Sasaki *et al.*, 1987b), fetal human liver cells (Pekonen *et al.*, 1987), anemic mouse spleen cells (Mufson and Gesner, 1987; Fraser *et al.*, 1988c), and fetal rat liver cells (Mayeux *et al.*, 1987b) since these cells are explanted from sources having high levels of erythropoiesis and therefore significant levels of EP. Contributing to this problem of endogenous EP being bound to the cell surface is the observation in FVA cells (Sawyer and Koury, 1987), mouse CFU-E (Landschulz *et al.*, 1989), and human CFU-E (Sawada *et al.*, 1989) that high-affinity receptors are rapidly downregulated after exposure to EP for as short a period as 2 hr and that only lower-affinity receptors are expressed during the maturation to reticulocytes. Therefore, erythroid cells explanted from tissues rich in EP may express only lower-affinity receptors because the higher-affinity class have been downregulated before the cells are harvested or the cells may be erythroblasts that have lost high-affinity receptors. In this regard, the FVA cell is well suited for study of the mechanism of actions of EP, including the receptor EP, since FVA proerythroblasts are isolated from mice that have low to normal levels of EP.

It should be noted that the number of receptors and affinities of the receptors for EP vary widely in different laboratories when binding of ^{125}I-EP to the same cells was measured. In Table I, note that a single class of receptors was found in FVA cells, fetal mouse liver cells, and TSA8 cells in the study by Sasaki *et al.* (1987b), yet other studies (Sawyer *et al.*, 1987b,c; Fukamachi *et al.*, 1987a,b) found two classes of higher- and lower-affinity receptors in these same cells. The dissociation constant for EP binding in Sasaki's study are either intermediate between higher- and lower-affinity receptors or less than the lower-affinity receptors—in particular the binding constants for fetal mouse liver cells. In this case, perhaps not enough data points were gathered to discriminate the existence of two classes of receptors.

Differences in the number of receptors and the binding constants between different laboratories are probably more the result of calculations than experimental conditions. This is particularly true for the earlier publications using AMGen EP, which is still only available in units of EP activity. Initially, the specific activity of AMGen EP was considered a trade secret. As discussed in an earlier section, reports of the specific activity and molecular weight of EP vary widely from urinary EP to recombinant EP such that up to a threefold difference in the dissociation constant and number of receptors can occur if one uses the molecular weight and specific activity for urinary EP to calculate the binding parameters for recombinant EP. In virtually all these reports on EP receptors, there is no information as to the molecular weight or specific activity of EP used for the calculations.

In cells having only the lower-affinity receptors for EP, the binding of EP to these receptors can still signal the cell to proliferate. This is also the case for FVA cells, mouse CFU-E, and human CFU-E, which quickly downregulate high-affinity receptors but still

Table I
Properties of the Receptor for EP

Source	Binding constant K_d (pM)	No. of receptors (higher affinity/lower affinity) per cell	Structure of receptor: No. of subunits	Structure of receptor: Mass (kDa)	Reference
Primary cells					
FVA cells	90/570	300/700	2	100/85	Sawyer *et al.* (1987b,c)
FVA cells[a]	240	180	2[b]	105/85	Sasaki *et al.* (1987b)
Mouse CFU-E	95/655	700/1000	2	104/89	Landschulz *et al.* (1989)
Human CFU-E	100/570	200/850	2	100/90	Sawada *et al.* (1988b); Sawyer *et al.* (1989)
Fetal mouse	410/3130	920/300	2	110/95	Fukamachi *et al.* (1987a); Tojo *et al.* (1987)
	140	420	2[b]	105/85	Sasaki *et al.* (1987b)
Fetal rat	180	517	2[b]	94/78	Mayeux *et al.* (1987b)
Mouse spleen	490	504	—	—	Mufson and Gesner (1987)
	70	450	—	—	Fraser *et al.* (1988c)
Fetal human	160	16	2	86/41	Pekonen *et al.* (1987)
Human bone marrow	~300	30	—	—	Fraser *et al.* (1988a)
EP-responsive MEL cells, precursorlike cell lines					
SKT6	150	470	3	109/94/63	Todokoro *et al.* (1987)
TSA8	250/6100	90/590	—	—	Fukamachi *et al.* (1987b)
Rauscher Red (5-1.5)	440	1690	2	105/95	Broudy *et al.* (1988)
Progenitorlike cell lines					
DA-1	540	131	—	—	Sakaguchi *et al.* (1988)
	1160	50	—	—	Tsao *et al.* (1988)
IC2	840	20	—	—	Tsao *et al.* (1988)
FDCP2	1390	20	—	—	Tsao *et al.* (1988)
HCD33; HCD57	550	4000	2	100/85	Sawyer and Hankins (1988)
EP-responsive human cell lines					
JK1	60/400	110/200	2[b]	105/85	Hitomi *et al.* (1988)
K562	270	4	—	—	Fraser *et al.* (1988b)
	—	0	—	—	Todokoro *et al.* (1988)

(Continued)

Table I (*Continued*)

Source	Binding constant K_d (pM)	No. of receptors (higher affinity/lower affinity) per cell	Structure of receptor: No. of subunits	Structure of receptor: Mass (kDa)	Reference
EP-unresponsive MEL cells					
745	490	504	—	—	Mayeux *et al.* (1987a); Sawyer *et al.* (1987b)
	800	760	2	100/85	
745A	210	113	2[b]	75/63	Sasaki *et al.* (1987b)
B8	1200	350	2	109/94	Tojo *et al.* (1988)
K-1	280	930	1	63	Todokoro *et al.* (1988)
T3Cl	620	690	1	63	Todokoro *et al.* (1988)
GM86(745)	270	110	1	63	Todokoro *et al.* (1988)
707	780	120	1	63	Todokoro *et al.* (1988)
Rauscher Red (5-1.5a)	550	2700	—	—	Broudy *et al.* (1988)
GM979	660	1635	—	—	Broudy *et al.* (1988)
EP-unresponsive human cell lines					
OCIM1	280	3000	—	—	Broudy *et al.* (1988)
HEL	290	34	—	—	Fraser *et al.* (1988a)
HEL	—	0	—	—	Todokoro *et al.* (1988)
Nonerythroid tissues					
Mouse placenta	1000	<100[c]	2	100/85	Sawyer *et al.* (1989)
Rat placenta	900	<100[c]	2	100/85	Sawyer *et al.* (1989)
Rat megakaryocytes	287	6500	—	—	Fraser *et al.* (1989)

[a]Unseparated mixture of cells.
[b]Claims of disulfide bridge between subunits.
[c]Binding to membranes only.

require EP for development when only the lower-affinity receptors are expressed. The effect of EP through the lower-affinity receptors is also shown in the HCD33 and HCD57 cells, which are totally dependent on EP for survival and proliferation and yet only express lower-affinity receptors (Sawyer and Hankins, 1988). Similarly, in DA-1, IC2, and FDCP2 cells, which only have the lower-affinity receptor, EP can substitute for the requirement of other hematological factors.

The number of receptors for EP on the surface of erythroid precursor cells most likely parallels the responsiveness of the cells to EP. CFU-E cells at the proerythroblast stage of development apparently have the greatest number of receptors (~ 1000 receptors/cell), and the number of receptors drops substantially as the cell matures to erythroblasts and mature erythroblasts while becoming less dependent on EP at each successive stage. In the FVA cell system, the total number of receptors drops to half when the cells become basophilic erythroblasts after culture for 24 hr. At this stage, these basophilic FVA erythroblasts are no longer dependent on EP for maturation (M. J. Koury and Bondurant, 1988). EP receptors are rapidly lost during the next 24 hr of culture as the basophilic FVA erythroblasts mature to reticulocytes, which have no receptors. This study was carried out by measuring total receptors and grain distribution by autoradiography (Sawyer and Koury, 1987). These results have also been seen in regenerating mouse CFU-E (Landschulz *et al.*, 1989) and human CFU-E (Sawada *et al.*, 1987). All of these findings in cell systems generally agree with the finding of Fraser *et al.* (1988b,c) who used autoradiography of bound ^{125}I-EP to estimate the relative distributions of receptors among erythroid cells at different stages of maturation in the cells from the spleens of anemic mice and normal human bone marrow.

It is not known how the number of receptors for EP change when the earliest erythroid progenitors differentiate into CFU-E and proerythroblasts. However, a study by Sawada *et al.* (1990) indicates very low numbers of EP receptors on purified BFU-E and an increase in receptor number as these BFU-E mature to CFU-E in culture. The low number of receptors for EP on cell lines that are multipotent such as DA-1, IC2, and FDCP2 (Sakaguchi *et al.*, 1987; Tsao *et al.*, 1988) reinforce the idea that the earliest erythroid cells have a low number of receptors. The finding that some MEL cells increase the number of receptors when induced by chemicals has been interpreted as the shift of an earlier progenitor to a more mature erythroid stage (Sasaki *et al.*, 1987b; Tojo *et al.*, 1988; Broudy *et al.*, 1988). This finding with MEL cells must be viewed with caution since the physiology of erythroleukemia cells and the normal erythroid precursors are quite distinct.

4. ENDOCYTOSIS OF ERYTHROPOIETIN AND METABOLISM OF THE RECEPTOR

Binding of ^{125}I-EP to FVA cells was observed to be twofold greater at 37°C than 0°C. Since this was typical for ligands that are internalized into the cell, the possibility that ^{125}I-EP was internalized by receptor-mediated endocytosis was pursued. Indeed, removal of surface-bound EP by a high-salt, pH 2.5 wash or by digestion with proteases demonstrated endocytosis of EP within 1 min of adding ^{125}I-EP to the cells. Initially, surface-bound EP accumulated within the first 15 min after treatment with ^{125}I-EP. After this time, surface-bound ^{125}I-EP declined slightly as radioactivity in the cell interior became greater than the surface-bound hormone. Internalized radioactivity continued to rise in a linear fashion from the initiation of the experiment until about 60–90 min when a steady-state distribution between surface bound EP and internalized radioactivity was reached.

^{125}I-EP was bound to the surface at 0°C, and the distribution of radioactivity between the cell surface, cell interior, and medium was examined at intervals after the FVA cells were washed and then warmed to 37°C. During the first 30 min, surface-bound ^{125}I-EP declined as radioactivity accumulated inside the cell. After 30 min, the radioactivity inside the cell declined as radioactivity was secreted into the medium. Analysis of the medium revealed that the majority of radioactivity was [^{125}I]iodotyrosine, which was the product of the degradation of ^{125}I-EP within the cell. The remaining radioactivity was free iodine and undegraded EP that had presumably been released intact from the cell surface. Inhibition of lysosomal function with NH_4Cl and chloroquine greatly reduced the degradation of internalized ^{125}I-EP, which suggested the degradation of EP in lysosomes after endocytosis. All of the above studies were performed by Sawyer *et al.* (1987b) and were repeated in human CFU-E by Sawada *et al.* (1988b). Reports from Mufson and Gesner (1987) documented the internalization of [^{35}S]-EP in erythroid cells from the spleens of anemic mice, and Fukamachi *et al.* (1987b) demonstrated internalization of ^{125}I-EP in TSA8 MEL cells and fetal mouse liver cells. Mufson and Gesner (1987) reported that ^{125}I-EP bound to the receptor for EP but the ^{125}I-EP was not endocytosed; however, this is completely contradictory to the other reports using ^{125}I-EP as well as our studies showing that ^{125}I-EP, [^{3}H]-EP, and [^{35}S]-EP were equally endocytosed.

The fate of the receptor for EP following the receptor-mediated endocytosis of ^{125}I-EP into the cell is not known. By analogy to the metabolism of other growth factors and hormones, the EP receptor will probably follow one or both of two pathways within the cell. In the first pathway, both the receptor and the hormone are degraded in the lysosomes following endocytosis into the cell. This is the fate of the epidermal growth factor receptor (Carpenter and Cohen, 1979). In the second pathway, the receptor can recycle to the surface to bind another ligand after the initial endocytosis of the receptor–ligand complex. In the case of the transferrin receptor, both the ligand and the receptor are recycled back to the surface of the cell following the internalization of the receptor and iron-loaded transferrin complex. Shortly after the endocytosis, the receptor–transferrin complex enters an acidic compartment distinct from the lysosomes (endosome) that has a pH low enough to dissociate the iron from transferrin. Apotransferrin has a higher affinity for the receptor at this low pH and remains bound to the receptor when the latter recycles to the cell surface (Dautry-Varsat *et al.*, 1983; Klausner *et al.*, 1983). Insulin and low-density lipoprotein are degraded when endocytosed but the receptor is recycled to the cell surface (reviewed by Goldstein *et al.*, 1985). The acidic nature of endosomes typically results in dissociation of peptide hormones and growth factors from the receptors. Thus, the receptor can recycle when the ligand is degraded in the lysosomes. Since bound EP is released from the receptor at or below pH 4.0, it is possible that the fate of EP and the receptor diverge following internalization into an acidic endosome, with the receptor recycling to the surface to the surface even though the EP is degraded.

In the case of a receptor that is degraded following endocytosis such as the receptor for epidermal growth factor, the number of surface receptors is downregulated to a much lower (10% to 20% of initial) steady-state level. An equilibrium is reached between new synthesis of receptors and degradation following treatment of the cells with the growth factor. When FVA cells, fetal mouse liver cells, or TSA8 cells were exposed to ^{125}I-EP, the ^{125}I-EP that bound to the surface of the cells increased during the first 30 min and then fell to 75% to 50% of the maximum by 2 hr. These experiments are inconclusive since the downregulation is small; and in the case of FVA and fetal mouse liver cells, erythroid maturation could account for the loss of receptors. In the case of FVA cells, mouse regenerating CFU-E, and fetal mouse liver cells, downregulation of higher-affinity receptors occurs to a greater extent than lower-affinity receptors for EP (Sawyer *et al.*, 1987b; Sawyer and Koury, 1987; Fukamachi *et al.*, 1987b; Landschulz *et al.*, 1989). Since the lower-affinity receptors are endocytosed equally well as the higher-affinity receptors (Sawyer *et al.*, 1989), these studies suggest the possibility of recycling of lower-affinity receptors and the degradation of higher-affinity receptors.

Our recent experiments in this laboratory show evidence of EP receptor degradation as the mechanism of downregulation of the receptor in HCD33 and HCD57 cells exposed to EP. These cells have only the lower-affinity receptors for EP but require EP for survival and proliferation. HCD33 and HCD57 cells do not differentiate in response to EP. When these cells are taken from cultures that contain EP, they have 400 to 600 receptors for EP per cell. However, the receptor number increases five- to tenfold when EP is removed from the culture. Upon subsequent reculture in the presence of ^{125}I-EP, the receptor number is downregulated to 10–20% by 24 hr in culture. The recovery of EP receptors after the downregulation requires protein synthesis. This result is consistent with the downregulation of EP receptors through degradation and the subsequent upregulation of receptors when EP is removed from the environment through synthesis of new receptors (Sawyer and Hankins, 1988).

More work is required to further understand the metabolism of both the receptor for EP and EP itself. We are beginning studies on the localization of ^{125}I-EP in cells using autoradiography at the electron microscope level. Localization of the receptor will await the production of antibodies that recognize the receptor. As the mechanism of action of EP is not known, these studies are of interest in determining the possibility of a direct translocation of EP or the receptor or a proteolytic fragment of either EP or the receptor to the nucleus of the cell.

5. STRUCTURE OF THE ERYTHROPOIETIN RECEPTOR

Cross-linking of ^{125}I-EP bound to plasma membranes from FVA cells using bifunctional cross-linkers first identified two proteins as the receptor for EP (Sawyer *et al.*, 1987c). These two ^{125}I-EP cross-linked proteins are bands of 140 and 125 kDa on SDS–PAGE. Subtraction of the molecular mass of EP in this SDS–PAGE system (40 kDa) from the molecular mass of the cross-linked protein results in apparent molecular masses of 100 and 85 kDa for the receptor proteins. The molecular mass of EP by SDS–PAGE is higher, 40 kDa, than the true molecular mass, 30 kDa, due to the extensive glycosylation. Therefore, the determination of molecular mass of the receptor for EP by cross-linking is subject to uncertainty due to the anomalous migration of ^{125}I-EP and possible anomalous migration of the cross-linked receptors on SDS–PAGE. Our recent results with a radiolabeled, photoactivatable, cleavable cross-linker derivatized to EP have shown that the true molecular masses of these two proteins are 100 and 90 kDa (Hosoi *et al.*, 1989). This experiment involved the transfer of the radiolabeled, photoactive group to the receptor proteins followed by the cleavage of the cross-linker molecule so that EP was no longer cross-linked to the radiolabeled receptor. Therefore, the initial cross-linking studies with bifunctional cross-linkers gave a very close estimation of the molecular mass of the EP receptor. Figure 2 shows the identical structure of the ^{125}I-EP cross-linked receptors in FVA cells, FVP cells, MEL clone 745 cells, and placentas from mice and rats.

The two receptor proteins labeled by cross-linking to ^{125}I-EP are not disulfide-bridged subunits of a larger receptor complex (Sawyer *et al.*, 1987c). Some workers have reported that these subunits are connected by disulfide bridges such that the receptor is so large a molecule that it does not enter the gel in SDS–PAGE (Sasaki *et al.*, 1987b; Mayeux *et al.*, 1987b). However, a number of reports show no disulfide bridging of these subunits. We initially reported in an abstract (Sawyer and Krantz, 1986b) the same apparent high-molecular-weight ^{125}I-EP cross-linked receptor when SDS–PAGE was carried out in the absence of reducing agents; however, this observation was an artifact of

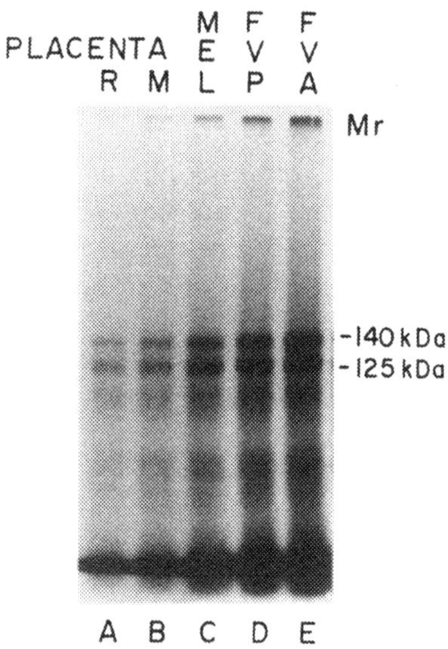

FIGURE 2. Cross-linking of ^{125}I-EP to the receptor in membranes from FVA cells, FVP cells, MEL clone 745 cells, and placentas from mice and rats. Plasma membrane protein (500 μg) was incubated with 10 U ^{125}I-EP/ml for 15 min at 37°C. Disuccinylmidyl suberate was added to a final concentration of 0.5 mM and incubated for 15 min at 0°C. The reaction was quenched with 150 mM Tris-HCl and unbound ^{125}I-EP was washed away. The membranes were dissolved in sample buffer and run on a 5% acrylamide, SDS–PAGE. The dried gel was exposed to X-ray film for 2 to 4 weeks. Lane A, rat placenta; B, mouse placenta; C, MEL clone 745 cells; D, FVP cells (immature proerythroblasts from the spleens of mice infected with FVP); E, FVA cells.

incomplete solubilization of the cross-linked membranes such that free ^{125}I-EP as well as cross-linked ^{125}I-EP were trapped at the top of the stacking gel and running gel (Sawyer, 1989). It seems virtually certain that the reports of a large EP receptor composed of disulfide-bridged subunits are the result of this same artifact.

Subsequent to our report of the receptor for EP in FVA cells as proteins of 100 and 85 kDa, others reported a similar structure in fetal mouse liver erythroid cells and cell lines (Sasaki *et al.*, 1987b; Tojo *et al.*, 1987, 1988; Hitomi *et al.*, 1988; Broudy *et al.*, 1988). Todokoro *et al.* (1987, 1988) reported the receptor in SKT6 cells as proteins of 109, 94, and 63 kDa with only the 63-kDa protein in T3C1, K-1, GM86 (MEL clone 745), and 707 cells, which are not responsive to EP. These authors claimed that the lack of EP responsiveness in these cells was due to a mutant EP receptor; however, EP receptors on MEL cells, clone 745, in our laboratory have the identical cross-linked structure as FVA cells, i.e., 100- and 85-kDa proteins. Broudy *et al.* (1988) and Tojo *et al.* (1988) have also reported the same structure for the EP receptor in MEL cells not responsive to EP. Mayeux *et al.* (1987b) reported the EP receptor as proteins of 94 and 78 kDa in rat fetal liver erythroid cells, while we found 100- and 85-kDa proteins in rat placenta (and mouse placenta) (see Figure 2) (Sawyer *et al.*, 1989). In the same study, the receptor of a normal human CFU-E was identified as proteins of 100 and 90 kDa, which is in close agreement with the report of Hitomi *et al.* (1988) on the receptor in JK1 human erythroleukemia cells. In contrast to all of these studies, Pekonen *et al.* (1987) found two proteins of 86 and 41 kDa cross-linked to ^{125}I-EP human fetal tissue. In view of the sensitivity of the EP receptor to endogenous proteinase activity (Sawyer *et al.*, 1987b) and the generation of lower-molecular-mass proteolytic fragments of the cross-linked receptors when rigorous steps are not taken to control proteinase activity (Sawyer *et al.*, 1988; Sawyer, 1989), the EP receptor is likely to exist as two proteins of approximately 100 to 90 kDa in all tissues. Reports of lower-molecular-mass EP receptors are most likely artifacts due to endogenous proteinase activities.

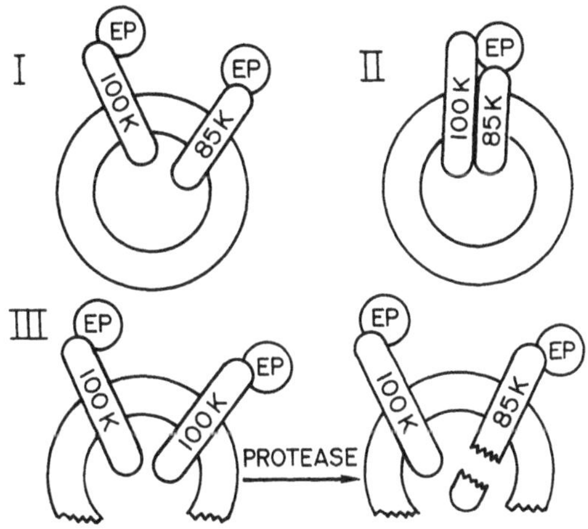

FIGURE 3. Models illustrating possible relationship of the 100- and 85-kDa proteins of the EP receptor. Models are described in detail in the text. Model I, two separate EP binding proteins; model II, dimer of 100- and 85-kDa proteins; model III, artifactual generation of 85-kDa protein by endogenous membrane proteinase activity.

Proteolytic peptide mapping by the Cleveland method showed that the 100- and 85-kDa proteins of the EP receptor have similar if not identical amino acid sequences (Sawyer *et al.*, 1988; Sawyer, 1989). Examination of the glycosylation of the two proteins demonstrated no detectable carbohydrate on either (Hosoi *et al.*, 1988; Sawyer *et al.*, 1988). Therefore, differential glycosylation of a single peptide cannot explain the differences in molecular mass of these two proteins. Rather, differential processing or posttranslational modification other than glycosylation by the cell may result in the expression of these two proteins from a common gene. Alternatively, the 85-kDa protein may be a laboratory artifact resulting from the cleavage of the 100-kDa receptor during the binding and cross-linking of ^{125}I-EP. However, the labeling of the 100- and 85-kDa proteins by ^{125}I-EP in cross-linking occurs in a 1 : 1 ratio when rigorous measures are taken to control proteinase activity. We have considered the possibility that the functional receptor for EP is a dimer of 100- and 85-kDa proteins. These three possibilities are considered in the models illustrated in Figure 3.

Model I shows the differential processing of a single gene or the expression of two closely related genes resulting in two distinct receptors for EP—one of 100 kDa and the other of 85 kDa. The existence of higher- and lower-affinity receptors for EP does not correlate with the presence of two receptor proteins. Cross-linking EP to these two proteins is not affected by varying the concentration of EP and consequently the ratio of higher- and lower-affinity receptors occupied. In addition, both proteins are equally labeled by ^{125}I-EP when cross-linking is carried out on cells that have only lower-affinity receptors (Sawyer *et al.*, 1987c, 1988, 1989).

Model II illustrates the possibility that a dimer of the 100- and 85-kDa proteins forms the functional receptor for EP. In this model, the 100- and 85-kDa proteins could arise from a common gene or two closely related genes. An analogous situation is the reverse transcriptase from avian retrovirus. The reverse transcriptase is composed of two subunits; one is a cleaved product of the other.

The third model appears to be the most likely possibility. In this model, the EP receptor is a single gene product of 100 kDa. During preparation of membranes, binding of ^{125}I-EP, cross-linking of ^{125}I-EP, or processing for SDS–PAGE, proteinase activity cleaves a 10- to 15-kDa fragment from the receptor. Similar cleavages have been found to occur in the receptors for epidermal growth factor, insulin, and others. It is equally possible that the cleaved protein (85 kDa) continues to bind EP with the same affinity or that only the intact 100-kDa protein can bind EP. The evidence supporting this model are the identical peptide maps from both 100- and 85-kDa proteins and the increased ratio of 100- to 85-kDa protein when additional inhibitors of proteases are included (Sawyer, 1989).

Alan D'Andrea, working in Harvey Lodish's laboratory at MIT, recently reported the cloning of an EP receptor (D'Andrea *et al.*, 1989). The cloning strategy was to make a cDNA library from MEL clone 745 cells that express receptors for EP, transfect COS cells with the cDNA, and screen and transfected COS cells for ^{125}I-EP binding. The approach requires that the receptor be encoded by a single gene, and the apparent success of the cloning eliminates the two-gene hypothesis from consideration. Screening of the COS cells led to the selection of cells that expressed thousands of both higher- and lower-affinity EP receptors of similar binding constants as the FVA cells. It is interesting that the cDNA is from a cell line that expresses only lower-affinity receptors (Mayeux *et al.*, 1987a; Sawyer *et al.*, 1987b; Sasaki *et al.*, 1987b). Cross-linking of ^{125}I-EP to the surface of these transfected COS cells results in cross-linked bands of 140 and 100 kDa, which is similar to but distinct from the receptor in the MEL 745 cells. Using the cDNA as a probe, mRNA of 2.1 kb was found expressed in erythroid cells but not nonerythroid tissues. Very surprisingly, the cDNA codes for a 55-kDa protein when mRNA is translated in an *in vitro* expression system. Sequencing the cDNA results in a putative amino acid sequence characteristic of a membrane protein: 23 hydrophobic amino acids apparently make up the membrane-spanning domain near the middle of the protein such that 213 amino acids make up the extracellular domain and 236 amino acids make up the cytoplasmic domain. Only a single asparagine is found in the extracellular domain as a possible site of N-linked glycosylation. The limited possible sites for glycosylation and our finding of no detectable carbohydrate in the EP receptor (Hosoi *et al.*, 1988; Sawyer *et al.*, 1988) seems to eliminate extensive glycosylation from the list of posttranslational modifications that might increase the size of the receptor from 55 kDa to 100 kDa.

No relation has been found between the 55-kDa putative receptor for EP and the 100-kDa protein that is cross-linked to ^{125}I-EP. Until biochemical or immunological methods demonstrate this relation, it is conceivable that the cloned 55-kDa protein acts as a factor that induces the expression of the EP receptor in COS cells. It is more likely that the 55-kDa protein is posttranslationally modified into a larger protein, perhaps by being covalently linked into a dimer with another 55-kDa protein or another protein expressed in COS cells. This linkage does not appear to be a disulfide bridge as all cross-linking studies carried out in the presence of reducing agents such as β-mercaptoethanol and dithiothreitol have demonstrated the EP receptor as the 100-kDa protein. In addition, the cross-linking of two closely situated 55-kDa subunits of the receptor to ^{125}I-EP simultaneously cannot explain the results. Cross-linking is limited such that less than 1% of bound ^{125}I-EP is cross-linked and therefore a similar percentage would be expected for neighboring proteins. It is expected that only 0.01% of bound ^{125}I-EP would be cross-linked to two proteins simultaneously. In addition, the use of a monofunctional cross-linker derivatized to EP also results in only the 100- and 90-kDa proteins labeled. It is impossible for this molecule to cross-link two 55-kDa proteins together.

It is interesting that the cloned putative EP receptor has some homology with the IL-2, IL-4, and IL-6 receptors. This would seemingly eliminate protein kinase activity in the receptor as a possible mechanism through which EP signals the cell to maintain or begin erythroid differentiation and suggests that EP may act through a similar but as yet unknown mechanism as those other receptors.

6. MECHANISM OF ACTION OF ERYTHROPOIETIN

EP interaction with its target cell occurs through surface receptors. It is possible that the binding of EP to the receptor on the cell surface results in a signal transduced to the interior of the cell. By analogy with other hormones, this signal might be the activation of nucleotide cyclases resulting in elevated levels of cAMP or cGMP; hydrolysis of phosphatidylinositol lipids resulting in increased intracellular calcium and activation of protein kinase C by diacylglycerol; or activation of a protein kinase such that a class of regulatory proteins become phosphorylated. In these cases, the endocytosis and degradation of EP may be to remove the signal at the cell surface. Since EP is required continuously in the cell's environment to maintain proliferation and erythroid maturation from the mature BFU-E stage until the basophilic erythroblast stage, the cell may depend on signals from freshly bound EP since previously bound EP is subsequently degraded and no longer can transmit a signal. In this way, fluctuations in EP levels are modulated as increased EP over an extended period is required for the maximum effect.

As for most other hormone receptors, an excess of receptors for EP are found on the cell surface. As illustrated in Figure 4, the maximum biological response to EP occurs with only a fraction of total receptors occupied (30%). Even the higher-affinity receptors for EP are only partially occupied with EP concentrations sufficient for the complete maturation of erythroid precursor cells. In the FVA cells, an EP concentration that results in greater than 90% of the maximum hemoglobin synthesis is only sufficient to occupy 50 to 100 surface receptors when the binding is carried out at 0°C. It is possible that a subpopulation of even higher-affinity receptors exist at 37°C but have not been detected

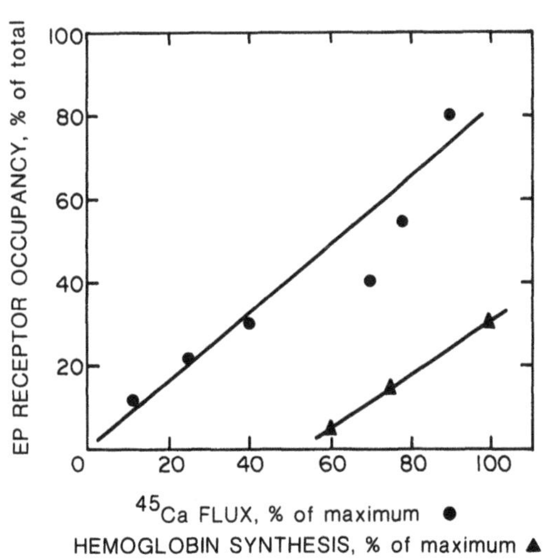

FIGURE 4. Relationship of the occupancy of the receptor for EP on FVA cells with erythroid differentiation as measured by hemoglobin synthesis and ^{45}Ca flux across the plasma membrane. ^{59}Fe incorporation into hemoglobin (Sawyer *et al.*, 1987a) and ^{45}Ca uptake (Sawyer and Krantz, 1984) were measured by FVA cells treated with a range of concentrations of EP. The results are plotted against the number of surface receptors occupied when the same concentration of ^{125}I-EP was used to determine the binding of the hormone.

with the available reagents. These results suggest that the EP signal to the cell interior may be one that is amplified since the number of receptors occupied is very small.

The percentage of EP receptors occupied on the erythroid precursor cells has no effect on the kinetics of erythroid maturation. The number of cells that successfully complete maturation depends on the concentration of EP in the environment and presumably on the number of molecules of EP bound to surface receptors. Heterogeneity of EP receptor number in the population of cells or changes in receptor density with the cell cycle may determine which cells mature under normal physiological levels of EP. The physiological EP concentrations of 10–30 mU/ml plasma result in only 10–20% maturation of proerythroblasts in culture (M. J. Koury and Bondurant, 1988). When FVA proerythroblasts are cultured in very low levels of EP or in the absence of EP, the cells lose their ability to respond to EP before the complete loss of cell viability. Under these conditions, the loss of ability to respond to EP correlates with the loss of higher-affinity receptors. FVA cells that survive 24 hr in the absence of EP have half the total number of EP receptors as the starting cells but no detectable higher-affinity receptors for EP (Sawyer and Koury, 1987). Only 10% of FVA cells deprived of EP for 24 hr can reach maturity when EP is restored to the culture (M. J. Koury and Bondurant, 1988). It is not understood why erythroid precursor cells die in the absence of EP; however, the first observable effect on FVA cells deprived of EP is the breakdown of DNA after 3–4 hr followed by a decline in metabolic activity between 6 and 18 hr later. The breakdown of DNA results in ineffective synthesis such that [^{3}H]thymidine incorporation in cells deprived of EP exceeds the incorporation of this labeled compound into DNA in cells proliferating in response to EP (M. J. Koury and Bondurant, 1988).

EP has been reported to stimulate RNA synthesis in a number of models. In FVA proerythroblasts, the first effect of EP seen is a change in the flux of ^{45}Ca within 1 min (Sawyer and Krantz, 1984) followed by increased glucose uptake between 2 and 4 hr (M. J. Koury *et al.*, 1987b) and increased synthesis of all classes of RNA by 6 hr. Synthesis of protein and DNA are not substantially increased by EP but are maintained at a steady level for the initial hours of culture (M. J. Koury and Bondurant, 1988). Erythroid-specific events such as the initiation of globin gene transcription (Bondurant *et al.*, 1985) and increased synthesis of transferrin receptors (Sawyer and Krantz, 1986a) begin after 6 hr in the presence of EP. These data do not provide a great deal of insight into the mechanism of action of EP.

The observation of increased ^{45}Ca uptake in FVA cells (Sawyer and Krantz, 1984) suggested that EP might influence the cell through elevation of the intracellular concentration of calcium. However, EP also stimulated the efflux of ^{45}Ca from FVA cells. Investigations on the intracellular calcium concentration in the FVA cells using the fluorescent chelators Quin-2 and Fura showed no effect of EP. Only in the presence of an analogue of cAMP did exposure to EP lower the intracellular calcium concentration (Thompson *et al.*, 1987). This probably indicates the activation of a calcium pump by EP and suggest a mechanism for the increased flux of ^{45}Ca in these cells that was observed previously. Figure 4 illustrates that the effect of EP on the calcium flux across the membrane directly correlates with the binding of EP to receptors, and that this flux of calcium requires much more binding of EP than that required for erythroid maturation. Moreover, no effect of EP was found on the hydrolysis of phosphorylated phosphatidylinositol phospholipids in FVA cells. The head group metabolites, inositol phosphates IP_3 and IP_4, have been shown to

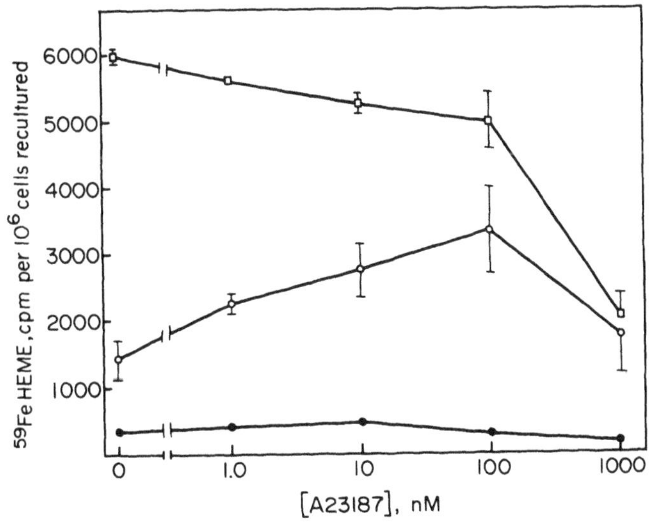

FIGURE 5. Enhancement of EP-dependent FVA cell erythroid maturation by calcium ionophore A23187. FVA cells were cultured in the presence of the indicated concentration of A23187 and no EP (●), 0.02 U EP/ml (○), and 1.0 U EP/ml (□) for 48 hr. Erythroid maturation was quantitated by hemoglobin synthesis. ^{59}Fe incorporation into hemoglobin was determined during the interval from 24 hr to 48 hr of culture.

release calcium from internal stores, and diacylglycerol activates protein kinase C activity. In this regard, phorbol esters, which directly activate protein kinase C activity, had no influence on the erythroid maturation of FVA cells, although these agents blocked chemical induction of MEL cell differentiation.

Calcium ion requirements have been reported for the normal development of CFU-E (Misti and Spivak, 1979). These authors also reported that calcium ionophores increase the number of CFU-E that develop by 25% although the ionophores had no effect in the absence of EP. Figure 5 shows that ionophore A23187 had no effect on hemoglobin synthesis in FVA cells in the absence of EP or in the presence of a high level of EP that produced a maximum response. However, A23187 enhanced the hemoglobin synthesis when a suboptimal level of EP was present in the culture. This result suggests that increased intracellular calcium may enhance the effect of EP but cannot substitute for the mechanism of action of EP.

In MEL cells, chemical inducers of differentiation such as DMSO apparently act at least partially through an increased intracellular calcium concentration. Extracellular calcium is required during the exposure to an inducer for commitment to differentiation to occur; and the lag between induction and commitment can be manipulated by altering the calcium in the medium (Bridges *et al.*, 1981). Inhibition of ion pumps and exchangers has resulted in the hypothesis that DMSO treatment damages the sodium/potassium pump such that calcium leaks into the cell from the medium through the sodium/calcium antiport system (Smith *et al.*, 1982; Levenson *et al.*, 1980). In addition, DMSO may damage mitochondria such that calcium is released into the cytoplasm from mitochondrial stores (Levenson *et al.*, 1982). These results do not imply that the mechanism by which EP acts necessitates the rise in intracellular calcium. Indeed, MEL cells, the majority of

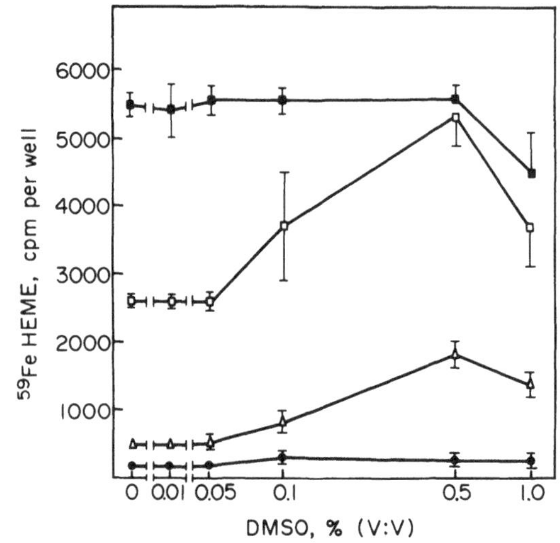

FIGURE 6. Enhancement of EP-dependent FVA cell erythroid maturation by dimethylsulfoxide (DMSO). FVA cells were cultured with the indicated concentration of DMSO and no EP (●), 0.005 U EP/ml (△), 0.05 U EP/ml (□), and 1.0 U EP/ml (■) for 48 hr. Erythroid maturation was quantitated by hemoglobin synthesis. ^{59}Fe incorporation into hemoglobin was determined during the interval from 24 hr to 48 hr of culture.

which are not responsive to EP, may have the normal mechanism by which EP acts fully activated at all times. Since differentiating cells are removed from the cell population, only cells that are blocked in differentiation survive. The intracellular level of calcium seems to be the rate-limiting step of differentiation of these MEL cells. FVA cells could have a higher intracellular concentration of calcium than MEL cells and thus the mechanism by which EP acts does not include a rise in intracellular calcium.

The effect of DMSO on FVA cell maturation is illustrated in Figure 6. DMSO had no effect on erythroid maturation in the absence of EP or in the presence of a concentration of EP that gives a maximum response. However, DMSO enhanced hemoglobin synthesis in FVA cells exposed to less than optimal levels of EP. These results together with the effect of ionophore A23187 shown in Figure 5 suggest that increased intracellular calcium or other changes in physiology due to DMSO can enhance the EP signal to maintain erythroid differentiation but they cannot substitute for the action of EP.

EP has been reported to increase the intracellular level of calcium using fluorescent chelators Quin-2 and Fura-2 in total human bone marrow cells depleted of red cells (Mladenovic and Kay, 1988) and partially purified erythroid progenitor cells from human cord blood (Miller *et al.*, 1988). In the first study, the average intracellular calcium concentration of bone marrow mononuclear cells rose from 138 nM to 450 nM after 4 min of exposure to EP. This result is difficult to reconcile with the fact that the EP-responsive proerythroblasts make up 4% of total bone marrow cells, and later erythroblasts, which have only a fraction of EP receptors and are essentially not responsive to EP, make up 13% of total marrow cell (Fraser *et al.*, 1988b). In the second study (Miller *et al.*, 1988), the entire population of partially pure progenitor cells responded to EP and 60% of the population responded to GM-CSF when increased intracellular calcium was investigated by fluorescence microscopy. In this study, no indication of the purity of the erythroid cells was presented. In addition, unpurified recombinant EP was used (approximately 10% EP). The conclusion of this study that BFU-E respond to both EP and GM-CSF by increased intracellular calcium may not be accurate since the response could be that of nonerythroid cells to a factor other than EP in the crude EP preparation (CHO cell supernatant). It is possible that the EP acts on the human erythroid progenitor cells in a different way than the action of EP on murine precursor cells; however, Linch *et al.* (1987) failed to detect an increase in intracellular calcium in proerythroblasts from human fetal liver when these cells were treated with EP. In other studies in press, EP failed to raise intracellular calcium in mouse CFU-E and erythroblasts as well as in EP-responsive murine cell lines HCD33, HCD57, and DA-1.

Very little is known about the mechanism of action of EP. More is known about second messengers of other hormones, which have been demonstrated to have no role in the stimulation of erythroid cells by EP. As of this time, cAMP, cGMP, tyrosine-specific protein kinase, phosphatidylinositol metabolites, diacylglycerol, and protein kinase C as well as intracellular calcium appear to have no major role in the action of EP in erythroid progenitor cells.

The early work on the mechanism of action of EP was done with impure mixtures of cells and impure EP. This combination resulted in erroneous reports of cGMP being involved in the mechanism of action of EP (Graber *et al.*, 1977; Rodgers *et al.*, 1976) when the agent that stimulated cGMP was endotoxin in the EP preparations (Graber *et al.*, 1979). Even recent work reporting that EP stimulated adenylate cyclase in membranes from rabbit bone marrow cells was carried out with EP that was approximately 1% pure (Bonanou-Tzedaki *et al.*, 1986). This conflicts with reports that EP had no effect on cAMP (Graber *et al.*, 1974). It is rather clear from recent work with the recombinant EP that the levels of cAMP and cGMP do not change in EP-responsive primary cells such as FVA cells or cell lines (Fukumoto *et al.*, 1989; Tsuda *et al.*, 1989) when the cells are treated with EP. Exogenously added analogues of cAMP or cGMP and agents that result in increased levels of these compounds in the cell have been reported to have effects on erythroid differentiation in EP-responsive cell lines (Fukumoto *et al.*, 1989; Tsuda *et al.*, 1989). However, these results have little meaning considering the number of possible effects these agents can have on the physiology of the cell. In FVA cells, addition of exogenous cAMP analogues or agents that increase cAMP had no effect whatsoever on the EP-dependent erythroid maturation or on the cells in absence of the hormone (Thompson *et al.*, 1987).

Protein phosphorylation as a result of the binding of EP to its receptor has not been documented as important in the mechanism of action of EP. The cloned putative receptor shows no homology to receptors belonging to the family of tyrosine-specific protein kinases. This makes the direct phosphorylation of proteins by the receptor unlikely. It is interesting that the tyrosine-specific protein kinase erb B, a truncated EGF receptor, is the oncogene responsible for erythroleukemias resulting from avian erythroblastosis virus infection (reviewed by Yarden and Ullrich, 1988). Examination of the protein phosphorylation in ^{32}P-labeled FVA cells by two-dimensional PAGE showed no changes as a result of exposure to EP during the first 6 hr. Treatment of the gels with sodium hydroxide, which enhances the phosphorylation of tyrosine residues (through removal of phosphates on serine and threonine residues), showed a number of labeled phosphotyrosyl proteins but no effect of EP during the first 6 hr. After 6 hr exposure to EP, a number of changes in protein phosphorylation were observed but these changes occur in conjunction with new protein synthesis. It is not clear that the change in protein phosphorylation after 6 hr is related to the mechanism of action of EP or is rather due to erythroid maturation.

In vitro phosphorylation using [γ-^{32}P]-ATP and FVA cell plasma membranes showed no effect of EP. Others have reported a reduction in the *in vitro* phosphorylation of a 43-kDa protein on serine residues after exposure of membranes from an EP-responsive cell line to EP (Choi *et al.*, 1987). However, the significance of this effect is not clear since the observation has not been made in the intact cells following EP stimulation. It is difficult to envision the mechanism through which EP binding to its receptor is isolated membranes without an energy source could lead to the inhibition of the protein kinase responsible for the phosphorylation of pp43.

Another remote possibility for the mechanism of action of EP is the release of a second messenger protein into the cytoplasm following binding of EP to its receptor. Chang and Goldwasser (1973) reported the existence of a cytoplasmic protein following the stimulation of bone marrow cells with EP which could stimulate RNA synthesis in isolated nuclei. However, impure cells were used and no other reports have confirmed this observation.

ACKNOWLEDGMENTS. The author thanks Mary Jane Rich for typing the manuscript, Judy Luna for assistance in proofreading, Drs. Mark Koury and Stephen Koury for helpful conversations regarding many aspects of this chapter, and Alan D'Andrea for sharing unpublished data on the cloning of the EP receptor. The author is supported by National Institutes of Health Grant DK-39781.

7. REFERENCES

Axelrad, A. A., McLeod, D. L., Shreeve, M. M., and Heath, D. S., 1974, Properties of cells that produce erythrocytic colonies in vitro, in: *Hemopoiesis in Culture* (W. A. Robinson, ed.), pp. 226–234, U.S. Government Printing Office, Washington, D.C.

Berridge, M. V., Fraser, J. K., Carter, J. M., and Lin, F. K., 1988, Effects of recombinant human erythropoietin on megakaryocytes and on platelet production in the rat, *Blood* **72:**970–977.

Beru, N., McDonald, J., Lacombe, C., and Goldwasser, E., 1986, Expression of the erythropoietin gene, *Mol. Cell. Biol.* **6:**2571–2575.

Bonanou-Tzedaki, S. A., Setchenska, M. S., Setchenska, M. A., and Arnstein, H. R. V., 1986, Stimulation of adenylate cyclase activity of rabbit bone marrow immature erythroblasts by erythropoietin and haemin, *Eur. J. Biochem.* **155:**363–370.

Bondurant, M. C., and Koury, M. J., 1986, Anemia induces accumulation of erythropoietin mRNA in the kidney and liver, *Mol. Cell. Biol.* **6:**2731–2732.

Bondurant, M., Koury, M., Krantz, S., Blevins, T., and Duncan, D., 1983, Isolation of erythropoietin-sensitive cells from Friend virus-infected marrow cultures: Characterization of the erythropoietin response, *Blood* **61:**751–758.

Bondurant, M. C., Lind, R. L., Koury, M. J., and Ferguson, M. E., 1985, Control of globin gene transcription by erythropoietin in erythroblasts from Friend virus-infected mice, *Mol. Cell. Biol.* **5:**675–683.

Branch, D. R., Turc, J. M., and Guilbert, L. J., 1987, Identification of an erythropoietin-sensitive cell line, *Blood* **69:**1782–1785.

Bridges, K., Levenson, R., Housman, D., and Cantley, L., 1981, Calcium regulates the commitment of MEL cells to terminal erythroid, *J. Cell Biol.* **90:**542–544.

Broudy, V. C., Lin, N., Egrie, J., DeHaen, C., Weiss, T., Papayannopoulo, T., and Adamson, J. W., 1988, Identification of the receptor for erythropoietin on human and murine erythroleukemia cells and modulation by phorbol ester and dimethyl sulfoxide, *Proc. Natl. Acad. Sci. USA* **85:**6513–6517.

Browne, J. K., Cohen, A. M., Egrie, J. C., Lai, P. H., Lin, F. K., Strickland, T., Watson, E., and Stebbins, N., 1986, Erythropoietin: Gene cloning, protein structure, and biological properties, *Cold Spring Harbor Symp. Quant. Biol.* **51:**693–699.

Carnot, P., and Deflandré, C., 1906, Sur l'activite hemopoietique des differents organes au cours de la regeneration du sang, *C.R. Acad. Sci.* **143**:432–435.

Carpenter, G., and Cohen, S., 1979, Epidermal growth factor, *Annu. Rev. Biochem.* **48**:193–216.

Chang, C.-S., and Goldwasser, E., 1973, On the mechanism of erythropoietin-induced differentiation. XII. A cytoplasmatic protein mediating induced nuclear RNA synthesis, *Dev. Biol.* **34**:246–254.

Chang, S. C.-S., Sikkema, D., and Goldwasser, E., 1974, Evidence for an erythropoietin receptor protein on rat bone marrow cells, *Biochem. Biophys. Res. Commun.* **57**:399–405.

Choi, H. S., Wojchowski, D. M., and Sytkowski, A. J., 1987, Erythropoietin rapidly alters phosphorylation of pp43, an erythroid membrane protein, *J. Biol. Chem.* **262**:2933–2936.

D'Andrea, A. D., Lodish, H. F., and Wong, G., 1989, Expression cloning of the murine erythropoietin receptor, *Cell* **57**:277–285.

Dautry-Varsat, A., Ciechanover, A., and Lodish, H. F., 1983, pH and the recycling of transferrin during receptor-mediated endocytosis, *Proc. Natl. Acad. Sci. USA* **80**:2258–2262.

Davis, J. M., Arakawa, T., Strickland, T. W., and Yphantis, D. A., 1987, Characterization of recombinant human erythropoietin produced in Chinese hamster ovary cells, *Biochemistry* **26**:2633–2638.

Dawson, P. J., Dresler, S. L., and Fieldsteel, A. H., 1979, Erythroid leukemia induced by Friend lymphatic leukemia virus in T-cell-depleted mice, *Cancer Res.* **39**:1611–1615.

DeBoth, N. J., Vermey, M., Van't Hull, E., Klootwijk-van-Dijke, E., van Griensven, L. J. L. D., Mol, J. N. M., and Stoof, T. J., 1978, A new erythroid cell line induced by Rauscher murine leukaemia virus, *Nature* **272**:626–628.

Dessypris, E. N., and Krantz, S. B., 1984, Effect of pure erythropoietin on DNA-synthesis by human marrow day 15 erythroid burst forming units in short-term liquid culture, *Br. J. Haematol.* **56**:295–306.

Dessypris, E. N., Gleaton, J. H., and Armstrong, O. L., 1987, Effect of human recombinant erythropoietin on human marrow megakaryocyte colony formation in vitro, *Br. J. Haematol.* **65**:265.

Dessypris, E. N., Graber, S. E., and Krantz, S. B., 1988, Effects of recombinant erythropoietin on the concentration and cycling status of human marrow hematopoietic progenitor cells in vivo, *Blood* **72**:2060–2062.

Dukes, P. P., and Einbinder, M., 1975, Response to marrow cell cultures to glass bound erythropoietin, *Blood* **45**:1024A.

Fraser, J. K., Lin, F. K., and Berridge, M. V., 1988a, Expression of high affinity receptors on human bone marrow and HEL cells, *Exp. Hematol.* **16**:836–842.

Fraser, J. K., Lin, F. K., and Berridge, M. V., 1988b, Expression and modulation of specific, high affinity binding sites for erythropoietin on the human erythroleukemia cell line, K562, *Blood* **71**:104–109.

Fraser, J. K., Nicholls, J., Coffey, C., Lin, F. K., and Berridge, M. V., 1988c, Down-modulation of high-affinity receptors for erythropoietin on murine erythroblasts by IL-3, *Exp. Hematol.* **16**:769–773.

Fraser, J. K., Tan, A. S., Lin, F. K., and Berridge, M. V., 1989, Expression of specific high affinity binding sites for erythropoietin on rat and mouse megakaryocytes, *Exp. Hematol.* **17**:10–16.

Fried, W., 1972, The liver as a source of extrarenal erythropoietin production, *Blood* **40**:671–677.

Friend, C., 1957, Cell-free transmission in adult Swiss mice of a disease having the character of a leukemia, *J. Exp. Med.* **105**:307–318.

Friend, C., Scher, W., Holland, J. G., and Sato, T., 1971, Hemoglobin synthesis in murine virus-induced leukemic cells in vitro: Stimulation of erythroid differentiation by dimethyl sulfoxide, *Proc. Natl. Acad. Sci. USA* **68**:378–382.

Fukamachi, H., Saito, T., Tojo, A., Kitamura, T., Urabe, A., and Takaku, F., 1987a, Binding of erythropoietin to CFU-E derived from fetal mouse liver cells, *Exp. Hematol.* **15**:833–837.

Fukamachi, H., Tojo, A., Saito, T., Kitamura, T., Nakata, M., Urabe, A. O., and Takaku, F., 1987b, Internalization of radioiodinated erythropoietin and the ligand-induced modulation of its receptor in murine erythroleukemia cells, *Int. J. Cell Cloning* **5**:209–219.

Fukumoto, H., Matsui, Y., and Obinata, M., 1989, Mechanism of erythropoietin action on the erythroid progenitor cells induced from murine erythroleukemia cells (TSA8), *Development* **105**:109–114.

Goldberg, M. A., Glass, G. A., Cunningham, J. M., and Bunn, H. F., 1987, The regulated expression of erythropoietin by two hepatoma cell lines, *Proc. Natl. Acad. Sci. USA* **84**:7972–7977.

Goldberg, M. A., Dunning, S. P., and Bunn, H. F., 1988, Regulation of the erythropoietin gene: Evidence that the oxygen sensor is a heme protein, *Blood* **72**:118a.

Goldstein, J. L., Brown, M. S., Anderson, R. G. W., Russell, D. W., and Schneider, W. J., 1985, Receptor-mediated endocytosis, *Annu. Rev. Cell Biol.* **1**:1–39.

Goldwasser, E., 1975, Erythropoietin and the differentiation of red blood cells, *Fed. Proc.* **34:**2285–2292.

Goldwasser, E., 1981, Erythropoietin and red cell differentiation, in: *Control of Cellular Division and Development, Part A* (D. Cunningham, E. Goldwasser, J. Watson, and D. F. Fox, eds.), pp. 487–494, Liss, New York.

Graber, S. E., Carrillo, M., and Krantz, S. B., 1974, Lack of effect of erythropoietin on cyclic adenosine 3′,5′-monophosphate levels in rat fetal liver cells, *J. Lab. Clin. Med.* **83:**288–295.

Graber, S. E., Bomboy, J. D., and Salmon, W. D., 1977, Effect of erythropoietin preparations on cyclic AMP and cyclic GMP levels in rat fetal liver cell cultures, *J. Lab. Clin. Med.* **90:**162–169.

Graber, S., Bomboy, J. D., and Salmon, W. D., 1979, Evidence that endotoxin is the cyclic 3′ : 5′-GMP-promoting factor in erythropoietin preparations, *J. Lab. Clin. Med.* **93:**25–31.

Gregory, C. J., and Evans, A. C., 1977, Human marrow cells capable of erythropoietic differentiation in vitro: Definition of three erythroid colony responses, *Blood* **49:**855–864.

Hankins, W. D., and Troxler, D., 1980, Polycythemia- and anemia-inducing erythroleukemia viruses exhibit differential erythroid transforming effects in vitro, *Cell* **22:**693–699.

Hankins, W. D., Kost, T. A., Koury, M. J., and Krantz, S. B., 1978, Erythroid bursts produced by Friend leukemia virus in vitro, *Nature* **276:**506–508.

Hankins, W. D., Chin, K., Dons, R., and Szabo, J., 1982, Isolation of erythropoietin-dependent cell lines suggest viability role for developmental hormones, *Blood* **70:**173a.

Hara, K., Suda, T., Suda, J., Ihle, J. N., Equachi, M., Nagata, S., Miura, Y., and Saito, M., 1988, Biopotential murine hemopoietic cell line (NFS-60) that is responsive to IL-3, GM-CSF and erythropoietin, *Exp. Hematol.* **16:**256–261.

Heath, D. S., Axelrad, A. A., McLeod, D. L., and Shreeve, M. M., 1976, Separation of the erythropoietin-responsive progenitors BFU-E and CFU-E in mouse bone marrow by unit gravity sedimentation, *Blood* **47:**777–792.

Hitomi, K., Fujita, K., Sasaki, R., Chiba, H., Okuno, Y., Ichiba, S., Takanashi, T., and Imura, H., 1988, Erythropoietin receptor of a human leukemic cell line with erythroid characteristics, *Biochem. Biophys. Res. Commun.* **154:**902–909.

Hitomi, K., Masuda, S., Ito, K., Ueda, M., and Sasaki, R., 1989, Solubilization and characterization of erythropoietin receptor from transplantable mouse erythroblastic leukemic cells, *Biochem. Biophys. Res. Commun.* **160:**1140–1148.

Hofer, E., Hofer-Warbinek, R., and Darnell, J. E., 1982, Globin RNA transcription: A possible termination site and demonstration of transcriptional control correlated with altered chromatin structure, *Cell* **29:**887–893.

Hosoi, T., Sawyer, S. T., and Krantz, S. B., 1988, The receptor for erythropoietin lacks detectable glycosylation, *Exp. Hematol.* **16:**118.

Hosoi, T., Sawyer, S. T., and Krantz, S. B., 1989, Identification of erythropoietin receptor in a ligand-free form with ^{125}I-labeled, photoreactive, cleavable cross-linker (Denny-Jaffe Reagent), *Exp. Hematol.* **17:**224a.

Imagawa, S., Smith, B. R., Palmer-Crocker, R., and Bunn, H. F., 1989, The effect of recombinant erythropoietin on intracellular calcium in erythropoietin-responsive cells, *Blood* **73:**1452–1457.

Iscove, N. N., 1977, The role of erythropoietin in regulation of population size and cell cycling of early and late erythroid precursors in mouse bone marrow, *Cell Tissue Kinet.* **10:**323–334.

Iscove, N. N., and Guilbert, L. J., 1978, Erythropoietin-independence of early erythropoiesis and a two-regulator model of proliferative control in the hematopoietic system, in: *In Vitro Aspects of Erythropoiesis* (M. J. Murphy, ed.), pp. 3–7, Springer, Berlin.

Iscove, N. N., and Sieber, F., 1975, Erythroid progenitors in mouse bone marrow detected by macroscopic colony formation in culture, *Exp. Hematol.* **3:**32–43.

Jacobs, K., Shoemaker, C., Rudersdorf, R., Neill, S. D., Kaufman, R. J., Mufson, A., Seehra, J., Jones, S. S., Hewick, R., Fritsch, E. F., Kawakita, M., Shimizu, T., and Miyake, T., 1985, Isolation and characterization of genomic and cDNA clones of human erythropoietin, *Nature* **313:**806–810.

Jacobson, L. O., Goldwasser, E., Fried, W., and Plzak, L., 1957, Role of the kidney in erythropoiesis, *Nature* **179:**633–634.

Jacobson, L. O., Marks, E. K., Gaston, E. O., and Goldwasser, E., 1959, Studies on erythropoiesis. XI. Reticulocyte response of transfusion-induced polycythemic mice to anemic plasma from nephrectomized mice and to plasma from nephrectomized rats exposed to low oxygen, *Blood* **14:**635–643.

Keppel, F., Allet, B., and Eisen, H., 1977, Appearance of a chromatin protein during the erythroid differentiation of Friend virus-transformed cell, *Proc. Natl. Acad. Sci. USA* **74:**653–657.

Klausner, R. D., Ashwell, G., Van Renswoude, J., Harford, J. B., and Bridges, K. R., 1983, Binding of apotransferrin to K562 cells: Explanation of the transferrin cycle, *Proc. Natl. Acad. Sci. USA* **80:**2263–2266.

Koury, M. J., and Bondurant, M. C., 1988, Maintenance by erythropoietin on viability and maturation of murine erythroid precursor cells, *J. Cell. Physiol.* **137:**65–74.

Koury, M. J., Bondurant, M. C., Duncan, D. T., Krantz, S. B., and Hankins, W. D., 1982, Specific differentiation events induced by erythropoietin in cells infected in vitro with the anemia strain of Friend virus, *Proc. Natl. Acad. Sci. USA* **79:**635–639.

Koury, M. J., Sawyer, S. T., and Bondurant, M. C., 1984, Splenic erythroblasts in anemia-inducing Friend disease: A source of cells for studies of erythropoietin-mediated differentiation, *J. Cell. Physiol.* **121:**526–532.

Koury, M. J., Bondurant, M. C., and Mueller, T. J., 1986, The role of erythropoietin in the production of principal erythrocyte proteins other than hemoglobin during terminal erythroid differentiation, *J. Cell. Physiol.* **126:**259–265.

Koury, M. J., Bondurant, M. C., and Rana, S. S., 1987a, Changes in erythroid membrane proteins during erythropoienin-mediated terminal differentiation, *J. Cell. Physiol.* **133:**438–448.

Koury, M. J., Bondurant, M. C., and Atkinson, J. B., 1987b, Erythropoietin control of terminal erythroid differentiation: Maintenance of cell viability, production of hemoglobin, and development of the erythrocyte membrane, *Blood Cells* **13:**217–226.

Koury, M. J., Bondurant, M. C., Graber, S. E., and Sawyer, S. T., 1988, Erythropoietin messenger RNA levels in developing mice and transfer of ^{125}I-erythropoietin by the placenta, *J. Clin. Invest.* **82:**154–159.

Koury, S. T., Bondurant, M. C., and Koury, M. J., 1988a, Localization of erythropoietin synthesizing cells in murine kidneys by *in situ* hybridization, *Blood* **71:**524–528.

Koury, S. T., Koury, M. J., and Bondurant, M. C., 1988b, Morphological changes in erythroblasts during erythropoietin-induced terminal differentiation in vitro, *Exp. Hematol.* **16:**758–763.

Koury, S. T., Koury, M. J., and Bondurant, M. C., 1988c, Distribution and function of cytoskeletal elements during enucleation of mammalian erythroblasts, *J. Cell Biol.* **107:**63a.

Koury, S. T., Koury, M. J., Bondurant, M. C., Caro, J., and Graber, S. E., 1989, Quantitation of erythropoietin-producing cells in kidneys of mice by *in situ* hybridization: Correlation with hematocrit, renal erythropoietin mRNA, and serum erythropoietin concentration, *Blood* **74:**645–651.

Krantz, S. B., and Goldwasser, E., 1965, On the mechanism of erythropoietin-induced differentiation. II. The effect on RNA synthesis, *Biochim. Biophys. Acta* **103:**325–333.

Krantz, S. B., and Goldwasser, E., 1984, Specific binding of erythropoietin to spleen cells infected with the anemia strain of Friend virus, *Proc. Natl. Acad. Sci. USA* **81:**7574–7578.

Krantz, S. B., and Jacobson, L. O., 1970, *Erythropoietin and the Regulation of Erythropoiesis,* University of Chicago Press, Chicago.

Krantz, S. B., Gallien-Lartigue, O., and Goldwasser, E., 1963, The effect of erythropoietin upon heme synthesis by marrow cells in vitro, *J. Biol. Chem.* **238:**4085.

Krantz, S. B., Sawyer, S. T., and Sawada, K.-I., 1988, The role of erythropoietin in erythroid cell differentiation, *Contrib. Nephrol.* **66:**25–37.

Krystal, G., 1983, A simple microassay for erythropoietin based on ^{3}H-thymidine incorporation into spleen cells from phenylhydrazine treated mice, *Exp. Hematol.* **11:**649–660.

LaCombe, C., DaSilva, J. L., Bruneval, P., Fournier, J. G., Wendling, F., Casadevall, N., Camilleri, J. P., Bariety, J., Varet, B., and Tambourin, P., 1988, Peritbular cells are the site of erythropoietin synthesis in the murine hypoxic kidney, *J. Clin. Invest.* **81:**620–623.

Landschulz, K. T., Noyes, A. V., Rodgers, D., and Boyer, S., 1989, Erythropoietin receptors on murine colony-forming units; natural history, *Blood* **72:**92a.

Levenson, R., Housman, D., and Cantley, L., 1980, Amiloride inhibits MEL cell differentiation: Evidence for Ca^{+2} requirement for commitment, *Proc. Natl. Acad. Sci. USA* **77:**5948–5952.

Levenson, R., Macara, I. G., Smith, R. L., Cantley, L., and Housman, D., 1982, Role of mitochondrial membrane potential in the regulation of murine erythroleukemia cell differentiation, *Cell* **28:**855–863.

Lin, F.-K., Suggs, S., Lin, C.-H., Browne, J. K., Smalling, R., Egrie, J. C., Chen, K. K., Fox, G. M., Martin, F., Stabinsky, Z., Badrawi, S. M., Lai, P.-H., and Goldwasser, E., 1985, Cloning and expression of the human erythropoietin gene, *Proc. Natl. Acad. Sci. USA* **82:**7580–7584.

Linch, D. C., Jones, H. M., Tidman, N., and Roberts, P. J., 1987, The effects of erythropoietin on primitive human erythroid cells, *Blood* **70:**177a.

McDonald, J. D., Lin, F.-K., and Goldwasser, E., 1986, Cloning, sequencing, and evolutionary analysis of the mouse erythropoietin gene, *Mol. Cell. Biol.* **6**:842–848.

McDonald, T. P., Cottrell, M. B., Clift, R. E., Cullen, C. W., and Lin, F. K., 1987, High doses of recombinant erythropoietin stimulate platelet production in mice, *Exp. Hematol.* **15**:719–724.

Marks, P. A., and Rifkind, R. A., 1978, Erythroleukemic differentiation, *Annu. Rev. Biochem.* **47**:419–448.

Marks, P. A., Sheffery, M., and Rifkind, R. A., 1987, Induction of transformed cells to terminal differentiation and the modulation of gene expression, *Cancer Res.* **47**:659–666.

Mayeux, P., Billat, C., and Jacquot, R., 1987a, Murine erythroleukemia cells (Friend cells) possess high-affinity binding sites for erythropoietin, *FEBS Lett.* **211**:229–233.

Mayeux, P., Billat, C., and Jacquot, R., 1987b, The erythropoietin receptor of rat erythroid progenitor cells, *J. Biol. Chem.* **262**:13985–13990.

Means, R. T., Krantz, S. B., Sawyer, S. T., and Gilbert, H. S., 1989, Erythropoietin receptors in polycythemia vera, *J. Clin. Invest.* **84**:1340–1344.

Melloni, E., Pontremoli, S., Michetti, M., Sacco, O., Cakiroglu, A. G., Jackson, J. F., Rifkind, R. A., and Marks, P. A., 1987, Protein kinase C activity and hexamethyl bisacetamide induced erythroleukemia cell differentiation, *Proc. Natl. Acad. Sci. USA* **84**:5282–5286.

Miller, B. A., Scaduto, R. C., Jr., Tillotson, D. L., Botti, J. L., and Cheung, J. Y., 1988, Erythropoietin stimulates a rise in intracellular free calcium concentration in single early human erythroid precursors, *J. Clin. Invest.* **82**:309–315.

Mirand, E. A., Steeves, R. A., Lange, R. D., and Grace, J. T., 1968, Virus-induced polycythemia in mice: Erythropoiesis without erythropoietin, *Proc. Soc. Exp. Biol. Med.* **128**:844–849.

Mishina, Y., and Obinata, M., 1985, Induction of commitment of murine erythroleukemia cell (TSA8) to CFU-E with DMSO, *Exp. Cell Res.* **162**:319–325.

Misti, J., and Spivak, J. L., 1979, Erythropoiesis in vitro. Role of calcium, *J. Clin. Invest.* **64**:1573–1579.

Miyake, T., Kung, C. K. H., and Goldwasser, E., 1977, Purification of human erythropoietin, *J. Biol. Chem.* **252**:5558–5564.

Mladenovic, J., and Kay, N. E., 1988, Erythropoietin induces rapid increases in intracellular free calcium in human bone marrow cells, *J. Lab. Clin. Med.* **112**:23–27.

Mufson, R. A., and Gesner, T. G., 1987, Binding and internalization of recombinant human erythropoietin in murine erythroid precursor cells, *Blood* **69**:1485–1490.

Nijhof, W., and Wierenga, P. K., 1983, Isolation and characterization of the erythroid progenitor cell: CFU-E, *J. Cell Biol.* **96**:386–392.

Nijhof, W., Wierenga, P. K., Pietens, J., and Bloem, R., 1984, Cell kinetic behavior of a synchronized population of erythroid precursor cells in vitro, *Cell Tissue Kinet.* **17**:629–639.

Ogawa, M., Porter, P. N., and Nakahata, T., 1983, Renewal and commitment to differentiation of hematopoietic stem cells (an interpretive review), *Blood* **61**:823–829.

Orkin, S. H., Swan, D., and Leder, P., 1975, Differential expression of α- and β-globin genes during differentiation of cultured erythroleukemia cells, *J. Biol. Chem.* **250**:8753–8759.

Patel, V. P., and Lodish, H. F., 1987, A fibronectin matrix is required for differentiation of murine erythroleukemia cells into reticulocytes, *J. Cell Biol.* **105**:3105–3118.

Pekonen, F., Rosenlof, K., Rutanen, E.-M., and Fyhrquist, F., 1987, Erythropoietin binding sites in human foetal tissues, *Acta Endocriniol.* (*Copenhagen*) **116**:561–567.

Peschle, C., and Condorelli, M., 1975, Biogenesis of erythropoietin: Evidence for pro-erythropoietin in a subcellular fraction of kidney, *Science* **190**:910–912.

Pfeffer, S. R., Huima, T., and Redman, C. M., 1986, Biosynthesis of spectrin and its assembly into the cytoskeletal system of Friend erythroleukemic cells, *J. Cell Biol.* **103**:103–113.

Recny, M. A., Scoble, H. A., and Kim, Y., 1987, Structural characterization of natural human urinary and recombinant DNA-derived erythropoietin, *J. Biol. Chem.* **262**:17156–17161.

Reissman, K. R., 1950, Studies on the mechanism of erythropoietic stimulation in parabiotic rats during hypoxia, *Blood* **5**:372–377.

Rich, I. N., Heit, W., and Kubanek, B., 1982, Extrarenal erythropoietin production by macrophages, *Blood* **60**:1007–1018.

Rodgers, G. M., Fisher, J. W., and George, W. J., 1976, Elevated cyclic GMP concentration in rabbit bone marrow culture and mouse spleen following erythropoietin stimulation, *Biochem. Biophys. Res. Commun.* **70**:287–291.

Roodman, G. D., Spivak, J. L., and Zanjani, E. D., 1981, Stimulation of erythroid colony formation in vitro by erythropoietin immobilized on agarose-bound lectins, *J. Lab. Clin. Med.* **98:**684–690.

Ruscetti, S., and Wolff, L., 1984, Spleen focus-forming virus: Relationship of an altered envelope gene to the development of a rapid erythroleukemia, *Curr. Top. Microbiol. Immunol.* **112:**21–44.

Sakaguchi, M., Koishihara, Y., Tsuda, H., Fujimoto, K., Shibuya, K., Kawakita, M., and Takatsuki, K., 1987, The expression of functional erythropoietin receptors on an interleukin-3 dependent cell line, *Biochem. Biophys. Res. Commun.* **146:**7–12.

Sasaki, R., Bothner, B., and Dell, A., 1987a, Carbohydrate structure of erythropoietin expressed in Chinese hamster ovary cells by a human erythropoietin cDNA, *J. Biol. Chem.* **262:**12059–12064.

Sasaki, R., Yanagawa, S., Hitomi, K., and Chiba, H., 1987b, Characterization of erythropoietin receptor of murine erythroid cells, *Eur. J. Biochem.* **168:**43–48.

Sassa, S., Takaku, F., and Nakao, K., 1968, Regulation of erythropoiesis in the Friend leukemia mouse, *Blood* **31:**758–765.

Sawada, K., Krantz, S. B., Kans, J. S., Dessypris, E. N., Sawyer, S. T., Glick, A. D., and Civin, C. I., 1987, Purification of human erythroid colony-forming units and demonstration of specific binding of erythropoietin, *J. Clin. Invest.* **80:**357–366.

Sawada, K., Dessypris, E. N., and Krantz, S. B., 1988a, CFU-E do not require accessory cells and hormones other than insulin (I) and erythropoietin (EP) for erythroid development, *J. Clin. Invest.* **83:**1701–1709.

Sawada, K., Krantz, S. B., Sawyer, S. T., and Civin, C. I., 1988b, Quantitation of specific binding of erythropoietin to human erythroid colony-forming cells, *J. Cell. Physiol.* **137:**337–345.

Sawada, K., Krantz, S. B., Dai, C.-H., Koury, S. T., Horn, S. T., Glick, A. D., and Civin, C. I., 1990, Purification of human blood burst-forming units-erythroid and demonstration of the evolution of erythropoietin receptors, *J. Cell. Physiol.*, in press.

Sawyer, S. T., 1989, The two proteins of the erythropoietin receptor are structurally similar, *J. Biol. Chem.* **264:**13343–13347.

Sawyer, S. T., and Hankins, W. D., 1988, Metabolism of erythropoietin in erythropoietin-dependent cell lines, *Blood* **72:**440.

Sawyer, S. T., and Koury, M. J., 1987, Erythropoietin requirements during terminal erythroid differentiation: The role of surface receptors for erythropoietin, *J. Cell Biol.* **105:**1077.

Sawyer, S. T., and Krantz, S. B., 1984, Erythropoietin stimulates $^{45}Ca^{2+}$ uptake in Friend virus-infected erythroid cells, *J. Biol. Chem.* **259:**2769–2774.

Sawyer, S. T., and Krantz, S. B., 1986a, Transferrin receptor number, synthesis and endocytosis during erythropoietin-induced maturation of Friend virus-infected erythroid cells, *J. Biol. Chem.* **261:**9187–9195.

Sawyer, S. T., and Krantz, S. B., 1986b, Identification of the receptor of erythropoietin on mouse erythroid cells by cross-linking, *J. Cell Biol.* **103**(Nov. 5, Part 2)**:**213a.

Sawyer, S. T., Koury, M. J., and Bondurant, M. C., 1987a, Large-scale procurement of erythropoietin-responsive erythroid cells: Assay for biological activity of erythropoietin, *Methods Enzymol.* **147:**340–352.

Sawyer, S. T., Krantz, S. B., and Goldwasser, E., 1987b, Binding and receptor-mediated endocytosis of erythropoietin in Friend virus infected erythroid cells, *J. Biol. Chem.* **262:**5554–5562.

Sawyer, S. T., Sawada, K.-I., and Krantz, S. B., and Luna, J., 1987c, Identification of the receptor for erythropoietin by cross-linking to Friend-virus-infected erythroid cells, *Proc. Natl. Acad. Sci. USA* **84:**3690–3694.

Sawyer, S. T., Hosoi, T., and Krantz, S. B., 1988, Structure of the erythropoietin receptor, *Blood* **72:**441.

Sawyer, S. T., Krantz, S. B., and Sawada, K.-I., 1989, Receptors for erythropoietin in mouse and human erythroid cells and placenta, *Blood* **74:**103–109.

Schuster, S. J., Badiavas, E. V., Costa-Giomi, P., Weinmann, R., Erslev, A. J., and Caro, J., 1989, Stimulation of erythropoietin gene transcription during hypoxia and cobalt exposure, *Blood* **73:**13–16.

Semenza, G. L., Traystamn, M. D., Gearhart, J. D., and Antonarakis, S. E., 1989, Polycythemia in transgeneic mice expressing the human erythropoietin gene, *Proc. Natl. Acad. Sci. USA* **86:**2301–2305.

Shibuya, T., and Mak, T. W., 1983, Isolation and induction of erythroleukemia cell lines with properties of erythroid progenitor burst-forming cells (BFU-E) and erythroid precursor cells (CFU-E), *Proc. Natl. Acad. Sci. USA* **80:**3721–3725.

Shoemaker, C. B., and Mistock, L. D., 1986, Murine erythropoietin gene: Cloning, expression, and human gene homology, *Mol. Cell. Biol.* **6:**849–858.

Smith, R. L., Macara, I. G., Leveson, R., Housman, D., and Cantley, L., 1982, Evidence that a Na^+/Ca^{++} antiport system regulates MEL cell differentiation, *J. Biol. Chem.* **257**:773–780.

Sytkowski, A. J., Salvado, A. J., Smith, G. M., and McIntyre, C. J., 1980, Erythroid differentiation of cloned Rauscher erythroleukemia cells in response to dimethyl sulfoxide, erythropoietin, *Science* **210**:74–76.

Tambourin, P. E., 1978, Haemopoietic stem cells and murine viral leukaemogenesis in: *Stem Cells and Tissue Homeostasis* (B. I. Lord *et al.*, eds.), pp. 259–316, Cambridge University Press, London.

Tambourin, P., Casadevall, N., Choppin, J., Lacombe, C., Heard, J. M., Fichelson, S., Wendling, F., Hankins, W. D., and Varet, B., 1983, Production of erythropoietin-like activity by a murine erythroleukemia cell line, *Proc. Natl. Acad. Sci. USA* **80**:6269–6273.

Tepperman, A. D., Curtis, J. E., and McCulloch, E. A., 1974, Erythropoietic colonies in cultures of human marrow, *Blood* **44**:659–669.

Thompson, L. P., Sawyer, S. T., Blackmore, P. F., and Krantz, S. B., 1987, A search for the second messenger of erythropoietin, *FASEB J.* **2**:A813.

Till, J. E., and McCulloch, E. A., 1961, A direct measurement of the radiation sensitivity of normal mouse bone marrow cells, *Radiat. Res.* **14**:213–222.

Till, J. E., McCulloch, E. A., and Siminovitch, L., 1964, A stochastic model of stem cell proliferation, based on the growth of spleen-colony forming cells, *Proc. Natl. Acad. Sci. USA* **51**:29–36.

Todokoro, K., Kanazawa, S., Amanuma, H., and Ikawa, Y., 1987, Specific binding of erythropoietin to its receptor on responsive mouse erythroleukemia cells, *Proc. Natl. Acad. Sci. USA* **84**:4126–4130.

Todokoro, K., Kanazawa, S., Amanuma, H., and Ikawa, Y., 1988, Characterization of erythropoietin receptor on erythropoietin-unresponsive mouse erythroleukemia cells, *Biochem. Biophys. Acta* **943**:326–330.

Tojo, A., Fukamachi, H., Kasuga, M., Urabe, A., and Takaku, F., 1987, Identification of erythropoietin receptors on fetal liver erythroid cells, *Biochem. Biophys. Res. Commun.* **148**:443–448.

Tojo, A., Kukamachi, H., Saito, T., Kasuga, M., Urabe, A., and Takaku, F., 1988, Induction of the receptor for erythropoietin in MEL cells after DMSO treatment, *Cancer Res.* **48**:1818–1822.

Troxler, D. H., and Scolnick, E. M., 1978, Rapid leukemia induced by cloned Friend strain of replicating murine type-C virus. Association with induction of xenotropic-related RNA sequences contained in spleen focus-forming virus, *Virology* **85**:17–27.

Tsao, C. J., Tojo, A., Fukamachi, H., Kitamura, T., Saito, T., and Urabe, A., 1988, Expression of the functional erythropoietin receptors on interleukin 3-dependent murine cell lines, *J. Immunol.* **140**:89–93.

Tsuda, H., Sawada, T., Sakaguchi, M., Kawakita, M., and Takatsuki, K., 1989, Mode of action of erythropoietin in Epo dependent murine cell line. I. Involvement of adenosine 3′ : 5′-cyclic monophosphate not as a second messenger but as a regulator of cell growth, *Exp. Hematol.* **17**:211–217.

Volloch, V., and Housman, D., 1982, Terminal differentiation of murine erythroleukemia cells: Physical stabilization of end-stage cells, *J. Cell Biol.* **93**:390–394.

Weiss, T. L., Kung, C. K. H., and Goldwasser, E., 1985, Quantitation of erythropoietin binding to bone marrow cells, *J. Cell. Biochem.* **27**:57–65.

Weiss, T. L., Barker, M. E., Selleck, S. E., and Wintroub, B. U., 1989, Erythropoietin binding and induced differentiation of Rauscher erythroleukemia cell line red 5-1.5, *J. Biol. Chem.* **264**:1804–1810.

Yarden, Y., and Ullrich, A., 1988, Growth factor receptor tyrosine kinases, *Annu. Rev. Biochem.* **57**:443–478.

Zanjani, E. D., Mclaurin, W. D., and Gordon, A. S., 1971, Biogenesis of erythropoietin: Role of the substrate for erythrogenin, *J. Lab. Clin. Med.* **77**:751–756.

Zanjani, E. D., Poster, J., and Burlington, H., 1977, Liver as the primary site of erythropoietin formation in the fetus. *J. Lab. Clin. Med.* **89**:640–644.

Chapter 13

The Transferrin Receptor and Iron Accumulation in Erythroid Cells

Klas Forsbeck

1. INTRODUCTION

Developing erythroid cells need large amounts of iron mainly for heme formation but also for other essential biochemical reactions. Several obstacles have to be overcome in order to get the iron distributed to the right place in sufficient quantity at the right time. In the ionic state, iron is reactive, potentially toxic, and poorly soluble. This imposes the need for appropriate carrier functions both outside and inside cells and during passage through hydrophobic membranes. The interval during erythroid differentiation with intense heme synthesis is limited, and creates a demand for an efficient iron extraction mechanism that can be turned on or shut off according to the actual iron requirement.

This chapter describes the cellular response to these demands and the central role played by the receptor for the iron-carrying plasma protein, transferrin. The iron moiety is accumulated in a pH- and reduction-dependent cyclic process that reutilizes both the receptor and the transferrin. The accumulation rate is governed by an iron-dependent negative feedback on the receptor synthesis in combination with the degenerative changes occurring during erythroid development.

2. GENERAL BACKGROUND

Erythroid development, which normally takes place in the bone marrow, consists of a number of stages from the multipotent stem cell to the mature erythrocyte. Two types of committed erythroid progenitors can be distinguished by *in vitro* cultures in methylcellulose, namely the relatively immature burst-forming units (BFU-E) and the more mature,

Klas Forsbeck Research and Development, Kabi, S-112 87 Stockholm, Sweden.

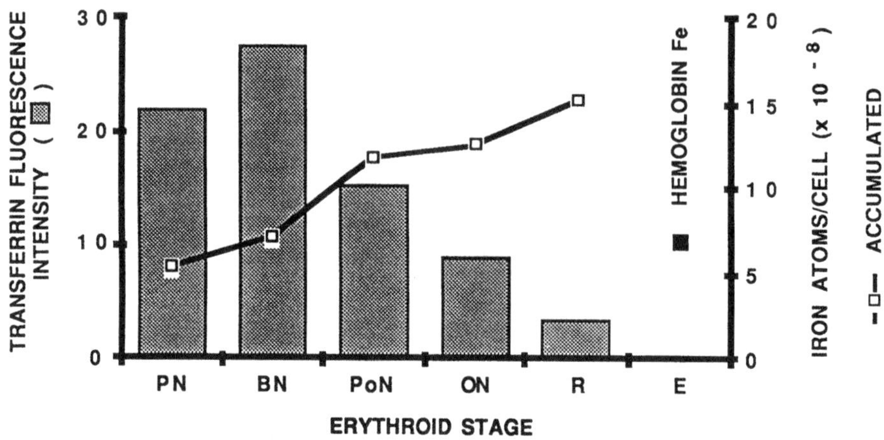

FIGURE 1. Cellular binding of transferrin and iron accumulation as a function of rat erythroid development. The cell-associated fluorescence-conjugated transferrin was estimated by microfluorometry (bars). The fluorescence of erythrocytes is at background level (compiled from Iacopetta *et al.*, 1983). The calculated total iron accumulation (right axis) is based on measurements of iron uptake rates by quantitative electron microscope autoradiography and cell maturation times. Values for all stages of development are shown. The final level of hemoglobin-associated iron (hemoglobin Fe) in erythrocytes is shown (compiled from Iacopetta and Morgan, 1983a). PN, pronormoblast; BN, basophilic normoblast; PoN, polychromatic normoblast; ON, orthochromatic normoblast; R, reticulocyte; E, erythrocyte.

erythropoietin-sensitive colony-forming units (CFU-E) (Iscove and Sieber, 1975). Committed progenitors develop by cell divisions into the pronormoblasts, basophilic, polychromatic, and orthochromatic normoblasts with concomitant cell shrinking and degenerative changes. Cell division ceases, the nucleus is expulsed, and the cells enter the circulation as reticulocytes and finally mature to erythrocytes (Arnstein *et al.*, 1987).

Iron accumulates most rapidly in the least mature erythroid stages (Figure 1; Iacopetta and Morgan, 1983a; Papayannopoulou and Finch, 1975), thereby paralleling the iron requirements for heme synthesis (Glass *et al.*, 1975). The total iron accumulated is about twice the amount present in hemoglobin of the mature rat erythrocyte, and excess iron is extruded (Iacopetta and Morgan, 1983a).

Several cellular models have been employed in iron metabolism studies. Due to their accessibility in the circulation, reticulocytes were originally used. However, many cellular functions are fading or already deranged in the reticulocyte, making it a poor representative of the earlier proliferative and differentiating cell forms in the erythroid series (Arnstein *et al.*, 1987). Improved cell culturing and fractionation, and new reagents, e.g., monoclonal antibodies, have recently made it possible to study these cells, which exist with low frequency mixed with other cell types in the bone marrow.

Established monoclonal tumor cell lines, finally, provide convenient and reproducible cellular models of the earlier erythroid stages. Examples are the human K562 (Andersson *et al.*, 1979; Forsbeck and Nilsson, 1983) and the murine Friend leukemias (Marks and Rifkind, 1978). The transferrin receptor and its general role in iron accumulation can also be studied in human cell lines of nonerythroid origin, such as HeLa (Bleil and Bretscher, 1982), KB carcinoma (Hanover *et al.*, 1984), HepG2 hepatoma (Ciechanover *et al.*, 1983), A431 epidermoid carcinoma (Davis *et al.*, 1987; Hopkins and Trowbridge, 1983), and various lymphoma and leukemia cells (Forsbeck and Nilsson, 1985). Cells of nonhuman origin are also used, an example being mouse teratocarcinoma (Karin and Mintz, 1981).

3. TRANSFERRIN RECEPTORS

It was found early that the number of transferrin receptors of a cell was influenced by the intracellular iron concentration, the growth rate of the cell, and the erythroid differentiation level. Recent advances of molecular biology techniques and their application on clonal tumor cell models clarify the regulatory mechanisms involved, as discussed below and summarized in Table I.

3.1. The Transferrin Receptor Gene

The human transferrin receptor gene is localized on chromosome 3 (Goodfellow *et al.*, 1982). The gene contains at least 19 coding regions distributed over a 31-kb DNA segment, as determined by heteroduplex analysis (McClelland *et al.*, 1984). No enhancer elements interacting with the gene have been found (Owen and Kühn, 1987).

3.2. Transcription

The transferrin receptor gene is transcriptionally blocked in nonproliferating cells, as shown in T-lymphocyte (Pelosi *et al.*, 1986; Pelosi-Testa *et al.*, 1988) and fibroblast (Owen and Kühn, 1987) models. Transcription is initiated when the cell is stimulated to proliferate, possibly by lowering the intracellular iron concentration (Pelosi-Testa *et al.*, 1988). The transferrin receptor expression is solely regulated by the proliferative status, e.g., in murine erythroid progenitors with no heme synthesis (Lesley *et al.*, 1984).

Miskimins *et al.* (1986) studied the molecular biology of transcriptional initiation

Table I
Effect of Various Factors on the Cellular Transferrin Receptors

Factor	Level	Mechanism	Effect on transferrin receptor numbers
Proliferative status[a]	Gene	Transcription initiation	↑
Iron concentration[b]	Gene	Lowered transcription rate	↓
Iron concentration[c]	mRNA	Lowered T_{50}	↓
Differentiation level[d]	Protein	Receptor shedding	↓

[a]Lesley *et al.* (1984), Miskimins *et al.* (1986), Owen and Kühn (1987), Pelosi *et al.* (1986), Pelosi-Testa *et al.* (1988).
[b]Casey *et al.* (1988a), Müllner and Kühn (1988), Rao *et al.* (1986).
[c]Müllner and Kühn (1988), Rao *et al.* (1986).
[d]Harding *et al.* (1983, 1985), Pan *et al.* (1985), Pan and Johnstone (1983).

using mouse fibroblasts. Four proteins interact with the transferrin receptor gene 5′ promoter region. Of these, three are DNA-binding proteins of 120, 105, and 95 kDa that are induced to higher levels in the cell, peaking at 3 hr after serum stimulation. An additional 88-kDa protein, not binding to DNA in the native state, is induced to binding with a maximum after about 6 hr. The levels of each protein declined after their respective maximum binding was reached. Transcription of the gene reaches a maximum 12 hr after stimulation. At this time DNA synthesis starts, and peaks at 24 hr.

The transcription rate of the transferrin receptor gene in the erythroid leukemia cell line K562 is inversely related to the iron concentration of the cell, as shown by the *in vitro* nuclear transcription assay (Rao *et al.*, 1986). Quantitatively, the transferrin receptor gene's promoter confers at the most a 3-fold iron-dependent regulation (Müllner and Kühn, 1988; Casey *et al.*, 1988a). Additional control mechanisms are operating on the mRNA level since the range observed for the iron-dependent regulation of the mRNA steady state is about 20-fold (Rao *et al.*, 1986).

3.2.1. Cytoplasmic mRNA Characteristics: Iron-Dependent Instability

The size of human transferrin receptor mRNA is around 5 kb, and varies depending on cell type (Kühn *et al.*, 1984; Schneider *et al.*, 1984; Müllner and Kühn, 1988), indicating that alternative splicing occurs. The mRNA has a 5′ coding region consisting of 2280 bases and two noncoding regions, one 5′ of approximately 280 bases and one 3′ of 2650 bases (Schneider *et al.*, 1984; McClelland *et al.*, 1984). The latter region contains at least one iron-responsive element (Müllner and Kühn, 1988) with a stem-loop structure that mediates an iron-dependent instability to the mRNA. Another study identified five such possible stem-loop structures confined to a nucleotide fragment of 678 bases within the 3′ untranslated region and demonstrated the function of two of these (Casey *et al.*, 1988b). The latter five iron-responsive elements are similar to the one originally identified in the 5′ untranslated region of ferritin mRNA (Aziz and Munro, 1987), but they differ from the one described above by Müllner and Kühn (1988).

Increased cellular iron concentrations reduce the half lives of the mRNAs in K562 (T_{50} = 2.5 hr) and mouse L cells (T_{50} = 30 hr) 5- and 20-fold, respectively (Rao *et al.*, 1986; Müllner and Kühn, 1988), which makes mRNA stability changes more important than the promoter as a transferrin receptor regulatory factor (Table I).

3.3. Synthesis and Turnover of the Transferrin Receptor

The transferrin receptor monomeric peptide contains no N-terminal signal sequence for membrane passage and is therefore not processed in microsomes (Schneider *et al.*, 1983, 1984; McClelland *et al.*, 1984). Instead, there is an internal transbilayer sequence, which possibly performs the dual function of signaling and anchoring the receptor in the membrane (Figure 2). These structural features fit with the model proposed by Sabatini *et al.* (1982), where cotranslational membrane insertion of the distal C-terminal part leads to an orientation with the extracellular C-terminus and cytoplasmic N-terminus.

The transferrin receptor protein has a long half-life in the cell, e.g., T_{50} values of 60 hr in a T-cell line (Omary and Trowbridge, 1981) and 15 hr in the erythroid K562 (Kohno *et al.*, 1986). The latter authors demonstrated an increased disappearance rate of receptors when the cells were subjected to phorbol ester treatment, reducing the T_{50} to 5 hr.

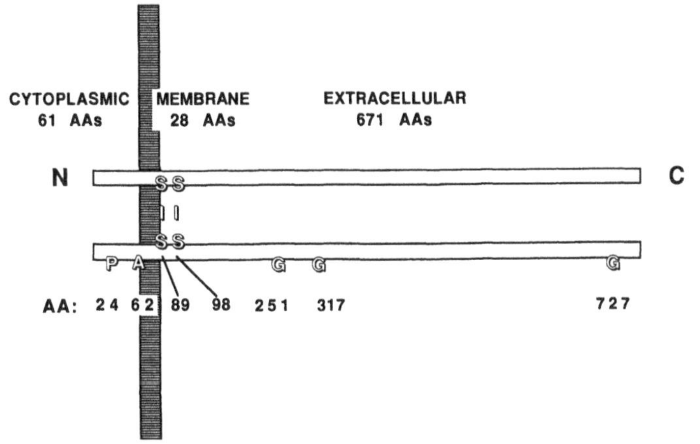

FIGURE 2. Proposed structure and orientation of the human transferrin receptor in the cell membrane. The receptor consists of two 90-kDa subunits (white bars) held together by two disulfide bonds (S–S) involving Cys-89 on the extracellular face of the membrane and Cys-98 in the extracellular, transferrin binding domain. The N-terminal is in the cytoplasm and the C-terminal is extracellular. The numbers of amino acids (AAs) of the cytoplasmic, membrane, and extracellular domains are shown at the top (according to McClelland *el al.*, 1984). Posttranslational modifications take place with fatty acylation of Cys-62 (A) on the cytoplasmic face of the membrane in the transmembrane domain, and Asp-linked glycosylation (G) in three sites in the extracellular domain. The cytoplasmic domain contains a Ser phosphorylation site (P). The amino acid positions are shown at the bottom (AA:).

3.4. Biochemistry of the Transferrin Receptor

The functional transferrin receptor, schematically shown in Figure 2 with its main biochemical features, is a membrane-spanning dimer of 180 kDA, consisting of two 90-kDa subunits (Hamilton *et al.*, 1979; Schneider *et al.*, 1982) that are linked by intermolecular disulfide bonds involving Cys-89 at the border between the extracellular and the membrane-spanning domains, and Cys-98 in the extracellular domain (Jing and Trowbridge, 1987). Each subunit binds one molecule of transferrin in its extracellular domain (Schneider *et al.*, 1982).

The complete sequence of the 760-amino-acid-long transferrin receptor monomer has been deduced from cDNA sequences by McClelland *et al.* (1984) and Schneider *et al.* (1984). From these data both groups conclude that the cytoplasmic N-terminus comprises 61 residues. The extracellular C-terminus is about 670 amino acids, and the connecting transmembrane stretch is slightly below 30 residues. The two groups arrive at minor differences concerning the length of the transmembrane domain, the former claiming 28 and the latter 26 residues.

The receptor is posttranscriptionally modified by one complex chain and two mannose-rich oligosaccharides at three N-Asp glycosylation sites in the extracellular part (Omary and Trowbridge, 1981; McClelland *et al.*, 1984; Schneider *et al.*, 1984) and a fatty acylation on Cys-62 adjacent to the transmembrane region on the cytoplasmic side (Jing and Trowbridge, 1987). Recent data also indicate the presence of O-linked glycans (Neefjes *et al.*, 1988).

The transferrin receptor is phosphorylated on Ser-24 in the intracellular domain by the phospholipid- and calcium-dependent protein kinase C (Davis *et al.*, 1986).

3.5. Developmental Influence on Cellular Transferrin Receptors

Transferrin receptors are already present at the erythroid progenitor stage with a low and rather variable expression on BFU-E, and increase with the developmental stage over CFU-E to recognizable normoblasts. The majority of the latter cells are equipped with the receptor, as detected by fluorescence-activated cell sorting using monoclonal anti-transferrin receptor antibodies or direct transferrin labeling (Lebman *et al.*, 1982; Lesley *et al.*, 1984; Sieff *et al.*, 1982).

The number of transferrin receptors is highest in the early normoblast stages (Figure 1; Nunez *et al.*, 1977; Iacopetta and Morgan, 1983a; Iacopetta *et al.*, 1983). With further maturation, normoblasts and reticulocytes lose receptors and the erythrocyte has none (Iacopetta *et al.*, 1982, 1983; Jandl and Katz, 1963; Lebman *et al.*, 1982). Iacopetta *et al.* (1983) calculate a tenfold decrease of receptors in developing rat erythroid cells using quantitative fluorometry of cell-associated transferrin. Since rat reticulocytes have approximately 100,000 transferrin receptors per cell, the authors calculate that basophilic normoblasts have about one million receptors and the pronormoblasts only slightly less, polychromatic normoblasts about 500,000 receptors, and orthochromatic normoblasts around 300,000 receptors per cell.

3.6. The Transferrin Ligand

Transferrin is by definition the main ligand of the transferrin receptor. The main characteristics of this protein including its physiologically important interaction with iron are described below.

Changes of pH play a major role in the overall iron accumulation process, and many of the individual steps take place in acidified cellular compartments. Thus, special emphasis is put on the influence of pH.

3.6.1. General Characteristics

Transferrin, the most important iron-carrying plasma protein, has been reviewed by Morgan (1981), Huebers and Finch (1987), and others. Thanks to the transferrin carrier, the normal Fe^{3+} iron concentration in plasma is approximately 20 μM (Bothwell *et al.*, 1979). The Fe^{3+} ion alone is poorly soluble with a maximum concentration of 10^{-18} M at pH 8 (Forth and Rummel, 1973).

The human transferrin gene is situated on chromosome 3 (Yang *et al.*, 1984), as are the genes for the transferrin receptor (Section 3.1) and the melanoma-specific iron-binding antigen p97 (Plowman *et al.*, 1983). The main site of transferrin synthesis is the liver, where transcription of the gene is inversely related to iron supply, as found in the rat system by Idzerda *et al.* (1986). Potential regulatory elements of the transferrin gene transcription, conferring sensitivity to heavy metals and glucocorticoids, have been identified in the 5′ end of the gene (Adrian *et al.*, 1986). The liver cell-specific expression control resides in the 620 bp upstream of the cap site in the transferrin gene (Brunel *et al.*, 1988). The latter authors detected at least five different nuclear factors that interacted with different sites of this region.

The complete sequence of the 679 amino acids in transferrin is known both from protein sequencing and from cDNA cloning (MacGillivray *et al.*, 1983; Yang *et al.*, 1984). The protein consists of a single polypeptide chain with two domains that result from gene duplication (Park *et al.*, 1985).

Transferrin contains two branched glycans in the C-terminal domain (Spik *et al.*, 1979). The functional role of these carbohydrate moieties is still unclear.

3.6.2. Transferrin Iron Binding Characteristics

Each of the two transferrin domains possesses a site capable of binding a ferric ion in a carbonate-dependent manner (Chasteen, 1983). Circulating transferrin is about 30% saturated with iron (Aisen, 1982). Huebers *et al.* (1981) reported a random iron distribution between the monoferric and diferric transferrin species, whereas Lane (1975) found a slight excess of iron in the two monoferric species.

The iron is strongly bound to transferrin in circulating blood, with a $K_a \approx 1\text{–}6 \times 10^{22}\ M^{-1}$ for the two types of sites (Aisen and Leibman, 1978). Dissociation gradually commences with pH decreasing below 6.5, being complete at pH 4.5 in solution (Surgenor *et al.*, 1949).

A closer analysis reveals that the two iron binding sites are not quite equal with regard to pH-dependent dissociation (Princiotto and Zapolski, 1976). The iron at the N-terminal site is completely released at about pH 6.0, whereas the C-terminal iron is more resistant to acid pH (Veldman *et al.*, 1986). Iron release from transferrin is a result of protonation of the carbohydrate anion (Chasteen, 1983).

Data presented by Dautry-Varsat *et al.* (1983) show that the iron dissociates from transferrin at pH ≈ 5.5 when the carrier is attached to the transferrin receptor.

In conclusion, a moderately acid pH is sufficient for removing the iron moiety from the transferrin carrier both in solution and attached to the transferrin receptor.

3.7. Iron Extraction from the Transferrin–Transferrin Receptor Complex: A Cyclic Process Regulated by pH

The transferrin interaction with the transferrin receptor strongly depends on the degree of iron saturation of the transferrin and the local pH value. This, together with the acid sensitivity of the iron–transferrin bond, is utilized by the cell in the iron accumulation process.

Under physiological conditions, the interaction between iron-laden transferrin and the receptor is characterized by a $K_d \approx 10^{-8}$ to 10^{-9} M (e.g., Hamilton *et al.*, 1979; Karin and Mintz, 1981; Ciechanover *et al.*, 1983; Klausner *et al.*, 1983a). There is a competitive advantage of diferric over monoferric transferrin for the transferrin receptor, and apotransferrin is the least competitive form (Huebers *et al.*, 1983; Young *et al.*, 1984). The transferrin concentration in plasma is about 30 μM as calculated from the total iron binding capacity (Bothwell *et al.*, 1979). At a physiological iron saturation, more than 15% or 4.5 μM of the transferrin is present in the diferric form (Lane, 1975), which is sufficient to saturate all transferrin receptors.

The transferrin–transferrin receptor complex rapidly encounters an acid pH when processed by the cell (van Renswoude *et al.*, 1982). This causes release of the iron moiety from transferrin.

Below pH 5.5 the apotransferrin formed binds as strongly to the transferrin receptor as does diferric transferrin at neutral pH (Klausner *et al.*, 1983a; Dautry-Varsat *et al.*, 1983), and this provides the basis for separation of the iron moieties from the stable, membrane-attached protein complex (Dautry-Varsat *et al.*, 1983; Klausner *et al.*, 1983a). Finally, when the complex once again is subjected to a neutral pH, apotransferrin is rapidly dissociated from the receptor (Klausner *et al.*, 1983a; Dautry-Varsat *et al.*, 1983), and the latter is ready to reenter the cycle.

Sipe and Murphy (1987) demonstrated the full sequence of exposure of diferric transferrin and transferrin receptor to acid pH followed by neutralization. Exposure to a pH of 6.0 for 10 min caused a 50% release of transferrin upon neutralization, and approximately 80% is released after being subjected to pH 5.5. The authors did not investigate the iron saturation of the released transferrin, but it is safe to assume that the release is due to iron losses.

The binding properties of transferrin to its receptor contrast with those of other ligands like asialoglycoprotein and insulin (Dautry-Varsat *et al.*, 1983), epidermal growth factor (Willingham *et al.*, 1984), α_2-macroglobulin (Yamashiro *et al.*, 1984), and low-density lipoprotein (Basu *et al.*, 1978) that all dissociate from their receptors at low pH. In this respect they are similar to the iron moiety.

The basic physicochemical steps operating in the transferrin receptor-mediated concentration and extraction of iron can be summarized as follows. Diferric transferrin reacts with the high-affinity transferrin receptor. Iron is released by exposure to acid pH and leaves the apotransferrin bound to the membrane receptor. When neutralized, apotransferrin is rapidly replaced by another diferric transferrin, and the cycle can start again. The next two sections describe how the process is organized at the cellular level.

4. FUNCTIONING OF THE TRANSFERRIN RECEPTOR SYSTEM DURING ERYTHROID DEVELOPMENT

During the 1960s the basic principles of transferrin uptake were clarified. Jandl and Katz (1963) showed that transferrin is utilized in a cyclic manner during iron uptake in a receptor-dependent process in the reticulocyte. Morgan and Appleton (1969) demonstrated that transferrin is internalized into the cell.

The presentation below contrasts current knowledge about the transferrin–transferrin receptor recycling pathways of the nucleated erythroid cells with the shedding mechanism of the developed erythroid cells. The central role played by endosomal pH in the receptor–ligand processing merits a special subsection. The influence of other factors, e.g., hormones, in experimental situations is also treated.

4.1. The Transferrin Receptor Recycling Pathways

The transferrin receptors move through the nucleated cell via the normal receptor-mediated endocytosis pathways as shown in Figure 3. A minority of the receptors are exposed on the cell surface (10–40%), and most are situated in intracellular compartments (e.g., Bleil and Bretscher, 1982; Klausner *et al.*, 1983b; Harding *et al.*, 1983; Iacopetta and Morgan, 1983b; Forsbeck and Nilsson, 1985; Lamb *et al.*, 1983).

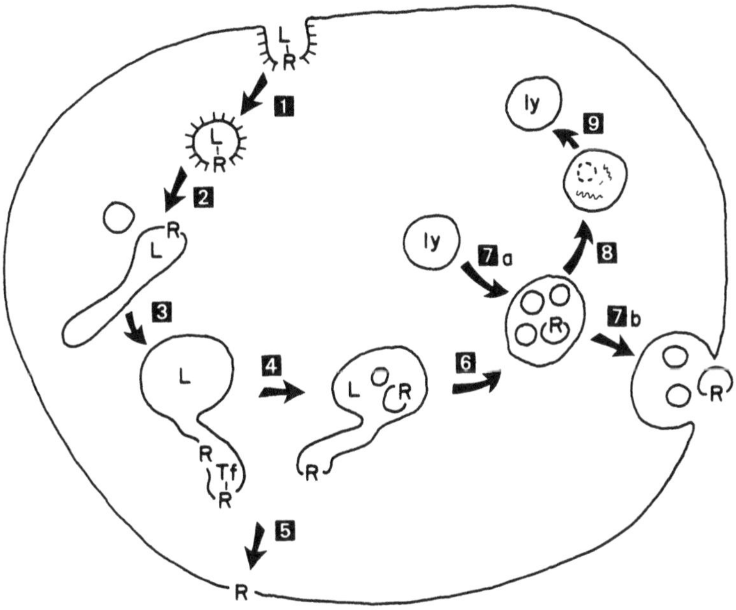

FIGURE 3. Pathways of receptor–ligand sorting. Receptor–ligand (L–R) complexes are internalized via coated pits and vesicles (1) and delivered by loss of coat and fusion to small uncoated vesicles and tubules (2). After transfer to large, sometimes multivesicular vesicles, analogous to CURL, some ligands dissociate in a pH-dependent manner (3, 4). These ligands (L) remain in the lumen of the vesicles, while receptors (R) are recycled to the plasma membrane (5). Transferrin (Tf), however, remains bound to its receptor and is also recycled. After sequestration in inclusion vesicles (4), receptors to be downregulated are concentrated in multivesicular bodies (6). In reticulocytes, developmental transferrin receptor loss is mediated by fusion of the multivesicular bodies to the plasma membrane, and resultant shedding to the extracellular space (7b). In macrophages, lysosomes (ly) may fuse with multivesicular bodies (7a), causing degradation of the receptors and inclusion bodies. This produces a degenerate multivesicular body (8), which eventually becomes a mature, densely staining lysosome (9). (After Harding *et al.*, 1985.)

The transferrin receptors are exposed on the cell surface with a T_{50} in the range of 2–5 min (Davis *et al.*, 1987; Karin and Mintz, 1981; Bleil and Bretscher, 1982; Hopkins and Trowbridge, 1983; Watts, 1985), and during this period iron-laden transferrins attach. The complexes cluster at clathrin-coated pits and are internalized (Bleil and Bretscher, 1982; Harding *et al.*, 1983; Hanover *et al.*, 1984; Willingham *et al.*, 1984) via coated vesicles (Figure 3, step 1; Figure 4A; Pearse, 1982; Choe *et al.*, 1987). Internalization also takes place in the absence of bound transferrins (Young *et al.*, 1984; Ajioka and Kaplan, 1986; Stein and Sussman, 1986; Watts, 1985).

The internalized transferrin receptors soon, within less than 5 min, localize in smooth vesicles, which via fusion events become larger (Figure 3, step 2; Figure 4A), and frequently connect with tubular extensions into the surrounding cytoplasm (Figure 3, step 3; Willingham *et al.*, 1984; Harding *et al.*, 1985). The larger endosomal compartments, which besides transferrin–transferrin receptor complexes contain an enrichment of other ligands with their receptors, have been termed receptosomes (Pastan and Willingham,

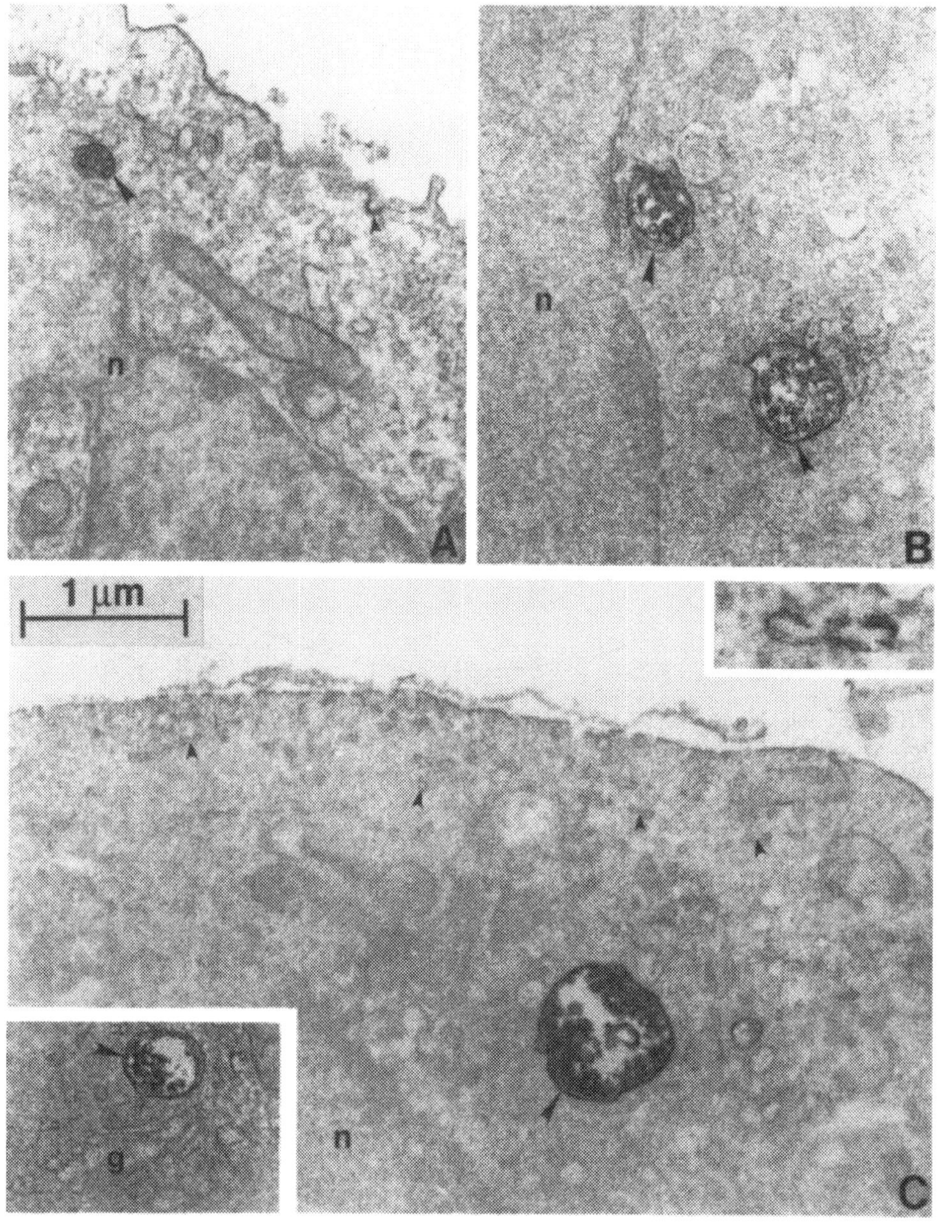

FIGURE 4. Immunoelectron microscopic localization of the transferrin cycle. Cell-surface human transferrin (bound at 4°C) was chased with bovine transferrin at 37°C in K562 cells for various time intervals. Human transferrin was then localized by electron microscopy employing immunoperoxidase staining after cells were fixed and permeabilized. (A) One-minute chase; reaction product is seen in plasma membrane invaginations (small arrowhead) and putative endocytic vesicles (large arrowhead); (B) 3-min chase; reaction product is seen in multivesicular bodies (arrowheads); (C) 5-min chase; reaction product is seen in multivesicular body (large arrowhead) and putative recycling vesicles (small arrowheads). Upper right inset is a high magnification of a dumbbell-shaped recycling vesicle. Lower left inset shows reaction product in a multivesicular body (arrowhead) closely associated with the membranes of a Golgi apparatus. n, nucleus; g, Golgi apparatus. (After Stein *et al.*, 1984.) One-micrometer bar inserted for A–C by the author. Further magnification of insets: upper right, × 3.3; lower left, × 1.25.

1983; Willingham *et al.*, 1984) or "compartments of uncoupling of receptor–ligand complex" (Geuze *et al.*, 1983; Harding *et al.*, 1985), widely named CURLs. In rat hepatocytes, solubilized asialoglycoproteins destined for lysosomal degradation remain in the vesicular part of CURLs, while the membrane receptors accumulate in the tubular extensions and return to the cell surface.

The acid pH in the endosomes (Helenius *et al.*, 1983; Anderson and Orci, 1988) plays an active role in the extraction of iron from the carrier transferrins and in the differential processing of various ligands (Sections 3.6.2 and 3.7). Iron release is so rapid that only apotransferrin could be detected in the endosomes of mouse cells by O'Keefe and Draper (1988). The stable apotransferrin–transferrin receptor complexes move into compartments separate from those containing ligands that are released from their receptors, e.g., epidermal growth factor (Dickson *et al.*, 1983b) and asialoglycoprotein (Stoorvogel *et al.*, 1987). The separation process has a T_{50} of approximately 2 min as studied in hepatoma cells using analysis by Percoll density gradients (Stoorvogel *et al.*, 1987). The separation of the released iron moieties is not known in such great detail, but presumably it is just as rapid.

The CURL structures represent the first stages of a series of morphologically developing endosomes, which gradually accumulate various electron-dense inclusions along with vesicles and transform into multivesicular bodies (MVBs) (Figure 3, steps 4 and 6; Figure 4B,C; Geuze *et al.*, 1983; Harding *et al.*, 1985). Two different developmental pathways have been observed for the MVBs, namely a degrading path with gradual transformation to secondary lysosomes as found in different cell types (Figure 3, steps 8–9; Geuze *et al.*, 1983; Harding *et al.*, 1985), and another path occurring in differentiated erythroid cells involving MVB fusion to the plasma membrane and subsequent release of the inclusion bodies (Figure 3, step 7b; Harding *et al.*, 1983).

Physical isolation of the main transferrin-containing endosomes by gradient centrifugation and gel filtration results in a population of vesicles with diameters between 0.2 and 0.4 μm. These vesicles are 30-fold enriched in transferrin receptors compared to plasma membrane and 9-fold depleted of lysosomal markers compared to lysosomes (Dickson *et al.*, 1983a; van Renswoude *et al.*, 1982).

During formation and development, the endosomes are actively transported toward the Golgi region of the cell (Willingham *et al.*, 1984), where a large number of transferrin–transferrin receptor-containing vesicles are gathered (Figure 4C, lower left inset). This concentration of transferrin and transferrin receptors in a juxtanuclear position can be easily detected by fluorescence-labeled transferrin and by immunocytochemical methods in nucleated cells, e.g., early normoblasts (Iacopetta *et al.*, 1983) and K562 cells (van Renswoude *et al.*, 1982; Forsbeck and Nilsson, 1985). Besides the usual constituents of the Golgi apparatus, this region also contains numerous mitochondria, frequently a centriole pair in the central part, but no ribosomes (Iacopetta *et al.*, 1983). Microtubules often emanate from the complex.

After passage through the immature MVBs, the apotransferrin–transferrin receptor complexes enter the exocytic pathway. This process is partly controlled by the Golgi apparatus as shown in the K562 cell line. Treatment with the carboxylic ionophore monensin, which causes major disruptions of Golgi-controlled processes (Ledger and Tanzer, 1984), results in an accumulation of transferrin in dilated MVBs and a 50% reduction of the surface transferrin receptor expression (Stein *et al.*, 1984). A non-Golgi-mediated recycling operates in parallel (Stein and Sussman, 1986). Two pathways have been demonstrated ultrastructurally in A431 cells (Hopkins, 1983) and by selection of a variant of mouse LMTK cells (O'Keefe and Draper, 1988). The recycling transferrin receptors never enter the *cis*-Golgi (Neefjes *et al.*, 1988) or the Golgi cisternae (Willingham *et al.*, 1984; Forsbeck *et al.*, 1986; Stoorvogel *et al.*, 1988) but connect to the exocytic path in the *trans*-Golgi reticulum (Stoorvogel *et al.*, 1988).

The exocytic transport takes place via tubular elements that are associated with microtubules, and closer to the cell membrane via small dumbbell-shaped vesicles (Figure 4C; Stein *et al.*, 1984; Willingham *et al.*, 1984). Fusion of the exocytic vesicles with the cell membrane leads to release of the intact apotransferrin from the transferrin receptor, and the cycle can start again (Figure 3, step 5).

The fastest transferrin passage through the cell is completed in 3–7 min (Karin and Mintz, 1981; Hopkins and Trowbridge, 1983; Iacopetta and Morgan, 1983b; Klausner *et al.*, 1983b). Due to the existence of the second pathway, which is slower, several hours is needed for a complete chase of a round of transferrin label (Karin and Mintz, 1981; O'Keefe and Draper, 1988; Stein *et al.*, 1984; Sorokin *et al.*, 1987). In the K562 cell line, neuraminidase-treated transferrin receptors are resialylated with such slow kinetics, presumably in the *trans*-Golgi region (Snider and Rogers, 1985). In a detailed study of the main rate constants for the transferrin recycling in human hepatoma cells, an average cycle time of 16 min was observed (Ciechanover *et al.*, 1983).

The intracellular transferrin processing is energy- and temperature-dependent (Jandl and Katz, 1963; Iacopetta and Morgan, 1983b), and requires an intact cytoskeleton (Hemmaplardh *et al.*, 1974; Galbraith and Galbraith, 1980), but is independent of protein synthesis (Iacopetta and Morgan, 1983b; Stein *et al.*, 1984).

4.2. Endosomal pH during the Transferrin–Transferrin Receptor Processing

The maintenance of low pH in the interior of both the endocytic and exocytic vesicles is a general phenomenon (Helenius *et al.*, 1983; Anderson and Orci, 1988). The different endosomes, like CURL and later multivesicular structures, are acidic as shown morphologically by the method of postembedding detection of acidic compartments (Anderson and Orci, 1988) in fibroblasts (Anderson *et al.*, 1984) and hepatocytes (Schwartz *et al.*, 1985). Interestingly, the tubular protrusions mediating the recycling are less acidic than the vesicle part.

The acid pH is maintained by energy- (Yamashiro *et al.*, 1983; Saermark *et al.*, 1985; Bakkeren *et al.*, 1987) and Mg^{2+}-dependent (Bakkeren *et al.*, 1987) proton pumps.

The acid interior of the endosomes can be neutralized, e.g., by addition of weak lysosomotropic amines (Ohkuma and Poole, 1978). Also, monensin exerts part of its effects by raising the endosomal pH (Ledger and Tanzer, 1984; Saermark *et al.*, 1985). This leads to improper routing of both ligands and receptors, and may even totally block recycling of the receptors to the surface (Tycko and Maxfield, 1982; Stoorvogel *et al.*, 1987). An observed effect is suppression of the steady-state levels of transferrin that are handled by the cells (Forsbeck and Nilsson, 1983).

In erythroid cells of different species and developmental stages, internalized transferrin–transferrin receptor complexes are subjected to pH $\approx$ 5.4–5.7 (Paterson *et al.*, 1984; van Renswoude *et al.*, 1982; Veldman *et al.*, 1986), whereas a pH above 6 has been reported in cells of nonerythroid origin (Sorkin *et al.*, 1988; Sipe and Murphy, 1987; Yamashiro *et al.*, 1984). This indicates a cell-type-specific difference in the acidification.

Full kinetic tracings of pH during the transferrin recycling were published by Sipe and Murphy (1987) for mouse Balb/c 3T3 cells. After a lag of 2 min upon binding to the functional receptor, transferrin is subjected to a transient acidification to pH 6.1 peaking at 5 min. A rise to external pH occurs with a T_{50} of 5 min. The entire cycle takes 15 min, which agrees with the recycling time measured by other methods.

4.3. The Shedding Mechanism

The number of transferrin receptors of erythroid cells decreases during development (Figure 1). Harding *et al.* (1985), studying the maturation of rat reticulocytes, demonstrated that transferrin receptors accumulate in the membranes of inclusion vesicles in MVBs (Figure 3, step 6). Subsequently, the MVBs fuse with the plasma membrane, and the inclusion vesicles are released (Figure 3, step 7b; Pan and Johnstone, 1983; Harding *et al.*, 1983, 1985; Pan *et al.*, 1985).

The shedded transferrin receptors still have the capacity to bind transferrin (Beguin *et al.*, 1988). Adam and Johnstone (1987) found that these receptors are modified in such a manner that they are no longer susceptible to phosphorylation by protein kinase C. The authors suggest that this modification provides the signal for externalization. This hypothesis needs further experimental evidence.

The occurrence of shedded transferrin receptors in the blood circulation is a positive indicator of normal as well as pathological erythropoiesis in the bone marrow as found in man (Kohgo *et al.*, 1987) and rat (Beguin *et al.*, 1988).

It is not clear if the lowering of the number of transferrin receptors also during earlier stages of erythroid development is due to the shedding mechanism. Decreased receptor synthesis in combination with cell division could contribute. Another possibility is the lysosomal pathway illustrated in Figure 3, steps 8 and 9.

4.4. Differentiation-Related Morphology

The morphology of the transferrin processing system in rat erythroid cells is strongly influenced by the degree of differentiation as revealed by the processing of fluorescence-labeled transferrin. Iacopetta *et al.* (1983) showed that in erythroid cells of all stages the transferrin label is located in discrete patches representing vesicles. Furthermore, early normoblasts contain a strongly labeled juxtanuclear area resulting from the gathering of transferrin-containing vesicles in the Golgi region (see Section 4.1).

The juxtanuclear labeled area becomes less accentuated with erythroid maturation for two reasons. First, the label is fainter due to receptor losses, and second, it is more spread around the nucleus due to the cell's inability to concentrate internalized vesicles when the Golgi structures and microtubules are involuted (Iacopetta *et al.*, 1983).

4.5. Distribution of the Transferrin Receptors

The relative distribution of the transferrin receptor population between the cell surface and the intracellular compartments is at a dynamic equilibrium governed by the flow rates between the different pools of receptors in the cell. These rates in turn can be influenced by hormonal signals as shown in epidermoid carcinoma cells (Davis *et al.*, 1987). Thus, both epidermal growth factor and insulinlike growth factor I redistribute the transferrin receptors from intracellular compartments to the cell surface. The former accomplishes this solely by increasing the rate of exocytosis and the latter by slightly increasing the rate of exocytosis and decreasing the rate of endocytosis.

The phospholipid- and calcium-dependent protein kinase C, a key enzyme in the transmission of extracellular signals into the cell (Ashendel, 1985), causes cell-type-dependent alterations in the distribution of the transferrin receptor pool with either increased (Buys *et al.*, 1984) or decreased (Forsbeck and Nilsson, 1985; Iacopetta *et al.*, 1986) surface expression, possibly by influencing the general rate of membrane trafficking (McGraw *et al.*, 1988). The cytoplasmic domain of the transferrin receptor is needed for receptor internalization, whereas the protein kinase C-mediated Ser-24 phosphorylation signal has no effect on the redistribution (Davis and Meisner, 1987; Rothenberger *et al.*, 1987; Zerial *et al.*, 1987).

In the rabbit and smaller rodents, heme causes a lowered internalization rate of transferrin receptors with resultant increased cell surface expression (Cox *et al.*, 1985). However, in human erythroid cells no such mechanism has been found (Gardner and Cox, 1988).

Alterations of the cytoplasmic calcium concentration play no major role in the partitioning of transferrin receptors. The only effect observed is a slight decrease of the internalization rate at a high cytoplasmic calcium concentration (Iacopetta *et al.*, 1986).

The carbohydrate moieties of the transferrin receptor have not been found to exert a regulatory role (Snider and Rogers, 1985; Neefjes *et al.*, 1988).

During mitosis, general rearrangements of the Golgi region lead to alterations in the vesicle traffic with a block of both endocytosis and exocytosis in the dividing cell as shown in epidermal carcinoma (Warren *et al.*, 1984) and HeLa cells (Sager *et al.*, 1984). Surface transferrin receptor numbers decrease and stay low until telophase, when the receptors reappear at higher levels than during interphase. The depletion is due to an immediate block in the exocytosis but continued internalization for a shorter period. At telophase, exocytosis resumes before endocytosis, resulting in the temporary excess of receptors (Warren *et al.*, 1984).

5. CELLULAR IRON METABOLISM

Even though ferric iron, bound to the transferrin–transferrin receptor complexes, is internalized in an endosomal compartment, it still has to be reduced to the ferrous state and traverse the membrane in order to reach the cytoplasm. The main factors operating in the iron accumulation process are discussed in Section 5.1 and summarized in Table II. In the cytoplasm, iron is partitioned between the different compartments via an intermediate transport pool, experimentally defined by its ability to form chelates. This pool regulates the number of transferrin receptors by a negative feedback mechanism and positively regulates the size of the iron-storing ferritin pool. The latter processes are schematically shown in Figure 5 and discussed in Section 5.2.

Table II
Factors Influencing the Cellular Accumulation of Iron

Factor	Stage of iron accumulation
Transferrin receptor numbers[a]	Concentration and entry of iron bound to transferrin
Endosomal acidification[b]	Iron release from transferrin in endosomes
Endosomal reduction[c]	Membrane passage from endosome to cytoplasm

[a]Muta *et al.* (1987), Nunez *et al.* (1977), Iacopetta *et al.* (1982), Iacopetta and Morgan (1983a).
[b]Bakkeren *et al.* (1987), Surgenor *et al.* (1949).
[c]Flatmark and Khan (1988), Thorstensen and Romslo (1988).

5.1. The Iron Accumulation Process

Iron accumulates at twice the rate at which diferric transferrin is internalized in reticulocytes (Iacopetta and Morgan, 1983b) and other cell types (Sorokin *et al.*, 1987). The entry takes place with iron still bound to transferrin, and evidences an intracellular site of iron release. Iron delivery to the cell is more rapid from diferric transferrin than

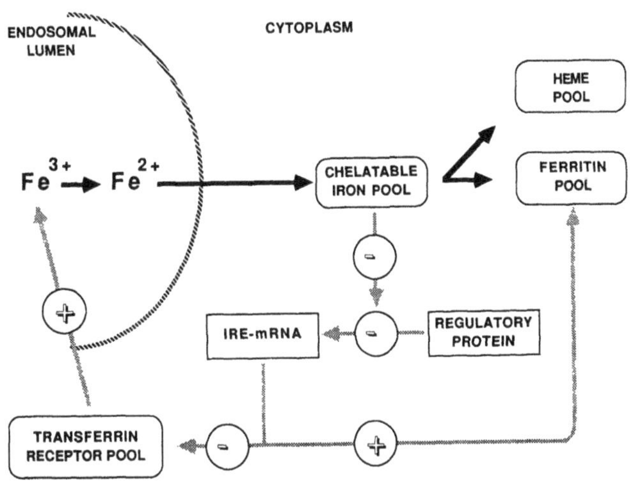

FIGURE 5. Feedback mechanisms controlling the cellular iron accumulation. The supply of ferric iron (Fe^{3+}) to the endosome is proportional to the size of the transferrin receptor pool. Ferric iron is released from the carrier transferrin and reduced (Fe^{2+}). After passage through the membrane into the cytoplasm, the iron is distributed mainly to the heme and ferritin pools via a chelatable transport pool. Iron-responsive elements in cellular mRNA (IRE-mRNA) take part in a regulatory mechanism by conferring an instability to the mRNA of the transferrin receptor and permitting full translation of the existing mRNA of ferritin. The cells contain a regulatory protein, which counteracts these effects by binding to the IRE in the absence of iron. The net outcome of such a binding is an increase in the transferrin receptor pool and a decrease in the ferritin pool.

Iron from the chelatable iron pool counteracts the binding of the regulatory protein to the IRE, thus causing a decrease in the transferrin receptor pool and a parallel increase in the ferritin pool. The mechanisms described above work in consort to keep a low level of chelatable iron.

Black arrows represent conservative iron flows, and gray arrows mark flows of information or regulatory events. ⊕, positive effect; ⊖, negative effect.

from the two monoferric species, the latter delivering their iron at similar rates (Huebers *et al.*, 1981).

The number of transferrin receptors is a major rate-determining factor in iron accumulation during the different stages of erythroid differentiation (Figure 1; Muta *et al.*, 1987; Nunez *et al.*, 1977; Iacopetta *et al.*, 1982). The rate is proportional to the steady-state level of transferrin associated with the cell as shown in, e.g., differentiating erythroid cells (Iacopetta and Morgan, 1983a) and K562 cells after various treatments (Forsbeck and Nilsson, 1985).

Whereas Iacopetta *et al.* (1982) found that each receptor extracts iron at an approximately fixed rate of 36 iron atoms per receptor per hour, irrespective of the differentiation stage within the erythroid series, there is evidence that the efficiency increases slightly with the differentiation stage in murine erythroid cells of both normal (Nunez *et al.*, 1977) and tumor origin (Hradilek and Neuwirt, 1987). The increases recorded are about 50 and 75%, respectively, and may depend on the degree of iron extraction from the carrier transferring, as found by the latter authors in erythroleukemia cells of various differentiation stages. It has been suggested that cell types with higher iron requirements use a more acidic pH in the transferrin-processing vesicles than cells with lower needs (Sipe and Murphy, 1987). The increase may also be due to a selective loss of the more slowly recycling transferrin receptors.

Low concentrations of lysosomotropic agents depress iron accumulation in the cell with no effects on the receptor recycling mechanism (Ciechanover *et al.*, 1983; Forsbeck and Nilsson, 1983; Iacopetta and Morgan, 1983b), whereas higher concentrations also interfere with the receptor recycling, demonstrating that the release of iron is more pH dependent than the receptor recycling.

A light and a heavy transferrin-containing endosome population can be isolated by Percoll density gradient centrifugation (Forsbeck *et al.*, 1986). By 3 to 5 min after uptake, iron is extracted from the transferrin and passes the endosome membrane in the heavy endosomes (Bakkeren *et al.*, 1987). The authors found that the iron is released into a low-molecular-weight, chelatable form in the acid endosomal interior. Addition of reducing equivalents from NADH is needed for membrane passage to occur (Flatmark and Khan, 1988). The inorganic Fe^{3+} ion has a maximum limited solubility of 10^{-18} M at pH 8, whereas the Fe^{2+} has a maximum of 1.6×10^{-2} M (Forth and Rummel, 1973), and several workers have pointed to the role of reductive systems in the transport and membrane passage of iron (for a review, see Crane *et al.*, 1985). There has been a tendency to put the mechanism of surface reduction against that of receptor-mediated endocytosis, but now they are considered to work in consort (Thorstensen and Romslo, 1988). Ferric iron, liberated in the acid endosomal milieu, is readily reduced by the transmembrane NADH diferric transferrin reductase in a transferrin receptor-dependent process (Löw *et al.*, 1986, 1987; Thorstensen and Romslo, 1988). The reductase produces protons, which add to the ones produced by the endosomal proton pumps. Experimental data favor reduction occurring at the luminal side of the membrane, since membrane-impermeable reagents like bathophenanthroline disulfonate react with the Fe^{2+} formed, and the membrane passage process shares many features of uptake of inorganic Fe^{2+} (Thorstensen and Romslo, 1988).

The morphological correlates to the physically isolated iron-extracting endosomes remain to be formally identified, but are most likely the receptosomes or CURLs as discussed in Section 4.1 (van Renswoude *et al.*, 1982; Anderson *et al.*, 1984; Schwartz *et al.*, 1985). Much work remains to clarify the molecular details of the iron release and passage through the endosome membrane and the roles of the transferrin receptor, if any, in this context.

5.2. Intracellular Iron Pools and Their Participation in Regulatory Mechanisms

Ferritin is the universal intracellular iron storage protein in all cells. It has an outer shell built of 24 subunits with varying proportions between the two H (heavy) and L (light) chain types of apoferritin and has a central content of inorganic ferric oxyhydroxide polymer with some phosphate (reviewed, e.g., by Chrichton and Charloteaux-Wauters, 1987).

Heme is the major iron-containing structure in mature erythroid cells. Fe^{2+} is incorporated in protoporphyrin IX by the enzyme ferrochelatase situated in the inner mitochondrial membrane (Sassa, 1983; Harbin and Dailey, 1985). During a 90-min pulse period, human bone marrow erythroid cells and reticulocytes use approximately 50% of the accumulated iron in heme groups, and rabbit reticulocytes use approximately 90% (Gardner and Cox, 1988).

The ferritin protein levels are regulated in proportion to the iron supply as demonstrated in K562 cells by Mattia *et al.* (1986). This supply directly governs the iron storage capacity, since the latter is positively related to the actual ferritin level. Preformed, inactive cytoplasmic ferritin mRNA is converted to an active form upon iron administration (Aziz and Munro, 1987) via an iron-responsive element, a stem-loop structure located in the 5′ untranslated region of the ferritin mRNA (Figure 5) of the same kind as the iron-responsive elements in the transferrin receptor mRNA (Casey *et al.*, 1988b). In addition to this strong positive translational control, iron-dependent transcription (Cairo *et al.*, 1985) and translation (Rouault *et al.*, 1987) have been demonstrated.

Low-molecular-weight intracellular transport compounds have been postulated to distribute the cytoplasmic iron as reviewed e.g. by Jacobs (1977). Application of various cell-penetrating iron chelators has demonstrated a chelatable iron pool that presumably represents such a transport intermediate (e.g., Bottomley *et al.*, 1985; Morgan, 1983; White *et al.*, 1976). Numerical estimates of the intermediate iron pool are scarce. Pulse–chase experiments in erythroid K562 cells have shown that less than 10% of the cellularly incorporated iron is in compartments not yet identified (Bottomley *et al.*, 1985). Higher fractions are found in hepatocytes, with a steady state of more than 20% of total iron and measuring 23 pg/10^6 cells (Young *et al.*, 1985). Mattia *et al.* (1986), using K562 cells, estimated the dwelling time of iron in this pool to be several hours.

The cell has a sensory system, which monitors the chelatable iron pool and governs the biosynthesis of the transferrin receptors and ferritins (Rogers and Munroe, 1987; Mattia and van Renswoude, 1988). Removal of iron from the pool by treatment with various chelators increases the synthesis and expression of transferrin receptors (Bottomley *et al.*, 1985; Forsbeck *et al.*, 1987; Mattia *et al.*, 1984; Rao *et al.*, 1986; Rudolph *et al.*, 1985), and an increase of the cellular iron content via iron-saturated media (Rao *et al.*, 1986; Rudolph *et al.*, 1985) or hemin (Bottomley *et al.*, 1985; Rao *et al.*, 1986) decreases the transferrin receptor expression. Reverse results are obtained for ferritin, where the iron exerts a positive effect on the biosynthesis (Rogers and Munroe, 1987).

The detailed molecular biology of the sensing and regulatory mechanisms between the iron and the transferrin receptor and the ferritin genes and mRNA is being explored. Rouault *et al.* (1988) showed that a regulatory protein binds to the iron-responsive element on ferritin mRNA when iron is scarce and dissociates when iron is abundant. In preliminary experiments they found that this regulatory protein also interacts with the iron-responsive element on the transferrin receptor mRNA, thus being a part of a universal regulatory pathway. Data obtained by Müllner and Kühn (1988) suggest that the regulatory protein factor has a T_{50} of around 40 min. When the regulatory protein binds to the 3′ elements in the transferrin receptor transcripts, the latter are protected from degradation, and when binding takes place to the 5′ iron-responsive element in the case of ferritin mRNA, attenuation of translation takes place (Casey *et al.*, 1988b). These regulatory mechanisms are schematically shown in Figure 5.

The ferritin iron stores are not easily mobilized for biosynthetic purposes in the living erythroid cell (Forsbeck and Nilsson, 1983), but represent a surplus (Zail *et al.*, 1964) that is actively excreted (Deiss and Cartwright, 1970; Mattia *et al.*, 1986). Iacopetta and Morgan (1983a) calculated that the rat erythroid cell takes up about twice as much iron as is needed for the hemoglobin synthesis. A strong negative influence on the cellular ferritin content is exerted by the erythroid differentiation process. The most immature erythroblasts contain about 2000 times more ferritin than erythrocytes (Ali *et al.*, 1983).

6. CONCLUDING COMMENTS

The transferrin receptor participation in the erythroid iron metabolism has many facets. The cellular development imposes a changing environment, in which receptor-mediated endocytosis of the iron carrier transferrin provides the synthesized hemes with iron. Beside this nutritional aspect, the processing of the receptor and its ligand is also a part of the vast field of cellular communication with the surroundings.

Two main phases can be distinguished concerning the operation of transferrin receptors during erythroid differentiation. The immature, nucleated cell efficiently recycles the synthesized transferrin receptors within a well-developed organelle system. The available iron, passing through a chelatable transport pool, regulates the synthesis of receptors by negative feedback. The iron concentration stays low when the immature cell synthesizes heme. Surplus iron is incorporated into ferritin, which is synthesized in increasing amounts when iron is available.

Later in erythroid development, heme synthesis decreases and less iron is needed. This signals a stop in transferrin receptor synthesis. A number of degenerative changes occur, some of them directly affecting the endocytosis mechanism. Instead of being recycled, the set of transferrin receptors is shed from the cell. These factors, possibly together with lysosomal action, result in disappearance of the receptors from the differentiated erythrocyte. During development, the iron accumulated is about twice as much as that finally associated with hemoglobin. The surplus, bound to ferritin, is actively extruded from the cell.

The outlines of receptor-mediated endocytosis are known. The iron-carrying transferrins bind to receptors exposed on the cell surface. The complexes enter acid endosomes where iron is released and accumulated in the cell, whereas the apotransferrin–receptor complexes enter one of two alternative exocytosis pathways, one fast and one slow. Finally, apotransferrins are released from the receptors on the cell surface when subjected to neutral pH, and a new functional cycle can start.

The endosomes involved in the receptor–ligand processing go through a development from immature smooth vesicles to increasingly multivesicular forms, MVBs. With time, primary lysosomes fuse to the MVBs, which develop into secondary lysosomes. In the immature erythroid cell, recycling receptors like the one for transferrin enter an exocytosis pathway from the early forms of MVBs. In the more mature erythroid cell, however, a new path leading to receptor shedding is followed. The receptors are gathered in the membranes of the inclusion vesicles of the MVBs. The MVBs fuse to the cell membrane and release the content of vesicles with the transferrin receptors. It is also possible that lysosomal degradation contributes to the observed loss of receptors in mature erythroid cells.

The detailed molecular mechanisms governing the functions of the transferrin receptor are poorly understood. Application of molecular biology increases our understanding of how the inorganic ionized iron exerts its control on the gene and mRNA levels. There is a fair understanding of the mechanisms at the mRNA level, but the regulatory proteins involved are only about to be characterized. The signals governing receptor movement both during normal iron accumulation and switch of paths during erythroid differentiation are totally unknown. Also, the molecular details concerning the release, reduction, and transport of iron are unknown. It even remains to be established up to which developmental stage the transferrin receptor is still actively involved.

ACKNOWLEDGMENTS. The author thanks F. Berglund and B. Sjöberg for critical reading of the manuscript and A. Tangen for secretarial aid.

REFERENCES

Adam, A., and Johnstone, R. M., 1987, Protein kinase C does not phosphorylate the externalized form of the transferrin receptor, *Biochem. J.* **242:**151–161.

Adrian, G. S., Korinek, B. W., Bowman, B. H., and Yang, F., 1986, The human transferrin gene: 5′ region contains conserved sequences which match the control elements regulated by heavy metals, glucocorticoids and acute phase reaction, *Gene* **49:**167–175.

Aisen, P., 1982, Current concepts in iron metabolism, *Clin. Haematol.* **11:**241–257.

Aisen, P., and Leibman, A., 1978, Thermodynamic and accessibility factors in the specific binding to human transferrin, in: *Transport by Proteins* (G. Blauer and H. Sund, eds.), pp. 277–294, de Gruyter, Berlin.

Ajioka, R. S., and Kaplan, J., 1986, Intracellular pools of transferrin receptors result from constitutive internalization of unoccupied receptors, *Proc. Natl. Acad. Sci. USA* **83:**6445–6449.

Ali, F. M. K., May, A., Jones, B. M., and Jacobs, A., 1983, Enrichment of erythroblasts from human bone marrow using complement-mediated lysis: Measurements of ferritin, *Br. J. Haematol.* **53:**227–235.

Anderson, R. G. W., and Orci, L., 1988, A view of acidic intracellular compartments, *J. Cell Biol.* **106:**539–543.

Anderson, R. G. W., Falck, J. R., Goldstein, J. L., and Brown, M. S., 1984, Visualization of acidic organelles in intact cells by electron microscopy, *Biochemistry* **24:**366–370.

Andersson, L. C., Nilsson, K., and Gahmberg, C. G., 1979, K562—A human erythroleukemic cell line, *Int. J. Cancer* **23:**143–147.

Arnstein, H. R. V., Langstaff, J. M., Ong, G., Threadgill, G. J., and Bonanou-Tzedaki, S. A., 1987, Control of macromolecular synthesis and degradation during terminal erythroid cell development, *Biomed. Biochem. Acta* **46:**S115–S119.

Ashendel, C. L., 1985, The phorbol ester receptor: A phospholipid-regulated protein kinase, *Biochim. Biophys. Acta* **822:**219–242.

Aziz, N., and Munro, H. N., 1987, Iron regulates ferritin mRNA translation through a segment of its 5′ untranslated region, *Proc. Natl. Acad. Sci. USA* **84:**8478–8482.

Bakkeren, D. L., de Jeu-Jaspars, C. M. H., Kroos, M. J., and van Eijk, H. G., 1987, Release of iron from endosomes is an early step in the transferrin cycle, *Int. J. Biochem.* **19:**179–186.

Basu, S. K., Goldstein, J. L., and Brown, M. S., 1978, Characterization of the low density lipoprotein receptor in membranes prepared from human fibroblasts, *J. Biol. Chem.* **253:**3852–3856.

Beguin, Y., Huebers, H. A., Josephson, B., and Finch, C. A., 1988, Transferrin receptors in rat plasma, *Proc. Natl. Acad. Sci. USA* **85:**637–640.

Bleil, J. D., and Bretscher, M. S., 1982, Transferrin receptor and its recycling in HeLa cells, *EMBO J.* **1:**351–355.

Bothwell, T. H., Charlton, R. W., Cook, J. D., and Finch, C. A., 1979, *Iron Metabolism in Man,* Blackwell, Oxford.

Bottomley, S. S., Wolfe, L. C., and Bridges, K. R., 1985, Iron metabolism in K562 erythroleukemic cells, *J. Biol. Chem.* **260:**6811–6815.

Brunel, F., Ochoa, A., Schaeffer, E., Boissier, F., Guillou, Y., Cereghini, S., Cohen, G. N., and Zakin, M. M., 1988, Interactions of DNA-binding proteins with the 5′ region of the human transferrin gene, *J. Biol. Chem.* **263:**10180–10185.

Buys, S. S., Keogh, E. A., and Kaplan, J., 1984, Fusion of intracellular membrane pools with cell surfaces of macrophages stimulated by phorbol esters and calcium ionophores, *Cell* **38:**569–576.

Cairo, G., Bardella, L., Schiaffoniata, L., Arosio, P., Levi, S., and, Bernelli-Zazzera, A., 1985, Multiple mechanisms of iron-induced ferritin synthesis in HeLa cells, *Biochem. Biophys. Res. Commun.* **133:**314–321.

Casey, J. L., di Jeso, B., Rao, K., Klausner, R. D., and Harford, J. B., 1988a, Two genetic loci participate in the regulation by iron of the gene for the human transferrin receptor, *Proc. Natl. Acad. Sci. USA* **85:**1787–1791.

Casey, J. L., Hentze, M. W., Koeller, D. M., Caughman, S. W., Rouault, T. A., Klausner, R. D., and Harford, J. B., 1988b, Iron-responsive elements: Regulatory RNA sequences that control mRNA levels and translation, *Science* **240:**924–928.

Chasteen, N. D., 1983, The identification of the probable locus of iron and anion binding in the transferrins, *Trends Biochem. Sci.* **8:**272–275.

Choe, H. R., Moseley, S. T., Glass, J., and Nunez, M. T., 1987, Rabbit reticulocyte coated vesicles carrying the transferrin–transferrin receptor complex: I. Purification and partial characterization, *Blood* **70:**1035–1039.

Chrichton, R. R., and Charloteaux-Wauters, M., 1987, Iron transport and storage, *Eur. J. Biochem.* **164:**485–506.

Ciechanover, A., Schwartz, A. L., Dautry-Varsat, A., and Lodish, H. F., 1983, Kinetics of internalization and recycling of transferrin and the transferrin receptor in a human hepatoma cell line. Effects of lysosomotropic agents, *J. Biol. Chem.* **258:**9681–9689.

Cox, T. M., O'Donnell, M. W., Aisen, P., and London, I. M., 1985, Hemin inhibits internalization of the membrane transferrin receptor, *Proc. Natl. Acad. Sci. USA* **82:**5170–5174.

Crane, F. L., Sun, I. L., Clark, M. G., Grebing, C., and Löw, H., 1985, Transplasma-membrane redox systems in growth and development, *Biochim. Biophys. Acta* **811:**233–264.

Dautry-Varsat, A., Ciechanover, A., and Lodish, H. F., 1983, pH and the recycling of transferrin during receptor-mediated endocytosis, *Proc. Natl. Acad. Sci. USA* **80:**2258–2262.

Davis, R. J., and Meisner, H., 1987, Regulation of transferrin receptor cycling by protein kinase C is independent of receptor phosphorylation at serine 24 in Swiss 3T3 fibroblasts, *J. Biol. Chem.* **262:**16041–16047.

Davis, R. J., Johnson, G. L., Kelleher, D. J., Anderson, J. K., and Czech, M. P., 1986, Identification of serine 24 as the unique site on the transferrin receptor phosphorylated by protein kinase C, *J. Biol. Chem.* **261:**9034–9041.

Davis, R. J., Faucher, M., Racaniello, L. K., Carruthers, A., and Czech, M. P., 1987, Insulin-like growth factor I and epidermal growth factor regulate the expression of transferrin receptors at the cell surface by distinct mechanisms, *J. Biol. Chem.* **262:**13126–13134.

Deiss, A., and Cartwright, G. E., 1970, Ferritin metabolism in reticulated-siderocytes, *J. Clin. Invest.* **49:**517–523.

Dickson, R. B., Beguinot, L., Hanover, J. A., Richert, N. D., Willingham, M. C., and Pastan, I., 1983a, Isolation and characterisation of a highly enriched preparation of receptosomes (endosomes) from a human cell line, *Proc. Natl. Acad. Sci. USA* **80:**5335–5339.

Dickson, R. B., Hanover, J. A., Willingham, M. C., and Pastan, I., 1983b, Prelysosomal divergence of transferrin and epidermal growth factor during receptor-mediated endocytosis, *Biochemistry* **22:**5667–5674.

Flatmark, T., and Khan, M. R., 1988, The release of iron from a subfraction of rat liver highly enriched in endosomal organelles requires both a functional H^+-ATPase and NADH, in: *Plasmamembrane Oxidoreductase in Control of Animal and Plant Growth* (F. L. Crane, D. J. Morré, and H. Löw, eds.), pp. 117–126, Plenum Press, New York.

Forsbeck, K., and Nilsson, K., 1983, Iron metabolism of established human hematopoietic cell lines in vitro, *Exp. Cell Res.* **144:**323–332.

Forsbeck, K., and Nilsson, K., 1985, The dynamic morphology of the transferrin–transferrin receptor system in human leukemia/lymphoma cell lines and its relation to iron metabolism and cell proliferation, *Scand. J. Haematol.* **35:**145–154.

Forsbeck, K., Ericsson, J., Birgegård, G., Malmgren, M., and Nilsson, K., 1986, Subcellular characterization of the transferrin–transferrin receptor and iron accumulating system of established human erythroid and monoblastoid tumour cell lines, *Acta Pathol. Microbiol. Immunol. Scand. Sect. A* **94:**245–252.

Forsbeck, K., Nilsson, K., and Kontoghiorghes, G. J., 1987, Variation in transferrin membrane binding and DNA synthesis in the K-562 and U-937 cell lines induced by chelators and their iron complexes, *Eur. J. Haematol.* **39:**318–325.

Forth, W., and Rummel, W., 1973, Iron absorption, *Physiol. Rev.* **53:**724–792.

Galbraith, G. M. P., and Galbraith, R. M., 1980, Metabolic and cytoskeletal modulation of transferrin receptor mobility in mitogen-activated human lymphocytes, *Clin. Exp. Immunol.* **42:**285–293.

Gardner, L. C., and Cox, T. M., 1988, Biosynthesis of heme in immature erythroid cells. The regulatory step for heme formation in the human erythron, *J. Biol. Chem.* **263:**6676–6682.

Geuze, H. J., Slot, J. W., and Strous, G. J. A. M., 1983, Intracellular site of asialoglycoprotein receptor–ligand uncoupling: Double-label immunoelectron microscopy during receptor-mediated endocytosis, *Cell* **32:**277–287.

Glass, J., Lavidor, L. M., and Robinson, S. H., 1975, Studies of murine erythroid cell development. Synthesis of heme and hemoglobin, *J. Cell Biol.* **65:**298–308.

Goodfellow, P. N., Banting, G., Sutherland, R., Greaves, M., Solomon, E., and Povey, S., 1982, Expression of human transferrin receptor is controlled by a gene on chromosome 3: Assignment using species specificity of a monoclonal antibody, *Somat. Cell. Mol. Genet.* **8:**197–206.

Hamilton, T. A., Wada, H. G., and Sussman, H. H., 1979, Identification of transferrin receptors on the surface of human cultured cells, *Proc. Natl. Acad. Sci. USA* **76:**6406–6410.

Hanover, J. A., Willingham, M. C., and Pastan, I., 1984, Kinetics of transit of transferrin and epidermal growth factor through clathrin-coated membranes, *Cell* **39:**283–293.

Harbin, B. M., and Dailey, H. A., 1985, Orientation of ferrochelatase in bovine liver mitochondria, *Biochemistry* **24:**366–370.

Harding, C., Heuser, J., and Stahl, P., 1983, Receptor-mediated endocytosis of transferrin and recycling of transferrin and recycling of the transferrin receptor in rat reticulocytes, *J. Cell Biol.* **97:**329–339.

Harding, C., Levy, M. A., and Stahl, P., 1985, Morphological analysis of ligand uptake and processing: The role of multivesicular endosomes and CURL in receptor–ligand processing, *Eur. J. Cell Biol.* **36:**230–238.

Helenius, A., Mellman, I., Wall, D., and Hubbard, A., 1983, Endosomes, *Trends Biochem. Sci.* **8:**245–250.

Hemmaplardh, D., Kailis, S. G., and Morgan, E. H., 1974, The effects of inhibitors of microtubule and microfilament function on transferrin and iron uptake by rabbit reticulocytes and bone marrow, *Br. J. Haematol.* **28:**53–65.

Hopkins, C. R., 1983, Intracellular routing of transferrin and transferrin receptors in epidermoid carcinoma A431 cells, *Cell* **35:**321–330.

Hopkins, C. R., and Trowbridge, J. S., 1983, Internalization and processing of transferrin and the transferrin receptor in human carcinoma A431 cells, *J. Cell Biol.* **97:**508–521.

Hradilek, A., and Neuwirt, J., 1987, Iron uptake and transferrin endocytosis in undifferentiated and differentiated erythroid cells, *Biomed. Biochem. Acta* **46:**S141–S145.

Huebers, H. A., and Finch, C. A., 1987, The physiology of transferrin and transferrin receptors, *Physiol. Rev.* **67**:520–582.

Huebers, H. A., Josephson, B., Huebers, E., Csiba, E., and Finch, C. A., 1981, Uptake and release of iron from human transferrin, *Proc. Natl. Acad. Sci. USA* **78**:2572–2576.

Huebers, H. A., Csiba, E., and Finch, C. A., 1983, Competitive advantage of diferric transferrin in delivering iron to reticulocytes, *Proc. Natl. Acad. Sci. USA* **80**:300–304.

Iacopetta, B. J., and Morgan, E. H., 1983a, Transferrin endocytosis and iron uptake during erythroid cell development, *Biomed. Biochem. Acta* **42**:S182–S186.

Iacopetta, B. J., and Morgan, E. H., 1983b, The kinetics of transferrin endocytosis and iron uptake from transferrin in rabbit reticulocytes, *J. Biol. Chem.* **258**:9108–9115.

Iacopetta, B. J., Morgan, E. H., and Yeoh, G. C. T., 1982, Transferrin receptors and iron uptake during erythroid cell development, *Biochim. Biophys. Acta* **687**:204–210.

Iacopetta, B. J., Morgan, E. H., and Yeoh, G. C. T., 1983, Receptor-mediated endocytosis of transferrin by developing erythroid cells from the fetal rat liver, *J. Histochem. Cytochem.* **31**:336–344.

Iacopetta, B. J., Carpentier, J. L., Pozzan, T., Lew, D. P., Gorden, P., and Orci, L., 1986, Role of intracellular calcium and protein kinase C in the endocytosis of transferrin and insulin by HL60 cells, *J. Cell Biol.* **103**:851–856.

Idzerda, R. L., Huebers, H., Finch, C. A., and McKnight, G. S., 1986, Rat transferrin gene expression: Tissue-specific regulation by iron deficiency, *Proc. Natl. Acad. Sci. USA* **83**:3723–3727.

Iscove, N. N., and Sieber, F., 1975, Erythroid progenitors in mouse bone marrow detected by macroscopic colony formation in culture, *Exp. Hematol.* **3**:32–43.

Jacobs, A., 1977, Low molecular weight intracellular iron transport compounds, *Blood* **50**:433–439.

Jandl, J., and Katz, J. H., 1963, The plasma-to-cell cycle of transferrin, *J. Clin. Invest.* **42**:314–326.

Jing, S., and Trowbridge, I. S., 1987, Identification of the intermolecular disulfide bonds of the human transferrin receptor and its lipid-attachment site, *EMBO J.* **6**:327–331.

Karin, M., and Mintz, B., 1981, Receptor-mediated endocytosis of transferrin in developmentally totipotent mouse teratocarcinoma stem cells, *J. Biol. Chem.* **256**:3245–3252.

Klausner, R. D., Ashwell, G., van Renswoude, J., Harford, J. B., and Bridges, K. R., 1983a, Binding of apotransferrin to K562 cells: Explanation of the transferrin cycle, *Proc. Natl. Acad. Sci. USA* **80**:2263–2266.

Klausner, R. D., van Renswoude, J., Ashwell, G., Kempf, C., Schechter, A. N., Dean, A., and Bridges, K. R., 1983b, Receptor-mediated endocytosis of transferrin in K562 cells, *J. Biol. Chem.* **258**:4715–4724.

Kohgo, Y., Niitsu, Y., Kondo, H., Kato, J., Tsushima, N., Sasaki, K., Hirayama, M., Numata, T., Nishisato, T., and Urushizaka, I., 1987, Serum transferrin receptor as a new index of erythropoiesis, *Blood* **70**:1955–1958.

Kohno, H., Taketani, S., and Tokunaga, R., 1986, Phorbol ester-induced regulation of transferrin receptors in human leukemia K562 cells, *Cell. Struct. Funct.* **11**:181–190.

Kühn, L. C., McClelland, A., and Ruddle, F. H., 1984, Gene transfer, expression, and molecular cloning of the human transferrin receptor gene, *Cell* **37**:95–103.

Lamb, J. E., Ray, F., Ward, J. H., Kushner, J. P., and Kaplan, J., 1983, Internalization and subcellular localization of transferrin and transferrin receptors in HeLa cells, *J. Biol. Chem.* **258**:8751–8758.

Lane, R. S., 1975, Differences between human Fe_1-transferrin molecules, *Br. J. Haematol.* **29**:511–520.

Lebman, D., Trucco, M., Bottero, L., Lange, B., Pessano, S., and Rovera, G., 1982, A monoclonal antibody that detects expression of transferrin receptor in human erythroid cells, *Blood* **59**:671–678.

Ledger, P. W., and Tanzer, M. L., 1984, Monensin—A perturbant of cellular physiology, *Trends Biochem. Sci.* **9**:313–314.

Lesley, J., Domingo, D. L., Schulte, R., and Trowbridge, I. S., 1984, Effect of an anti-murine transferrin receptor–ricin A conjugate on bone marrow stem and progenitor cells treated in vitro, *Exp. Cell Res.* **150**:400–407.

Löw, H., Sun, I. L., Navas, P., Grebing, C., Crane, F. L., and Morré, D. J., 1986, Transplasmalemma electron transport from cells is part of a diferric transferrin reductase system, *Biochem. Biophys. Res. Commun.* **139**:1117–1123.

Löw, H., Grebing, C., Lindgren, A., Tally, M., Sun, I. L., and Crane, F. L., 1987, Involvement of transferrin in the reduction of iron by the transplasma membrane electron transport system, *J. Bioenerg. Biomembr.* **19**:535–549.

McClelland, A., Kühn, L. C., and Ruddle, F. H., 1984, The human transferrin receptor gene: Genomic organization, and complete primary structure of the receptor deduced from a cDNA sequence, *Cell* **39**:267–274.

MacGillivray, R. T. A., Mendez, E., Shewale, J. G., Sinha, S. K., Lineback-Zins, J., and Brew, K., 1983, The primary structure of human serum transferrin. The structures of seven cyanogen bromide fragments and the assembly of the complete structure, *J. Biol. Chem.* **258**:3543–3553.

McGraw, T. E., Dunn, K. W., and Maxfield, F. R., 1988, Phorbol ester treatment increases the exocytic rate of the transferrin receptor recycling pathway independent of serine-24 phosphorylation, *J. Cell Biol.* **106**:1061–1066.

Marks, P. A., and Rifkind, R. A., 1978, Erythroleukemic differentiation, *Annu. Rev. Biochem.* **47**:419–448.

Mattia, E., and van Renswoude, J., 1988, The pivotal role of ferritin in cellular iron homeostasis, *Bioessays* **8**:107–111.

Mattia, E., Rao, K., Shapiro, D. S., Sussman, H. H., and Klausner, R. D., 1984, Biosynthetic regulation of the human transferrin receptor by desferrioxamine in K562 cells, *J. Biol. Chem.* **259**:2689–2692.

Mattia, E., Josic, D., Ashwell, G., Klausner, R., and van Renswoude, J., 1986, Regulation of intracellular iron distribution in K562 human erythroleukemia cells, *J. Biol. Chem.* **261**:4587–4593.

Miskimins, W. K., McClelland, A., Roberts, M. P., and Ruddle, F. H., 1986, Cell proliferation and expression of the transferrin receptor gene: Promoter sequence homologies and protein interactions, *J. Cell Biol.* **103**:1781–1788.

Morgan, E. H., 1981, Transferrin: Biochemistry, physiology and clinical significance, *Mol. Aspects Med.* **4**:1–123.

Morgan, E. H., 1983, Chelator-mediated iron efflux from reticulocytes, *Biochim. Biophys. Acta* **733**:39–50.

Morgan, E. H., and Appleton, T. C., 1969, Autoradiographic localization of ^{125}I-labelled transferrin in rabbit reticulocytes, *Nature* **223**:1371–1372.

Müllner, E. W., and Kühn, L. C., 1988, A stem-loop in the 3′ untranslated region mediates iron-dependent regulation of transferrin receptor mRNA stability in the cytoplasm, *Cell* **53**:815–825.

Muta, K., Nishimura, J., Ideguchi, H., Umemura, T., and Ibayashi, H., 1987, Erythroblast transferrin receptors and transferrin kinetics in iron deficiency and various anemias, *Am. J. Hematol.* **25**:155–163.

Neefjes, J. J., Verkerk, J. M. H., Broxterman, H. J. G., van der Marel, G. A., van Boom, J. H., and Ploegh, H. L., 1988, Recycling glycoproteins do not return to cis-Golgi, *J. Cell Biol.* **107**:79–87.

Nunez, M. T., Glass, J., Fischer, S., Lavidor, L. M., Lenk, E. M., and Robinson, S. H., 1977, Transferrin receptors in developing murine erythroid cells, *Br. J. Haematol.* **36**:519–526.

Ohkuma, S., and Poole, B., 1978, Fluorescence probe measurement of intralysosomal pH in living cells and the perturbation of pH by various agents, *Proc. Natl. Acad. Sci. USA* **75**:3327–3331.

O'Keefe, D. O., and Draper, R. K., 1988, Two pathways of transferrin recycling evident in a variant of mouse LMTK-cells, *Somat. Cell Mol. Genet.* **14**:473–487.

Omary, M. B., and Trowbridge, I. S., 1981, Biosynthesis of the human transferrin receptor in cultured cells, *J. Biol. Chem.* **256**:12888–12892.

Owen, D., and Kühn, L. C., 1987, Noncoding 3′ sequences of the transferrin receptor gene are required for mRNA regulation by iron, *EMBO J.* **6**:1287–1293.

Pan, B. T., and Johnstone, R. M., 1983, Fate of the transferrin receptor during maturation of sheep reticulocytes in vitro: Selective externalization of the receptor, *Cell* **33**:967–977.

Pan, B. T., Teng, K., Wu, C., Adam, M., and Johnstone, R. M., 1985, Electron microscopic evidence for externalization of the transferrin receptor in vesicular form in sheep reticulocytes, *J. Cell Biol.* **101**:942–948.

Papayannopoulou, T., and Finch, C. A., 1975, Radioiron measurements of red cell maturation, *Blood Cells* **1**:535–546.

Park, I., Schaeffer, E., Sidoli, A., Baralle, F. E., Cohen, G. N., and Zakin, M. M., 1985, Organization of human transferrin gene: Direct evidence that it originated by gene duplication, *Proc. Natl. Acad. Sci. USA* **82**:3149–3153.

Pastan, I., and Willingham, M. C., 1983, Receptor-mediated endocytosis: Coated pits, receptosomes and the Golgi, *Trends Biochem. Sci.* **8**:250–254.

Paterson, S., Armstrong, N. J., Iacopetta, B. J., McArdle, H. J., and Morgan, E. H., 1984, Intravesicular pH and iron uptake by immature erythroid cells, *J. Cell. Physiol.* **120**:225–232.

Pearse, B. M. F., 1982, Coated vesicles from human placenta carry ferritin, transferrin, and immunoglobulin G, *Proc. Natl. Acad. Sci. USA* **79**:451–455.

Pelosi, E., Testa, U., Louache, F., Thomopoulos, P., Salvo, G., Samoggia, P., and Peschle, C., 1986, Expression of transferrin receptors in phytohemagglutinin-stimulated human T-lymphocytes, *J. Biol. Chem.* **261**:3036–3042.

Pelosi-Testa, E., Samoggia, P., Gianella, G., Montesoro, E., Caravita, T., Salvo, G., Camagana, A., Isacchi, G., Testa, U., and Peschle, C., 1988, Mechanisms underlying T-lymphocyte activation: Mitogen initiates and IL-2 amplifies the expression of transferrin receptors via intracellular iron level, *Immunology* **64**:273–279.

Plowman, G. D., Brown, J. P., Enns, C. A., Schröder, J., Nikinmaa, B., Sussman, H. H., Hellström, K. E., and Hellström, I., 1983, Assignment of the gene for human melanoma-associated antigen p97 to chromosome 3, *Nature* **303**:70–72.

Princiotto, J. V., and Zapolski, E. J., 1976, Functional heterogeneity and pH-dependent dissociation properties of human transferrin, *Biochim. Biophys. Acta* **428**:766–771.

Rao, K., Harford, J. B., Rouault, T., McClelland, A., Ruddle, F. H., and Klausner, R. D., 1986, Transcriptional regulation by iron of the gene for the transferrin receptor, *Mol. Cell. Biol.* **6**:236–240.

Rogers, J., and Munro, H., 1987, Translation of ferritin light and heavy subunit mRNAs is regulated by intracellular chelatable iron levels in rat hepatoma cells, *Proc. Natl. Acad. Sci. USA* **84**:2277–2281.

Rothenberger, S., Iacopetta, B. J., and Kühn, L. C., 1987, Endocytosis of the transferrin receptor requires the cytoplasmic domain but not its phosphorylation site, *Cell* **49**:423–431.

Rouault, T. A., Hentze, M. W., Dancis, A., Caughman, S. W., Harford, J. B., and Klausner, R. D., 1987, Influence of altered transcription on the translational control of human ferritin expression, *Proc. Natl. Acad. Sci. USA* **84**:6335–6339.

Rouault, T. A., Hentze, M. W., Caughman, S. W., Harford, J. B., and Klausner, R. D., 1988, Binding of a cytosolic protein to the iron-responsive element of human ferritin messenger RNA, *Science* **241**:1207–1210.

Rudolph, N. S. Ohlsson-Wilhelm, B. M., Leary, J. F., and Rowley, P. T., 1985, Regulation of K562 cell transferrin receptors by exogenous iron, *J. Cell. Physiol.* **122**:441–450.

Sabatini, D. D., Kreibach, G., Morimoto, T., and Adesnik, M., 1982, Mechanisms for the incorporation of proteins in membranes and organelles, *J. Cell Biol.* **92**:1–22.

Saermark, T., Flint, N., and Evans, W. H., 1985, Hepatic endosome fractions contain an ATP-driven proton pump, *Biochem. J.* **225**:51–58.

Sager, P. R., Brown, P. A., and Berlin, R. D., 1984, Analysis of transferrin recycling in mitotic and interphase HeLa cells by quantitative fluorescence microscopy, *Cell* **39**:275–282.

Sassa, S., 1983, Heme biosynthesis in erythroid cells: Distinctive aspects of the regulatory mechanism, in: *The Regulation of Hemoglobin Biosynthesis* (E. Goldwasser, ed.), pp. 359–383, Elsevier, Amsterdam.

Schneider, C., Sutherland, R., Newman, R., and Greaves, M., 1982, Structural features of the cell surface receptor for transferrin that is recognized by the monoclonal antibody OKT9, *J. Biol. Chem.* **257**:8516–8522.

Schneider, C., Asser, U., Sutherland, D. R., and Greaves, M. F., 1983, In vitro biosynthesis of human cell surface receptor for transferrin, *FEBS Lett.* **158**:259–264.

Schneider, C., Owen, M. J., Banville, D., and Williams, J. G., 1984, Primary structure of human transferrin receptor deduced from the mRNA sequence, *Nature* **311**:675–678.

Schwartz, A. L., Strous, G. J. A. M., Slot, J. W., and Geuze, H. J., 1985, Immunoelectron microscopic localization of acidic intracellular compartments in hepatoma cells, *EMBO J.* **4**:899–904.

Sieff, C., Bicknell, D., Caine, G., Robinson, J., Lam, G., and Greaves, M. F., 1982, Changes in cell surface antigen expression during hemopoietic differentiation, *Blood* **60**:703–713.

Sipe, D. M., and Murphy, R. F., 1987, High-resolution kinetics of transferrin acidification in BALB/c 3T3 cells: Exposure to pH 6 followed by temperature-sensitive alkalinization during recycling, *Proc. Natl. Acad. Sci. USA* **84**:7119–7123.

Snider, M. D., and Rogers, O. C., 1985, Intracellular movement of cell surface receptors after endocytosis: Resialylation of asialotransferrin receptor in human erythroleukemia cells, *J. Cell Biol.* **100**:826–834.

Sorkin, A. D., Teslenko, L. V., and Nikolsky, N. N., 1988, The endocytosis of epidermal growth factor in A431 cells: A pH of microenvironment and the dynamics of receptor complex dissociation, *Exp. Cell Res.* **175**:192–205.

Sorokin, L. M., Morgan, E. H., and Yeoh, G. C. T., 1987, Transferrin receptor numbers and transferrin and iron uptake in cultured chick muscle at different stages of development, *J. Cell. Physiol.* **131**:342–353.

Spik, B., Fournet, B., Cheron, A., Strecker, G., Montreuil, J., Dorland, L., and Vliegenthart, J. F. G., 1979, Comparative study of the structure of the glycans isolated from different transferrins, *International Symposium on Glycoconjugates 5*, Kiel, pp. 21–22.

Stein, B. S., and Sussman, H. H., 1986, Demonstration of two distinct transferrin receptor recycling pathways and transferrin-independent receptor internalization in K562 cells, *J. Biol. Chem.* **261**:10319–10331.

Stein, B. S., Bensch, K. G., and Sussman, H. H., 1984, Complete inhibition of transferrin recycling by monensin in K562 cells, *J. Biol. Chem.* **259**:14762–14772.

Stoorvogel, W., Geuze, H. J., and Strous, G. J., 1987, Sorting of endocytosed transferrin and asialoglycoprotein occurs immediately after internalization in HepG2 cells, *J. Cell Biol.* **104**:1261–1268.

Stoorvogel, W., Geuze, H. J., Griffith, J. M., and Strous, G. J., 1988, The pathways of endocytosed transferrin and secretory protein are connected in the trans-Golgi reticulum, *J. Cell Biol.* **106**:1821–1829.

Surgenor, D. M., Koechlin, B. A., and Strong, L. E., 1949, Chemical, clinical and immunological studies on the products of human plasma fractionation. XXXVII. The metal-combining globulin of human plasma, *J. Clin. Invest.* **28**:73–96.

Thorstensen, K., and Romslo, I., 1988, Uptake of iron from transferrin by isolated rat hepatocytes, *J. Biol. Chem.* **263**:8844–8850.

Tycko, B., and Maxfield, F. R., 1982, Rapid acidification of endocytic vesicles containing α_2-macroglobulin, *Cell* **28**:643–651.

van Renswoude, J., Bridges, K. R., Harford, J. B., and Klausner, R. D., 1982, Receptor-mediated endocytosis of transferrin and the uptake of Fe in K562 cells: Identification of a nonlysosomal acidic compartment, *Proc. Natl. Acad. Sci. USA* **79**:6186–6190.

Veldman, A., Van der Heul, C., Kroos, M. J., and Van Eijk, H. G., 1986, Fluorescence probe measurements of the pH of the transferrin microenvironment during uptake by rat bone marrow erythroid cells, *Br. J. Haematol.* **62**:155–162.

Warren, G., Davoust, J., and Cockcroft, A., 1984, Recycling of transferrin receptors in A431 cells is inhibited during mitosis, *EMBO J.* **3**:2217–2225.

Watts, C., 1985, Rapid endocytosis of the transferrin receptor in the absence of bound transferrin, *J. Cell Biol.* **100**:633–637.

White, G. P., Bailey-Wood, R., and Jacobs, A., 1976, The effect of chelating agents on cellular iron metabolism, *Clin. Sci.* **50**:145–152.

Willingham, M. C., Hanover, J. A., Dickson, R. B., and Pastan, I., 1984, Morphologic characterization of the pathway of transferrin endocytosis and recycling in human KB cells, *Proc. Natl. Acad. Sci. USA* **81**:175–179.

Yamashiro, D. J., Fluss, S. R., and Maxfield, F. R., 1983, Acidification of endocytic vesicles by an ATP-dependent proton pump, *J. Cell Biol.* **97**:929–934.

Yamashiro, D. J., Tycko, B., Fluss, S. R., and Maxfield, F. R., 1984, Segregation of transferrin to a mildly acidic (pH 6.5) para-Golgi compartment in the recycling pathway, *Cell* **37**:789–800.

Yang, F., Lum, J. B., McGill, J. R., Moore, C. M., Naylor, S. L., van Bragt, P. H., Baldwin, W. D., and Bowman, B. H., 1984, Human transferrin: cDNA characterization and chromosomal localization, *Proc. Natl. Acad. Sci. USA* **81**:2752–2756.

Young, S. P., Bomford, A., and Williams, R., 1984, The effect of the iron saturation of transferrin on its binding and uptake by rabbit reticulocytes, *Biochem. J.* **219**:505–510.

Young, S. P., Roberts, S., and Bomford, A., 1985, Intracellular processing of transferrin and iron by isolated rat hepatocytes, *Biochem. J.* **232**:819–823.

Zail, S. S., Charlton, R. W., Torrance, J. D., and Bothwell, T. H., 1964, Studies on the formation of ferritin in red cell precursors, *J. Clin. Invest.* **43**:670–680.

Zerial, M., Suomalainen, M., Zanetti-Schneider, M., Schneider, C., and Garoff, H., 1987, Phosphorylation of the human transferrin receptor by protein kinase C is not required for endocytosis and recycling in mouse 3T3 cells, *EMBO J.* **6**:2661–2667.

Chapter 14

Iron-Mediated Oxidative Stress and Erythrocytes

Catherine Rice-Evans

1. INTRODUCTION

1.1. Why Is Iron Important?

Iron in its labile, chelatable form has been implicated in the generation of radical species such as the hydroxyl radical, ferryl species [an Fe(IV)–oxygen complex], perferryl ($Fe^{2+}-O_2 \leftrightarrow Fe^{3+}-O_2^{-}\cdot$), and $Fe^{2+}/Fe^{3+}/O_2$ complexes (for reviews see Halliwell and Gutteridge, 1986; Czapski and Goldstein, 1986; Aust and White, 1986; Aust and Minotti, 1987; Gutteridge and Halliwell, 1989) and in the propagation of oxidative events.

The majority of iron in the human body is localized within the intraerythrocytic compartment in the divalent state involved in the oxygen-transporting function of hemoglobin. The rest is, in the main, distributed within heme (e.g., myoglobin, cytochromes, cyclooxygenase) and nonheme proteins (e.g., ferritin, transferrin, lipoxygenases) safely compartmentalized from participating in the initiation of damaging, free-radical-mediated reactions. A small "transit pool" of low-molecular-weight iron chelates is present within cells (Jacobs, 1977) other than normal mature erythrocytes. The exact chemical nature of this pool is not clear, but it probably represents iron ions attached to ATP, ADP, or GTP, to organic acids, and perhaps to the polar head groups of DNA or lipids.

Hemoglobin may be involved in injurious reactions in pathological situations in which it becomes delocalized as membrane-bound inclusions or as a result of microbleeding in the tissues. For example, in abnormal erythrocytes the intracellular accumulation of breakdown products of the oxidative denaturation of hemoglobin or of iron in a low-molecular-weight pool, residual from the reticulocyte stage, can induce cellular damage. In addition, evidence is accumulating that when hemoglobin is released during micro-

Catherine Rice-Evans Department of Biochemistry and Chemistry, Royal Free Hospital School of Medicine, London NW3 2PF, England.

bleeding processes it becomes toxic: specific observations have been made in the brain (Panter *et al.*, 1985), in retinal degeneration in the eye (Doly *et al.*, 1986), and in bleeding at sites of inflammation (Yoshino *et al.*, 1985). Microbleeding in certain tumours (Rowland and Symons, 1988) has also been detected.

At sites of tissue damage, phagocytes are activated and microbleeding often occurs across the damaged endothelium. Stimulation of chemoattractants may exacerbate the injury. The generation of hydrogen peroxide from the superoxide radical formed at such sites may interact with methemoglobin to form the ferryl hemoglobin species (George and Irvine, 1952). In this iron(IV) state, the iron is in an oxidation state one higher than that in methemoglobin. Depending on the relative concentrations of hydrogen peroxide and methemoglobin, iron can be mobilized from the hemoglobin (Whitburn, 1987; Gutteridge, 1986, 1987; Halliwell, 1988). An intermediate stage in the mechanism for release of iron from these hemoproteins may be via the formation of the reactive ferryl radical species (Petersen *et al.*, 1989; Rice-Evans *et al.*, 1989).

As a result of the increased iron availability, the possibility arises for the initiation of free radical damage mediated by hydroxyl radical formation (Puppo and Halliwell, 1988; Aruoma *et al.*, 1989) via the iron-catalyzed Haber–Weiss reaction or the formation of iron–ferryl radicals or other iron–oxygen complexes (Minotti and Aust, 1987a,b).

Superoxide radical is an effective reducing agent for iron(III):

$$Fe^{3+} + O_2^{-}\cdot \rightarrow Fe^{2+} + O_2$$

Other reducing agents also perform a similar function such as ascorbate (Winterbourn, 1979, 1981; Rowley and Halliwell, 1983), reduced glutathione (Rowley and Halliwell, 1982a; Searle and Tomasi, 1982), and reduced nicotinamide complexes (Rowley and Halliwell, 1982b). Superoxide radical can be efficiently converted to hydrogen peroxide by superoxide dismutase:

$$2O_2^{-}\cdot + 2H^+ \rightarrow H_2O_2 + O_2$$

Hence, the interactions of iron(II) with hydrogen peroxide can predispose cells and tissues to the deleterious effects of the extremely reactive hydroxyl radical via the Fenton reaction:

$$Fe^{2+} + H_2O_2 \rightarrow Fe^{3+} + OH^{\cdot} + OH^-$$

The ferryl hemoglobin radical can induce similar damage. The propagation of oxidative events in erythrocyte membranes is also amplified by the decomposition of lipid hydroperoxides by iron complexes (O'Brien, 1969; Labeque and Marnett, 1988) such as methemoglobin, heme, hemin generating further radical species (see Section 1.4.1).

1.2. Susceptibility of Erythrocytes to Oxidative Stress

The erythrocyte is at increased risk from oxidative processes for a variety of reasons. It is continually exposed to high oxygen tensions, hemoglobin is susceptible to autoxidation and can function as an oxidase and a peroxidase (Chung and Wood, 1971; Peisach *et al.*, 1975; Goldberg *et al.*, 1976), it is unable to repair damaged components by resynthesis, and the membranes are composed of components that are vulnerable to peroxidative decomposition such as the polyunsaturated fatty acid side chains and the specific amino acyl side chains that become oxidatively modified, undergoing fragmentation or aggregation.

The binding of oxygen in hemoglobin involves a substantial migration of charge from the heme iron to oxygen, effectively forming a superoxide anion, as the sixth coordinating ligand in oxyhemoglobin, and low-spin ferric iron(III) (Weiss, 1964; Peisach *et al.*, 1968; Wittenberg *et al.*, 1970). On deoxygenation the shared electron is normally returned to the iron when oxygen is released. The location of the oxygen binding site in the highly hydrophobic environment of the globin chain folding and the complexing of the iron to the imidazole side chains in the globin chains stabilize the functional form of hemoglobin and the tendency of reduced heme iron to undergo autoxidation is decreased (Williams, 1971; Aisen, 1977). The distal histidine protects the ferrous heme iron by acting as a proton trap (Perutz, 1989), autoxidation being catalyzed by protons. When the hydrophobic region about the heme moiety is disturbed, slight modifications in the heme pocket may allow access of small anions or water. Consequently, the heme iron loses an electron and is oxidized to the high-spin ferric state, methemoglobin, releasing the superoxide radical (Goldberg *et al.*, 1976).

Within the erythrocyte there normally develops a balance between the spontaneous production of methemoglobin and superoxide radical, on autoxidation of hemoglobin (Misra and Fridovich, 1972), and restoration of the hemoglobin to its normal functional state controlled by the antioxidant defenses.

Any pathological situation that increases the turnover of this cycle, whether in-

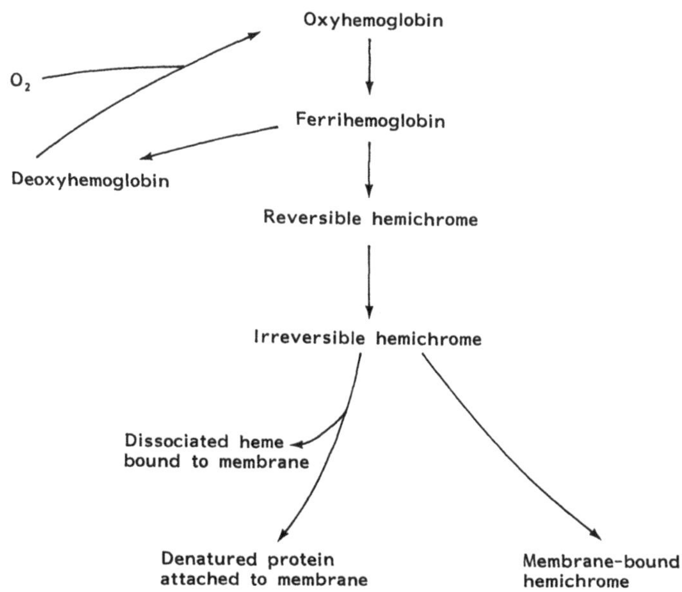

FIGURE 1. Denaturation of oxidized hemoglobin.

creased oxidative stress or impaired antioxidant defenses will enhance production of oxidized hemoglobin and generation of active oxygen species. During the oxidative denaturation of hemoglobin (Figure 1), methemoglobin is converted to reversible hemichromes, which are reducible to functional hemoglobin. In due course, reversible hemichromes are denatured further, separation of α and β chains may occur, and irreversible hemichromes form. Hemichromes are also generated as intermediate denaturation products in the oxidation of isolated α or β hemoglobin subunits (as in the thalassemias) and in some abnormal hemoglobin variants. These species cannot be reconstituted to functional hemoglobin and precipitate (Rachmilewitz *et al.*, 1971) to form inclusion bodies, which may bind to the membrane in some pathological states. Under normal physiological conditions the link between heme and globin is very stable (Brunori *et al.*, 1965) but in abnormal situations, e.g., in sickle cells, increased heme dissociation occurs (Bunn and Jandl, 1966; Liu *et al.*, 1988). Heme, when released from its natural globin anchor, incorporates into the hydrophobic lipid area of the membrane and binds to cytoskeletal proteins (Shaklai *et al.*, 1985). Residual globin is also thought to become membrane-bound. The formation of inclusion bodies exerts considerable oxidative stress as a result of increased production of oxygen radicals and may be involved in membrane disruption, decreasing the deformability and ultimate survival of the erythrocyte.

1.3. Antioxidant Defenses

The erythrocyte is well endowed with antioxidant defenses, free radical scavengers and enzymatic systems that maintain the normal state of the cell (Figure 2). NADH-dependent methemoglobin reductase converts the autoxidized hemoglobin back to deoxyhemoglobin; Cu-Zn superoxide dismutase converts the superoxide (McCord and Fridovich, 1969) formed from hemoglobin autoxidation to hydrogen peroxide, which has limited reactivity; catalase and glutathione-dependent selenium-containing glutathione peroxidase detoxify the hydrogen peroxide (Cohen and Hochstein, 1963). In addition, glutathione peroxidase removes lipid hydroperoxides after cleavage from membranes. Maintenance of the reduced glutathione levels is essential for sustaining the reduced state of the thiol side chains of the cellular enzymes, and of the intramembranous and cytoskeletal proteins; the cell is thus equipped with NADPH-dependent glutathione reductase to reduce oxidized glutathione. α-Tocopherol is one of the most important antioxidants

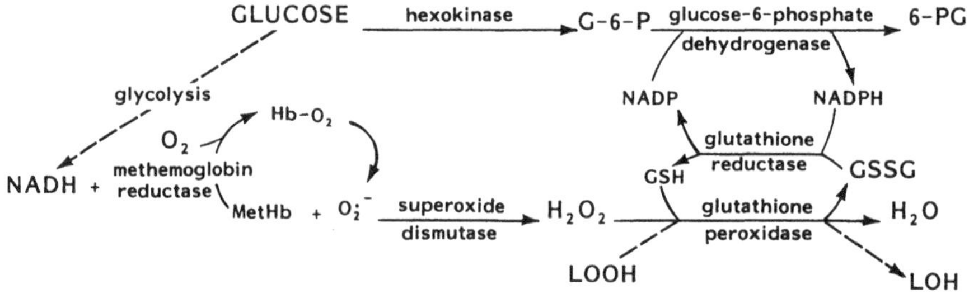

FIGURE 2. The intracellular antioxidant defenses of the human erythrocyte.

residing in the membrane where it can function as a chain-breaking antioxidant, interrupting the propagation of oxygen radical mechanisms (Tappel, 1962; Diplock, 1983).

The ascorbate level within the erythrocyte is low (0.043 mM) (Moser, 1987) but it features as one of the major extracellular plasma antioxidant mechanisms working synergistically with tocopherol. Extracellularly, ceruloplasmin, which acts as a ferroxidase (Osaki *et al.*, 1966), may function in facilitating iron binding to transferrin (Chidambaram *et al.*, 1983), the latter normally being in excess. In plasma, superoxide dismutase (Marklund, 1984) and selenium glutathione peroxidase activities are very low, and catalase is essentially absent. Albumin binds copper strongly and iron weakly, thus behaving as a site-specific sacrificial antioxidant (Halliwell and Gutteridge, 1986). Haptoglobin and hemopexin have been suggested to inhibit radical reactions by binding to free hemoglobin or heme, thereby preventing the promotion of decomposition of lipid hydroperoxides. Uric acid is also an important physiological antioxidant (Ames *et al.*, 1981) as an efficient chelator of metal ions (K. J. A. Davies *et al.*, 1986).

1.4. Consequences of Free Radical Damage to Membrane Lipids and Proteins

Red cell damage by oxidant stress is generally thought to be the end result of two processes: the oxidation of hemoglobin followed by denaturation of methemoglobin to hemichromes, and free radical attack on the membrane components: the polyunsaturated fatty acid side chains of the membrane lipids, the reduced thiol groups and other susceptible amino acid side chains of the membrane proteins.

1.4.1. Membrane Lipids and Reactive Oxygen Species

The peroxidation of the membrane lipids is a consequence of many types of cellular injury in which free radical intermediates are produced in excess of local defense mechanisms and has been extensively reviewed (Girotti, 1985; Kappus, 1985; Halliwell and Gutteridge, 1984). Lipid peroxidation may be initiated by any primary reactive free radical species that has sufficient reactivity to abstract a hydrogen atom (Figure 3), e.g., such species as hydroxyl radical $OH^{\cdot}$, ferryl hemoglobin radical, peroxy and alkoxy radical. The lipid radical takes up oxygen to form a peroxy radical and propagation occurs leading to the formation of lipid monohydroperoxides. This propagation phase can be repeated many times. Thus, an initial event triggering lipid peroxidation can be amplified.

Lipid hydroperoxides are fairly stable molecules at physiological temperatures but their decomposition is catalyzed by transition metals and metal complexes such as low-molecular-weight iron complexes, methemoglobin, and heme. Both iron(II) and iron(III) are effective catalysts for hydroperoxide degradation but the former more so. Reduced metal complexes react with lipid peroxides to give alkoxy radicals:

$$\mathrm{LOOH} + \mathrm{Fe}^{2+}\text{-complex} \rightarrow \mathrm{LO}^{\cdot} + \mathrm{Fe}^{3+}\text{-complex} + \mathrm{OH}^{-}$$

Oxidized iron complexes react more slowly to produce alkoxy and peroxy radicals (M. J. Davies and Slater, 1987) and, under certain conditions, ferryl radicals (Labeque and Marnett, 1988):

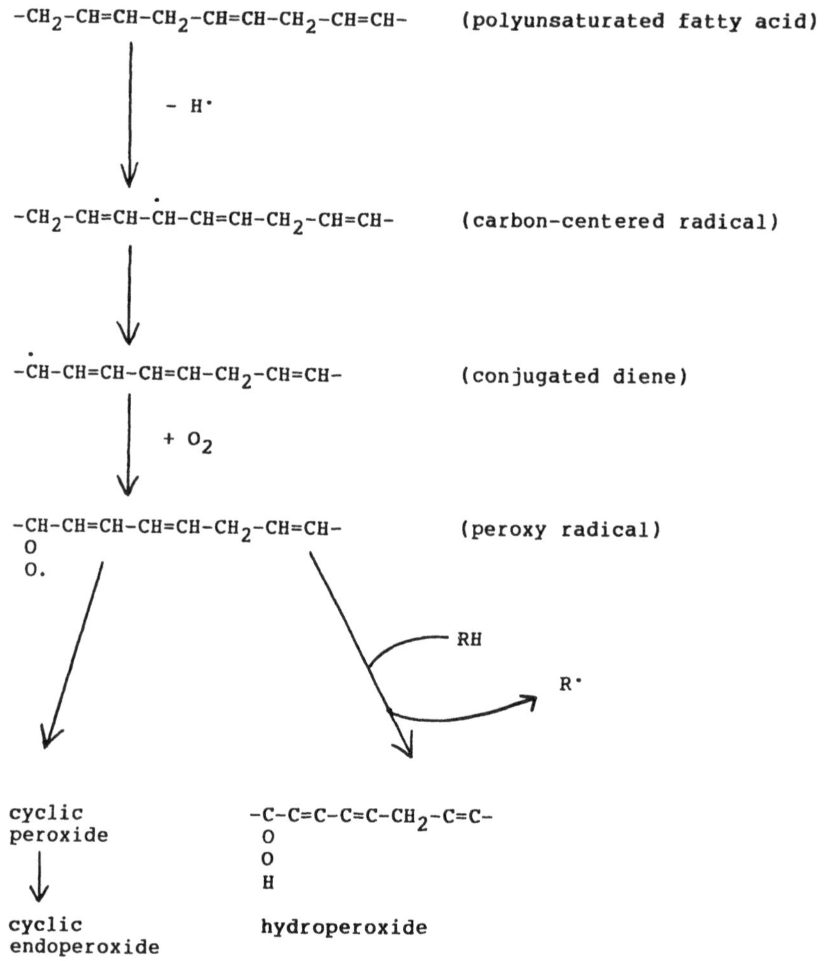

FIGURE 3. Mechanism of peroxidation of a polyunsaturated fatty acyl chain.

$$LOOH + Fe^{3+}\text{-complex} \rightarrow LO^{\bullet} + [Fe(IV){=}O\ \text{complex}]^{2+} + H^{+}$$

$$\searrow$$

$$LOO^{\bullet} + Fe^{2+}\text{-complex} + H^{+}$$

These radical species stimulate the chain reaction of lipid peroxidation by abstracting further hydrogen atoms (Figure 3). Cleavage of carbon bonds during lipid peroxidation reactions (Figure 4) results in the formation of aldehydic metabolites such as malonaldehyde (Tappel, 1980) or alkenals, such as 4-hydroxynonenal (Esterbauer, 1985).

The production of a highly reactive free radical leads to primary reactions and damage in the immediate surroundings of where the radical is produced; it will not diffuse far before it interacts within its microenvironment, which in a membrane would include the initiation of lipid peroxidation. Secondary products of lipid peroxidation such as lipid

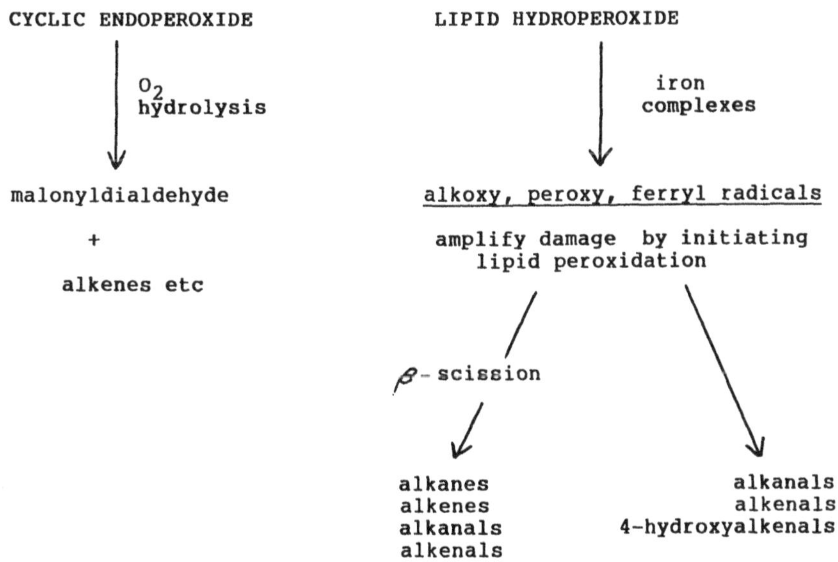

FIGURE 4. Cleavage of carbon bonds during lipid peroxidation.

peroxy radicals and lipid hydroperoxides may diffuse in the plane of the membrane before reacting further, thereby spreading the biochemical lesion. Such processes therefore not only affect the structural and functional integrity of the membrane and its fluidity, for example, but also the breakdown of products of lipid peroxidation can further damage cellular function. For example, alkenals, alkanals, or lipid hydroperoxides may undergo degradative reactions and be metabolized rapidly; some of these, like the lower-molecular-weight hydroperoxides, aldehydes, and 4-hydroxyalkenals, can escape from the membrane and cause disturbances at a distance. Therefore, a reaction that originally produces a radical that interacts within its own microenvironment, may produce a sequence of later events that direct disturbances throughout the cell, its membrane, and in some instances into the extracellular domain.

1.4.2. Vulnerability of Proteins to Oxidative Stress

Proteins may also be critical targets of free radical attack as they are present both inside and outside cells in very high concentrations and, because many are catalytic, modifications by free radicals may have an amplified effect. The consequences of such damage may be protein aggregation and cross-linking or protein degradation and fragmentation, depending on the nature of the vulnerable protein component and the attacking radical species (Wolff *et al.*, 1986; Wolff and Dean, 1986). Oxidized proteins become increasingly susceptible to proteolytic attack (Levine *et al.*, 1981), so in pathological states radical damage to proteins may lead to intracellular accumulation of denatured proteins, of which erythrocytes have no means of disposing.

Several amino acids crucial for protein and membrane function are particularly susceptible to radical damage (Figure 5) (Sies, 1986; Singh *et al.*, 1982; Roshchupkin *et*

Arginine	→	glutamic semialdehyde + NO
Lysine	→	glutamic semialdehyde
Proline	→	glutamic semialdehyde → pyroglutamate → glutamate
Histidine	→	histidine endoperoxides
	→	aspartate, asparagine
Cysteine	→	Cys-disulphides mixed-disulphides
	→	sulphenic acid, sulphinic acid, sulphonic acid (via alkyl thioradicals)
Methionine	→	methionine sulphoxide methionine sulphone
Tryptophan	→	5-hydroxy tryptophan
	→	kynurenine
	→	N-formyl kynurenine
Tyrosine	→	bityrosine (not in the presence of O_2, $O_2^{\cdot -}$)
Phenylalanine	→	tyrosine (in presence of $^{\cdot}OH$)

FIGURE 5. Modifications of amino acid side chains by oxygen radicals.

al., 1979). Proteins are susceptible to either direct attack of radical species such as $OH^{\cdot}$ generated from hydrogen peroxide accumulated in the presence of low-molecular-weight chelates of iron, or lipid alkoxy and peroxy radicals as a consequence of the formation of radical intermediates of lipid peroxidation. These may react with proteins closely associated with the peroxidizing lipids. Lysine, for example, may be modified by stable products of lipid peroxidation such as malonaldehyde or 4-hydroxynonenal. Methionine oxidation to methionine sulfoxide (Sies, 1986) and cysteine to cysteic acid may be mediated by superoxide radicals; oxidation of tryptophan to kynurenine, *N*-formyl kynurenine, 5-hydroxy tryptophan, and other species (Butler *et al.*, 1988) may reflect direct attack by hydroxyl radicals or by peroxy radicals formed as metabolites of adjacent lipid hydroperoxides in the membrane.

Radical-mediated protein breakdown in some cellular membranes has been shown to be an earlier event than lipid peroxidation (Davies and Goldberg, 1987).

2. SICKLE ERYTHROCYTES, MEMBRANE-BOUND IRON, AND OXIDATIVE DAMAGE

2.1. Membrane Abnormalities

Even though the development of the membrane abnormalities in sickle-cell erythrocytes is due to the presence of a mutant gene product of hemoglobin, sickle-cell anemia is in a sense a membrane disease. The relationship between the induction of the membrane modifications in sickle erythrocytes and the pathology of the disease is, however, not clearly understood. It has been suggested that one of the significant factors in the induction of sickling may be the development of abnormal membrane–hemoglobin interactions (Fischer *et al.*, 1975; Asakura *et al.*, 1977; Lessin *et al.*, 1978; Shaklai *et al.*, 1981) especially in homozygous sickle-cell disease with less than 10% fetal hemoglobin (Kim *et al.*, 1980). The consequences of such interactions may be amplified by the involvement of such iron complexes and their denaturation products in the promotion of oxidative events in the membrane (Rice-Evans *et al.*, 1986c).

When sickle erythrocytes pass through the deoxygenation phase of the sickling–unsickling cycle, the hemoglobin undergoes a drastic reduction in solubility, during which polymerized fibers of deoxyhemoglobin S adhere to each other and to the internal surface of the membrane (White, 1974). This distorts the cells into the characteristic drepanocytic morphology, which returns to normal on reoxygenation in the first instance. During the process of reversible sickling, membrane damage is induced which may be due to membrane–hemoglobin interaction, culminating in the formation of irreversibly sickled cells. Irreversibly sickled cells, even under reoxygenation conditions, retain the sickled morphology (Figure 6) and fail to function normally, implicating the membrane as an important locus of secondary cellular damage. These cells become dense and dehydrated in the circulation (Clark *et al.*, 1980), generating a variable fraction of cells with decreased deformability (reviewed in Palek, 1977; Rice-Evans and Chapman, 1981) and an increased mean cell hemoglobin concentration. Observations of membrane abnormalities have also been related to the molecular defect in hemoglobin, in particular increased calcium permeability (Eaton *et al.*, 1974), altered phospholipid asymmetry (Lubin *et al.*, 1981), change in surface charge distribution (Rice-Evans *et al.*, 1978; Hebbel *et al.*, 1980).

During repeated sickling–unsickling cycles, not only has the gross morphology of

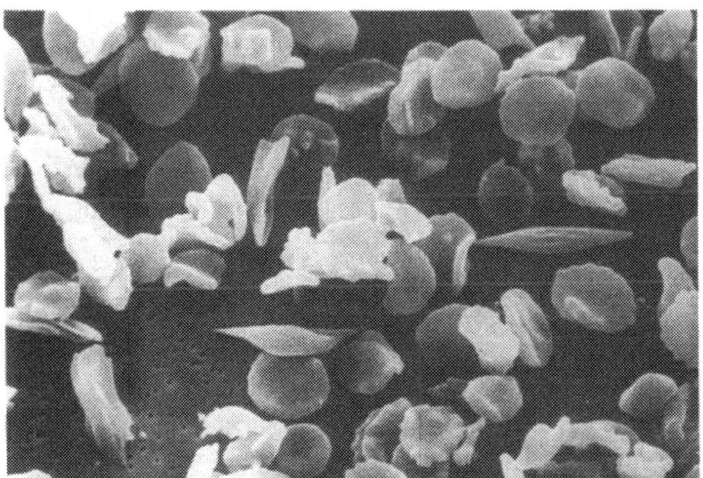

FIGURE 6. Erythrocytes (oxygenated) from a patient with homozygous sickle-cell anemia.

the cell undergone a structural transition, but the cytoskeletal network of proteins underlying the erythrocyte membrane, consisting of spectrin, actin, and protein 4.1, has also modified its structural arrangement in an identical manner (Lux *et al.*, 1976; Platt *et al.*, 1985). The erythrocyte membrane cytoskeletal network connects to the intramembranous proteins through ankyrin binding sites on spectrin. It has been shown using immunofluorescence spectroscopy (Waugh *et al.*, 1986) that at most sites where hemichrome aggregates themselves are firmly attached to the membrane, band 3 is also clustered; in addition, ankyrin and glycophorin also frequently are observed to colocalize with inclusion body sites on the membrane. Others have suggested that interactions between the hemoglobin and proteins of the cytoskeleton contribute to the increased membrane rigidity in sickle cells (Fortier *et al.*, 1988) and that oxidative cross-linking may be involved in such interactions. Current research is beginning to suggest that the cytoskeletal modifications are induced by oxidative damage to the membrane mediated by iron-containing denaturation products of hemoglobin.

2.2. Displaced Iron in Sickle Erythrocytes

In sickle-cell anemia, in contrast to thalassemia, intraerythrocytic hemoglobin precipitation is not extensive, the iron from premature erythrocyte destruction becoming available for other developing red cells. Increased levels of membrane-bound iron in sickle erythrocyte membranes have been calculated to correspond to 0.8% cellular hemoglobin equivalent (Rice-Evans *et al.*, 1986d; Rice-Evans and Omorphos, 1986), total iron being determined by atomic absorption spectroscopy. Asakura *et al.* (1977) had previously proposed the identification of this component as membrane-bound oxidative denaturation products of hemoglobin, in the form of irreversible hemichromes. Nonheme iron complexes compatible with iron in ferritin or hemosiderin have also been detected in sickle erythrocytes (Bauminger *et al.*, 1979). The essential questions that are currently being addressed are: what are the precise forms of iron bound to the sickle erythrocyte membrane, what damage is it causing, and how can it be controlled?

Earlier studies had revealed membrane changes consistent with oxidative damage to the sickle erythrocyte membrane (Rice-Evans *et al.*, 1978; Lubin *et al.*, 1981). Such membrane defects were particularly significant in blood samples with a higher proportion of irreversibly sickled cells. In particular, decreased lipid fluidity (Rice-Evans *et al.*, 1978) and diminished availability of reduced thiol groups of the membrane proteins (Rice-Evans *et al.*, 1978; Rank *et al.*, 1985) were demonstrated (Table I), which could not be accounted for by oxidative cross-linking of the type demonstrated in Hb-Koln disease, for example. Recent studies (Rice-Evans and Hartley, 1989) investigating oxidative protein constituents by HPLC analysis (Griffiths *et al.*, 1988) of oxidized amino acid components of sickle-cell membranes after pronase digestion have revealed other oxidation products of proteins including those of tryptophan, mainly identified as *N*-formyl kynurenine, kynurenine, and 5-hydroxy tryptophan, consistent with the involvement of iron-mediated hydroxyl radicals or propagating lipid peroxy radicals.

The modified levels or activities of the scavengers and enzymes normally involved in the antioxidant defenses of the erythrocyte are well documented. Limited antioxidant defenses create a situation in which the reduction mechanisms are not able to attenuate efficiently the consequences of increased hemoglobin autoxidation. These modifications

Table I
Membrane Abnormalities in Sickle Erythrocytes[a] Consistent with Oxidative Damage

	Sickle	Normal
Endogenous lipid peroxidation: TBAR products (A_{532}/mg protein)	0.150 ± 0.03[b] (n = 19)[c]	0.072 ± 0.01 (n = 21)
Membrane protein oxidation: % tryptophan oxidized	11% (n = 4)	1.3% (n = 4)
Iron levels: membrane-bound iron (nmol iron/mg protein)	35.4 ± 17 (n = 16)	1.5 ± 1 (n = 11)
Reduced thiol groups (nmol/mg protein)	65 ± 5 (n = 14)	84 ± 9 (n = 24)

[a]Samples from patients with elevated proportions of ISCs.
[b]Values are means ± SD.
[c]*n*, number of samples.

include decreased activities of glutathione peroxidase and catalase (Das and Nair, 1980), decreased levels of vitamin E in plasma and erythrocytes (Chiu and Lubin, 1979), and increased levels of superoxide dismutase (Das and Nair, 1980). Similar modifications in antioxidant levels have been reported in other erythrocyte disorders in the pathophysiology of which oxidative mechanisms have been implicated. Such a situation creates a scenario for increased oxidative stress and the formation of reactive oxygen species in the sickle erythrocytes, thereby increasing the vulnerability of the membrane lipids to peroxidation and the specific amino acid constituents on the proteins become susceptible to oxidation.

2.3. Free Radical Damage in Sickle-Cell Membranes

Footprints of free-radical-mediated oxidative damage in sickle erythrocytes have been recognized for a number of years. Excessive spontaneous generation of superoxide by sickle erythrocytes has been reported (Hebbel *et al.*, 1982). Increased endogenous levels of lipid peroxides and their secondary metabolites have been identified in sickle erythrocyte membranes (Rice-Evans *et al.*, 1982; Rice-Evans and Omorphos, 1983; Rice-Evans and Baysal, 1987a) prepared from samples containing a higher proportion of irreversibly sickled cells, without exogenous stress. These samples retained elevated levels of membrane-associated iron (Hartley and Rice-Evans, 1989a). The bioavailability of membrane-associated iron was assessed by incubating membranes for 5 hr at 37°C and the generation of aldehydic secondary metabolites of lipid peroxidation was enhanced. The implication of membrane-bound species as mediators of the amplification of peroxidative lipid damage was confirmed by the incorporation of scavengers and an iron chelator during the incubation (Table II). The incubation-induced propagation of oxidative damage was inhibited by ascorbate and vitamin E. The iron chelator desferrioxamine was also effective in suppressing the propagation of the peroxidation of membrane lipids: this finding was unexpected since desferrioxamine will not bind iron in heme or hemichromes

Table II
Effects of Radical-Scavenging Systems on Oxidative Damage in Sickle-Cell Membranes

Conditions	TBAR products (A_{532}/mg membrane protein)
Sickle-cell membranes	
Control, no incubation	0.150 ± 0.03 (n = 10)
Control, incubated[a]	0.260 ± 0.03 (n = 5)
+ Desferrioxamine (0.4 mM)	0.111 ± 0.02 (n = 3)
+Ascorbate (5 mM)	0.141 ± 0.01 (n = 4)
+ Ascorbate (5 mM) + desferrioxamine (0.4 mM)	0.094 ± 0.03 (n = 3)
+ Vitamin E (0.1 mg/ml)	0.104 ± 0.02 (n = 4)
Normal erythrocyte membranes	
Control, no incubation	0.070 ± 0.01 (n = 10)
Control, incubated[a]	0.069 ± 0.01 (n = 5)
+ Ascorbate (5 mM)	0.085 ± 0.01 (n = 5)
+Desferrioxamine (0.4 mM)	0.074 ± 0.01 (n = 5)

[a]Incubation was for 5 hr at 37°C.

or methemoglobin, leading to the questions: is chelatable iron dissociating from such species, do the membrane-bound iron species originate from ferritin or hemosiderin or low-molecular-weight iron residual from the sickle reticulocytes that has become decompartmentalized via intracellular oxidation processes, or is desferrioxamine acting in some other capacity?

These data suggest that oxidative damage is related, at least in part, to mechanisms involving the increased oxidative denaturation of hemoglobin. In order to identify the nature of membrane-bound iron species, several studies have been undertaken. Direct spectrofluorimetric and spectrophotometric assays for heme iron and nonheme iron associated with the sickle erythrocyte membrane indicate that the majority is nonheme iron (Table III) (Hartley and Rice-Evans, 1989a,b). The levels of heme iron bound to sickle membranes are low and correspond to about 2.1 nmol heme iron/mg membrane protein (Hartley and Rice-Evans, 1989a; Kuross *et al.*, 1988). This is only a minor proportion of the total membrane-associated iron in erythrocytes from patients with a higher proportion of endogenous lipid hydroperoxides. The majority of the iron seems to be nonheme iron of which about half can be identified as a labile form of chelatable iron, capable of catalyzing radical reactions, mainly localized in the vicinity of the cytoskeletal proteins (Hartley *et al.*, 1990; Hartley and Rice-Evans, 1989a,b). The association of such iron species close to

Table III
Elevated Levels of Membrane-Bound Iron Species in Sickle Erythrocyte Membranes

	Sickle-cell membrane	Normal erythrocyte membrane
Total iron (nmol/mg membrane protein)	25.4 ± 23 (n = 39)	1.5 ± 1 (n = 14)
Heme iron (nmol/mg membrane protein)	2.1 ± 1 (n = 12)	1.9 ± 1 (n = 8)
Nonheme iron (nmol/mg membrane protein)	23.3 ± 22 (n − 11)	0 (n = 5)

the cytoskeleton may be relevant to the observed oxidation of the protein 4.1 (Schwartz *et al.*, 1987) and the development of other cytoskeletal abnormalities in sickle erythrocytes (Lux *et al.*, 1976; Rank *et al.*, 1985). This localization of the more rapidly reacting component of the membrane-associated nonheme iron is in contrast with the reported findings of Hebbel, who implicated binding to the polar head group of the membrane phospholipid (Kuross and Hebbel, 1988). The remainder is in a form less available for mediation of injurious radical reactions, possibly in ferritin, consistent with the findings of Kuross and Hebbel (1988), or in hemosiderin. ESR spectroscopic investigations have been applied to clarify the mechanism by which desferrioxamine can suppress amplification of peroxidative membrane damage. The results (Figure 7) (Hartley *et al.*, 1990) show that, at lower concentrations, desferrioxamine functions mainly as chelator of the more accessible component of the nonheme iron, thereby suppressing the propagation of oxidative events in the membrane (Rice-Evans *et al.*, 1986c). At higher desferrioxamine concentrations, this trihydroxamate acts as a chain-breaking antioxidant, as well as an iron chelator, possibly via its electron transfer properties (Rice-Evans *et al.*, 1989; Morehouse *et al.*, 1987; M. J. Davies *et al.*, 1987).

The dual mechanism of action of desferrioxamine, as an iron chelator and as a chain-breaking antioxidant, may account for its effectiveness in inhibiting the propagation of oxidative events in the sickle erythrocyte membrane, catalyzed by the membrane-associated iron complexes.

3. IRON OVERLOAD, OXIDATIVE DAMAGE, AND THALASSEMIA

The thalassemias are a group of inherited disorders of hemoglobin synthesis in which the globin chain synthesis is impaired. In β-thalassemia major, the suppressed formation of β-globin chains, while α-chain synthesis proceeds normally, leads to an accumulation of unpaired α chains. These precipitate and form intracellular inclusions that can be observed even in young nucleated erythroid cells (Polliack and Rachmilewitz, 1973). Such inclusion bodies may interact with the membrane and form localized areas of membrane rigidity contributing toward a decrease in deformability of thalassemic erythro-

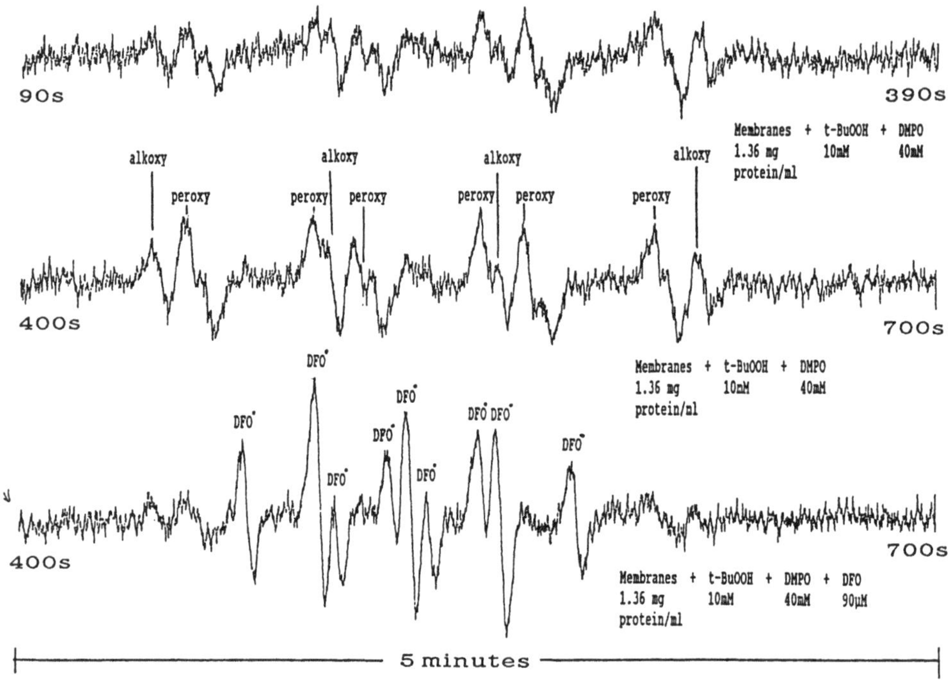

FIGURE 7. Alkoxyl-DMPO and peroxyl-DMPO radical adduct signals as detected by ESR produced by the interaction of sickle erythrocyte membranes with t-butyl hydroperoxide with and without desferrioxamine. Spectra were recorded at pH 7.4 and 22°C, under air-saturated conditions, 90 and 400 sec after mixing. Sickle erythrocyte membranes were mixed with t-butyl hydroperoxide (10 mM) and DMPO (40 mM) in the absence and presence of 90 μM desferrioxamine. Spectra were identified by comparison of hyperfine coupling constants with previously reported data.

cytes (for reviews see Polliack and Rachmilewitz, 1973; Palek, 1977; Tillman and Schroter, 1979; Rice-Evans and Chapman, 1981; Rice-Evans and Dunn, 1982).

3.1. Iron Excess in Thalassemia

In hemolytic anemias such as sickle-cell disease or hemoglobin H disease where the hemoglobin does not extensively precipitate, the iron from erythrocyte destruction is available for other developing red cells. In unstable hemoglobins such as thalassemia major, in which globin chains are deficient or absent, the chains are directly oxidized to the hemichrome state prior to the formation of intracellular inclusions, via oxidation of isolated chains.

The formation of inclusion bodies exerts considerable oxidative stress as a result of increased production of oxygen radicals. Isolated globin chains autoxidize much faster than the natural tetramer (Brunori *et al.*, 1975), producing breakdown products of oxidized hemoglobin chains and superoxide radicals. This indicates that the association of α- and β-globin chains to form tetramers in hemoglobin acts as a protective mechanism against the attack of oxidants present in red cells, such as hydrogen peroxide or superoxide, a protection that is defective in diseases such as thalassemia.

The iron from the α-chain inclusions digested within the reticuloendothelial cells seems to be moved preferentially into iron stores. This abnormal handling of iron from premature erythrocyte destruction, as well as the regular blood transfusion to combat the severe anemia and the abnormally high iron absorption (Pearson and O'Brien, 1975; Risdon *et al.*, 1975), cause iron overload. The latter route is particularly important in milder forms of thalassemia, thalassemia intermedia, in which iron loading occurs without blood transfusion. There is no efficient protective mechanism to eliminate excess body iron, hence patients are treated for iron overload. The iron chelator desferrioxamine forms an iron(III) ferrioxamine complex, which is rapidly excreted in the urine.

The major cause of the cellular pathology in thalassemia major is the α-chain imbalance. Chain imbalance is actually comparable in hemoglobin H disease and in thalassemia major, yet the degree of ineffective erythropoiesis is much less in the former disease because only a small proportion of unpaired chains precipitate in the marrow—instead, they associate to form soluble though nonfunctional β_4 hemoglobin H). These differences lead to hemolytic anemia in hemoglobin H disease rather than the ineffective erythropoiesis in thalassemia major. In β-thalassemia it is the timing of precipitation and the nature of the precipitates that determine the timing of cell injury and death (Modell and Berdoukas, 1984).

Untreated patients with thalassemia major die in infancy unless regularly transfused. Regular blood transfusion causes iron loading, leading to the appearance of low-molecular-weight iron complexes in the blood serum (Hershko *et al.*, 1978) and iron deposition mainly in the liver, reticuloendothelial system, and the endocrine glands. Apoferritin synthesis is greatly enhanced in thalassemic reticulocytes and relatively large amounts of iron in the form of the storage proteins, ferritin and hemosiderin, are present in the erythropoietic cells obtained from patients with this disease (Eylar and Matioli, 1965; Polliack *et al.*, 1974).

The ways in which iron may cause tissue and cellular damage in such pathological conditions as thalassemia are complex but in the blood such toxicity may arise from several sources: the excess availability of non-protein-bound iron when plasma transferrin becomes saturated (Worwood, 1983), from the accumulation of breakdown products of oxidized hemoglobin in the form of hemichrome close to the membrane surface of the thalassemic erythrocyte (Rachmilewitz, 1985), or from heme released from its natural globin anchor (Shaklai *et al.*, 1985) and from the presence of ferritinlike iron residual from the thalassemic reticulocyte (Bauminger *et al.*, 1979). Several interrelated factors, therefore, contribute toward the altered rheological behavior and accelerated destruction of the thalassemic erythrocyte. Biochemical abnormalities of these cells include: elevated levels of intracellular calcium (Rachmilewitz, 1985) (especially in thalassemia intermedia), a lowered level of potassium ions with a consequent loss of water and reduction in cell volume and osmotic fragility (Kahane and Rachmilewitz, 1976). The final outcome is the shrinkage of the cell, which would contribute toward the increased rigidity of the membrane. When individual Heinz bodies are present in cells, they can be selectively removed during their passage through the spleen by physical pitting or plucking of the inclusion as it passes through the splenic cords. The resulting loss of surrounding membrane material that occurs as the Heinz bodies are removed leads to a decreased surface-

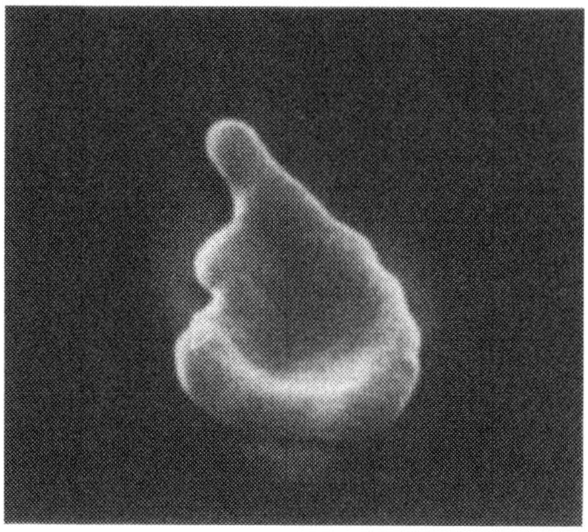

FIGURE 8. A thalassemic dacryocyte. (From Bessis *et al.*, 1973.)

area-to-volume ratio and therefore contributes further to reduced deformability and decreased cellular survival. The typical thalassemic dacryocyte is shown in Figure 8. The appearance of the red cells in the peripheral blood of patients with severe forms of thalassemia is unique in terms of the variety of pathological changes seen: target cells, anisocytosis, poikilocytosis are just some of the markedly distorted shapes observed.

3.2. Oxidative Damage in Thalassemic Erythrocytes

The formation of Heinz bodies exerts considerable oxidative stress as a result of the increased production of oxygen radicals (Winterbourn *et al.*, 1976). The increased susceptibility of the thalassemic erythrocyte to oxidative stress is well documented (Stocks *et al.*, 1972; Rachmilewitz *et al.*, 1976; Chiu *et al.*, 1982; Poli *et al.*, 1986). Modified polyunsaturated fatty acid composition reflects lipid peroxidation in the membranes, resulting in decreased lipid fluidity (Rice-Evans *et al.*, 1980), e.g., decreased arachidonic acid and increased palmitic acid (Kahane and Rachmilewitz, 1976). Lowered levels of free thiol groups in the membrane proteins (Rice-Evans *et al.*, 1980) also implicate oxidative damage.

Free-radical-induced peroxidation of unsaturated fatty acids in the membrane phospholipids leads to the formation of aldehydic secondary lipid breakdown products, some of which, e.g., malonaldehyde, may become involved in the polymerization of the membrane proteins with important consequences for the mechanical properties of the cell and its survival time (Hochstein *et al.*, 1981); 4-hydroxynonenal has been implicated in the hemolysis of thalassemic erythrocytes (Poli *et al.*, 1986).

In thalassemia major and intermedia, the major threat to the erythrocyte from iron-mediated oxidative damage seems to be from the iron in the intracellular inclusions in the thalassemic erythrocyte, and from the possible presence of iron complexes which may be residual from the thalassemic reticulocyte. It is possible that the increased production of active oxygen species from isolated hemoglobin chains may lead to the accumulation of levels of peroxide that can interact with the hemoglobin, producing available iron complexes or reactive higher oxidation states of hemoglobin that are able to promote oxidative damage. It is also conceivable that superoxide radicals can interact with residual ferritin from the thalassemic reticulocyte, again releasing free iron. This iron would then be available to catalyze oxidative damage both intracellularly and in the membrane. Modifications to the cytoskeletal assembly as well as to the lipid structure and organization would result and, via modifications to the interactions with the intramembranous proteins, may alter the distribution of charge at the membrane surface.

3.3. Bioavailability of Membrane-Bound Iron Complexes

We have assessed the bioavailability of membrane-bound iron complexes and their role in the oxidative degeneration of thalassemic membranes (Rice-Evans and Hartley, 1989) by studying total iron retention of the membranes, the ability of membrane-bound iron complexes to stimulate lipid peroxidation via t-butyl hydroperoxide-mediated radical production, and the plasma vitamin E levels (Table IV).

The erythrocytes were divided into three groups according to the vitamin E status: normal erythrocytes, thalassaemic Group I (from thalassemic patients with vitamin E levels at the lower end of the normal range), and thalassemic Group II (from thalassemic patients, deficient in vitamin E). A correlation is shown between patients whose erythrocyte membranes contained elevated membrane-bound iron levels and the ability to propagate peroxidative damage to the membrane lipids mediated by t-butyl hydroperoxide; those effects are more pronounced in the vitamin E-deficient samples.

4. HEMOGLOBIN AS A PROTECTIVE DEVICE

Depending on the circumstances, hemoglobin may function in the erythrocyte in the protection against oxidative damage (Rice-Evans, 1990), rather than in the promotion of such damage.

The ability of iron-containing proteins to redox cycle enables them to serve as additional oxidant sinks protecting the membrane from the deleterious effects of hydrogen peroxide and other hydroperoxides.

Table IV
Vitamin E Status, Membrane-Bound Iron Levels, and Iron-Induced Lipid Peroxidation in Thalassemic Erythrocytes

	Thalassemic erythrocytes		Normal erythrocytes
	Group I	Group II	
Plasma vitamin E (μg/ml)	12.12 ± 4.6 (n = 26)	5.82 ± 2.1 (n = 25)	13.03 ± 5.6 (n = 13)
Membrane-bound iron (nmol/mg protein)	5.4 ± 4 (n = 3)	14.5 ± 10 (n = 5)	2.7 ± 2 (n = 6)
TBAR products (nmol/mg protein)	1.4 ± 0.6 (n = 5)	2.1 ± 1 (n = 12)	0.6 ± 0.06 (n = 8)

Table V
The Influence of the Oxidation State of Hemoglobin on the Response of Membrane Lipids to Oxidative Stress

	Methemoglobin		TBAR-reactive compounds	
	5 hr	24 hr	5 hr	24 hr
Erythrocytes	(%)		(nmol/10^{10} cells)	
− ATP	6 ± 2 (n = 5)	74 ± 8 (n = 5)	6.2 ± 1.0 (n = 4)	9.4 ± 30 (n = 7)
+ ATP	0 (n = 5)	15 ± 4 (n = 5)	22.3 ± 3.0 (n = 3)	41.3 ± 8.0 (n = 3)
			(nmol/mg protein)	
Membranes (hemoglobin-free)	—	—	11.1 ± 2.0 (n = 6)	—
Hemolysate (membrane-free)	2 ± 1 (n = 13)	43 ± 7 (n = 13)	—	—

Exposing intact erythrocytes to extracellular iron-mediated oxidative stress *in vitro* induces high levels of hemoglobin oxidation intracellularly but membrane peroxidation only occurs to a small extent (Rice-Evans and Baysal, 1987b; Rice-Evans, 1987) (Table I). Hemoglobin seems to be acting as an oxidant sink in protecting the membrane from radical attack and, in the process, itself being converted to methemoglobin. If during the course of exposing the cells to iron-mediated oxidative stress the energy requirements of the cell are maintained, thereby substantially suppressing the formation of, and any secondary effects due to, methemoglobin, the peroxidation of the membrane is considerably enhanced (Table V). This implies that when methemoglobin production is minimized under conditions of oxidative stress, the membranes become more vulnerable to oxidative damage and increased lipid peroxidation ensues. This concept is substantiated by exposure of hemoglobin-free membranes and membrane-free hemolysates to oxidative stress under the same conditions (Table V). In the former case, high levels of membrane peroxidation are observed in the absence of the hemoglobin; in the latter case, the induction of hemoglobin oxidation was considerably less extensive than in the intact erythrocyte. These observations support the role of hemoglobin in intercepting the mediator of the oxidative damage to the membrane in the intact erythrocyte exposed to oxygen radicals. In other words, the mechanism of scavenging propagating oxidative species in the membrane lipids is balanced by the oxidation of oxyhemoglobin to methemoglobin within the erythrocyte in its capacity to act as an oxidant sink.

Consistent with this, others (Snyder *et al.*, 1988) have demonstrated that pretreatment of cells with carbon monoxide, thus preventing its action as an oxidant sink, prior to exposure to hydrogen peroxide, inhibited hemoglobin–spectrin cross-linking and other related oxidative membrane alterations but failed to inhibit lipid peroxidation.

As described previously, during oxidation of hemoglobin within erythrocytes under oxidative stress, or when hemoglobin is released via microbleeding processes at sites of tissue damage, the methemoglobin formed may be activated by hydrogen peroxide to form the ferrylhemoglobin species.

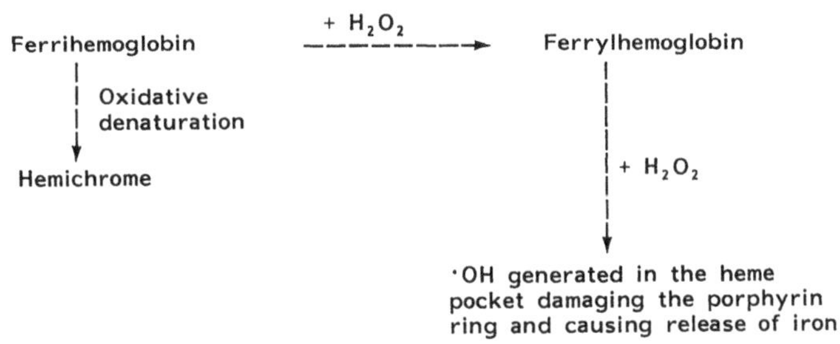

FIGURE 9. The interaction of hemoglobin with hydrogen peroxide.

The intermediate oxidation of iron within these proteins to a higher valence state may be an important protective device and this concept is enhanced by experiments that illustrate the capacity of a variety of electron donors to reduce rapidly the iron of these heme proteins, e.g., the reduction of ferryl myoglobin to metmyoglobin by desferrioxamine (Rice-Evans *et al.*, 1989). It has been proposed that the functional site of myoglobin, for example, is protected from oxidative damage by the proximity of several aromatic groups. The transfer of the radical site to a tyrosine or histidine on the surface of the globin allows it to be reduced by local electron-donating substrates. Higher concentrations of hydrogen peroxide can also interact with ferrylhemoglobin or ferrylmyoglobin as shown (Petersen *et al.*, 1989) in Figure 9. The subsequent generation of the hydroxyl radical within the heme pocket will damage the porphyrin ring and, after several rounds of attack, leads to the release of iron from hemoglobin or myoglobin. It is this released iron that can stimulate hydroxyl radical formation away from the heme group and that, in addition to the ferrylhemoglobin radical, has the potential to induce membrane, cell, and tissue damage (Puppo and Halliwell, 1988; Rice-Evans *et al.*, 1989).

5. CONCLUSIONS

In situations under which erythrocytes are placed under oxidative stress or in conditions of tissue damage and microbleeding in the tissues, hemoglobin oxidation and delocalization of iron complexes occur. The presence of iron-containing denaturation products of hemoglobin and decompartmentalized iron complexes in sickle-cell and thalassemic membranes provides a source of continued oxidative stress that possibly underlies the membrane abnormalities and contributes toward premature erythrocyte destruction. These findings in conjunction with the ability of desferrioxamine to not only complex with low-molecular-weight iron chelators but also intercept radical species (Rice-Evans *et al.*, 1989; Darley-Usmar *et al.*, 1989) might seem to emphasize the relevance of desferrioxamine as a therapeutic approach in sickle-cell anemia and thalassemia. However, desferrioxamine, while reported to clear iron from hepatocytes, does not apparently enter normal erythrocytes (Rice-Evans, 1989) within a time scale longer than its lifetime in plasma.

Furthermore, this iron chelator, currently in therapeutic use for the treatment of iron overload disorders, is not entirely nontoxic at the high doses applied for the treatment of iron overload (Jackson *et al.*, 1983; S. C. Davies *et al.*, 1983; Arden *et al.*, 1984; Romeo *et al.*, 1984; Guerini *et al.*, 1985; Blake *et al.*, 1985). *In vitro* studies have demonstrated oxidation of hemoglobin in hemolysates by desferrioxamine and its suppression in the presence of ascorbate (Rice-Evans *et al.*, 1986a,b).

The efficacy of other specific electron-donating drugs that are able to enter erythrocytes is currently under investigation in the author's laboratory for the suppression of intraerythrocytic iron-mediated oxidative stress.

6. REFERENCES

Aisen, P., 1977, Some physicochemical aspects of iron metabolism, *CIBA Found. Symp.* **51**:1–17.

Ames, B. N., Cathcart, R., Schwiers, E., and Hochstein, P., 1981, Uric acid provides an antioxidant defence in humans against oxidant- and radical-caused aging and cancer: A hypothesis, *Proc. Natl. Acad. Sci. USA* **78**:6858–6862.

Arden, G. B., Wonke, B., Kennedy, C., and Huehns, E. R., 1984, Ocular changes in patients undergoing long-term desferrioxamine treatment, *Br. J. Ophthalmol.* **68**:873–877.

Aruoma, O., Halliwell, B., Laughton, M. J., Quinlan, G. J., and Gutteridge, J. M. C., 1989, The mechanism of initiation of lipid peroxidation. Evidence against a requirement for iron(II)–iron(III) complex, *Biochem. J.* **258**:617–620.

Asakura, T., Minakata, K., Adachi, K., Russell, M. O., and Schwartz, E., 1977, Denatured haemoglobin in sickle erythrocytes, *J. Clin. Invest.* **59**:633–640.

Aust, S. D., and White, B. C., 1986, The role of iron in lipid peroxidation, in: *Free Radicals, Cell Damage and Disease* (C. Rice-Evans, ed.), pp. 15–27, Richelieu Press, London.

Aust, S. D. and Minotti, G., 1987, The role of iron in the inhibition of lipid peroxidation, *Chem. and Physics of Lipids* **44**:141–207.

Bauminger, E. R., Cohen, S. G., Offer, S., Racmilewitz, E. A., 1979, Quantitative studies of ferritin-like iron in erythrocytes of thalassaemia, sickle cell anemia and haemoglobin Hammersmith by Mossbauer spectroscopy, *Proc. Natl. Acad. Sci. USA* **76**:939–943.

Blake, D. R., Winyard, P., and Lunec, J., 1985, Cerebral and ocular toxicity induced by desferrioxamine, *Q. J. Med.* **219**:345–355.

Brunori, M., Wyman, J., Antonini, E., and Rossi-Fanelli, A., 1965, Studies on the oxidation–reduction potentials of haem proteins, *J. Biol. Chem.* **240**:3317–3324.

Brunori, M., Falcioni, G., Fioreti, E., Giardina, B., and Rotilio, G., 1975, Formation of superoxide in the autoxidation of isolated α and β chains of human haemoglobin and its involvement in haemichrome precipitation, *Eur. J. Biochem.* **53**:99–104.

Bunn, F., and Jandl, J., 1966, Exchange of haem among haemoglobin molecules, *Proc. Natl. Acad. Sci. USA* **56**:974–978.

Butler, J., Hoey, B. M., Lea, J. S., 1988, The measurement of radicals by pulse radiolysis, in: *Free Radicals, Methodology and Concepts* (C. Rice-Evans, and B. Halliwell, eds.), pp. 457–479, Richelieu Press, London.

Bessis, M., 1983, Red cell shapes, in *Red Cell Shape* (M. Bessis, R. I. Weed, P. F. Leblond, eds.), Springer, New York.

Chidambaram, M. V., Barnes, G., and Frieden, E., 1983, Caeruloplasmin and the reactions forming diferric transferrin, *FEBS Lett.* **159**:137–140.

Chiu, D., and Lubin, B., 1979, Abnormal vitamin E and glutathione peroxidase levels in sickle cell anaemia: Evidence for increased susceptibility to lipid peroxidation in vivo, *J. Lab. Clin. Med.* **94**:542–548.

Chiu, D., Lubin, B., and Shohet, S., 1982, Peroxidative reactions in red cell biology, in: *Free Radicals in Biology* (W. A. Pryor, ed.), Volume V, pp. 115–160, Academic Press, New York.

Chung, J., and Wood, J. L., 1971, Oxidation of thiocyanate to cyanide catalysed by haemoglobin, *J. Biol. Chem.* **246**:555–560.

Clark, M. R., Mohandas, N., and Shohet, S. B., 1980, The deformability of oxygenated irreversibly sickled cells, *J. Clin. Invest.* **65**:185–196.

Cohen, G., and Hochstein, P., 1963, Glutathione peroxidase, the primary agent for the elimination of hydrogen peroxide in erythrocytes, *Biochemistry* **2:**1420–1428.

Czapski, G., and Goldstein, S., 1986, When do metal complexes protect biological systems from superoxide toxicity and when do they enhance it? *Free Radical Res. Commun.* **1:**157–161.

Darley-Usmar, V. M., Hersey, A., and Garland, L. G., 1989, A method for the comparative assessment of antioxidants as peroxyl radical scavengers, *Biochem. Pharmacol.* **38:**1465–1467.

Das, S. K., and Nair, R. C., 1980, Superoxide dismutase, glutathione peroxidase, catalase and lipid peroxidation of normal and sickled erythrocytes, *Br. J. Haematol.* **44:**87–92.

Davies, K. J. A., and Goldberg, A. L., 1987, Oxygen radicals stimulate intracellular proteolysis and lipid peroxidation by independent mechanisms in erythrocytes, *J. Biol. Chem.* **262:**8220–8226.

Davies, K. J. A., Sevanian, A., Mukkassah-Kelly, S. F., and Hochstein, P., 1986, Uric acid–iron ion complexes, *Biochem. J.* **235:**747–754.

Davies, M. J., and Slater, T. F., 1987, Studies on the metal ion and lipoxygenase-catalysed breakdown of hydroepoxides using electron spin resonance spectroscopy, *Biochem J.* **245:**167–173.

Davies, M. J., Donkor, R. E., Dunster, A., Gee, C. A., Jonas, S., and Willson, R. L., 1987, Desferrioxamine and superoxide free radicals: Formation of an enzyme-damaging nitroxide, *Biochem. J.* **246:**725–729.

Davies, S. C., Marcus, R. E., Hungerford, J. L., Miller, M. H., Arden, G. B., and Huehns, E. R., 1983, Ocular toxicity of high dose intravenous desferrioxamine, *Lancet* **2:**181–184.

Diplock, A. T., 1983, The role of vitamin E in biological membranes, *CIBA Found. Symp.* **101:**45–53.

Doly, M., Bonhomme, B., and Vennat, J. C., 1986, Experimental study of the retinal toxicity of haemoglobin iron, *Ophthalmic Res.* **18:**21–27.

Eaton, J. W., Skelton, T. O., Swofford, H. S., Kaplin, C. E., and Jacobs, H. S., 1974, Elevated calcium in sickle cell disease, *Nature* **246:**105–106.

Esterbauer, H., 1985, Lipid peroxidation products: Formation, chemical properties and biological activities, in: *Free Radicals in Liver Injury* (G. Poli, K. Cheeseman, M. U. Dianzani, and T. F. Slater, eds.), pp. 29–47, IRL Press, Oxford, England.

Eylar, E. H., and Matioli, G., 1965, Apoferritin synthesis in human erythroid cells in thalassaemia, *Nature* **208:**661–664.

Fischer, S., Nagel, R. L., Bookchin, R. M., Roth, E. F., and Tellez-Nagel, I., 1975, The binding of haemoglobin to membranes of normal and sickle erythrocytes, *Biochim. Biophys. Acta* **375:**422–433.

Fortier, N. L., Snyder, L. M., Ganver, F., Kiefer, C., McKenney, J., and Mohandas, N., 1988, The relationship between in vivo generated haemoglobin skeletal protein complex and increased cell and membrane rigidity, *Blood* **71:**1427–1431.

George, P., and Irvine, D. H., 1952, The interaction between metmyoglobin and hydrogen peroxide, *Biochem. J.* **52:**511–515.

Girotti, A. W., 1985, Mechanism of lipid peroxidation, *J. Free Radical Biol. Med.* **1:**87–95.

Goldberg, B., Stern, A., and Peisach, J., 1976, The mechanism of superoxide generation by the interaction of phenylhydrazine with haemoglobin, *J. Biol. Chem.* **251:**3045–3051.

Griffiths, H., Unsworth, J., Blake, D. R., and Lunec, J., 1988, Oxidation of amino acids within serum proteins, in: *Free Radicals, Chemistry, Pathology and Medicine* (C. Rice-Evans and T. Dormandy, eds.), pp. 439–454, Richelieu Press, London.

Guerini, A., London, G., Marchais, S., Metivier, F., and Relisse, J. M., 1985, Acute deafness and desferrioxamine, *Lancet* **2:**39–40.

Gutteridge, J. M. C., 1986, Iron promoters of the Fenton reaction and lipid peroxidation can be released from haemoglobin by peroxides, *FEBS Lett.* **201:**291–295.

Gutteridge, J. M. C., 1987, The antioxidant activity of haptoglobin towards haemoglobin-stimulated lipid peroxidation, *Biochim. Biophys. Acta* **917:**219–223.

Gutteridge, J. M. C., and Halliwell, B., 1989, Iron toxicity and oxygen radicals, *Bailliere's Clin. Haematol.* **2:**195–256.

Halliwell, B., and Gutteridge, J. M. C., 1984, Oxygen toxicity, oxygen radicals, transition metals and disease, *Biochem. J.* **219:**1–14.

Halliwell, B., and Gutteridge, J. M. C., 1986, Oxygen free radicals and iron in relation to biology and medicine. Some problems and concepts, *Arch. Biochem. Biophys.* **246:**501–514.

Hartley, A., and Rice-Evans, C., 1989a, Membrane-associated iron species and membrane oxidation in sickle cell disease, *Biochem. Soc. Trans.* **17:**116–118.

Hartley, A., and Rice-Evans, C., 1989b, The nature of membrane-bound iron species involved in radical-mediated damage to sickle erythrocytes, *Biochem. Soc. Trans.* **17**:488–489.

Hartley, A., Davies, M. J., and Rice-Evans, C., 1990, Desferrioxamine as a lipid chain-breaking antioxidant in sickle erythrocyte membranes, *FEBS Lett.* (in press).

Hebbel, R. P., Yamada, O., Moldow, C. F., Jacob, H. S., White, J. G., and Eaton, J. W., 1980, Abnormal adherence of sickle erythrocytes to cultured vascular endothelium. Possible mechanisms for microvascular occlusion in sickle cell disease, *J. Clin. Invest.* **65**:154–158.

Hebbel, R. P., Eaton, J. W., Balasingham, M., Steinberg, M. H., 1982, Spontaneous oxygen radical generation by sickle erythrocytes, *J. Clin. Invest.* **70**:1253–1259.

Hershko, C., Graham, G., Bates, G. W., and Rachmilewitz, E. A., 1978, Nonspecific serum iron in thalassaemia: An abnormal serum iron fraction of potential toxicity, *Br. J. Haematol.* **49**:255–263.

Hochstein, P., Jain, S. K., and Rice-Evans, C., 1981, The physiological significance of oxidative perturbations in erythrocyte membrane lipids and proteins, in: *The Red Cell* (G. Brewer, ed.), pp. 449–459, Liss, New York.

Jackson, M. E., Brenton, D. P., and Modell, B., 1983, DTPA in management of iron overload in thalassaemia, *J. Inher. Metab. Dis.* **6**:97–98.

Jacobs, A., 1977, Low molecular weight intracellular iron transport compounds, *Blood* **50**:433–439.

Kahane, I., and Rachmilewitz, E. A., 1976, Alteration of the red blood cell membrane and the effect of vitamin E on osmotic fragility in thalassaemia major, *Isr. J. Med. Sci.* **12**:11–15.

Kappus, H., 1985, Lipid peroxidation: Mechanisms, analysis, enzymology and biological relevance, in: *Oxidative Stress* (H. Sies, ed.), pp. 273–310, Academic Press, New York.

Keberle, H., 1964, The biochemistry of desferrioxamine and its relation to iron metabolism, *Ann. N.Y. Acad. Sci.* **119**:758–768.

Kim, H. C., Friedman, S., Asakura, T., and Schwartz, E., 1980, Inclusions in red blood cells containing HbS or HbC, *Br. J. Haematol.* **44**:547–554.

Kuross, S. A., and Hebbel, R. P., 1988, Nonhaem iron in sickle erythrocyte membranes: Association with phospholipids and potential role in lipid peroxidation, *Blood* **72**:1278–1285.

Kuross, S. A., Rank, B. H., and Hebbel, R. P., 1988, Excess haem in sickle erythrocyte inside-out membranes: Possible role in thiol oxidation, *Blood* **71**:876–882.

Labeque, R., and Marnett, L., 1988, Reaction of haematin with allylic fatty acid hydroepoxides, *Biochemistry* **27**:7060–7070.

Lessin, L. S., Kuranstin-Mills, J., Wallas, C., and Weems, H., 1978, Membrane alterations in irreversibly sickled cells: Haemoglobin–membrane interactions, *J. Supramol. Struct.* **9**:537–543.

Levine, R. L., Oliver, C. N., Fulks, R. M., and Stadtman, E. R., 1981, Turnover of bacterial glutamate synthetase: Oxidative inactivation precedes proteolysis, *Proc. Natl. Acad. Sci. USA* **78**:2120–2124.

Liu, S. C., Zhai, S., and Palek, J., 1988, Detection of haemin release during haemoglobin S denaturation, *Blood* **71**:1755–1758.

Lubin, B., Chiu, D., Bastacky, J., Roelofsen, B., and Van Deenen, L. L. M., 1981, Abnormalities in the membrane phospholipid organisation in sickled erythrocytes, *J. Clin. Invest.* **67**:1643–1649.

Lux, S. E., John, K. M., and Karnovsky, M. J., 1976, Irreversible deformation of the spectrin–actin lattice in irreversibly sickled cells, *J. Clin. Invest.* **58**:955–963.

McCord, J. M., and Fridovich, I., 1969, Superoxide dismutase an enzymic function for erythrocuprein (haemocuprein), *J. Biol. Chem.* **244**:6049–6055.

Marklund, S. L., 1984, Extracellular superoxide dismutase and other superoxide dismutase isoenzymes in tissues from nine mammalian species, *Biochem. J.* **222**:649–655.

Minotti, G., and Aust, S. D., 1987a, Role of iron in the initiation of lipid peroxidation, *Chem. Phys. Lipids* **44**:191–208.

Minotti, G., and Aust, S. D., 1987b, The requirement for iron (IV) in the initiation of lipid peroxidation by iron (II) and hydrogen peroxide, *J. Biol. Chem.* **262**:1098–1104.

Misra, H. P., and Fridovich, I., 1972, The role of superoxide anion in the autoxidation of epinephrine and a simple assay for superoxide dismutase, *J. Biol. Chem.* **247**:3170–3175.

Modell, B., and Berdoukas, V., 1984, *The Clinical Approach to Thalassaemia*, Grune & Stratton, New York.

Morehouse, K. M., Flitter, W. D., and Mason, R. P., 1987, The enzymatic oxidation of desferal to a nitroxide radical, *FEBS Lett.* **222**:246–250.

Moser, U., 1987, Uptake of ascorbic acid by leukocytes, *Ann. N.Y. Acad. Sci.* **498**:200–215.
O'Brien, P. J., 1969, Intracellular mechanisms for the decomposition of a lipid hydroperoxide, *Can. J. Biochem.* **47**:485–492.
Osaki, S., Johnson, D. A., and Frieden, E., 1966, Possible significance of the ferrous oxidase activity of caeruloplasmin in normal human serum, *J. Biol. Chem.* **241**:2746–2751.
Palek, J., 1977, Red cell membrane injury in sickle cell anaemia, *Br. J. Haematol.* **35**:1–9.
Panter, S. S., Sadrzadeh, S. M., Hallaway, P. E., Haines, J. L., Anderson, V. E., and Eaton, J. W., 1985, Hypohaptoglobinaemia associated with familial epilepsy, *J. Exp. Med.* **161**:748–754.
Pearson, H. A., and O'Brien, R. T., 1975, The management of thalassaemia major, *Semin. Haematol.* **12**:255–265.
Peisach, J., Blumberg, W. E., Wittenberg, B. A., and Wittenberg, J. B., 1968, The electronic structure of protohaem proteins. III. The configuration of the haem and its ligands, *J. Biol. Chem.* **243**:1871–1880.
Peisach, J., Blumberg, W. E., and Rachmilewitz, E. A., 1975, The demonstration of ferrihaemochrome intermediates in Heinz body formation following the oxidation of oxyhaemoglobin by acetylphenylhydrazine, *Biochem. Biophys. Acta* **393**:404–418.
Perutz, M., 1989, Myoglobin and haemoglobin: Role of distal residues in reactions with haem ligands, *Trends Biochem. Sci.* **14**:42–44.
Petersen, R. I., Symons, M. C. R., and Taiwo, F. A., 1989, Application of irradiation and ESR spectroscopy to the study of ferryl haemoglobin and myoglobin, *J. Chem. Soc. Faraday Trans.* **85**:2435–2443.
Platt, O. S., Falcone, J. F., and Lux, S. E., 1985, Molecular defect in the sickle erythrocyte cytoskeleton. Abnormal spectrin binding to sickle inside-out vesicles, *J. Clin. Invest.* **75**:266–271.
Poli, G., Ramenghi, U., David, O., Biasi, F., Cecchini, G., Carini, R., Chiarpotto, E., and Dianzani, M. U., 1986, Lipid peroxidation in red blood cells obtained from thalassaemic subjects and exposed to oxidative stress, in: *Free Radicals, Cell Damage and Disease* (C. Rice-Evans, ed.), pp. 449–459, Richelieu Press, London.
Polliack, A., and Rachmilewitz, E. A., 1973, Ultrastructural studies in thalassaemia major, *Br. J. Haematol.* **24**:319–326.
Polliack, A., Yataganas, X., Thorell, B., and Rachmilewitz, E. A., 1974, An electron microscopic study of the nuclear abnormalities in erythroblasts in beta-thalassaemia major, *Br. J. Haematol.* **26**:201–204.
Puppo, A., and Halliwell, B., 1988, Formation of hydroxyl radicals from hydrogen peroxide in the presence of iron: Is haemoglobin a biological Fenton reagent? *Biochem. J.* **249**:185–190.
Rachmilewitz, E. A., 1985, Erythrocyte membrane alterations in beta-thalassaemia, *Clin. Haematol.* **14**:163–182.
Rachmilewitz, E. A., Peisach, J., and Blumberg, W. E., 1971, Studies on the stability of oxyhaemoglobin A and its constituent chains and their derivatives, *J. Biol. Chem.* **246**:3356–3366.
Rachmilewitz, E. A., Lubin, B. H., and Shohet, S. B., 1976, Lipid membrane peroxidation in beta-thalassaemia major, *Blood* **47**:495–505.
Rank, B. H., Carlsson, J., and Hebbel, R. P., 1985, Abnormal redox status of membrane: Protein thiols in sickle erythrocytes, *J. Clin. Invest.* **75**:1531–1537.
Rice-Evans, C., 1987, Oxidative modifications in erythrocytes induced by iron, in: *Free Radicals, Oxidant Stress and Drug Action* (C. Rice-Evans, ed.), pp. 307–330, Richelieu Press, London.
Rice-Evans, C., 1989, Iron chelators and the suppression of oxidative damage in erythrocytes: Extracellular and intracellular responses, in: *Free Radicals, Metal Ions and Biopolymers* (P. Beaumont, D. Deeble, B. Parsons, and C. Rice-Evans, eds.), pp. 283–301, Richelieu Press, London.
Rice-Evans, C., 1990, Decompartmentalised, microbleeding and membrane oxidation, in: *Free Radicals, Lipoproteins and Membrane Lipids* (A. Crastes de Paulet, L. Douste Blazy, and R. Paoletti, eds.), in press.
Rice-Evans, C., and Baysal, E., 1987a, Role of membrane-bound haemoglobin products in oxidative damage in sickle cell membranes, *Acta Haematol.* **78**:105–108.
Rice-Evans, C., and Baysal, E., 1987b, Iron-mediated oxidative stress in erythrocytes, *Biochem. J.* **244**:191–196.
Rice-Evans, C., and Chapman, D., 1981, Red blood cell biomembrane structure and deformability, *Scand. J. Clin. Lab. Invest.* **156**:99–110.
Rice-Evans, C., and Dunn, M. J., 1982, Erythrocyte deformability and disease, *Trends Biochem. Sci.* **7**:282–286.
Rice-Evans, C., and Hartley, A., 1989, Free radicals erythrocyte disorders and iron decompartmentalisation, in:

Medical, Biochemical and Chemical Aspects of Free Radicals (O. Hayaishi, E. Niki, M. Kondo, and T. Yoshikawa, eds.), Elsevier, Amsterdam, pp. 1185–1194.

Rice-Evans, C., and Omorphos, S., 1983, Hydrazine and sickle cells, *Biochem. Soc. Trans.* **11:**180–181.

Rice-Evans, C., and Omorphos, S. C., 1986, Free radical-induced membrane damage in sickle erythrocytes, in: *Approaches to the Therapy of Sickle Cell Anaemia* (Y. Beuzard, S. Charache, and F. Galacteros, eds.), pp. 329–336, INSERM, Paris.

Rice-Evans, C., Bruckdorfer, K. R., and Dootson, G., 1978, Studies on the altered membrane characteristics of sickle cells, *FEBS Lett.* **94:**81–86.

Rice-Evans, C., Johnson, A., and Flynn, D. M., 1980, Red cell membrane abnormalities in beta-thalassaemia major, *FEBS Lett.* **119:**53–57.

Rice-Evans, C., Omorphos, S., and White, J. M., 1982, The effects of hydrazine on sickle cells, *Biochim. Biophys. Acta* **691:**367–371.

Rice-Evans, C., Baysal, E., Flynn, D., and Kontoghiorghes, G., 1986a, Iron-mediated free radical effects on erythrocytes: The role of desferrioxamine, *Biochem. Soc. Trans.* **14:**368–369.

Rice-Evans, C., Baysal, E., Flynn, D., and Kontoghiorghes, G., 1986b, The effects of desferrioxamine and ascorbate on haemoglobin from thalassaemic erythrocytes, in: *Superoxide and Superoxide Dismutase in Chemistry, Pathology and Medicine* (G. Rotilio, ed.), pp. 662–666, Elsevier, Amsterdam.

Rice-Evans, C., Omorphos, S. C., and Baysal, E., 1986c, Sickle cell membranes and oxidative damage, *Biochem. J.* **237:**265–269.

Rice-Evans, C., Omorphos, S. C., and Baysal, E., 1986d, Sickle cell pathology: Is the membrane important? in: *Free Radicals, Cell Damage and Disease* (C. Rice-Evans, ed.), pp. 149-166, Richelieu Press, London.

Rice-Evans, C., Okunade, G., and Khan, R., 1989, The suppression of iron release from activated myoglobin by physiological electron donors and by desferrioxamine, *Free Radical Res. Commun.* **7:**45–54.

Risdon, R. A., Barry, M., and Flynn, D., 1975, Transfusional iron overload: Hepatic fibrosis in thalassaemia, *J. Pathol.* **116:**83–95.

Romeo, M. A., DiGregorio, F., and Schiliro, G., 1984, Allergy to desferrioxamine, *J. Inher. Metab. Dis.* **7:**121–124.

Roshchupkin, D. I., Talitsky, V. V., and Peleritsyn, A. N., 1979, Fluorescence study of tryptophan photolysis, *Photochem. Photobiol.* **30:**635–641.

Rowland, I., and Symons, M. C. R., 1988, Tumours and iron: The use of electron spin resonance, in: *Free Radicals, Methodology and Concepts* (C. Rice-Evans and B. Halliwell, eds.), pp. 169–184, Richelieu Press, London.

Rowley, D. A., and Halliwell, B., 1982a, Superoxide-dependent formation of hydroxyl radicals in the presence of thiol compounds, *FEBS Lett.* **138:**33–36.

Rowley, D. A., and Halliwell, B., 1982b, Superoxide-dependent formation of hydroxyl radicals from NADH and NADPH in the presence of iron salts, *FEBS Lett.* **142:**39–41.

Rowley, D. A., and Halliwell, B., 1983, Formation of hydroxyl radicals from hydrogen peroxide and iron salts by superoxide- and ascorbate-dependent mechanisms: Relevance to the pathology of rheumatoid disease, *Clin. Sci.* **64:**649–653.

Schwartz, R., Rybicki, A. C., Heath, R. H., and Lubin, B. H., 1987, Protein 4.1 in sickle erythrocytes, *J. Biol. Chem.* **262:**15666–15672.

Searle, A., and Tomasi, A., 1982, Hydroxyl free radical production in iron-cysteine solutions and protection by zinc, *J. Inorg. Chem.* **17:**161–166.

Shaklai, N., Sharma, V. S., and Ranney, H. M., 1981, Interaction of sickle cell haemoglobin with erythrocyte membranes, *Proc. Natl. Acad. Sci. USA* **78:**65–68.

Shaklai, N., Shviro, Y., Rabizadeh, E., and Kirscher-Zilbe, I., 1985, Accumulation and drainage of haemin in the red cell membrane, *Biochim. Biophys. Acta* **821:**355–356.

Sies, H., 1986, Biochemistry of oxidative stress, *Angew. Chem. Int. Ed. Engl.* **25:**1058–1071.

Singh, A., Koroll, G. W., and Cundall, R. B., 1982, Pulse radiolysis of aqueous solutions of sodium azide: Reactions of azide radical, tryptophan and tyrosine, *Radiat. Phys. Chem.* **19:**137–146.

Snyder, L. M., Fortier, N. L., Leb, L., McKenney, J., Trainor, J., Sheerin, H., and Mohandas, N., 1988, The role of membrane protein sulphydryl groups in hydrogen peroxide-mediated membrane damage in human erythrocytes, *Biochim. Biophys. Acta* **937:**229–240.

Stocks, J., Offerman, E. L., Modell, C. B., and Dormandy, T. L., 1972, The susceptibility to autoxidation of human red cell lipids in health and disease, *Br. J. Haematol.* **23:**713–724.

Tappel, A. L., 1962, Vitamin E as the biological lipid antioxidant, *Vitam. Horm.* (N.Y.) **20**:493–510.

Tappel, A. L., 1980, Measurement of and protection from in vivo lipid peroxidation, in: *Free Radicals in Biology* (W. A. Pryor, ed.), Volume IV, pp. 1–45, Academic Press, New York.

Tillman, W., and Schroter, W., 1979, Rheological properties of erythrocytes in homozygous and heterozygous thalassaemia, *Br.J. Haematol.* **41**:401–411.

Waugh, S. M., Willardson, B. M., Kannan, R., Labotka, R. J., and Low, P. S., 1986, Heinz bodies induce clustering of band 3, glycophorin and ankyrin in sickle cell erythrocytes, *J. Clin. Invest.* **78**:1155–1160.

Weiss, J., 1964, The nature of iron bound in oxyhaemoglobin, *Nature* **202**:83–84.

Whitburn, K., 1987, The interaction of oxymyoglobin with hydrogen peroxide: The formation of ferryl myoglobin at moderate excess of hydrogen peroxide, *Arch. Biochem. Biophys.* **253**:419–430.

White, J. G., 1974, Ultrastructural features of erythrocyte and haemoglobin sickling, *Arch. Intern. Med.* **133**:545–562.

Williams, D. R., 1971, *The Metals of Life,* Van Nostrand–Rheinhold, Princeton, N.J.

Winterbourn, C. C., 1979, Comparison of superoxide with other reducing agents in the biological production of hydroxyl radicals, *Biochem. J.* **182**:625–628.

Winterbourn, C. C., 1981, Hydroxyl radical production in body fluids. Roles of metal ions, ascorbate and superoxide, *Biochem. J.* **198**:125–131.

Winterbourn, C. C., McGrath, B. N. M., and Carrell, R. W., 1976, Reactions involving superoxide and normal and unstable haemoglobins, *Biochem. J.* **155**:493–502.

Wittenberg, J. B., Wittenberg, B. A., Peisach, J., and Blumberg, W. E., 1970, On the state of iron and the nature of the ligand in oxyhaemoglobin, *Proc. Natl. Acad. Sci. USA* **67**:1846–1853.

Wolff, S. P., and Dean, R. T., 1986, Fragmentation of proteins by free radicals and its effect on their susceptibility to enzymic hydrolysis, *Biochem. J.* **234**:399–403.

Wolff, S. P. Garner, A., and Dean, R. T., 1986, Free radicals, lipids and protein degradation, *Trends Biochem. Sci.* **11**:27–31.

Worwood, M. J., 1983, Iron and haemochromatosis, *J. Inher. Metab. Dis.* **6**:63–69.

Yoshino, S., Blake, D. R., Hewitt, S., Morris, C., and Bacon, P. A., 1985, Effect of blood on the activity and persistence of antigen reduced inflammation in the rat air pouch, *Ann. Rheum. Dis.* **44**:495–490.

Chapter 15

Erythrocyte Membrane Damage in Hemolytic Anemias

Augusta Brovelli

1. INTRODUCTION

Molecular defects directly affecting a membrane component can modify the shape of the red cell and reduce its survival capacity; the same effect can result from molecular lesions at the level of a cytoplasmic protein or enzyme. In some diseases where the primary defect does not involve a membrane component, a secondary membrane lesion develops, contributing to or causing red cell death. For instance, in some hemoglobinopathies, a less stable anomalous hemoglobin interacts with the membrane and modifies lipid and proteins; in erythroenzymopenias, such as glucose-6-phosphate dehydrogenase (G6PD) deficiency and pyruvate kinase (PK) deficiency, the altered structure of a key enzyme involved in glycolysis or the pentose pathway, impairs the ability of red cells to withstand stress conditions, and there are clues that lead us to suspect a membrane lesion secondary to the enzyme deficiency. In this review, membrane processes found in the three above-mentioned groups of diseases will be described: in particular, we will deal with two hemoglobinopathies, thalassemia and sickle-cell anemia; with the enzymopathies glucose-6-phosphate dehydrogenase and pyruvate kinase deficiency; and with the hereditary membrane skeleton diseases spherocytosis, elliptocytosis, and pyropoikilocytosis. A summary of membrane lesions found in each disease will be presented, more detailed reviews concerning each of these topics being available [see Wagner *et al.* (1985), Hebbel

Abbreviations used in this chapter: G6PD$^+$, glucose-6-phosphate dehydrogenase deficiency; PK$^+$, pyruvate kinase deficiency; HS, hereditary spherocytosis; HE, hereditary elliptocytosis; HPP, hereditary pyropoikilocytosis; ISC, irreversibly sickled cells; PE, phosphatidylethanolamine; PS, phosphatidylserine; PC, phosphatidylcholine; PI, phosphatidylinositide; TPI, triphosphoinositide; MDA, malonyldialdehyde; Hb, hemoglobin.

Augusta Brovelli Department of Biochemistry, Faculty of Science, University of Pavia, 27100 Pavia, Italy.

et al. (1985), and Hebbel (1986) for sickle cell anemia; Rachmilewitz *et al.* (1985) for thalassemia; Palek and Lux (1983), Becker and Lux (1985), Palek (1985) for hereditary defects of membrane skeleton; Beutler (1983), Valentine *et al.* (1983) for G6PD and PK deficiencies]. Membrane damage associated with an impaired survival capacity of pathological red cells will be compared with membrane properties of senescent normal red cells (see Bartosz, this volume). The general aim of the review is to analyze which membrane lesions in erythrocyte pathology shorten red cell life span, and to find insights into membrane processes that would be worth investigating in those diseases, where indirect evidence suggests the presence of membrane damage. Structural integrity of the membrane allows the erythrocyte to perform its biological role of oxygen transport from the lungs throughout the body. The ability to deform reversibly during flow is crucial to the red cell for performing its function. This property, known as cellular deformability, is determined by membrane material characteristics, cell surface-area-to-volume ratio, and cytoplasmic viscosity, and is strictly related to the red cell survival capacity (Mohandas *et al.*, 1983). Since structure and function of the membrane are highly involved in determining cell deformability (through the maintenance of ion gradients and water distribution, and through the transport of nutrients and catabolites), the life span of a red cell ultimately depends on the integrity of its membrane.

2. THE HUMAN ERYTHROCYTE MEMBRANE

The red cell membrane is composed of 41% lipid, 52% protein, and 7% carbohydrate by weight (for a review see Haest, 1982), and the organization of these components is critical for its function. Lipids are arranged in a bilayer, with polar head groups facing the aqueous phase of the plasma and cytoplasm and the hydrophobic acyl chains are oriented toward the interior of the bilayer. Phospholipids and unesterified cholesterol are present in approximately equimolar amounts, and account for nearly 70% and 23% by weight of the total lipid fraction, respectively (3% is represented by glycolipids). Phospholipids and glycolipids are distributed asymmetrically in the bilayer: glycolipids are present in the outer leaflet of the membrane, together with 75% of the phosphatidylcholine (PC), 80% of the sphingomyelin, and 20% of the phosphatidylethanolamine (PE). Phosphatidylserine (PS) and phosphatidylinositides (PI) are confined to the inner side of the bilayer. All classes of phospholipids are heterogeneous, depending on the properties of their acyl chains, which differ in length and number of double bonds. Asymmetry of the bilayer derives from the distribution of different polar heads in the two leaflets and of unsaturated fatty acids in the two monolayers: unsaturated fatty acids are more abundant in the inner half of the bilayer, which makes this side more fluid than the external one. Cholesterol is detectable on both sides of the bilayer, since it can rapidly move from one side to the other.

Proteins present on the erythrocyte membrane are classified as integral or peripheral, depending on the procedure required to isolate them, and on their localization with respect to the lipid bilayer: they can be identified by their mobility in SDS–PAGE, and named by a progressive number increasing with the electrophoretic mobility. The band 3 protein, glucose transporter, and sialoglycoproteins, belonging to the family of glycophorins (A, B, C, D) are the integral proteins best characterized from the molecular point of view. All these

Table I
Molecular Properties of Integral Proteins of the Human Erythrocyte Membrane

	M_r (kDa)	Copies/cell	Assembly state	Interactions	Function
Band 3	93	10^6	Dimer	Ankyrin, spectrin, 4.1, 4.2, glyceraldehyde 3-phosphate dehydrogenase, aldolase	Anion transport
Glycophorin					
A (α)	36	$5–9 \times 10^5$	Dimer		M, N antigens
B (δ)	20	3×10^5			Ss, U antigens
C (β)	32	10^5		Protein 4.1	Membrane skeleton–integral domain connection
D (γ)	23	2×10^4			Membrane skeleton–integral domain connection
Band 4.5	56				Glucose transport
Decay-accelerating factor	75				Inhibits C3 convertase assembly
CR_1	200				Complement receptor

proteins are glycosylated, to different extents, by N-linked and or O-linked saccharide chains, and their extracellular domain carries many of the erythrocyte antigens. Molecular properties of integral proteins of the human erythrocyte membrane are summarized in Table I. Among the integral proteins represented in lower quantities we should mention the ion-dependent ATPases (Na^+/K^+ and Ca^{2+}), ion channels, transporters, some enzymes and receptors, and some minor constituents, known only by their antigenic properties. In human erythrocyte membrane, peripheral proteins (see Table II) are only present on the cytoplasmic side of the bilayer. Some of them are organized into a supramolecular net of filaments, the membrane skeleton, conferring to the erythrocyte its shape and deformability (Marchesi, 1985; Woods and Lazarides, 1988): the principal component (by mass) of this structure is the heterodimeric protein αβ-spectrin, which self-associates into $\alpha_2\beta_2$ tetramers and higher-order oligomers, which are cross-linked by short oligomers (10–20 globular monomers) of actin: the affinity of spectrin for actin is low, but the formation of a ternary complex with protein 4.1 increases the affinity by 10^6-fold. Membrane skeleton stability seems to depend on the organization of spectrin in tetramers and higher oligomers. Sites connecting membrane skeleton components (spectrin, band 4.1 protein, actin, band 4.9 protein, tropomyosin, myosin, adducin) to give rise to the supramolecular structure, and sites of interaction between the membrane skeleton and the integral domain [band 3 protein–ankyrin–spectrin, band 3–4.1 protein, glycophorin–triphosphoinositide (TPI)–protein 4.1 connections] are crucial for the maintenance of erythrocyte shape and deformability. Many proteins are involved in "the building up" of these sites, and a lesion in any one of these components may dramatically modify the membrane supramolecular structure locally. Regulatory factors producing the supramolecular organization have not been elucidated, although phosphorylation could be one of these factors, since most of the membrane proteins are phosphorylated at multiple sites by kinases under hormonal, Ca^{2+}–calmodulin, and cAMP control.

Table II
Molecular Properties of Peripheral Proteins of the Erythrocyte Membrane

	M_r (kDa)	Copies/cell	Assembly state	Interactions	Function
Spectrin α (1)	240	10^5	$(\alpha\beta)_2$	Band 3, ankyrin, protein 4.1	Membrane skeleton
β (2)	225				
Ankyrin	215	10^5	Monomer	Band 3, spectrin, protein 4.2	Membrane skeleton–integral domain connection
Band 4.1	78	2×10^5	Dimer(?)	Spectrin, actin, PS, glycophorins C and D	Membrane skeleton
Band 4.2	72	2×10^5	Tetramer	Ankyrin, band 3	
Band 4.9	45	2×10^5			Membrane skeleton
Actin	43	5×10^5	Oligomers	Spectrin, 4.1	Membrane skeleton
Tropomyosin	29	7×10^4	Dimer	Actin	Actin stabilizing
Adducin α	103	5×10^5	αβ	Spectrin, actin	Actin cross-linking
β	97				
Band 6	35	4×10^5	Tetramer	Band 3	Glyceraldehyde 3-phosphate dehydrogenase
Band 7	29	5×10^5			
Band 8	23	10^5			Ca^{2+} binding

Throughout its life span the red cell is continually exposed to oxidant stress. This is in part due to its iron content and to the spontaneous oxidation of hemoglobin to methemoglobin, with subsequent generation of oxygen radicals, and partly to the contact with macrophages and activated granulocytes, which can induce oxidative injury by the toxic substances they produce. Since it is anucleate, the erythrocyte is not able to resynthesize the damaged structures; it has only two ways of reacting to the structural insults: to engage its metabolism in the detoxification and, if the challenge is too much, to degrade the damaged cellular components, thus losing part of its ability to survive. Enzyme activities responsible for oxygen radical detoxification are localized in the cytoplasm: antioxidant systems specifically involved in surveillance on the membrane (vitamin E, phospholipases) act on lipids. It can be supposed that reduced glutathione serves to detoxify oxidative damages undergone by membrane proteins; when oxidized glutathione accumulates, it is exported outside the cell in an ATP-dependent process (Kondo *et al.*, 1987), and new molecules of reduced glutathione have to be synthesized. Moreover, the red cell contains a proteolytic system, partly localized in the cytoplasm, partly on the membrane: heavy oxidized proteins may undergo conformational changes, become recognizable by intracellular proteases, and be degraded. Some *in vitro* experiments suggest this possibility (Davies and Goldberg, 1987), and the reduction of cell surface that the erythrocyte undergoes during its life span in the circulation could in part be due to this mechanism. Ankyrin and protein 4.1, which build the connections between the integral domain and membrane skeleton, are substrates for Ca^{2+}-proteases (Pontremoli and Melloni, 1986). The damage to the connections may locally destabilize the membrane supramolecular structure and produce loss of membrane fragments as vesicles.

3. HEMOGLOBINOPATHIES

As hemoglobin (Hb) is the most important cytoplasmic component of the red cell, membrane lesions secondary to diseases due to a structural defect of Hb, such as sickle-cell anemia, or an unbalanced regulation of globin chain biosynthesis, such as thalassemia, will be considered.

3.1. Sickle-Cell Anemia

Sickle-cell anemia represents a disease where a more detailed study of the membrane lesions has been carried out; the disease is due to the substitution of β6 valine for glutamic acid. This mutant Hb has been called HbS. Hb exists in two conformations, oxy- and deoxyhemoglobin. When the pO_2 decreases, molecules of deoxyhemoglobin S aggregate into filaments and these crystalline precipitates cause the red cell to have the typical sickle shape. On oxygenation, oxyhemoglobin S becomes soluble again. When the cell sickles and unsickles repeatedly, the membrane is affected and the cell becomes irreversibly sickled. This results in reduced red cell survival and microvascular occlusion, and the clinical manifestations are anemia and vasoocclusive episodes. While the relative insolubility and deoxygenation-induced polymerization of HbS are well understood, the relationship between this event and the development of membrane abnormalities is less known. An attractive hypothesis proposed by Hebbel (1986) is that autoxidation of the red cell occurs, consequent to iron decompartmentalization, resulting in oxidative lesions to the membrane.

Sickle-cell membrane can be damaged by lipid peroxidation, which modifies molecular characteristics in particular of the inner leaflet of the membrane, and leads to accumulation of by-products, aldehydes and dialdehydes; these could affect aminophospholipid and protein conformation, because of their reactivity toward amino and thiol groups, and protein distribution on the cell surface, acting as cross-linking agents. Malonyldialdehyde (MDA) appears to be the major dialdehyde by-product. The fact that sickle erythrocytes undergo lipid peroxidation is shown by the increased amount of MDA they contain (Das and Nair, 1980) and generate during incubation under aerobic conditions (Hebbel, 1986). MDA generation leads to the formation of fluorescent chromolipids, found in increased amounts in sickle cells (Das and Nair, 1980), as well as of cross-linked aminophospholipids (Jain and Shohet, 1984). A partial disruption of the asymmetric distribution of phospholipids in the bilayer, with exposure of PS and PE on the outer surface of the membrane, was shown in sickle red cells and in irreversibly sickled cells (ISC) by Lubin *et al.* (1981). This disruption of asymmetry may be important for the genesis of sickle-cell behavior. Loss of asymmetry in some areas of the membrane makes normal red cells, like sickle cells, more adherent to endothelial cells (Williamson and Schlegel, 1984; Hebbel *et al.*, 1985); moreover, it can enhance erythrocyte binding to macrophages (Tanaka and Schroit, 1983). Finally, exposure of PS and PE provides a surface capable of activating clotting factors (Zwaal, 1978), contributing to the hypercoagulable status of the patients. Deoxygenation and reoxygenation of sickle cells *in vitro* results in the release of spectrin-free spicules (Allan *et al.*, 1982), similar to those present in the circulation of patients with sickle-cell anemia and having, like ISC, procoagulant activity *in vitro;* these spicules contain the region where phospholipid asymmetry was lost

(Choe *et al.*, 1986), and probably these are sites where bundles of HbS protrude through the membrane skeleton, uncoupling it from the bilayer, thus enhancing the transbilayer mobility of aminophospholipids (Middelkoop *et al.*, 1988).

Lipid peroxidation and PS exposure are associated with structural modifications of membrane proteins. Although for some years evidence of membrane protein oxidation has been sought, such as the finding of an abnormal thiol state of some membrane skeletal or connecting components [spectrin, protein 4.1, and ankyrin (Rank *et al.*, 1985)], only recently has direct evidence for an increased oxidative state of protein 4.1 been obtained (Schwartz *et al.*, 1987). Since on protein 4.1 a high-affinity binding site for PS (Rybicki *et al.*, 1988; Schiffer *et al.*, 1988) has been identified, localized on the cystein-rich 30-kDa chymotryptic peptide (Cohen *et al.*, 1988), and since the effect of its oxidation on the capacity to bind to membrane vesicles has been shown (Schwartz *et al.*, 1987), protein 4.1 oxidation could play an important role in determining the PS exposure in the sickle cell. The increased oxidative state of spectrin and ankyrin can also be taken into account due to the location of these proteins and their interaction with band 3 protein: interestingly, the cytoplasmic domain of band 3 is one of the sites where Hb and hemichromes bind to the membrane (Waugh and Low, 1985). The presence of oxidative insults to proteins localized in the environment of Hb binding sites supports the role of Hb in initiating membrane damage (Hebbel, 1986; Rice-Evans *et al.*, 1986). An increased binding of denatured Hb (hemicromes) to the membrane is documented in the sickle cell (Campwala and Desforges, 1982); moreover, HbS binds more tightly to the membrane than HbA, and this could favor the transfer of the heme moiety (Kuross *et al.*, 1988) to the lipid bilayer and the start of the oxidative cascade. Glycoprotein topography and membrane skeleton supramolecular structure seem to be modified in these cells. Heinz bodies, found associated with the cytoplasmic surface of the membrane, induce clustering of glycophorin, band 3, and ankyrin in the sickle cell (Waugh *et al.*, 1986). Triton-extracted membrane skeletons from ISC are fixed in the sickle configuration (Lux *et al.*, 1976). This situation probably derives from abnormal interactions between membrane skeleton constituents. Evidence of abnormal interaction has been obtained for normal spectrin, which binds inefficiently to ankyrin of inside-out vesicles prepared from sickled cells (Platt *et al.*, 1985); this effect could be related to abnormal thiol redox status found for some proteins like ankyrin, spectrin, and protein 4.1 (Rank *et al.*, 1985; Schwartz *et al.*, 1987). Thiol redox status of spectrin moreover influences its self-association (Streichman *et al.*, 1988) and its interaction with ankyrin (Bennet, 1978) and protein 4.1 (Wolfe *et al.*, 1982). Abnormal redox status of membrane thiols may produce the abnormal homeostasis of cations observed in the sickle-cell membrane, which accumulates Ca^{2+} (Eaton *et al.*, 1973; Palek, 1977) and loses K^+ and water (Clark *et al.*, 1978). The reason for this abnormal intracellular Ca^{2+} concentration in sickle cells is unknown. However, an increased intracellular Ca^{2+} level was also observed in G6PD-deficient cells (see below), concomitant to the membrane thiol oxidation produced by fava bean glycosides. Moreover, normal red cell treatment with thiol reagents (Rothstein, 1981) promotes the loss of K^+. PI-turnover abnormalities (Raval and Allan, 1986) of these cells could also be related to the increased intracellular Ca^{2+} concentration. Finally, microrheologic measurements have detected an abnormal intrinsic rigidity of the ISC membrane (Nash *et al.*, 1984; Evans *et al.*, 1984).

All membrane lesions produced in the sickle cell may derive from an increased autoxidation produced by Hb molecules denatured and bound to the membrane, generating twice the normal amount of activated oxygen species [superoxide, peroxide, and hydroxyl radicals (Hebbel, 1986)]. Most of the antioxidant mechanisms employed by the red cell to detoxicate radicals are located in the cytoplasm, and since the sickle cell needs to engage its antioxidant system to overcome the increased production of radicals, its capacity would not be sufficient to avoid the contact between harmful intermediates and membrane components. The last event leading to the removal of ISC from the circulation seems to be the binding of autoantibodies, promoting extravascular hemolysis of the densest subpopulations of sickle cells (Petz *et al.*, 1984; Hebbel *et al.*, 1985).

3.2. Thalassemia

The major abnormality of the red cell in patients affected by the severe form of thalassemia results from the precipitation of an unstable Hb chain, which is present in excess: the other chain is synthesized only in small amounts or not at all. The excess precipitating globin chains are already present at the level of the nucleated erythroid cells in homozygous α and β thalassemia; the excess α-globin chains disintegrate rapidly into unstable monomers; the excess β chains form a tetramer (β_4), also known as HbH (Rachmilewitz *et al.*, 1969). As a result of the precipitation of unstable unpaired Hb subunits, a chain of events starts, leading to the early removal of thalassemic red cells in bone marrow and from the peripheral blood. Lesions at the level of the membrane are present in these cells, and are evident also from their abnormal morphological and rheological properties (Tillman and Schroeter, 1979). Although less extensively characterized than sickle-cell membranes, thalassemic cell membranes present signs of oxidative injury partly superimposable on those found in the sickle cell (Rachmilewitz *et al.*, 1985). Phospholipid and cholesterol amounts are twofold increased in these cells. As a consequence, membrane surface increases and the cells become less dense (Kahane and Rachmilewitz, 1976). Clues of lipid peroxidation in this membrane are represented by the percent decrease of unsaturated fatty acids, such as arachidonic, in phospholipids and by a corresponding increase of saturated fatty acids, such as palmitic (Rachmilewitz *et al.*, 1985), and by the presence of MDA in the membrane after exposure of thalassemic cells to exogenous peroxide (Stocks *et al.*, 1972). The decreased content of PE was also related to lipid peroxidation, since this phospholipid is particularly vulnerable (Jacob and Lux, 1968).

The oxidative damage at the level of membrane proteins is documented only by the presence of aggregates in ghosts subjected to SDS–PAGE: these aggregates cannot be dissociated by reduction, and therefore they could derive from protein cross-linking induced by MDA (Kahane *et al.*, 1978a). The increased permeability to cations indirectly suggests that the redox state of membrane thiol groups can be altered; Ca^{2+} accumulates intracellularly (Shalev *et al.*, 1984) and K^+ is lost (Gunn *et al.*, 1972); however, direct evidence about the thiol group redox status in thalassemia is not available.

The membrane skeleton situation has not been studied in detail; this fraction contains four times more Hb than the normal counterpart. Its anomalous architecture can be explained by spectrin properties, which could be partly derived from the interaction with Hb. A decreased α-spectrin content and defective self-association of dimers to tetramers (Lamchiagdhase *et al.*, 1987) have been found in thalassemia. These membrane skeleton abnormalities could be responsible for the patchy, instead of uniform, distribution of

Table III
Properties of Membrane Proteins and Lipids in Aged and Pathological Red Cells[a]

	SCA	Th	A	HS	HE	HPP	G6PD$^+$	PK$^+$
Phospholipids/cell	↓[b]	↑	↓	↓		↓		
Phospholipid composition		≠	=	≠			≠	
PS exposure	+		+	+				
Flip-flop	↑							
PI turnover	↑							↑
Saturated and unsaturated fatty acid ratio		↑						
Lipid peroxidation products	++	++	+				+	
Protein composition	=	αSp$^+$	=	Sp$^+$			b3$^+$(f)	=
Membrane skeleton shape	≠	≠	≠?	≠	≠	≠		
Spectrin self-assembly		↓			↓	↓		
Hb on the membrane	++	++	+	+	+	+	+(f)	
Glycoprotein topography-mobility	≠	≠		≠				
Thiols	↓	↓		↓			↓	?
Protein aggregates	−	+	−				+	
Oxidation targets	1, 2.1, 4.1	Sp?					Calpain(f) Ca^{2+} pump(f) spectrin	

[a]Abbreviations: A, aging; b3$^+$, band 3 deficiency; f, fabic crisis; Hb, hemoglobin and derivatives; SCA, sickle-cell anemia; SP$^+$, spectrin deficiency; Th, thalassemia.
[b]+, presence; ++, presence in abnormal amount; −, absence; =, unmodified; ≠, modified; ↑, increased; ↓, decreased.

glycophorin on the membrane (Kahane *et al.*, 1978b). The reduced sialic acid content of glycophorin (Kahane *et al.*, 1980) may also contribute to its different topography on the membrane and to the exposure of glycolipid α-galactosyl sites, which are recognized by autologous antibodies (Galili *et al.*, 1983). Perturbation of the integral component topography and of membrane skeleton architecture may also derive from the increased proteolysis of membrane proteins (Ballas *et al.*, 1982), perhaps triggered by the oxidative lesion promoted by unpaired globin chains interacting with the membrane. The oxidant challenge produced by accumulation of unpaired globin chains is documented by a twofold increase of superoxide, peroxide, and hydroxyl radicals; an increased activity of all systems involved in detoxication of these radicals has also been demonstrated (Rachmilewitz *et al.*, 1985). Levels of vitamin E are dramatically low in thalassemic subjects, and the reason for this has not been elucidated. Administration of vitamin E does not correct membrane anomalies (Rachmilewitz *et al.*, 1985), probably because of a defect at the level of the vitamin E-binding protein. Since vitamin E represents the most significant defense the membrane can engage against oxidative processes, its deficiency may potentiate the damage produced by the presence of an increased amount of Hb on the membrane.

Tables III and IV summarize membrane lesions described in senescent (Clark, 1988), sickled, and thalassemic red cells. The comparison suggests a common pathway, perhaps with differences in the quantitative expression, in the genesis of lesions leading to the impairment of the survival capacity of these red cells.

Table IV
Membrane Processes in Aged and Pathological Red Cells[a]

	SCA	Th	A	HS	HE	HPP	G6PD^{+}	PK^{+}
Cation perturbations	+[b]	+	+	+	+	+	+	+
Ca^{2+}	↑	↑				↑	↑ (f)	
K^{+}	↓	↓	↓				↓ (f)	
Na^{+} fluxes				≠				
Cell water	↓		↓					
Mean corpuscular volume	↓	↑	↓	↓		↓	≃	≃
Deformability	↓	↓	↓	↓	↓	↓		↓
Loss of membrane fragments	+		+	+	−	+		
IgG binding	+	+	+					
Interactions								
Endothelium	+							
Macrophages	+	+	+					+
Proteolysis		+	+				≠	=
Phosphorylation				≠				
Sialylation		↓						

[a]Abbreviations: A, aging; f, fabic crisis; SCA, sickle-cell anemia; Th, thalassemia.
[b]+, presence; −, absence; =, unmodified; ≃, nearly modified; ≠, modified; ↑, increased; ↓, decreased.

4. ENZYMOPATHIES

Enzymopathies such as G6PD and PK deficiency deserve attention because they can throw light on the relationship between the two main metabolic pathways of the red cell and membrane structure.

4.1. G6PD Deficiency

Membrane properties of G6PD-deficient red cells can in particular help to throw light on processes taking place when the capacity to maintain the functional efficiency of the glutathione system is impaired. The first report about a membrane abnormality in G6PD deficiency is from Allen *et al.* (1978). These authors demonstrated the presence of protein aggregates in G6PD-deficient red cell membranes, by SDS–PAGE and gel filtration, and interpreted them as early and sensitive indicators of oxidative damage to red cells. Johnson *et al.* (1979) and Coetzer and Zail (1980) reported similar observations, and the involvement of spectrin in the formation of these aggregates. Rice-Evans *et al.* (1981), examining membrane lipid abnormalities in G6PD deficiency of the Mediterranean and A type, found a decreased cholesterol/phospholipid ratio, increased lipid fluidity, and elevated lipid peroxidation level. Kosower *et al.* (1982) described the membrane thiol–disulfide status of G6PD-deficient red cells upon treatment with a thiol-oxidizing reagent, diamide, and found a correlation between the thiol state of the membrane and the intracellular glutathione content. The oxidative state of individual membrane proteins has not been systematically investigated in these cells; however, membrane functional studies have shown an impairment of the shape-recovery mechanism in G6PD-deficient erythrocytes, which seems to correlate with the impairment of the shunt pathway (Alhanaty *et al.*, 1984). Diamide treatment produces an impairment of Ca^{2+}-ATPase activity (Shalev *et al.*, 1985), and this observation is useful in explaining the properties of G6PD-deficient cells during fava bean hemolysis. Turrini *et al.* (1985) studied the cell Ca^{2+} level, Ca^{2+}-ATPase activity, GSH intracellular levels, and membrane protein pattern by SDS–PAGE in red cell subpopulations of different density during fabic crisis, and found that the densest fraction contained the most damaged red cells, in terms of Ca^{2+} accumulation, Ca^{2+}-ATPase impairment, the presence of protein aggregates and decreased recovery of some membrane proteins (particularly of band 3 protein). De Flora *et al.* (1985), besides an accumulation of intracellular Ca^{2+}, concomitant to a decreased Ca^{2+}-ATPase activity, also described a decrease in the intraerythrocytic K^+ content in fabic red cells during crises. These authors were able to reproduce the same events *in vitro,* by incubating G6PD-deficient erythrocytes with divicine, the substance present in fava beans and strongly implicated in the genesis of favism. More recently, the same group (Morelli *et al.*, 1987) described in the densest subpopulations of circulating red cells during fabic crisis the presence of reducible protein aggregates, and an impairment of the red cell proteolytic system, shown by the abnormally low procalpain activity and by an inversion in the subcellular distribution of acid endopeptidase, which in this situation are mainly found in the cytosol, resulting in unrestrained Heinz body formation. These observations demonstrate that proteins damaged by oxidation are cleared by red cell proteolytic systems, as previously suggested by Davies and Goldberg (1987). Extensive oxidation can inactivate the thiol enzyme calpain, usually implicated in the clearance of oxidized proteins, and in this case partially degraded polypeptides, including Hb, bind to the membrane. During fabic crisis these events concern the densest cells, i.e., those cells where an impairment of detoxication systems (both antioxidant and proteolytic) becomes more evident as a consequent of red cell senescence (reviewed by Clark, 1988). During fava bean hemolysis, red cell morphology is modified and membrane cross bonding, adhesion between opposing areas of the cytoplasmic face of the red cell membrane, takes place (Fischer *et al.*, 1985). Since a role for spectrin in determining this phenomenon can be hypothesized, an altered distribution of this molecule on the membrane seems to be induced by oxidative processes triggered by fava bean ingestion.

4.2. PK Deficiency

Deficiency of erythrocyte PK was first shown by Valentine *et al.* (1961) to cause hereditary nonspherocytic hemolytic anemia. PK is a key enzyme in glycolysis and at present no quantitative relationship between the residual red cell PK activity and the severity of clinical manifestations has been documented. About 5% of the enzyme was found on the membrane (Schroeter and Tillmann, 1975), and this finding led to the hypothesis that a membrane abnormality could be associated with the enzyme defect. While cytoplasmic normal enzyme is an allosteric enzyme, exhibiting cooperative substrate binding, the enzyme bound to the membrane shows nonsigmoidal behavior in normal and in PK-deficient subjects as well (Lakomek *et al.*, 1983). The first membrane dysfunction documented in PK-deficient erythrocytes was a decreased red cell deformability, thought to be secondary to ATP depletion (Schroeter and Tillmann, 1975; Shohet and Ness, 1976; Bernard, 1977). The study of the PK-deficient red cell membrane was then approached by Zanella *et al.* (1979) who found a decreased glycoprotein stability during nutrient deprivation and an increased sensitivity of these cells to the lytic action of mouse macrophages. These abnormalities were more evident in young than in mature and old erythrocytes. Allen *et al.* (1983) found increased absorption of cytoplasmic proteins to the erythrocyte membrane in ATP-depleted normal and PK-deficient mature cells and reticulocytes, as shown by the increased protein/phospholipid ratio in membrane preparations and by the appearance of a 50-kDa polypeptide in SDS–PAGE. The increased protein absorption to the membrane is not associated with increased Ca^{2+} uptake, thiol oxidation, or altered phosphorylation of membrane proteins. The authors were able to reproduce all the abnormalities found for the pathological cells by depleting ATP from reticulocyte-enriched normal red cells, and excluded the possibility that the cytoplasmic protein found on the membrane could be catalase or the PK monomer. Marik *et al.* (1988) found that membrane skeleton behavior in PK-deficient cells can be related to the high amount of reticulocytes present, and measured a normal membrane fluidity. Zanella *et al.* (1976) obtained a qualitative and quantitative correction of PK abnormalities by *in vitro* and *in vivo* treatment of PK-deficient red cells with thiol compounds. These findings suggest the possibility of a generalized abnormality of red cell thiol groups in this disease. An impairment at the level of the antioxidant system could be derived from their lower ATP content, and in fact these cells show an impairment of NAD synthesis (Zerez and Tanaka, 1987). Lastly, the PI-phosphate-head-group turnover is increased in PK-deficient reticulocytes (Schwarzer *et al.*, 1987). Cation flux abnormalities have been described in PK deficiency (Bernard, 1977), and this finding points to a perturbation of thiol group redox state in membrane proteins.

5. MEMBRANE SKELETON PATHOLOGY

This final section deals with membrane properties of erythrocytes in which a primary defect of a membrane protein is the cause of anemia. Several authoritative reviews are available on the topics of hereditary spherocytosis (HS), elliptocytosis (HE), and pyropoikilocytosis (HPP) (Palek and Lux, 1983; Palek, 1985; Becker and Lux, 1985). In this review we will consider red cell membrane properties found in hereditary membrane skeleton diseases, in order to illustrate the role directly played by individual membrane proteins in maintaining the functional efficiency of the red cell. It is worth considering the possibility that perturbations of membrane properties similar to those produced by an inherited molecular defect could also derive from structural insults or degenerative processes.

5.1. Hereditary Spherocytosis

It has long been recognized that the inheritance pattern and clinical manifestations of HS are heterogeneous. The phenotype of a spherocytic red cell can be inherited as an autosomal dominant or recessive trait, and the clinical manifestations can range from a mild compensated hemolytic state to severe hemolytic anemia. Generally, modifications found in HS result in altered shape, with microcytosis, decreased deformability, increased microviscosity, and decreased fluidity of the membrane. Spectrin deficiency was first observed in mouse mutants with autosomal recessive hemolytic anemia, characterized by spherocytosis and having some clinical manifestations reminiscent of those presented by human HS (Greenquist *et al.*, 1978). In these mouse erythrocytes the molecular defect appeared to be the spectrin deficiency, since reconstitution of the membranes with spectrin led to a decrease in fragmentation and vesiculation, characteristic of untreated spherocytes (Shohet, 1979). Some years later, studies on human HS also showed that membranes from all patients with HS are deficient in spectrin (Agre *et al.*, 1985), and that the degree of deficiency correlates with the clinical severity. Recent studies have begun to delineate the molecular basis for heterogeneity in clinical manifestations. A moderate decrease in spectrin was noted in erythrocytes from individuals with the dominantly inherited form of the disorder, whereas more marked spectrin deficiencies [HS(Sp$^+$) according to the Palek and Lux (1983) nomenclature] were found in a group of families with non-dominantly inherited disease (Agre *et al.*, 1985). Most of these individuals were found to be homozygous for a variation in the spectrin αI domain tryptic peptide (Winkelman *et al.*, 1986), and in all of them the increase in osmotic fragility and the rate of hemolysis appeared to correlate with the spectrin content of the membrane (Agre *et al.*, 1986). Electron microscopy studies indicate that the membrane skeleton disorder is proportional to the extent of spectrin deficiency (Liu *et al.*, 1986). To establish that in HS(Sp$^+$) spectrin deficiencies are directly responsible for generation of spherocytes with reduced surface area, Chasis *et al.* (1988) have tried to document by ectacytometry that a quantitative relationship exists between membrane surface area, mechanical stability, and membrane spectrin content. These authors showed this correlation in the 18 HS subjects studied, thus concluding that spectrin deficiency is a major factor determining the pathophysiology of HS. Besides spectrin deficiency, other HS subgroups show a defective binding between spectrin and protein 4.1 (***Sp***–4.1) (Goodman *et al.*, 1982; Wolfe *et al.*, 1982), due to an increased oxidant sensitivity of the spectrin β chain (Becker *et al.*, 1987). Direct oxidation of normal spectrin with diamide can produce a defective spectrin–actin–4.1 interaction, similar to that present in an ***Sp***–4.1 variant of HS. Furthermore, reduction by dithiothreitol of the HS (***Sp***–4.1) spectrin restores it to 85% of the normal spectrin–actin–4.1 interaction (Becker and Lux, 1985). An increased sensitivity to oxidation has been generally reported in HS. This property could be present in different spectrin mutants, as a consequence of an altered protein conformation, and could also play a role in determining the spectrin deficiency, since proteolysis of oxidized molecules may occur. An increased sensitivity to proteolysis of spectrin and a concomitant decreased sensitivity of band 3 has been reported in HS (Gaczynska *et al.*, 1987). It could be supposed that Hb participates in these oxidative processes, since additional Hb has been found in membrane preparations of HS erythrocytes (Becker and Lux, 1985). Spectrin abnormalities apparently do not play a role in all individuals with HS. Recently, spherocytosis with ankyrin deficiency (Coetzer *et al.*, 1988a), protein 4.2 deficiency (Rybicki *et al.*, 1986), and with membrane surface loss without spectrin deficiency (Chasis *et al.*, 1988) have been described. Although the biochemical basis for membrane instability in spectrin-deficient HS erythrocytes seems to be clear, there could be a variety of mechanisms leading to spectrin deficiency. Among them we can hypothesize defects in the synthesis and stability of spectrin and defects in synthesis and stability of other membrane proteins, which are essential for the assembly (Woods and Lazarides, 1988) and maintenance of a normal membrane skeleton, as suggested by the characteristics of the rare HS variants. Membrane skeleton abnormalities consequent to altered spectrin structure produce several other structural and functional abnormalities in the spherocytic red cell, which were recognized before the biochemical identification of the primary defect (see Tables III and IV).

The lipidic component of the HS membrane is modified in composition, in absolute amount, perhaps as a reflection of microcytosis, and in the supramolecular organization, with transfer of PS to the outer leaflet of the bilayer (Becker and Lux, 1985).

The oxidative state of membrane proteins increases, probably because the decreased deformability lengthens the transit time through the spleen, where neutrophils and monocytes could oxidize red cells. cAMP-dependent and -independent phosphorylation is modified, but not against exogenous substrates. This indicates a normal protein kinase activity and suggests a different exposure of phosphorylation sites of the substrates (Becker and Lux, 1985). Cation homeostasis, generally concomitant with an increased oxidative state of membrane proteins, is perturbed. ATP consumption by Na^+/K^+-ATPase is twice that found in normal cells, intracellular K^+ concentration is decreased, and the HS cell is dehydrated. Mean corpuscular Hb concentration increases, probably as a consequence of the loss of membrane fragments, water, and K^+. This may play a role in producing a decreased cell deformability (Mohandas *et al.*, 1983).

5.2. Hereditary Elliptocytosis and Pyropoikilocytosis

Elliptocytosis is a very heterogeneous disease in terms of inheritance, severity of hemolysis, red cell morphology, and the underlying molecular defect. Nine clinical variants have been identified, ranging from an asymptomatic carrier state to a very severe hemolytic anemia, with elliptocytes, microspherocytes, and poikilocytes, and to the unusual elliptocytosis with dyserythropoiesis. The clinical heterogeneity of the HE/HPP syndrome is paralleled by biochemical heterogeneity, and several different molecular defects have been described in HE and HPP (Coetzer *et al.*, 1988b). The possibility that the molecular defect underlying this group of disorders affects the red cell membrane skeleton was first suggested by morphological and structural studies of the membrane skeletal residues of red cell ghosts extracted with Triton X-100. While skeleton preparations from normal cells have a discoidal shape, those prepared from HE red cells have an elliptical shape (Tomaselli *et al.*, 1981). Studies on the mechanical stability of these skeletons revealed that membrane skeletons from patients with HE and a closely related disorder, HPP, were unstable when exposed to mechanical and thermal stress (Liu *et al.*, 1981), suggesting a defect in interactions between proteins constituting the skeleton network. The major structural elements of the membrane skeleton are spectrin α and β heterodimers (SpD): this protein is multifunctional, and molecular properties of structural and functional domains of spectrin can be analyzed by limited proteolytic or chemical fragmentation of the molecule, and separation of the resulting peptides by two-dimensional electrophoresis (Knowles *et al.*, 1983). This technique has been a powerful tool in identifying the localization of the molecular defect in specific subsites of the spectrin molecule (Knowles *et al.*, 1983; Marchesi *et al.*, 1986). Among these, the 80-kDa domain, derived from tryptic digestion of the spectrin α chain, and designated the αI domain, appears to be the most involved in HE and HPP pathogenesis. This domain represents the contact site where α and β chains belonging to two spectrin dimers interact, to give rise, through this head-to-head assembly, to tetramers or higher-order oligomers, essential for the building up of the normal membrane skeleton network (Knowles *et al.*, 1983). In some HE and HPP patients the αI domain abnormality is associated with an enrichment in spectrin dimers (SpD),1; according to the Palek nomenclature, these defects are named HE and HPP ($Sp\alpha^{I/74}$), ($Sp\alpha^{I/46}$), ($Sp\alpha^{I/65}$), since tryptic digestion of spectrin dimers produces peptides having M_r of 74, 46, and 65 kDa, respectively, instead of the αI 80-kDa peptide. The corresponding domain on the β chain of spectrin (βI) contains the major phosphorylation site of the molecule, and this domain has been found to be defective in HE (***Sp***Dβ–SpD) (Dhermy *et al.*, 1982). The distal end of the spectrin dimer contains the binding site for protein 4.1. While defects of this interaction are present in a subclass of spherocytosis, partial (HE 4.1^+) or total (HE 4.1^0) deficiency of protein 4.1 (Tchernia *et al.*, 1981), or a shortened form of this protein (Alloisio *et al.*, 1982), produce elliptocytosis. In this case the anomaly involves contacts between this protein and the integral domain of the membrane: this defect may be associated also with a defect of minor glycophorins (Anstee *et al.*, 1984) and protein 4.9. The HE phenotype also appears when another connecting site between the membrane skeleton and an integral domain is deficient; the junction spectrin–ankyrin–band 3 is functionally impaired, as a consequence of defective spectrin–ankyrin (HE ***Sp***–2.1) or ankyrin–band 3 (HE 2.1–***3***) interactions (Palek, 1985). It is interesting to observe that spectrin deficiency, recognized as the main determinant of the HS phenotype, occurs also in α-spectrin variants of HS and HPP; in these cases, membranes are less unstable than spectrin-deficient HS membranes with comparable spectrin density. The location of the molecular lesion explains why HE erythrocytes have little loss of surface area, whereas HS erythrocytes undergo a consistent loss of surface area. In HE and HPP, molecular abnormalities affect the region of protein–protein association, whereas HS involves a change of density of the protein network underlaying the lipid bilayer. It is worth noting that lesions at the level of proteins involved in the membrane skeleton–integral domain interaction result in the ellipto or poikilo phenotype, and lesions leading to an impairment of the supramolecular assembly of membrane skeleton occur at the level of multisite junctions.

Unlike in HS, in HE and HPP membrane properties secondary to the membrane skeleton defect have been little characterized. HPP cells show an elevated Ca^{2+} content and an increased Ca^{2+} flux (Palek *et al.*, 1981; Wiley and Gill, 1976). A comparable increased Ca^{2+} permeability can be induced in normal cells by thermal injury (Palek *et al.*, 1981). The accumulation of Ca^{2+} may contribute to the adherence of Hb and cytoplasmic proteins to the membrane (Palek *et al.*, 1981) and can explain rigidity and fragility of the membrane. Elliptical deformation and an increased membrane permeability to cations, on the other hand, can be induced in normal cells by high shear stress (Larsen *et al.*, 1981).

6. CONCLUDING REMARKS

Characteristics of the erythrocyte membrane in the above-mentioned hemolytic anemias, and comparison of membrane processes taking place during physiological aging and in hemoglobinopathies, enzymopathies, and membrane skeleton diseases, show that an oxidative lesion is present, to different quantitative levels, in all the erythrocytes where an impairment of the survival capacity is evident. For this reason one could be tempted to say that all structural and functional damage to the membrane, such as membrane skeleton structural alterations and cation homeostasis perturbations, are the consequence of an oxidative lesion. Although one cannot ignore that any defect to a molecule could alter its conformation and increase its sensitivity to oxidation, the presence of an oxidative lesion in PK deficiency and in most of the variants of HS, HE, and HPP is still speculative. On the contrary, in sickle-cell anemia, thalassemia, and G6PD deficiency, and during aging of normal erythrocytes, the presence of oxidative insults to membrane components seems to be well documented. It would be worth investigating whether membrane processes as shown in aging, sickle-cell anemia, and thalassemia are also present in diseases where membrane characteristics have been less extensively investigated.

ACKNOWLEDGMENTS. The author thanks Professor Alberto Zanella for his critical comments, Professor Cesare Balduini for helpful discussions and suggestions, and Anna Vai and Katharine Dyne for their help in preparing the manuscript. Work supported by Ministero Pubblica Istruzione.

7. REFERENCES

Agre, P., Casella, J. F., Zinkham, W. H., McMillan, C., and Bennett, V., 1985, Partial deficiency of erythrocyte spectrin in hereditary spherocytosis, *Nature* **314**:380–383.

Agre, P., Asimos, A., Casella, J. F., and McMillan, C., 1986, Inheritance pattern and clinical response to splenectomy as a reflection of erythrocyte spectrin deficiency in hereditary spherocytosis, *N. Engl. J. Med.* **315**:1579–1583.

Alhanaty, E., Snyder, M., and Sheetz, M. P., 1984, G6PD-deficient erythrocytes have an impaired shape recovery mechanism, *Blood* **63**:1198–1202.

Allan, D., Limbrick, A. R., Thomas, P., and Westerman, M. P., 1982, Release of spectrin-free spicules on reoxygenation of sickled erythrocytes, *Nature* **295**:612–613.

Allen, D. W., Johnson, G. J., Cadman, S., and Kaplan, M. E., 1978, Membrane polypeptide aggregates in G6PD-deficient and in vitro aged red blood cells, *J. Lab. Clin. Med.* **91**:321–327.

Allen, D. W., Groat, J. D., Finkel, B., Rank, B. H., Wood, P. A., and Eaton, J. W., 1983, Increased adsorption of cytoplasmic proteins to the erythrocyte membrane in ATP-depleted normal and PK-deficient mature cells and reticulocytes, *Am. J. Hematol.* **14**:11–25.

Alloisio, N., Dorleac, E., Delauney, J., Girot, R., Galand, C., and Boivin, P., 1982, A shortened variant of red cell membrane protein 4.1, *Blood* **60**:265–267.

Anstee, D. J., Parsons, S. F., Ridgwell, K., Tanner, M. J. A., Merry, A. H., and Thomson, E. E., 1984, Two individuals with elliptocytic red cells apparently lack three minor erythrocyte membrane sialoglycoproteins, *Biochem. J.* **218**:615–619.

Ballas, S. K., Burka, E. R., and Gill, F. M., 1982, Abnormal red cell membrane proteolytic activity in severe heterozygous β-thalassemia, *J. Lab. Clin. Med.* **99**:263–274.

Becker, P. S., and Lux, S. E., 1985, Hereditary spherocytosis and related disorders, *Clin. Haematol.* **14**:15–43.

Becker, P. S., Morrow, J. S., and Lux, S. E., 1987, Abnormal oxidant sensitivity and β-chain structure of spectrin in hereditary spherocytosis associated with defective spectrin–4.1 binding. *J. Clin. Invest.* **80**:557–565.

Bennet, V., 1978, Purification of an active proteolytic fragment of the membrane attachment site for human erythrocyte spectrin, *J. Biol. Chem.* **253**:2292–2299.

Bernard, J. F., 1977, Anomalies des transferts cationiques transmembranaires au cours des anémies hémolitiques congenitals, *Nouv. Rev. Fr. Hematol.* **18**:117–137.

Beutler, E., 1983, Erythrocyte disorders: Anemias due to increased destruction of erythrocytes with enzyme deficiencies, in: *Hematology*, 3rd ed. (W. J. Williams, E. Beutler, A. J. Erslev, and M. A. Lichtman, eds.), pp. 561–582, McGraw–Hill, New York.

Campwala, H. Q., and Desforges, J. F., 1982, Membrane-bound hemichrome in density-separated cohorts of normal (AA) and sickled (SS) cells, *J. Lab. Clin. Med.* **99:**25–28.
Chasis, J. A., Agre, P., and Mohandas, N., 1988, Decreased membrane mechanical stability and in vivo loss of surface area reflect spectrin deficiencies in hereditary spherocytosis, *J. Clin. Invest.* **82:**617–623.
Choe, H. R., Schlegel, R. A., Rubin, E., Williamson, P., and Westerman, M. P., 1986, Alterations of red cell membrane organization in sickle cell anemia, *Br. J. Haematol.* **63:**761–773.
Clark, M. R., 1988, Senescence of red blood cells: Progress and problems, *Physiol. Rev.* **68:**503–554.
Clark, M. R., Morrison, C. E., and Shohet, S. B., 1978, Monovalent cation transport in irreversibly sickled cells, *J. Clin. Invest.* **62:**329–337.
Coetzer, T., and Zail, S., 1980, Membrane protein complexes in GSH- depleted red cells, *Blood* **56:**159–167.
Coetzer, T. L., Lawler, J., Liu, S. C., Prchal, J. T., Gualtieri, R. J., Brain, M. C., Dacie, J. V., and Palek, J., 1988a, Partial ankyrin and spectrin deficiency in severe, atypical hereditary spherocytosis, *N. Engl. J. Med.* **318:**230–234.
Coetzer, T., Lawler, J., Prchal, J. T., and Palek, J., 1988b, Molecular determinants of clinical expression of hereditary elliptocytosis, *Blood* **70:**766–772.
Cohen, A. M., Liu, S. C., Lawler, J., Derick, L., and Palek, J., 1988, Identification of the protein 4.1 binding site to phosphatidylserine vesicles, *Biochemistry* **27:**614–619.
Das, S. K., and Nair, R. C., 1980, Superoxide dismutase, glutathione peroxidase, catalase and lipid peroxidation of normal and sickled erythrocytes, *Br. J. Haematol.* **44:**87–92.
Davies, K. J. A., and Goldberg, A. L., 1987, Proteins damaged by oxygen radicals are rapidly degraded in extracts of red blood cells, *J. Biol. Chem.* **262:**8227–8234.
De Flora, A., Benatti, U., Guida, L., Forteleoni, G., and Meloni, T., 1985, Favism: Disordered erythrocyte calcium homeostasis, *Blood* **66:**294–297.
Dhermy, D., Lecomte, M. C., Garbarz, M., Bournier, O., Galand, C., Gautero, H., Feo, C., Alloisio, N., Delaunay, J., and Boivin, P., 1982, Spectrin β chain variant associated with hereditary elliptocytosis, *J. Clin. Invest.* **70:**707–715.
Eaton, J. W., Skelton, T. D., Swofford, H. S., Kolpin, C. E., and Jacob, H. S., 1973, Elevated erythrocyte calcium in sickle cell disease, *Nature* **246:**105–106.
Evans, E., Mohandas, N., and Leung, A., 1984, Static and dynamic rigidities of normal and sickle erythrocytes. Major influence of cell hemoglobin concentration. *Blood* **73:**477–488.
Fischer, T. M., Meloni, T., Pescarmona, G. P., and Arese, P., 1985, Membrane cross bonding in red cells in favic crisis: A missing link in the mechanism of extravascular hemolysis, *Br. J. Haematol.* **59:**159–169.
Gaczynska, M., Bartosz, G., Judkiewicz, L., and Rosin, J., 1987, Abnormal degradation of erythrocyte membrane proteins in hereditary spherocytosis, *Clin. Chim. Acta* **168:**7–11.
Galili, U., Korkesh, A., Kahane, I., and Rachmilewitz, E. A., 1983, Demonstration of a natural antigalactosyl Ig antibody on thalassemic red blood cells, *Blood* **61:**1258–1264.
Goodman, S. R., Shiffer, K. A., Casoria, L. A., and Eyster, M. E., 1982, Identification of the molecular defect in the erythrocyte membrane skeleton of some kindreds with hereditary spherocytosis, *Blood* **60:**772–784.
Greenquist, A. C., Shohet, S. B., and Berstein, S. E., 1978, Marked reduction of spectrin in hereditary spherocytosis in the common house mouse, *Blood* **51:**1149–1155.
Gunn, R. B., Silvers, D. N., and Rosse, W. F., 1972, Potassium permeability in β-thalassemia minor red blood cells, *J. Clin. Invest.* **51:**1044–1049.
Haest, C. W. M., 1982, Interaction between membrane skeleton proteins and the intrinsic domain of the erythrocyte membrane, *Biochim. Biophys. Acta* **694:**331–352.
Hebbel, R. P., 1986, Autoxidation and the sickle erythrocyte membrane: A possible model of of iron decompartmentalization, in: *Free Radicals, Aging and Degenerative Diseases* (J. E. Johnson, D. Harman, R. Walford, and J. Miguel, eds.), pp. 395–424, Liss, New York.
Hebbel, R. P., Schwartz, R. S., and Mohandas, N., 1985, The adhesive sickle erythrocyte: Cause and consequence of abnormal interactions with endothelium, monocytes/macrophages and model membranes, *Clin. Haematol.* **14:**141–161.
Jacob, H. S., and Lux, S. E., 1968, Degradation of membrane phospholipids and thiols in peroxide hemolysis: Studies on vit E deficiency, *Blood* **32:**549–568.
Jain, S. K., and Shohet, S. B., 1984, A novel phospholipid in irreversibly sickled erythrocytes: Evidence for in vivo peroxidative membrane damage in sickle cell disease, *Blood* **63:**362–367.
Johnson, G. J., Allen, D. W., Cadman, S., Fairbanks, V. F., White, J. G., Lampkin, B. C., and Kaplan, M. E.,

1979, Red cell membrane polypeptide aggregates in G6PD mutants with chronic hemolytic disease, *N. Engl. J. Med.* **301**:522–527.

Kahane, I., and Rachmilewitz, E. A., 1976, Alterations in the red blood cell membrane and the effect of vit E on osmotic fragility in thalassemia major, *Isr. J. Med. Sci.* **12**:11–15.

Kahane, I., Shifter, A., and Rachmilewitz, E. A., 1978a, Cross linking of red blood cell membrane proteins induced by oxidative stress in β-thalassemia, *FEBS Lett.* **85**:267–271.

Kahane, I., Polliack, A., Rachmilewitz, E. A., Bayer, E. A., and Skutelsky, E., 1978b, Distribution of sialic acids on the red blood cell membrane in β-thalassemia, *Nature* **271**:674–675.

Kahane, I., Ben-Chetrit, E., Shifter, A., and Rachmilewitz, E. A., 1980, The erythrocyte membranes in β thalassemia. Lower sialic acid levels in glycophorin, *Biochim. Biophys. Acta* **596**:10–17.

Knowles, W. J., Morrow, J. S., Speicher, D. W., Zarkowsky, H. S., Mohandas, N., Mentzer, W. C., Shohet, S. B., and Marchesi, V. T., 1983, Molecular and functional changes in spectrin from patients with hereditary pyropoikilocytosis, *J. Clin. Invest.* **71**:1867–1877.

Kondo, T., Kawakami, Y., Taniguchi, N., and Beutler, E., 1987, Glutathione disulfide stimulated Mg^{2+}-ATPase of human erythrocyte membranes, *Proc. Natl. Acad. Sci.* **84**:7373–7377.

Kosower, N. S., Zipser, Y., and Faltin, Z., 1982, Membrane thiol–disulfide status in G6PD-deficient red cells: Relationship to cellular glutathione, *Biochim. Biophys. Acta* **691**:345–352.

Kuross, S. A., Rank, B. H., and Hebbel, R. P., 1988, Excess heme in sickle erythrocyte inside-out membranes: Possible role in thiol oxidation, *Blood* **71**:876–882.

Lakomek, M., Tillmann, W., Scharnetzky, M., Schroter, W., and Winkler, H., 1983, Erythrocyte pyruvate kinase deficiency. A kinetic study of the membrane-localized and cytoplasmatic enzyme from six patients, *Enzyme* **29**:189–197.

Lamchiagdhase, P., Wilairat, P., Sahaphong, S., Bunyaratvej, A., and Fucharoen, S., 1987, Defective spectrin dimer self-association in thalassemic red cells, *Eur. J. Haematol.* **38**:246–250.

Larsen, R. L., Katz, S., Roufogalis, B. D., and Brooks, D. E., 1981, Physiological shear stresses enhance the Ca^{2+} permeability of human erythrocytes, *Nature* **294**:667–668.

Liu, S. C., Palek, J., Prchal, J., and Castleberry, R. P., 1981, Altered spectrin dimer–dimer association and instability of erythrocyte membrane skeletons in hereditary pyropoikilocytosis, *J. Clin. Invest.* **68**:597–605.

Liu, S. C., Derick, L. H., Agre, P., and Palek, J., 1986, Alteration of the red cell membrane skeletal ultrastructure in hereditary spherocytosis, *Blood* **68**:56a.

Lubin, B., Chiu, D., Bastacky, J., Roelofsen, B., and van Deenen, L. L. M., 1981, Abnormalities in membrane phospholipid organization in sickled erythrocytes, *J. Clin. Invest.* **67**:1643–1649.

Lux, S. E., John, K. M., and Karnovsky, M. J., 1976, Irreversible deformation of spectrin–actin lattice in irreversibly sickled cells, *J. Clin. Invest.* **58**:955–963.

Marchesi, S. L., Knowles, W. J., Morrow, J. S., Bologna, M., and Marchesi, V. T., 1986, Abnormal spectrin in hereditary elliptocytosis, *Blood* **67**:141–151.

Marchesi, V. T., 1985, Stabilizing infrastructure of cell membranes, *Annu. Rev. Cell Biol.* **1**:531–561.

Marik, T., Brabec, V., Kodicek, M., and Jarolim, P., 1988, Pyruvate kinase-deficiency anemia: Membrane approach, *Biochem. Med. Metab. Biol.* **39**:55–63.

Middelkoop, E., Lubin, B. H., Bevers, E. M., Op Den Kamp, J., Comfurius, P., Chiu, D. T. Y., Zwaal, R. F. A., van Deenen, L. L. M., and Roelofsen, B., 1988, Studies on sickled erythrocytes provide evidence that the asymmetric distribution of phosphatidylserine in the red cell membrane is maintained by both ATP-dependent translocation and interaction with membrane skeletal proteins, *Biochim. Biophys. Acta* **937**:281–288.

Mohandas, N., Chasis, J. A., and Shohet, S. B., 1983, The influence of membrane skeleton on red cell deformability, membrane material properties, and shape, *Semin. Hematol.* **20**:225–242.

Morelli, A., Grasso, M., Meloni, T., Forteleoni, G., Zocchi, E., and De Flora, A., 1987, Favism: Impairment of proteolytic systems in red blood cells, *Blood* **69**:1753–1758.

Nash, G. B., Johnson, C. S., and Meiselman, H. J., 1984, Mechanical properties of oxygenated red blood cells in sickle cell (HbSS) disease, *Blood* **63**:73–82.

Palek, J., 1977, Red cell calcium content and transmembrane calcium movements in sickle cell anemia, *J. Lab. Clin. Med.* **89**:1365–1374.

Palek, J., 1985, Hereditary elliptocytosis and related disorders, *Clin. Haematol.* **14**:45–87.

Palek, J., and Lux, S. E., 1983, Red cell membrane skeletal defects in hereditary and acquired hemolytic anemias, *Semin. Hematol.* **20**:189–224.

Palek, J., Liu, S. C., Liu, P. Y., Prehal, J., and Castleberry, R. P., 1981, Altered assembly of spectrin in red cell membranes in hereditary pyropoikilocytosis, *Blood* **57**:130–139.

Petz, L. D., Yam, P., Wilkinson, L., Garatty, G., Lubin, B., and Mentze, W., 1984, Increased IgG molecules bound to the surface of red blood cells of patients with sickle cell anemia, *Blood* **64**:301–304.

Platt, O. S., Falcone, J. F., and Lux, S. E., 1985, Molecular defect in the sickle erythrocyte skeleton: abnormal spectrin binding to sickle inside-out vesicles, *J. Clin. Invest.* **75**:266–271.

Pontremoli, S., and Melloni, E., 1986, Extralysosomal protein degradation, *Annu. Rev. Biochem.* **55**:455–481.

Rachmilewitz, E. A., Pleisach, J., Bradley, T. B., and Blumberg, W., 1969, Role of hemicromes in the formation of inclusion bodies in haemoglobin H disease, *Nature* **222**:248–250.

Rachmilewitz, E. A., Shinar, E., Shalev, O., Galili, U., and Schrier, S. L., 1985, Erythrocyte membrane alterations in β-thalassemia, *Clin. Haematol.* **14**:163–182.

Rank, B. H., Carlsson, J., and Hebbel, R. P., 1985, Abnormal redox status of membrane–protein thiols in sickle erythrocytes, *J. Clin. Invest.* **75**:1531–1537.

Raval, P. J., and Allan, D., 1986, Changes in membrane polypeptides, polyphosphoinositides and phosphatidate in dense fractions of sickle cells, *Biochim. Biophys. Acta* **856**:595–601.

Rice-Evans, C., Rush, J., Omorphos, S. C., and Flynn, D. M., 1981, Erythrocyte membrane abnormalities in G6PD-deficiency of the Mediterranean and A-types, *FEBS Lett.* **136**:148–152.

Rice-Evans, C., Omorphos, S. C., and Baysal, E., 1986, Sickle cell pathology: Is the membrane important? in: *Free Radicals, Cell Damage and Disease* (C. Rice-Evans, ed.), Richelieu Press, London.

Rothstein, A., 1981, Mercurials and red cell membranes, in: *The Function of Red Blood Cells: Erythrocyte Pathobiology* (D. F. H. Wallach, ed.), pp. 105–131, Liss, New York.

Rybicki, A. C., Heath, R., Wolf, J. L., Lubin, B., and Schwartz, R. S., 1986, Deficiency of protein 4.2 in erythrocytes from a patient with a Coombs negative hemolytic anemia, *J. Clin. Invest.* **81**:893–901.

Rybicki, A. C., Russel, H., Lubin, B., and Schwartz, R. S., 1988, Human erythrocyte protein 4.1 is a phosphatidylserine binding protein, *J. Clin. Invest.* **81**:255–260.

Schiffer, K. A. Goerke, J., Duzgunes, N., Fedor, J., and Shohet, S. B., 1988, Interaction of erythrocyte protein 4.1 with phospholipids. A monolayer and liposome study, *Biochim. Biophys. Acta* **937**:269–280.

Schroeter, W., and Tillmann, W., 1975, Membrane-localized pyruvate kinase of red blood cells in hemolytic anemia associated with PK-deficiency, *Klin. Wochenschr.* **53**:1101–1106.

Schwartz, R. S., Rybicki, A. C., Heath, R. H., and Lubin, B. H., 1987, Protein 4.1 in sickle erythrocytes. Evidence for oxidative damage, *J. Biol. Chem.* **262**:15666–15672.

Schwarzer, E., Reimann, B., Jacobasch, G., and Maretzki, D., 1987, Metabolism of polyphosphositides of G6PD- and PK-deficient red blood cells, *Biomed. Biochim. Acta* **46**:182–183.

Shalev, O., Mogilner, S., Shinar, E., Rachmilewitz, E. A., and Schrier, S. L., 1984, Impaired erythrocyte calcium homeostasis in β-thalassemia, *Blood* **64**:564–566.

Shalev, O., Lavi, V., Hebbel, R. P., and Eaton, J. W., 1985, Erythrocyte (Ca^{2+} + Mg^{2+})-ATPase activity: Increased sensitivity to oxidative stress in G6PD-deficiency, *Am. J. Hematol.* **19**:131–136.

Shohet, S. B., 1979, Reconstitution of spectrin deficient spherocytic mouse erythrocyte membranes, *J. Clin. Invest.* **64**:483–494.

Shohet, S. B., and Ness, P. M., 1976, Hemolytic anemias. Failure of the red cell membrane, *Med. Clin. North Am.* **60**:913–932.

Stocks, J., Offerman, E. L., Modell, C. B., and Dormandy, T. L., 1972, The susceptibility to autooxidation of human red cell lipids in health and disease, *Br. J. Haematol.* **23**:713–724.

Streichman, S., Hertz, E., and Tatarsky, I., 1988, Direct involvement of spectrin thiols in maintaining erythrocyte membrane thermal stability and spectrin dimer self association, *Biochim. Biophys. Acta* **942**:333–340.

Tanaka, Y., and Schroit, A. J., 1983, Insertion of fluorescent phosphatidylserine into the plasma membrane of red blood cells. Recognition by autologous macrophages, *J. Biol. Chem.* **258**:11335–11343.

Tchernia, G., Mohandas, N., and Shohet, S. B., 1981, Deficiency of skeletal membrane protein band 4.1 in homozygous hereditary elliptocytosis. Implications for erythrocyte membrane stability, *J. Clin. Invest.* **68**:454–460.

Tillmann, W., and Schroeter, W., 1979, Rheological properties of erythrocytes in heterozygous and homozygous β-thalassemia, *Br. J. Haematol.* **43**:401–411.

Tomaselli, M. B., John, K. M., and Lux, S. E., 1981, Elliptical erythrocyte membrane skeletons and heat-sensitive spectrin in hereditary elliptocytosis. Implications for red cell membrane stability, *Proc. Natl. Acad. Sci. USA* **78**:1911–1915.

Turrini, F., Naitana, A., Mannuzzu, L., Pescarmona, G. P., and Arese, P., 1985, Increased red cell calcium, decreased calcium adenosine triphosphatase, and altered membrane proteins during fava bean hemolysis in G6PD-deficient (Mediterranean variant) individuals, *Blood* **66**:302–305.

Valentine, W. N., Tanaka, K. R., and Miwa, S., 1961, A specific erythrocyte glycolytic enzyme defect (pyruvate kinase) in three subjects with congenital non-spherocytic hemolytic anemia, *Trans. Assoc. Am. Physicians.* **74**:100–110.

Valentine, W. N., Tanaka, K. R., and Paglia, D. E., 1983, Pyruvate kinase and other enzyme deficiency disorders of the erythrocyte, in: *The Metabolic Basis of Inherited Disease* (J. B. Stanbury, J. B. Wyngaarden, D. S. Fredrickson, J. L. Goldstein, and M. S. Brown, eds.), pp. 1606–1628, McGraw–Hill, New York.

Wagner, G. M., Schwartz, R. S., Tsun-Yee Chiu, D., and Lubin, B. H., 1985, Membrane phospholipid organization and vesiculation of erythrocytes in sickle cell anemia, *Clin. Haematol.* **14**:183–200.

Waugh, S. M., and Low, P. S., 1985, Hemichrome binding to band 3: Nucleation of Heinz bodies on the erythrocyte membrane, *Biochemistry* **24**:34–39.

Waugh, S. M., Willardson, B. M., Kannan, R., Labotka, R. J., and Low, P. S., 1986, Heinz bodies induce clustering of band 3, glycophorin, and ankyrin in sickle cell erythrocytes, *J. Clin. Invest.* **78**:1155–1160.

Wiley, J. S., and Gill, F. M., 1976, Red cell calcium leak in congenital hemolytic anemia with extreme microcytosis, *Blood* **47**:197–210.

Williamson, P., and Schlegel, R., 1984, Maintenance of phospholipid asymmetry and its role in erythrocyte pathology, in: *Erythrocyte Membranes: Recent Clinical and Experimental Advances* (W. C. Kruckeberg, J. W. Eaton, J. Aster, and G. J. Brewer, eds.), pp. 123–136, Liss, New York.

Winkelman, J. C., Marchesi, S. L., Watkin, P., Linnenback, A. J., Agre, P., and Forget, B. G., 1986, Recessive hereditary spherocytosis is associated with an abnormal alpha-spectrin subunit, *Clin. Res.* **34**:474A.

Wolfe, L. C., John, K. M., Falcone, J. C., Byrne, A. M., and Lux, S. E., 1982, A genetic defect in the binding of protein 4.1 to spectrin in a kindred with hereditary spherocytosis, *N. Engl. J. Med.* **307**:1367–1373.

Woods, C. M., and Lazarides, E., 1988, The erythroid membrane skeleton: Expression and assembly during erythropoiesis, *Annu. Rev. Med.* **39**:107–122.

Zanella, A., Rebulla, P., Giovanetti, A. M., Curzio, M., Pescarmona, G. P., and Sirchia, G., 1976, Effects of sulphydryl compounds on abnormal red cell pyruvate kinase, *Br. J. Haematol.* **32**:373–385.

Zanella, A. Brovelli, A., Mantovani, A., Izzo, C., Rebulla, P., and Balduini, C., 1979, Membrane abnormalities of pyruvate kinase deficient red cells, *Br. J. Haematol.* **42**:101–108.

Zerez, C. R., and Tanaka, K. R., 1987, Impaired nicotinamide adenine dinucleotide synthesis in PK-deficient human erythrocytes: A mechanism for decreased total NAD content and a possible secondary cause of hemolysis, *Blood* **69**:999–1005.

Zwaal, R. F. A., 1978, Membrane and lipid involvement in blood coagulation, *Biochim. Biophys. Acta* **515**:163–205.

Chapter 16

Action of Drugs on the Erythrocyte Membrane

Bernhard Deuticke, Reinhard Grebe, and Cees W. M. Haest

1. INTRODUCTION

Drugs in the sense of pharmacologically active compounds have been important tools in the analysis of the structure and function of the erythrocyte (RBC) membrane. Conversely, the RBC membrane is also a well-established experimental model for the pharmacologically and medically relevant targets of many membrane-active drugs.

For these reasons a large amount of data are available concerning the interaction of drugs with the RBC membrane and their effects on membrane properties. In the following, this material is reviewed and categorized, in particular according to three important features:

1. Dynamics of membrane constituents
2. Membrane mechanics and cell shape
3. Permeability and transport systems

To facilitate the presentation, the essentials of these three features of the membrane are outlined in a prefatory section.

Like all plasma membranes, the RBC membrane consists of lipids and proteins, part of the latter being glycosylated (Viitala and Järnefelt, 1985), and some being acylated (Staufenbiel, 1987). The lipids, essentially phospholipids ($\approx$ 4.3 μmol/ml cells), glycolipids ($\approx$ 0.25 μmol/ml cells), and cholesterol ($\approx$ 3.5 μmol/ml cells) (Broekhuyse, 1974; Viitala and Järnefelt, 1985), are arranged as a bilayer with a mean thickness of 4 nm. Besides a characteristic variation in the head group and interfacial domain, phospholipids and glycolipids vary in the nature of their apolar chains with respect to length (16–24

Bernhard Deuticke, Reinhard Grebe, and Cees W. M. Haest Department of Physiology, Medical Faculty, Technical University, D-5100 Aachen, Federal Republic of Germany.

carbons) and extent of unsaturation (0–6 double bonds). Phospholipids and glycolipids are distributed asymmetrically between the two leaflets of the bilayer (Op den Kamp, 1979). Patterns of lipids, but not the principles of their transbilayer distribution, vary among mammalian species (Deuticke, 1977). Cholesterol may be distributed symmetrically (Lange and Slayton, 1982), but direct evidence on this point is lacking.

Due to the particular structure of the individual lipids and their fractional contents in the bilayer, these constituents provide the membrane with physical properties that are usually characterized in terms of the following parameters:

1. Order and mobility of the acyl chains as quantified by analyzing motions or collisions of spectroscopic probes (NMR, ESR), which often bear close structural similarity to native membrane lipids (Schubert, 1987; Yeagle, 1987; Lentz, 1989)
2. Membrane polarity, as characterized by dielectric properties, usually determined by fluorescent probes (Griffith *et al.*, 1974)

The environment of a drug is highly dependent on its localization in the membrane. The lipid segment at the center of the bilayer has a low order (nearly isotropic), a low polarity, and a high mobility. The region at the membrane interface has a high order (highly anisotropic), a high polarity, and a low mobility. Moreover, the dynamics of the lipids in the inner and outer leaflet may differ. Thus the term lipid "fluidity" or its inverse, "microviscosity," which are often used to describe lipid dynamics, have no precise meaning.

RBC membrane proteins, which contribute about 50% of the membrane mass ($\approx$ 4.5 mg protein/ml packed cells), are usually divided, on an operational basis, into integral and peripheral species, although this "classic" dualism is becoming less appropriate as more of the minor protein fractions are studied (Lande *et al.*, 1982; Wold, 1986; Low, 1987).

The refinement of analytical techniques has begun to close the gap between the fairly small number of polypeptides (about 20) that could be classified by conventional separation techniques (Goodman *et al.*, 1988) and the considerably larger number of proteins whose presence has to be taken for granted on the basis of functional membrane properties. This progress is particularly due to the advent of (1) immunological strategies for the detection and separation of very small peptide fractions (Whitfield *et al.*, 1985; Latron *et al.*, 1987; Denker *et al.*, 1988) and (2) the improvement of strategies of unequivocal assignment of functions to membrane proteins, mainly by the techniques of membrane reconstitution (Madden, 1986; Scheuring *et al.*, 1986) and affinity labeling with specific inhibitors, often derived from drugs (see Section 4).

Functionally, RBC membrane proteins are conveniently divided into three classes:

1. Peripheral proteins forming a subbilayer network serving as a membrane support (Goodman *et al.*, 1988; Bennett, 1989; Carraway and Carothers Carraway, 1989). Their established and putative functions for membrane stability, for area-, shear-, and bending-elasticity, for viscosity, for cell shape and deformability will be outlined in Section 3.
2. Integral, mostly membrane-spanning proteins, many of them serving as transport pathways (Section 4). These pathways may be subclassified into pores or channels (for water, nonelectrolytes, and certains ions), carrier-mediated net movements of nonelectrolytes, cotransport or ion exchange of anions and cations, and finally primary active transport systems, i.e., ion pumps. Certain integral proteins

provide for the attachment of the skeletal network (Goodman *et al.*, 1988; Bennett, 1989). These attachment poles and other proteins also form the glycocalyx, crucially involved in surface properties of the cell, e.g., blood group antigenicity and surface charge (Viitala and Järnefelt, 1985; Seaman, 1983; Marcus, 1981).
3. Exoenzymes (Low, 1987), receptors (e.g., Bree *et al.*, 1984; Kelleher *et al.*, 1987), and proteins of unknown function (Lande *et al.*, 1982; Denker *et al.*, 1988) complete the list of integral membrane proteins, which may well add up to more than 100 different species, the total number of copies of integral protein monomers amounting to at least about 2×10^6 per cell.

The lateral organization of these lipid-intercalated proteins is not well established, except for what is known from electron microscopy and some cross-linking studies. Lateral mobility of the integral proteins is very low (Peters, 1988), mainly due to restrictions imposed by the skeletal network (Ridgwell *et al.*, 1984; Tsuji and Ohnishi, 1986).

The interaction of surrounding lipids with the integral proteins is thought to modulate protein activity due to changes of their native conformation, of the rate of conformational changes, and, probably, electrostatic effects (Deuticke and Haest, 1987; Yeagle, 1989). Drugs intercalating into the lipid domain may thus affect protein function in this indirect way (Goldstein, 1984), in addition to direct interactions with the protein. Distinction between these two mechanisms is one of the major forthcoming tasks in the study of drug action on the RBC membrane.

2. PARTITIONING OF DRUGS AND THEIR EFFECTS ON MEMBRANE STRUCTURE AND DYNAMICS

2.1. Partitioning of Drugs into Membranes

Most membrane-active drugs are amphiphilic substances containing apolar domains with high affinity for the hydrophobic interior of the lipid bilayer as well as polar domains that prevent full penetration of the drug into the hydrophobic core and force the drug to orient at the bilayer interface. Besides such a nonspecific interaction with the membrane lipid, drugs may undergo either specific, i.e., site-directed, or nonspecific, i.e., hydrophobic or electrostatic, interactions with membrane proteins.

Partition coefficients for a membrane and the adjacent aqueous phase can be defined in principle either as ratios of the *mole fractions* of solute in the membrane to those in the medium or as ratios of *concentrations* in the membrane to those in the medium. Both definitions are not unambiguous in the heterogeneous biomembrane due to uncertainties concerning the exact membrane volume and the fraction of the total membrane volume accessible to the drug. Other complications in the quantitation of partitioning concern the possible perturbation of the membrane by the drug which affects binding of other molecules. Moreover, drugs may aggregate in the membrane or in the aqueous phase. These complications can be avoided by measurements at low drug concentrations. However, in this case the investigator should be aware of significant binding of apolar drugs to the vessel wall.

A number of methods are available to measure partitioning of drugs into membrane.

Following equilibrium dialysis or equilibration of cells (isolated membranes) with a drug and either centrifugation or hygroscopic desorption filtration (Conrad and Singer, 1979), the extent of drug binding to membranes may be measured by gas chromatography (Miller *et al.*, 1977), ultraviolet (Welti *et al.*, 1984; Rogers *et al.*, 1986), spin resonance (Gaffney *et al.*, 1983), or fluorescence (Pjura *et al.*, 1984) spectroscopy, or by using radioactively labeled compounds (Roth and Seeman, 1972; Korten and Miller, 1979; Korten *et al.*, 1980; Bondy and Remien, 1981; Antunes-Madeira and Madeira, 1984, 1985; Luxnat and Galla, 1986; Zachowski and Durand, 1988). Drug binding can also be detected or quantified by measuring changes of the cell electrophoretic mobility (Isomaa *et al.*, 1979; Tamura *et al.*, 1982), by effects on cell shape (Deuticke, 1968; Fujii *et al.*, 1979), and by the increase of the critical hemolytic volume due to an increase of the membrane surface area to cell volume ratio (Seeman, 1972; Deuticke and Ruska, 1976; Haest *et al.*, 1981). Insertion of foreign molecules into the RBC membrane has been confirmed by photoaffinity labeling of membrane components by phenothiazine derivatives (Wildenauer and Zeeb-Wälde, 1983), by transverse resonance energy transfer from membrane-bound fluorescent fatty acids to intracellular hemoglobin (Eisinger and Flores, 1982), by analyzing the shift of the ultraviolet or fluorescence spectrum of chlorpromazine and fluorescent fatty acids upon membrane binding (Welti *et al.*, 1984; Pjura *et al.*, 1984), and by determining the orientation of spin-labeled fatty acids in membranes (Gaffney *et al.*, 1983).

Simple partitioning of drugs between RBC membrane and medium, as indicated by a concentration independence of the partition coefficient, has been observed for the anionic anesthetic pentobarbital and for neutral anesthetics (benzyl alcohol, alkanols, phenols) up to high membrane concentrations (Seeman *et al.*, 1971; Machleidt *et al.*, 1972; Roth and Seeman, 1972). At very high concentrations of benzyl alcohol, its partition coefficient even increases due to perturbation of the bilayer (Metcalfe *et al.*, 1968). Partitioning of the cationic β-adrenergic antagonist propranolol is also only slightly affected by concentration (Rogers *et al.*, 1986). On the other hand, partitioning of other cationic drugs such as chlorpromazine and imipramine (Roth *et al.*, 1972; Roth and Seeman, 1972; Bondy and Remien, 1981; Luxnat and Galla, 1986; Zachowski and Durand, 1988) as well as of negatively charged drugs like fatty acids phloretin and merocyanine (Pjura *et al.*, 1984; Jennings and Solomon, 1976; Allan *et al.*, 1989), exhibits saturation. Saturating partitioning of charged drugs can be described in terms of a Langmuir adsorption isotherm, which takes into account the membrane expansion and the increase of surface charge density by membrane-inserted drug (Seelig *et al.*, 1988; Barthel *et al.*, 1988). Partition coefficients of various drugs for RBC have been reviewed by Seeman (1972). These partition coefficients agree well with more recent data. Established partition coefficients (concentration in membrane/concentration in medium) vary from about 5000 for chlorpromazine and imipramine (Welti *et al.*, 1984; Zachowski and Durand, 1988) to 0.05 and 0.5 for methanol and barbital (Seeman, 1972).

From studies on lipid vesicles, partitioning of drugs into the membrane may be expected to depend on the dielectric energy difference for the drug between the aqueous and the membrane compartment as well as on the dipole potential of the bilayer phospholipid head group. Distribution of charged drugs will additionally depend on the surface charge of the lipid phase, which influences the drug concentration at the membrane surface (Surewicz and Leyko, 1981; Schlieper and Steiner, 1983; De Young and Dill,

1988). Primary adsorption of a drug to the membrane surface is followed by intercalation of its nonpolar part between the fatty acid chains of the lipids. Drug molecules will partition into the hydrophobic region of the membrane according to their lipid solubility (Seeman, 1972; Diamond and Katz, 1974). However, partitioning into the membrane bilayer does not resemble dissolution into bulk solvent, since the solvent properties of the bilayer are dominated by its anisotropic and interfacial characteristics (Simon *et al.*, 1979). Therefore, the smaller sized drugs will not uniformly distribute in the bilayer but rather according to a gradient of drug concentration. For different depths in the bilayer the probability of localization will differ. The stability of orientation of a drug between the acyl chains will increase with increasing length (axial ratio) of the drug, while its freedom of transverse mobility will decrease. Moreover, when the drug is permeable or can undergo transbilayer reorientation (flip-flop), its distribution between the inner and outer membrane leaflet is not necessarily symmetric as has been derived from differential effects of drugs on cell shape (Sheetz and Singer, 1974; Fujii *et al.*, 1979, see also Section 3). Finally, studies on model systems show that the partition of drugs between a membrane and the adjacent aqueous phase and their lateral and transverse distribution in the membrane depend on (1) the physical state of the lipid (i.e., the lipid-crystalline or gel state; not relevant for the RBC membrane), (2) the surface packing density of the phospholipids, as influenced by the acyl chain ordering, their degree of unsaturation, the fatty acyl chain length, the head group structure of the phospholipid, and the cholesterol content, and (3) lipid–protein interactions (Colley and Metcalfe, 1972; Miller *et al.*, 1977; Antunes-Madeira and Madeira, 1984, 1985; Luxnat and Galla, 1986; Müller *et al.*, 1986; De Young and Dill, 1988; Zachowski and Durand, 1988).

In line with these results of model studies, the partition coefficient for thiopental between the RBC membrane and the medium was reported to increase 1.5-fold from 24 to 37 after cholesterol depletion of the RBC membrane (Korten *et al.*, 1980). The drug partition coefficients for membranes of various origin correlate well with the cholesterol content (Antunes-Madeira and Madeira, 1984, 1985). Bilayers prepared from RBC lipids had a higher partition coefficient for thiopental and fatty acids than the native RBC membrane (Korten *et al.*, 1980; Gaffney *et al.*, 1983). This might indicate restriction of the drug to the lipid domain of the RBC membrane (i.e., exclusion from the integral protein domain). On the other hand, the partition coefficient for the uncharged insecticide lindane was 2-fold higher in RBC membranes than in bilayers prepared from membrane lipids (Antunes-Madeira and Madeira, 1985).

In contrast to a claim of Conrad and Singer (1979), the partition coefficients of drugs in the RBC membrane have been confirmed to correspond to those for artificial lipid membranes containing cholesterol (Luxnat and Galla, 1986; Antunes-Madeira and Madeira, 1984, 1985). The lower partitioning of drugs in biomembranes had been postulated to be due to a high "internal pressure" in biological membranes (Conrad and Singer, 1979). Indeed, the insertion of the anesthetic dibucaine into lipid monolayers and bilayers decreases with increasing the surface pressure. At 39.5 mN/m the drug is extruded from the lipid phase (Seelig, 1987). However, there is evidence that the surface pressure of the lipid phase in biomembranes is only 31–35 mN/m (Demel *et al.*, 1975). Possible errors in the quantification of partition coefficients for drugs due to membrane microvesiculation and formation of mixed micelles of drug and membrane components as suggested by Conrad and Singer (1981), have been checked (Gaffney *et al.*, 1983; Luxnat and Galla,

1986; Zachowski and Durand, 1988). At low concentrations of drugs, neither membrane solubilization nor micelle formation could be detected.

2.2. Nature of Drug-Binding Forces

Information on the forces dominating the binding of drugs to the membrane comes from studies on the effects of ionic strength, temperature, and pH. An increase of ionic strength decreases binding of cationic drugs, increases binding of anionic drugs, and has no effect on binding of neutral drugs (Seeman, 1972; Roth *et al.*, 1972). Ca^{2+} ions are weak antagonists to the binding of cationic drugs (Koblin *et al.*, 1975). At 10 mM Ca^{2+}, membrane binding of chlorpromazine decreases only by 33% (Zachowski and Durand, 1988). Ca^{2+} also displaces bound chlorpromazine from the membrane (Seeman, 1972).

An increase of pH from 6 to 8 decreases the partitioning of fatty acids into the RBC membrane (Gaffney *et al.*, 1983). In contrast, partitioning of chlorpromazine increases upon an increase of pH from 6.0 to 8.5 (Bondy and Remien, 1981). These changes indicate preferential binding of the uncharged form of the drugs, in line with pH effects on drug binding to artificial lipid vesicles (Westman *et al.*, 1982; Gaffney *et al.*, 1983; Schreier *et al.*, 1984; Eftink *et al.*, 1985). For the same reason, the partition coefficient of the anionic form of barbiturates is almost zero, but partition occurs upon charge neutralization by lowering the pH (Korten and Miller, 1979).

An increase of temperature slightly diminishes the partitioning of chlorpromazine and of the insecticide parathion into the RBC membrane (Seeman, 1972; Antunes-Madeira and Madeira, 1984). The partitioning of lindane is biphasic with a minimum (Antunes-Madeira and Madeira, 1985). On the other hand, partitioning into lipid vesicles has a negative temperature dependence over the whole temperature range for lindane and alkanes (Jones and Lee, 1985; Simon *et al.*, 1979). Such behavior is expected for a predominantly hydrophobic interaction (Tanford, 1980). The hydrophobic nature of RBC membrane binding of drugs has also been derived from the close correlation between the partition coefficients for drugs between apolar solvents and water on the one hand, and RBC or artificial membranes and their suspending medium on the other hand (Seeman, 1972; Diamond and Katz, 1974; Stein, 1986). The average environment of drugs in the membrane is hydrophobic albeit not as hydrophobic as the center of the bilayer. Moreover, hydrophobicity of binding was derived from the free energies of adsorption of alcohols and chlorpromazine.

For polar substituent groups on drugs, the contributions to the free energy change ($dF = dH - TdS$), driving their transfer from the aqueous phase to the lipid bilayer, are mostly enthalpy dominated (positive dH) since the positive entropy (dS) for insertion of these groups leads to a negative contribution ($-TdS$) to dF, which does not exceed, however, the positive contribution of dH. On the other hand, the free energy of transfer of the apolar CH_2 groups is negative (−450 to −650 kcal/mol) as a result of a negative enthalpy due to the interaction of CH_2 groups and the acyl chains, despite the negative entropy caused by the loss of mobility of these groups upon lipid insertion. Thus, entropically apolar drugs favor the aqueous phase whereas enthalpically they prefer the bilayer (Katz and Diamond, 1974; Simon *et al.*, 1979; Smith *et al.*, 1981).

2.3. Localization of Drugs in the Membrane

From experiments and molecular mean field calculations (Gruen and Haydon, 1981; Pope *et al.*, 1989) on lipid bilayers, it is likely that short alkanes will orient in the disordered region of the bilayer center, whereas alkanes with 16–19 carbons probably align parallel to the acyl chains in the ordered region of the bilayer. At higher alkane to lipid ratios (0.1 : 1) the alkanes C_{16} and C_6, but not C_{12} are extruded from the bilayer periphery and produce an isotropic phase in the bilayer center. Cholesterol reduces solubility in the ordered region (Pope *et al.*, 1989, and references therein).

Short alkanols (up to C_3) may favor hydrophilic interactions at the bilayer/water interface; alkanols C_{4-8} have been claimed to accumulate in the peripheral part of the hydrocarbon domain and have the tendency to cluster (Brasseur *et al.*, 1985). On the other hand, alkanols up to C_6 have been claimed to accumulate in the bilayer interior and increase its dielectric constant (Orme *et al.*, 1988, and references therein). Alkanols $\geq C_8$ probably align parallel to the phospholipid acyl chains with the OH group anchored to the membrane surface. The depth of penetration of charged drugs may depend on the packing density of the acyl chains, since cholesterol incorporation in the bilayer decreases the depth of penetration (Auger *et al.*, 1988). Penetration of charged drugs into artificial bilayers is also pH-dependent. At pH values below the pKa the cationic group of the anesthetic tetracaine lies in between the polar head groups of the phospholipids, whereas above the pKa the deprotonated form of the molecule completely inserts into the apolar domain (Boulanger *et al.*, 1981).

Insertion into the membrane decreases the pKa of the cationic group of chlorpromazine (Schreier *et al.*, 1984) and increases the pKa of fatty acids (Rooney *et al.*, 1983) due to the lower dielectric constant of the bilayer as compared to the aqueous phase.

2.4. Effects of Drugs on Membrane Lipids

Penetration of the apolar part of a drug into the hydrophobic membrane domain changes lipid–lipid interactions. According to studies on lipid vesicles, drug insertion between the acyl chains decreases their packing density, produces membrane expansion, membrane thinning, changes of the surface potential and the polarity gradient across the bilayer, changes of the phospholipid head group conformation, and changes of the physical state of the lipid phase, i.e., transitions from the liquid-crystalline to the gel phase or from bilayer to nonbilayer phases (Jain and Wu, 1977; Cullis and Verkleij, 1979; Singer and Jain, 1980; Boulanger *et al.*, 1981; Hornby and Cullis, 1981; Nicolay *et al.*, 1985; Forrest and Rodham, 1985; Hanpft and Mohr, 1985; Haydon and Elliott, 1986; Frezzatti *et al.*, 1986; Orme *et al.*, 1988; Seelig *et al.*, 1988). Moreover, drugs may change the order and the rate of anisotropic motions of the lipid molecules in the membrane. Such anisotropic motions of lipids include *trans–gauche* isomerization of the acyl chains, lateral diffusion in the plane of the membrane, rotation about their long axis, and transbilayer reorientation (flip-flop). Any of these motions could be changed by drugs (Goldstein, 1984).

Compared to our knowledge of drug effects on pure lipid systems, data for the RBC membrane are scarce. The neutral drugs benzyl alcohol and ethanol disorder membrane lipids and increase their mobility (Metcalfe *et al.*, 1968; Chin and Goldstein, 1977; Giraud *et al.*, 1981). The effect is stronger in the core of the bilayer than near the surface and the head group is not affected (Giraud *et al.*, 1981). By using permeable and impermeable lipid spin labels and comparing ghost and erythrocyte data, evidence was obtained for a preferential decrease of the order of inner layer lipids by benzyl alcohol (Chabanel *et al.*, 1985). Alkanols C_{5-7} and cannabinol (Pang and Miller, 1978; Kutchai *et al.*, 1980) decrease the order of the membrane lipids. The C_7 alcohol is most effective. Whether alkanols $> C_{16}$ have the opposite, ordering effect, as in postsynaptic membranes (Miller *et al.*, 1989), remains to be seen.

Cationic drugs such as tetracaine, chlorpromazine, xylocaine, imipramine, primaquine, chloroquine, ticlopidine, propranolol, and diazepam, the nonionic detergent Triton X-100, as well as anions such as oleate and benzenesulfonate have a disordering effect on membrane lipids (Feinstein *et al.*, 1975; Giraud *et al.*, 1981; Suda *et al.*, 1981; Janoff *et al.*, 1981; Surewicz, 1982; Daveloose *et al.*, 1982; Noji *et al.*, 1982). Lipid mobility is increased (Giraud *et al.*, 1981; Ogiso *et al.*, 1981). Chlorpromazine and trifluoperazine were reported to disorder lipids in whole RBC membranes, but not in membranes depleted of the cytoskeletal protein spectrin or ghosts resealed with antispectrin antibodies (Minetti and Di Stasi, 1987). These drugs were therefore postulated to exert their effect on membrane lipids due to their calmodulin antagonistic effects discussed in Section 4. Calmodulin and Ca^{2+} have been claimed to destabilize the membrane skeletal network (Takakuwa and Mohandas, 1988; Bennett, 1989), which in turn is thought to influence the stability of the lipid domain (Haest, 1982). Interestingly, a number of membrane binding drugs—procaine, isoxsuprine, lindane, and salicylate—do not seem to affect lipid order (Suda *et al.*, 1981; Noji *et al.*, 1982; Antunes-Madeira and Madeira, 1989).

Besides effects of drugs on the mobility within one leaflet of the bilayer, there is also evidence that drugs affect transverse lipid mobility between the two leaflets of the bilayer. Tetracaine and nupercaine, the channel-forming antibiotics amphotericin and gramicidin as well as bacterial cytotoxins have been shown to enhance passive, nonmediated transbilayer mobility of phospholipids (Schneider *et al.*, 1986; Classen *et al.*, 1987). Chlorpromazine inhibits the recently discovered carrier-mediated translocation of aminophospholipids and produces a transient fast transbilayer movement of a small fraction of membrane phospholipids across the bilayer (Rosso *et al.*, 1988). Enhancement of transverse mobility in some cases results in a decrease of the asymmetric distribution of phospholipids between the inner and outer membrane leaflet (Schneider *et al.*, 1986; Rosso *et al.*, 1988).

There is some evidence that drugs may affect the physical state of RBC membrane lipids. Long-chain alcohols ($\geq C_{14}$) inserted into the membrane produce rod-shaped projections that contain more alcohol and more sphingomyelin than the cell, indicating lipid phase separation (Grunze *et al.*, 1982). Moreover, the channel-forming antibiotic gramicidin at high concentrations produces nonbilayer lipid phases in RBC membranes (Tournois *et al.*, 1987). Finally, it should be mentioned that at high concentrations drugs can induce fusion of RBC membranes (Hope and Cullis, 1981; Lang *et al.*, 1984), thought to require local formation of nonbilayer phases (Ellens *et al.*, 1989).

2.5. Effects of Drugs on Membrane Proteins

Modifications of membrane protein function may conceptually result either from primary alterations of the lipid domain leading to secondary modification of lipid–protein interactions or from direct nonspecific or specific binding of the drug to protein (Dodson and Moss, 1984). In some cases the extent of interaction of drugs with the membrane is crucially dependent on minor structural details of the drug. This is usually taken as evidence for specific interaction with proteins. However, a similar structural dependence was also found for effects of the drugs on lipid phases (Makriyannis *et al.*, 1986). Evidence for a tight binding of chlorpromazine to membrane proteins was derived from quenching of tryptophan fluorescence by the anesthetic in RBC membranes (Yamaguchi *et al.*, 1985). Moreover, the rotational mobility of spin-labeled local anesthetics diminishes after decreasing the number of positive charges on membrane proteins by modification with amino reagents (Koblin and Wang, 1976). Further evidence for an interaction between drugs and membrane proteins comes from changes of the mobility of protein sulfhydryl groups labeled with maleimide nitroxide spin label. Labeling produces an ESR spectrum indicative of strongly and weakly immobilized sites (McConnell and McFarland, 1970). Propranolol, imipramine, desipramine, and phenothiazines were found to decrease the fraction of mobile sites (Holmes and Piette, 1970; Leterrier *et al.*, 1973; Surewicz, 1982; Benga *et al.*, 1983). This is similar to the stabilizing effect of cholesterol on membrane protein (Rooney *et al.*, 1985). However, the disordering effect of the drugs on membrane lipids is opposite to the increase of lipid order and decrease of lipid mobility by cholesterol (Yeagle, 1985).

On the other hand, the antiviral drugs adamantine, rimandantine, and adamantane carboxylic acid, the anthracycline antibiotics adriamycin, daunomycin, and acridines as well as actinomycin D and Pt compounds increase the mobile fraction of maleimide spin-labeled protein side chains (Sinha and Chignell, 1979; Donath *et al.*, 1987).

All these effects of drugs on protein dynamics, however, do not prove a direct modification of membrane proteins since the drug could induce lipid-mediated changes of protein structure. The best evidence for direct interaction of a drug with membrane proteins can be derived from drug inhibition of protein-mediated transport processes (see below). Two features of such inhibition can be used as an argument for direct interaction between inhibitor and transport protein. (1) In many cases the drugs are inhibitory at concentrations at which, even assuming high partition coefficients, their concentration in the lipid domain would be too low to induce changes of its "bulk" properties (mobility, polarity, surface charge). (2) Photolabeling techniques can be used to demonstrate that binding to protein is the basis of the inhibitory effect of a drug. Examples for both are given in Section 4.

2.6. Membrane Expansion by Drugs

Conceptually, the insertion of drugs and other exogenous compounds should expand the RBC membrane unless the drugs occupy free volume space. This concept was first put forward and verified experimentally by Seeman and his associates (Seeman, 1972). Due to an increase of cell surface area resulting from such a membrane-expanding effect, drugs are able to stabilize RBC against hypotonic lysis in addition to their hemolytic effect at higher concentrations. Originally, the membrane expansion estimated from this anti-hemolytic effect and the observed increase of size of ghosts after insertion of drug into the membrane seemed to be much higher than the increase of surface area calculated from the total volume of the inserted drug molecules assuming uniform expansion of the membrane in all directions (Seeman, 1972). Later it was recognized that an increase in area of a lipid bilayer must go along with some bilayer thinning to keep its volume constant (Trudell, 1977). Under these premises, expected and observed values were found to agree much better. The predicted decrease of membrane thickness accompanying drug insertion was, however, small. This anisotropic membrane expansion was experimentally confirmed by Bull *et al.* (1982) and Kita and Miller (1982).

A further uncertainty in calculations of membrane expansion by drugs concerns the volume of the drug in the membrane. The volume of the membrane after drug insertion is not simply the sum of the volume of the native membrane and that of the inserted molecules. Upon transfer of the drug from an aqueous medium into the membrane, there is an excess of volume increase (Smith *et al.*, 1981; Kita *et al.*, 1981; Kita and Miller, 1982). The volume a drug assumes in the membrane is close to its molecular volume in apolar solvents (Kita and Miller, 1982). In case the total number of drug molecules inserted into the RBC membrane can be reliably quantified, and its molecular volume in the membrane is known, the calculated membrane expansion can be correlated with the concomitant increase of the critical hemolytic cell volume as long as the drug does not increase membrane leak permeability and thereby abolishes the osmometer properties of the RBC (Ponder, 1948).

The considerations presented so far tacitly assumed symmetric distribution of an inserted drug between the two leaflets of the bilayer. Effects of a selective or preferential interaction of drugs with one of the leaflets of the membrane will be a subject of Section 3.

3. DRUG EFFECTS ON MEMBRANE MECHANICS, CELL SHAPE, AND DEFORMABILITY

3.1. General Aspects

The microscopic observation of suspended erythrocytes is in principle one of the simplest methods (available since the invention of the microscope 300 years ago) to analyze the interaction of drugs with the RBC membrane by their influence on RBC shape. Although there are now numerous refined methods available to investigate such interactions, detailed analysis of shape and shape changes of RBC remains important because the incident shape of nonnucleated RBC is the global expression of the sum of changes in local membrane properties on a molecular level, i.e., of drug action. Furthermore, normal cell shape can now be quantified unequivocally by using a set of geometrical parameters such as volume, surface area, and mean mean curvature (MMC) (Grebe and Schmid-Schönbein, 1985). This has become possible because new computer-based techniques for image enhancement and processing with subsequent data processing are now available.

The RBC is well suited for analysis of the response of the cell and its membrane to external influences because of its simple structure. It consists of a viscoelastic membrane and a Newtonian solution of 0.31 kg/liter of hemoglobin (Cokelet and Meiselman, 1968) as the liquid interior. Thence, all of its elastic characteristics reside in the components of the cell membrane; when at rest, the cell shape is determined by the overall elastic behavior of the membrane. Its ability to change shape is primarily based on an excess of surface area compared to an isoplane sphere, a fact that can be expressed as the sphericity index SI given by

$$\text{SI} = (4.836 \times \text{volume}^{2/3})/\text{surface area}$$

with a mean value of 0.792 (Fung *et al.*, 1981).

Since 40 to 45% of the blood volume consists of RBC, the ability of the cells to deform when exposed to the shear forces generated in flow largely determines the macroscopic flow properties of blood through all vessels of the body. The mode of cell deformation at constant volume is dominated by their membrane properties and will thereby be influenced by drugs acting on them.

3.2. RBC Shape

3.2.1. General Characteristics

The RBC at rest is characterized by its well-known biconcave shape to which it transforms from the essentially globular shape of its precursor, the reticulocyte, shortly after its delivery into circulation (Lichtman and Santillo, 1986) by a process that alters the SI. The average human RBC maintains a discocytic shape with a volume V of 98 μm^3, a surface area A of 140 μm^2, a diameter of 8.0 μm, and a thickness at the minimum of 1.1 μm and at the maximum of 2.6 μm (Hochmuth, 1987). These values depend, however, on cellular age. Due to loss of volume, the hemoglobin concentration increases in RBC aging *in vivo*. This causes an increase in cell density (Clark, 1988).

Even without changes of volume and very small changes in surface area, RBC shape can be changed in two different ways, one resulting in membrane protrusions or crenation, one in membrane invaginations, termed echinocytosis and stomatocytosis, respectively (Deuticke, 1968; Bessis, 1977).

3.2.2. Quantification (Measurement Methods)

A variety of procedures to quantify shape changes have been introduced. Best known are those based on subjective classification according to the shape sequence introduced by Bessis (1977) (Fujii *et al.*, 1979). Objective methods include the recording of changes in light scattering, used also to record the time course of shape changes (Nagasawa-Fujimori *et al.*, 1981; Artmann, 1986). Coakley and Deeley (1980) introduced a method that evaluates fragmentation patterns of heated cells.

Our own approach to this field was based on the fact that besides volume and surface area the so-called mean curvature lends itself as a geometric parameter describing the appearance of a body. Mean curvature of a point of a surface is defined as the inverse of the radius of the best fitting sphere. To obtain a parameter describing the shape of the whole cell, the average of all local mean curvatures has to be determined, the entity already used, termed mean mean curvature (MMC) (DeHoff and Rhines, 1968).

Using standardized surface parts or adapting arbitrary closed mathematical descriptions of the cell geometry, several authors have indirectly used the MMC to estimate the difference in area of the two leaflets of the bilayer for the typical RBC shape sequence (Svetina *et al.*, 1985; Leitmannova and Glaser, 1977; Beck, 1978). By numerical optimization of the free energy of RBC shapes (Grebe *et al.*, 1988b), evidence has been provided that in the transition from spherostomatocytes via discocytes up to spheroechinocytes there is an increase in the difference between the areas of the outer and the inner layer, which coincides with a respective increase of the MMC.

3.2.3. Computation of Difference of Layer Area

The differences in layer area (dA) needed to produce different RBC shapes have usually been calculated by geometrical models constructed from parts of spheres, cones, and tori (Svetina *et al.*, 1982; Beck, 1978; Ferrell *et al.*, 1985), assuming that the membrane components behave like incompressible fluids. If the MMC of the RBC is known, the formula given below provides a much simpler way to calculate this difference for the RBC standard shapes:

$$dA = 2 \times d \times \mathrm{MMC} \times A$$

where d is the thickness of the membrane and A is the surface area of the RBC.

For the shapes of the sequence exhibiting rotational symmetry, the values for the MMC earlier reported (Grebe *et al.*, 1988b) have recently been confirmed by numerical optimization (Grebe and Zuckermann, 1989). They vary from 0.32/μm for the discocyte down to 0.27/μm for the most stomatocytic cell (stomatocyte III according to the Bessis classification). Thence, the difference in area between the two bilayer surfaces is given for the discocyte as

$$dA(134\ \mu\mathrm{m}^2) = 0.343\ \mu\mathrm{m}^2$$

and for the stomatocyte III as

$$dA(134\ \mu\mathrm{m}^2) = 0.289\ \mu\mathrm{m}^2$$

To transform, e.g., a discocyte into a stomatocyte III, the surface area of the inside layer has to be increased by 0.054 μm^2. Assuming that vertical displacement of cone-shaped molecules does not take place, one can calculate, using 0.6 nm^2 (Cevc and Marsh, 1987) as the mean cross-sectional area of a typical phospholipid molecule, that an additional 9×10^4 intercalated phospholipid molecules in the inner layer would be required for this transformation. For some phospholipids, e.g., rod-shaped molecules, this calculation may be too simple, because it neglects the induced increase of intramembranous pressure in the middle of the bilayer. For drugs the situation is even more complex, because shape and localization in the bilayer have to be taken into account.

These theoretically predicted area differences leading to defined shape changes are accessible to experimental testing.

3.3. RBC Membrane Models and Properties

3.3.1. Mechanical Aspects

When composite objects some orders of magnitude larger than their components, such as a membrane with a large variety of structural elements, are to be investigated, it becomes difficult to derive information on the "macroscopic" behavior only from studies of properties on the molecular level. Thence, besides the above-outlined molecular membrane model a global continuums model can be used. Both approaches have their own set of experimental tools and provide different types of information that have to be related to each other. Some of the properties of the molecular model have been described above (Section 2).

The RBC membrane can only be treated as a continuum in the plane of the membrane, thus compressing classical three-dimensional continuum mechanics into two dimensions (Hochmuth and Waugh, 1987). Most commonly, a theoretical approach is applied, which considers local membrane properties to be constant over the whole membrane, at least in the time range of a few minutes (Evans and Skalak, 1979a,b). This assumption, however, is in conflict with our knowledge about the high lateral mobility of the membrane lipids (Peters, 1988). If, on the other hand, a local and global fluidity of the membrane components is taken into account, properties cannot be assumed to be constant over the entire membrane for the simple reason that their components are able to change position thereby varying local composition, and thence local membrane properties. Considering, in addition, that we are confronted with the finite size of the cell, new constraints have to be introduced due to the fact that the sum of membrane components and their properties are finite and constant. Therefore, global properties must be investigated in order to describe the mechanical behavior of the whole cell membrane, taking into account the free mobility of membrane components.

As an alternative, a membrane model treating the membrane as a composite of fluids exhibiting different viscosities has been proposed (Schmid-Schönbein *et al.*, 1983). Numerical optimization of free energy on the basis of this model leads to discocytic and stomatocytic shapes as the configurations that exhibit the global free energy minima (Grebe and Zuckermann, 1989).

If deformed briefly by external forces, RBC exhibit a local memory of their previous shape. After removal of the force, they recover with time constants of about 0.1 sec (Chien *et al.*, 1978), which highly depend on temperature (Hochmuth *et al.*, 1980). The elastic response to such deformations is due to three fundamental and independent elastic components each of which can be characterized by its own related elastic modulus, namely that of shear/bending and area expansion. If the forces are applied for more than a few minutes, persistent viscous deformations occur (Evans and LaCelle, 1975; Markle *et al.*, 1983). This may be related, on the molecular level, to the spectrin dimer/tetramer association reaction, a weak linkage that is characterized by a half-time of approximately 7 min (Ungewickell and Gratzer, 1978).

In this context, one must keep in mind that a completely different type of deformation can be induced, namely the (more or less permanent) formation of tethers under shear, long tubules with a typical radius of about 15 nm (Hochmuth *et al.*, 1982). The tethers are assumed to be free of spectrin so their generation is due to a process separating the spectrin network from the bilayer. This extreme local shape change caused by external forces is in principle due to the same mechanism as the exovesiculation at the final stage of echinocytotic shape change driven by internal forces of the membrane (Lutz, 1978).

3.3.1.1. Lipids. Present concepts about the shape of the resting cell can be summarized as follows: overall configuration of the cell is a function of the molecular shape and the electrical properties of its membrane components, their lateral and their transverse arrangement in the bilayer. It is generally accepted that membrane phospholipids can have either the shape of rods or that of cone sections, where positive and negative inclines can be assumed (Kleinfeld, 1987).

Due to their individual geometry, all the phospholipids and the cholesterol together will constitute a basic curvature of the membrane (Cevc and Marsh, 1987). The phospholipids in the RBC membrane are distributed asymmetrically between the two layers (Op den Kamp, 1979). Phosphatidylcholine, phosphatidylserine, sphingomyelin, and glycolipids are rod-shaped, thus *per se* cannot introduce a finite curvature. Phosphatidylethanolamine (representing 10% of the phospholipids of the outer leaflet and 20% of the inner one) is cone-shaped and may thereby introduce an overall positive curvature (Kleinfeld, 1987; Kirkpatrick, 1979). The phosphatidylserine molecules, found exclusively in the inner leaflet of the bilayer, carry net negative charges and thereby contribute to the surface tension.

Bending of the membrane by external forces will induce a difference in surface pressure and thereby a difference in surface area between the two layers. This will result in effects complementary to the insertion of drugs, which according to the "bilayer couple hypothesis" formulated by Sheetz and Singer (1974), leads to shape changes. In addition, however, the difference in surface pressure acts as a driving force for a redistribution of membrane constituents capable of high transbilayer mobility. According to present views, the major candidate for such a redistribution (squeezing out of the compressed into the expanded leaflet) is cholesterol, which seems to flip-flop rapidly (Lange, 1981). This redistribution will lead to a new energy minimum characterized by partial compensation of the differences in surface pressure.

3.3.1.2. Proteins: Glycocalyx. In the past, the role of membrane proteins in determining resting membrane shape has been largely neglected in comparison to the attention given to membrane phospholipids. Recent data show that this is not justified. The external membrane surface is known to be covered mainly by the extramembranous parts of the intrinsic glycoproteins (e.g., glycophorins, band 3, band 4.5) and of glycolipids, forming the glycocalyx (Viitala and Järnefelt, 1985).

The glycophorins carry most of the cells sialic acid residues (about 2.5×10^7/cell), which are fully ionized at physiological pH (Viitala and Järnefelt, 1985) and provide for a negative net surface charge. Recent developments of the methods used to measure the electrophoretic mobility of cells have provided evidence that the effective negative surface charge of the RBC (their so-called zeta potential) is influenced not only by the density of these charges but also by their steric arrangement (Donath and Voigt, 1986; McLaughlin, 1985). The exofacial surface charge influences bilayer properties by increasing the surface tension due to the coulombic forces acting between the charge-carrying molecules.

Change of these coulombic forces by enzymatic removal (Seaman, 1983; Grebe *et al.*, 1988a), chemical modification (Tenforde, 1970), or noncovalent shielding (Isomaa *et al.*, 1979) of the effective surface charge should change the previously established equilibrium between inside and outside surface tension and thereby induce changes in MMC and RBC shape. This was recently substantiated by Grebe and Schmid-Schönbein (1985).

In vivo, the outer membrane surface is covered by a layer of plasma proteins, mainly albumin, long known to influence the cell surface properties (Jay, 1975). This molecule, known to be strongly stomatocytogenic, may also exert an influence on RBC shape due to diminution of the effective surface charge (Grebe and Schmid-Schönbein, 1985).

3.3.1.3. Proteins: Membrane Skeleton. The most important factor stabilizing cell shape at the inner membrane face is a protein layer formed mainly by the densely crowded peripheral proteins spectrin and actin (band 5), connected via ankyrin (band 2.1) and band 4.1 protein to the intrinsic proteins band 3 and glycophorin (Marchesi, 1985; Bennett, 1989) or directly to other integral proteins (Denker *et al.*, 1988). The resulting "membrane skeleton" can be isolated by treatment of red cells or ghosts with Triton X-100 (Shen, 1986; Liu *et al.*, 1987). It shows the shape of its precursor (Lange and Slayton, 1982) and undergoes similar dynamic shape changes. Although it is attached to the intrinsic domain by high-affinity bonds, dynamic processes of dissociation and reassociation occur.

Different models have been developed to describe the properties of the spectrin–band 3–ankyrin–actin hetero-complex, e.g., as entropic springs (Evans and Skalak, 1979a,b; Stokke *et al.*, 1986) or as a material structured by weak, reversible intramolecular self-associations (Shen *et al.*, 1986).

The long, flexible spectrin molecule carries a negative net charge, although it also has positively charged surface domains, which may provide for its planar electrostatic adhesion to the anionic phospholipids of the inner bilayer surface. Compared to the above binding forces in the hetero-complex, these coulombic interactions seem to be very weak (Bennett, 1985; Bitbol *et al.*, 1989). An alternative spatial arrangement of spectrin might be the formation of free loops reaching into the cytoplasm (Scheven and Stibenz, 1983).

The negative net charge of the skeleton can be regarded as the direct counterpart of the negative net charge of the glycocalyx. It also induces a lateral pressure and may contribute to the interfacial free energy of the inner membrane surface. It also contributes to the potential profile over the membrane (Heinrich *et al.*, 1982; Ohshima *et al.*, 1986).

If one considers the tight coupling between the two leaflets of the bilayer, the integral membrane glycoproteins, and the spectrin hetero-complex, it follows that all the components could play a role in the action of many echinocytogenic and stomatocytogenic drugs. Thus, it seems reasonable to extend the former bilayer couple not only to a trilayer (Morrow and Anderson, 1986) but even to a "quadrilayer couple hypothesis" of structures that determine RBC shape change.

3.3.2. Extramembrane Factors

In analyzing the mechanical membrane properties, it is also essential to take into account the adjacent aqueous phases as well as their physical state. In principle, the membrane properties may physiologically be influenced by endo- and exofacial proteins, mono- and polyvalent ions, pH, and temperature.

Although experimental data for these physiological modulations of membrane properties and cell shape are available (e.g., see Johnson, 1985), most of the mechanistic details are entirely unsettled. It is also for the latter reason that modification of membrane properties by drugs meets so much interest, since it is anticipated that the analysis of perturbations will give insight into the unknown mechanisms.

Hemoglobin can adsorb to the inner membrane surface (Lilley and Fung, 1987). Static properties (shear elastic modulus and bending elasticity) are not influenced by hemoglobin (Evans *et al.*, 1984) while dynamic parameters such as the recovery time after sudden removal of a force of deformation increase with hemoglobin concentration (Linderkamp and Meiselman, 1982).

3.4. RBC Deformability

The RBC consists of a membrane having the above-reported mechanical properties and a fluid interior. In normal resting cells there is no influence of the homogeneous, viscous hemoglobin solution in the interior on the static cell properties. In contrast, the behavior of the cell in flow is influenced by the internal viscosity as well. The dynamic properties of the whole cell are often described by a parameter called "deformability," which should not be envisaged as a mechanical property of the cell or its membrane in a strict physical sense. It refers mainly to elastic shape change and is quantitated by different methods providing different magnitudes describing the response of the cell to deforming forces. The main purpose of these methods is to predict the behavior of the cell in the microcirculation. For a refined definition of the term "deformability," see Teitel and Schmid-Schönbein (1990).

Normal RBC "deformability" is dominated by the membrane elastic behavior characterized by the elastic moduli of extension, shear, and bending. The smaller these moduli, the more deformable the RBC becomes. During the deformation process, energy is dissipated by the viscoelasticity of the RBC membrane as well as by the internal viscosity of the cell. Deformability decreases as the viscosity (of cytosol and membrane) increases.

The most common methods to measure deformability comprise (Chien, 1987):

- Shear-induced single-cell elongation in the rheoscope (Fischer and Schmid-Schönbein, 1977)
- Micropipette aspiration and relaxation methods (Chien *et al.*, 1978)
- Floor attachment methods evaluating shape changes of superfused, point-attached cells (Artmann, 1986)
- Viscometry of RBC suspensions (Dintenfass, 1985)
- Filtration methods using filters of varying materials and geometry (Bull, 1986)

3.5. Drug Effects on RBC Shape

Deuticke (1968) formulated the rule that echinocytosis of RBC is induced by anionic or nonionized drugs, stomatocytosis by cationic ones. Drugs in these two classes were shown to be true antagonists since they are able to reverse each other's effect.

This basic concept, which conceptualized a considerable number of observations scattered in the older literature, has essentially been confirmed by many studies published in recent years and rationalized by Sheetz and Singer (1974) on the basis of the charge asymmetry of the RBC membrane phospholipids. Sheetz and Singer claimed that anionic drugs insert with some preference into the outer leaflet of the bilayer, containing the zwitterionic choline phospholipids, due to their lower affinity for the inner leaflet containing the anionic aminophospholipids. Cationic drugs were proposed to prefer the inner leaflet for analogous electrostatic reasons. Preferential expansion of the respective layers was assumed to lead to the shape changes observed.

Direct evidence for this concept could be adduced using exogenously added charged compounds such as diacyl- or lysophospholipids. For such compounds a selective insertion into the outer membrane leaflet may be assumed since their transbilayer reorientation rates are very low (half-times of several hours; Bergmann *et al.*, 1984; Middelkoop *et al.*, 1986; Daleke and Huestis, 1989). The consecutive selective expansion of the outer leaflet indeed transforms the biconcave erythrocytes into echinocytes (Mohandas *et al.*, 1978; Haest, 1982). Ultimately, such echinocytosis will result in exovesiculation upon a sufficient increase of the amount of inserted phospholipid.

Conceptually, the leaflet-expanding effect of a drug could result from its geometric contribution to the total surface area of the leaflet, which is determined by the drug's mean cross-sectional area. In addition, ionized drugs will increase or decrease the net charge density in the leaflet and thereby augment or diminish the surface pressure in this leaflet.

The above concept of a coupled quadrilayer instead of a bilayer couple furthermore necessitates the postulate of drug effects on shape via a modification of the organization or charge density of the exofacial glycocalyx or of the endofacial membrane skeleton.

The concentrations of various drugs required to produce shape changes vary quite considerably (by two orders of magnitude), presumably not only as a consequence of different partitioning into the membrane, but also as a result of different tendencies of the drugs to distribute asymmetrically between the two leaflets.

Drugs complying with the original bilayer concept include (Deuticke, 1968; Fujii *et al.*, 1979):

- Among echinocytogenic drugs: alkanoates, alkanesulfonates; simple and substituted aromatic and other cyclic anions including barbiturates, nonsteroidal anti-inflammatory drugs; anionic phenothiazines
- Among stomatocytogenic drugs: alkylamines; local anesthetics; cationic phenothiazines; antihistamines

Apparently contradictory findings, e.g., of an echinocytogenic action of amphiphilic cations with a quaternary nitrogen could also be reconciled with the original concept by assuming that there is no transbilayer reorientation of such compounds. The echinocytogenic effect of zwitterionic lysophospholipids and detergents (Isomaa *et al.*, 1987) was explained by the same assumption.

On the other hand, the original concept of Deuticke and Sheetz and Singer cannot account for the shape effects of readily permeable nonionic detergents, e.g., Triton X-100. Moreover, alcohols (Fujii *et al.*, 1979) and some other membrane-inserting uncharged drugs, for which there is no *a priori* reason to assume an asymmetric insertion on electrostatic grounds, also induce shape changes. These inconsistencies may eventually be resolved by demonstrating interactions of these shape-changing compounds with the glycocalyx or the membrane skeleton.

Finally, a number of other conditions under which shape changes are observed, clearly demonstrate that differences in area between the inner and outer leaflet of the membrane (each comprising two layers affecting curvature) can be induced by procedures which are not yet reconcilable with the "bilayer couple" concept. To mention a few examples: albumin induces stomatocytes (Jay, 1975) and metabolic depletion produces echinocytes (Feo and Mohandas, 1977). Drug-induced echinocytes revert to disco- or stomatocytes upon stimulation of the hexosemonophosphate shunt by methylene blue (Alhanaty and Sheetz, 1984) or in the presence of thiols (Truong *et al.*, 1986). Compensatory stomatocytogenesis is inhibited by vanadate (Schrier *et al.*, 1986). Oxidative stressing of erythrocytes in the presence of inosine, vanadate, and ferricyanide produces stomatocytes (Heller *et al.*, 1987; Heller and Deuticke, unpublished data).

Factors adduced to explain these shape changes include:

- Spectrin phosphorylation/dephosphorylation (Bennett, 1985)
- Phosphoinositide phosphorylation/dephosphorylation (Ferrell and Huestis, 1984)

Some of the stomatocytogenic cationic drugs are known to be calmodulin antagonists (see also Section 4.6.3.2). Since calmodulin has been shown on the other hand to interact with major membrane skeletal proteins (Bennett, 1989), an involvement of calmodulin in RBC shape control and its perturbation by calmodulin antagonists were postulated by Nelson *et al.* (1983). Recent observations of Isomaa and Engblom (1988), however, have failed to support this hypothesis.

3.6. Drugs and RBC Deformability

Rheological properties of whole blood strongly depend on rheological properties of the RBC. Drugs used to improve these properties could act by influencing the parameters constituting membrane "deformability" or altering the cytoplasmic viscosity, as discussed above.

The host of drugs for which such therapeutic effects are claimed has recently been reviewed (Ernst, 1987; Lowe, 1988). Their compilation contains, e.g.,

- Pentoxifylline (xanthine derivative)
- Buflomedil, isoxsuprine, bencyclane (vasoactive drugs)
- Cinnarizine, flunarizine, nifedipine (Ca^{2+} entry blockers)
- Eicosapentaenoic acid (a component of fish oil)
- Flavonoids such as hydroxyethyl-rutosides
- Dipyridamole
- Local anesthetics
- Ketanserin (serotonin antagonist)

Suggested mechanisms for an improvement of RBC "deformability" comprise effects on ATP, membrane plasma protein interactions, RBC surface properties and charge, and modification of transmembrane electrolyte transport processes, e.g., those involved in cell volume control. None of the above drugs can unambiguously be associated with one of these mechanisms at present (Ernst, 1987). It is not even clear whether such direct mechanisms are involved at all.

Eicosapentaenoic acid and administration of insulin to diabetics have been claimed to affect "deformability" due to changes of RBC membrane "fluidity" resulting from altered lipid composition (see Lowe, 1988, for references). Since the relationship between membrane "fluidity" as derived from measurements of lipid order and mobility, and the continuums mechanical viscoelasticity of the membrane is far from clear, such correlations still await their mechanistic interpretation. Chabanel *et al.* (1985) have demonstrated that benzyl alcohol, which causes an increase in the fluidity of membrane lipids, has no effect on membrane viscoelastic properties as tested by micropipette aspiration. Thus, RBC "deformability" cannot directly be related to lipid order and mobility.

4. DRUG EFFECTS ON PERMEABILITY AND TRANSPORT SYSTEMS

The third major type of drug effects on the RBC membrane concerns its unspecific permeability and specific transport systems. These effects cannot be covered comprehensively in the available space. We have therefore selected a number of processes, which seemed of particular interest either in terms of physiological aspects or in the light of the relevance of information derived from studies with inhibitory drugs.

4.1. Ground Permeability

Solutes can permeate the lipid phase of the RBC membrane as a function of their membrane/water partition coefficient. Evidence for this concept mainly stems from studies using mono- and polyhydroxy alcohols, and small amides, but also organic acids (Deuticke, 1977; Stein, 1986). The permeation barrier in the lipid domain seems to behave like hexadecane with respect to the partition of polar nonelectrolytes (Stein, 1986).

In addition to partition, permeability depends on intramembrane diffusion coefficients. The unexpectedly steep mass dependence of these diffusion coefficients suggests (Stein, 1986) that the lipid domain should not be envisaged as a homogeneous solvent (liquid hydrocarbon) in which solutes move according to the Stokes–Einstein relationship, $D \approx 1/r$. Rather, it might be regarded as a "soft polymer" in which the number and probability of formation of holes (free volume) between the hydrocarbon chains play a decisive role for transverse diffusion.

Within the framework of this solution-diffusion model for the "ground permeability," factors influencing the polarity (dielectric constant) of the lipid domain and the order (packing density) and mobility of its constituents will affect permeability. Indeed, the simple permeability of the RBC membrane to small nonelectrolytes or the nonionized form of organic acids increases with a decreasing cholesterol content and an increasing mean number of double bonds of the phospholipid acyl chains (Deuticke, 1977). These effects are probably due to an increase of the intramembrane diffusion coefficient and of the partition coefficient (see Section 2.1).

In analogy, insertion into the lipid domain of drugs that alter the diffusion coefficient of permeants by changing order and mobility, or the partition coefficients, e.g., by altering the polarity of the membrane barrier or the packing of its acyl chains, will affect the ground permeability. Effects of this type are documented in particular for alkanols (C_{2-8}), which have been studied as model anesthetics (Seeman, 1972). Their effects on the ground permeability depend on the concentration range tested and on the permeant considered. At nerve blocking concentrations (Seeman, 1972), they enhance the permeability of polar uncharged solutes such as polyols or nonionized organic acid (Deuticke, unpublished results). At these concentrations, RBC membrane lipids are slightly disordered (Kutchai *et al.*, 1980). A causal relationship between the two phenomena, occurring at a molar ratio alkanol/membrane lipid of about 0.1, is likely. Interestingly, the diffusional permeability of water and small alcohols is affected only very little by these treatments.

In addition to such permeability enhancement via "fluidization," narcotic levels of the lower alcohols ($\leq C_6$) seem to increase ion permeability by increasing the polarity of the membrane interior. This has been concluded from an enhancement, by these alcohols, of the permeability to large hydrophobic ions (Orme *et al.*, 1988). The extent of this enhancement, which depends crucially on the radius of the permeant, markedly exceeds the enhancement of nonelectrolyte permeability, at comparable sizes of the permeants.

At high concentrations of alkanols (alkanol/lipid ratio > 0.3), these permeability changes are superimposed by a conspicuous increase of the permeability to small inorganic ions, and also to sucrose (Wieth *et al.*, 1974), which is normally impermeant. These effects have been known for a long time as "prelytic changes" because they initiate the sequence of events known as "colloid-osmotic hemolysis" (Wilbrandt, 1941). Similar increases of ion permeability were also observed in artificial lipid membranes (Johnson and Bangham, 1969; Gutknecht and Tosteson, 1970; Barchfeld and Deamer, 1985). This type of leakiness is assigned by some authors to a decrease in the dielectric energy difference for ions between water and membrane phase, with an additional contribution from changes of the dipole potential at the membrane interface (Orme *et al.*, 1988). Others have inferred that the alcohols may produce (transient) defects in the bilayer acting as aqueous pores for ions and polar nonelectrolytes (Barchfeld and Deamer, 1985). In the latter case, these alcohol effects might be comparable to those of detergents (see below). The two concepts are implicitly based on different assumptions concerning the localization of the alcohol in the membrane. Dielectric changes require alcohol accumulation in the membrane core; the pore concept presumes an alignment in-between the phospholipid chains, the hydroxyl groups being localized in the interfacial region.

Detergents, which differ from alkanols by being charged or more bulky in their hydrophilic domain, are well known for their lytic action on RBC at concentrations below those required for membrane solubilization. A selective enhancement of nonelectrolyte permeability, reflecting changes of membrane "fluidity," is not well established for such amphiphiles. They induce, however, aqueous membrane leaks (Isomaa *et al.*, 1986). Generalizing, their leak-forming potency increases with (1) increasing length of the alkyl chain, (2) decreasing bulkiness of their polar moiety, and (3) in the order anionic > cationic > zwitterionic (Isomaa *et al.*, 1986; Zaslavsky *et al.*, 1978). These differences are partly related to variations in partitioning since they diminished when the effects were related to equal membrane-stabilizing capacities (see Section 2.6), which reflect equal intramembrane concentrations.

Leakiness already occurs at detergent concentrations that stabilize cells against osmotic lysis (Isomaa and Hägerstrand, 1988). The original distinction between an antihemolytic and the hemolytic concentration range for amphiphiles is thus merely due to the experimental conditions used in the work establishing this dualism (Seeman, 1972). Lysis is a secondary event, resulting from colloid-osmotic swelling, the rate of which is determined by the induced leak permeability. Size and selectivity of detergent-induced leaks are unknown and their structural basis is a matter of conjecture. Formation of nonbilayer structures in the lipid bilayer has been proposed (Isomaa *et al.*, 1986). It will be interesting to see how much these leaks have in common with those produced by chemical modification of RBC membrane proteins (Deuticke *et al.*, 1983, 1989).

Many cationic drugs, e.g., local anesthetics and phenothiazines, are unspecific membrane perturbants, resembling charged detergents. They are lytic to RBC at concentrations somewhat higher than the membrane-stabilizing range (Seeman, 1972). Lysis is based on the formation of ion leaks. A radius of 0.6–0.7 nm could be assigned to these leaks in chlorpromazine-doped cells (Lieber *et al.*, 1984). Tetracaine also increases nonelectrolyte permeability, which may be related to the well-established increase of "membrane fluidity" or the enhancement of transbilayer lipid mobility by such local anesthetics (Noji *et al.*, 1982; Schneider *et al.*, 1986).

Besides alkanols and compounds with detergent properties, the diphenol compound phloretin and probably other flavonoid aglycones affect RBC ground permeability. These compounds, which are inhibitors of many mediated transport processes (see below), enhance nonelectrolyte permeability not only in artificial lipid membranes (Andersen *et al.*, 1976), but also in RBC (Owen and Solomon, 1972; Deuticke, 1974). This may be a consequence of an increase of membrane "fluidity." On the other hand, these compounds also affect the permeability for hydrophobic ions including fatty acids, presumably by changing the dipole potential at the membrane interface (Orme *et al.*, 1988). Indirect evidence further suggests that flavonoids (Biesendorfer *et al.*, 1981) and their aglycones (Deuticke, unpublished results) suppress the formation of various types of aqueous membrane leaks or lower the accessibility of such leaks to polar permeants.

4.2. Channels or Low-Affinity Carriers for Water, Small Polyols, and Amides

Proteinaceous pathways for water, small polyols (ethylene glycol, glycerol), and small amides (urea and some of its derivatives and analogues) seem to be present in the RBC membrane, in addition to the ground permeability for these solutes. The concept stems from the inhibition of their transport by protein-reactive agents (mercurials, Cu^{2+}) as well as saturation kinetics and competition phenomena (Deuticke, 1977; Brahm, 1983). The nature and molecular identity of these pathways and their possible interrelationships are a matter of continuous debate (Brahm, 1983; Chasan and Solomon, 1985; Solomon, 1986; Brahm and Galey, 1987). The transport system for urea ($K_m \approx$ 300–400 mM) and other small amides is strongly inhibited by phloretin (K_I = 18–25 μM) in its uncharged form (Toon and Solomon, 1987) and by many highly lipophilic urea analogues (Mayrand and Levitt, 1983). The transfer system for glycerol (in human and some other types of RBC), which also has a very low affinity for its substrate, is highly sensitive to inhibition by H^+ and Cu^{2+} (Carlsen and Wieth, 1976; Deuticke, 1977). Hexanol inhibits noncompetitively (K_I = 0.8 mM) without affecting inactivation by H^+ (Deuticke, unpublished). Inhibition by other alkanols (C_{1-5}) has been reported (Stein and Danielli, 1956). Phloretin inhibits only weakly (I_{max} = 75%) (Carlsen and Wieth, 1976). Anionic drugs (salicylate, phenopyrazone), like strongly adsorbing inorganic anions (e.g., SCN^-), inhibit by shifting the H^+ inactivation curve to higher pH values (Deuticke, unpublished).

The water permeability of RBC comprises a considerable component involving membrane proteins, which somehow serve as a "pore" (Macey, 1984). The major biochemical evidence for this component is its inhibition by mercurials (Macey, 1984; Toon and Solomon, 1986). While this is a feature shared by many other protein pathways, the water channel is unique in that neither the most widely effective noncovalent inhibitors of protein-mediated transports in RBC—alcohols and phloretin (Brahm, 1982; Toon and Solomon, 1987)—nor other drugs have unequivocally acceptable inhibitory effects.

4.3. Carrier-Mediated Transport of Nonelectrolytes

4.3.1. Glucose Transport

4.3.1.1. General Properties. Passive transmembrane movements of glucose and many other monosaccharides (LeFevre, 1967) are catalyzed in human RBC by a transporter (1.5–3 × 10^5 copies/cell). Recent reviews should be consulted for its kinetics, molecular identity, and structural features (Wheeler and Hinkle, 1985; Krupka and Devés, 1986; Stein, 1986; Walmsley, 1988). Molecular biological evidence (Mueckler *et al.*, 1985; Sarkar *et al.*, 1988) strongly supports that a 55-kDa glycoprotein is involved, although some immunological (Langdon and Holman, 1988) and affinity labeling (e.g., May, 1987) data still point to a hitherto unclear role of a 100-kDa (band 3 region) protein.

Most of the kinetic data indicate that the transport system operates as a mobile carrier, to be envisaged as a gated channel (Krupka and Devés, 1986), exposing a single substrate binding site alternately to the *cis* or the *trans* interface of the membrane. Reorientation of the binding site, equivalent to a (minor) change of conformation, occurs as a consequence of the binding not only of substrate or substrate analogues, but also of inhibitory drugs (see below). Objections against this "simple" model (Naftalin, 1988; Carruthers, 1988; Stein, 1986) are still a matter of controversy. More recently, refined experimentation seems to have refuted at least some of these objections (Wheeler and Whelan, 1988). The drugs presented below have provided important information on the glucose transporter.

4.3.1.2. Diphenol Inhibitors. Phloretin is one of the oldest known inhibitors of hexose transport (see LeFevre, 1961, and Krupka, 1985, for references). K_I values reported vary between 0.25 and 9 μM, depending on experimental conditions. The inhibitory potency crucially depends on both of the aromatic rings and the presence of some of the substituent hydroxyl groups (LeFevre, 1961; Kotyk *et al.*, 1965). Phloridzin, the glycoside of phloretin, also inhibits, although with a K_I about 80 times higher. Both, the permeant (Jennings and Solomon, 1976) phloretin and the impermeant phloridzin compete with glucose in net entry and equilibrium exchange experiments (increase of K_m) but inhibit noncompetitively in net exit experiments (decrease of V_{max}). This surprising result was shown (Krupka, 1985; Krupka and Devés, 1986) to be the consequence of a selective binding of these inhibitors to the outward-facing form of the transporter and a consecutive shift, in transmembrane distribution, of the transporter in favor of its outward-facing orientation, a "recruitment." Exofacial binding of inhibitor thus gives rise not only to competition with external substrate but also to a decrease of the number of transporters available to internal substrate under net exit conditions (lowering of V_{max} = noncompetitive inhibition). These data provide evidence for the actual existence of two conformations of the transporter. In spite of its competitive mechanism, phloretin probably does not interact with the glucose binding site but with an (adjacent?) region on the protein. This follows from different effects of phloretin and maltose, an impermeable substrate analogue, on the rate of reaction of covalent inhibitors of glucose transport with the transporter (Krupka and Devés, 1980a). Such modifying effects of substrates or inhibitory drugs on the rate of binding of covalent inhibitors may be taken as further indicators of conformational changes in a transporter.

Among other diphenol compounds tested for their inhibitory action (LeFevre, 1961), stilbestrol, hexestrol, and their alkyl derivatives proved even more effective than phloretin (I_{50} 0.5–1 μM), producing mixed-type inhibition under net exit conditions.

4.3.1.3. Cytochalasin B. This fungal metabolite is a widely used high-potency inhibitor of glucose transfer ($I_{50} < 1$ μM) (see Krupka and Devés, 1986, and Walmsley, 1988, for references). It competes selectively with substrate for the inward-facing form of the transporter, as indicated by competitive inhibition of exit as compared to noncompetitive inhibition of glucose entry. Due to its high binding affinity ($K_D \approx 0.1$ μM) (Jung and Rampal, 1977), cytochalasin B has served to identify the glucose transporter and to determine its number per cell (e.g., Basketter and Widdas, 1978). Its binding site could be located to the inner membrane surface, but is probably not identical with the substrate binding site (see Krupka and Devés, 1986, for references). Cytochalasin B binding also modifies the covalent binding of various inhibitors (Krupka, 1985; Pinkofsky and Jung, 1985) indicative of conformational changes induced by its binding, which manifest themselves in a recruitment of substrate binding sites to the inner membrane surface (Wang *et al.*, 1986). Cytochalasin B binds covalently to the transporter upon *in situ* photoactivation, presumably of a tryptophan in the transport protein (see Walmsley, 1988, for references). Its binding site on the transporter could thus be localized with some confidence.

4.3.1.4. Forskolin. This established activator of adenyl cyclase inhibits hexose transport by direct interaction with the transporter (Sergeant and Kim, 1985; Lavis *et al.*, 1987). Noncompetitive inhibition of hexose uptake (Sergeant and Kim, 1985) and mutual exclusion with cytochalasin B suggest binding to the inward-facing form of the transporter. A photoreactive derivative of forskolin (Wadzinski *et al.*, 1987) is the most efficient inhibitor of hexose transport presently known ($I_{50} \approx 0.05$ μM), besides photoreactive derivatives of phloretin and phloridzin (Fannin *et al.*, 1981; Wyse *et al.*, 1989). It may prove to be a useful tool for detecting and characterizing the transporter even at high dilution, e.g., in reconstituted systems.

4.3.1.5. Miscellaneous Inhibitory Drugs. Information on the sidedness and asymmetry of the transporter has also been obtained by analyzing inhibition by C_{19} and C_{21} steroids (I_{50} 20–300 μM) (Lacko *et al.*, 1975). This class of compounds exhibits variable preference for the two orientations of the transporter (Krupka and Devés, 1980b). The relevance of hydrogen-bonding groups for ligand binding to the substrate site has been evaluated using pyridine derivatives (Hershfield and Richards, 1976).

In addition, inhibition of glucose transport has been reported for numerous classes of drugs. Many of them [local anesthetics, $I_{50} > 0.05$ μM (Lacko *et al.*, 1977, 1979), benzoic acid esters representing part of a local anesthetic (Lacko *et al.*, 1981), numerous benzodiazepines (K_I 12–316 μM) (Lacko and Wittke, 1984), as well as cannabis alkaloids (Schurr *et al.*, 1974)] inhibit glucose entry competitively. Aliphatic and aromatic alcohols produce mixed-type inhibition (Lacko *et al.*, 1974, 1978). Since there is no structural similarity between all these competitive inhibitors, "competition" must be due either to a merely partial overlap between the binding sites for inhibitor and substrate, or to a true allosteric competition between distant but conformationally coupled binding sites, or to an indirect effect mediated by alteration of the lipids surrounding the transport protein. An influence of the lipid environment on the glucose transporter has been demonstrated (Deuticke and Haest, 1987; Carruthers and Melchior, 1988).

Phenothiazines, on the other hand, inhibit noncompetitively (Owen and Gunn, 1983). Anionic and cationic forms have similar potencies (K_I 20–400 μM) A noncompetitive mechanism has also been reported for furosemide ($K_I \approx 2$ mM; Jung and Mookerjee, 1976), which inhibits glucose self-exchange like other anthranilate derivatives [niflumate ($I_{50} \approx$ 100 μM), meclofenamate (I_{50} 70 μM) (Cousin and Motais, 1979)]. Last but not least, dipyridamole deserves interest as an inhibitor of glucose transport (I_{50} = 30 μM) (Plagemann and Woffendin, 1987) in view of its other transport inhibitory effects (see below).

4.3.2. Nucleoside Transport

4.3.2.1. General Properties. Passive nucleoside transport, involved in the salvage of purine and pyrimidine nucleosides, is mediated in human RBC by a glycoprotein ($M_r \approx 55$ kDa, about $1–1.5 \times 10^4$ copies/cell). Available information on this system up to 1987 has been reviewed by Plagemann *et al.* (1988), where references for subsequent statements not further verified by a reference may be found. The transporter, which differs in number per cell, kinetics, and other properties between mammalian RBC, has a broad substrate specificity, but affinities vary considerably. Kinetic evidence (Stein, 1986; Plagemann *et al.*, 1988) indicates that in fresh cells it operates as a simple mobile carrier.

4.3.2.2. Inhibitory Nucleoside Analogues. Much has been learned about this transporter using lipophilic nucleoside analogues, e.g., nitrobenzyl mercaptopurine ribosides (NBMPR), which are highly potent transport inhibitors ($K_I < 0.1$ nM in sheep, $I_{50} \approx$ 5–10 nM in human RBC). They are tightly bound to the transporter. K_D values (≈ 0.1 nM) are lower by orders of magnitude than K_m's of the physiological substrates (50–500 μM), indicating a major relevance of hydrophobic interactions for inhibitor binding. Suppression of binding by other compounds serves as a screening test for inhibitors of nucleoside transport. Moreover, NBMPR are used for photolabeling the transporter.

NBMPRs are preferentially bound to the outward-oriented form of the transporter, as indicated by competition for influx and noncompetitive effects on efflux of uridine in sheep RBC (Jarvis *et al.*, 1982). In human RBC the kinetic situation is somewhat ambiguous (Plagemann *et al.*, 1988). While these inhibitors clearly compete with substrates for binding, it is not unequivocally settled whether their binding involves the substrate binding site (completely or partly) or is allosteric (Jarvis *et al.*, 1983; Jarvis, 1987; Plagemann *et al.*, 1988).

4.3.2.3. Vasodilators, Calcium Antagonists, and Other Inhibitors. Besides the potent substrate-analogue inhibitors, two classes of drugs have proven inhibitory: vasodilators (dipyridamole, dilazep, hexobendine, lidoflazine), and Ca^{2+} channel antagonists (e.g., nimodipine and other 1,4-dihydropyridines, phenylalkylamines, and diltiazem). Dipyridamole and many of its congeners (Gerlach *et al.*, 1965; Plagemann and Woffendin, 1987b) inhibit nucleoside transport. Binding involves the outward-oriented conformation as indicated by a pattern of inhibition resembling that for NBMPR. K_I values vary among different species (5–40 nM; Jarvis *et al.*, 1982; Jarvis, 1986; Plagemann *et al.*, 1988). [^{3}H]-Dipyridamole binds with high affinity ($K_D \approx 10$ nM) to about 5 $\times 10^5$ sites/cell (Woffendin and Plagemann, 1987b), which exceeds the number of transporters as defined by binding of NBMPR (Jarvis, 1986). The inhibitor has been used for photolabeling the transporter with some specificity, although low-affinity binding sites are also present in the membrane, which is not surprising in view of its inhibitory effects on other transporters (see Sections 4.3.1.5 and 4.4.1.3).

Two other inhibitory vasodilators, dilazep and hexobendine, are about equally effective (I_{50} 10–30 nM; Striessnig *et al.*, 1985; Plagemann and Kraupp, 1986). The mechanism of inhibition is not known. Both suppress the binding of NBMPR (K_I 0.2 and 3 nM; Hammond *et al.*, 1985; Plagemann *et al.*, 1988). Dilazep also suppresses binding of dipyridamole, indicative of partly overlapping or allosterically linked binding sites. The latter would seem plausible in view of the rather divergent structure of these inhibitors. On the other hand, lidoflazine is a potent ($K_I = 12$ nM) competitive inhibitor (Plagemann and Woffendin, 1987b), which also inhibits NBMPR binding. The interplay between these potent inhibitors is presently not understood.

The same is true for the surprising inhibitory influence on nucleoside transport of Ca^{2+} antagonists of varying chemical constitution. According to studies by Ford *et al.* (1985), Striessnig *et al.* (1985), Hammond *et al.* (1985), and Plagemann and Woffendin (1987b), 1,4-dihydropyridines inhibit nucleoside transport though with lesser effectiveness than the vasodilators, and suppress binding of NBMPR. Nimodipine, the most potent inhibitor of this group ($I_{50} \approx 0.6$ μM, $K_D \approx 50$ nM), is reported to inhibit transport competitively, while its specific binding to the membrane is suppressed noncompetitively by NBMPR and the vasodilators. These observations are hard to reconcile with any simple model. The other dihydropyridines as well as diltiazem and the phenylalkylamines (e.g., verapamil) are less effective. Their K_I values (10–160 μM) in this system correspond to the I_{50} values reported by Engelmann and Duhm (1989) for inhibition of Ca^{2+} entry into RBC. Interaction of these potent calcium antagonists with the RBC membrane exhibits stereoselectivity, opposite to that characteristic for their interaction with electrically excitable tissues (Striessnig *et al.*, 1985).

Some other drugs related to membrane–Ca^{2+} interactions (pimozide, trifluoperazine, papaverine, TMB 8) also inhibit the site-specific binding of NBMPR, while propranolol and flunarizine are ineffective (Hammond *et al.*, 1985). Benzodiazepines (diazepam and many others) affect both nucleoside transport and NBMPR binding, at concentrations exceeding those of the potent inhibitors by orders of magnitude. The kinetic pattern of inhibition suggests substrate-competitive binding to the transporter in both of its orientations (Hammond *et al.*, 1983).

Considering the structural heterogeneity of these inhibitors and the ambiguities concerning their mechanisms, no consistent model for the relationship between substrate and inhibitor binding sites on the nucleoside transporter can yet be envisaged.

An intriguing problem concerns the relationship between nucleoside and monosaccharide transporter. As reviewed by Plagemann and Woffendin (1987a), the two proteins are likely to have common structural features. Such similarities are also indicated by some overlap in their inhibitor sensitivities. (1) Adenosine inhibits glucose transport ($I_{50} \approx 3$ μM) although hexoses do not inhibit nucleoside transport at all. (2) Both cytochalasin B and dipyridamole not only block "their" system with high potency but also act on the complementary one, though with a 100-fold lower potency. (3) Nucleosides inhibit cytochalasin B binding to the glucose transporter with higher affinity than glucose, presumably by competition for the binding site on the cytoplasmic face of the membrane.

4.4. Anion Movements

4.4.1. Anion Exchange via Band 3

4.4.1.1. General Characteristics. The rapid exchange of Cl^- for HCO_3^- across the RBC membrane that contributes to the respiratory transport of CO_2 in blood is mediated by a 96-kDa integral membrane protein "band 3," of which, however, only a C-terminal domain of about 60 kDa takes part in transport. Human RBC contain about 1.2×10^6 copies/cell and the primary sequence has been elucidated (Tanner *et al.*, 1988); many details of its molecular organization and its kinetics are available and have been reviewed extensively up to the present (e.g., Knauf, 1979, 1986; Fröhlich and Gunn, 1986; Passow, 1986; Jennings, 1989; Cabantchik and Eidelman, this volume, where references to the following can be found).

In its predominant mode of operation, band 3 catalyzes an obligatory, electrically silent exchange of anions by a "Ping-Pong" mechanism, to be envisaged as a conformational change of the protein upon binding a single substrate ion to a "transfer site," which leads to a change of its orientation from one side of the barrier to the other one, or else to a shifting of the barrier from the *trans* to the *cis* side of the binding region. There seems to be no "slippage" of the unloaded carrier. The exchange process is modulated by ligands (anions, H^+) that bind to a number of operationally defined modifier sites for which either an allosteric linkage to the transfer site (Passow, 1986) or interference with the approach of the substrate to the rate-limiting barrier (Tanford, 1985) is claimed.

A second mode of operation of band 3, conductive electrodiffusion, occurs to a very limited extent and seems to reflect channel-like properties of the exchanger (Fröhlich, 1984).

In the exchange mode, band 3 accepts a wide variety of mono- and divalent inorganic and organic anions, including even monovalent complexes between alkali cations (Na^+, Li^+) and certain divalent anions of the carbonate type (Becker and Duhm, 1978). This "anionic cation transport" may be regarded as a third mode of operation of the transporter.

The anion exchanger exhibits considerable conformational flexibility, which is reflected in a variety of allosteric interactions between the binding sites of noncovalent and covalent ligands of the transporter (substrate and modulator ions, reversibly binding drugs, chemical modifiers). These ligand/band 3 interactions have been the object of numerous reviews and compilations (Passow and Schnell, 1969; Deuticke, 1970, 1977; Cabantchik *et al.*, 1978; Knauf, 1979, 1986; Passow, 1986; Jennings, 1989).

4.4.1.2. Bimodal Inhibitors. Major progress in the analysis of anion exchange has come from "bimodal" inhibitors, which either can bind noncovalently only or can subsequently be linked covalently to their target site. Stilbene disulfonate derivatives (SDs), e.g., 4,4′-dinitrostilbene-2,2′-disulfonate (DNDS), are the most widely used probes of this type (Cabantchik *et al.*, 1978). As reversibly binding impermeable anionic probes, they inhibit with varying efficiencies, I_{50} values ranging over three orders of magnitude between 1.2 μM and 1.3 mM. High inhibitory potency requires localization of the sulfonate groups at appropriate positions and increases with increasing lipophilicity and electrophilicity of the inhibitor, as quantified by the Hansch factor π and the Hammet factor σ. Inhibitor binding is competitive with substrate binding as revealed by transport studies, NMR spectroscopic analyses of Cl^- binding to band 3, and binding studies with labeled analogues of the inhibitors (see Passow, 1986, for references). These data suggest binding of substrate and inhibitor to at least overlapping binding sites, although more complicated allosteric mechanisms cannot be excluded (Passow, 1986; Fröhlich and Gunn, 1987). The SDs are very selective inhibitors of band 3-mediated anion transport. Only at much higher concentrations do they affect transport via the monocarboxylate transporter (see Deuticke, 1982; Jennings and Adams-Lackey, 1982) and K^+/Cl^- cotransport (see section 4.5.2.2).

Introduction of amino-reactive isothiocyano groups converts reversibly binding SDs into inhibitors [e.g., 4,4'-diisothiocyano stilbene-2,2'-disulfonate (DIDS)] binding covalently to the site of their primary noncovalent binding. These compounds have provided much of the available information on the molecular identity of the exchanger, its number per cell, the localization of the SD binding site in the primary sequence, and its microenvironment. The binding site, only accessible from the outer membrane surface but probably located in an intrabilayer domain of the protein (see Jennings, 1989), takes part in conformational changes. It can be recruited to the outer surface, dependent on the substrate concentration at the inner surface of the membrane. Ligand association with the site, in turn, induces conformational changes (Passow, 1986; Knauf, 1986).

A photoreactive N-(4-azido-nitrophenyl) derivative of taurine (NAP-taurine) has also proven to be a valuable bimodal inhibitor. Being transported ($K_m \approx 700$ μM) it inhibits competitively when present on the inside and may be used as a label of the substrate binding site. From the outside, NAP-taurine inhibits much more effectively ($K_I \approx 20$ μM) by binding to an exofacial "modifier site" closely adjacent to the substrate and DIDS binding site. It can also be used to monitor and induce conformational changes in the protein (Knauf, 1979, 1986; Passow, 1986).

4.4.1.3. Reversible Noncovalent Inhibitors. Besides these bimodal inhibitors, many other groups of chemically unrelated drugs inhibit anion exchange. Within many of these groups the inhibitory potency is basically provided by an anionic group and increases with the hydrophobicity of the molecule—indicating the involvement of hydrophobic domains (clefts, pockets, surface domains) in binding, and its electrophilicity—indicating the significance of charge transfer complexes in inhibitor binding. The contribution of the two parameters may vary.

Substituted benzene sulfonates (Cabantchik *et al.*, 1978) inhibit competitively but only at considerably higher concentrations (K_I 0.2–60 mM) than their "dimeric analogues," the SDs. Motais and associates have contributed extensive data for chloro- and nitro-substituted phenol compounds (Motais *et al.*, 1978), benzoates (I_{50} 0.6 μM–12 mM), salicylates (I_{50} 1.8 μM–1.3 mM), phenoxyacetates, including ethacrynic acid (I_{50} 0.7 μM) (Motais and Cousin, 1978), sulfonamido compounds known as inhibitors of carbonic anhydrase or as diuretics (Cousin and Motais, 1976; Motais and Cousin, 1976) as well as anthranilate compounds [niflumate (I_{50} 0.6 μM), flufenamate (Cousin and Motais, 1979, 1982a,b)]. Inhibitors of the anthranilate type have been analyzed in more detail. Inhibition is noncompetitive [niflumate (Cousin and Motais, 1979)] or of mixed type [furosemide K_I 50 μM, bumetanide K_I 360 μM (Brazy and Gunn, 1976; Gunn, 1985)] and seems to involve an outward-facing modifier site, which may overlap or interfere allosterically to some extent with the SD binding site (Knauf and Mann, 1984; Passow, 1986).

Phloridzin inhibits ($I_{50} \approx 0.2$ mM) from outside and noncompetitively. Phloretin produces mixed inhibition ($I_{50} \approx 1$–3 μM) predominantly in its protonated form ($pK_a \approx 4.6$) by binding to an exofacial site (Fröhlich and Gunn, 1987; Fröhlich and Mayer, 1989). Its interference with SD binding, on the other hand, has been assigned to the anionic form (Forman *et al.*, 1982). The inhibitory potency of phloretin and other acetophenone derivatives depends on lipophilicity and dipolar properties, but not on electrophilicity (Cousin and Motais, 1978). Inhibitory effects of simple alkyl derivatives including detergents (Deuticke, 1970, 1977) are not well characterized except for the case of alkanols C_{2-10} (Forman *et al.*, 1985). They inhibit noncompetitively (I_{50} 10–100 mM within the membrane). The short-chain members also interfere with SD equilibrium binding, while all of them enhance the rate of that binding, taken to indicate an alkanol-mediated facilitation of conformational changes following ligand binding.

Local anesthetics (Deuticke, 1970; Feinstein *et al.*, 1977; Gunn and Cooper, 1975) and cationic phenothiazines (Motais *et al.*, 1980) inhibit (probably noncompetitively) at concentrations not far from, or in the lytic range (> 0.5 mM). For these compounds as for the alkyl derivatives, an indirect inhibitory mechanism—via perturbation of the lipid domain—may be considered. In contrast, anionic phenothiazine carboxylates are potent inhibitors (I_{50} 2–10 μM) even below their membrane-stabilizing range of concentrations (Motais *et al.*, 1980).

Dipyridamole, a cationic (pK ≈ 6.3) pyrimido-pyrimidine [and many of its congeners (Gerlach *et al.*, 1965)], has long been known as a highly effective inhibitor of anion transport ($I_{50} \approx 5$ μM; Deuticke, 1970). Its competitive (divalent anions) or mixed-type (monovalent anions) inhibition (Renner *et al.*, 1988) and its interference with binding of other ligands to the exchanger were recently shown to depend on the presence of Cl^- whose binding is obviously required to force the protein into a conformation able to accept dipyridamole (Renner *et al.*, 1988; Legrum and Passow, 1989).

4.4.2. Monocarboxylate/H^+ Cotransport

Most of the inhibitors reviewed in Section 4.4.1 are by no means specific for the anion exchanger. In addition to their effects on other transport processes, as covered in this review, most of them are also inhibitors of the second anion transporter of the RBC membrane, the monocarboxylate/H^+ cotransporter. This subject including inhibitors has recently been covered in two reviews (Deuticke, 1982, 1989).

4.5. Transport of Monovalent Cations

4.5.1. The Na^+/K^+-Pumping ATPase

4.5.1.1. General Characteristics. RBC of man but not all other species (Duggan *et al.*, 1965) establish and maintain transmembrane gradients of alkali cations, Na^+ and K^+, by the ATP-dependent Na^+/K^+ pump. It consists of two proteins, the catalytic α subunit forming several transmembrane helical loops ($M_r \approx 100$ kDa) and the glycosylated β subunit ($M_r \approx 45$ kDa) whose function is not well characterized (see Ovchinnikov *et al.*, 1987, for references to structural data). Human RBC contain about 200–400 pump sites per cell, the number of subunits per pump site being uncertain. The available information on pump and ATPase activity has been compiled and reviewed extensively (Hoffman and Forbush, 1983; Stein, 1986) or concisely (Skou, 1986; Duhm, 1989). Available material for RBC was recently summarized by Hoffman (1986).

In its normal, active transport, mode of operation the enzyme mediates the uphill exchange of 3 Na_i^+ for 2 K_e^+, fueled by the hydrolysis of 1 ATP. Analysis of the system, and of its inhibition, can thus be carried out under vectorial transport (Na^+/K^+ pump) or scalar enzymatic (Na^+/K^+-ATPase) aspects. According to the most widely accepted model, transport is essentially based on a conformational transition of the protein between an E1 form releasing K^+ and binding Na^+ at the cytoplasmic face of the membrane and an E2 form releasing Na^+ and binding K^+ at the outer surface. The alternating reversible changes in affinity required for this process are coupled to a cycle of phosphorylation (Na^+ dependent) and dephosphorylation (K^+ dependent) of the enzyme by ATP at the cytoplasmic surface. The energy for the conformational changes underlying the changes in cation affinity is derived from the transition of the protein-phosphoryl bond from a high-energy state of loose covalent binding (E1 ~ P) to a low-energy state of tight binding (E2–P) (e.g., see Tanford, 1983). Numerous intermediate steps of the cycle, e.g., non-covalent binding of the substrates and products of ATP cleavage, occlusion of transported cations, have been postulated and experimentally supported (Jorgensen and Andersen, 1988). The entire pump cycle is reversible, producing ATP from the dissipation of cation gradients through the pump. Besides this physiological mode of operation, additional transport modes (K_e^+/K_i^+ and Na_i^+/Na_e^+ exchange as well as uncoupled movements) have been established, some of them requiring ATP (and ADP) without its hydrolysis (Hoffman, 1986; Duhm, 1989).

Due to the diversity of ligands of the transport system and the allosteric linkage between their binding sites, considerable complexity of inhibitory mechanisms and a large body of inhibitory drugs might b expected. Surprisingly, this is not the case. Besides the group of cardiac glycosides, only a very limited number of inhibitors have been established in studies on RBC, as will become evident below (see also Schwartz *et al.*, 1975).

4.5.1.2. Glycoside Inhibitors. The highly selective inhibition of the Na^+/K^+ pump by cardiac glycosides (Schatzmann, 1953) has contributed decisively to the analysis of the molecular details of the transporter. Ouabain, most frequently used in RBC work, and other cardiotonic steroids inhibit pump and ATPase activity at concentrations that markedly depend on experimental conditions, due to a very slow rate of their binding (see Forbush, 1983, for a general summary). It requires several hours of incubation to reach equilibrium (e.g., see Joiner and Lauf, 1978). I_{50} and K_D values for pump inhibition and glycoside binding obtained under equilibrium conditions in human RBC are on the order of 2–5 nM for the most effective glycosides (proscillaridine and its esters and ethers, ouabain). Data for less effective glycosides ($I_{50} > 5$ nM) vary among different investigators (Belz, 1981; Brown and Erdmann, 1984; Senn *et al.*, 1988). Without preincubation, considerably higher concentrations are required [$I_{50} \geq 1$–2 μM for the above-mentioned glycosides (Senn *et al.*, 1988)]. Binding and inhibition are virtually irreversible due to the slow dissociation of the glycoside–enzyme complex (Forbush, 1983).

The glycosides and their aglycones (genins) bind exofacially at a 1 : 1 ratio to the α subunit, with high preference for the E2–P state. The kinetics of association are markedly affected by the concentrations, concentration ratios, and sidedness of the transported cations (and of nucleotide) (Hoffman, 1986; Joiner and Lauf, 1978; Forbush, 1983), in line with the general concept that the binding characteristics also reflect the allosteric conformation changes of the α subunit and the Na^+/K^+ competition dominating the rates of ion transport (Forbush, 1983; Anner, 1985). Competition between glycoside and K_e^+ (which induces dephosphorylation of E2–P) is a particularly well-studied feature. Additional allosteric cation binding sites different from the cation transfer site(s), however, also seem to interfere with glycoside binding, and its nucleotide (ATP) requirement may arise from noncovalent binding sites in addition to the phosphorylation at the high-affinity site for ATP.

Structural requirements for the high affinity of cardiac glycosides for their "receptor," mostly derived from studies on non-RBC ATPases (Fullerton *et al.*, 1983; Dittrich *et al.*, 1983; Forbush, 1983) but also on the RBC pump (e.g., see Belz, 1981, and Brown and Erdmann, 1984, for references), comprise in particular the steric arrangement of the steroid ring, crucial for the binding rate, its OH substituents, and the nature of the lactone ring as well as its orientation relative to the steroid moiety. Aglycones may be equally or even more effective than their corresponding glycosides (Senn *et al.*, 1988), while the glucuronides are generally less effective (Belz and Heinz, 1977). The number and nature of the sugar groups mainly modulates the dissociation rates of the glycoside receptor complex.

The marked selectivity of the glycosides for the Na^+/K^+ pump has stimulated their use as affinity labels. Photoactivatable azido derivatives have proven particularly useful (Forbush, 1983) and are now available at sufficiently high specific activity for sensitive labeling (Lowndes *et al.*, 1987). Yet, the glycoside binding site has so far not been identified.

In spite of the intricate structural requirements for glycoside effectivity, rather unrelated glycoside analogues and other polycyclic compounds also inhibit the pump. This was shown for the erythrophleum alkaloids, cassaine ($I_{50} \approx 50$ μM) and couminginine ($I_{50} \approx 0.1$ μM), which lack many of the structurally important features summarized above, but also inhibit competitive with K_e^+ (Kahn, 1955). Canrenone, a spironolactone metabolite, is a partial inhibitor ($I_{50} \approx 90$ μM) that decreases V_{max} and substrate affinity and suppresses ouabain binding ($K_I \approx 200$ μM) (Garay *et al.*, 1985). Sanguinarine, a plant alkaloid, inhibits at $I_{50} \approx 30$ μM (Cala *et al.*, 1982).

4.5.1.3. Further Inhibitors. In contrast to the abundant material on glycoside inhibition, most other drugs inhibiting the Na^+/K^+ pump are not very well characterized. Vanadate inhibits pumping and ATP hydrolysis at micromolar concentrations on the cytoplasmic side, presumably by competing with ATP and locking the system in one of the E2 conformations (Beaugé *et al.*, 1980; Hoffman and Forbush, 1983). Eosin Y, under nonphotooxidative conditions, inhibits RBC Na^+/K^+-ATPase reversibly in mixed-type kinetics involving competition with ATP binding (Blaiklock and Green, 1971), possibly to a modifier site (Skou and Esmann, 1981). Moreover, oligomycin inhibits the pump ($I_{50} \approx$ 1–2 μg/ml) (Askari and Rao, 1970). In analogy with ATPases from other sources, this probably results from a slowing of the major conformation changes (E1 ~ P/E2–P, E1/E2, i.e., of the ion translocation step) (Robinson *et al.*, 1984). Inhibitory effects of flavonoids and their aglycones on the Na^+/K^+ pump of RBC are not well established except for quercetin ($K_I \approx 60$ μM; Wüthrich and Schatzmann, 1980), although they are anticipated and potentially revealing in view of studies on the enzyme from other sources (Robinson, 1969; Kuriki and Racker, 1976; Robinson *et al.*, 1984).

Local anesthetics (e.g., tetracaine $\leq$ 1.3 mM, dibucaine $\leq$ 0.8 mM) and related drugs [chlorpromazine ($I_{50} \approx$ 0.5 mM), imipramine ($I_{50} \approx$ 0.2 mM)] inhibit pumping (Andersen, 1968; Askari and Rao, 1970). The simultaneous occurrence of leakiness (see Section 4.1) has precluded a more detailed analysis. ATPase activity is inhibited (mixed-type) by tetracaine ($I_{50} \approx$ 1.8 mM), going along with a lowering of the affinities for both transported ions (Bond and Hudgins, 1976). Conversely, these cations (and ATP) protect the enzyme against irreversible inactivation by tetracaine (Bond and Hudgins, 1977). Inhibitory effects of ethacrynic acid ($I_{50} \approx$ 0.3 mM) (Sachs and Welt, 1968; Askari and Rao, 1970) and furosemide (I_{50} 2 mM) (Askari and Rao, 1970) are not particularly well characterized. Interference with ATP binding has been suggested. Finally, suramine, a large, impermeable polyaromatic hexasulfonate anion, inhibits ATPase ($I_{50} \approx$ 50 μM) and pump activity from the cytoplasmic face (Fortes *et al.*, 1973). The polyanionic character of the inhibitor suggests interference with nucleotide binding. In contrast, the impermeable polycation, compound 48/80, one of the most potent inhibitors of the Ca^{2+} pump, has very little effect on the Na^+/K^+ pump (DiJulio *et al.*, 1989).

4.5.2. Passive Cation Transporters

4.5.2.1. General Aspects. Besides being transported actively, the major cations of the RBC (Na^+, K^+) move passively across the membrane using a number of pathways defined more or less operationally by their insensitivity to ouabain and their sensitivity to certain other inhibitors. In line with the requirements of electroneutrality, these cation movements involve a coupling of fluxes, i.e., either a cation/Cl^- cotransport (symport) or a cation/cation exchange (antiport). Since the coupling is not electrical but "molecular" in nature, these essentially passive transporters can mediate uphill transport of one of their substrates, for which the chemical gradient is low, by dissipating the higher chemical potential gradient of the other, "driver" substrates (see Stein, 1986; Lauf, 1986; Geck and Heinz, 1986; Duhm, 1989). Transports are saturable and exhibit *trans* effects. Some of them are controlled and seem to be involved in the regulation of cell cation composition and volume (see Hoffman and Simonsen, 1989, for a recent review), others are functionally not well understood.

4.5.2.2. K^+/Cl^- Cotransport. After its original establishment in avian, fish, and sheep RBC (see McManus and Schmidt, 1978; Lauf, 1985, 1986; Lauf *et al.*, 1987), good evidence for an electroneutral, Na^+-independent K^+/Cl^- cotransport J_K (Cl) is now also available for human and porcine RBC (see Lauf, 1986; Kaji, 1986; O'Neill, 1987; Duhm, 1987; Kim *et al.*, 1989; Brugnara *et al.*, 1989, where further references may be found). This transport pathway is, however, essentially silent (except in avian RBC) under native, isosmotic conditions. Ouabain-insensitive, Cl^--coupled efflux (or influx) of K^+ becomes evident upon activation by (1) hypotonic and other types of cell swelling (Lauf, 1986; Kaji, 1986; O'Neill, 1987; Brugnara *et al.*, 1989) and (2) modification or oxidation of crucial membrane SH groups (Lauf, 1985; Duhm, 1987; Haas and Harrison, 1989). The former mode of activation (see Lauf, 1985, for a model) may be the physiologically relevant stimulus inducing a compensatory loss of KCl and water from swollen cells (Dunham and Ellory, 1981; Borgese *et al.*, 1987; Lytle and McManus, 1987). Activation via SH groups may play a pathophysiological role in hemoglobin-related RBC anomalies [sickle cell anaemia, HbCC (Canessa *et al.*, 1986; Berkowitz and Orringer, 1987)] and other conditions going along with oxidative stress (Haas and Harrison, 1989).

J_K (Cl) is moderately sensitive to classical loop diuretics of the 5-sulfamoyl-benzoate (anthranilate) type. For bumetanide, I_{50} values vary between 50 and 200 μM in various types of RBC (Lauf, 1986; Kaji, 1986; O'Neill, 1987). Furosemide is either less (Kaji, 1986) or more (Lytle and McManus, 1987) effective, piretanide essentially ineffective. This comparatively low sensitivity to loop diuretics distinguishes J_K (Cl) from $J_{Na,K}$ (Cl) (see Section 4.5.2.3). Loop diuretics of other classes (muzolimine, cycletanide, ethacrynic acid) do not inhibit (Garay *et al.*, 1988). Moreover, neither amiloride nor notoriously membrane-active drugs (quinine, quinidine, phenothiazines, TMB8, phloretin) are inhibitory (Kaji, 1986; Garay *et al.*, 1988). SDs inhibit only at considerably higher concentrations than required for inhibition of anion exchange (Lytle and McManus, 1987; Borgese *et al.*, 1987; O'Neill, 1987; Haas and Harrison, 1989; Kim *et al.*, 1989).

The lack of sensitive and more specific inhibitors of K^+/Cl^- cotransport has recently been overcome by Garay *et al.* (1986, 1988). They found some indane analogues of ethacrynic acid (dihydroindenyl alkanoates) to inhibit J_K (Cl) at I_{50} values of 10–30 μM without inhibition of $J_{Na,K}$ (Cl). The site of attack of bumetanide, furosemide, and these new inhibitors is not clear. Furosemide induces mixed-type inhibition in low-K^+ sheep RBC (Lauf, 1984). Interference with Cl^- binding seems not to be involved (Lauf, 1984), while enhancement of the inhibitory efficiency by external cations has been demonstrated (Lytle and McManus, 1987; Lauf, 1984; Duhm, 1987) and may be taken to reflect binding of the inhibitors to a substrate-independent site on the outward-facing, K^+-loaded conformation of the transporter.

4.5.2.3. $Na^+/K^+/2Cl^-$ Cotransport. Na^+ and K^+ can penetrate the membrane of nucleated and nonnucleated RBC in an ouabain-resistant, Cl^--dependent, coupled mode of transport (see Chipperfield, 1986, Lauf, 1986, Lauf *et al.*, 1987, and Haas, 1989, for references). Elegant work on duck (McManus, in Lauf *et al.*, 1987) and human (Duhm, 1987) RBC has provided good evidence that these movements [$J_{Na,K}$(Cl)] are mediated by an electroneutral cotransport of 1 Na^+ + 1 K^+ + 2 Cl^-, exhibiting "glide symmetry" (first on–first off) in the ordered binding and debinding of its four substrate ions and no "slippage" of partially loaded carrier forms. In avian RBC this cotransporter is activated not only by β-adrenergic catecholamines and their second messenger, cAMP, but also by osmotic shrinking, hypoxia, and other factors (compiled by Haas, 1989) and may effect a regulatory volume increase. Shrinkage also activates in certain types of nonnucleated mammalian RBC (rat, ferret). In human RBC the transporter may contribute to the setting of volume but seems to function in a KCl (and water)-extruding mode (Duhm and Göbel, 1984). This cotransporter is inhibited by very low concentrations of loop diuretics, particularly of the 5-sulfamoyl-benzoate type. I_{50} values increase in the order bumetanide (0.6 μM) < piretanide (1.2 μM) < furosemide (9 μM) in human RBC. Differences in reported values may arise from species differences or the effects of the external cation level on inhibitory potency (Palfrey *et al.*, 1980; Garay *et al.*, 1986; Chipperfield, 1986; Duhm, 1987).

Comparative studies with various drugs of this type have revealed I_{50}'s as low as 0.05 μM (Palfrey *et al.*, 1980) and demonstrated the particular relevance, beside the anionic group, of the substituents at the 3 position of the benzoate ring, in contrast to substituents at position 4. Inhibition is also reported for ethacrynic acid after transformation to its cysteyl adduct (I_{50} = 0.4 μM) (Ellory and Stewart, 1982) while potent inhibitors of J_K (Cl), dihydroindenyl alkanoates (Garay *et al.*, 1986, 1988), have little or no effect. This is also true for unrelated diuretics such as thiazides, tienilic acid, triamterene, and amiloride (Ellory and Stewart, 1982; Garay *et al.*, 1986). Phloretin inhibits weakly (80% at 100 μM) (Chipperfield, 1981).

The requirement of an anionic group for inhibition suggested inhibitor binding to one of the chloride sites on the transporter. This was confirmed by kinetic and inhibitor binding studies (Haas and McManus, 1983; Haas and Forbush, 1986). Since binding is stimulated by low and inhibited by high Cl concentrations, but stimulated by cations at all concentrations, the inhibitor is thought to bind at the chloride site assumed to become occupied last on the tetravalent carrier–substrate complex (McManus, in Lauf *et al.*, 1987; Duhm, 1987). Catecholamine-stimulated, saturable, and loop-diuretic-sensitive binding of labeled bumetanide to dog RBC ($K_D \approx 0.1$ μM) has also provided a number of 10^3 (duck) to 10^4 (ferret) copies per cell of the transporter (Haas and Forbush, 1986; Mercer and Hoffman, 1985). The label binds to a 150-kDa peptide in RBC membranes, which thus may be involved in $J_{Na,K}$ (Cl).

4.5.2.4. Na^+/Li^+ (Na^+/Na^+) Exchange. Observations made in patients undergoing lithium treatment of affective disorders have led to the establishment, in the RBC membrane, of an ouabain-resistant transporter mediating a 1 : 1 coupled exchange of Li^+ against Na^+, or Na^+ for Na^+ (see Haas *et al.*, 1975; Duhm *et al.*, 1976; Duhm and Becker, 1978, 1979; Pandey *et al.*, 1978; Sarkadi *et al.*, 1978, for details). The transporter, which is highly selective (25 : 1) for Li^+ over Na^+, has no established physiological function, since it does not accept other monovalent cations. It is expressed in RBC with considerable interindividual and interspecific differences concerning capacity but not affinity (Duhm and Becker, 1979; Hannaert and Garay, 1986), operates independently of ATP by a consecutive (Ping-Pong) exchange mechanism, and may exhibit asymmetry (Hannaert and Garay, 1986) in the conformational changes occurring upon cation binding (Becker and Duhm, 1979).

Inhibitors of other types of cation movements in RBC (bumetanide, amiloride, chlorobutanol, oligomycin, dipyridamole, harmine) do not affect this transporter (Duhm and Becker, 1977, 1979; Pandey *et al.*, 1978; Morgan *et al.*, 1989). Data for ethacrynic acid, furosemide, quinine, and quinidine are conflicting (Pandey *et al.*, 1978; Duhm and Becker, 1979), but there is reason to regard them as ineffective. Phloretin and some of its congeners (Duhm and Becker, 1978) inhibit to a species-dependent extent [I_{50} 25 μM, I_{max} 100% in human RBC (Pandey *et al.*, 1978)]. The mechanistic details are still obscure. Stimulation of the transporter (increase of V_{max}) by internal Ca^{2+} has been reported (Agam *et al.*, 1987). This stimulation was suppressed by calmidazolium (40% at 5 μM) possibly indicative of a role for calmodulin. A 2.5-kDa peptide from leukemic promyelocytes, "inhibitin" (Morgan and Mir, 1984), has been shown to be a very potent (K_I 0.2 μM) and highly selective, strictly noncompetitive inhibitor of this transporter, which may facilitate its further analysis (Morgan *et al.*, 1989).

4.5.2.5. Na^+/H^+ Exchange. A Na^+/H^+ exchanger, present in many types of cells, also occurs in nucleated RBC from amphibia and fish (Siebens and Kregenow, 1985; Kregenow *et al.*, 1985; Borgese *et al.*, 1986; Mahé *et al.*, 1985) as well as in mammalian RBC (rabbit, dog, man) (Jennings *et al.*, 1986; Escobales and Rivera, 1987;

Parker, 1986; Escobales and Canessa, 1986; Semplicini *et al.*, 1989). In fish (trout) it is activated by catecholamines in a cAMP-involving signal transmission chain leading to cell swelling caused by the uptake of Na_e^+ (in exchange for H_i^+), followed by a secondary uptake of Cl^- (via the anion exchanger) and water. In amphibia (*Amphiuma*) the system is activated by osmotic cell shrinking and thus provides for regulatory volume increase. In mammalian RBC the ouabain- and bumetanide-resistant Na^+/H^+ antiporter becomes operative after an increase of the chemical gradients for its substrates (H_i^+, Na_e^+). It is also activated by cell shrinking (dog, rabbit) or by elevated Ca_i^{2+} (human), H_i^+ or Li_i^+ (rabbit, dog), while increasing H_e^+ is inhibitory, indicative of an external H^+-binding modifier site.

Evidence for the presence of this antiporter also stems from its sensitivity to the pyrazine carboxamides, amiloride and derivatives. Gradient-induced and Ca_i^{2+}-stimulated Na^+/H^+ exchange is partly (30–60%) inhibited by amiloride ($I_{50} \approx 17$ μM) (Escobales and Canessa, 1985; Semplicini *et al.*, 1989) and its more potent analogue, dimethylamiloride ($I_{50} \approx 0.9$ μM) (Escobales and Rivera, 1987). Shrinking-activated Na^+/H^+ exchange is fully suppressed in amphibian ($I_{50} \approx 1$–3 μM), rabbit ($I_{50} \approx 50$ μM), and dog RBC (Siebens and Kregenow, 1985; Jennings *et al.*, 1986; Parker, 1986). Inhibition is probably competitive. Catecholamine stimulation of the antiporter in fish is sensitive to drugs interfering with the formation of cAMP (Mahé *et al.*, 1985). Amiloride also acts in this way on Na^+/H^+ exchange ($I_{50} \approx 6$ μM) in addition to its direct but less pronounced ($I_{50} \approx 60$ μM) effect on the antiporter. Quinidine inhibits the antiporter in dog RBC ($I_{50} = 10$ μM) (Parker, 1983b) while drugs affecting other carrier-mediated cation movements (bumetanide, phloretin, DIDS, trifluoperazine) are ineffective (Escobales and Canessa, 1985).

4.5.2.6. Ca_e^{2+}/Na_i^+ Exchange. In some mammalian RBC (dog, ferret) lacking an efficient Na^+/K^+ pump and therefore having a low or no transmembrane gradient for K^+ and Na^+, volume regulation involves a Ca^{2+}/Na^+ exchange mechanism that upon osmotic swelling initiates a regulatory volume decrease by driving Na^+ (+ Cl^- and water) outward in dissipating the inward Ca^{2+} gradient (Parker, 1978, 1983a; Milanick, 1989). Inhibition of this exchanger has been demonstrated so far only for quinidine ($I_{50} \approx$ 20 μM) (Parker, 1978, 1983a), while a cardiotonic dipyridine derivative, amrinone, is stimulatory at low millimolar concentrations (Parker and Harper, 1980).

4.5.2.7. Na^+/Mg^{2+} Exchange. In analogy to other membrane systems, a transporter exchanging Na^+ (or Mn^{2+}) for Mg^{2+} is present in chicken, human, and rat RBC (Günther and Vormann, 1985; Féray and Garay, 1986, 1987; Lüdi and Schatzmann, 1987). It is not clear whether it operates as a normal antiporter (at a stoichiometry of 3 Na_e^+ for 1 Mg_i^{2+}) requiring ATP for unknown reasons (Féray and Garay, 1986, 1987) or as a Mg^{2+} extrusion pump driven by metabolic energy directly, and transporting Na_e^+ inward passively (Lüdi and Schatzmann, 1987). The system is insensitive to ouabain, bumetanide, and DIDS. Phloretin ($I_{50} \approx 150$ μM), amiloride ($K_I \approx 400$ μM), and quinidine ($I_{50} \approx 50$ μM) are inhibitory (Féray and Garay, 1986, 1987; Lüdi and Schatzmann, 1987). Féray and Garay (1988) have recently shown some tricyclic antidepressants to be rather selective inhibitors. Imipramine, carpipramine, and dothiepine are most effective (I_{50} 25–40 μM). Nontricyclic antidepressants and neuroleptics (e.g., chlorpromazine, pimozide) are almost ineffective.

4.6. Calcium-Related Transport Systems

4.6.1. Passive Calcium Influx

RBC fully deprived of calcium pump activity (by inhibition) exhibit a considerable passive Ca^{2+} uptake (10–50 μmol/liter per hour at 37°C) mediated by a saturable system (K_m 0.8 mM) (McNamara and Wiley, 1986). The system may comprise several components, a basal one and additional ones operationally defined by sensitivity to K_e^+, membrane potential, and cell volume (Engelmann and Duhm, 1989). These additional pathways are inhibited to varying extents by Ca^{2+} entry blockers (verapamil, diltiazem, nifedipine, nitrendipine, flunarizine) (Varecka and Carafoli, 1982; Neyses *et al.*, 1985; Engelmann and Duhm, 1989).

In addition, quinidine inhibits indirectly by preventing the hyperpolarizing increase of g_K (Ca) following Ca^{2+} influx (Varecka and Carafoli, 1982). Propranolol (1 mM) stimulates passive Ca^{2+} entry, by enhancing the hyperpolarization via g_K (Ca) (see Section 4.6.2). Salicylate stimulates Ca^{2+} (and Mg^{2+}) uptake by acting as an ionophore, although much less effective than A23187 (Bürgin and Schatzmann, 1979).

4.6.2. Calcium-Activated K^+ Channel

4.6.2.1. General Properties. Accumulation of Ca^{2+} (1–5 μM) in RBC induces a rapid and highly selective efflux of K^+ [g_K (Ca), P_K (Ca)]. This effect is normally prevented by the continuous outward pumping of Ca^{2+}. Ca^{2+} accumulates when pumping is inhibited or Ca^{2+} influx is enhanced by ionophore (e.g., A23187). This K^+ efflux (for reviews see Lew and Ferreira, 1978; Schwarz and Passow, 1983), which can also be elicited by Pb^{2+}, is activated by K_e^+ (2 mM) and modulated by Na_i^+ and H_i^+ (Stampe and Vestergaard-Bogind, 1985). Bulk electrophysiological and single-cell patch-clamp analyses have revealed the underlying pathway as a channel with a conductance of about 20 pS, present in human RBC at 15–200 copies/cell (Freedman and Novak, 1983; Grygorczyk and Schwarz, 1985; Vestergaard-Bogind *et al.*, 1985; Alvarez and Garcia-Sancho, 1987). It is likely to be similar to Ca^{2+}-activated K^+ channels (SK) in other cells (Latorre *et al.*, 1989). Channel opening produces hyperpolarization of the cell but also salt loss and cell shrinkage in normal extracellular environment. The function of the channel for RBC is still a matter of conjecture.

4.6.2.2. Activators. Propranolol and chemically related β-adrenergic agonists stimulate g_K (Ca) (ED_{50} 2–10 μM) by sensitizing the channel protein for gating by Ca^{2+} or Pb^{2+} (Schwarz *et al.*, 1989). In addition, displacement of Ca^{2+} from membrane stores may contribute (Porzig, 1975). At higher levels, the drugs inhibit P_K (Ca) by reducing the probability of channel opening (Schwarz *et al.*, 1989). Activation of P_K (Ca) has also been reported for alkanols and general anesthetics (Yamamoto and Harris, 1983a; Caldwell and Harris, 1985).

Channel gating is somehow related to the metabolic and redox state of the cell as indicated by stimulating effects of electron donor systems present externally (phenazine methosulfate + ascorbate) or internally (NADH, NADPH, GSH) (Garcia-Sancho *et al.*, 1979). This stimulation is based on an enhanced open probability of the channel (Fuhrmann *et al.*, 1985) and may involve a membrane-bound oxidoreductase (see below). A role for redox equilibria is also indicated by stimulating effects of menadione and vanadate on channel gating (Fuhrmann *et al.*, 1985).

4.6.2.3. Inhibitors. Inhibition of P_K (Ca) occurs at various levels. Besides blockers of normal Ca^{2+} influx (see above), K^+-channel blockers such as TEA, quinine, and quinidine [I_{50} 5–100 μM, depending on K_e^+ and membrane potential (Reichstein and Rothstein, 1981; Simons, 1985)] are inhibitory (see Grygorczyk and Schwarz, 1985, for an electrophysiological analysis). K^+-channel blockage may also underlie the strong inhibitory effects of potential-sensitive carbocyanine dyes (Simons, 1985) and of venoms from scorpion (Castle and Strong, 1986) or snakes (Alvarez and Garcia-Sancho, 1989), which act from outside at nanomolar concentrations. Apamin, on the other hand, an established blocker of K^+ channels in various tissues (Latorre *et al.*, 1989), does not inhibit P_K (Ca).

Cationic drugs (phenothiazines, butyrophenones, cetiedil) known as "calmodulin antagonists" (see Section 4.6.3.2), also inhibit P_K (Ca) at the low concentrations required for binding to calmodulin (Hoffman *et al.*, 1980; Lackington and Orrego, 1981; Berkowitz and Orringer, 1984). This may indicate control of the channel by calmodulin in addition to its activation by a recently described 23-kDa cytoplasmic protein, calpromotin (Plishker *et al.*, 1988). Intramembrane levels of the apolar calmodulin inhibitors, on the other hand, are very high (about 10 mol% of the membrane lipids) even at the low aqueous concentrations required for their effect on calmodulin (Luxnat and Galla, 1986; Welti *et al.*, 1984). Therefore, direct effects of these drugs on the membrane-intercalated channel protein also have to be considered, as in the case of the Ca^{2+}-pumping ATPase (see Section 4.6.3.2). Similar considerations hold for the inhibitory effect of antihistamines (Gardos and Szasz, 1968), alkanols C_{2-7} (Szasz *et al.*, 1978), and anticonvulsants (Caldwell and Harris, 1985).

Surprisingly, inhibitors of Na^+/K^+-ATPase [ouabain (I_{50} = 9 nM) and oligomycin ($I_{50} \approx$ 2 μg/ml)] inhibit P_K (Ca) (Hoffman *et al.*, 1980). How the Na^+/K^+ pump is related to the Ca^{2+}-activated K^+ channel is not clear. A similar uncertainty concerns the basis of the inhibitory effects on P_K (Ca) of compounds known to act as inhibitors of membrane-bound oxidoreductase [atebrin (= mepacrine), chlorpromazine, probucol] (Garcia-Sancho *et al.*, 1979; Howland *et al.*, 1984). In spite of striking parallelisms of the two inhibitions (Fuhrmann *et al.*, 1985; Fehlau *et al.*, 1989), an indirect nature of the blockage of the K^+ channel via inhibition of the oxidoreductase still remains to be proven.

4.6.3. The Calcium-Pumping ATPase

4.6.3.1. General Properties. Cytoplasmic free Ca^{2+} in RBC is kept at its normal low level (≈ 20 nM; Lew *et al.*, 1982) by the activity of a calcium pump fueled by ATP. This pump (400–2000 copies/cell; Agre *et al.*, 1983) consists of a 140-kDa protein that catalyzes outward transport, presumably in a Ca^{2+} : ATP stoichiometry of 1 : 1, according to the E1–E2 type of operation (for reviews see Schatzmann, 1983; Vincenzi and Hinds, 1988; Jorgensen and Andersen, 1988). Its transport capacity and substrate affinity are considerably enhanced (Volpi *et al.*, 1981) by the ubiquitous cytoplasmic Ca^{2+}-activated protein calmodulin (M_r = 16.7 kDa) present in RBC at about 2.5 μM. At Ca_i^{2+} levels exceeding 10 nM (Agre *et al.*, 1983), calmodulin (CaM) assumes a conformation that activates the pump with a Ca^{2+}-dependent affinity (Scharff and Foder, 1982), thus contributing to the feedback control of the low cytoplasmic Ca^{2+} levels. This activation, which can in part be mimicked by anionic amphiphiles or mild tryptic digestion of the pump (see Gietzen *et al.*, 1982a, and Vincenzi, 1982, for references), goes along with an increase of the V_{max} and a lowering of the K_m for Ca^{2+} (Volpi *et al.*, 1981).

4.6.3.2. Calmodulin-Related Inhibitors. Suppression of CaM binding to the pump is probably the predominant basis of action of most of the hitherto described inhibitors of the Ca^{2+} pump or its ATPase function, usually studied in inside-out vesicles (transport) or broken membrane preparations (ATPase). For a recent review, see Vincenzi and Hinds (1988). These inhibitors, which bind to CaM and thereby block its activating interaction with the pump protein (see also Levin and Weiss, 1979; Vogel, 1988), can be ranked according to their inhibitor constants (I_{50}):

1. Very-high-affinity inhibitors ($I_{50} < 1$ μM): Calmidazolium, a disubstituted imidazole derived from an antimycotic, and compound 48/80, a mixture of hydrophobic polycations ($M_r \approx 1300$ Da), long known as a histamine liberator, are the most potent CaM-antagonistic inhibitors of the calcium pump presently known (Gietzen *et al.*, 1981, 1983; DiJulio *et al.*, 1989). Compound 48/80, being impermeable, is likely to be ineffective in native cells, in contrast to calmidazolium
2. High-affinity inhibitors ($I_{50} < 10$ μM): Penfluridol and pimozide, drugs known for their antipsychotic activity (Gietzen *et al.*, 1980).
3. Intermediate-affinity inhibitors ($I_{50} < 100$ μM): Phenothiazines (e.g., trifluoperazine, chlorpromazine; Gietzen *et al.*, 1980), naphthalene sulfonamides (Kobayashi *et al.*, 1979), microtubule inhibitory vinca alkaloids (e.g., vinblastine, vincristine) (Gietzen *et al.*, 1982b), and the vasodilators cetiedil and bepridil (Agre *et al.*, 1984).
4. Low-affinity inhibitors ($I_{50} > 100$ μM): Local anesthetics (e.g., dibucaine, tetracaine), atebrin, propranolol (Volpi *et al.*, 1981).

4.6.3.3. "Direct" Inhibitors. Besides their CaM-related effects, at least a number of these cationic inhibitors also exert direct effects on the pump function, as indicated by studies of the basal pump rate in the absence of CaM and on the pump activated by trypsin or anionic amphiphiles (Vincenzi, 1982; Gietzen *et al.*, 1982a; DiJulio *et al.*, 1989).

These findings agree with the observation of CaM-independent inhibitors of the Ca^{2+} pump. Besides potent inorganic inhibitors such as La^{3+} (acting as a Ca^{2+} analogue, $I_{100} < 1$ μM) (Sarkadi *et al.*, 1980), ruthenium red (I_{50} 4–6 μM) (Hinds *et al.*, 1981), and vanadate (acting as a phosphate analogue) (Rossi *et al.*, 1981), the flavonoid aglycones quercetin (I_{50} 4–6 μM) and phloretin ($I_{100} < 100$ μM) produce noncompetitive inhibition in inside-out vesicles, possibly by preventing the conformation change of the phosphorylated pump protein ($E_1 \sim P/E_2$–P) responsible for the translocation step (Wüthrich and Schatzmann, 1980). Suramine, an aromatic hexasulfonate ($I_{50} \approx 50$ μM) (Schatzmann, 1983), may interfere with ATP binding.

Stimulation of the Ca^{2+} pump has been reported for ethanol and pentobarbital (Yamamoto and Harris, 1983b). These effects may be related to the activation of the pump by acidic amphiphiles (phospholipids, fatty acids), which seem to mimic the effect of CaM (see Gietzen *et al.*, 1982a, for references).

ACKNOWLEDGMENTS. Work from the authors' laboratories quoted in this review was supported by the Deutsche Forschungsgemeinschaft. We are deeply indebted to Charlotte Wigge for her invaluable help.

5. REFERENCES

Agam, G., Hatzav, P., Abekasis, S., and Livne, A., 1987, Elevated intracellular Ca^{2+} affects Li_i–Na_o countertransport in human red blood cells, *Biochim. Biophys. Acta* **904:**207–215.

Agre, P., Gardner, K., and Bennett, V., 1983, Association between human erythrocyte calmodulin and the cytoplasmic surface of human erythrocyte membranes, *J. Biol. Chem.* **258:**6258–6265.

Agre, P., Virshup, D., and Bennett, V., 1984, Bepridil and cetiedil. Vasodilators which inhibit Ca^{2+}-dependent calmodulin interactions with erythrocyte membranes, *J. Clin. Invest.* **74:**812–820.

Alhanaty, E., and Sheetz, M. P., 1984, Cell membrane shape control—Effects of chloromethyl ketone peptides, *Blood* **63:**1203–1208.

Allan, D., Hagelberg, C., Kallen, K.-J., and Haest, C. W. M., 1989, Echinocytosis and microvesiculation of human erythrocytes induced by insertion of merocyanine 540 into the outer membrane leaflet, *Biochim. Biophys. Acta* **986:**115–122.

Alvarez, J., and Garcia-Sancho, J., 1987, An estimate of the number of Ca^{2+}-dependent K^+ channels in the human red cell, *Biochim. Biophys. Acta* **903:**543–546.

Alvarez, J., and Garcia-Sancho, J., 1989, Inhibition of red cell Ca^{2+}-dependent K^+ channels by snake venoms, *Biochim. Biophys. Acta* **980:**134–138.

Andersen, N. B., 1968, The effect of local anesthetic and pH on sodium and potassium flux in human red cells, *J. Pharmacol. Exp. Ther.* **163:**393–406.

Andersen, O. S., Finkelstein, A., Katz, I., and Cass, A., 1976, Effect of phloretin on the permeability of thin lipid membranes, *J. Gen. Physiol.* **67:**749–771.

Anner, B. M., 1985, The receptor function of the Na^+, K^+-activated adenosine triphosphatase system, *Biochem. J.* **227:**1–11.

Antunes-Madeira, M. C., and Madeira, V. M. C., 1984, Partition of parathion in synthetic and native membranes, *Biochim. Biophys. Acta* **778:**49–56.

Antunes-Madeira, M. C., and Madeira, V. M. C., 1985, Partition of lindane in synthetic and native membranes, *Biochim. Biophys. Acta* **820:**165–172.

Antunes-Madeira, M. C., and Madeira, V. M. C., 1989, Membrane fluidity as affected by the insecticide lindane, *Biochim. Biophys. Acta* **982:**161–166.

Artmann, G., 1986, A microscopic photometric method for measuring erythrocyte deformability, *Clin. Hemorheol.* **6:**617–627.

Askari, A., and Rao, S. N., 1970, Drugs affecting sodium transport in human erythrocyte ghosts, *J. Pharmacol. Exp. Ther.* **172:**211–223.

Auger, M., Jarrell, H. C., and Smith, I. C. P., 1988, Interactions of the local anesthetic tetracaine with membranes containing phosphatidylcholine and cholesterol: A ^2H NMR study, *Biochemistry* **27:**4660–4667.

Barchfeld, G. L., and Deamer, D. W., 1985, The effect of general anesthetics on the proton and potassium permeabilities of liposomes, *Biochim. Biophys. Acta* **819:**161–169.

Barthel, D., Zschoernig, O., Lange, K., Lenk, R., and Arnold, K., 1988, Interaction of electrically charged drug molecules with phospholipid membranes, *Biochim. Biophys. Acta* **945:**361–366.

Basketter, D. A., and Widdas, W. F., 1978, Asymmetry of the hexose transfer system in human erythrocytes. Comparison of the effects of cytochalasin B, phloretin and maltose as competitive inhibitors, *J. Physiol. (London)* **278:**389–401.

Beaugé, L. A., Cavieres, J. J., Glynn, I. M., and Grantham, J. J., 1980, The effects of vanadate on the fluxes of sodium and potassium ions through the sodium pump, *J. Physiol. (London)* **301:**7–23.

Beck, J. S., 1978, Relations between membrane monolayers in some red cell shape transformations, *J. Theor. Biol.* **75**:487–501.

Becker, B. F., and Duhm, J., 1978, Evidence for anionic cation transport of lithium, sodium and potassium across the human erythrocyte membrane induced by divalent anions, *J. Physiol. (London)* **282**:149–168.

Becker, B. F., and Duhm, J., 1979, Studies on lithium transport across the red cell membrane. VI. Properties of a sulfhydryl group involved in ouabain-resistant Na^+–Li^+ (and Na^+–Na^+) exchange in human and bovine erythrocytes, *J. Membr. Biol.* **51**:287–310.

Belz, G. G., 1981, Rubidium uptake in erythrocytes, in: *Cardiac Glycosides* (K. Greeff, ed.), pp. 95–113, Springer-Verlag, Berlin.

Belz, G. G., and Heinz, N., 1977, The influence of polar and nonpolar digoxin and digitoxin metabolites on the ^{86}Rb-uptake of human erythrocytes and the contractility of guinea pig papillary muscles, *Arzneim. Forsch.* **27**:653–655.

Benga, G., Ionescu, M., Popescu, O., and Pop, V. I., 1983, Effect of chlorpromazine on proteins in human erythrocyte membranes as inferred from spin labeling and biochemical analysis, *Mol. Pharmacol.* **23**:771–778.

Bennett, V., 1985, The membrane skeleton of human erythrocytes and its implications for more complex cells, *Annu. Rev. Biochem.* **54**:273–285.

Bennett, V., 1989, The spectrin–actin junction of erythrocyte membrane skeletons, *Biochim. Biophys. Acta* **988**:107–121.

Bergmann, W. L., Dressler, V., Haest, C. W. M., and Deuticke, B., 1984, Reorientation rates and asymmetry of distribution of lysophospholipids between the inner and outer leaflet of the erythrocyte membrane, *Biochim. Biophys. Acta* **772**:328–336.

Berkowitz, L. R., and Orringer, E. P., 1984, An analysis of the mechanism by which cetiedil inhibits the Gardos phenomenon, *Am. J. Hematol.* **17**:217–223.

Berkowitz, L. R., and Orringer, E. P., 1987, Cell volume regulation in hemoglobin CC and AA erythrocytes, *Am. J. Physiol.* **252**:C300–C306.

Bessis, P. M., 1977, Erythrocyte form and deformability for normal blood and some hereditary hemolytic anemias, *Nouv. Rev. Fr. Hematol.* **18**:75–94.

Biesendorfer, H., Felix, W., and Wildenauer, D. B., 1981, Studies on the haemolytic action of amphiphilic substances in vitro, *Biochem. Pharmacol.* **30**:2287–2292.

Bitbol, M., Dempsey, C., Watts, A., and Devaux, P. F., 1989, Weak interaction of spectrin with phosphatidylcholine–phosphatidylserine multilayers. A ^2H and ^{31}P NMR study, *FEBS Lett.* **244**:217–222.

Blaiklock, R. G., and Green, J. W., 1971, Nonphotooxidative eosin Y inhibition of Na^+-K^+ ATPase activity in the human erythrocyte membrane, *Arch. Biochem. Biophys.* **145**:43–49.

Bond, G. H., and Hudgins, P. M., 1976, Inhibition of ATPase activity in human red cell membranes by tetracaine, *Biochem. Pharmacol.* **25**:267–270.

Bond, G. H., and Hudgins, P. M., 1977, Irreversible inactivation of human red cell ATPase activity by tetracaine, *Biochem. Pharmacol.* **26**:2241–2245.

Bondy, B., and Remien, J., 1981, Differential binding of chlorpromazine to human blood cells: Application of the hygroscopic desorption method, *Life Sci.* **28**:441–449.

Borgese, F., Garcia-Romeu, F., and Motais, R., 1986, Catecholamine-induced transport systems in trout erythrocyte. Na^+/H^+ countertransport or NaCl cotransport? *J. Gen. Physiol.* **87**:551–566.

Borgese, F., Garcia-Romeu, F., and Motais, R., 1987, Control of cell volume and ion transport by β-adrenergic catecholamines in erythrocytes of rainbow trout, Salmo gairdneri, *J. Physiol. (London)* **382**:123–144.

Boulanger, Y., Schreier, S., and Smith, I. C. P., 1981, Molecular details of anesthetic–lipid interaction as seen by deuterium and phosphorus-31 nuclear magnetic resonance, *Biochemistry* **20**:6824–6830.

Brahm, J., 1982, Diffusional water permeability of human erythrocytes and their ghosts, *J. Gen. Physiol.* **79**:791–819.

Brahm, J., 1983, Urea permeability of human red cells, *J. Gen. Physiol.* **82**:1–23.

Brahm, J., and Galey, W. R., 1987, Diffusional solute flux during osmotic water flow across the human red cell membrane, *J. Gen. Physiol.* **89**:703–716.

Brasseur, R., Chatelain, P., Goormaghtigh, E., and Ruysschaert, J.-M., 1985, A semi-empirical conformational analysis of the interaction of n-alkanols with dipalmitoylphosphatidylcholine, *Biochim. Biophys. Acta* **814**:227–236.

Brazy, P. C., and Gunn, R. B., 1976, Furosemide inhibition of chloride transport in human red blood cells, *J. Gen. Physiol.* **68:**583–599.

Bree, F., Gault, I., D'Athis, P., and Tillement, J. P., 1984, Beta adrenoceptors of human red blood cells, determination of their subtypes, *Biochem. Pharmacol.* **33:**4045–4050.

Broekhuyse, R. M., 1974, Improved lipid extraction of erythrocytes, *Clin. Chim. Acta* **51:**341–343.

Brown, L., and Erdmann, E., 1984, Comparison of the affinity of human, beef and cat heart (Na^+ + K^+)-ATPase for different digitalis derivatives, *Arzneim. Forsch.* **34:**1314–1318.

Brugnara, C., Van Ha, T., and Tosteson, D. C., 1989, Role of chloride in potassium transport through a K–Cl cotransport system in human red blood cells, *Am. J. Physiol.* **256:**C994–C1003.

Bull, B. S., 1986, Report: JCSH expert panel on blood rheology: Guidelines for measurement of blood viscosity and erythrocyte deformability, *Clin. Hemorheol.* **6:**439–453.

Bull, M. H., Brailsford, J. D., and Bull, B. S., 1982, Erythrocyte membrane expansion due to the volatile anesthetics, the 1-alkanols, and benzyl alcohol, *Anesthesiology* **57:**399–403.

Bürgin, H., and Schatzmann, H. J., 1979, The relation between net calcium, alkali cation and chloride movements in red cells exposed to salicylate, *J. Physiol. (London)* **287:**15–32.

Cabantchik, Z. I., Knauf, P. A., and Rothstein, A., 1978, The anion transport system of the red blood cell. The role of membrane protein evaluated by use of "probes," *Biochim. Biophys. Acta* **515:**239–302.

Cala, P. M., Norby, J. G., and Tosteson, D. C., 1982, Effects of the plant alkaloid sanguinarine on cation transport by human red blood cells and lipid bilayer membranes, *J. Membr. Biol.* **64:**23–31.

Caldwell, K. K., and Harris, R. A., 1985, Effects of anesthetic and anticonvulsant drugs on calcium-dependent efflux of potassium from human erythrocytes, *Eur. J. Pharmacol.* **107:**119–125.

Canessa, M., Spalvins, A., and Nagel, R. L., 1986, Volume-dependent and NEM-stimulated K^+, Cl^- transport is elevated in oxygenated SS, SC and CC human red cells, *FEBS Lett.* **200:**197–202.

Carlsen, A., and Wieth, J. O., 1976, Glycerol transport in human red cells, *Acta Physiol. Scand.* **97:**501–513.

Carraway, K. L., and Carothers Carraway, C. A., 1989, Membrane–cytoskeleton interactions in animal cells, *Biochim. Biophys. Acta* **988:**147–171.

Carruthers, A., 1988, The glucose transporter reconsidered, *Trends Biochem. Sci.* **13:**426–427.

Carruthers, A., and Melchior, D. L., 1988, Effect of lipid environment on membrane transport: The protein/lipid bilayer system, *Annu. Rev. Physiol.* **50:**257–271.

Castle, N. A., and Strong, P. N., 1986, Identification of two toxins from scorpion (*Leiurus quinquestriatus*) venom which block distinct classes of calcium-activated potassium channel, *FEBS Lett.* **209:**117–121.

Cevc, G., and Marsh, D., 1987, *Phospholipid Bilayers: Physical Principles and Models,* Wiley, New York.

Chabanel, A., Abbott, R. E., Chien, S., and Schachter, D., 1985, Effects of benzyl alcohol on erythrocyte shape, membrane hemileaflet fluidity and membrane viscoelasticity, *Biochim. Biophys. Acta* **816:**142–152.

Chasan, B., and Solomon, A. K., 1985, Urea reflection coefficient for the human red cell membrane, *Biochim. Biophys. Acta* **821:**56–62.

Chien, S., 1987, Red cell deformability and its relevance to blood flow, *Annu. Rev. Physiol.* **49:**177–192.

Chien, S., Sung, K.-L. P., Skalak, R., Usami, S., and Tözeren, A., 1978, Theoretical and experimental studies on viscoelastic properties of erythrocyte membrane, *Biophys. J.* **24:**463–487.

Chin, J. H., and Goldstein, D. B., 1977, Effects of low concentrations of ethanol on the fluidity of spin-labeled erythrocyte and brain membranes, *Mol. Pharmacol.* **13:**435–441.

Chipperfield, A. R., 1981, Chloride dependence of furosemide- and phloretin-sensitive passive sodium and potassium fluxes in human red cells, *J. Physiol. (London)* **312:**435–444.

Chipperfield, A. R., 1986, The (Na^+-K^+-Cl^-) cotransport system, *Clin. Sci.* **71:**465–476.

Clark, M. R., 1988, Senescence of red blood cells: Progress and problems, *Physiol. Rev.* **68:**503–554.

Classen, J., Haest, C. W. M., Tournois, H., and Deuticke, B., 1987, Gramicidin-induced enhancement of transbilayer reorientation of lipids in the erythrocyte membrane, *Biochemistry* **26:**6604–6612.

Coakley, W. T., and Deeley, J. O. T., 1980, Effects of ionic strength, serum protein and surface charge on membrane movements and vesicle production in heated erythrocytes, *Biochim. Biophys. Acta* **602:**355–375.

Cokelet, G. R., and Meiselman, H. J., 1968, Rheological comparison of hemoglobin solutions and erythrocyte suspensions, *Science* **162:**275–277.

Colley, C. M., and Metcalfe, J. C., 1972, The localisation of small molecules in lipid bilayers, *FEBS Lett.* **24:**241–246.

Conrad, M. J., and Singer, S. J., 1979, Evidence for a large internal pressure in biological membranes, *Proc. Natl. Acad. Sci. USA* **76:**5202–5206.

Conrad, M. J., and Singer, S. J., 1981, The solubility of amphipathic molecules in biological membranes and lipid bilayers and its implications for membrane structure, *Biochemistry* **20:**808–818.

Cousin, J. L., and Motais, R., 1976, The role of carbonic anhydrase inhibitors on anion permeability into ox red blood cells, *J. Physiol. (London)* **256:**61–80.

Cousin, J. L., and Motais, R., 1978, Effect of phloretin on chloride permeability: A structure–activity study, *Biochim. Biophys. Acta* **507:**531–538.

Cousin, J. L., and Motais, R., 1979, Inhibition of anion permeability by amphiphilic compounds in human red cells: Evidence for interactions of niflumic acid with the band 3 protein, *J. Membr. Biol.* **46:**125–153.

Cousin, J. L., and Motais, R., 1982a, Inhibition of anion transport in the red blood cells by anionic amphiphilic compounds. I. Determination of the flufenamate-binding site by proteolytic dissection of the band 3 protein, *Biochim. Biophys. Acta* **687:**147–155.

Cousin, J. L., and Motais, R., 1982b, Inhibition of anion transport in the red blood cell by anionic amphiphilic compounds. II. Chemical properties of the flufenamate-binding site on the band 3 protein, *Biochim. Biophys. Acta* **687:**156–164.

Cullis, P. R., and Verkleij, A. J., 1979, Modulation of membrane structure by Ca^{2+} and dibucaine as detected by ^{31}P NMR, *Biochim. Biophys. Acta* **552:**546–551.

Daleke, D. L., and Huestis, W. H., 1989, Erythrocyte morphology reflects the transbilayer distribution of incorporated phospholipids, *J. Cell Biol.* **108:**1375–1385.

Daveloose, D., Sablayrolles, M., Molle, D., and Leterrier, F., 1982, Interaction of ticlopidine with the erythrocyte membrane, *Biochem. Pharmacol.* **31:**3949–3954.

DeHoff, R. T., and Rhines, F. N., 1968, *Quantitative Microscopy*, McGraw–Hill, New York.

Demel, R. A., Geurts Van Kessel, W. S. M., Zwaal, R. F. A., Roelofsen, B., and van Deenen, L. L. M., 1975, Relation between various phospholipase actions on human red cell membranes and the interfacial phospholipid pressure in monolayers, *Biochim. Biophys. Acta* **406:**97–107.

Denker, B. M., Smith, B. L., Kuhajda, F. P., and Agre, P., 1988, Identification, purification, and partial characterization of a novel Mr 28,000 integral membrane protein from erythrocytes and renal tubules, *J. Biol. Chem.* **263:**15634–15642.

Deuticke, B., 1968, Transformation and restoration of biconcave shape of human erythrocytes induced by amphiphilic agents and changes of ionic environment, *Biochim. Biophys. Acta* **163:**494–500.

Deuticke, B., 1970, Anion permeability of the red blood cell, *Naturwissenschaften* **57:**172–179.

Deuticke, B., 1974, Acetate transfer across mammalian red cell membrane: Evidence for two different pathways, in: *Comparative Biochemistry and Physiology of Transport* (L. Bolis, K. Bloch, S. E. Luria, and F. Lynen, eds.), pp. 338–345, North-Holland, Amsterdam.

Deuticke, B., 1977, Properties and structural basis of simple diffusion pathways in the erythrocyte membrane, *Rev. Physiol. Biochem. Pharmacol.* **78:**1–97.

Deuticke, B., 1982, Monocarboxylate transport in erythrocytes, *J. Membr. Biol.* **70:**89–103.

Deuticke, B., 1989, Monocarboxylate transport in red blood cells: Kinetics and chemical modification, *Methods Enzymol.* **173:**300–329.

Deuticke, B., and Haest, C. W. M., 1987, Lipid modulation of transport proteins in vertebrate cell membranes, *Annu. Rev. Physiol.* **49:**221–235.

Deuticke, B., and Ruska, C., 1976, Changes of nonelectrolyte permeability in cholesterol-loaded erythrocytes, *Biochim. Biophys. Acta* **433:**638–653.

Deuticke, B., Lütkemeier, P., and Sistemich, M., 1984, Ion selectivity of aqueous leaks induced in the erythrocyte membrane by crosslinking of membrane proteins, *Biochim. Biophys. Acta* **775:**150–160.

Deuticke, B., Poser, B., Lütkemeier, P., and Haest, C. W. M., 1983, Formation of aqueous pores in the human erythrocyte membrane after oxidative cross-linking of spectrin by diamide, *Biochim. Biophys. Acta* **731:**196–210.

Deuticke, B., Henseleit, U., Haest, C. W. M., Heller, K. B., and Dubbelman, T. M. A. R., 1989, Enhancement of transbilayer mobility of a membrane lipid probe accompanies formation of membrane leaks during photodynamic treatment of erythrocytes, *Biochim. Biophys. Acta* **982:**53–61.

De Young, L. R., and Dill, K. A., 1988, Solute partitioning into lipid bilayer membranes, *Biochemistry* **27:**5281–5289.

Diamond, J. M., and Katz, Y., 1974, Interpretation of nonelectrolyte partition coefficients between dimyristoyl lecithin and water, *J. Membr. Biol.* **17**:121–154.

DiJulio, D., Hinds, T. R., and Vincenzi, F. F., 1989, Inhibition of basal and calmodulin-activated Ca^{2+}-pump ATPase by fractionated compound 48/80, *Biochim. Biophys. Acta* **981**:337–342.

Dintenfass, L., 1985, *Blood Viscosity, Hyperviscosity and Hyperviscosaemia,* MTP Press, Lancaster.

Dittrich, F., Berlin, P., Köpke, K., and Repke, K R. H., 1983, Stereoelectronic interaction between cardiotonic steroids and Na, K-ATPase: Molecular mechanism of digitalis action, *Curr. Top. Membr. Transp.* **19**:251–255.

Dodson, B. A., and Moss, J., 1984, Molecular mechanism of action of general anesthetics, *Mol. Cell. Biochem.* **64**:97–103.

Donath, E., and Voigt, A., 1986, Electrophoretic mobility of human erythrocytes. On the applicability of the charged layer model, *Biophys. J.* **499**:493–499.

Donath, E., Herrmann, A., Coakley, W. T., Groth, T., Egger, M., and Taeger, M., 1987, The influence of the antiviral drugs amantadine and rimantadine on erythrocyte and platelet membranes and its comparison with that of tetracaine, *Biochem. Pharmacol.* **36**:481–487.

Duggan, D. E., Bäer, J. E., and Noll, R. M., 1965, Membrane adenosinetriphosphatase and cation composition of mammalian erythrocytes, *Naturwissenschaften* **10**:264.

Duhm, J., 1987, Furosemide-sensitive K^+ (Rb^+) transport in human erythrocytes: Modes of operation, dependence on extracellular Na^+, kinetics, pH dependency and the effect of cell volume and N-ethylmaleimide, *J. Membr. Biol.* **98**:15–32.

Duhm, J., 1989, Na^+ and K^+ transport in human and rat erythrocytes: Features complicating the interpretation of data, in: *Salt and Hypertension* (R. Rettig, D. Ganten, and F. C. Luft, eds.), pp. 35–51, Springer-Verlag, Berlin.

Duhm, J., and Becker, B. F., 1977, Studies on the lithium transport across the red cell membrane. II. Characterization of ouabain-sensitive and ouabain-insensitive Li^+ transport. Effects of bicarbonate and dipyridamole, *Pfluegers Arch.* **367**:211–219.

Duhm, J., and Becker, B. F., 1978, Studies on Na^+-dependent Li^+ countertransport and bicarbonate-stimulated Li^+ transport in human erythrocytes, in: *Cell Membrane Receptors for Drugs and Hormones: A Multidisciplinary Approach* (R. W. Straub and L. Bolis, eds.), pp. 281–295, Raven Press, New York.

Duhm, J., and Becker, B. F., 1979, Studies on lithium transport across the red cell membrane. V. On the nature of the Na^+-dependent Li^+ countertransport system of mammalian erythrocytes, *J. Membr. Biol.* **51**:263–286.

Duhm, J., and Göbel, B. O., 1984, Role of furosemide-sensitive Na^+/K^+ transport system in determining the steady-state Na^+ and K^+ content and volume of human erythrocytes in vitro and in vivo, *J. Membr. Biol.* **77**:243–254.

Duhm, J., Eisenried, F., Becker, B. F., and Greil, W., 1976, Studies on lithium transport across the red cell membrane. I. Li^+ uphill transport by the Na^+-dependent Li^+ countertransport system of human erythrocytes, *J. Membr. Biol.* **364**:147–155.

Dunham, P. B., and Ellory, J. C., 1981, Passive potassium transport in low potassium sheep red cells: Dependence upon cell volume and chloride, *J. Physiol. (London)* **318**:511–530.

Eftink, M. R., Puri, R. K., and Ghahramani, M. D., 1985, Local anesthetic–phospholipid interactions. The pH dependence of the binding of dibucaine to dimyristoylphosphatidylcholine vesicles, *Biochim. Biophys. Acta* **813**:137–140.

Eisinger, J., and Flores, J., 1982, The relative locations of intramembrane fluorescent probes and of the cytosol hemoglobin in erythrocytes, studied by transverse resonance energy transfer, *Biophys. J.* **37**:6–7.

Ellens, H., Siegel, D. P., Alford, D., Yeagle, P. L., Boni, L., Lis, L. J., Quinn, P. J., and Bentz, J., 1989, Membrane fusion and inverted phases, *Biochemistry* **28**:3692–3703.

Ellory, J. C., and Stewart, G. W., 1982, The human erythrocyte Cl-dependent Na-K cotransport system as a possible model for studying the action of loop diuretics, *Br. J. Pharmacol.* **75**:183–188.

Engelmann, B., and Duhm, J., 1989, Distinction of two components of passive Ca^{2+} entry blockers, *Biochim. Biophys. Acta* **981**:36–42.

Ernst, E., 1987, Hemorheological treatment, in: *Clinical Hemorheology* (S. Chien, J. Dormandy, E. Ernst, and A. Matrai, eds.), pp. 229–373, Nijhoff, The Hague.

Escobales, N., and Canessa, M., 1985, Ca^{2+}-activated Na^+ fluxes in human red cells. Amiloride sensitivity, *J. Biol. Chem.* **260**:11914–11923.

Escobales, N., and Canessa, M., 1986, Amiloride-sensitive Na^+ transport in human red cells: Evidence for a Na/H exchange system, *J. Membr. Biol.* **90**:21–28.

Escobales, N., and Rivera, A., 1987, Na^+ for H^+ exchange in rabbit erythrocytes, *J. Cell. Physiol.* **132**:73–80.

Evans, E. A., and LaCelle, P. L., 1975, Intrinsic material properties of the erythrocyte membrane indicated by mechanical analysis of deformation, *Blood* **45**:29–43.

Evans, E. A., and Skalak, R., 1979a, Mechanics and thermodynamics of biomembranes: Part 1, *CRC Crit. Rev. Bioeng.* **3**:181–323.

Evans, E. A., and Skalak, R., 1979b, Mechanics and thermodynamics of biomembranes: Part 2, *CRC Crit. Rev. Bioeng.* **3**:331–418.

Evans, E., Mohandas, N., and Leung, A., 1984, Static and dynamic rigidities of normal and sickle erythrocytes, *J. Clin. Invest.* **73**:477–488.

Fannin, F. F., Evans, J. O., Gibbs, E. M., and Diedrich, D. F., 1981, Phloretinyl-3′-benzylazide: A high affinity probe for the sugar transporter in human erythrocytes. I. Hexose transport inhibition and photolabelling of mutarotase, *Biochim. Biophys. Acta* **649**:189–201.

Fehlau, R., Grygorczyk, R., Fuhrmann, G. F., and Schwarz, W., 1989, Modulation of the Ca^{2+}- or Pb^{2+}-activated K^+-selective channels in human red cells. II. Parallelisms to modulation of the activity of a membrane-bound oxidoreductase, *Biochim. Biophys. Acta* **978**:37–42.

Feinstein, M. B., Fernandez, S. M., and Sha'afi, R. I., 1975, Fluidity of natural membranes and phosphatidylserine and ganglioside dispersions. Effects of local anesthetics, cholesterol and protein, *Biochim. Biophys. Acta* **413**:354–370.

Feinstein, M. B., Volpi, M., Perrie, S., Makriyannis, A., and Sha'afi, R. I., 1977, Mechanisms of local anesthetic action on the permeability of erythrocytes, leukocytes, and liposomes containing the erythrocyte anion channel protein, *Mol. Pharmacol.* **13**:840–851.

Feo, C., and Mohandas, N, 1977, Clarification of the role of ATP in red cell morphology and function, *Nature* **265**:166.

Féray, J.-C., and Garay, R., 1986, An Na^+-stimulated Mg^{2+}-transport system in human red blood cells, *Biochim. Biophys. Acta* **856**:76–84.

Féray, J.-C., and Garay, R., 1987, A one-to-one Mg^{2+} : Mn^{2+} exchange in rat erythrocytes, *J. Biol. Chem.* **262**:5763–5768.

Féray, J.-C., and Garay, R., 1988, Demonstration of a Na^+ : Mg^{2+} exchange in human red cells by its sensitivity to tricyclic antidepressant drugs, *Naunyn-Schmiedeberg's Arch. Pharmacol.* **338**:332–337.

Ferrell, J. E., Jr., and Huestis, W. H., 1984, Phosphoinositide metabolism and the morphology of human erythrocytes, *J. Cell Biol.* **98**:1992–1998.

Ferrell, J. E., Lee, K.-J., and Huestis, W. M., 1985, Membrane bilayer balance and erythrocyte shape: A quantitative assessment, *Biochemistry* **24**:2849–2857.

Fischer, T., and Schmid-Schönbein, H., 1977, Tank tread motion of red cell membranes in viscometric flow: Behavior of intracellular and extracellular markers (with film), *Blood Cells* **3**:351–365

Forbush, B., III, 1983, Cardiotonic steroid binding to Na,K-ATPase, *Curr. Top. Membr. Transp.* **19**:167–201.

Ford, D. A., Sharp, J. A., and Rovetto, M. J., 1985, Erythrocyte adenosine transport: Effects of Ca^{2+} channel antagonists and ions, *Am. J. Physiol.* **248**:H593–H598.

Forman, S. A., Verkman, A. S., Dix, J. A., and Solomon, A. K., 1982, Interaction of phloretin with the anion transport protein of the red blood cell membrane, *Biochim. Biophys. Acta* **689**:531–538.

Forman, S. A., Verkman, A. S., Dix, J. A., and Solomon, A. K., 1985, n-Alkanols and halothane inhibit red cell anion transport and increase band 3 conformational change rate, *Biochemistry* **24**:4859–4866.

Forrest, B. J., and Rodham, D. K., 1985, An anesthetic-induced phosphatidylcholine hexagonal phase, *Biochim. Biophys. Acta* **814**:281–288.

Fortes, P. A. G., Ellory, J. C., and Lew, V. L., 1973, Suramin: A potent ATPase inhibitor which acts on the inside surface of the sodium pump, *Biochim. Biophys. Acta* **318**:262–272.

Freedman, J. C., and Novak, T. S., 1983, Membrane potentials associated with Ca-induced K conductance in human red blood cells: Studies with a fluorescent oxonol dye, WW 781, *J. Membr. Biol.* **72**:59–74.

Frezzatti, W. A., Jr., Toselli, W. R., and Schreier, S., 1986, Spin label study of local anesthetic–lipid membrane interactions. Phase separation of the uncharged form and bilayer micellization by the charged form of tetracaine, *Biochim. Biophys. Acta* **860**:531–538.

Fröhlich, O., 1984, How channel-like is a biological carrier? Studies with the erythrocyte anion transporter, *Biophys. J.* **45:**93–94.

Fröhlich, O., and Gunn, R. B., 1986, Erythrocyte anion transport: The kinetics of a single-site obligatory exchange system, *Biochim. Biophys. Acta* **864:**169–194.

Fröhlich, O., and Gunn, R. B., 1987, Interactions of inhibitors on anion transporter of human erythrocyte, *Am. J. Physiol.* **252:**C153–C162.

Fröhlich, O., and Mayer, S., 1989, Mechanism of phloretin inhibition of erythrocyte anion exchange, *Biophys. J.* **55:**145a.

Fuhrmann, G. F., Schwarz, W., Kersten, R., and Sdun, H., 1985, Effects of vanadate, menadione and menadione analogs on the Ca^{2+}-activated K^+ channels in human red cells. Possible relations to membrane-bound oxidoreductase activity, *Biochim. Biophys. Acta* **820:**223–234.

Fujii, T., Sato, T., Tamura, A., Wakatsuki, M., and Kanaho, Y., 1979, Shape changes of human erythrocytes induced by various amphipathic drugs acting on the membrane of the intact cells, *Biochem. Pharmacol.* **28:**613–620.

Fullerton, D. S., Kitatsuji, E., and Deffo, T., 1983, Use of prophet and MMS-X computer graphics in the study of the cardiac steroid receptor site of Na,K-ATPase, *Curr. Top. Membr. Transp.* **19:**257–264.

Fung, Y. C., Tsang, W. C. O., and Patitucci, P., 1981, High-resolution data on the geometry of red blood cells, *Biorheology* **18:**369–385.

Gaffney, B. J., Willingham, G. L., and Schepp, R. S., 1983, Synthesis and membrane interactions of spin-label bifunctional reagents, *Biochemistry* **22:**881–892.

Garay, R. P., Diez, J., Nazaret, C., Dagher, G., and Abitol, J. P., 1985, The interaction of canrenone with the Na^+,K^+ pump in human red blood cells, *Naunyn-Schmiedeberg's Arch. Pharmacol.* **329:**311–315.

Garay, R. P., Hannaert, P. A., Nazaret, C., and Cragoe, E. J., Jr., 1986, The significance of the relative effects of loop diuretics and anti-brain edema agents on the Na^+,K^+,Cl^- cotransport system and the $Cl^-/NaCO_3^-$ anion exchanger, *Naunyn-Schmiedeberg's Arch. Pharmacol.* **334:**202–209.

Garay, R. P., Nazaret, C., Hannaert, P. A., and Cragoe, E. J., Jr., 1988, Demonstration of a $[K^+,Cl^-]$-cotransport in human red cells by its sensitivity to [(dihydroindenyl)oxy]alkanoic acids: Regulation of cell swelling and distinction from the bumetanide-sensitive $[Na^+,K^+,Cl^-]$-cotransport system, *Mol. Pharmacol.* **33:**696–701.

Garcia-Sancho, J., Sanchez, A., and Herreros, B., 1979, Stimulation of monovalent cation fluxes by electron donors in the human red cell membrane, *Biochim. Biophys. Acta* **556:**118–130.

Gardos, G., and Szasz, I., 1968, The mechanism of ion transport in human erythrocytes. II. The role of histamine in the regulation of cation transport, *Acta Biochim. Biophys. Acad. Sci. Hung.* **3:**13–27.

Geck, P., and Heinz, E., 1986, The Na-K-2Cl cotransport system, *J. Membr. Biol.* **91:**97–105.

Gerlach, E., Deuticke, B., and Koss, F. W., 1965, Einfluβ von Pyrimidopyrimidin- und Pteridin-Derivaten auf Phosphat- und Adenosin-Permeabilität menschlicher Erythrozyten, *Arzneim. Forsch.* **15:**558–563.

Gietzen, K., Mansard, A., and Bader, H., 1980, Inhibition of human erythrocyte Ca^{2+}-transport ATPase by phenothiazines and butyrophenones, *Biochem. Biophys. Res. Commun.* **94:**674–681.

Gietzen, K., Wüthrich, A., and Bader, H., 1981, R 24571: A powerful inhibitor of red blood cell Ca^{++}-transport ATPase and of calmodulin-regulated functions, *Biochem. Biophys. Res. Commun.* **101:**418–425.

Gietzen, K., Sadorf, I., and Bader, H., 1982a, A model for the regulation of the calmodulin-dependent enzymes erythrocyte Ca^{2+}-transport ATPase and brain phosphodiesterase by activators and inhibitors, *Biochem. J.* **207:**541–548.

Gietzen, K., Wüthrich, A. and Bader, H., 1982b, Effects of microtubular inhibitors on plasma membrane calmodulin-dependent Ca^{2+}-transport ATPase, *Mol. Pharmacol.* **22:**413–420.

Gietzen, K., Adamczyk-Engelmann, P., Wüthrich, A., Konstantinova, A., and Bader, H., 1983, Compound 48/80 is a selective and powerful inhibitor of calmodulin-regulated functions, *Biochim. Biophys. Acta* **736:**109–118.

Giraud, F., Claret, M., Bruckdorfer, K. R., and Chailley, B., 1981, The effects of membrane lipid order and cholesterol on the internal and external cationic sites of the Na^+-K^+ pump in erythrocytes, *Biochim. Biophys. Acta* **647:**249–258.

Goldstein, D. B., 1984, The effects of drugs on membrane fluidity, *Annu. Rev. Pharmacol. Toxicol.* **24:**43–64.

Goodman, S. R., Krebs, K. E., Whitfield, C. F., Riederer, B. M., and Zagon, I. S., 1988, Spectrin and related molecules, *CRC Crit. Rev. Biochem.* **23:**171–234.

Grebe, R., and Schmid-Schönbein, H., 1985, Tangent counting for objective assessment of erythrocyte shape changes, *Biorheology* **22:**455–469.

Grebe, R., and Zuckermann, M., 1989, Erythrocyte shape simulation by numerical optimization, *Biorheology* **26:**505.

Grebe, R., Wolff, H., and Schmid-Schoenbein, H., 1988a, Influence of red cell surface charge on red cell membrane curvature, *Pfluegers Arch.* **413:**77–83.

Grebe, R., Zuckermann, M., and Schmid-Schönbein, H., 1988b, Erythrocyte shape is influenced by free electric and chemical energy, in: *Electromagnetic Fields and Biomembranes* (M. Markov and M. Blank, eds.), pp. 141–144, Plenum Press, New York.

Griffith, O. H., Dehlinger, P. J., and Van, S. P., 1974, Shape of the hydrophobic barrier of phospholipid bilayers (evidence for water penetration in biological membranes), *J. Membr. Biol.* **15:**159–192.

Gruen, D. W. R., and Haydon, D. A., 1981, A mean-field model of the alkane-saturated lipid bilayer above its phase transition, *Biophys. J.* **33:**167–188.

Grunze, M., Haest, C. W. M., and Deuticke, B., 1982, Lateral segregation of membrane lipids and formation of stable rod-shaped membrane projections in erythrocytes treated with long-chain alcohols, *Biochim. Biophys. Acta* **693:**237–245.

Grygorczyk, R., and Schwarz, W., 1985, Ca^{2+}-activated K^+ permeability in human erythrocytes: Modulation of single-channel events, *Eur. Biophys. J.* **12:**57–65.

Gunn, R. B., 1985, Bumetanide inhibition of anion exchange in human red blood cells, *Biophys. J.* **47:**326a.

Gunn, R. B., and Cooper, J. A., Jr., 1975, Effect of local anesthetics on chloride transport in erythrocytes, *J. Membr. Biol.* **25:**311–326.

Günther, I., and Vormann, J., 1985, Mg^{2+} efflux is accomplished by an amiloride sensitive Na^+/Mg^{2+} antiport, *Biochem. Biophys. Res. Commun.* **130:**540–545.

Gutknecht, J., and Tosteson, D. C., 1970, Ionic permeability of thin lipid membranes. Effects of n-alkyl alcohols, polyvalent cations, and a secondary amine, *J. Gen. Physiol.* **55:**359–374.

Haas, M., 1989, Properties and diversity of (Na–K–Cl) cotransporters, *Annu. Rev. Physiol.* **51:**443–457.

Haas, M., and Forbush, B., III, 1986, [^{3}H]bumetanide binding to duck red cells. Correlation with inhibition of (Na + K + 2Cl) cotransport, *J. Biol. Chem.* **261:**8434–8441.

Haas, M., and Harrison, J. H., Jr., 1989, Stimulation of K–Cl cotransport in rat red cells by a hemolytic anemia-producing metabolite of dapsone, *Am. J. Physiol.* **256:**C265–C272.

Haas, M., and McManus, T. J., 1983, Bumetanide inhibits (Na + K + 2Cl) cotransport at a chloride site, *Am. J. Physiol.* **245:**C235–C240.

Haas, M., Schooler, J., and Tosteson, D. C., 1975, Coupling of lithium to sodium transport in human red cells, *Nature* **258:**425–427.

Haest, C. W. M., 1982, Interactions between membrane skeleton proteins and the intrinsic domain of the erythrocyte membrane, *Biochim. Biophys. Acta* **694:**331–352.

Haest, C. W. M., Plasa, G., and Deuticke, B., 1981, Selective removal of lipids from the outer membrane layer of human erythrocytes without hemolysis, *Biochim. Biophys. Acta* **649:**701–708.

Hammond, J. R., Jarvis, S. M., Paterson, A. R. P., and Alexander, S., 1983, Benzodiazepine inhibition of nucleoside transport in human erythrocytes, *Biochem. Pharmacol.* **32:**1229–1235.

Hammond, J. R., Williams, E. F., and Clanachan, A. S., 1985, Affinity of calcium channel inhibitors, benzodiazepines, and other vasoactive compounds for the nucleoside transport system, *Can. J. Pharmacol.* **63:**1302–1307.

Hannaert, P. A., and Garay, R. P., 1986, A kinetic analysis of Na–Li countertransport in human red blood cells, *J. Gen. Physiol.* **87:**353–368.

Hanpft, R., and Mohr, K., 1985, Influence of cationic amphiphilic drugs on the phase-transition temperature of phospholipids with different polar headgroups, *Biochim. Biophys. Acta* **814:**156–162.

Haydon, D. A., and Elliott, J. R., 1986, Surface potential changes in lipid monolayers and the "cut-off" in anesthetic effects of N-alkanols, *Biochim. Biophys. Acta* **863:**337–340.

Heinrich, R., Gaestel, M., and Glaser, R., 1982, The electric potential profile across the erythrocyte membrane, *J. Theor. Biol.* **96:**211–231.

Heller, K. B., Jahn, B., and Deuticke, B., 1987, Peroxidative membrane damage in human erythrocytes induced by a concerted action of iodoacetate, vanadate and ferricyanide, *Biochim. Biophys. Acta* **901:**67–77.

Hershfield, R., and Richards, F. M., 1976, Reversible inhibition of glucose transport in human erythrocytes by a series of pyridine derivatives, *J. Biol. Chem.* **251:**5141–5148.

Hinds, T. R., Raess, B. U., and Vincenzi, F. F., 1981, Plasma membrane Ca^{2+} transport: Antagonism by several potential inhibitors, *J. Membr. Biol.* **58:**57–65.

Hochmuth, R. M., 1987, Properties of red blood cells, in: *Handbook of Bioengineering* (R. Skalak and S. Chien, eds.), pp. 12.1–12.17, McGraw–Hill, New York.
Hochmuth, R. M., and Waugh, R. E., 1987, Erythrocyte membrane elasticity and viscosity, *Annu. Rev. Physiol.* **49:**209–219.
Hochmuth, R. M., Buxbaum, K. L., and Evans, E. A., 1980, Temperature dependence of the viscoelastic recovery of red cell membrane, *J. Biophys.* **29:**177–182.
Hochmuth, R. M., Wiles, H. C., Evans, E. A., and McCown, J. T., 1982, Extensional flow of erythrocyte membrane from cell body to elastic tether. II. Experiment, *Biophys. J.* **39:**83–89.
Hoffman, E. K., and Simonsen, L. O., 1989, Membrane mechanisms in volume and pH regulation in vertebrate cells, *Physiol. Rev.* **69:**315–382.
Hoffman, J. F., 1986, Active transport of Na^+ and K^+ by red blood cells, in: *Membrane Transport Disorders,* 2nd ed. (T. Andreoli, J. F. Hoffman, S. G. Schultz, and D. D. Fanestil, eds.), pp. 221–234, Plenum Press, New York.
Hoffman, J. F., and Forbush, B., III (eds.), 1983, *Structure, Mechanism, and Function of the Na/K Pump, Curr. Top. Membr. Transp.* **19.**
Hoffman, J. F., Yingst, D. R. Goldinger, J. M., Blum, R. M., and Knauf, P. A., 1980, On the mechanism of Ca-dependent K transport in human red blood cells, in: *Membrane Transport in Erythrocytes, Alfred Benzon Symposium 14* (U. V. Lassen, H. H. Ussing, and J. O. Wieth, eds.), Munksgaard, Copenhagen.
Holmes, D. E., and Piette, L. H., 1970, Effects of phenothiazine derivatives on biological membranes: Drug-induced changes in electron spin resonance spectra from spin-labeled erythrocyte ghost membranes, *J. Pharmacol. Exp. Ther.* **173:**78–84.
Hope, M. J., and Cullis, P. R., 1981, The role of nonbilayer lipid structures in the fusion of human erythrocytes induced by lipid fusogens, *Biochim. Biophys. Acta* **640:**82–90.
Hornby, A. P., and Cullis, P. R., 1981, Influence of local and neutral anesthetics on the polymorphic phase preferences of egg yolk phosphatidylethanolamine, *Biochim. Biophys. Acta* **647:**285–292.
Howland, J. L., Daughtey, J. N., Donatelli, M., and Theofrastous, J. P., 1984, Inhibition of the erythrocyte calcium-sensitive potassium channel by probucol, *Pharmacol. Res. Commun.* **16:**1057–1064.
Isomaa, B., and Engblom, A. C., 1988, Is calmodulin inhibition involved in shape transformations induced by amphiphiles in erythrocytes? *Biochim. Biophys. Acta* **940:**121–126.
Isomaa, B., and Hägerstrand, H., 1988, Effects of nonionic amphiphiles at sublytic concentrations on the erythrocyte membrane, *Cell Biochem. Funct.* **6:**183–190.
Isomaa, B., Bergman, H., and Sandberg, P., 1979, The binding of CTAB, a cationic surfactant, to the rat erythrocyte membrane, *Acta Pharmacol. Toxicol.* **44:**36–42.
Isomaa, B., Hägerstrand, H., Paatero, G., and Englbom, A. C., 1986, Permeability alterations and anti-haemolysis induced by amphiphiles in human erythrocytes, *Biochim. Biophys. Acta* **860:**510–524.
Isomaa, B., Hägerstrand, H., and Paatero, G., 1987, Shape transformations induced by amphiphiles in erythrocytes, *Biochim. Biophys. Acta* **899:**93–103.
Jain, M. K., and Wu, N. M., 1977, Effect of small molecules on the dipalmitoyl lecithin liposomal bilayer. III. Phase transition in lipid bilayer, *J. Membr. Biol.* **34:**157–201.
Janoff, A. S., Mazorow, D. L., Coughlin, R. T., Bowdler, A. J., Haug, A., and McGroarty, E. J., 1981, The modification of human erythrocyte membrane structure by membrane stabilizers: An electron spin resonance study, *Am. J. Hematol.* **10:**171–179.
Jarvis, S. M., 1986, Nitrobenzylthioinosine-sensitive nucleoside transport system: Mechanism of inhibition by dipyridamole, *Mol. Pharmacol.* **30:**659–665.
Jarvis, S. M., 1987, Kinetics and molecular properties of nucleoside transporters in animal cells, in: *Topics and Perspectives in Adenosine Research* (E. Gerlach and B. F. Becker, eds.), pp. 102–117, Springer-Verlag, Berlin.
Jarvis, S. M., McBride, D., and Young, J. D., 1982, Erythrocyte nucleoside transport: Asymmetrical binding of nitrobenzylthioinosine to nucleoside permeation sites, *J. Physiol. (London)* **324:**31–46.
Jarvis, S. M., Janmohamed, S. N., and Young, J. D., 1983, Kinetics of nitrobenzylthioinosine binding to the human erythrocyte nucleoside transporter, *Biochem. J.* **216:**661–667.
Jay, A. W. L., 1975, Geometry of the human erythrocyte. I. Effect of albumin on cell geometry, *Biophys. J.* **15:**205–222.
Jennings, M. L., 1989, Structure and function of the red blood cell anion transport protein, *Annu. Rev. Biophys. Biophys. Chem.* **18:**397–430.

Jennings, M. L., and Adams-Lackey, M., 1982, A rabbit erythrocyte membrane protein associated with l-lactate transport, *J. Biol. Chem.* **257**:12866–12871.

Jennings, M. L., and Solomon, A. K., 1976, Interaction between phloretin and the red blood cell membrane, *J. Gen. Physiol.* **67**:381–397.

Jennings, M. L., Douglas, S. M., and McAndrew, P. E., 1986, Amiloride-sensitive sodium–hydrogen exchange in osmotically shrunken rabbit red blood cells, *Am. J. Physiol.* **251**:C32–C40.

Johnson, R. M., 1985, pH effects on red cell deformability, *Blood Cells* **11**:317–321.

Johnson, S. M., and Bangham, A. D., 1969, The action of anaesthetics on phospholipid membranes, *Biochim. Biophys. Acta* **193**:92–104.

Joiner, C. H., and Lauf, P. K., 1978, The correlation between ouabain binding and potassium pump inhibition in human and sheep erythrocytes, *J. Physiol. (London)* **283**:155–175.

Jones, O. T., and Lee, A. G., 1985, Interactions of hexachlorocyclohexanes with lipid bilayers, *Biochim. Biophys. Acta* **812**:731–739.

Jorgensen, P. L., and Andersen, J. P., 1988, Structural basis for E_1–E_2 conformational transitions in Na,K pump and Ca-pump proteins, *J. Membr. Biol.* **103**:95–120.

Jung, C. Y., and Mookerjee, B. K., 1976, Inhibitory effect of furosemide on glucose transport, *J. Lab. Clin. Med.* **87**:960–966.

Jung, C. Y., and Rampal, A. L., 1977, Cytochalasin B binding sites and glucose transport carrier in human erythrocyte ghosts, *J. Biol. Chem.* **252**:5456–5463.

Kahn, J. B., Jr., 1955, Effects of two erythrophleum alkaloids on potassium transfer in human erythrocytes, *Proc. Soc. Exp. Biol. Med.* **110**:412–414.

Kaji, D., 1986, Volume-sensitive K transport in human erythrocytes, *J. Gen. Physiol.* **88**:719–738.

Katz, Y., and Diamond, J. M., 1974, Thermodynamic constants for nonelectrolyte partition between dimyristoyl lecithin and water, *J. Membr. Biol.* **17**:101–120.

Kelleher, R. S., Murray, E. F., and Peterson, S. W., 1987, Insulin causes insulin-receptor internalization in human erythrocyte ghosts, *Biochem. J.* **241**:93–97.

Kim, H. D., Sergeant, S., Forte, L. R., Hwan Sohn, D., and Hyok, Im, J., 1989, Activation of a Cl-dependent K flux by cAMP in pig red cells, *Am. J. Physiol.* **256**:C772–C778.

Kirkpatrick, F. H., 1979, New models of cellular control: Membrane cytoskeletons, membrane curvature potential, and possible interactions, *Biosystems* **11**:93–109.

Kita, Y., and Miller, K. W., 1982, Partial molar volumes of some 1-alkanols in erythrocyte ghosts and lipid bilayers, *Biochemistry* **21**:2840–2847.

Kita, Y., Bennett, L. J., and Miller, K. W., 1981, The partial molar volumes of anesthetics in lipid bilayers, *Biochim. Biophys. Acta* **647**:130–139.

Kleinfeld, A. M., 1987, Current views of membrane structure, *Curr. Top. Membr. Transp.* **29**:1–27.

Knauf, P. A., 1979, Erythrocyte anion exchange and the band 3 protein: Transport kinetics and molecular structure, *Curr. Top. Membr. Transp.* **12**:249–363.

Knauf, P. A., 1986, Anion transport in erythrocytes, in: *Membrane Transport Disorders,* 2nd ed. (T. Andreoli, J. F. Hoffman, S. G. Schultz, and D. D. Fanestil, eds.), pp. 191–220, Plenum Press, New York.

Knauf, P. A., and Mann, N. A., 1984, Use of niflumic acid to determine the nature of the asymmetry of the human erythrocyte anion exchange system, *J. Gen. Physiol.* **83**:703–725.

Kobayashi, R., Tawata, M., and Hidaka, H., 1979, Ca^{2+} regulated modulator protein interacting agents: Inhibition of Ca^{2+}-Mg^{2+}-ATPase of human erythrocyte ghost, *Biochem. Biophys. Res. Commun.* **88**:1037–1045.

Koblin, D. D., and Wang, H. H., 1976, The effect of chemical modifiers on the interaction of a spin-labeled local anesthetic with human erythrocyte membranes, *Biochem. Pharmacol.* **25**:1405–1413.

Koblin, D. D., Pace, W. D., and Wang, H. H., 1975, The penetration of local anesthetics into the red blood cell membrane as studied by fluorescence quenching, *Arch. Biochem. Biophys.* **171**:176–182.

Korten, K., and Miller, K. W., 1979, Erythrocyte ghost–buffer partition coefficients of phenobarbital, pentobarbital, and thiopental support the pH-partition hypothesis, *Can. J. Physiol. Pharmacol.* **57**:325–328.

Korten, K., Sommer, T. J., and Miller, K. W., 1980, Membrane composition modulates thiopental partitioning in bilayers and biomembranes, *Biochim. Biophys. Acta* **599**:271–279.

Kotyk, A., Kolinska, J., Verse, K., and Szammer, J., 1965, Inhibition by phloretin and phlorizin derivatives of sugar transport in different cells, *Biochem. Z.* **342**:129–138.

Kregenow, F. M., Caryk, T., and Siebens, A. W., 1985, Further studies of the volume-regulatory response of

Amphiuma red cells in hypertonic media. Evidence for amiloride-sensitive Na/H exchange, *J. Gen. Physiol.* **86:**565–584.

Krupka, R. M., 1985, Asymmetrical binding of phloretin to the glucose transport system of human erythrocytes, *J. Membr. Biol.* **83:**71–80.

Krupka, R. M., and Devés, R., 1980a, Evidence for allosteric inhibition sites in the glucose carrier of erythrocytes, *Biochim. Biophys. Acta* **598:**127–133.

Krupka, R. M., and Devés, R., 1980b, Asymmetric binding of steroids to internal and external sites in the glucose carrier of erythrocytes, *Biochim. Biophys. Acta* **598:**134–144.

Krupka, R. M., and Devés, R., 1986, Looking for probes of gated channels: Studies of the inhibition of glucose and choline transport in erythrocytes, *Biochem. Cell Biol.* **64:**1099–1107.

Kuriki, Y., and Racker, E., 1976, Inhibition of (Na^+,K^+) adenosine triphosphatase and its partial reactions by quercetin, *Biochemistry* **15:**4951–4956.

Kutchai, H., Chandler, L. H., and Geddis, L. M., 1980, Effects of anesthetic alcohols on membrane transport processes in human erythrocytes, *Biochim. Biophys. Acta* **600:**870–881.

Lackington, I., and Orrego, F., 1981, Inhibition of calcium-activated potassium conductance of human erythrocytes by calmodulin inhibitory drugs, *FEBS Lett.* **133:**103–106.

Lacko, L., and Wittke, B., 1984, The affinities of benzodiazepines to the transport protein of glucose in human erythrocytes, *Drug Res.* **34:**403–407.

Lacko, L., Wittke, B., and Geck, P., 1974, Interaction of alcohols with the transport system of glucose in human erythrocytes, *J. Cell. Physiol.* **83:**267–274.

Lacko, L., Wittke, B., and Geck, P., 1975, Interaction of steroids with the transport system of glucose in human erythrocytes, *J. Cell. Physiol.* **86:**673–680.

Lacko, L., Wittke, B., and Lacko, I., 1977, Interaction of local anesthetics with the transport system of glucose in human erythrocytes, *J. Cell. Physiol.* **92:**257–264.

Lacko, L., Wittke, B., and Lacko, I., 1978, Inhibition of glucose transport in human erythrocytes by benzylalcohol, *J. Cell. Physiol.* **96:**199–202.

Lacko, L., Wittke, B., and Lacko, I., 1979, The effect of homologous local anesthetics of the 4-alkoxy- and 4-alkylamino-benzoic acid-diethylamino-ethylester-hydrochloride-series on the glucose transport in human erythrocytes, *J. Cell. Physiol.* **100:**169–174.

Lacko, L., Wittke, B., and Zimmer, G., 1981, Interaction of benzoic acid derivatives with the transport system of glucose in human erythrocytes, *Biochem. Pharmacol.* **30:**1425–1431.

Lande, W. M., Thiemann, P. V. W., and Mentzen, W. C., Jr., 1982, Missing band 7 membrane protein in two patients with high Na, low K erythrocytes, *J. Clin. Invest.* **70:**1273–1280.

Lang, R. D. A., Wickenden, C., Wynne, J., and Lucy, J. A., 1984, Proteolysis of ankyrin and of band 3 protein in chemically induced cell fusion, *Biochem. J.* **218:**295–305.

Langdon, R. G., and Holman, V. P., 1988, Immunological evidence that band 3 is the major glucose transporter of the human erythrocyte membrane, *Biochim. Biophys. Acta* **945:**23–32.

Lange, Y., and Slayton, J. M., 1982, Interaction of cholesterol and lysophosphatidylcholine in determining red cell shape *Lipid Res.* **23:**1121–1127.

Lange, Y., Dolde, J., and Steck, T. L., 1981, The rate of transmembrane movement of cholesterol in the human erythrocyte, *J. Biol. Chem.* **256:**5321–5323.

Latorre, R., Oberhauser, A., Labarca, P., and Alvarez, O., 1989, Varieties of calcium-activated potassium channels, *Annu. Rev. Physiol.* **51:**385–399.

Latron, F., Blanchard, D., and Carton, J. P., 1987, Immunochemical characterization of the human blood cell membrane glycoprotein recognized by the monoclonal antibody 12E7, *Biochem. J.* **247:**757–764.

Lauf, P. K., 1984, Thiol-dependent passive K/Cl transport in sheep red cells. IV. Furosemide inhibition as a function of external Rb^+, Na^+, and Cl^-, *J. Membr. Biol.* **77:**57–62.

Lauf, P. K., 1985, $K^+ : Cl^-$ cotransport: Sulfhydryls, divalent cations, and the mechanism of volume activation in a red cell, *J. Membr. Biol.* **88:**1–13.

Lauf, P. K., 1986, Chloride-dependent cation cotransport and cellular differentiation: A comparative approach, *Curr. Top. Membr. Transp.* **27:**89–125.

Lauf, P. K., McManus, T. J., Haas, M., Forbush, B., Duhm, J., Flatman, P. W., Saier, M. H., Jr., and Russell, J. M., 1987, Physiology and biophysics of chloride and cation cotransport across cell membranes, *Fed. Proc.* **46:**2377–2394.

Lavis, V. R., Lee, D. P., and Shenolikar, S., 1987, Evidence that forskolin binds to the glucose transporter of human erythrocytes, *J. Biol. Chem.* **262:**14571–14575.

LeFevre, P. G., 1961, Sugar transport in the red blood cell: Structure–activity relationships in substrates and antagonists, *Pharmacol. Rev.* **13**:39–70.

Legrum, B., and Passow, H., 1989, Inhibition of inorganic anion transport across the human red blood cell membrane by chloride-dependent association of dipyridamole with a stilbene disulfonate binding site on the band 3 protein, *Biochim. Biophys. Acta* **979**:193–207.

Leitmannova, A., and Glaser, R., 1977, Mathematical modelling of human echinocytes and the membrane bending of disocytes, stomatocytes and echinocytes, *Stud. Biophys.* **64**:123–141.

Lentz, B. R., 1989, Membrane "fluidity" as detected by diphenylhexatriene probes, *Chem. Phys. Lipids* **50**:171–190.

Leterrier, F., Rieger, F., and Mariaud, J. F., 1973, Comparative study of the action of phenothiazine and parafluorobutyrophenone derivates on rat brain membranes using the spin label technique, *J. Pharmacol. Exp. Ther.* **186**:609–615.

Levin, R. M., and Weiss, B., 1979, Selective binding of antipsychotics and other psychoactive agents to the calcium-dependent activator of cyclic nucleotide phosphodiesterase, *J. Pharmacol. Exp. Ther.* **208**:454–459.

Lew, V., and Ferreira, H. G., 1978, Ca^+ -transport and the properties of a Ca-sensitive K channel in red cell membranes, *Curr. Top Membr. Transp.* **10**:217–271.

Lew, V. L., Tsien, R. Y., and Miner, C., 1982, Physiological $[Ca^{2+}]_i$ level and pump-leak turnover in intact red cells measured using an incorporated Ca chelator, *Nature* **298**:478–481.

Lichtman, M. L., and Santillo, P., 1986, Red cell egress from the marrow:? vis-a-tergo, *Blood Cells* **12**:11–19.

Lieber, R. M., Lange, Y., Weinstein, R. S., and Steck, T. L., 1984, Interaction of chlorpromazine with the human erythrocyte membrane, *J. Biol. Chem.* **259**:9225–9234.

Lilley, G. L., and Fung, L. W. M., 1987, Hemoglobin–membrane interaction at physiological ionic strength and temperature, *Life Sci.* **41**:2429–2439.

Linderkamp, O., and Meiselman, H. J., 1982, Geometric, osmotic, and membrane mechanical properties of density-separated human red cells, *Blood* **59**:1121–1127.

Liu, S. C., Derick, L. H., and Palek, J., 1987, Visualization of the hexagonal lattice in the erythrocyte membrane skeleton, *J. Cell Biol.* **104**:527–536.

Low, M. G., 1987, Biochemistry of the glycosyl-phosphatidylinositol membrane protein anchors, *Biochem. J.* **244**:1–13.

Lowe, G. D. O., 1988, Rheological therapy, in: *Clinical Blood Rheology* Volume 2 (D. Gordon and O. Lowe, eds.), pp. 1–22, CRC Press, Boca Raton.

Lowndes, J. M., Hokin-Neaverson, M., and Ruoho, A. E., 1987, Photoaffinity labeling of erythrocyte membrane (Na^+ + K^+)-ATPase with high specific activity [^{125}I]iodoazidogalactosyl digitoxigenin, *Biochim. Biophys. Acta* **904**:154–158.

Lüdi, H., and Schatzmann, H. J., 1987, Some properties of a system for sodium-dependent outward movement of magnesium from metabolizing human red blood cells, *J. Physiol. (London)* **390**:367–382.

Lutz, H. U., 1978, Vesicles isolated from ATP-depleted erythrocytes and out of thrombocyte-rich plasma, *J. Supramol. Struct.* **8**:375–389.

Luxnat, M., and Galla, H.-J., 1986, Partition of chlorpromazine into lipid bilayer membranes: The effect of membrane structure and composition, *Biochim. Biophys. Acta* **856**:274–282.

Lytle, C., and McManus, T. J., 1987, Effect of loop diuretics and stilbene derivatives on swelling-induced K–Cl cotransport, *J. Gen. Physiol.* **90**:28a.

McConnell, H. M., and McFarland, B. G., 1970, Physics and chemistry of spin labels, *Q. Rev. Biophys.* **3**:91–136.

McLaughlin, S., 1985, New experimental models for the electrokinetic properties of biological membranes: The location of fixed charges affects the electrophoretic mobility of model membranes, *Stud. Biophys.* **110**:25–28.

McManus, T. J., and Schmidt, W. F., 1978, Ion and co-ion transport in avian red cells, *Membr. Transp. Processes* **1**:79–106.

McNamara, M. K., and Wiley, J. S., 1986, Passive permeability of human red blood cells to calcium, *Am. J. Physiol.* **250**:C26–C31.

Macey, R. I., 1984, Transport of water and urea in red blood cells, *Am. J. Physiol.* **246**:C195–C203.

Machleidt, H., Roth, S., and Seeman, P., 1972, The hydrophobic expansion of erythrocyte membranes by the phenol anesthetics, *Biochim. Biophys. Acta* **255**:178–189.

Madden, T. D., 1986, Current concepts in membrane protein reconstitution, *Chem. Phys. Lipids* **40**:207–222.

Mahé, Y., Garcia-Romeu, F., and Motais, R., 1985, Inhibition by amiloride of both adenylate cyclase activity and the Na^+/H^+ antiporter in fish erythrocytes, *Eur. J. Pharmacol.* **116:**199–206.

Makriyannis, A., Siminovitch, D. J., Das Gupta, S. K., and Griffin, R. G., 1986, Studies on the interaction of anesthetic steroids with phosphatidylcholine using ^2H and ^{13}C solid state NMR, *Biochim. Biophys. Acta* **859:**49–55.

Marchesi, V. T., 1985, Stabilizing infrastructure of cell membranes, *Annu. Rev. Cell Biol.* **1:**531–561.

Marcus, D. M. (ed.), 1981, Blood group immunochemistry and genetics, *Semin. Hematol.* **17:**1–71.

Markle, D. R., Evans, E. A., and Hochmuth, R. M., 1983, Force relaxation and permanent deformation of erythrocyte membrane, *Biophys. J.* **42:**91–98.

May, J. M., 1987, Labeling of human erythrocyte band 3 with maltosylisothiocyanate, *J. Biol. Chem.* **262:**3140–3145.

Mayrand, R. R., and Levitt, D. G., 1983, Urea and ethylene glycol-facilitated transport system in the human red cell membrane, *J. Gen. Physiol.* **81:**221–237.

Mercer, R. W., and Hoffman, J. F., 1985, Bumetanide-sensitive Na/K cotransport in ferret red blood cells, *Biophys. J.* **47:**157a.

Metcalfe, J. C., Seeman, P., and Burgen, A. S. V., 1968, The proton relaxation of benzyl alcohol in erythrocyte membranes, *Mol. Pharmacol.* **4:**87–95.

Middelkoop, E., Lubin, B. H., Op den Kamp, J. A. F., and Roelofsen, B., 1986, Flip–flop rates of individual molecular species of phosphatidylcholine in the human red cell membrane, *Biochim. Biophys. Acta* **855:**421–424.

Milanick, M. A., 1989, Na–Ca exchange in ferret red blood cells, *Am. J. Physiol.* **256:**C390–C398.

Miller, K. W., Hammond, L., and Porter, E. G., 1977, The solubility of hydrocarbon gases in lipid bilayers, *Chem. Phys. Lipids* **20:**229–241.

Miller, K. W., Firestone, L. L., Alifimoff, J. K., and Streicher, P., 1989, Nonanesthetic alcohols dissolve in synaptic membranes without perturbing their lipids, *Proc. Natl. Acad. Sci. USA* **86:**1084–1087.

Minetti, M., and Di Stasi, A. M. M., 1987, Involvement of erythrocyte skeletal proteins in the modulation of membrane fluidity by phenothiazines, *Biochemistry* **26:**8133–8137.

Mohandas, N., Greenquist, A. C., and Shohet, S. B., 1978, Bilayer balance and regulation of red cell shape changes, *J. Supramol. Struct.* **9:**453–458.

Morgan, K., and Mir, M. A., 1984, Isolation of a sodium transport inhibitory factor, inhibitin, from cultured leukemic promyelocytes, *J. Clin. Invest.* **74:**1132–1142.

Morgan, K., Spurlock, G., Collins, P. A., and Mir, M. A., 1989, Interaction of inhibitin with the human erythrocyte $Na^+(Li^+)_i/Na_o^+$ exchanger, *Biochim. Biophys. Acta* **979:**53–61.

Morrow, J. S., and Anderson, R. A., 1986, Shaping the too fluid bilayer, *Lab. Invest.* **54:**237–240.

Motais, R., and Cousin, J. L., 1976, The inhibitor effect of probenecid and structural analogues on organic anions and chloride permeabilities in ox erythrocytes, *Biochim. Biophys. Acta* **419:**309–313.

Motais, R., and Cousin, J. L., 1978, A structure activity study of some drugs acting as reversible inhibitors of chloride permeability in red cell membranes: Influence of ring substituents, in: *Cell Membrane Receptors for Drugs and Hormones: A Multidisciplinary Approach* (R. W. Straub and L. Bolis, eds.), Raven Press, New York. pp. 219–225.

Motais, R., Sola, F., and Cousin, J. L., 1978, Uncouplers of oxidative phosphorylation. A structure–activity study of their inhibitory effect on passive chloride permeability, *Biochim. Biophys. Acta* **510:**201–207.

Motais, R., Baroin, A., Motais, A., and Baldy, S., 1980, Inhibition of anion and glucose permeability by anesthetics in erythrocytes. The mechanisms of action of positively and negatively charged drugs, *Biochim. Biophys. Acta* **599:**673–688.

Mueckler, M., Caruso, C., Baldwin, S., Panico, M., and Blench, I., 1985, Sequence and structure of a human glucose transporter, *Science* **229:**941–945.

Müller, H.-J., Luxnat, M., and Galla, H.-J., 1986, Lateral diffusion of small solutes and partition of amphipaths in defect structures of lipid bilayers, *Biochim. Biophys. Acta* **856:**283–289.

Naftalin, R. J., 1988, Pre-steady-state uptake of D-glucose by the human erythrocyte is inconsistent with a circulating carrier mechanism, *Biochim. Biophys. Acta* **946:**431–438.

Nagasawa-Fujimori, H., Hiromi, K., Moriwaki, N., and Fujii, T., 1981, Application of light scattering stopped-flow method to the observation of rapid processes of shape changes of human erythrocytes induced by various drugs, *Biochem. Int.* **2:**129–135.

Nelson, G. A., Andrews, M. L., and Karnovsky, M., 1983, Control of erythrocyte shape by calmodulin, *J. Cell Biol.* **96:**730–735.

Neyses, L., Locher, R., Stimpel, M., Streuli, R., and Vetter, W., 1985, Stereospecific modulation of the calcium channel in human erythrocytes by cholesterol and its oxidized derivatives, *Biochem. J.* **227**:105–112.

Nicolay, K., van der Neut, R., Fok, J. J., and de Kruijyff, B., 1985, Effects of adriamycin on lipid polymorphism in cardiolipin-containing model and mitochondrial membranes, *Biochim. Biophys. Acta* **819**:55–65.

Noji, S., Takahashi, T., and Kon, H., 1982, A spin-label study of the correlation between stomatocyte formation and membrane fluidization of erythrocytes, *Biochem. Pharmacol.* **31**:3173–3180.

Ogiso, T., Iwaki, M., and Mori, K., 1981, Fluidity of human erythrocyte membrane and effect of chlorpromazine on fluidity and phase separation of membrane, *Biochim. Biophys. Acta* **649**:325–335.

Ohshima, H., Makino, K., and Kondo, T., 1986, Potential distribution across a membrane with surface charge layers: Effects of nonuniform charge distribution, *J. Colloid Interface Sci.* **113**:369–374.

O'Neill, W. C., 1987, Volume-sensitive Cl-dependent K transport in human erythrocytes, *Am. J. Physiol.* **253**:C883–C888.

Op den Kamp, J. A. F., 1979, Lipid asymmetry in membranes, *Annu. Rev. Biochem.* **48**:47–71.

Orme, F. W., Moronne, M. M., and Macey, R. I., 1988, Modification of the erythrocyte membrane dielectric constant by alcohols, *J. Membr. Biol.* **104**:57–68.

Ovchinnikov, Y., A., Arzamazova, N. M., Arystarkhove, E. A., Gevondyan, N. M., Aldanova, N. A., and Modyanov, N. N., 1987, Detailed structural analysis of exposed domains of membrane-bound Na^+,K^+-ATPase. A model of transmembrane arrangement, *FEBS Lett.* **217**:269–274.

Owen, J. D., and Solomon, A. K., 1972, Control of nonelectrolyte permeability in red cells, *Biochim. Biophys. Acta* **290**:414–418.

Owen, N. E., and Gunn, R. B., 1983, Kinetic mechanism of chlorpromazine inhibition of erythrocyte 3-O-methylglucose transport, *Biochim. Biophys. Acta* **727**:213–216.

Palfrey, H. C., Feit, P. W., and Greengard, P., 1980, cAMP-stimulated cation cotransport in avian erythrocytes: Inhibition by "loop" diuretics, *Am. J. Physiol.* **238**:C139–C148.

Pandey, G. N., Sarkadi, B., Haas, M., Gunn, R. B., Davis, J. M., and Tosteson, D. C., 1978, Lithium transport pathways in human red blood cells, *J. Gen. Physiol.* **72**:233–247.

Pang, K.-Y. Y., and Miller, K. W., 1978, Cholesterol modulates the effects of membrane perturbers in phospholipid vesicles and biomembranes, *Biochim. Biophys. Acta* **511**:1–9.

Parker, J. C., 1978, Sodium and calcium movements in dog red blood cells, *J. Gen. Physiol.* **71**:1–17.

Parker, J. C., 1983a, Passive calcium movements in dog red blood cells: Anion effects, *Am. J. Physiol.* **244**:C318–C323.

Parker, J. C., 1983b, Volume-responsive sodium movements in dog red blood cells, *Am. J. Physiol.* **244**:C324–C330.

Parker, J. C., 1986, Interactions of lithium and protons with the sodium–proton exchanger of dog red blood cells, *J. Gen. Physiol.* **87**:189–200.

Parker, J. C., and Harper, J. R., Jr., 1980, Effects of amrinone, a cardiotonic drug, on calcium movements in dog erythrocytes, *J. Clin. Invest.* **66**:254–259.

Passow, H., 1986, Molecular aspects of band 3 protein-mediated anion transport across the red blood cell membrane, *Rev. Physiol. Biochem. Pharmacol.* **103**:61–203.

Passow, H., and Schnell, K. F., 1969, Chemical modifiers of passive ion permeability of the erythrocyte membrane, *Experientia* **25**:460–468.

Peters, R., 1988, Lateral mobility of proteins and lipids in the red cell membrane and the activation of adenylate cyclase by β-adrenergic receptors, *FEBS Lett.* **234**:1–7.

Pinkofsky, H. B., and Jung, C. Y., 1985, Accessibility of sulfhydryl residues induced by cytochalasin B binding and conformational dynamics in the human erythrocyte glucose transporter, *Arch. Biochem. Biophys.* **240**:94–101.

Pjura, W. J., Kleinfeld, A. M., and Karnovsky, M. J., 1984, Partition of fatty acids and fluorescent fatty acids into membranes, *Biochemistry* **23**:2039–2043.

Plagemann, P. G. W., and Kraupp, M., 1986, Inhibition of nucleoside and nucleobase transport and nitrobenzylthioinosine binding by dilazep and hexobendine, *Biochem. Pharmacol.* **35**:2559–2567.

Plagemann, P. G. W., and Woffendin, C., 1987a, Comparison of the equilibrium exchange of nucleoside and 3-O-methylglucose in human erythrocytes and of the effects of cytochalasin B, phloretin and dipyridamole on their transport, *Biochim. Biophys. Acta* **899**:295–301.

Plagemann, P. G. W., and Woffendin, C., 1987b, Effects of Ca^{2+}-channel antagonists on nucleoside and nucleobase transport in human erythrocytes and cultured mammalian cells, *Biochim. Biophys. Acta* **928**:243–250.

Plagemann, P. G. W., Wohlhueter, R. M., and Woffendin, C., 1988, Nucleoside and nucleobase transport in animal cells, *Biochim. Biophys. Acta* **947**:405–443.

Plishker, G. A., Pope, C. N., and Seinsoth, L. S., 1988, Calpromotin, a high molecular weight oligomer involved in calcium-dependent potassium transport, *Biophys. J.* **53**:533a.

Ponder, E., 1948, *Hemolysis and Related Phenomena,* Grune & Stratton, New York.

Pope, J. M., Littlemore, L. A., and Westerman, P. W., 1989, Chain-length dependence of n-alkane solubility in phosphatidylcholine bilayers: A ^{2}H-NMR study, *Biochim. Biophys. Acta* **980**:69–76.

Porzig, H., 1975, Comparative study of the effects of propranolol and tetracaine on cation movements in resealed human red cell ghosts, *J. Physiol. (London)* **249**:27–49.

Reichstein, E., and Rothstein, A., 1981, Effects of quinine on Ca^{++}-induced K^{+} efflux from human red blood cells, *J. Membr. Biol.* **59**:57–63.

Renner, M., Dietl, M., and Schnell, K. F., 1988, Chloride mediated inhibition of the phosphate and the sulfate transport by dipyridamole in human erythrocyte ghosts, *FEBS Lett.* **238**:77–81.

Ridgwell, K., Tanner, M. J. A., and Anstee, D. J., 1984, The rhesus (D) polypeptide is linked to the human erythrocyte cytoskeleton, *FEBS Lett.* **174**:7–12.

Robinson, J. D., 1969, Effects of phlorizin on membrane cation-dependent adenosine triphosphatase and p-nitrophenyl phosphatase activities, *Mol. Pharmacol.* **5**:584–592.

Robinson, J. D., Robinson, L. J., and Martin, N. J., 1984, Effects of oligomycin and quercetin on the hydrolytic activities of the (Na^{+} + K^{+})-dependent ATPase, *Biochim. Biophys. Acta* **772**:295–306.

Rogers, J. A., Cheng, S., and Betageri, G. V., 1986, Association and partitioning of propranolol in model and biological membranes, *Biochem. Pharmacol.* **35**:2261–2264.

Rooney, E. K., East, J. M., Jones, O. T., McWhirter, J., Simmonds, A. C., and Lee, A. G., 1983, Interaction of fatty acids with lipid bilayers, *Biochim. Biophys. Acta* **728**:159–170.

Rooney, M. W., Yachnin, S., Kucuk, O., Lis, L. J., and Kauffman, J. W., 1985, Oxygenated cholesterols synergistically immobilize acyl chains and enhance protein helical structure in human erythrocyte membranes, *Biochim. Biophys. Acta* **820**:33–39.

Rossi, J. P. F. C., Garrahan, P. F., and Rega, A. F., 1981, Vanadate inhibition of active Ca^{2+} transport across human red cell membranes, *Biochim. Biophys. Acta* **648**:145–150.

Rosso, J., Zachowski, A., and Devaux, P. F., 1988, Influence of chlorpromazine on the transverse mobility of phospholipids in the human erythrocyte membrane: Relation to shape changes, *Biochim. Biophys. Acta* **942**:271–279.

Roth, S., and Seeman, P., 1972, The membrane concentrations of neutral and positive anesthetics (alcohols, chlorpromazine, morphine) fit the Meyer–Overton rule of anesthesia; negative narcotics do not, *Biochim. Biophys. Acta* **255**:207–219.

Roth, S., Seeman, P., Akerman, S. B. A., and Chau-Wong, M., 1972, The action and adsorption of local anesthetic enantiomers on erythrocyte and synaptosome membranes, *Biochim. Biophys. Acta* **255**:199–206.

Sachs, J. R., and Welt, L. G., 1968, Concentration dependence of active potassium transport in the human red blood cell in the presence of inhibitors, *J. Clin. Invest.* **47**:949.

Sarkadi, B., Alifimoff, J. K., Gunn, R. B., and Tosteson, D. C., 1978, Kinetics and stoichiometry of Na-dependent Li transport in human red blood cells, *J. Gen. Physiol.* **72**:249–265.

Sarkadi, B., Szasz, I., and Gardos, G., 1980, Characteristics and regulation of active calcium transport in inside-out red cell membrane vesicles, *Biochim. Biophys. Acta* **598**:326–338.

Sarkar, H. K., Thorens, B., Lodish, H. F., and Kaback, H. R., 1988, Expression of the human erythrocyte glucose transporter in Escherichia coli, *Proc. Natl. Acad. Sci. USA* **85**:5463–5467.

Scharff, O., and Foder, B., 1982, Rate constants for calmodulin binding to Ca^{2+}-ATPase in erythrocyte membranes, *Biochim Biophys. Acta* **691**:133–143.

Schatzmann, H. J., 1953, Herzglykoside als Hemmstoffe für den aktiven Kalium and Natrium-Transport durch die Erythrocytenmembran, *Helv. Physiol. Pharmacol. Acta* **11**:346.

Schatzmann, H. J., 1983, The red cell calcium pump, *Annu. Rev. Physiol.* **45**:303–312.

Scheuring, U., Kollewe, K., Haase, W., and Schubert, D., 1986, A new method for the reconstitution of the anion transport system of the human erythrocyte membrane, *J. Membr. Biol.* **90**:123–135.

Scheven, C. H., and Stibenz, D., 1983, Die Elastizität der Erythrozytenmembran: Übersicht und Versuch einer Deutung auf Grund neuer Daten zur Ultrastruktur des Membranskeletts, *Morph. Jahrb.* **129**:287–298.

Schlieper, P., and Steiner, R., 1983, The effect of different surface chemical groups on drug binding to liposomes, *Chem. Phys. Lipids* **34**:81–92.

Schmid-Schönbein, H., Grebe, R., and Heidtmann, H., 1983, A new membrane concept for viscous RBC deformation in shear: Spectrin oligomer complexes as a Bingham-fluid in shear and a dense periodic colloidal system in bending, *Ann. N.Y. Acad. Sci.* **416**:225–254.

Schneider, E., Haest, C. W. M., Plasa, G., and Deuticke, B., 1986, Bacterial cytotoxins, amphotericin B and local anesthetics enhance transbilayer mobility of phospholipids in erythrocyte membranes. Consequences for phospholipid asymmetry, *Biochim. Biophys. Acta* **855**:325–336.

Schreier, S., Frezzatti, W. A., Jr., Araujo, P. S., Chaimovich, H., and Cuccovia, I. M., 1984, Effect of lipid membranes on the apparent pK of the local anesthetic tetracaine. Spin label and titration studies, *Biochim. Biophys. Acta* **769**:231–237.

Schrier, S. L., Junga, I., and Ma, L., 1986, Studies on the effect of vanadate on erythrocyte and shape changes in human red blood cells and ghosts, *Blood* **68**:1008–1014.

Schubert, D., 1987, Biophysical approaches to the study of biological membranes, in: *Biological Membranes: A Practical Approach* (J. B. C. Findlay and W. H. Evans, eds.), pp. 241–280, IRL Press, Oxford.

Schurr, A., Sheffer, N., Graziani, Y., and Livne, A., 1974, Inhibition of glucose efflux from human erythrocytes by hashish components, *Biochem. Pharmacol.* **23**:2005–2009.

Schwartz, A., Lindenmayer, G. E., and Allen, J. C., 1975, The sodium–potassium adenosine triphosphatase: Pharmacological, physiological and biochemical aspects, *Pharmacol. Rev.* **27**:3–134.

Schwarz, W., and Passow, H., 1983, Ca^{2+}-activated K^+ channels in erythrocytes and excitable cells, *Annu. Rev. Physiol.* **45**:359–374.

Schwarz, W., Keim, H., Fehlau, R., and Fuhrmann, G. F., 1989, Modulation of the Ca^{2+}- or Pb^{2+}-activated K^+-selective channels in human red cells. I. Effects of propranolol, *Biochim. Biophys. Acta* **978**:32–36.

Seaman, G. V. F., 1983, Electrochemical properties of the peripheral zone of erythrocytes, *Ann. N.Y. Acad. Sci.* **416**:176–189.

Seelig, A., 1987, Local anesthetics and pressure: A comparison of dibucaine binding to lipid monolayers and bilayers, *Biochim. Biophys. Acta* **899**:196–204.

Seelig, A., Allegrini, P. R., and Seelig, J., 1988, Partitioning of local anesthetics into membranes: Surface charge effects monitored by the phospholipid head-group, *Biochim. Biophys. Acta* **939**:267–276.

Seeman, P., 1972, The membrane actions of anesthetics and tranquilizers, *Pharmacol. Rev.* **24**:583–655.

Seeman, P., Roth, S., and Schneider, H., 1971, The membrane concentrations of alcohol anesthetics, *Biochim. Biophys. Acta* **225**:171–184.

Semplicini, A., Spalvins, A., and Canessa, M., 1989, Kinetics and stoichiometry of the human red cell Na^+/H^+ exchanger, *J. Membr. Biol.* **107**:219–228.

Senn, N., Lelièvre, L. G., Braquet, P., and Garay, R., 1988, High sensitivity of the Na^+,K^+-pump of human red blood cells to genins of cardiac glycosides, *Br. J. Pharmacol.* **93**:803–810.

Sergeant, S., and Kim, H. D., 1985, Inhibition of 3-O-methylglucose transport in human erythrocytes by forskolin, *J. Biol. Chem.* **260**:14677–14682.

Sheetz, M. P., and Singer, S. J., 1974, Biological membranes as bilayer couples. A molecular mechanism of drug–erythrocyte interactions, *Proc. Natl. Acad. Sci. USA* **71**:4457–4461.

Shen, B. W., Josephs, R., and Steck, T. L., 1986, Ultrastructure of the intact skeleton of the human erythrocyte membrane, *J. Cell Biol.* **102**:997–1006.

Siebens, A. W., and Kregenow, F. M., 1985, Volume-regulatory response of Amphiuma red cells in anisotonic media. The effect of amiloride, *J. Gen. Physiol.* **86**:527–564.

Simon, S. A., Stone, W. L., and Bennett, P. B., 1979, Can regular solution theory be applied to lipid bilayer membranes? *Biochim. Biophys. Acta* **550**:38–47.

Simons, T. J. B., 1985, Influence of lead ions on cation permeability in human red cell ghosts, *J. Membr. Biol.* **84**:61–71.

Singer, M. A., and Jain, M. K., 1980, Interaction of four local anesthetics with phospholipid bilayer membranes: Permeability effects and possible mechanisms, *Can. J. Biochem.* **58**:815–821.

Sinha, B. K., and Chignell, C. F., 1979, Interaction of antitumor drugs with human erythrocyte ghost membranes and mastocytoma P815: A spin label study, *Biochem. Biophys. Res. Commun.* **86**:1051–1057.

Skou, J. C., 1986, The sodium, potassium pump, *Scand. J. Clin. Lab. Invest.* **46**:11–23.

Skou, J. C., and Esmann, M., 1981, Eosin, a fluorescent probe of ATP binding to the $(Na^+ + K^+)$-ATPase, *Biochim. Biophys. Acta* **647**:232–240.

Smith, R. A., Porter, E. G., and Miller, K. W., 1981, The solubility of anesthetic gases in lipid bilayers, *Biochim. Biophys. Acta* **645**:327–338.

Solomon, A. K., 1986, On the equivalent pore radius, *J. Membr. Biol.* **94**:227–232.

Stampe, P., and Vestergaard-Bogind, B., 1985, The Ca^{2+}-sensitive K^+-conductance of the human red cell membrane is strongly dependent on cellular pH, *Biochim. Biophys. Acta* **815:**313–321.

Staufenbiel, M., 1987, Ankyrin-bound fatty acid turns over rapidly at the erythrocyte plasma membrane, *Mol. Cell. Biol.* **7:**2981–2984.

Stein, W. D., 1986, *Transport and Diffusion across Cell Membranes*, Academic Press, New York.

Stein, W. D., and Danielli, J. F., 1956, Structure and function in red cell permeability, *Discuss. Faraday Soc.* **21:**238–251.

Stokke, B. T., Mikkelsen, A., and Elgsaeter, A., 1986, The human erythrocyte membrane skeleton may be an ionic gel. I. Membrane mechano-chemical properties, *Eur. Biophys. J.* **13:**203–218.

Striessnig, J., Zernig, G., and Glossmann, H., 1985, Human red-blood-cell Ca^{2+}-antagonist binding sites. Evidence for an unusual receptor coupled to the nucleoside transporter, *Eur. J. Biochem.* **150:**67–77.

Suda, T., Shimizu, D., Maeda, N., and Shiga, T., 1981, Decreased viscosity of human erythrocyte suspension induced by chlorpromazine and isoxsuprine, *Biochem. Pharmacol.* **30:**2057–2064.

Surewicz, W. K., 1982, Propranolol-induced structural changes in human erythrocyte ghost membranes. A spin label study, *Biochem. Pharmacol.* **31:**691–694.

Surewicz, W. K., and Leyko, W., 1981, Interaction of propranolol with model phospholipid membranes. Monolayer, spin label and fluorescence spectroscopy studies, *Biochim. Biophys. Acta* **643:**387–397.

Svetina, S., Ottova-Leitmannova, A., and Glaser, R., 1982, Membrane bending energy in relation to bilayer couples concept of red blood cell shape transformations, *J. Theor. Biol.* **94:**13–23.

Svetina, S., Brumen, M., and Zeks, B., 1985, Lipid bilayer elasticity and the bilayer couple interpretation of red cell shape transformations and lysis, *Stud. Biophys.* **110:**177–184.

Szasz, I., Sarkadi, B., and Gardos, G., 1978, Effect of drugs on the calcium-dependent rapid potassium transport in calcium-loaded intact red cells, *Acta Biochim. Biophys. Acad. Sci. Hung.* **13:**133–141.

Takakuwa, Y., and Mohandas, N., 1988, Modulation of erythrocyte membrane material properties by Ca^{2+} and calmodulin, *J. Clin. Invest.* **82:**394–400.

Tamura, A., Morita, K., Fujii, T., and Kojima, K., 1982, Detection of the electrical surface charge induced by treatment of the membrane lipid bilayer of human erythrocytes, *Cell Struct. Funct.* **7:**21–27.

Tanford, C., 1980, *Hydrophobic Effect*, Wiley, New York.

Tanford, C., 1983, Mechanism of free energy coupling in active transport, *Annu. Rev. Biochem.* **52:**379–409.

Tanford, C., 1985, Simple model can explain self-inhibition of red cell anion exchange, *Biophys. J.* **47:**15–20.

Tanner, M. J. A., Martin, P. G., and High, S., 1988, The complete amino acid sequence of the human erythrocyte membrane anion-transport protein deduced from the cDNA sequence, *Biochem. J.* **256:**703–712.

Teitel, P., and Schmid-Schönbein, H., 1990, Towards standardization of terminology in clinical hemorheology: Let us call a spade a spade, *Clin. Hemorheol.* in press.

Tenforde, T., 1970, Microelectrophoretic studies on the surface chemistry of erythrocytes, *Adv. Biol. Med. Phys.* **13:**43–105.

Toon, M. R., and Solomon, A. K., 1986, Control of red cell urea and water permeability by sulfhydryl reagents, *Biochim. Biophys. Acta* **860:**361–371.

Toon, M. R., and Solomon, A. K., 1987, Modulation of water and urea transport in human red cells: Effects of pH and phloretin, *J. Membr. Biol.* **99:**157–164.

Tournois, H., Leunissen-Bijvelt, J., Haest, C. W. M., de Gier, J., and de Kruijyff, B., 1987, Gramicidin-induced hexagonal H_{II} phase formation in erythrocyte membranes, *Biochemistry* **26:**6613–6621.

Trudell, J. R., 1977, The membrane volume occupied by anesthetic molecules: A reinterpretation of the erythrocyte expansion data, *Biochim. Biophys. Acta* **470:**509–510.

Truong, H.-T. N., Ferrell, J. E., and Huestis, W. H., 1986, Sulfhydryl reducing agents and shape regulation in human erythrocytes, *Blood* **67:**214–221.

Tsuji, A., and Ohnishi, S., 1986, Restriction of the lateral motion of band 3 in the erythrocyte membrane by the cytoskeletal network: Dependence on spectrin association state, *Biochemistry* **25:**6133–6139.

Ungewickell, E., and Gratzer, W., 1978, Self-association of human spectrin. A thermodynamic and kinetic study, *Eur. J. Biochem.* **88:**379–385.

Varecka, L., and Carafoli, E., 1982, Vanadate-induced movements of Ca^{2+} and K^+ in human red blood cells, *J. Biol. Chem.* **257:**7414–7421.

Vestergaard-Bogind, B., Stampe, P., and Christophersen, P., 1985, Single-file diffusion through the Ca^{2+}-activated K^+ channel of the human red cells, *J. Membr. Biol.* **88:**67–75.

Viitala, J., and Järnefelt, J., 1985, The red cell surface revisited, *Trends Biochem. Sci.* **10**:392–395.

Vincenzi, F. F., 1982, Pharmacological modification of the Ca^{2+}-pump ATPase activity of human erythrocytes, *Ann. N.Y. Acad. Sci.* **402**:368–380.

Vincenzi, F. F., and Hinds, T. R., 1988, Drug effects on plasma membrane calcium transport, in: *Calcium in Drug Actions* (P. F. Baker, ed.), pp. 147–162, Springer-Verlag, Berlin.

Voge, H. J., 1988, Ligand-binding sites on calmodulin, in: *Calcium in Drug Actions* (P. F. Baker, ed.), pp. 57–87, Springer-Verlag, Berlin.

Volpi, M., Sha'afi, R. I., and Feinstein, M. B., 1981, Antagonism of calmodulin by local anesthetics. Inhibition of calmodulin-stimulated calcium transport of erythrocyte inside-out membrane vesicles, *Mol. Pharmacol.* **20**:363–370.

Wadzinski, B. E., Shanahan, M. F., and Ruoho, A. E., 1987, Derivatization of the human erythrocyte glucose transporter using a novel forskolin photoaffinity label, *J. Biol. Chem.* **262**:17683–17689.

Walmsley, A. R., 1988, The dynamics of the glucose transporter, *Trends Biochem. Sci.* **13**:226–231.

Wang, J.-F., Falke, J. J., and Chan, S. T., 1986, A proton NMR study of the mechanism of the erythrocyte glucose transporter, *Proc. Natl. Acad. Sci. USA* **83**:3277–3281.

Welti, R., Mullikin, L. J., Yoshimura, T., and Helmkamp, G. M., Jr., 1984, Partition of amphiphilic molecules into phospholipid vesicles and human erythrocyte ghosts: Measurements by ultraviolet difference spectroscopy, *Biochemistry* **23**:6086–6091.

Westman, J., Boulanger, Y., Ehrenberg, A., and Smith, I. C. P., 1982, Charge and pH dependent drug binding to model membranes. A ^{2}H-NMR and light absorption study, *Biochim. Biophys. Acta* **685**:315–328.

Wheeler, T. J., and Hinkle, P. C., 1985, The glucose transporter of mammalian cells, *Annu. Rev. Physiol.* **47**:503–517.

Wheeler, T. J., and Whelan, J. D., 1988, Infinite-cis kinetics support the carrier model for erythrocyte glucose transport, *Biochemistry* **27**:1441–1450.

Whitfield, C. F., Coleman, D. B., Kay, A. M. B., Shiffer, K. A., Miller, J., and Goodman, S. R., 1985, Human erythrocyte membrane proteins of zone 4.5 exist as families of related proteins, *Am. J. Physiol.* **240**:C70–C79.

Wieth, J. O., Funder, J., Gunn, R. B., and Brahm, J., 1974, Passive transport pathways for chloride and urea through the red cell membrane, in: *Comparative Biochemistry and Physiology of Transport* (L. Bolis, K. Bloch, S. E. Luria, and F. Lynen, eds.), pp. 317–338, North-Holland, Amsterdam.

Wilbrandt, W., 1941, Osmotische Natur sogenannter nicht osmotischer Hämolysen (Kolloidomotische Hämolyse), *Pfluegers Arch. Gesamte Physiol. Menschen Tiere* **245**:23–52.

Wildenauer, D. B., and Zeeb-Wälde, B. C., 1983, Solubility of phenothiazines in red blood cell membranes as evidenced by photoaffinity labeling, *Biochem. Biophys. Res. Commun.* **116**:469–477.

Woffendin, C., and Plagemann, P. G. W., 1987, Interaction of [^{3}H]dipyridamole with the nucleoside transporters of human erythrocytes and cultured animal cells, *J. Membr. Biol.* **98**:89–100.

Wold, F., 1986, Fatty acylation of proteins (Keep fit with fat?), *Trends Biochem. Sci.* **11**:58–59.

Wüthrich, A., and Schatzmann, H. J., 1980, Inhibition of the red cell calcium pump by quercetin, *Cell Calcium* **1**:21–35.

Wyse, J. W., Blanl, M. E., Maynard, C. L., Diedrich, D. F., and Butterfield, D. A., 1989, Electron spin resonance investigation of the interaction of the anion and glucose transport inhibitor, p-azidobenzylphlorizin, with the human red cell membrane, *Biochim. Biophys. Acta* **979**:127–131.

Yamaguchi, T., Watanabe, S., and Kimoto, E., 1985, ESR spectral changes induced by chlorpromazine in spin-labeled erythrocyte ghost membranes, *Biochim. Biophys. Acta* **820**:157–164.

Yamamoto, H.-A., and Harris, R. A., 1983a, Calcium-dependent ^{86}Rb efflux and ethanol intoxication: Studies of human red blood cells and rodent brain synaptosomes, *Eur. J. Pharmacol.* **88**:357–363.

Yamamoto, H.-A., and Harris, R. A., 1983b, Effects of ethanol and barbiturates on Ca^{2+}-ATPase activity of erythrocyte and brain membranes, *Biochem. Pharmacol.* **32**:2787–2791.

Yeagle, P. L., 1985, Cholesterol and the cell membrane, *Biochim. Biophys. Acta* **822**:267–287.

Yeagle, P. L., 1987, *The Membranes of Cells,* pp. 83–111, Academic Press, New York.

Yeagle, P. L., 1989, Lipid regulation of cell membrane structure and function, *FASEB J.* **3**:1833–1842.

Zachowski, A., and Durand, P., 1988, Biphasic nature of binding of cationic amphipaths with artificial and biological membranes, *Biochim. Biophys. Acta* **937**:411–416.

Zaslavsky, B. Y., Ossipow, N. N., Krivich, V. S., Baholdina, L. P., and Rogozhin, S. V., 1978, Action of surface-active substances on biological membranes. II. Hemolytic activity of nonionic surfactants, *Biochim. Biophys. Acta* **507**:1–7.

Index

www.ingramcontent.com/pod-product-compliance
Lightning Source LLC
LaVergne TN
LVHW062243200926
848484LV00006B/1603
9780306434624